Advances in Manufacturing Technology XII

Advances in Manufacturing Technology XII

Proceedings of the Fourteenth National Conference on Manufacturing Research
University of Derby, UK
7–9 September 1998

Edited by

Professor R W Baines
Dean of the School of Engineering

Professor A Taleb-Bendiab
Head of the Division of Design and Manufacture

Dr Z Zhao
Reader in Computer Integrated Manufacturing

School of Engineering
University of Derby
UK

Professional Engineering Publishing

Professional Engineering Publishing Limited
Bury St Edmunds and London, UK.

First Published 1998

ISBN 1 86058 172 2

A CIP catalogue record for this book is available from the British Library.

Printed and bound in Great Britain by Bookcraft (Bath) Limited.

About the Editors

Professor Ray Baines is the Conference Chairman, Dean of the School of Engineering, University of Derby and a member of the University Governing Council. He is a Professor of Manufacturing Systems and the Director of the Integrated Manufacturing Research Centre (IMRC). His research interests include manufacturing systems, technology, and enterprise modelling.

Professor Azzelrabe Taleb-Bendiab is currently Head of Division of Design and Manufacture in the School of Engineering, University of Derby. He is a Professor of integrated product development and the Deputy Director of the Integrated Manufacturing Research Centre (IMRC). His research interests are in computer-supported collaborative design and manufacture and intelligent manufacturing systems.

Dr Zhengxu Zhao is a reader in Computer Integrated Manufacturing within the School of Engineering, University of Derby. He spent several years, first in manufacturing industry, then in higher education in the People's Republic of China, before starting his academic career at Staffordshire University where he gained his PhD. His research interests include computer-aided process planning, CAD modelling and three-dimensional simulation, virtual manufacturing, sensors and control.

Related Titles of Interest

Title	Editor/Author	ISBN
Managing Enterprises – Stakeholders, Engineering, Logistics, and Achievement	D T Wright, M M Rudolph V Hanna, D Gillingwater, and N D Burns	1 86058 066 1
Developments in Rapid Prototyping and Tooling	G Bennett	1 86058 048 3
Advanced Manufacturing Processes, Systems, and Technologies (AMPST '96)	M K Khan, C S Wright, and R Whalley	0 85298 989 X
Computer-Aided Production Engineering	IMechE Conference 1995–3	0 85298 949 0

For the full range of titles published by Professional Engineering Publishing contact:

Sales Department
Professional Engineering Publishing Limited
Northgate Avenue
Bury St Edmunds
Suffolk
IP32 6BW
UK

Tel: 01284 724384
Fax: 01284 718692

Contents

Manufacturing Systems

Production and Inventory Control

Process, Planning, and Scheduling

Virtual Reality

Robots, Automation, and CAD

Modelling and Simulation

Manufacturing Strategy/Management

Business Process Re-engineering

Design Management

Knowledge-Based Systems

Quality Control Management

Acknowledgements

The organization of the conference has been carried out within the School of Engineering at the University of Derby, UK. The editors wish to record special thanks to Linda Marshall and Rita Malpass for conference arrangements.

The papers have been rigorously refereed and the editors wish to express their thanks to the referees below:

Professor R W Baines	University of Derby
Professor G Blount	Coventry University
Dr A Boyle	Liverpool John Moores University
Professor A Bramley	University of Bath
Professor K Case	Loughborough University
Dr G Colquhoun	Liverpool John Moores University
Dr P Corcoran	University of Derby
Dr R Gill	Middlesex University
Professor D Harrison	Glasgow Caledonian University
Mr B Ion	University of Strathclyde
Professor N Juster	University of Strathclyde
Professor J Kay	Cranfield University
Professor E Lo	Sheffield Hallam University
Professor T Mileham	University of Bath
Professor B Mills	Liverpool John Moores University
Professor T Medland	University of Bath
Professor B Rowe	Liverpool John Moores University
Professor A Taleb-Bendiab	University of Derby
Professor D Wilcock	Sunderland University
Dr M Wu	University of Derby
Dr V Vitanov	University of Portsmouth
Dr Z Zhao	University of Derby

Preface

The Consortium is an independent body and was established at a meeting held at Loughborough University on 17 February 1978. Its main aim is to promote manufacturing engineering education, training, and research. To achieve this the Consortium maintains a close liaison with those Government Departments and other bodies concerned with the initial and continuing education and training of professional engineers and responds to appropriate consultative and discussion documents and other initiatives. It organizes and supports national manufacturing engineering education research conferences and symposia. The Institution of Electrical Engineers, with whom there is a close working arrangement, undertakes the secretarial duties.

The Consortium consists of heads of those university departments or sections whose first priority is to manufacturing engineering and who have a direct responsibility for running honours degree courses in the field of manufacturing engineering.

COMEH decided in 1984 that a national forum was needed in which the latest research work in the field of manufacturing engineering and manufacturing management could be disseminated and discussed. The result of this initiative was that an annual series of these national conferences on manufacturing research (NCMR) was started. The first NCMR was held at the University of Nottingham in 1985 and subsequently at Napier College (Edinburgh), Nottingham University, Sheffield Polytechnic, Huddersfield Polytechnic, Strathclyde University, Hatfield Polytechnic, University of Central England, University of Bath, Loughborough University of Technology, De Montfort University, University of Bath, and Glasgow Caledonian University. The fourteenth conference, of which this is the published proceedings, was held at the University of Derby, 7–9 September 1998.

Professor Ray Baines (Conference Chairman and Editor), Professor Azzelrabe Taleb-Bendiab, and Dr Zhengxu Zhao (Editors), School of Engineering, University of Derby, UK.

Manufacturing Technology

Human mismatches in machining

M R ABDUL RANI
Faculty of Mechanical Engineering, Universiti Teknologi Malaysia, Malaysia
K CASE
Department Manufacturing Engineering, Loughborough University, UK
M A SINCLAIR
Department of Human Sciences, Loughborough University, UK

ABSTRACT

This research highlights human factors issues in machining by examining mismatches in machining tasks and their relationships with various human characteristics. Mismatches refer to incompatibilities, inappropriateness, unsuitabilities or inconsistencies which, if not addressed, would lead to errors. Knowledge of the rate of mismatches and an understanding of the causes is invaluable in the design of new working environments, machines and tasks.

The Human Task-Mismatch Matching Method was developed to study these issues in manual turning operations using experimental and questionnaire techniques on groups of 16 skilled and 12 unskilled operators. The skilled subjects were drawn from local industry and university technical staff. Unskilled subjects were engineering students all of whom had some experience and knowledge of machining through periods of industrial placement.

With a single group of skilled and unskilled operators statistically significant relationships were established between mismatches and many of the human characteristics studied, but for skilled operators the only significant relationship was between self-confidence and trust.

The general conclusion is that whilst studying operators in their own workplace provides invaluable information for the design and operation of future workplaces, the relationships between performance and human characteristics remain difficult to formally establish.

1. INTRODUCTION

Human factors issues in machining tasks are studied by examining the problems of mismatches and their relationships with various human characteristics including age, skill, work experience, self-confidence and trust. Mismatches refer to incompatibilities, inappropriateness, unsuitabilities or inconsistencies (1), and are considered here with an emphasis on turning operations. Human characteristics may influence the occurrence of mismatches and hence a deeper understanding is useful to reinforce and extend existing knowledge especially of the requirements of the human-machine interface.

The allocation of function for manual machining operations is investigated by focusing on mismatches between the expected and actual operator actions. An understanding of the elements of expertise provides the principles for more effective training of machinists and the design of machine tools with human-centred features properly based on human psychological, technical and physical characteristics. The specific objective then was to establish relationships between (i) *mismatches* and skill, self-confidence, level of trust, work experience and age, (ii) *self-confidence* and skill, level of trust, work experience and age and (iii) *level of trust* and skill, work experience and age.

Most operator errors arise from a mismatch between the properties of the system as a whole and the characteristics of human information processing (4), and thus mismatches between operators and equipment need to be identified by task. Research into trust and self-confidence of operators on machine systems has been focused on automated systems (2,5,6) but an understanding of trustworthiness and self-confidence on manually-operated machines provides the basis for development of those characteristics on automated systems.

2. MISMATCHES

The problems of mismatches in machining operations are centred on people, and the actions of individual operators result in matches or mismatches between tasks and actions. In this research inappropriate, incompatible, unsuitable or inconsistent actions are considered mismatches and consist of *Intrusion* (help required in proceeding), *Omission* (step omitted), *Commission* (step performed incorrectly), *Reversal* (repeated steps due to earlier omission), *Wrong Request* (for tools, etc.), *Wrong Components, Repetition, Misapplication* (incorrect execution of a method) and *Violations* (standard procedures contravened).

3. THE HYPOTHESES

The twelve hypotheses below (H1 - H12) were developed after an extensive review of literature on human problems in machining operations (3).

Mismatches
- H1: More skilled operators commit fewer mismatches.
- H2: Operators having high self-confidence commit fewer mismatches.
- H3: Operators having a high level of trust commit fewer mismatches.
- H4: Operators with greater experience commit fewer mismatches.
- H5: Older operators commit fewer mismatches.

Self-confidence.
- H6: More skilled operators have higher self-confidence.
- H7: The higher the self-confidence, the higher the level of trust.
- H8: Operators with greater experience have more self-confidence.
- H9: Older operators have more self-confidence.

Level of trust.
- H10: More skilled operators have more trust.
- H11: Operators with greater experience have more trust.
- H12: Older operators have more trust.

Figure 1 depicts the relationships between the variables of these hypotheses.

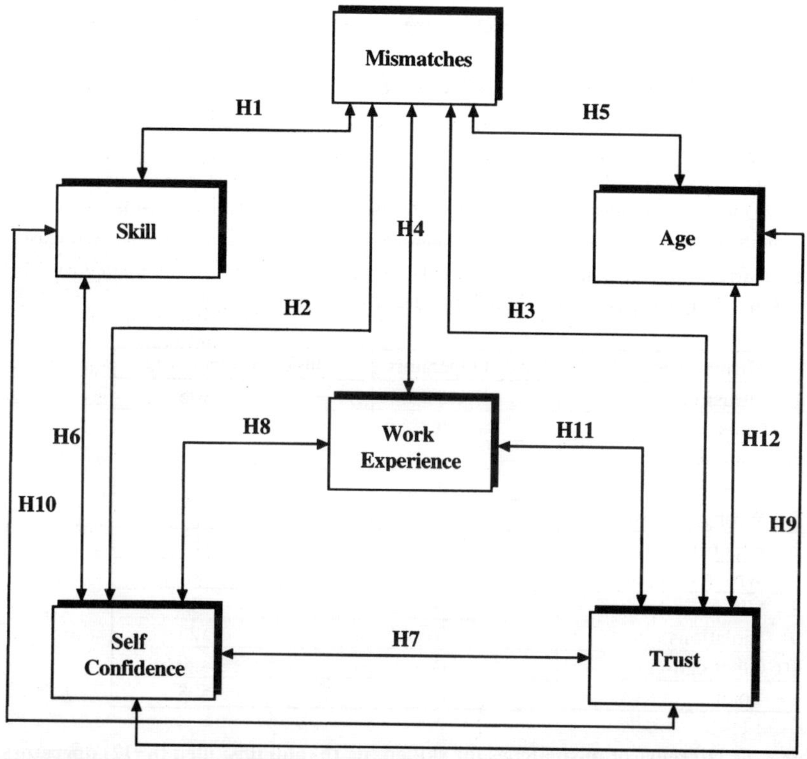

Figure 1. A model of hypotheses depicting the relationships between variables.

4. THE HUMAN TASK-MISMATCH MATCHING METHOD

The Human Task-Mismatch Matching Method (HTMM) was developed to establish the relationships between human characteristics and mismatches and resembles the Task Difficulty and Criticality Rating (TDCR) method (7). The HTMM method takes its name from the way in which each sub-task is matched to the possible mismatches. Mismatches are quantified based on the performance of task elements in machining operations, and the total compared to the ideal performance. The method is unusual in that both simulated-field study and questionnaire survey techniques are used to generate reliable and valid results.

Machining operations are divided into the distinct categories of selection, setting-up and inspection. Selection functions are carried out to choose appropriate tools, set-up functions are carried out before cutting, and inspection concerns the acceptability of a machined product. In each category mismatches are prone to occur due to operator actions and machine design, and may result in time loss, re-work, scrap products and low productivity.

The simulated field study took the form of observing operators using familiar machines in their own workplace machining a simple standard part with some freedom in the operational methods employed. In addition to the experimental aspects, questionnaire studies were used as a subjective technique to evaluate the performance of operators in machining tasks based on their own experience, expert opinions and daily encounters at the workplace. The questionnaires were administered both before and after the machining experiment in the form of a structured interview with the subjects taking part in the experimental studies.

5. ANALYSIS AND INTERPRETATION OF RESULTS

Analysis of the data was carried using methods including Mann-Whitney U, Chi-square, Wilcoxon matched-pairs signed ranks tests and Spearman Correlation Coefficients. Objective performance data was obtained from the total number and categories of mismatches committed by subjects during the machining trials. Subjective data was collected from the questionnaire survey using psychophysical ranking (1 to 10) for self-confidence and trust.

The occurrence of mismatches (Table 1) shows that "repetition" predominates for skilled operators and "requiring assistance" predominates for unskilled operators.

Mismatches	Skilled Operators	Unskilled Operators
1. Intrusion	5	306
2. Omission	6	27
3. Commission	4	37
4. Reversal in sequence	8	7
5. Wrong request	1	3
6. Repetition	112	83
7. Wrong components	12	27
8. Misapplication	0	13
9. Violations	8	25
10. Other causes	0	0
Total	156	528

Table 1. Occurrence of mismatches for skilled (n=16) and unskilled (n=12) operators.

5.1. Mismatches and Skill

Skill is a product of extensive training, long-term exposure to manufacturing processes and frequent execution of similar machining tasks. Repetition constituted 72% of mismatches committed by skilled operators, and these resulted from mechanical failures including mechanical engagement (e.g. selecting speed by engaging levers), parts assembly (e.g. setting up tools on tool posts) and problems in manual aspects of machining processes (e.g. tool selection). The repetition mismatch is considered significant in the sense that the design and handling of mechanical parts requires improvements in terms of design, accuracy and the consideration of user aspects for better handling in set-up and selection functions. Unskilled operators committed more mismatches and this can be attributed to lower anticipation which is known to be important for skilled performance (8). Lower anticipation results from inadequate training, and a detailed study of the results should be useful in establishing training requirements.

5.2. Mismatches and Self-confidence

A significant relationship between mismatches and self-confidence for both skill categories of machine operators was found; the greater the self-confidence, the fewer the mismatches. Unlike unskilled operators, skilled operators have high self-confidence due to their skill and training and this allows them to commit fewer mismatches. Understanding the relationship between mismatches and self-confidence should be useful in the development of training programmes and machine or system design.

5.3. Mismatches and Trust

A relationship was found between the occurrence of mismatches and trust; the greater the trust, the fewer the mismatches, and this corresponds to previous findings (2) where it was suggested that system performance and occurrence of faults could affect trust. Trust can be developed through experience, training and familiarisation. This indicates the importance of careful planning of technology acquisition, and of attention to human resource programmes. Other parameters such as sociological aspects may also be significant and worthy of investigation in this context.

5.4. Mismatches and Work Experience

The results confirm that operators with longer work experience commit fewer mismatches. This finding formally establishes an important characteristic of skilled operators, but the challenge remains to achieve low mismatches with short working experience.

5.5. Mismatches and Age

The study established that older operators commit fewer mismatches, contradicting earlier opinions (9). Age is normally synonymous with maturity which is a factor in human capability. However, the findings here are limited to the working age group (20-55 years). The conclusion to be drawn from this is that machine design should include an analysis of the controls and job aids that would benefit younger operators.

5.6. Self-confidence and Skill

Skill has been shown to be related to self-confidence. Skilfulness may instil high self-confidence among machine operators, and thus it is necessary to maintain skilfulness simply for the sake of maintaining self-confidence. Skilfulness could be maintained by refresher courses and a reduced level of automation to ensure mental stimulation.(10).

5.7. Self-confidence and Trust

It has been suggested that mistrust would cause inappropriate task allocation strategies and influence operators' reliance (2). Therefore, operators' trust on machines needs to develop in parallel to their self-confidence. The analysis confirmed that operators having higher levels of self-confidence have correspondingly high levels of trust in machines.

5.8. Self-confidence and Working Experience

The informal view that working experience instils self-confidence is confirmed by the study, and agrees with (11) that self-efficacy would cause, and might be caused by, performance experiences. However, the extremes of the characteristic, lack of self-confidence and over-confidence for skilled or unskilled operators are detrimental to the execution of tasks.

5.9. Self-confidence and Age

Age and self-confidence were found to be related and this is in keeping with the earlier finding (the relationship between mismatches and age) and emphasises that age is an important contributor towards the design of tasks and machine designs. Learning and experience increase with age and instil self-confidence particularly for psychomotor skills.

5.10. Trust

Trust has always been associated with self-confidence (2). Positive relationships were found between trust and skill but no significant relationships were found with experience and age. This concurs with the previous finding that high trust corresponds to fewer mismatches, and also supports the suggestion that trust helps to reduce complexity and uncertainty (12). This should be exploited in manufacturing where machine designs need to cater for all types of operators. Training should be designed to enable operators, especially new and unskilled operators, gain an adequate level of trust. The absence of relationships between trust and work experience and trust and age suggest that trust does not naturally occur in older and more experienced workers, so retraining is justified if only to reinforce the level of trust.

6. CONCLUSIONS

Human needs, skill, creativity and potential should be the focus of human centred technological systems, and this calls for a critical analysis of human performance, including mismatches. This research has provided a part of this analysis and extended understanding of some important human characteristics as they relate to machining tasks. It is envisaged that this knowledge will be useful in designing better manual systems whilst at the same time providing insights into the human needs of automated systems.

7. REFERENCES

1. Urdang, L. The Oxford Thesaurus, Oxford University Press, 1991.
2. Lee, J.D. & Moray, N., Trust, Self-confidence and Operators' Adaptation to Automation. Int. J. of Human-Computer Studies. No1, pp153-184, 1994.
3. Abdul Rani, M.R. Human Mismatches in Machining. PhD Thesis, Loughborough University, 1997.
4. Reason, J. Cognitive Aids in Process Environments : Prostheses or Tools *? Int.J.of Man-Machine Studies*, Vol.27, pp463-470, 1987.
5. Muir, B.M. Trust in automation: Part I. Theoretical Issues in the Study and Human Intervention in Automated Systems. Ergonomics. Vol 37, No.11, pp1905-1922, 1994.
6. Muir, B.M. & Moray, N. Trust in Automation. Part II. Experimental Studies of Trust and Human Intervention in a Process Control Simulation. Ergonomics. Vol 39, No.3, pp429-460, 1996.
7. Williams, J.C. Human Factors Analysis of Automation Requirements- A methodology for Allocating Functions in *10th Advances in Reliability Technology Symposium* (Ed.) Libberton, G.P.1988. Elsevier, Applied Science, London, 1988.
8. Holding, D.H., Skill research. in *Human Skills* (Ed.) Holding, D.H.. John Wiley and Sons, 1981.
9. Welford, A.T. Ageing and Human Skill. The Nuffield Foundation, Oxford University Press, UK, 1958.
10. Ekkers, C.L., Pasmooij, C.K., Brouwers, A.A.F. & Janusch, A.J. Human Control Tasks: A Comparative Study in Different Man-Machine Systems. Proceedings of the IFAC Workshop, Enschede, Netherlands. Pergamon Press, England, 1979.
11. Silver, W., Mitchell, T.R. & Bist, M.E. Responses to Successful and Unsuccessful Performances: The Moderating Effect of Self-Efficacy on the Relationship between Performance and Attributions. Organisational Behaviour and Human Decision Process. Vol. 62, No. 3, pp286-299, 1995.
12. Luhman, N. Trust and Power, Wiley New York, 1980.

Investigation of parameters' influence and relationships in reaming using statistical design of experiments

S ARUNACHALAM and **J M O'SULLIVAN**
Department of Manufacturing Systems Engineering, University of Hertfordshire, Hatfield, UK

ABSTRACT

Accurate control of bore precision in superabrasive reaming is affected by many parameters. Previous research work in this area concentrated only on key parameters and hence failed to deliver reliable information to improve the process behaviour. The present work involves analysing the efficiency of the technique with 15 process variables arranged in a Taguchi L_{16} orthogonal array at two levels. The effect and significance of individual factors were assessed using the ANOVA method.

1 INTRODUCTION

SAR is an attractive technique for internal finishing cylindrical bores [1]. The process has potential for producing bores with size and cylindricity tolerance to submicron level. Figure 1 shows different designs of abrasive sleeves used in the process. Many manufacturers provide specifications for abrasive reamers, but most of the current specifications are for a static condition or a range of extreme conditions of performance, e.g. surface finish produced, the size tolerance etc. When a reamer performs machining, however, neither the manufacturer nor the user know its actual process capability characteristics. The manufacturer lacks an acceptable methodology and measuring technique to determine this capability. Companies using reamers often do not have the personnel or the measuring equipment needed to determine the process capability of superabrasive reaming (PCSAR). Unfortunately, many quality control objectives cannot be met if the PCSAR is unknown. A major concern of the reamer user is how to determine the best operating characteristics for optimising PCSAR. Therefore, determining PCSAR and optimising the process's operational parameters are important.

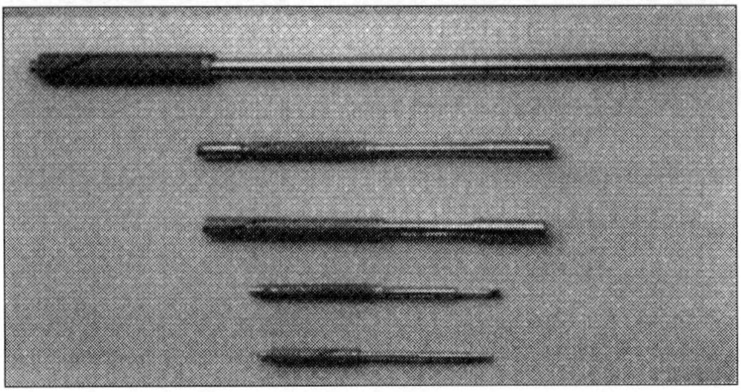

Figure 1: Diamond and cBN Precision Reamers for high precision bore finishing

2 SELECTION OF PROCESS PARAMETERS

SAR is prone to a number of bore-geometry related problems. From previous work [1][2], preliminary tests and experience, 15 variables and the variables' level were identified to ensure that they were chosen within practical limits (table 1). The floating fixture (gimbal mechanism) discussed in reference [3] has been designed, built and used in this experiment to eliminate the misalignment problem between the axes of the reamer and the bore and to provide a near-uniform stock removal around the bore circumference. The variable named 'gimbal centre' means whether the pivoting point of the floating system was set at entry or exit end of the specimen.

Table 1: Factors and level descriptions.

Process variables	Level 1	Level 2
Specimen-related		
Axial length (A)	50 mm	100 mm
Wall thickness (B)	5 mm	15 mm
External force (C)	0 N	5 N
External moment (D)	0 Nm	0.5 Nm
Gimbal centre (E)	entry	exit
Material (F)	aluminium	meehanite
External geometry (G)	round	square
Reamer-related		
Dressing (H)	CBN	ND
Grit size (J)	357 μm	64 μm
Sleeve parallel length (K)	19 mm	50 mm
Plain or grooved (L)	plain	grooved
Process-related		
Reamer rotational speed (M)	200 rpm	300 rpm
Axial feed rate (N)	0.076mm/rev	0.127mm/rev
Axial feed pattern (O)	5mm	20 mm
Stock removal (P)	10 μm	20 μm

3 EXPERIMENT PLAN

The first-order effect of these variables has been investigated using Taguchi's L_{16} design matrix [4]. Each trial was replicated twice, in a randomised manner, totalling 32 runs. Those variables that emerged as being important to the process quality have been further investigated using the ANOVA (analyse of variance) method to study the influences of factors and interactions. Interaction effects have not been considered at this stage. Measurements of the bore geometry after reaming was carried out by Taylor Hobson Ltd, UK and shown in table 2. ANOVA was calculated and significance of the variables is expressed as percentage contribution of each factor towards the bore geometry (table 3). The overall aim of the experiment is to identify the most significant process variables that will affect the bore quality.

4 PARAMETERS INFLUENCING BORE QUALITY

Bore quality is discussed in terms of internal diameter, roundness, cylindricity, straightness and surface finish. From table 3, it can be seen that external geometry of specimen, gimbal centre, grit size, depth of cut, reamer rotational speed and the dressing nature of the sleeve significantly affect the overall bore quality. Trials with feed rate at low setting produced a bore surface with smaller Ra value, indicating that the feed rate is also a governing factor of surface finish. It can be assumed that at lower feed, there may be improper reaming action due to insufficient normal force, while at higher feed, there may be unstable movement of the workpiece due to increased reaming forces [5]. Also at higher reamer rotational speed, dulling of the cutting edges of abrasive grains (Figure 2) occurs and results in the production of rough bores.

Table2: Measurements after reaming.

Trial no	Diameter	Roundness error	Cylindricity error	Straightness error	Surface finish
1	18.961	5.47	20.975	9.725	0.875
2	18.971	7.885	33.875	6.6	0.466
3	19.039	26.315	88.8	63.7	1.033
4	19.062	50.415	175.475	122.45	1.906
5	19.06	78.535	125.525	68.225	0.616
6	19	45.845	144.225	53.75	1.947
7	19.015	55.75	124.875	45.5	0.813
8	18.993	27.335	126.725	69.825	1.345
9	18.982	26.48	79.725	18.525	0.806
10	18.984	19.01	72.25	26.15	0.802
11	19.076	90.265	149.725	36.45	0.633
12	19.006	37.574	80.225	20.275	1.013
13	18.974	18.54	47.825	22.675	2.525
14	18.964	9.295	19.65	11.424	0.941
15	19.038	56.025	122.175	16.775	1.249
16	18.99	26.93	68	36.024	0.988

* Unit of measurement is μm except for diameter

Table 3: Factor influence on bore geometry.

	Factors	Percentage contribution				
		Diameter	Roundness error	Cylindricity error	Straightness	Surface finish
A	Axial length	0.000	9.083	0.000	0.901	0.000
B	Wall thickness	1.705	0.000	0.000	2.067	0.000
C	External force	13.278*	6.959*	14.835*	20.224*	0.000
D	External moment	31.816*	53.786*	40.350*	3.286*	0.000
E	Gimbals centre	5.074*	0.000	0.000	0.000	3.337*
F	Material	2.988*	0.000	0.000	0.721	0.367
G	External geometry	0.000	1.967	0.000	0.000	3.523*
H	Dressing	0.000	0.000	0.733	0.000	5.335*
J:	Grit size	1.288	0.000	14.007*	13.859*	36.451*
K	Sleeve parallel length	0.000	0.000	0.000	3.246*	1.440
L	Plain or grooved	2.454	0.000	0.000	0.000	1.796
M	Reamer rotational speed	0.000	0.000	0.000	0.000	4.686*
N	Axial feed rate	0.000	0.757	3.931*	0.715	3.847*
O	Axial feed increment	2.337	3.419*	0.000	0.000	0.000
P	Stock removal	7.878*	7.114*	0.590	0.000	16.927*

* Influencing variables at 90% confidence level

Deterioration of abrasives on the reamer sleeve can also lead to poor surface finish due to a decrease in the fraction of cutting points (Figure2). Lower mechanical strength of the cutting points would therefore contribute to rapid deterioration. This could result, in an unsatisfactory shape of the cutting point, a situation which is likely to produce a rough surface. The presence of defects, such as cracks, in the abrasive particles would also have a major influence on the strength of the cutting point.

Figure 3 shows that swarf removed during the reaming process fills all or part of the interstices between abrasive grits. This limits the cutting action. The effect would be felt especially for fine grits and presumably, for large sample diameters. Also insufficient space for the swarf to be accommodated between the interstices causes the swarf fragments to re-weld to the reamer surface. This occurs quite severely with aluminium specimens (Figure4). Such fragments cannot be removed easily. When the reamer is used again, this affects the cutting efficiency.

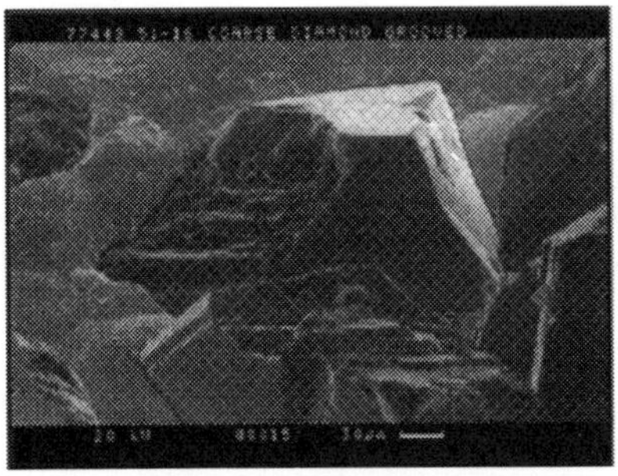

Figure 2: Fractured abrasive cutting point after reaming

Figure 3: Swarf inclusion- coarse grains

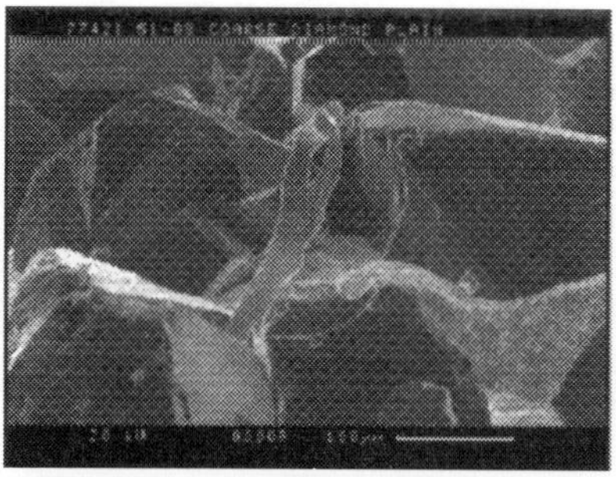

Figure 4: Aluminium swarf adhered to the abrasives

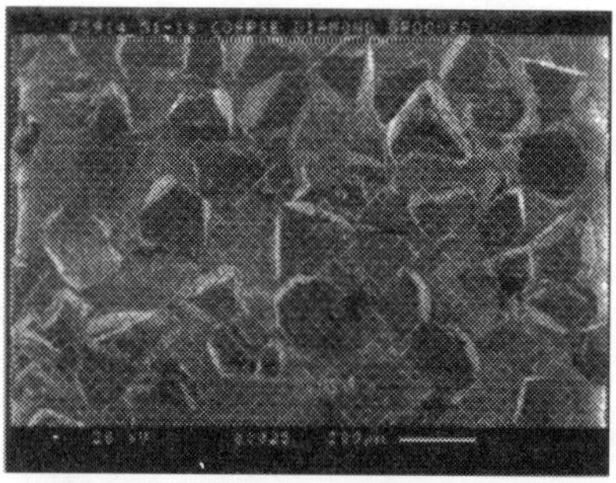

Figure 5: Spacing between coarse grains

Lal's [6] experiments in centerless grinding with four different sizes of abrasive grains, identified that in each case, effective abrasive grain radius decreased with increase in size. He concluded that the grain size will not only influence the transverse radius at the tip of a grain but also the spacing of effective cutting points (Figure 5) and these effects will be in opposite directions. The net effect of an increase in nominal grain diameter will be to increase surface roughness. Results indicate reamers with finer abrasive size produce smaller Ra values whereas coarser grits increase the surface roughness. It can also be said that surface topography or surface finish produced by SAR is a result of the reamer profile on to the bore surface which has a specific motion relative to the former. Hence the surface finish is governed by both the controllability of relative motion between reamer and workpiece and the transfer characteristic of the tool profile.

The only process variable significant for its effect on bore diameter is depth of cut and the pivoting point of the gimbal centre. Feed increment and plain or grooved nature of the sleeve offers just about 2 percent. Material type has a significance of 3 percent. Being a soft material, aluminium is prone to diametrical expansion just because the grits tend to chew off more material.

In trials where an external moment of force was introduced, the final bore diameter is slightly larger than expected. For example in trial no.6 the depth of cut set was 20 μm. Bore diameter before the reaming was 18.955 mm. In fact after reaming the bore diameter should be 18.975 (18.955 + 20μm) mm. But metrology shows 19.000 mm, 5 μm larger. This is because the introduction of external force and moment caused overcutting which increased the bore diameter. It could be argued that absence of external force or moment should produce better size tolerance. But this is not true from experimental results. For example in trial 16, where the external force and moment is zero the bore diameter produced was 27 μm larger. In trial 13, the diameter increased by 11 μm. Miller [2] argues that this is an indication that the dressing nature of the abrasives is an affecting factor of bore diameter. But this situation is also similar with abrasives dressed with ND. In trial 9, where CBN is dressed with CBN, bore diameter was 49 μm larger than what it should be. This is not clear why but it is suspected that torque produced by the external moment could be a reason. Analysis of main effects showed that dressing nature of abrasives produced only an average of 0.003 microns difference in bore diameter.

ANOVA identifies the major influencing factors on roundness as external geometry, external force and the grit size. When external force and moment were introduced, despite the floating characteristic of the reaming fixture, the specimen was forced to sway to one side causing the reamer to cut excessively on one side and less at the other side. Besides, the external moment introduced a torque which lead to aggressive reaming. This scenario affects the roundness and cylindricity and the straightness of the reamed bore. In trials 4,5,6,11 where both force and moment were introduced, the roundness and cylindricity error produced are worse in comparison to those produced in trials 1,2 and 3 where toque was nil.

Cutting force on an abrasive grain in SAR, like any other abrasive process, is usually very small, may be at sub Newton level or less (very difficult to measure). In principle, this is equivalent to that on a single abrasive grain in grinding [7]. The cutting force reflects clearly the chip removal process and is an important physical parameter for understanding cutting phenomena in SAR. When side load is introduced onto the specimen, the cutting force on the abrasive grain increases linearly when it reaches the area where the side load acts. Thus the contact pressure of the reamer with the workpiece is high on this area. Aggressive cutting takes place at this region whilst undercutting takes place on the remaining portion of the bore. Chip formation takes a different form and this is evident from figure 6, showing swarf collected from trial 16. Big chips indicate formation during the aggressive phase and small chips during normal phase.

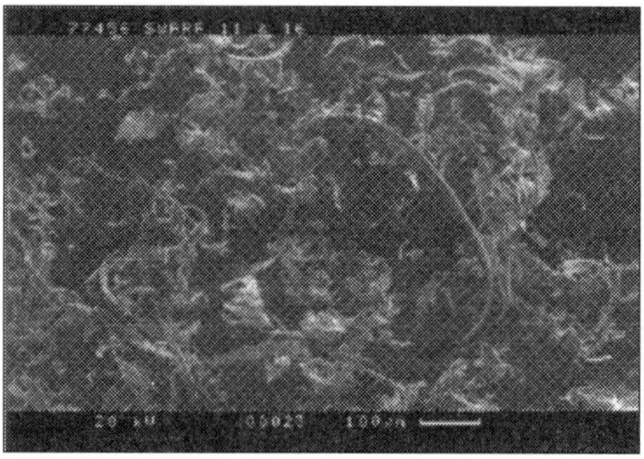

Figure 6: Swarf from irregular reaming

In order for controlled influence of the bore geometry it is therefore necessary to control the contact pressure. It can be justified that this kind of irregularity in the cutting behaviour of the reamer tends to seriously affect the roundness, cylindricity and the straightness. From analysis it is also identified that depth of cut is another influencing parameter. Combination of existence of the side load and bigger depth of cut tend to affect the bore geometry. It is quite unclear how the axial length of the specimen influences the roundness whilst effecting no other aspects of bore geometry. Poor roundness and cylindricity is obtained from many trials regardless the size of the specimens. This could be due to interaction effect between variables or uncontrollable noise factors.

5 CONCLUSION AND FURTHER WORK

From experimental results and ANOVA analysis, the major process variables which have strong influence on overall bore geometry have been identified. as the grit size, reamer rotational speed, feed rate, amount of stock removal, and axial length of specimen. The next stage of experiment is to perform a complete factorial experiment to study the effects of selected variables and interactions on bore geometry. Having confirmed from analysis that the torque produced during reaming severely affects the bore geometry, it has been decided to eliminate introduction of external forces on to the specimen in future investigations.

6 ACKNOWLEDGEMENT

This project is supported by the International Scientific Co-operation of the European Commission.

REFERENCES

1. S.Arunachalam and J.M.O'Sullivan, *Superabrasive Reaming - An Innovative Technique for Bore Finishing,* Industrial Diamond Review, Vol.57, No.574, pp.76-80, 1997
2. J.Miller, *Superabrasive Reaming of Bores In Cast Iron,* Industrial Diamond Review, Vol.56, No.2, pp.52-57, 1996
3. S.Arunachalam, J.M.O'Sullivan and J.Miller, *Design of Equipment for Experimental Investigation of Superabrasive Reaming Process,* 13th NCMR, Glasgow, pp.1-5, 1997J
4. K.R.Ranjit, *A Primer on the Taguch Method,* Van Nostrand Reinhold, 1990
5. S.S.Pande, A.R.Naik and S.Somasudaram, *Some Investigations on Plunge Feed Centreless Grinding Process,* International Journal of Production Research, Vol. 30, No.12, pp.2825-2837, 1992
6. G.K.Lal and M.C.Shaw, *On the Dressability of Abrasive Grains,* International Journal of . Machine Tool Design, Vol. 13, pp.131-140, 1973
7. T.Ueda and A.Yamamoto, *An Analytical Investigation of the Honing Mechanism,* Journal of Engineering for Industry, Vol. 106, pp.237-241, 1984

An investigation into the design flow problems in multi-chip module technology

R AZIZ and **J C BEAMISH**
Department Manufacturing Systems Engineering, University of Hertfordshire, Hatfield, UK

ABSTRACT

Currently, one of the most time-consuming tasks in a multichip module (MCM) technology design flow, is trying to find the necessary components in *known good die* (KGD) format. This task alone can take several months. This paper presents an overview of the MCM technology and illustrates the technological benefits which can be derived by the application of this technology. The paper further identifies the existing design problems within the MCM industry and highlights the methodology undertaken to identify the problems. The findings show the working relationship within the MCM supply chain needs to be analysed and enhanced so that the design-flow time within the MCM industry is substantially reduced.

1 INTRODUCTION

Semiconductor chips are the main hardware input for the majority of the electronic equipment and systems industry. Thus, they are widely used in industries such as consumer electronics, telecommunications, computers, aerospace, defence, instrumentation, medical equipment and automotive. However, the demand for increased performance and circuit densities of semiconductor chips has become an ever more important factor in the manufacture of commercially viable electronic equipment and systems. Furthermore, the increased performance and circuit densities of semiconductor chips require comparative technological advancements in chip interconnection. The conventional interconnection approaches such as printed circuit boards (PCB), hybrids and surface mount technology (SMT) cannot fulfil these requirements. Consequently, a new generation of interconnection packaging called Multichip Module (MCM) technology has emerged.

Multichip Module technology is a functional electronic module comprising of several bare semiconductor chips that are interconnected to a high density substrate to provide a single package. The nature of the substrate may be either ceramic, laminate, silicon or a combination of these (1).

The quantum leap from packaging single semiconductor chip onto PCBs to packaging bare semiconductor chips onto MCMs has lead to dramatic improvements in the performance and reliability of an electronic system. The performance gains are largely due to the reduction in signal delay, since the inteconnect distances between the bare semiconductor chips is much shorter than the distance between the chips on a printed circuit board. The reliability improves because there are fewer package levels. Moreover, the technology, due to its fabrication structure has lead to a reduction in the weight and size of the system. These reductions are due to the elimination of packages for the individual chips, and hence the closer placement of the chips within the MCMs. The reduction in size and weight and number of modules per system achieved with MCM technologies is the greatest ever achieved in the history of electronic packaging. Thus, it is widely anticipated that multichip module technology will be a major force in the electronic equipment and systems industry in the next millennium (2-3). The benefits of MCM technologies in comparison to the traditional technologies are summarised in Table 1.

Almost all efforts in the area of MCM technology have been devoted to technological issues such as establishing new fabrication processes and materials to improve the technology, with little having been done to improve the design flow process. Moreover, their is a question mark, as to whether the existing commercial electronic design automation tools can accommodate the physical design requirements of MCM technology. This paper will focus on MCM technologies from a system's perspective, and will address the problem of the current lack of a structured and automated approach to an optimised MCM design flow.

Table 1 - Benefits of MCM technology in comparison to traditional technologies

Benefits	**Characteristics**
Electrical Performance	In comparison to SMT and PCB solutions the circuits are maximised in smaller space due to closer placement of chips within the circuit
Reduced Size	Very compact at system level. Approximately 5-10 times smaller in area then SMT solution
Improved Reliability	Total number of interconnection and length of wiring are minimised compared to PCB and SMT solutions
Higher Density	The physical distances across which signals are transmitted is reduced in comparison to PCB and SMT solutions
Improved Packaging Efficiency	Significant increase in the ratio of silicon chip area to substrate area (30-80%). The ratio is 5-25% with traditional technologies

2 DESIGN ISSUES

This section highlights the operations that are perilous to the MCM design flow. These are; the availability of information on component description (4), and the ability of the existing commercial electronic design automation tools to accommodate the physical design requirements of MCM technology (4-5). These two issues are discussed below; and hence, constitute the major area in which research is undertaken by the authors.

2.1 Availability of Information on Component Description

One of the most time consuming and error-prone tasks during an MCM design flow is the availability of the information needed for MCM design and the data entry into the design framework (6). Information on each component in a design must be obtained. Great care must be taken to ensure that the data is correct and corresponds to the revision of the die being used in the die (7).

While such information on packaged parts for PCBs is readily available, bare semiconductor chip or known good die (KGD) data must be gathered from integrated circuit (IC) manufacturers or die distributors and entered from scratch. Unfortunately the design data provided by some IC manufacturers is still very poor. Often the design data are provided as facsimile copies of bonding diagrams, without indication of scale, pad co-ordinates and pad identification. However, this information is extremely important during the layout phase for the correct generation of the physical part and during the early design phases. Other important data such as die size and total number of pads also need to be considered, in order to compare different alternatives and take optimum design and manufacturing decisions. This clearly shows that the die information is currently specified to the CAD system manually, thus, increasing the time to market.

Consequently, the MCM design tools do not contain any physical part libraries nor electrical symbol libraries. Automatically generating such component libraries is a faster and more accurate approach than the manual technique mentioned above. The purpose of such MCM component model libraries would help develop the industry infrastructure to provide for the free flowing and efficient availability of component models and data needed to support the emerging MCM industry (8).

2.2 Commercial Electronic Design Automation Tools for MCMs

Computer Aided Design (CAD) tools are required to manage the design task and establish a comprehensive data base for design simulation, analysis, manufacturing and testing of MCMs. Currently, most of the available MCM CAD systems are adaptations from PCB CAD systems. However, practical experience with these MCM design tools has shown that the environment they provide to the designer is far from complete and solutions where vendors are combined only partly solve these problems. Table 2 summarises the features of PCB tools in comparison to MCM design requirements (9).

Table 2-Features of PCB Tools in Comparison to MCM Design Requirements

Design Issues	PCB Tools ability to cater MCM design Requirements
Physical Design hierarchy handling	No
Schematic design hierarchy handling	Yes
Placement and route	Has major problems
Graphical database ready for mask tooling	Problematic
Special purpose mask support	No
Transmission line support	Yes (requires substantial improvements)
Accurate transmitter/receiver modelling	No
Simultaneous switching noise modelling support	No
Substrate testing support	No
After assembly testing	No
Physical design verification	Yes (can't handle multilayer rules)

The physical design tools derived from PCB packages are often unlikely to have the fine database resolutions and flexibility necessary for MCM design. In general, physical design tools like routers need to be rules driven in order to ensure that manufacturable designs are created. The finer design rules associated with MCMs and the higher performance that usually accompanies their use require new routers which understand the rules and constraints required by MCM design. In addition to the general design issues expressed above, MCMs pose unique electrical and thermal design problems (10).

3 RESEARCH METHODOLOGY

This section describes the research method used to obtain the necessary information from MCM CAD tool suppliers regarding their ability to supply sufficient commercial CAD tools for MCM design requirements and the obstacles preventing the CAD tool suppliers from obtaining the necessary data in design format.

The principal organisational research designs are; experimental research, case study/qualitative research, triangulation, and survey research. However, the researchers felt that the triangulation method was the most suitable for this research project. Denzin (11) used the term triangulation, a term borrowed from navigation and military strategy, to argue for the combination of methodologies in the study of the same phenomenon. The concept of triangulation is based on the fact that any bias inherent in particular data sources, investigator, and method would be neutralised when used in conjunction with other data sources, investigators, and methods (12). In this study, a survey method, using semi-structured interviews, followed by a demonstration showing the capability of the CAD vendors' tools was deemed the most suitable method.

3.1 Research instrument

The surveys in the present research work were based on close-ended questions. The major advantages of this question style according to Adams and Schavendt (13), are the ease of completing the questions; brevity of response time; specification of the frame of reference for the subject and ease in scoring, coding, and tabulation. Furthermore, Bailey (14) suggests that some respondents may have a difficult time in writing an answer that reflects their feelings even if they are motivated and willing to participate, hence the need for close-ended questions. An instrument (or protocol) was developed to be used in the survey for the following reasons. First it operates as an *aide-memoire*, ensuring that the same aspects are covered with all the subjects, thus increasing the reliability of the research. Second, it helps focus the process of data collection on the relevant issues. Third, it makes it easier to register the information collected.

The main priority of the work was to attain an understanding of CAD tool suppliers ability to supply sufficient commercial CAD tools for MCM design requirements and the obstacles preventing the CAD tool suppliers from obtaining the necessary data in design format. With this in mind, it was thought that, if the researchers adopted the approach of asking the organisations to confirm or refute a particular design issue in question, the possibility would arise of inducing the organisations to accept the a priori ideas. This would increase the risk of jeopardising the internal reliability of the research. Consequently, in addition to undertaking the survey, each organisation was asked to demonstrate the capability of its commercial CAD tool, with reference to questions asked in the survey and design issues summarised in Table 2.

3.2 Research Procedure

The sample was drawn from a list of MCM CAD tool suppliers who were members of Printed Circuit Interconnection Federation (PCIF) Design Group or International Microelectronics and Packaging Society (IMAPS), two of the most significant trade federations for microelectronics, and electronic interconnection and packaging industries. It became apparent that only five companies were addressing commercial CAD tools for MCM design.

The initial approach to a company was by telephoning directly the individuals who were representing their company in IMAPS or PCIF design group, explaining the purpose of the interview and demonstration and asking the organisation to help. A date was then set for an interview and a demonstration with the companies willing to participate in the research. In all four organisations (out of a possible five) agreed to participate in the research.

4 FINDINGS

The findings from both, the semi-structured interviews and the demonstrations undertaken by the CAD tool vendors to assess the commercial capability of the tools to fulfil MCM design requirements, showed that the tools, in contradiction to much of the reported literature, were sufficient for designing MCM technologies. Thus, the design issues presented in Table 2 were easily accommodated by the participants' CAD tools. All the CAD tool suppliers have been updating the CAD packages so that they could accommodate MCM design requirements.

Several CAD tool suppliers reported that the problem lay in attracting customers (MCM manufacturers) to use their CAD tools. The CAD tool suppliers felt that the MCM manufacturers were reluctant to use their CAD tools because they were under the (false) impression that the tools were priced too highly and as yet could not cater for all their requirements.

However, the CAD tool suppliers were having severe problems in obtaining information on the availability and specifications of KGD. Thus, the CAD tools did not contain any automated libraries containing such information, thus increasing the design flow time. It became apparent that the die manufacturers were reluctant to expose information to the CAD tool suppliers about the die size, geometries, etc., for reasons unknown to the CAD tool suppliers. Furthermore, when the CAD tool suppliers were asked if they were aware that a database (Die Information Exchange) was being constructed to carry the required information, it became apparent that the CAD tool suppliers were not aware of this activity. This clearly shows that their is a lack of working relationship between the MCM supply chain, i.e. between MCM manufacturers, bare die suppliers and CAD tool suppliers. By working closely together, the entire supply chain can reduce the design flow time, and hence reduce the time taken for the MCM to enter the market.

5 FURTHER WORK

The interviews and demonstrations have clearly shown that the existing CAD tools are commercially capable of designing MCMs. The greatest obstacle remains in obtaining the information on bare dies from bare die suppliers/manufacturers. Moreover, the communication and trust between the groups in the MCM supply chain is extremely limited. Further work will be carried out to establish a framework to analyse and improve the relationship between the interfaces (MCM CAD tool suppliers and Die manufacturer/suppliers) in the MCM supply chain. The researchers have asked the members in the supply chain to participate in the project. The response has been extremely positive.

6 CONCLUSION

Whilst MCMs have the potential to supply the electronic packaging industry with the required performance, density, and size reductions; limitation imposed by the existing design flow hinders the time to market MCM technology and hence limits the widespread adoption of this technology. The authors have shown that lack of communication between MCM CAD tool suppliers and bare die suppliers is responsible for the time consuming MCM design flow. The essence of future work will be to analyse and improve the working relationship between the two groups in the supply chain.

REFERENCES

1. Aziz, R., Beamish, J.C., Spurgeon, D. C., Implementation of Multichip Module Technology in SMEs, *Advances in Manufacturing Technology X- Proceedings of the 12th NCMR*, pp 281-285, 1996
2. BPA (technology and management) Ltd. *Technology Roadmaps for Different Packaging Application Report*, 1995
3. Frost and Sullivan, *World Multichip Module Markets Report*, 1992
4. Ringoot, E.,Truzzi, C. and Beyne, E., A System Level approach to a Structured MCM Design Methodology, *Proceedings of IEEE Workshop on Chip Package Co-Design (CPD 98)*, 24-26 March, 1998
5. Mazullo, T., Introduction to MCM Design, *Proceeding of 3rd Annual PCB Design Conference*, 1994, pp 85-92
6. Truzzi, C., The Good -Die Design Database, *The Good-Die Newsletter*, Number2/March 1997
7. Ladd, C., Advanced MCM Design, *Printed Circuit Design*, August 1995, pp 10-16
8. Demmin, J.C., Faster, Cheaper MCM Development, *ICEMM Proceedings*, 1993, pp 119-124
9. Sherwani, N., Yu, Q., Badida, S., *Introduction to Multichip Modules*, John Wiley and Sons, Inc., New York, 1995
10. Licari, J.J., *Multichip Module Design, Fabrication, and Testing*, McGraw-Hill, Inc., New York, 1995
11. Denzin, N.K., *The research act: A theoretical introduction to sociological methods*, 2nd Edition, New York: John Wiley, 1978
12. Jick, T.D., Mixing qualitative and quantitative methods: Triangulation in action, *Administrative Science Quarterly*, 24, 1979, pp 602-611
13. Adams,, G.R., and Schvaneveldt, J.D., *Understanding Research Methods*, Longman Inc., New York, 1985
14. Bailey, K.D., *Methods of Social Research*, New York; The Free Press, 1978

Selection choices and applicability of manufacturing control systems in an injection moulding environment

I BANIK, A DAVIES, E G JOHN, and **M USENER**
School of Engineering, Cardiff University, UK

ABSTRACT

This paper examines the problems of selecting a manufacturing control system. It focuses on the criteria and cost benefits of selecting the various options and discusses the difficulties of implementation.

1. INTRODUCTION

Manufacturing control systems can vary from simple paper based systems to large integrated computer based systems. Selecting an appropriate system can be a laborious process which SME's limited by their resources, both in terms of capital and people, may struggle to complete satisfactorily.

This paper reports on a study undertaken to identify and select an appropriate manufacturing control system for a plastics manufacturing company. The company produces an extensive range of durable plastic items using both compression and injection moulding technologies. The company has an annual turnover of around £5.5 million and employs about 160 people. Products vary in both size and colour which creates a product range of approximately 1,500 items each of which has a fairly flat bill of material, typically 2 to 4 levels. The aim of this study is to highlight and discuss some of the problems of identifying and then selecting a manufacturing management control system.

2. OPTIONS

Before discussing the procedure of selection, the available options must first be identified and considered. The options available can be divided into three categories, each of which is detailed and their corresponding benefits and limitations examined. However it must be remembered that the benefits and limitations of any such system will be influenced by, and depend upon, the environment in which the implementation is undertaken.

The options available are :- 1. Upgrade the Existing System.
2. Invest in a Bought Out System.
3. Develop a system In-house.

2.1 Upgrade the Existing System

The computer system currently in use at the company, was installed well over ten years ago. It is written in COBOL and operates in a BOS (Business Operating System) environment. Its data libraries contain product, supplier and customer information. The system is fundamentally a database with the facility ('Reporter') to produce simple reports to meet the information needs of decision makers within the company. The system was not designed to create on-line production plans and therefore is unable to provide an adequate managerial control facility. The source code is not available to the company and therefore an upgrade of the existing computer system would only be possible with their input.

Benefits:

- Employees/users are familiar with the existing system, resulting in minimum re-training.
- Upgrading to an integrated system would permit those additional computer and paper based systems, which have been developed to supplement the original software, to be incorporated into the new system.

Limitations:

- As previously stated, the original software is still under licence. Given this situation it is first necessary, that the licence holder is prepared to enter into a contract to produce what would be, in effect, new bespoke software to meet the changed requirements of the company. Secondly they should be prepared to undertake this on a competitive cost basis.
- The existing data structures are no longer appropriate or representative of the changed manufacturing requirements and would require a complete restructuring.
- The current system, both in terms of user interface and the ability to modify output formats is not particularly user friendly and would thus need a total re-examination.

2.2. Invest in a Bought Out System

The current market availability of manufacturing management and control software, for in particular the plastics processing industry, can be divided broadly into two categories, industry specific and general purpose systems, each having its own benefits and limitations. If industry specific software were to be considered, it would have the advantage that the provider should have adequate knowledge of the needs of the end-user, to better tailor the package to requirement. Conversely a general purpose system would afford a much broader menu of modules, but may require a significant directive contribution from the purchaser.

The second consideration of any Bought out system would be the decision of having to identify and quantify the degree of complexity that is necessary in the package to enable the business to operate efficiently, without over complicating its business operation. Typically the decision will range between simple systems such as MRP to totally integrated business control software.

Benefits:

- The level of support and accountability for the effective and efficient operation of a bought out system can be assigned, by agreement, between the user and the provider of the system. This can vary from outright purchase of both the software and hardware to a lease or rental arrangement.
- Bought out systems should also have the advantage of proven software, by its widespread use by the many customers of the software supplier and which should thus be easily and successfully implemented within the company.

Limitations:

- The costs of a bought out package can be unpredictable, since unknown costs, such as training and implementation, can be extremely difficult to predict at the evaluation stage of the project.
- The actual acquisition costs of a bought out package have also increased significantly with the expansion of capability and there is evidence to suggest that over provisioning of specification can happen.
- It is also necessary to consider that, given the recent developments in software, it may not be possible to implement a given selected package without a significant upgrade in existing company hardware.
- A bought out package could also necessitate organisational, attitudinal and procedural changes which may not be well received by the end users.

2.3. Develop a System In-house.

The final option is to undertake, in-house, the development of a bespoke system. This can only be considered because of the extended availability of suitable general purpose database, spreadsheet and enquiry software packages, together with appropriate interconnecting software such as netware. Naturally this option could not be considered unless there is sufficient expertise, available within the organisation, to undertake the work. In this particular situation the company was in collaboration with a university on a Teaching Company programme and as such it would be possible to redirect sufficient resource to provide the time and technical development expertise required to undertake the work.

Benefits:

- A bespoke system, created under the foregoing conditions, would enable the program to be designed to exactly meet the current needs of the company and to provide for future requirement.
- In house development would ensure access to the program code should future expansion or application be required.
- An in-house development and implementation programme can be tailored to meet the existing management pressures and any funding availability problems the company may have.
- Since the system will be developed on site, whilst production continues, opportunity will exist for unexpected problems which will occur to be identified and resolved, even during

the later stages of the implementation phase. Such opportunity would not exist if either of the other strategies were to be adopted.

Limitations:

- As with any in-house programme, it is difficult to be precise in regard of the time necessary to design, develop and implement a given system. If not properly planned and managed, such imprecision may cause difficulties and conflicts within the organisation.
- Again, as with any in-house development programme, and although due care and attention would naturally be assigned to the project there would be no guarantee of success.

3 OPTION SELECTION

The selection of which option is the most appropriate is, in itself, a difficult decision making process. This study suggests a methodology which, by comparing the various options, determines the most appropriate solution according to a given set of performance and measurement attributes. To achieve this the methodology first considers those requirement issues which are paramount to the company, then undertakes a comparison and matching exercise of the options to the requirements, and finally determines which option is most suitable given the foregoing conditions.

3.1. Selection Attributes

There have been many contributions [1,2,3]which detail advice on how such planning and control systems should be selected and implemented. One such contribution "An executive guide to manufacturing control systems selection" [4] suggests that there are two facets of the problem which should bear mention, and which should be clearly defined prior to choosing a system.

1. The strategic requirements of the company, i.e. those requirements which will ultimately improve the image of the company.

2. The functional requirements of the system, i.e. the specification and performance required of any system which are designed and selected to meet the specific needs of the company.

In contrast to section 2, which details the benefits and limitations of the available options, this section defines the selection attributes of any given option. These are grouped under generic headings (1. and 2. above), and detail the paramount requirements, that is the selection attributes, of the company. Cost requirements have been incorporated to represent the financial limitations of the company.

Strategic Requirements

- Reduce the throughput order time to customer.
- Improve internal managerial control of production.
- Enable improved customer order status information.
- Enable reliable delivery schedules to be generated.

Functional Requirements

- User Requirements
    - Generate a comprehensive production operations plan.
    - Permit on-line data updates and status regeneration.
    - Operate via a user friendly interface.
    - Produce schedules which are compatible with the operational environment.
    - Enable variance scenarios to be modelled.
- Design and Implementation Requirements
    - The design and implementation of the system should be clearly defined and understood by all concerned.
    - The implementation time schedule should be in accord with that identified by the company.
    - System integrity should be proven by parallel operation prior to live operation
- Future Requirements
    - The system should be upwardly compatible in respect of projected software and operating environment developments and portable in respect of changing hardware availability.

Cost Requirements

- At all times during the project development the system must remain within the specified limits of capital and budget conditions.

3.2 Comparison of Available Options

The purpose and ultimate objective of the selection process is to decide upon which of the possible options is the most appropriate to the needs of the user i.e. to determine that best option relative to the others. The key word is 'relative', the absolute benefits are not crucial, the important consideration is which of the options will best meet the existing and future managerial and production control needs of the company and therefore ultimately provide the greatest contribution to long term profitability. Table 1 illustrates the comparisons that were made between the various option choices, the first is a comparison between the requirements themselves, and the second compares the available solutions.

Comparison of Selection Attributes (Weightings)

Of the performance criteria defined in section 3.1, it is evident that certain of the requirements are more important to the feasibility and ultimate success of the project than others. In an attempt to quantify the decision making process, it was decided that a weighting scheme should be devised which could then be applied uniformly to each of the identified attributes.

The weightings are such that the higher the weighting, the more important is considered the requirement. As with any such scheme, both the number of assignable categories and the weights assigned to each is arbitrary. For the purposes of this study it was decided that the number of categories should mirror the initially identified attributes and that there would be five categories. viz Cost, Strategic, User, Future and Implementation.

Protracted discussions with company management, suggested that the relative weights which should be assigned to each of these five categories would be:-

Cost	=	0.4
Strategic	=	0.2
User	=	0.2
Future	=	0.1
Implementation	=	0.1

Although this assignment of weights does permit each of the important attributes to be included in the analysis, it would be prudent to undertake a sensitivity analysis on the assigned weighting factors in order to determine how varying each affects the eventual decision outcomes.

Comparison of Available Solutions (Relative Value)
With certain of the identified attributes, such as Cost, comparison can be relatively easy to undertake since it can be assisted by the absolute values involved, however with other of the attributes such as Future requirements, the situation is more problematic. First there is the obvious difficulty of attempting to quantify the actual benefit of selecting a particular solution choice and secondly there is the difficulty of estimating the effects of present and future unpredictability. This problem is not restricted to the subjective attributes, but can also occur with certain of the objective attributes such as Cost. For example, it might be apparent that initially, one of the options is the least expensive, however what may not be so apparent is that another option may be more unpredictable in respect of implementation resulting in a change in the cost balance of the project.

Exact figures cannot be put to other criteria, however experience will allow a comparison to be made between the choices enabling the assignment of a value between 1 and 3 (relative value) in each case.

Explanation of Comparison Procedure (Table 1)
In table 1 a risk value has been used to compare the solutions, the lower the risk value the more suitable the solution. The risk value defines the element of risk of each solution for each selection criteria.

$$Risk\ Value = Weighting * Relative\ Value$$

For example, in Table 1, 1.2 is the highest risk value and corresponds to the cost requirement of a bought in system. This reflects the fact that cost is the most important requirement to be met and a bought in system is the least likely to meet it. The final column of Table 1 totals all risk values for each solution and in this case the development of a bespoke system is shown to be the most appropriate i.e. it has the lowest value.

Having decided which option to implement, the selection procedure can be repeated within that option. The development of a bespoke system could be tackled in different ways and the most appropriate method may not be obvious. If a bought in system had proved to be most appropriate then the defined selection procedure could be used to decide which of the numerous packages available best fit the company requirements.

Table 1 - Comparison of Options using defined Selection Criteria

Selection Criteria		Cost	Strategic	Functional			TOTAL
				User	Future	Implementation	
Solutions	Weighting	0.4	0.2	0.2	0.1	0.1	1
Current System	Relative Value	2	2	3	3	1	
	Risk value	0.8	0.4	0.6	0.3	0.1	2.2
Bought In	Relative Value	3	1	2	1	3	
	Risk value	1.2	0.2	0.4	0.1	0.3	2.2
Bespoke	Relative Value	1	3	1	2	2	
	Risk value	0.4	0.6	0.2	0.2	0.2	1.6

Key: Relative Values 1 = least risk ⟶ 3 = highest risk

4 CONCLUSION

Difficulties in selecting the most suitable manufacturing control system stem from an uncertainty of the real value of the benefits and limitations of the different options. A simple methodology, introducing the use of relative values and thus eliminating such uncertainty, has been used to successfully select a control system in an industrial injection moulding environment.

REFERENCES

1. Browne J., Haren J., Shivnan J,. "Production Management Systems", (1996), Addison -Wesley, ISBN 0-201-42297-2
2. Rembold, U., Nnaji, B. O., Storr, A., "Computer Integrated Manufacturing and Engineering" (1993), Addison - Wesley, ISBN 0-201-56541-2
3. Cummings R., Ezingeard J-N., Grieve R. J. James M., "Manufacturing Information Systems Implementation in SME's: A Case Study", Proc. XII National Conf. on Manufacturing Research, Ed. Bramley A. N., Mileham A. R., Owen G. W., Sept 1996.
4. *An Executive Guide to Manufacturing Control Systems Selection*, Manufacturing Information Systems Working Party, The Institution of Production Engineers., 1982.

A tool design methodology for precision forging

F BONNAVAND
Laboratoire 3S, Domaine Universitaire Grenoble, France
A N BRAMLEY and **D J MYNORS**
Department of Mechanical Engineering, University of Bath, UK

Synopsis/abstract

The paper proposes a numerical tool design methodology for the precision forging process. The procedure takes into account the temperature, stress, and displacement fields associated with the warm forging process using die insert methods. Whilst commercial forging modelling packages can enable this, a high overhead of computation time is incurred. This paper demonstrates how the time can be reduced dramatically by applying separately the elastic and plastic modelling tools. An example is given of the application of the technique using FORGE2 and ANSYS for the design of tools for a backward extrusion process.

1 INTRODUCTION

Because of the competitive environment, forging companies always try to improve the profitability of their products and reduce the time required for tool design. There are two main directions to improve this profitability: firstly to improve the tool's performances, and secondly to improve the predictability and therefore the precision of the final product in order to avoid or minimise subsequent operations of material-removal. The design of forging tools is often carried out by doing trials which could cost a lot of money and require a lot of time. Numerical simulations of forging processes are one solution for saving money, and they can lead to a significant reduction of the time required to design tools. Numerical simulations are often used to predict the deflection of the dies (1), or for die stress analysis (2) and can give very accurate results, but are not optimised to the needs of the tool designer.

For the axisymmetric forging process, the objective of the designer is to determine the configuration and the dimensions of the dies (see Fig 1). In this paper, the inner shape of the die is assumed to be similar to the external shape of the desired workpiece. At the end of the proposed methodology, this inner shape can be updated in order to compensate for the die deflection. Consequently, for axisymmetric forging processes, the parameters which have to be found are the following: die material data, number of die inserts, interference values, radii of the different inserts, etc... Thus, the tools are defined by a finite number of parameters and the designer has to find the best set of these parameter values in order to optimise tool performance and produce the required shape.

The purpose of this paper is not to suggest the criteria which characterise tool performance. These are typically based on the stress, displacement, and temperature fields. This paper proposes a methodology which is based on the separation of analysis of the workpiece and the dies, which enables the fast determination of all these different fields in the dies for different sets of parameters. Afterwards the designer can easily choose the optimum solution according to criteria such as *lowest possible maximum equivalent stress, minimum radial deflection of the die*, etc...

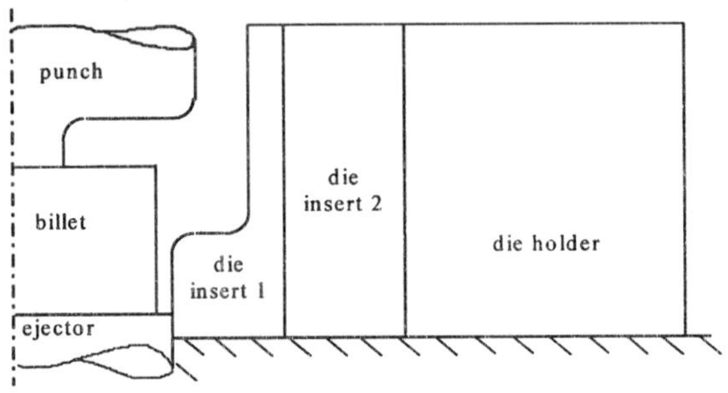

Fig 1: example of configuration of the dies for the axisymmetric forging process

2 GENERAL PRESENTATION OF THE METHODOLOGY

As seen above, the purpose of the following methodology is to find the temperature, stress and displacement fields in the dies for several sets of parameter values which define the die configuration. The computation time should be as short as possible in order to minimise the lead time of the industrial companies. The basic principle for designing the dies is to do one simulation with a forging modelling package for each set of values. However, this procedure is not applicable, because as the following example shows, it would require a very long time to apply it. For the simple case of the back extrusion process with one prestressed die insert and one die holder, the average computation time needed to run one simulation is about one hour. If the designer wants to determine three parameters (for example the outer radius of the die insert, the interference value, and the outer radius of die holder), and if each parameter can have ten values, one thousand simulations have to be run. Thus, the duration of the computation would in this case be at least forty days.

The proposed methodology is based on one main assumption that the results in the workpiece do not depend a lot on the configuration of the dies. Thus, it is not necessary to re-run the thermo-elastoplastic simulation of the workpiece (which needs a lot of time) each time a parameter value which defines the dies has been changed. In fact, it is possible to do the analysis of the workpiece and the analysis of the dies separately. At the beginning of the calculation, the user has to initiate the set of parameter values which define the dies. Afterwards, one simulation is done with a forging package using this configuration in order to determine the different fields in the workpiece. In fact, as shown in section 4, the choice of the initial values does not have a real influence on the results in the workpiece. Thus, the user can choose any reasonable set of values. Afterwards, for each set of parameter values which define the configuration of the dies, a thermo-elastic analysis of the dies is carried out. These simulations can be done with a classical finite element package. The boundary conditions, the thermal load, and the structural load are calculated by using the results of the analysis of the workpiece which is done at the beginning of the procedure. Fig 2 describes this methodology in the form of a flow chart showing the different operations which have to be done. At the end of the calculation, the designer knows the results in the dies for each configuration, and can choose the most appropriate one according to different criteria.

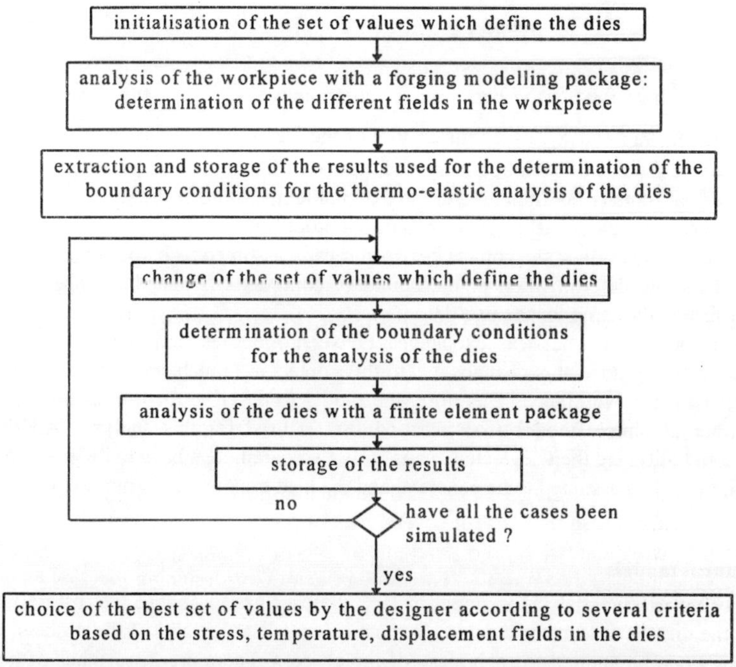

Fig 2: description of the methodology for designing forging tools

This methodology enables a lot of time to be saved. For example, the analysis of the dies with a finite element package requires about 2 minutes for the same case of back extrusion process as above. Thus, the computation time to run the one thousand simulations will not exceed one and half days.

The boundary conditions for the analysis of the dies are calculated by using the results of the analysis of the workpiece. The boundary conditions could have a great influence on the results in the dies. Therefore it is important to specify accurately the models used for their determination. Section 3 gives details of the models which are used, and explains the calculation procedure.

3 MODELS FOR THE ANALYSIS OF THE DIES

3.1 Thermals models

The analysis of the dies is thermo-elastic. Therefore, firstly the thermal analysis is done, and afterwards, the elastic analysis is done. The type of the thermal analysis is transient, because the temperature is time dependent. Therefore, all the boundary conditions have to be determined several times, and afterwards they are linearly interpolated between these times. There are three types of boundary conditions which correspond to three types of contact:

- Contact with the workpiece: the model used for the determination of these boundary conditions is the same as the one used in almost all forging modelling packages. The equation 1 gives the heat flux through the wall:

$$q = h(T_{workpiece} - T_{dies}) \qquad (1)$$

where q is the heat flux, h the heat transfer coefficient, $T_{workpiece}$ the temperature of the workpiece at the wall and T_{die} the temperature of the die at the wall. Most of finite elements packages enable the use of such boundary conditions. In fact, only two parameters have to be defined: the heat transfer coefficient, and the external temperature at the wall which is a result of the workpiece simulation at the beginning of the procedure.

- Contact between dies: the contact between dies is assumed to be perfect. Therefore the model used for the determination of the boundary conditions is that the temperature at the contact surface is the same for the two dies.
- Contact with air: the heat exchanges between the dies and the air are not very significant, because the heat exchanges due to this contact are much smaller than those due to the contact with the workpiece. Therefore, the model to describe them does not need to be very accurate. A simple model of convection flux is used for this analysis. All the other effects like radiation are then neglected. Thus, only two parameters have to be given: the room temperature which is assumed to be constant, and the heat transfer coefficient with the air.

3.2 Structural models

The type of the elastic analysis is static. In fact, the user has to choose the instant when the results in the dies have to be calculated. In most cases it is at the end of the process, but this methodology can be applied to each stage of the process. There are three types of boundary conditions:

- Contact with the workpiece: the friction between the dies and the workpiece is neglected. Therefore, only the normal pressure has to be defined. The normal pressure is read into the results of the analysis of the workpiece.
- Contact between dies: the contact between two dies is supposed to be without friction. Therefore the model just imposes a similar normal displacement at the wall for the two dies.
- Thermal load for thermal expansion calculation: the temperature field calculated in the previous analysis is input as a body load for the structural analysis in order to calculate the strain due to heat expansion. Most of the finite element packages enable this to be done automatically by transferring the results from the thermal analysis to the structural analysis.

All these models are used for the analysis of the dies, and it could take a long time to manually apply these boundary conditions. However most finite elements packages enable the user to write a small subroutine which can do that automatically.

4 VALIDITY OF THE METHODOLOGY

The validity of the methodology is characterised by the quality of the stress and temperature fields in the dies obtained by this method. In fact there are two things which can introduce some errors: the models used for the transfer of the data from the workpiece's analysis to the die's analysis, and the main assumption* that the different fields in the workpiece do not depend on the configuration of the die.

To estimate the quality of the results, the methodology was applied to a simple case of back extrusion with a single container. Two packages were used: FORGE2 for the elasto-plastic analysis and ANSYS for the elastic analysis. Three tests were carried out as described in Fig3.

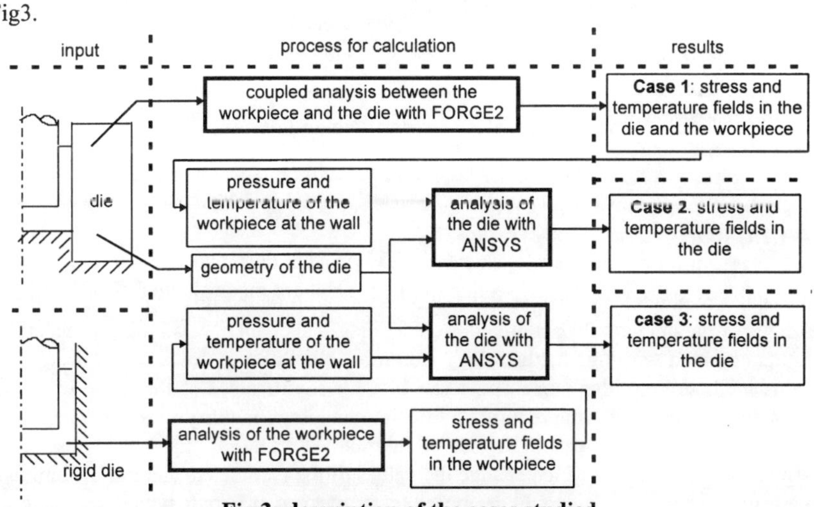

Fig 3: description of the cases studied

The results for case 1 are obtained by a coupled analysis between the workpiece and the die, with FORGE2. Thus, in this case there are no errors due to the transfer of the data and the main assumption*. Therefore, the results of the case 1 are considered as reference results, and the other cases will be compared against them. For the three cases, Fig 4 presents the equivalent stress along a horizontal line situated at the same height as the head of the punch.

For case 2, the configuration of the die used to determine the fields in the workpiece (and so the load for the die analysis with ANSYS) is the real configuration. Thus, even if the main assumption* (see beginning of the section) is not true, the results would be similar to the reference results. In fact, the errors are only due to the models used for the transfer of the data. As the equivalent stress for the case 2 is very close to reference one (it is impossible to see the difference on Fig 4), it can be assumed that the models used for this transfer of data does not introduce significant errors.

The results for case 3 are not so accurate. The maximum error occurs close to the die workpiece interface, and is about sixteen percent. This difference is due to the main assumption* on which the methodology is based. However, the configuration of this case is supposed to be the worst, because the die used to calculate the fields in the workpiece is rigid and at a constant temperature. If the characteristics of the die used for the analysis of the workpiece represent more accurately the true working conditions, the maximum error would be less than five percent. This accuracy is sufficient for design forging tools; and thus the proposed methodology can be used as an aid to optimise tool design.

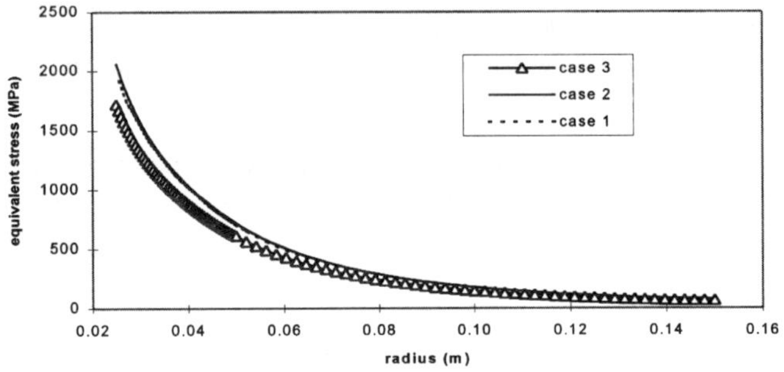

Fig 4: Equivalent stress along an horizontal

5 CONCLUSION

The presented methodology enables the fast determination of the different fields in the dies for each set of parameter values which define the dies. This paper shows that the accuracy of the results obtained with this methodology is compatible with the needs of the designer. By using this methodology, the gain in time can be very significant. For the given example in section 2, the computation time required by this methodology is thirty times shorter. Most of the finite element packages enable the use of subroutines written by the user. Thus, the user can write small subroutines which change the values of the parameters and run the simulation automatically. Therefore, it should be possible to make a program which can determine automatically the fields in the dies for different sets of parameter values.

REFERENCES

1. H. Kim and T. Yagu. Estimation of the dimensional differences between cold forgings and tools in backward cup extrusion of steel. Proceedings of 5th International Conference on Technology of Plasticity, pp: 433-436, 1996.

2. R Balendra and H Long. Prediction of geometrical accuracy of back-extruded components using FE simulation. Proceedings of Forging and Related Technology conference (IFCT'98), pp 183-190, 1998

Modelling surface roughness improvement in grinding

X CHEN and **W B ROWE**
School of Engineering, Liverpool John Moores University, UK

ABSTRACT

Grinding is a precision machining process used to improve workpiece surface roughness. This paper demonstrates that spark-out is the most important stage of a plunge grinding cycle for achievement of a low value of roughness. The nature of the spark-out process is investigated including the effects of machine stiffness, system deflection, wheel sharpness and plastic deformation on improvement of surface roughness. Experiments and a comparison with predictions from a computer simulation programme demonstrate the effects of the fundamental factors. It is demonstrated that the time constant for roughness is more than twice the time constant for system deflections.

1. INTRODUCTION

Grinding behaviour in a grinding cycle is strongly affected by the deflections, between the grinding wheel and the workpiece, resulting from grinding force and system flexibility. Selection of optimum grinding cycle parameters depends on the knowledge of deflection performance of a grinding cycle. A plunge grinding cycle consists of infeed and dwell stages. During the infeed stage, the wheel is fed into the workpiece at a constant rate. Due to the deflection of the grinding system, the reduction of workpiece size lags behind the indicated infeed position. This difference depends on the compliance of the grinding system and on the grinding force. In the dwell stage, the residual stock continues to be removed until the wheel is retracted. The roundness and roughness also improve during the dwell period. The deflection in a basic plunge grinding cycle is illustrated in figure 1. The grinding cycle can be divided into three stages including spark-in, steady grinding and spark-out. The spark-in and spark-out stages are transient processes. The grinding behaviour in these stages is indicative of the compliance of the grinding system. Spark-out is the most important stage for reduction of surface roughness in a plunge grinding cycle. Prediction of the period for the roughness to reach a minimum value requires better understanding of the mechanics of grinding performance during spark-out.

In order to describe the removal rate, the grinding system is often simplified as a first order linear system. The workpiece removal can be represented by [1]

$$\tau \cdot \ddot{r} + \dot{r} = v_f \qquad (1)$$

where r is the radius of the workpiece, v_f is the infeed rate, τ is the time constant of the grinding system removal rate. The removal rate time constant is a measure of the effects of the system stiffness k_e, the grinding force coefficient k_c and the workpiece rotational speed n_w.

$$\tau = \frac{k_c}{k_e \cdot n_w} \qquad (2)$$

This first order model of material removal can be used to control grinding power, workpiece size and roundness[1]. The effect of the grinding system compliance and the wheel sharpness are jointly represented by a single parameter, the removal rate time constant. To describe the improvement of surface roughness during spark-out may also be described approximately by a time constant. A commonly used model is

$$R(t) = (R_0 - R_\infty)e^{-t/\tau_R} + R_\infty \qquad (3)$$

where $R(t)$ is the roughness at the time t from commencement of spark-out, R_0 is the roughness at the commencement of spark-out, R_∞ is the asymptotic value of roughness during spark-out and τ_R is the time constant of the roughness decay curve. The roughness time constant is not usually equal to the removal rate time constant of the grinding system. Rowe[2] demonstrated that the roughness improvement curve depended on the stiffness of the grinding system. Salje, Mushardt and Scherf[3] also noticed the roughness improvement in spark-out did not follow an exponential decay as illustrated in figure 2. These facts imply that the surface roughness follows a non-linear relationship with the grinding depth. In this paper, the generation of the workpiece surface is investigated comparing experimental results with a computer simulation. The reduction of surface roughness during spark-out is discussed including the effect of machine stiffness.

2. GENERATION OF THE WORKPIECE SURFACE IN GRINDING

A ground workpiece surface is created when the abrasive grains on the wheel surface pass through the workpiece removing material. The mechanism of material removal consists of rubbing, ploughing and cutting. A grain transfers its shape onto the workpiece surface although only a proportion of the material in front of the grain is removed due to elastic deformation and plastic ploughing.

Local elastic deflection of a grain reduces the real cutting depth of a grain. Saini, Wager and Brown[4] revealed that the local deflections of the wheel when a grain is in contact with steel are of the same order of magnitude as the undeformed chip thickness. The variation of the deflection of the grain centre δ_c has a trend and scale similar to the total deflection. Therefore the deflection of the grain centre is considered as the local deflection in the analyses of both the dressing and the grinding process. Nakayama, Brecker and Shaw[5] described the deflection of the grain centre as following the form of a Hertz distribution,

$$\delta_c = C f_n^{2/3} \qquad (4)$$

where C is a constant and f_n is the normal force acting on the grain.

Ploughing displaces material to the side of the grain. The amount of displaced material depends on the workpiece hardness[6]. The proportions of rubbing, ploughing and cutting also depend on the grinding depth. Lortz[7] observed a dead zone in the area ahead of the grain. If the cutting depth of a grain is smaller than a critical value, no material is removed as the grain passes through the grinding zone. The critical depth of the dead zone depends on the grain shape and the friction between the grain and the workpiece. Due to the dead zone, the proportions of cutting ploughing and rubbing vary with the depth of cut. A large grinding depth increases the proportion of cutting.

During spark-out, the cutting depth decreases. The proportion of ploughing and rubbing increases. Because of the decrease of grinding depth, the ploughing function plays an important role in the spark-out stage. Ploughing squeezes the engaged material to the sides of the grain to form ridges. The ridges are removed by subsequent grains.

3. SIMULATION OF SURFACE GENERATION IN GRINDING

During grinding, each grain that cuts the workpiece removes a portion of the workpiece material and leaves a trace on the workpiece surface as shown in figure 3. The final workpiece surface is determined by accumulating the interactions of grains that pass the workpiece. When the workpiece passes through the grinding contact zone of length l_c, the distance travelled by the surface of the wheel is l_s.

$$l_s = \frac{v_s}{v_w} l_c \qquad (5)$$

where v_s is the grinding speed and v_w is the workspeed. This means that all the active grains over a length l_s engage with the cross section of the workpiece ABB'A'. When a cutting edge passes through a cross section of the grinding zone, less material is removed than implied by the undeflected cutting edge depth. Assuming the grain $G_{i,j,k}$ passes through the grinding zone at abb'a', the shaded area will be removed. After the grinding wheel passes through the cross section ABB'A', the remaining surface contour is the ground surface of the workpiece.

In grinding, only a proportion of the material in the interference zone is removed by a grain. The main influences are elastic deflection and plastic deformation. The elastic deflection is represented by equation 4. The remaining plastically deformed material piles up at the sides of the grain. It was assumed that the area of the material removed within a groove formed by a grain is proportional to the cutting area of the grain. The ratio of the volume of the material removed to the volume of the undeformed chip is defined as the cutting efficiency ratio β. The material piled up alongside the groove will be the proportion (1 - β) of the volume of the undeformed chip. Therefore the area A_p of the material remaining on each side of the groove is

$$A_p = \frac{(1-\beta)\cdot A}{2} \qquad (6)$$

where A is the cross sectional area of the undeformed chip. The shape of the displaced material was approximated by a parabola superimposed at each side of the groove.

The depth of cut decreases during spark-out. For a system represented by equations 1 and 2, the depth of cut in spark-out pass i will be[8]

$$a_i = \frac{a_{i-1}}{1 + \dfrac{1}{\tau \cdot n_w}} \qquad (7)$$

Grinding force can be estimated by considering the interactions of the grains that pass through the grinding zone[8,9]. The grinding force on each grain was estimated using a spherical grain model as illustrated in figure 4. The spherical grain model takes acount of grain sharpness with grain depth of cut.

$$f_{t,i} = \frac{3\pi}{4}\frac{b}{t}H(\frac{C'}{3})A(\sin\theta + \mu\cos\theta) \tag{8}$$

$$f_{n,i} = \frac{3\pi}{4}\frac{b}{t}H(\frac{C'}{3})A(\cos\theta + \mu\sin\theta) \tag{9}$$

where f_{ti} is the tangential force on the grain i, f_{ni} is the normal force on the grain i, b is the cutting width of the grain, t is the cutting depth of the grain, H is the hardness of the workpiece, C' is a constraint factor, A is the cross sectional area of the undeformed chip, μ is the friction coefficient and θ is the angle of the indentation force from the normal. The total grinding force is the sum of the individual forces on each of the grains that pass through the grinding zone to remove the required depth of cut.

$$F_t = \sum^{n} f_{t,i} \tag{10}$$

$$F_n = \sum^{n} f_{n,i} \tag{11}$$

4. RESULTS AND DISCUSSION

The basis of the computer program[9] to simulate the process of surface generation is illustrated in figure 5. Three functions are represented in the total grinding simulation. The first is the dressing process, which provides the location of the grains and the shape of the cutting edges. The second is the grinding process by the grains according to the kinematics and the deformations in the process. The third function is to represent the action of each individual grain as it contributes to the total grinding forces and the surface profile. The simulation demonstrates the grinding performance by accumulating the influences of the grain shape, the elastic and plastic deformation and the reduction of depth of cut during spark-out. In the results reported here, the function for wear of the grains is de-actived.

Figure 6 shows the comparison of simulated and experimental results during spark-out. The experiments were carried out on a Jones & Shipman Series 10 CNC grinding machine. Dressing lead f_d = 0.15 mm/r, dressing depth a_d = 0.015 mm. Wheel speed v_s = 33m/s, workspeed v_w = 0.25 m/s, infeed rate v_f = 10 μm/s. The computer simulation was undertaken with the same parameters. The simulation results can be seen to be similar in nature to the experimental results. If the improvement of surface roughness during spark-out is represented by an exponential decay, the average roughness time constant is 1.21 seconds. This value is more than double the removal rate time constant, which was 0.54 seconds. The difference between the two time constants for the removal rate and for the surface roughness demonstrates that surface roughness is not related proportionally to the system deflection. The rate at which the residual stock is removed is slowed down due to plastic pile-up and therefore slows down the improvement in surface roughness. The simulation results also show that the time constant for the surface roughness is larger than the removal rate time constant. Table 1 shows the effect of removal rate time constant on the improvement of surface roughness. A small removal rate time constant provides less improvement of surface roughness during spark-out. As indicated in equation 2, the removal rate time constant represents the effects of wheel sharpness and the machine system stiffness. For the same wheel sharpness, a small time constant represents a stiff machine, and less residual stock is left for spark-out to remove. This demonstrates that a small time constant leaves less margin for surface roughness improvement.

5. CONCLUSION

Investigation shows that the reduction of surface roughness does not follow the changes of the system deflection. To achieve best surface roughness requires longer spark-out than that for size and roundness control. The ratio of roughness time constant to the deflection time constant varies. Therefore for control of surface roughness, dwell time should be determined according to the roughness time constant.

In order to provide a comprehensive explanation of the improvement in surface roughness during spark-out, more investigations would be required to determine more precisely the effect of plastic deformation and non-linear performance of the grinding system.

REFERENCES

[1] Chen, X., Allanson, D., Thomas, A., Moruzzi, J. L., Rowe, W. B., Simulation of Feed Cycles for Grinding Between Centres, Int. J. Mach. Tools & Manufact., 1994, 34, 5, 603-616.

[3] Rowe, W. B., An Experimental Investigation of Grinding Machine Compliances and Improvements in Productivity, Proceedings of the 14th Int. MTDR Conference, Sept. 1973, 479-486.

[2] Salje, E., Mushardt, H., Scherf, E., Optimization of Short Grinding Cycles, The Annals of CIRP, 1980, 29, 2, 477-495.

[4] D. P. Saini, J. G. Wager and R. H. Brown, Practical Significance of Contact Deflections in Grinding, The Annals of CIRP, 1982, 31, 1, 215-219.

[5] K. Nakayama, J. Brecker, M. C. Shaw, Grinding Wheel Elasticity, Trans. of the ASME, Journal of Engineering for Industry, 1971, 93, 5, 609-614.

[6] Hamed, M. S., Grinding Mechanics - Single Grit Approach, PhD Thesis, Leicester Polytechnic, 1977.

[7] Lortz, W., A Model of the Cutting Mechanism in Grinding, Wear, 1979, v.53, 115-128.

[8] Chen, X., Strategy for the Selection of Grinding Wheel Dressing Conditions, PhD Thesis, Liverpool John Moores University, April, 1995.

[9] X. Chen, W. B. Rowe, B. Mills, D. R. Allanson, "Analysis and Simulation of the Grinding Process - Part IV, Effects of Wheel Wear", International Journal of Machine Tools & Manufacture, 1998, Vol.38, No.1-2, 41-49..

Table 1. Simulation results of improvement of surface roughness

System time constant	Roughness time constant	R_∞	$R_0 - R_\infty$
0.10 second	0.43 second	0.46 μm	0.15 μm
0.55 second	1.28 second	0.37 μm	0.32 μm
1.00 second	2.27 second	0.33 μm	0.35 μm

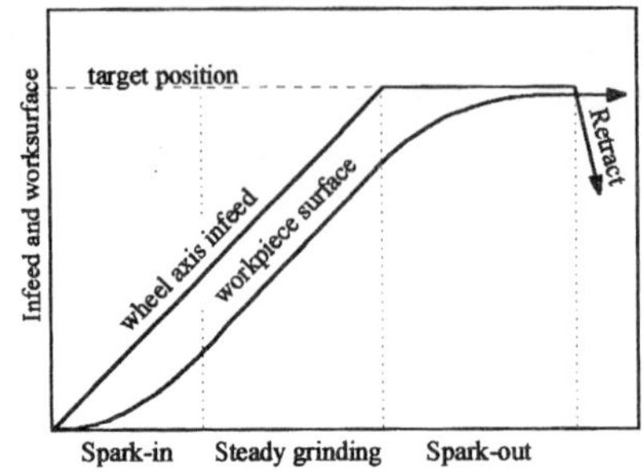

Figure 1. The deflection in a grinding cycle

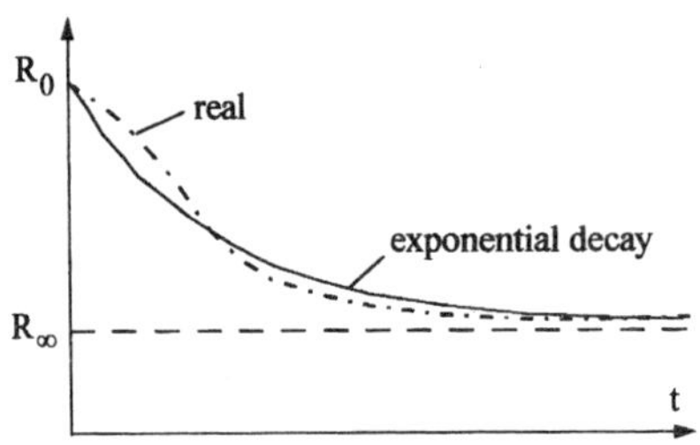

Figure 2. Reduction in workpiece roughness during spark-out

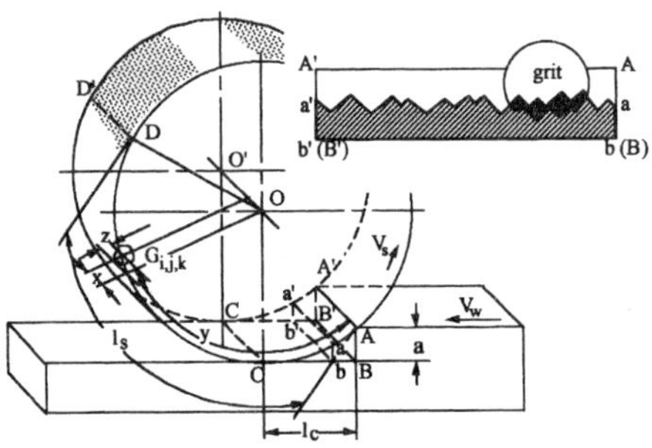

Figure 3. Generation of the workpiece surface

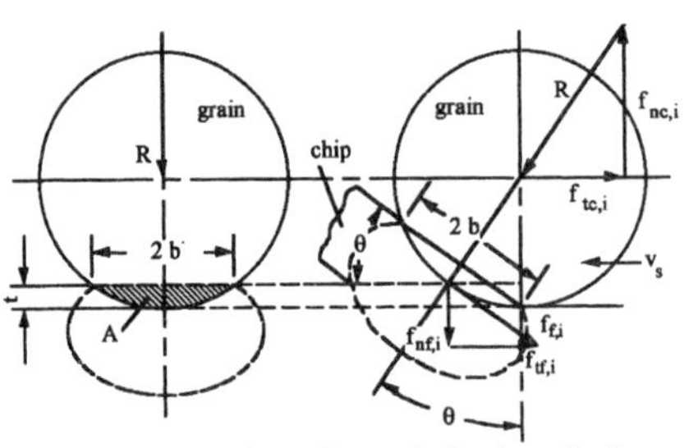

Figure 4. Penetration of a grain in the grinding process

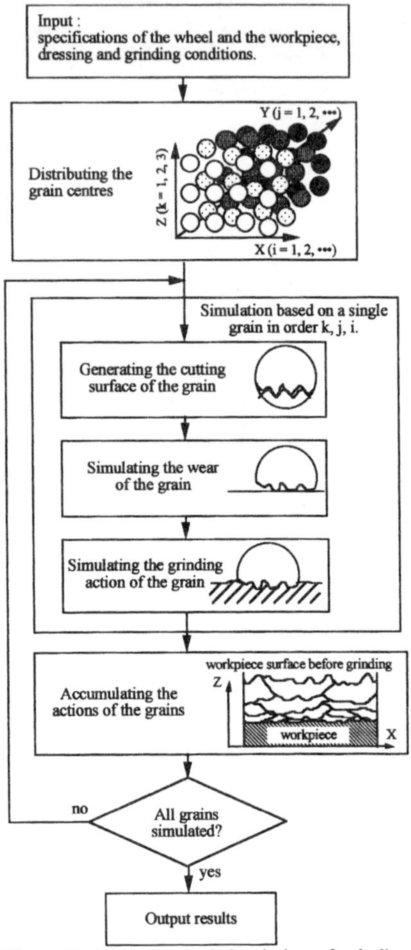

Figure 5. Flow chart of simulation of grinding process

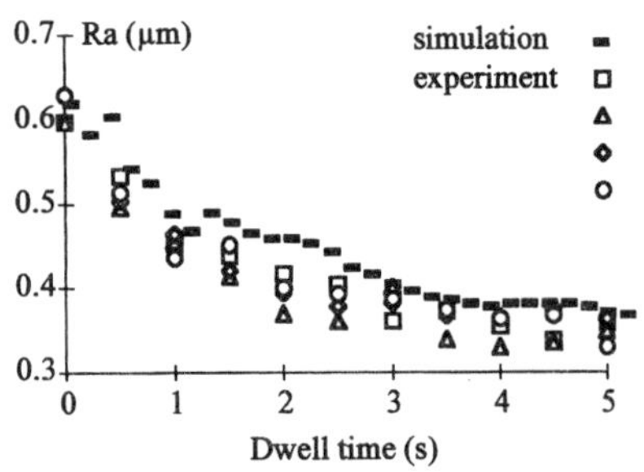

Figure 6. Roughness improvement during spark-out

Analysis of component forces in high-speed threading of Ni–Cr–Mo steel with coated carbide inserts

E O EZUGWU and **C I OKEKE**
School of Engineering Systems and Design, South Bank University, London, UK

ABSTRACT
Forces generated when threading a Ni-Cr-Mo (817M40T) steel with two grades (P20 - P30 and K05 -K20 types) of PVD coated carbide inserts under various cutting conditions were recorded and analysed. Other data obtained during the threading operation include tool life at specified cutting conditions, tool wear rate, the extent of damage on threaded surfaces and hardness variation at various locations of the threaded surfaces. The test results show that lower cutting forces were generated when machining with the single layer Al_2O_3 coated inserts consisting of a K05-K20 substrate and incorporating a honed edge. The edge geometry as well as the composition of the substrate tend to ensure lower tool wear rate, hence longer tool life in addition to minimum damage of the threaded surface with only slight deviation from the apparent hardness of the workpiece after machining.

KEY WORDS

Coated carbides, Steel, Threading, Component forces, Wear volume

1. INTRODUCTION

In metal cutting, each element of machining systems has a significant influence on the efficiency and reliability of the machining operations. Cutting speed, feed rate, infeed method (for threading), depth of cut, cutting forces and power comsumption as well as tool geometry are some of the parameters that can be controlled to justify the reliability and efficiency of cutting tool materials during machining operation. The cutting edge is the most stressed part of machining system because of the large force generated on a small area during machining. This makes the cutting tool the weakest part of the machining system, thus the need to re-inforce the cutting edge by way of tool geometry and coating to enhance performance and avoid premature fracture during machining operation. Variation in tool nose radius, chamfering, sharp corners and honed edges and different tool shapes (i.e. square, triangle, rhomboid and round shapes) and coating(s) are typical design alterations found in cutting tool materials. Each design has influence on tool performance in terms of tool life, quality of surface generated as well as the ''structure-sensitive'' properties of the work material after machining operation. Analyses of component forces are frequently based on orthogonal reference systems involving the machine tool and the tool material (1- 3). The resultant component of the x-y-z system directly characterises the cutting process as a whole.

In this paper, the cutting and feed forces during the threading operation were recorded and analysed with respect to other relevant machining parameters.

2. EXPERIMENTAL PROCEDURE

The machining test was carried out on previously cleaned, centred and trued workpiece bars (470 mm long and 150 diameter mm) using an electronically controlled centre lathe which has a manually selected three range gear box giving 18-200 rpm, 48-605 rpm and 140-1800 rpm respectively. This lathe is driven by a steplessly variable 11 kW motor drive, which provides a torque of 1411 Nm and a wider spread of power at the cutting surface than conventional direct drive machine tools. Coolant was not used for the threading tests in order to minimise tool failure relating to thermal shock.

The chemical composition (%) as specified by British Standard BS 970: Part 1: 1983 for the work material is given in Table 1. The steel bars were supplied in hot-rolled, hardened and tempered condition. Two grades of coated cemented carbide indexable threading tools were used for the machining trials. Insert type A consists of a P20-P30 substrate with sharp cutting edges and coated with a single TiN layer (3-4 μm) using the Physical Vapour Deposition (PVD) technique. Insert type B consists of a straight grade (K05-K20) substrate with honed edges and coated with a single Al_2O_3 layer (3-4μm) using the PVD technique. Tables 2a and 2b show the composition and properties of substrate materials respectively. Standard laydown triangular indexable inserts and a 25 mm shank toolholder with ISO designation Al25-3C RH, to provide a clearance angle of 6°, were used for the single-point thread cutting test.

3. EXPERIMENTAL RESULTS AND DISCUSSIONS

Tables 3 and 4 show recorded flank wear, wear volume, flank wear rate, component forces, tool life, and average hardness values taken at different regions of the thread surfaces at various cutting conditions. Analysis of the component forces at different flank wear values (average/maximum) during threading was based on the assumption that there were no external factors such as vibration and that only the length of wear along tool face (h) and width of wear land (2r) varies and also that the chip formation process remained constant. The variations of cutting forces at different thread passes as well as the average cutting forces after threading for 5 minutes at various conditions are shown in Figures 1a and 1b. These clearly show that insert type A gave higher cutting forces during the first pass (i.e at the start of cut). This suggests higher stresses and higher temperature per unit area hence higher flank wear rate, which increases the tool-workpiece contact length/area. Insert type B gave lower forces during the first pass. This increased during the second pass and rapidly droped in the third pass to a uniform level during the remaining passes (Figure 1b). This suggests a temporary restriction in the first pass and reduced localised macro plastic deformation of the cutting edge due to the honed edge profile hence mild wear, lower forces and longer tool life. Component forces recorded at the end of tool life (Table 4) generally increased with increasing cutting speed contrary to expection. This may be attributed to the perceived rubbing effect between the two contacting bodies, especially at the pullout stage as well as the effect of forces exerted on the tool nose and the clearance surface (2). These causes severe surface deterioration at the groove of the threaded surfaces, which worsen after machining with the insert type A (Figures

2a and 2b). Average cutting forces at the start of threading (Figure 1) are generally higher by about 10% than at the end of tool life, suggesting that prolonged machining, which also increases the tool-workpiece contact length/length of wear (h), leads to localised plastic deformation as the mean shear strength of the work material increases with decreasing infeed rate at the shear plane. The feed forces generally remained low with little fluctuation at higher cutting conditions, suggesting that the alternating radial infeed method used ensures even wear on both sides of the cutting edge. The test results also show that the cutting (tangential) forces are much higher (between 10% - 15% times) than the feed forces during threading and also increases at higher speed conditions (Table 4). A plausible cause for this may be as a result of the forces at the tool nose and clearance surface which tends to create additional 'parasitic' loading on the cutting edge due to increased rubbing action between the tool and the machined surface. An increase in feed rate from 0.40 mm/rev to 0.44 mm/rev gave only a slight increase (about 3%) in cutting forces. This clearly shows that the cutting speed had the most significant effect on the component forces when threading 817M40T with the two grades of carbide tools.

In previous work average flank wear (AFW) of 0.587 mm (half of the maximum flank wear, MFW) was used to formulate the relationship between the flank wear and the growth in thread root (GTR) as stated below: (4)

$$GTR = \frac{v_b . \tan 3^\circ \times 100}{h_3 . \sin 30^\circ} \quad (1)$$

where v_b is the measured average flank wear, h_3 is the thread height and GTR is 5.
This equation can be further expressed, to obtain the relationship between the flank wear and the height of the thread, as:

$$V_b = 0.48 h_3 \text{ (for AFW)} \quad \text{or} \quad 0.96 h_3 \quad \text{(for MFW)} \quad (2)$$

The significance of this formula is that once the height of the external thread is given, it will be possible to predict the length of wear (AFW/MFW) along the tool nose/face which is expected to give the V-type shapes, Figures 3a and 3b, when machining with the two grades of thread cutting inserts. Thus, the MFW was used to derive the formula for the wear volume based on the curvature/shape as this gives the full length of wear along the tool surface. The volume of a parabola for insert type A, is given as

$$V = \pi\vartheta \int r^2 dh , \quad (3)$$

where $r^2 = h$ and ϑ = flank wear rate i.e. $V = \frac{\pi\vartheta}{2}\left[h^2\right]_0^h , \Rightarrow V = \frac{\pi\vartheta}{2}h^2$

For the volume of a right cone for Tool B grade:

$$V = \pi\vartheta \int h^2 dr \quad (4)$$

where h= 0, r = h and ϑ = flank wear rate i.e. $V = \frac{\pi\vartheta}{3} r^2 . h$

Equations 3 and 4 were used for calculating the volume of material abraded from the tool surface as shown in Table 3. This shows that insert type A did not only give higher wear volume with respect to length of wear (h) as cutting conditions increases, but also higher component focres as well as shorter tool life than insert type B. It is worth mentioning that flank wear rate had been introduced in this derivation to represent both wearing action and wearability of the two mating bodies (i.e. tool material and the workpiece) which replaces the generalised wear coefficients as recommended by Rabinowicz (5). The nature of metallurgically transformed species (i.e. deposited oxide debris, BUE, etc.) present at the interface are generally influenced by the value of coefficient of wear, which in turn is also affected by the cutting conditions (6), and cannot be used for predicting wear volume. This is the reason for introducing flank wear rate as, from practical point of view, this does not seem to affect the oxidation characteristics at the cutting zone.

The V-shape wear observed on both tool types during threading, Figure 3a and 3b. This may be due to side flow of the continuous helical chips plus the effect of the trailing and leading edge grooving wear mechanism of the inserts. Evidence of more adhesion of oxide debris were obsevered on worn type A inserts, particularly after machining at higher cutting conditions. This suggest that the temperature attained during threading is sufficiently high to cause thermal softening of the cobalt binder in the substrate. As a result, insert type B with lower Co content (6 vol %) will probably have higher resistance to plastic deformation than insert type A due to the mean thickness of the cobalt binder between the carbide grains (7).

Insert type type B generally gave longer tool life during threading (Table 3). This suggests that Al_2O_3 coating on the honed edge have greater resistance to various wear mechanims likely to manifest themselves singly or in combination to undermine tool perfomance (8-9) since it exhibits excellent chemical stability, low thermal conductivity at elevated temperature and high hot hardness. The hardness variation along four different regions after threading (Table 4) suggests that the top of the thread is softer than the sides, root and base due perhaps to its thinner cross section. About 3-5% reduction in hardness was observed after threading in all the conditions investigated. This can be attributed to mild tempering effect caused by high temperature gradient associated with single point thread cutting operation.

4. CONCLUSIONS

The test results show that the cutting (tangential) forces are much higher (between 10-15% times) than the feed forces during threading. An increase in cutting speed led to an increase in component forces due to the rubbing effect of the tool-workpiece interface, particularly at the tool nose and clearance face during threading.

The Al_2O_3 coated type B inserts with honed edges generate lower component forces, longer tool life, minimum tool wear with less adhesion of flake-like oxide debris on the tool face compared with sharp edge type A inserts with TiN coating.

Two different types of wear curvature, parabolic and right-cone shapes for insert types A and B respectively, were identified. The V-shape type of wear scar can be associated with the side flow mechanism of the continuous helical chips which tends to abrade the tool face in a V-shape form because of the alternating cutting edge profile.

A formula for determining the volume of particles worn from the tool face was derived based on the shape of the wear which relates both the length of wear (h), the width of wear land (2r) and flank wear rate (ϑ). The wear volume increases as the cutting conditions increases and in all cases, insert type B gave lower wear volume, hence longer tool life and lower component forces.

The edge geometry as well as the composition of the substrate tend to ensure lower tool wear rate, hence longer tool life in addition to minimum damage of the threaded surface with only a slight deviation from the apparent hardness of the workpiece after machining.

5. ACKNOWLEDGEMENT

The authors acknowledge the support from Macreadys-Carbon and alloy steels and Damstahl Tooling which enabled this work to be carried out.

6. REFERENCES

1. **Kaczmarek. J.**, *Principles of metal machining*, Peter Peregrinus, Ltd., 1976.
2. **Wallace, P. M.** and **Boothroyd, G.**, *Tool forces and tool-chip friction in orthogonal machining*, Jnr Mechanical Engineering Science, vol.6, no.1, 1964, pp75-85
3. **Thomas, G. G.**, *Production Technology*, Oxford University Press, 1977, pp187-211.
4. **Ezugwu, E. O., C. I. Okeke and Machado, A. R.**, *High speed threading of Inclusion-modified steel using coated carbide inserts*, Accepted paper for publication in the Journal of Material Processing Technology
5. **Rabinowicz, E.**, *Friction and wear of materials*, New York, 1977, p.155-158
6. **Bhansali, K. J.** *Adhesive wear of nickel- and cobalt-base alloys*, Wear, 60(1980)95-100
7. **Ezugwu, E. O.** *Wear of coated carbide tools when machining an Ni-Cr-Mo steel*, Lubrication Engineering, Nov. 1997, pp18-24
8. **Stot, F. H. Lin, D. S.** and **Wood, G. C.**, *The structure and mechanism of formation of the glaze oxide layer produced on nickel-based alloys during wear at high temperatures*, Corrosions Science, 13(1973)449-469
9. **Bittes, G., Leroy, F,** and **Auclair, G.**, *The relationship between inclusionary deposits and the wear of cutting tools*, Jr of Materials Processing Technology 54(1995)88-96

7. TABLE AND FIGURE CAPTIONS

Table 1-Nominal chemical composition of 817M40T (En 24T)

C	Si	Mn	S	P	Cr	Mo	Ni	Ca	Fe	Hardness
0.36	0.10	0.45	0.4	0.40	1.10	0.20	1.30	0.001	Bal	285 HV
0.44	0.40	0.70	Max	Max	1.40	0.35	1.70	0.004	Bal	Measured average

Table 2a-Chemical composition of the substrate of the threading tools (vol.%)

Tool Type	WC	TiC + TaC	Co	Cr_2C_2	ISO Code
A	68.5	21	10.5	-	P20-30
B	93.5	-	6	0.5	K05-20

Table 2b-Properties of the tool materials

Tool type	Coating thickness (μm)	Grain size (μm)	TRS (N/mm^2)	Density (g/cm^3)	Hardness (HV)
A	TiN 3-4	2.0	1700	12.4	91.1
B	Al_2O_3 3-4	1.0	3000	14.9	94.0

Table 3: Flank wear rate (FWR) volume of wear (WV) for inserts types A and B when threading 817M40T steel at different cutting conditions

Insert Type	Speed (m/min	Feed (mm/rev	AFW (mm)	MFW (mm)	Wear width-2r	Tool life (min)	FWR (mm/min)	Wear vol. (mm^3)
	150	0.40	0.547	1.094	0.365	35	0.016	3.0×10^{-2}
	175	0.40	0.565	1.130	0.377	32	0.018	3.6×10^{-2}
	200	0.40	0.560	1.120	0.373	31	0.018	3.55×10^{-2}
A	225	0.40	0.694	1.388	0.463	2.5	0.278	8.4×10^{-2}
	150	0.44	0.546	1.092	0.364	17	0.032	6.0×10^{-2}
	175	0.44	0.562	1.124	0.375	15	0.037	7.4×10^{-2}
	200	0.44	0.574	1.148	0.383	13	0.044	9.1×10^{-2}
	225	0.44	0.72	1.440	0.48	2.3	0.313	1.0×10^{-2}
	150	0.40	0.577	1.152	0.364	41	0.014	6.0×10^{-3}
	175	0.40	0.538	1.124	0.375	39	0.014	6.0×10^{-3}
	200	0.40	0.563	1.148	0.382	36	0.016	7×10^{-3}
B	225	0.40	0.750	1.500	0.500	2.6	0.288	3×10^{-2}
	150	0.44	0.579	1.158	0.386	26	0.022	1.0×10^{-3}
	175	0.44	0.578	1.140	0.380	16	0.036	3.0×10^{-3}
	200	0.44	0.565	1.130	0.377	14	0.040	3.4×10^{-3}
	225	0.44	0.770	1.540	0.513	2.4	0.321	$3.4\ 10^{-2}$

Table 4-Average cutting forces, sub-surface hardness, flank wear and tool lives recorded after threading 817M40T steel at various conditions with insert types A and B

Insert Type	Speed (m/min)	Feed rate (mm/rev)	AFW (mm)	MFW (mm)	Tool life (min)	Cutting force (N)	Feed force (N)	Ave. hardness after threading (HV) Top	Side	Root	Base	Ave.
	150	0.40	0.544	1.094	35	459	30	271	268	288	281	277
A	175	0.40	0.565	1.130	32	641	50	269	276	279	281	277
	200	0.40	0.560	1.120	31	800	50	273	271	279	283	277
	225	0.40	0.694	1.388	2.5	840	50	276	273	282	282	278
	150	0.40	0.577	1.154	41	467	50	256	260	288	260	266
B	175	0.40	0.538	1.076	39	595	60	254	261	260	258	258
	200	0.40	0.563	1.126	36	684	50	255	270	256	270	263
	225	0.40	0.750	1.500	2.6	750	30	260	270	263	272	266
	150	0.44	0.546	1.092	17	473	33	258	257	268	270	263
A	175	0.44	0.562	1.124	15	660	51	256	260	270	273	265
	200	0.44	0.574	1.148	13	820	51	259	270	273	271	268
	225	0.44	0.720	1.440	2.3	870	60	257	273	268	273	268
	150	0.44	0.579	1.158	26	480	50	256	258	260	260	259
B	175	0.44	0.578	1.156	16	614	65	257	260	270	270	264
	200	0.44	0.565	1.135	14	705	50	255	270	271	274	268
	225	0.44	0.770	1.540	2.4	773	30	256	270	274	270	268

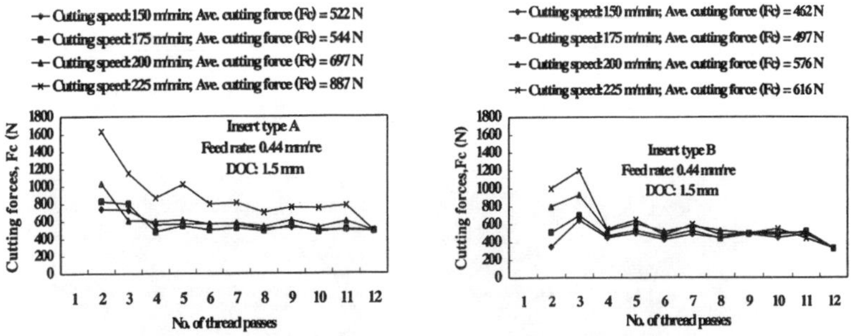

(a) (b)

Figure 1a-1b-Variation in cutting forces on thread passes and average cutting forces at the start of thread when threading 817M40T with the two grades of the cutting tools at various conditions

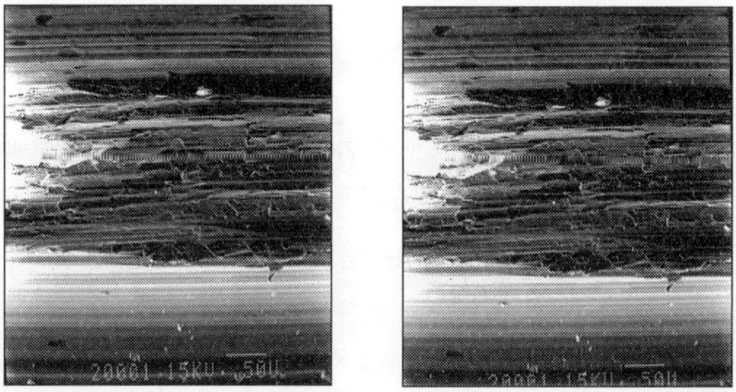

(a) (b)

Figure 2-Threaded surface produced after machining 817M40T steel with (a) insert type A for 13 minutes (b) insert types B for 14 minutes at a speed of 200 m/min, feed rate of 0.44 mm/rev and DOC of 1.5 mm

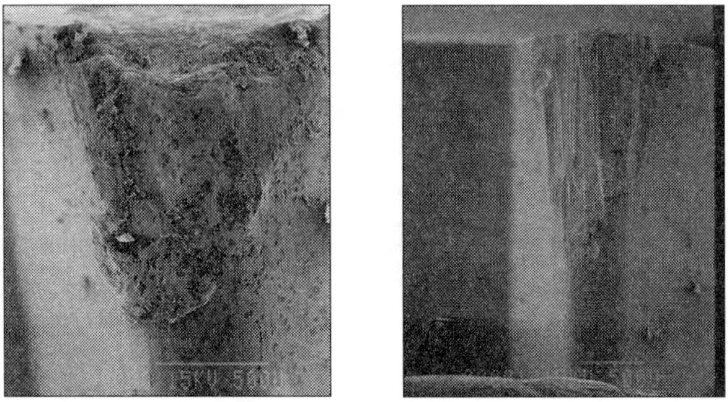

(a) (b)

Figure 3a-3b-Typical V-shape wear on the tool face showing (a) parabolic shape for insert type A with adhering debris and (b) right-cone shape for insert type B with minimum or no adhering debris after machining 817M40T

Investigations into local deformation in incremental forming operations

J FERREIRA and **F H OSMAN**
Department of Mechanical Engineering, University of Bath, UK

Open die forging is a process in which a large variety of simple tools, plain and profiled, are employed to produce complex shapes by incremental deformation. The process requires low energy machines where the forging step is performed on a small section at any one time. In its traditional form, the process is carried out manually thus requiring experience and intelligent manipulation of the workpiece material. The development of a new operating environment for open die forging will increase its competitiveness in the global market where small batch manufacturing mode has become an essential feature of the new production trends. This paper presents a new on-line computer controlled set-up for the manipulation of the workpiece material and to provide accurate interaction between workpiece orientation and the action provided by the tools. Also, an analytical approach to predict the mode of deformation after an incremental step effected by a semi-circular tool has been presented and the results compared with those obtained experimentally.

1. INTRODUCTION

Open die forging is one of the incremental forming processes where the material is allowed to flow freely without severe constraints as in closed die forging. There are two main steps involved in the process; these are the manipulation of the workpiece material and the action of the pressing tools. Both action are inter-related and must be synchronised accurately in order produce the desired profile. The invention and development of new machinery for forging has always been a milestone for the forging industry. Most current forging machines e.g. hammers and mechanical, screw, and hydraulic presses are designed for a single forming operation. The entire open die forging operation is usually carried out in such a way that the billet is deformed incrementally and progressively over its entire length. The level of deformation may vary for each step and the workpiece moves axially and/or rotationally around its longitudinal axis between every consecutive compression steps. Therefore the design of a new forging machine would need different approach in order to implement the level of automation and mechanisation required for a flexible open die forging manufacturing cell. Most recently, some special purpose machines have been developed for incremental forging[1,2]. The introduction of robots in forging has been limited to handling the workpiece from the furnace to the press, from one cavity to another or to perform a co-ordinated operation with press parameters[3,4]. Also, limited types of rail mounted manipulators have been used in the traditional open die forging for very large and heavy workpieces [5].

2. OPEN DIE FORGING PRESS

A new hydraulic press, for use in open die forging research, has been designed with full computer interface to control workpiece positioning and press stroke and speed. The hydraulic circuit consists of a double acting cylinder, a servo control mechanism and a directional control valve which sustains a constant ram speed irrespective of the press load. The stroke of the ram is measured using a linear encoder.

2.1 Workpiece manipulation

Manipulation and positioning of the workpiece is of vital importance in free open die forging. The main use of the manipulator is to adjust the orientation of the workpiece, to advance it along its length and to release it prior to the pressing stroke so as to enable free deformation. The workpiece must accurately be aligned otherwise the workpiece is likely to undergo undesirable bending[10].

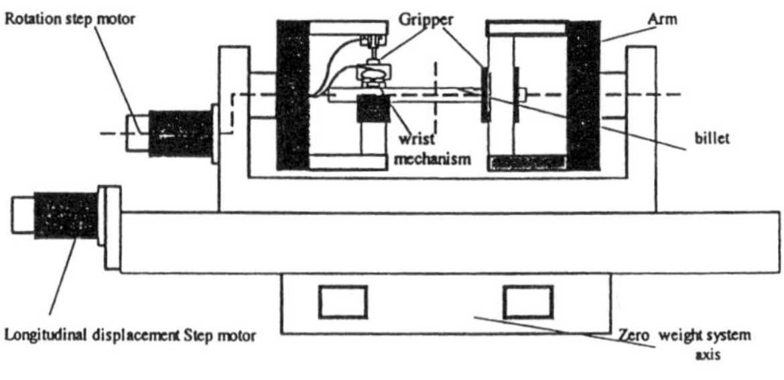

Fig. 1 : Workpiece manipulator

Fig. 1 Shows a schematic diagram of the workpiece manipulator main components which includes a zero weight system to ensure that the workpiece is free from any loading during the positioning process, i.e rotation or axial displacement. The set-up also shows a wrist mechanism which is used for the insertion or removal of the billet. The manipulator is powered and controlled by a pneumatic-electronic circuit with four small cylinders for the holding, releasing and rotating of the workpiece and two cylinders for the vertical positioning of the workpiece. The positioning tolerance is 10 microns and 0.015 degrees.

2.2 Press Control System

The overall set-up is governed by the press and the manipulator control systems and local sensing devices. The breakdown of the displacement control system is shown in Fig. 2.

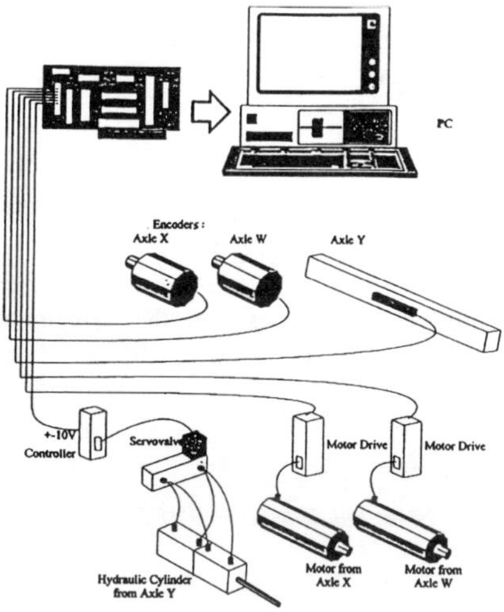

Fig. 2 : PC23 Control system

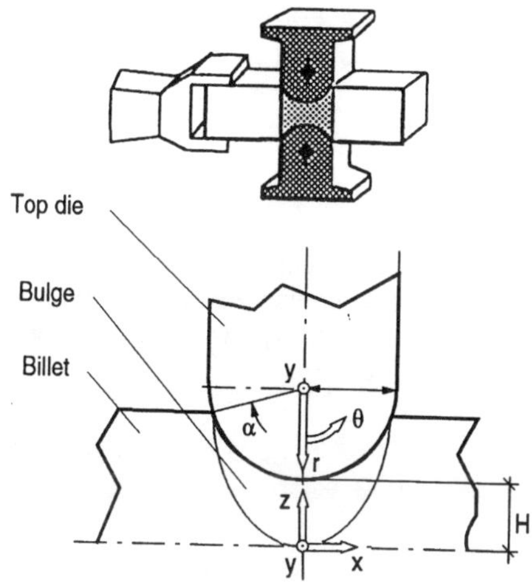

Fig. 3 : Analytical model

The overall control is achieved through an intelligent peripheral PC23 which slots into a computer bus expansion. Control functions are interfaced using C and BASIC high level languages. The PC23 card provides three control lines, two of which are used for the axial and rotational displacement while the third enables the interface with the press control parameters. The machine can be operated by computer pointing device e.g. keyboard, mouse and joystick. The machine is compact with an overall height of 110cm, width 110cm and depth of 60cm, and is easy to use. It was built as a prototype for research purposes only.

3. UPPER BOUND ANALYSIS

One of the early investigation, using the Upper Bound Method, to predict loading requirements in open die forging was carried out by Barya and Johnson[6]. Juneja[7] and, Yang and Kim[8] investigated spread and loading requirement in connection with forging of polygonal blocks. Adaptation of upper bound solutions to predict the mode of deformation in flat tools-open die forging was proposed by Osman et al [9]. They investigated the effect of the deformation cycle (compression followed by 90^0 rotation) on the final cross-sectional area of the material between the tools and provided a systematic approach to predict the optimum operating sequence for achieving the desired reduction in area. Very little work has been carried out using tools other than flat dies. Ferreira and Osman[10] investigated experimentally the effect of the billet aspect ratio on spread and elongation using semi-circular tools.

3.1 Velocity Field and Load Prediction

Fig. 3 shows a schematic diagram of a section being deformed. The velocity component in the direction of the tool movement may be expressed by,

$$\dot{u}_z = -\dot{u}_t \frac{r - r_t - H}{H} \tag{1}$$

where $\dot{u}_t$ and r_t are the tool velocity and radius. The velocity component in the axial direction is assumed to be a function of θ, thus;

$$\dot{u}_x = \varsigma \dot{u}_t \frac{\theta}{\alpha} \tag{2}$$

where ζ is a constant and α is the angle from the centre of the moving tool to the edge contact point with the deforming material. By transforming the velocities into the cylindrical coordinate system and applying the condition of volume constancy, the strain rate field becomes ;

$$\dot{\varepsilon}_{rr} = -\dot{u}_t \frac{1}{H} \cos\theta \tag{3}$$

$$\dot{\varepsilon}_{\theta\theta} = \varsigma \dot{u}_t \frac{1}{r\alpha} \cos\theta \tag{4}$$

$$\dot{\varepsilon}_{yy} = \dot{u}_t \cos\theta \left[\frac{1}{H} - \varsigma \frac{1}{r\alpha}\right] \tag{5}$$

$$\dot{\varepsilon}_{r\theta} = \dot{u}_t \sin\theta \left[\frac{1}{2H} - \varsigma \frac{1}{2r\alpha}\right] \tag{6}$$

$$\dot{\varepsilon}_{y\theta} = \dot{u}_t \frac{y}{2r} \sin\theta \left[\varsigma \frac{1}{r\alpha} - \frac{1}{H}\right] \tag{7}$$

$$\dot{\varepsilon}_{ry} = \varsigma \dot{u}_t \frac{y}{2r^2\alpha} \cos\theta \tag{8}$$

The power dissipated due to deformation, ($\dot{W}_d$) over the deforming volume V is;

$$\dot{W}_d = \frac{2}{\sqrt{3}} \sigma_o \int_V \sqrt{\frac{1}{2} \dot{\varepsilon}_{ij} \dot{\varepsilon}_{ij}} \, . dV \tag{9}$$

The power dissipated due to friction between the tool and the deforming material ($\dot{W}_f$) is,

$$\dot{W}_f = m . \frac{\sigma_o}{\sqrt{3}} \int_A |\Delta \dot{v}| . dA \tag{10}$$

where Δv is the relative velocity between the tool and deforming material, A is the area of contact and m is the friction factor. Adding equation (9) and (10) gives the total power required to effect the deformation, this in turn is equal to the power necessary to be exerted by the tools. Because the deformation pattern is assumed to consist of one zone with a uniform flow function and the material deforms in a free mode, shear losses between the deforming zone and the solid material are not taken into account.

4. COMPARISON WITH EXPERIMENTS

Experiments were carried on billets of commercially pure aluminium and using semi-cylindrical tool. Fig. 4 shows comparison between the load predicted by the above method and experiments. The billets used for this test were of 13mm width and an aspect ratio (height/width) of one. The cylindrical tool used in this experiments has an end radius of 15mm. The theoretical results obtained from equation (9) and (10) and optimised with respect to ζ are shown for friction factor of m=0.05 and 1.0. The results also show that the predicted load increases with the increased penetration and friction. It is also noticed that the experimental loads for 60% and 80% penetration are comparable. This may be attributed to the diminishing size of the dead metal zone, known to exist in simple compression, with the increased penetration.

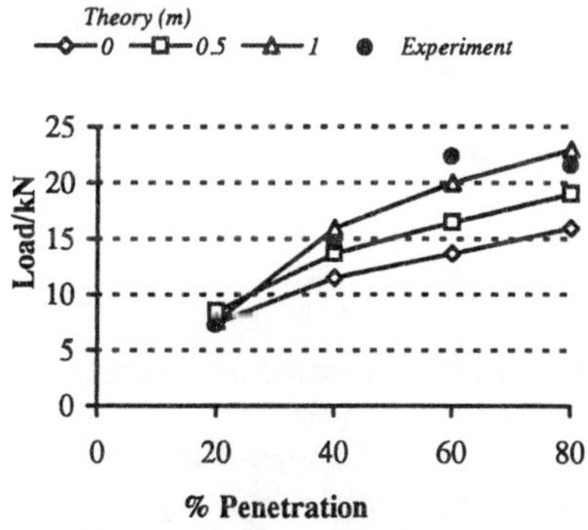

Fig. 4 : Load against penetration

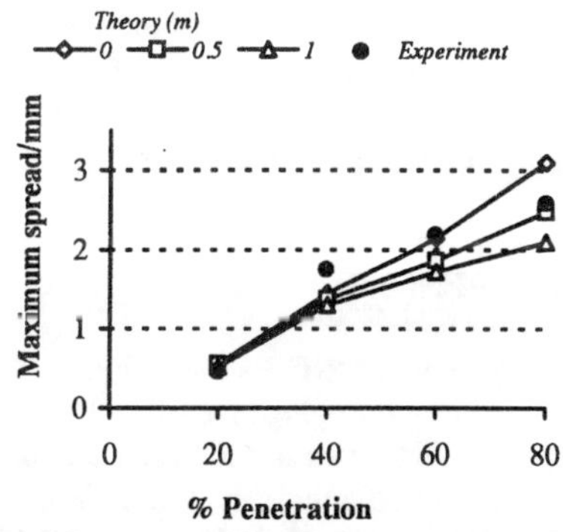

Fig. 5 : Spread at different penetration level

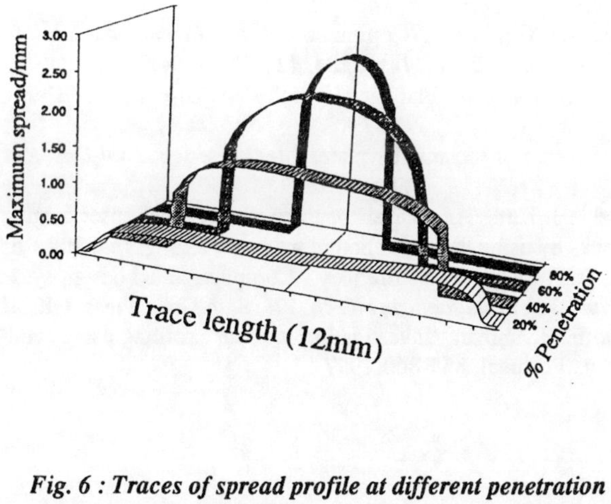

Fig. 6 : Traces of spread profile at different penetration

Fig. 5 shows the maximum spread or lateral flow along the central section of the billet. The actual values of spread are quite small and most of the material has actually deformed along the billet length. While maximum spread increases with the penetration its profile varies with the level of penetration. At small penetration levels maximum spread occurs nearer to the tool. As the penetration increases the profile becomes parabolic with the maximum spread occurring at the centre of the billet. However, for billets with a large aspect ratio it is possible for the material at the centre of the billet to fold inward rather than spreading outwards. Fig. 6 shows the trace of spread at the centre of the billet and along the z axis. Naturally, the span of the trace reduces as the penetration increases. It is clear that in the early stages of the process the bulge is not uniform.

5. CONCLUSIONS

A new machine specific for incremental open die forging and incorporating robot control technology have been presented. The workpiece manipulation system included axial and rotational displacement in addition to an interface which coordinates the link between press, robot and workpiece. Also, the process of compression of a rectangular bar with semi-cylindrical tool was analysed. Both experimental and predicted results showed that side spread produced by semi-cylidrical tools is much smaller than that of flat tools.

REFERENCES

[1] K. Lange, 'NC Radial forging - a new concept in flexible automated manufacturing of precision forging in small quantities', *25th MTDR Conf.*, 63-72, (1985).

[2] M. Wehrheim and R. Kopp 'Automated process to produce a near-net-shape workpiece by open die forging', *Proc. 4th Int. Conf. Tech. Plasticity*, **2**,1102-1107, (1993).

[3] B.W. Rooks, K.O. Okpere and R.M.H Cheng, 'Asynchronous sequential control of billet handling system for feeding a hot forging machine', *The Industrial Robot*, **1**, 66-71, (1973).

[4] E. Appleton, W.B. Heginbotham and M. Kohono, 'Design study and feasibility trial for a robot blacksmith', *Proc. 8th ISIR*, Stutgart, (1978).

[5] J.M. Thomas, A. Tomlinson 'Computer control for automation of heavy forging production', *J. of Iron and Steel Inst.*, **204**, 423-433, (1966).

[6] G.L. Barya and W. Johnson. 'Flat bar forging', *Proc. 5th Int. Tool Des. Conf.*, 449-469, (1964).

[7] B.L. Juneja, 'Forging of rectangular plates', *Int. J. Mach. Tool Des. Res.*, **13**, 141-153, (1973).

[8] D.J. Yang and J.H. Kim, 'An analysis of three dimensional upset forging of rectangular polygonal blocks by using the upper bound method', *J. Eng. Ind*, **109**, 155-160, (1987)

[9] F. H. Osman, B. Aksakal, A. N. Bramley, 'Computer aided design system for open die forging', *Advances in Manufacturing Tech. VII*, Bath Univ. Press, UK, 216-221,(1993).

[10] J. Ferreira and F. H. Osman, 'Investigation into bar profiling using semi-circular tools', *AMPT'97 Conf.*, Portugal, 855-860(1997).

The effects of organic coatings on heat transfer in underwater pipelines

J C HATCHMAN
Consultant, Upton, UK

SYNOPSIS

Until now the methods of calculating the thermal resistance of an underwater pipeline carrying an internal fluid has had to rely on the use of commonly indexed overall heat transfer coefficients or of a simplification which includes the resistance of a paint film in that of the pipe wall (additional insulation and fouling allowances are of course supplementary to this). These results show that for uninsulated steelwork the thermal resistance of the coating can be significant. In some circumstances it can even be the controlling medium in a heat transfer process. Concise data on the heat transfer properties of typical industrial and marine coatings is sparse or unavailable. The paper gives measured values of essential properties and illustrations of possible applications.

NOTATION

A_{ext}	surface area of external section of pipe (m^2)
A_{int}	surface area of internal section of pipe(m^2)
$A_{manti\text{-}fouling}$	logarithmic mean area of marine anti-fouling paint coat section (m^2)
$A_{mbase\text{-}coat}$	logarithmic mean area of base-coat of paint section (m^2)
A_{mpipe}	logarithmic mean area of pipe wall (m^2)
a.a.f.	marine ablative anti-fouling paint
c.a.f.	marine conventional anti-fouling paint
C_p	heat capacity (J/kg.K)
c.r.	chlorinated rubber base-coat industrial paint
c.t.e.	coal tar epoxy base-coat industrial paint
z.r.e.	zinc rich epoxy base-coat industrial paint
HFO	heavy residual fuel oil with a kinematic viscosity of 380 cSt at 50°C
h_{ext}	external convection heat transfer coefficient ($W/m^2.K$)
h_{int}	internal convection heat transfer coefficient ($W/m^2.K$)
IFO	light (intermediate) fuel oil with a kinematic viscosity of 40 cSt at 50°C
k	thermal conductivity (W/m.K)

$k_{anti\text{-}fouling}$	thermal conductivity of marine anti-fouling paint (W/m.K)
$k_{base\text{-}coat}$	thermal conductivity of industrial base-coat paint (W/m.K)
k_{pipe}	thermal conductivity of pipe wall material (W/m.K)
L	pipe section length (m)
R_{tot}	total thermal resistance of the coated pipe (K/W)
r_i	relevant section radius (m)
$x_{anti\text{-}fouling}$	radial film thickness of anti-fouling paint coat (m)
$x_{base\text{-}coat}$	radial film thickness of base-coat paint (m)
x_{pipe}	radial thickness of pipe wall (m)
ε	total normal emissivity

1 INTRODUCTION

For something so common as a paint an abundance of data on the heat transfer properties would be expected. However that is not the case and data is sparse or unavailable. In standard reference works such as the *TPRC Series* (**1**) or specialized works such as that by Mills (**2**) or Roff and Scott (**3**) there is no data for the thermal conductivity of paints. There is a limited amount of data on radiation emissivities of paints such as that given by Gubareff et al in (**4**) but it does not cover the common industrial and marine paints described here. Published data on heat capacities for commonly indexed polymers is available in technical bulletins such as (**5**) and data for paints was previously given by the author to the Institute (**6**). A similar situation exists with the data for the thermal conductivity and results and a detailed description of the experimental methodology has been submitted to the Institute (**7**). There is again a limited amount of data available for commonly indexed polymers but the measured paint data differs significantly from that.

For illustrative purposes, the effects of paint coatings on the heat transfer from uninsulated pipelines, underwater, have been analysed. The heat transfer scenario analysed was that for single-run pipelines carrying residual fuel oils and freshwater. The pipelines were taken to be sufficiently large so as not to require trenching and any weight coating would be intermittent. Predicted and measured heat transfer from a pipeline carrying a residual fuel oil were described by Stephenson (**8**). However that was for a completely different situation with the pipeline being mostly underground and the remainder being exposed to the atmosphere. The pipeline was also insulated but (perhaps for commercial reasons) no details were given of how. The situation analysed here is simply to evaluate the effect of how the application of different combinations of paint coatings can influence the overall heat transfer coefficient and heat transfer from steel work to seawater. This is seen to be relevant to short runs of uninsulated pipelines carrying residual fuel oils or crude oils across water where there is a serious risk of pumping failure due to a fall in temperature of the oil towards the pour point. If this occurs there is the additional problem of the behaviour of the oil becoming non-Newtonian and pseudoplastic. A further situation seen to be relevant is that of an underwater pipeline carrying freshwater in or below normal ambient freezing conditions where there is also a danger of pumping failure or pipe blockage.

With regard to the paint systems, normal procedure for underwater steelwork is to first apply a thin coat of adhesion primer, of the order of 15 - 30 micron dry film thickness to the bare metal. The thermal resistance of this film would be extremely small and the resistance associated with it was taken to be included in the terms for the additional coats which have a far greater thickness. A base coat is then applied on top of the primer. In this analysis it was assumed that a marine anti-fouling paint coat would be applied on top of the base coat to retard or prevent the growth and adhesion of marine fouling organisms. It is accepted that this is currently quite a rare practice in pipeline protection but it is useful for analytical purposes and for those rare cases where it is used. The paint system would, in effect, be similar to that used to protect the underwater areas of merchant ships and has been shown to afford maximum protection over prolonged periods. The present day cost of these paint systems is of the order of £8 - 15 per m^2 of metal surface and this determines that extreme care is taken during the application of the paint coats to achieve the required accuracy in film thickness.

2 MATERIALS THERMOPHYSICAL PROPERTIES

The pipe was taken to be manufactured from a carbon steel corresponding approximately to a Lloyds Category 1 Material and in the calculations its thermal conductivity was taken to be 50 W/m.K.

Thermophysical properties for the residual fuel oils were taken from the specifications described by Newberry (9) and a standard reference work on fuels (10).

Thermophysical properties for the freshwater were taken from standard thermodynamic tables (11).

Thermophysical properties for the seawater were taken from (12) and (13).

Relevant thermophysical properties of the organic coatings are shown in Table 1. These values are a general guide and the properties will vary with differing chemical compositions from individual paint manufacturers.

Paint type	C_p (J/kg.K)	k (W/m.K)	ε
Ablative anti-fouling	835	0.71	0.88
Chlorinated rubber	995	0.25	0.79
Coal tar epoxy	1350	0.46	0.86
Conventional anti-fouling	665	0.66	0.85
Zinc rich epoxy	550	3.75	0.79

Table 1 Thermophysical properties of typical industrial and marine paints at 30°C

3 PIPE FLOW EXAMPLES

In the examples the pipeline was taken to be a 800 mm nominal diameter pipe with a wall thickness of 10 mm. The outer surface of the pipe was taken to be protected with a base coat of chlorinated rubber or coal tar epoxy or zinc rich epoxy to a minimum dry film thickness of 200 micron and a maximum of 300 micron. On top of this a marine anti-fouling paint, of either the conventional or ablative type, would be applied to a minimum dry film thickness of 100 micron and a maximum of 300 micron. Thus the total dry film thickness could vary between 300 and 600 micron. For the purpose of the analysis no further external insulation was considered. The inner surface of the pipe was taken to be unprotected. Fouling resistances for either surface were not included in the analysis because they are a general term.

To illustrate the effect of the thermal resistance of a coating in the heat transfer process, the system was analysed for pipelines carrying three different fluids :

(i) A heavy residual fuel oil corresponding to a BS MA 100: 1982: grade M7 (ISO 8217: 1996: grade RMH35) marine fuel (9) with a kinematic viscosity of 380 cST at 50°C.

(ii) A light (intermediate) residual fuel oil corresponding to a BS MA !00: 1982 grade M4 (ISO 8217: 1996: grade RMA10) marine fuel (9) with a kinematic viscosity of 40 cSt at 50°C.

(iii) Freshwater.

In the analysis, all three fluids were allocated bulk temperatures in the range 45 - 60°C which is standard for transferring fuel oils having a minimum pour point of 30°C. The freshwater flow was analysed in the same temperature range for comparison. The mean velocity of the internal flow of the fuel oils was taken to be 1.9 m/s and that of the freshwater, 2.5 m/s. The pipes were taken as being fully immersed in and surrounded by shallow seawater in which the transport properties were not affected by pressure. The temperature and salinity of the seawater were taken to be 10°C and 35 mass parts per thousand, respectively. These conditions were taken as being typical for an average annual mean for the Western Europe continental shelf in the North Atlantic Ocean. The external flow was analysed with seawater velocities in the range 0.5 - 5 m/s which correspond approximately to seawater current flow in the range 1 - 10 knot.

A view of the pipes and internal and external flow is shown in Fig 1 and the heat transfer system for the purpose of this analysis was taken to be steady-state, one dimensional (i.e. temperature gradients in the radial direction only) conduction through a composite cylinder with internal and external convection. Using the thermal resistance analogy the resistance of the equivalent thermal circuit is

$$R_{tot} = \Sigma \frac{1}{h_{\text{int}} A_{\text{int}}} + \frac{x_{pipe}}{k_{pipe} A_{mpipe}} + \frac{x_{base-coat}}{k_{base-coat} A_{mbase-coat}} + \frac{x_{anti-fouling}}{k_{anti-fouling} A_{manti-fouling}} + \frac{1}{h_{ext} A_{ext}}$$

where A_{mi} are the logarithmic mean areas of the tubular sections defined by

$$A_{mi} = \frac{2\pi(r_{n+1} - r_n)L}{\ln(\frac{r_{n+1}}{r_n})}$$

The internal heat transfer coefficients for the heavy fuel oil flow were computed using both the Gnielinski correlation (14) for constant liquid properties and the Sleicher and Rouse correlation (15) for variable liquid properties. The indications were found to be that even though the flow at these conditions would be in the laminar and transitional regions, non-Newtonian pseudoplastic effects would be small.

The light fuel oil and freshwater flows at these conditions, would be well into the turbulent regime and the heat transfer coefficients were computed using the Gnielinski correlation.

The external flow of the seawater would be well into the turbulent regime throughout the range of flow rate investigated and the heat transfer coefficients were computed using the Zukauskas correlation (16).

4 RESULTS

The thermal resistances of the organic coatings against total film thickness are shown in Fig 2. The thermal resistance of the pipe wall was computed separately and is not included in these values. The extrema in total film thickness shown in Fig 2 refer to the maximum and minimum applications of the base-coat and anti-fouling paints i.e. a 300 micron total film thickness refers to a 200 micron film thickness of a base-coat paint plus a 100 micron film thickness of an anti-fouling paint, etc. It can be seen from Fig 2 that the thermal resistances of the coatings using the chlorinated rubber as a base coat are significantly higher than those of the coatings using a coal tar epoxy or a zinc rich epoxy paint for a base coat. The zinc rich epoxy base-coat paint systems, as would be expected, show the lowest thermal resistances, they are several times lower than that of the other paint systems. The coal tar epoxy base-coat systems show intermediate values of thermal resistance.

The external convective heat transfer coefficients against seawater velocity are shown in Fig 3. It can be seen that the external heat transfer coefficient is very sensitive to seawater velocity but is almost unaffected by the small changes in paint film thickness which could occur.

For each of the different types of internal fluid flow the results are summarized as follows:

1. Heavy residual fuel oil flow

The internal convective heat transfer coefficients for the HFO flow are shown in Fig 4. As would be expected they are low, of the order of 50 $W/m^2.K$ and show a small difference between the two correlations used to evaluate them. It can also be seen that they are significantly lower than the results for the IFO flow which are also shown in Fig 4. The total thermal resistances of the pipeline per m length against bulk oil temperature are shown in Fig 5 and it can be seen how sensitive the results are to the HFO bulk temperature. The results shown are for the extreme cases :

(i) maximum coating resistance and maximum external convective resistance
and
(ii) minimum coating resistance and minimum external convective resistance.

The thermal resistances of the coatings as a percentage of the total thermal resistances against HFO bulk temperature are shown in Fig 6. As would be expected the thermal resistances of the coatings are comparatively small, the maximum value being some 10% of the total resistance at a HFO temperature of 60°C. The paint system at this value would be a 300 micron film thickness of chlorinated rubber base-coat paint plus a 300 micron film thickness of conventional anti-fouling paint. The seawater velocity for these results would be the minimum, 0.5 m/s. The corresponding range in overall heat transfer coefficients is from 30 to 60 $W/m^2.K$.

At the other extreme the minimum thermal resistance of the coatings would be expected to be approximately 1% of the total resistance at a HFO temperature of 45°C. The paint system at this value would be a 200 micron film thickness of zinc rich epoxy base-coat paint plus a 100 micron film thickness of ablative anti-fouling paint. The seawater velocity for these results would be the maximum, 5.0 m/s. The corresponding range in overall heat transfer coefficients is from 32 to 69 $W/m^2.K$.

The range in overall heat transfer coefficient values compare to quoted values (17) for heavy oil to water heat transfer of 60 - 300 $W/m^2.K$.

2. Light residual fuel oil flow

The internal convective heat transfer coefficients for the IFO flow are shown on the same figure as the for the heavy residual fuel oil flow, Fig 4. It can be seen from the results that the heat transfer coefficient would be expected to be several times greater than that for the HFO flow, 150 - 220 $W/m^2.K$ cf. 50 $W/m^2.K$. The total thermal resistances of the pipeline per m length are again shown in Fig 5 and it can be seen that they are significantly lower than for the HFO flow and they are far less sensitive to changes in the bulk temperature. The results for the thermal resistances of the coatings as a percentage of the total thermal resistances against bulk IFO temperature are again shown in Fig 6. The results for the extreme cases of paint system composition and external flow show:

(i) The maximum thermal resistance of the coating can now account for some 20 - 25% of the total resistance in the heat transfer process. The corresponding range in overall heat transfer coefficients is from 118 to 144 $W/m^2.K$.

(ii) The minimum thermal resistance of the coatings in the total process is of the order of 3%. The corresponding range in overall heat transfer coefficients is from 144 to 186 $W/m^2.K$.

The same parameters in the analysis apply as for the paint systems in the previous case. The range in overall heat transfer coefficients compare to quoted values (17) of 350 - 900 $W/m^2.K$ for light oil to water heat transfer.

3. Freshwater flow

The internal convective heat transfer coefficients for the freshwater flow are shown in Fig 7. As would be expected these are more than an order of magnitude greater than for the residual oil flows, 7500 - 8200 $W/m^2.K$ cf. 47 - 220 $W/m^2.K$. The total resistances of the pipeline per m length are shown in Fig 5. The results for the thermal resistances of the coatings as a percentage of the total thermal resistances against bulk freshwater temperature are shown in Fig 6. From these results it can be seen:

(i) The total thermal resistances of the pipeline are almost unaffected by the bulk temperature of the freshwater.

(ii) The maximum thermal resistance of the coating can now account for slightly more than 70% of the total in the process. The corresponding overall heat transfer coefficient is some 435 $W/m^2.K$, there being no noticeable variation over the range in internal fluid bulk temperature.

(iii) The minimum thermal resistance of the coatings in the total process is slightly lower than 40%. The corresponding overall heat transfer coefficient is some 1955 $W/m^2.K$ and again there being no noticeable variation with change in internal fluid bulk temperature.

The same parameters in the analysis apply as in the two previous cases. However it can be appreciated that the thermal resistance of the coatings in this case is of the same order as that of the internal and external convective films. The range in overall heat transfer coefficients compare to quoted values (17) of 800 - 1500 $W/m^2.K$ for water to water heat transfer.

5 CONCLUSIONS

1. The analyses showed that overall heat transfer coefficients for pipelines coated with modern paints can vary outside of the ranges quoted for indexed coefficients.

2. In the case of a heavy oil flow the influence of the type of paint system used can account for a difference of approximately 5% in overall heat transfer coefficients.

3. In the case of a light oil flow the type of paint system can account for some 25% difference in overall heat transfer coefficients.

4. In the. case of a freshwater flow the type of paint system can account for a difference of some 350% in the overall heat transfer coefficients.

5. The relevance of these results is seen to be that in the knowledge of the internal and external fluid flow, the pipe wall material, the paint type and thickness, the overall heat transfer coefficients can be calculated with a reasonable certainty. Fouling factors on either or both surfaces can be introduced in a normal manner.

The results indicate that it is possible to take the heat transfer calculations of paint protected underwater steelwork into a true technological dimension with an associated impact on international manufacturing. Coincidentally it is a century since the publication of Lees' work **(18)** on poor conductors at Owens College.

ACKNOWLEDGEMENTS

The experimental work measuring the thermophysical properties of the paints was originally funded under the Total Technology Programme of the SERC and the Co-operating Body was Shell International Marine Ltd. The author wishes to acknowledge the contribution of the paint manufacturer, W & J Leigh & Co. Nor should it pass without mentioning the contributions of Professors A J Morton and J D Jackson to the experimental work. Full responsibility for all facts and opinions remains, however, with the author.

REFERENCES

1 *Thermophysical Properties of Matter, Volumes 2 and 5,* 1978, The TPRC Data Series, IFI/Plenum.

2 **Mills, W. G. B.** *Paint Chemist's Handbook,* 1962, Scott Greenwood.

3 **Roff, W. J. and Scott, J. R.** *Fibres, Films, Plastics and Rubbers,* 1971, Butterworth.

4 **Gubareff, G. G., Janssen, J. E. and Torborg, R. H.** *Thermal Radiation Properties Survey,* 1960, Honeywell Research Center.

5 *Physical and chemical properties of polymers and moulding compounds,* Shell Plastics Technical Bulletin No. T3003/8.82/700.

6 **Hatchman, J. C.** In Proceedings of Second UK National Conference on Heat Transfer, Glasgow, 14 - 16 September 1988, paper C208/88.

7 **Jackson, J. D. and Hatchman, J. C.** Paper submitted to IMechE, 1998.

8 **Stephenson, P. L.** In Proceedings of Second UK National Conference on Heat Transfer,

Glasgow, 14 - 16 September 1988, paper C149/88.
9 **Newberry, P. J.** *Trans IMarE,* 1996, **108**, 241.
10 **Rose, J. W. and Cooper, J. R.** *Technical Data on Fuel,* 1977, British National Committee World Energy Conference.
11 **Rogers, G. F. C. and Mayhew, Y. R.** *Thermodynamic and Transport Properties of Fluids - SI Units,* 1988, Basil Blackwell.
12 Editor **F. G. Walton Smith,** *Handbook of Marine Science,* 1974, CRC.
13 **Cox, R. A. and Smith, N. D.** *Proceedings of the Royal Society,* 1959, A, **252**, 51.
14 **Gnielinski, V.** *Int. Chem. Engng.* 1976, **16**, 359.
15 **Sleicher, C. A. and Rouse, M. W.** *Int. J. Heat Mass Transfer,* 1975, **18**, 677.
16 **Zukauskas, A.** In *Handbook of Single-Phase Convective Heat Transfer,* 1987, Wiley-Interscience.
17 Editors **Perry, R. H. and Green, D. W.** *Chemical Engineers Handbook,* 1997, McGraw-Hill.
18 **Lees, C. H.** *Phil. Trans. Roy. Soc. of London,* 1898, Series A, **191**, 399.

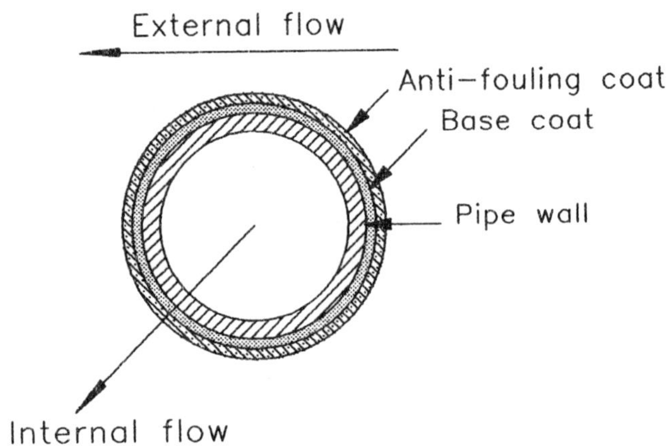

Fig 1 Exaggerated view of coated pipe system

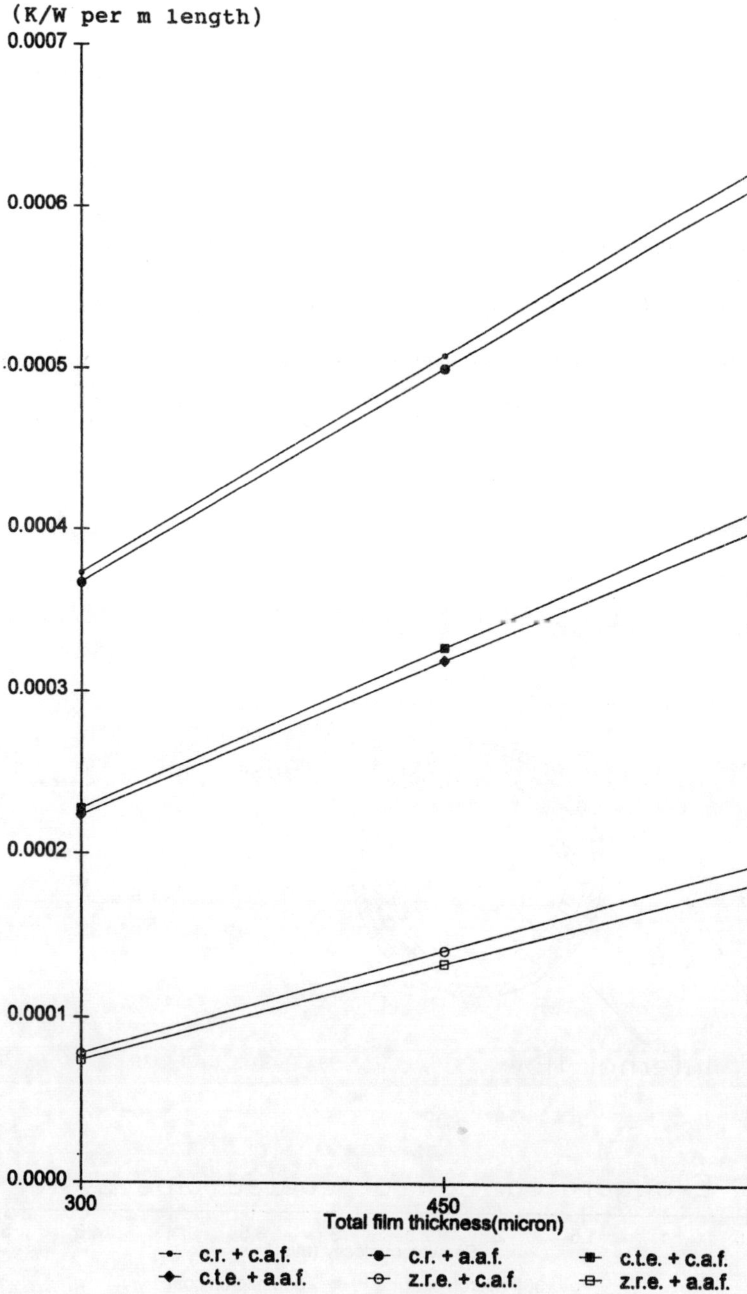

Fig 2 Coating thermal resistance against film thickness

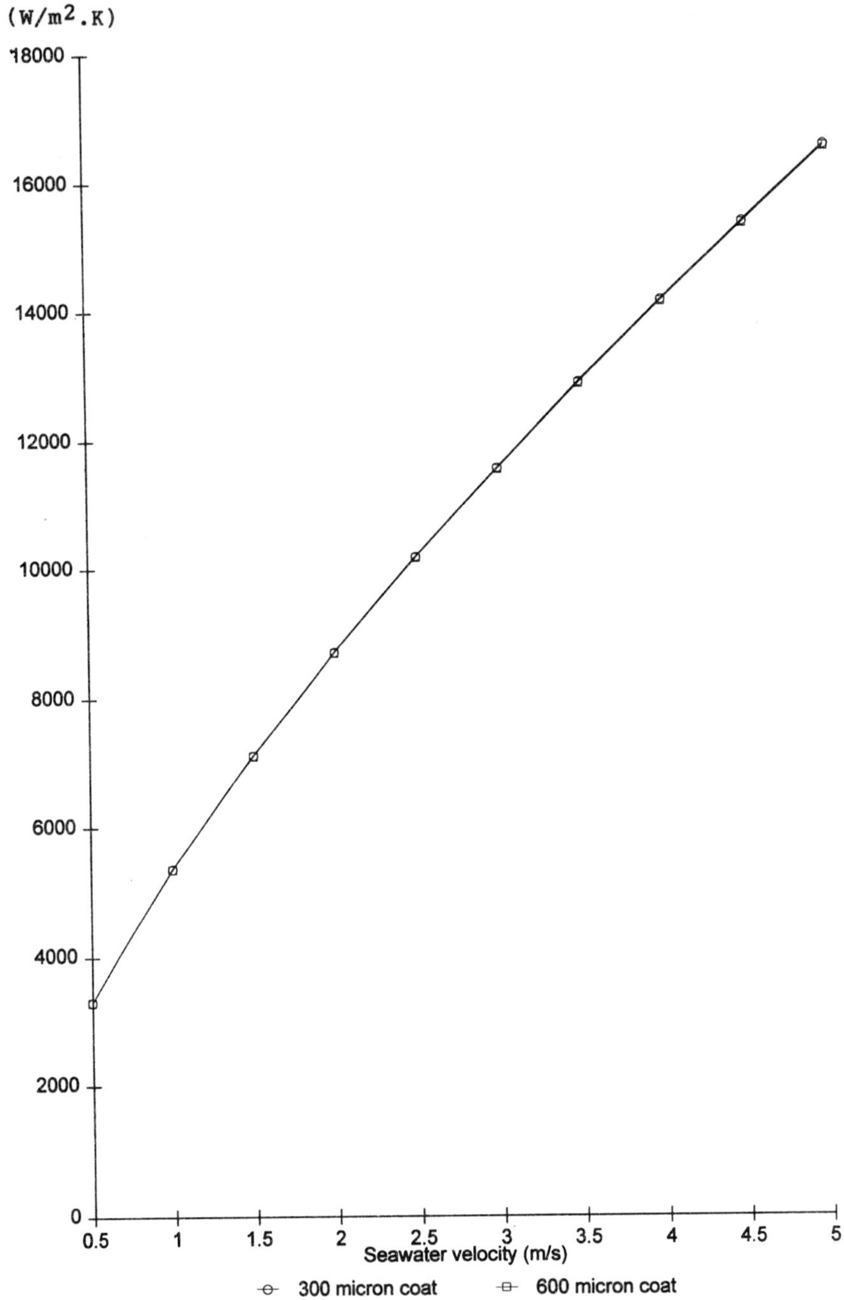

Fig 3 External heat transfer coefficients

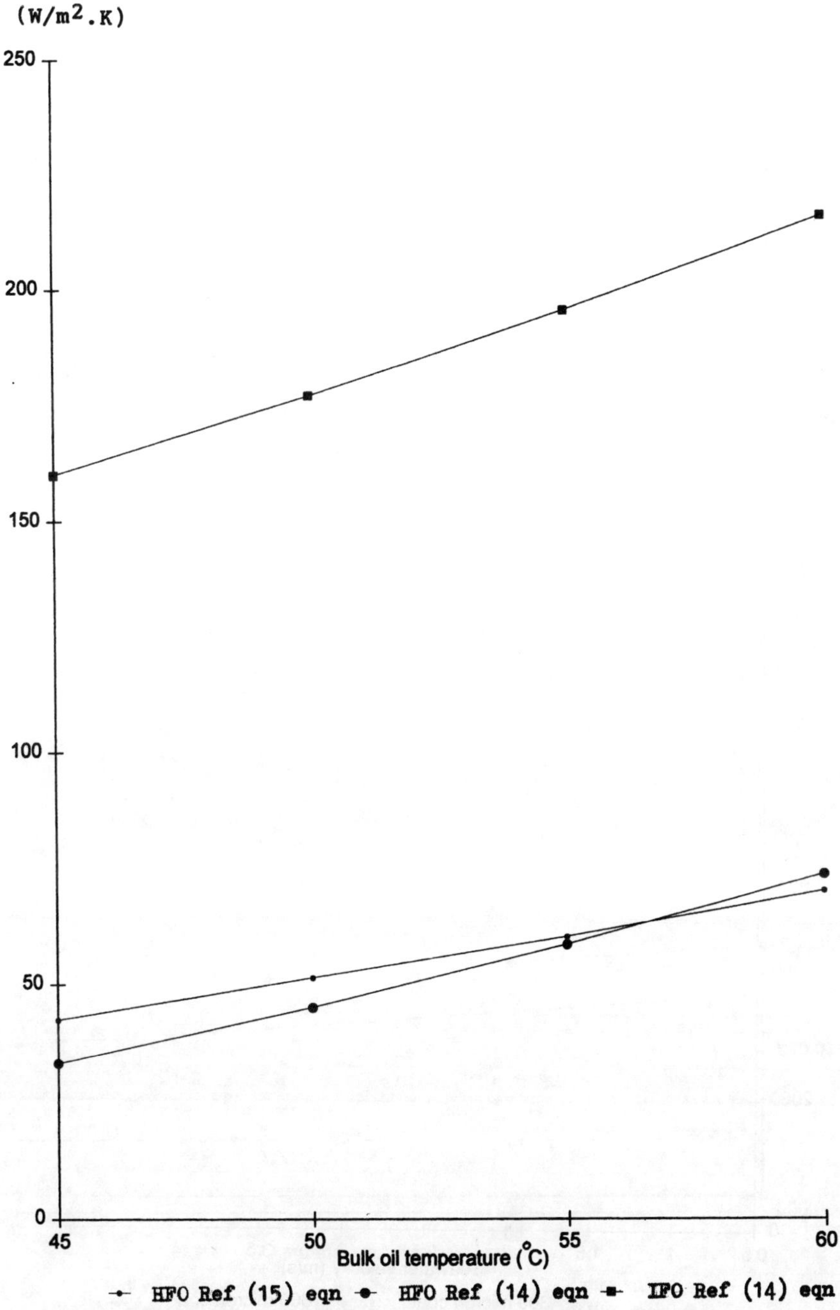

Fig 4 Internal heat transfer coefficients for oil flow

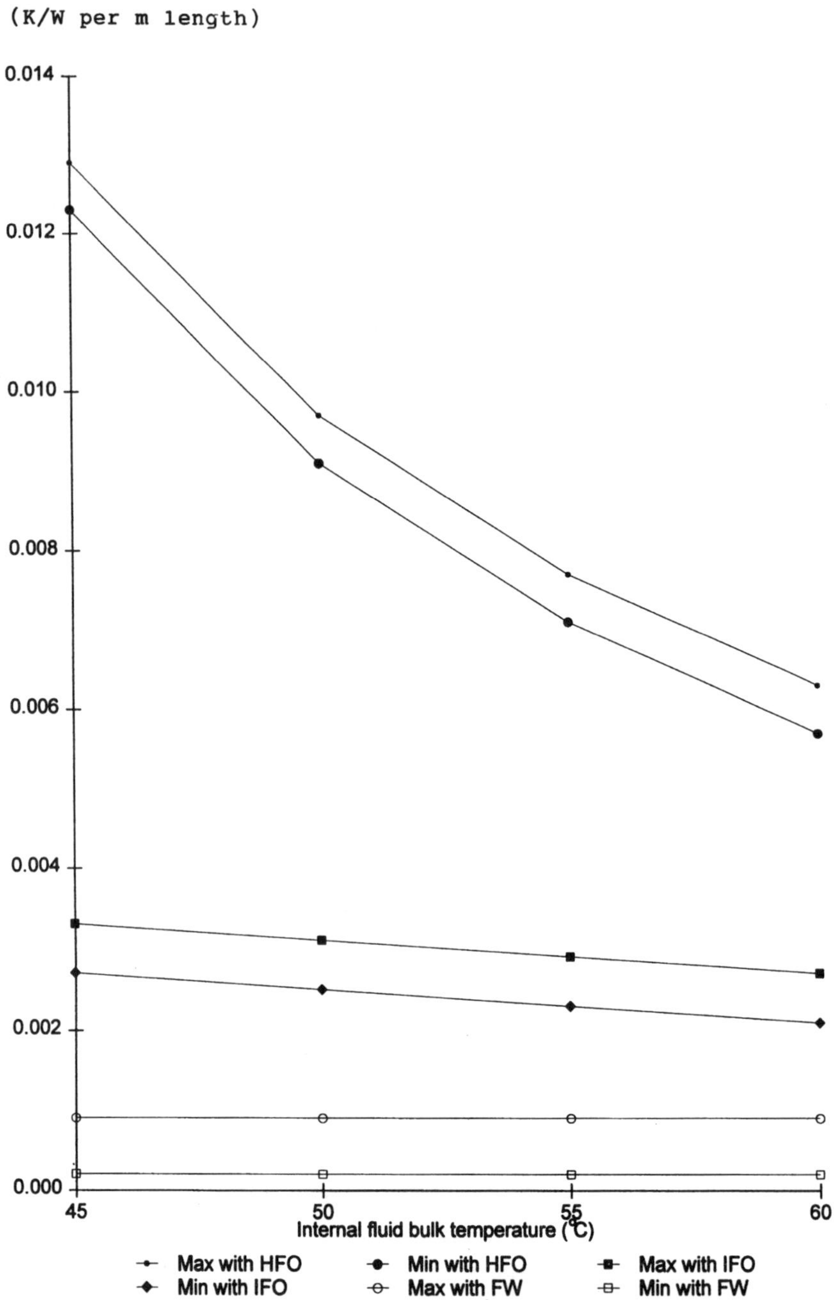

Fig 5 Total thermal resistance of the pipelines

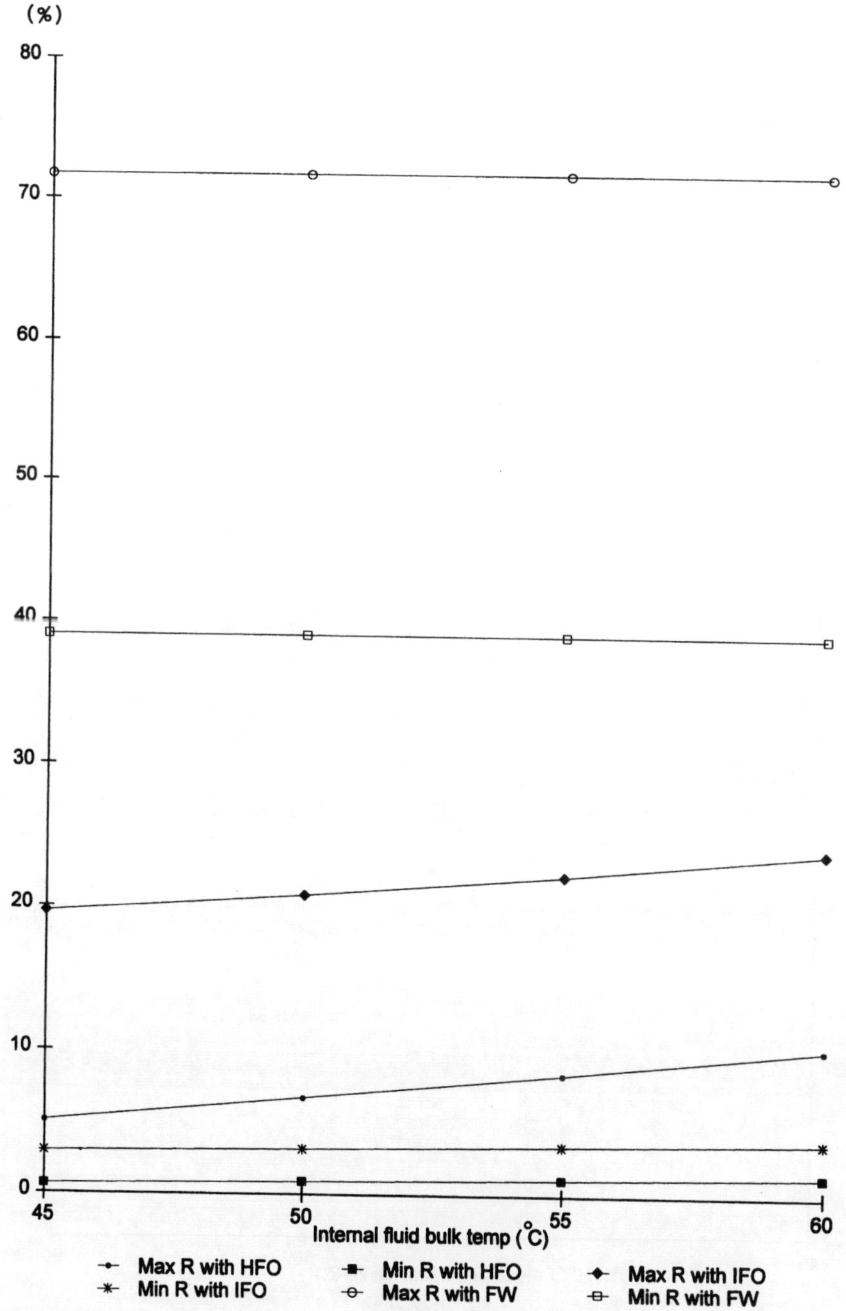

Fig 6 Coating thermal resistance as (%) of total resistance

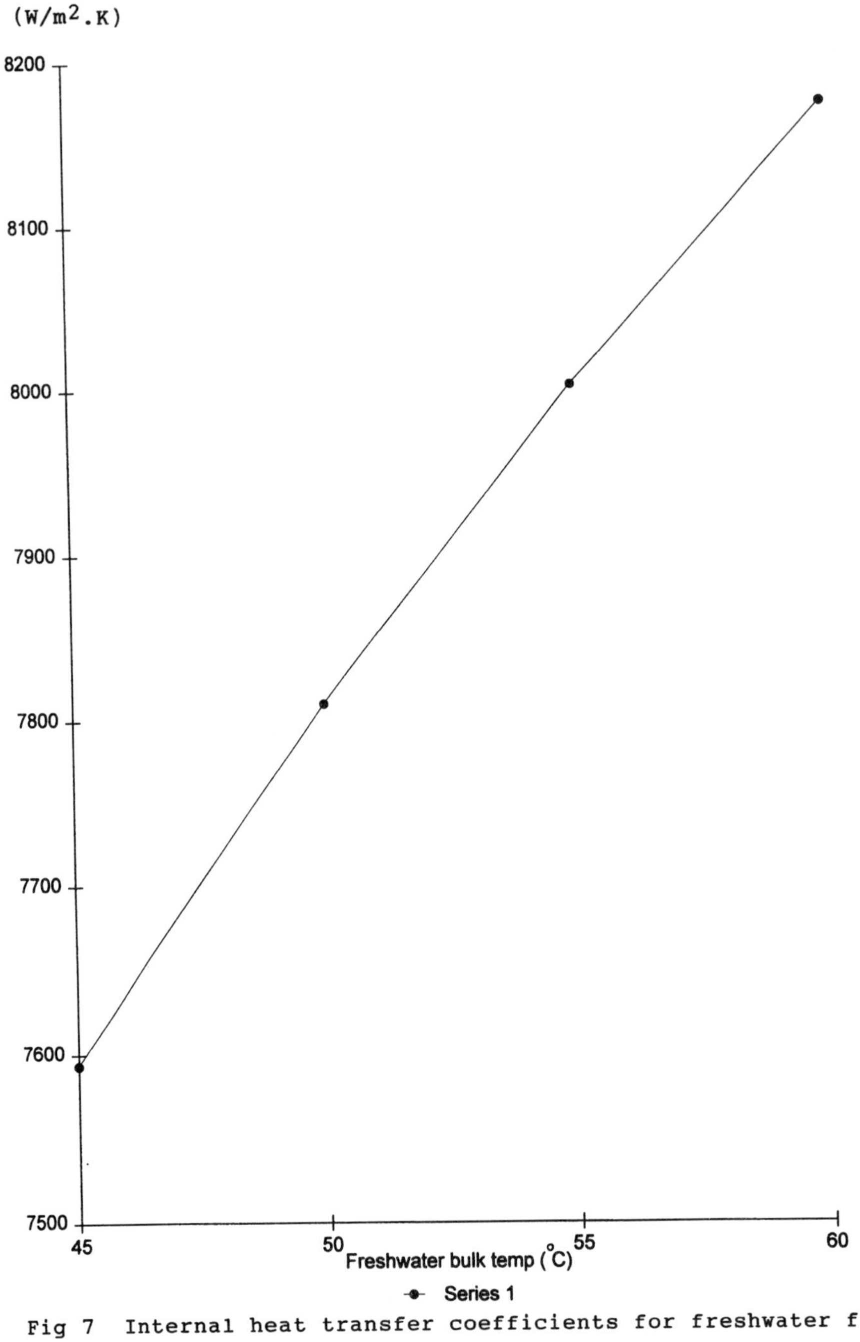

Fig 7 Internal heat transfer coefficients for freshwater flow

Case study on the manufacture of a low-pressure die-casting component used as a major structural member for the Jaguar XK8 sports car

J JACKSON
Alumasc Limited, Kettering, UK
J E KEMPSON
Faculty of Engineering and Computer Technology, University of Central England, Birmingham

ABSTRACT

The project in this case study involved the installation and commissioning of a (i) low pressure die casting machine, (ii) a 5 axis machining centre and (iii) a CNC Co-ordinate measuring machine. The paper describes how the production on this large equipment was initiated, proved to the customer and brought to a production level that was 40% beyond the maximum planned by the customer when the plant was being specified. Technology innovations were achieved in both the die casting process and the monitoring systems added to the casting machine. The process capability of the machining was monitored by the CMM. Significant changes in the methods of measuring this major structural component were required before the process could be described as within statistical control.

1.0 INTRODUCTION

The Case study describes the introduction of Advanced Manufacturing Technology Equipment and Systems. The first part of the project involved the running of an innovative low pressure die casting process together with implementing an improved PC computer based monitoring systems. The second part involved the installation, commissioning and operation of a five axis machining centre and a CNC CMM (Co-ordinate Measuring machine). The CMM was to be used as a gauge for 100% inspection.

PBM Components is a company within the UK group Birmid Holdings Ltd. This group is owned by the global German steel manufacturing company Thyssen Gmbh. The Birmid Group has a long history as a group of foundries which have latterly concentrated on non-ferrous components. PBM components has in turn concentrated on production of high volume automotive parts supplying primarily to global companies such as Ford, GM, Toyota, Nissan and Volvo. The company has a range of aluminium die-casting processes including conventional gravity casting and computerised High Pressure Diecasting machines. At the beginning of the project the company had just commissioned a GIMA low pressure die casting machine that could produce castings of some 18 kg in weight.

1.1 THE JAGUAR CONTRACT

The Jaguar XK8 Sports Car was the first new sports model since the introduction of the XKS some twenty years previously. For the company to make significant progress in the specification of the vehicle particularly in powered accessories such as air-conditioning, it was important to reduce weight in existing members wherever possible. The Jaguar engine is supported on the suspension via a sub-frame which is traditionally made from a welded steel fabrication. It is this sub-frame, whose length is the full width of the car, that has been made as an aluminium die-casting. The contract for the manufacture of this component was won by PBM Components after initial prove out of the concept was achieved at the German plant where the GIMA casting machine was manufactured. As this was an innovative design change, Jaguar also commissioned the production of a prototype steel fabrication as a fall back position in case the technology proved hard to run at production rates. It was soon found in testing that this fall back position was not in fact usable as the weight distribution changed the performance of the car too radically. At this time Ford Motor Company took over Jaguar on the expected successful introduction of the XK8.

One of the factors, reported by the MD, of the company's success in bidding for this contract was having the plans in place to run a Teaching Company Programme between UCE and PBM. This was seen as a mechanism to overcome some of the technological hurdles that would be posed. The TCAs would compliment the existing experienced team that had seen considerable retrenchment, as the automotive industry has regrouped and decreased in size in order to become competitive, and sufficiently agile to maintain world class standards of productivity. Other major factors in winning this contract were

(i) the close proximity of PBM to the assembly line some 17 miles away in Coventry.

(ii) the capability of the company to take on the casting, the heat treatment and the final machining of the component all on one site.

2.0 THE GIMA CASTING PROCESS

This low pressure die casting process is an innovative process that uses a vertical hydraulic closing of the dies together with the automatic provision of ladling of the molten aluminium. Significant development of the die cavity, the venting, the temperature control and the exact relative positioning of the two dies has been achieved which enables very large castings the width of the Jaguar sports car to be cast. However the control of these factors by limit switch for position and manual thumb wheels for temperature was not a simple task. The plant is operated on a four shift continental timing basis so that operator variations produced scrap castings only too easily. It was also found that the best way to return to the production of good castings was to cast out of any problems, thus scrapping a minimum of 5 castings each time a malfunction occurred. That minimum assumed the controls were reset properly. The opening time of the dies was critical, so that any slight problem with extraction tended to take the machine out of its desired process configuration. The solution to this process variability was to design a real time monitoring system that would prompt the operators to take the right corrective action such as manually intervene with a die warming gas jet. This was particularly successfully achieved by doing a trend analysis on the thermocouple outputs as shown graphically in fig 1. From the implementation of this controller the percentage of good components rose as much as 55%. The system was developed by using a commercial analysis package. The TCA developed a bespoke front end to display the results required to the company.

Further work was also needed on the archiving of data recorded from the automated X-ray facility. Each casting had four significant displays saved onto optical discs in a very hostile environment. The task of logging this information for a hot casting, which was to subsequently go through heat treatment, accurately was a significant task. As the component was a structural member of the vehicle an audit trail was necessary in case failure or apparent failure occurred, particularly as the vehicle is sold extensively in the United States of America.

3.0 THE MACHINING OF THE CASTINGS

A Huller Hille five axis machining centre had been purchased as part of a Flexible Manufacturing Cell(FMC) by Jaguar for installation into PBM together with the DEA Co-ordinate Measuring Machine. The main task for the TCA for this project was to demonstrate to the company and the customer that the process capability of the FMC was to the standard required for World Class Manufacturing. As reported by Eraifej(1), FMC commissioning has significant technological hurdles to overcome and measurement of performance needs to be organised systematically. For this purpose SPC software was purchased for use on the CMM. It however soon became apparent that there were a considerable number of process variables that were not under control. Eliminating these one by one was the programme for the two years, with the added constraint that production rates had to be maintained. This production initially consisted of machining the sub-frames for the prototype vehicles followed by those for the prove-out vehicles and eventually the full production run. There was even a period when the production required was 140% of the planned production of the plant. The cell had always been planned to run for 24 hours a day so there was no spare capacity available apart from making improvements in set up times, cycle times, maintenance response etc. Eventually a second five axis machine was obtained by Thyssen to respond to this pressure, though this machine had originally been planned to be used for a second component at PBM. The justification of the capital employed was in this case not a significant exercise as Thyssen had unused machines within its global resources, which was is very different from most Small and Medium sized Enterprises (SMEs) (Hayward 3).

Process capability constraints:- The inspection of this item was not simple, particularly as there were two features that set up the angle of the front suspension arms of the car (see fig 2). The fourth axis of the machine helped in the twist aspect. At the planning stage it was hoped that the fifth axis would take care of the remaining tilt requirement for the tracking. However the fifth axis was a C axis rather than an A axis so that this final rotation had to be achieved by having a manually operated fixture. This meant that the programme had to mid-cycle extract the pallet from the machine, wait for the operator to release clamps raise or lower a relatively heavy fixture, push in shot bolts and reclamp. Not an operation that is conducive to effective automatic 24 hour operation! Significant human failings were observed during the two years. These failings also included mis-positioning of the casting so that it was stressed as the clamps were tightened. The castings do vary considerably and the fixture allowed for this, but with frequent use operators tended to clamp in the wrong order, forget to release the spring loaded restraints or forget to tighten them after loading (see fig 3) such that significant dimensional variation was observed.

During the project considerable metrology problems had to be overcome, initially the DEA CMM appeared to give gross errors even on simple dimensions that could be checked with slip gauges. This was eventually found to be due to the way the datum for the component had been assigned on the drawing. There then ensued a lengthy and often heated discussion, that took a number of weeks, between the customer, the company and the CMM supplier with the Teaching Company Associate acting as communicator. The CMM company had provided the programme for their machine as part of a turnkey contract. In the end success was achieved but the conclusion was drawn that programmable machines such as CMMs need to be fully controlled by the user company as the requirements of customers do change almost weekly. Even when, as in this case, the machine was only measuring one component. Secondly it became apparent that one CMM programme was needed to give the company setting adjustments for the machining centres, but another programme was needed to ensure that the inspection results obtained on PBMs machine were equivalent to those produced on the customer's CMM. It should be noted that the previous fabricated steel component used for this application had not been inspected in this manner. The technology had not been available at the fabricators plant. One of the major dimensional variations was eventually tracked down to the tilting fixture previously described. Another was due to unintentional forced deflection of this large casting. This was particularly possible as the operators had the use of pneumatic drivers to tighten the many clamping nuts. As can be imagined obtaining process capability data that was at all acceptable was extremely difficult so that 100% inspection is still needed. This has associated dangers that are particularly relevant when operating on a 24 hour basis as described by Hallak (2)

4.0 HEAT TREATMENT AND OTHER PROCESSES.

The company had traditionally used a local sub-contractor to heat treat their die castings. However, as part of the plan for the production of this component, agreement had been reached with this company that new plant would be purchased and employees from the sub-contractor would be employed to run the new plant. In this way the logistics of the supply of the casting to Jaguar Cars was much improved. This was by no means the only new plant installed for the contract. Every casting was crack detected with the various work stations for application of dye penetrant and subsequent washing baths. A deburring drum of considerable size was also installed as part of the Flexible Manufacturing Cell commissioned with the machining centre to minimise the effects of flash. A straightening plant was also needed to compensate for the deflections caused as the hot casting was handled after its extraction from the metal dies.

5.0 CONCLUSIONS

The project enabled Jaguar to introduce the vehicle at a greater rate of delivery than had been planned. The commercial pressure to achieve this was enormous and the further introduction of an extra CNC machine and process equipment made the Engineering task even more exacting. During the two years the value of the capital plant introduced was greater than £2,000,000. This was an unusual situation where the availability of capital plant was almost greater than the capability of the company to install and run it. It meant that significant responsibility was given to everyone involved who had any technical expertise.

6.0 ACKNOWLEDGEMENTS

This case study comes from work done as part of a teaching company scheme between PBM Components Ltd and the University of Central England in Birmingham. The project was part financed by both EPSRC and dti.

7.0 REFERENCES

1. Erafej F.A., Kempson J.E., Oraifige I.O. 'The commissioning of flexible manufacturing cells' Proceedings of 4th IMCC, Beijing Institute of Technology, Beijing April 1989 pgs A197-212
2. Hallak H, Kempson J.E. 'The implementation of In-Process Measurement and Control Procedures to enhance process capability and reduce set up times' 12th NCMR Glasgow Caledonian University Sept 1997 pgs 265-271
3. Hayward A.S. Kempson J.E 'Strategic Influence of TCS on Capital Projects' TTI Conference, London July 1994 pgs 24-28

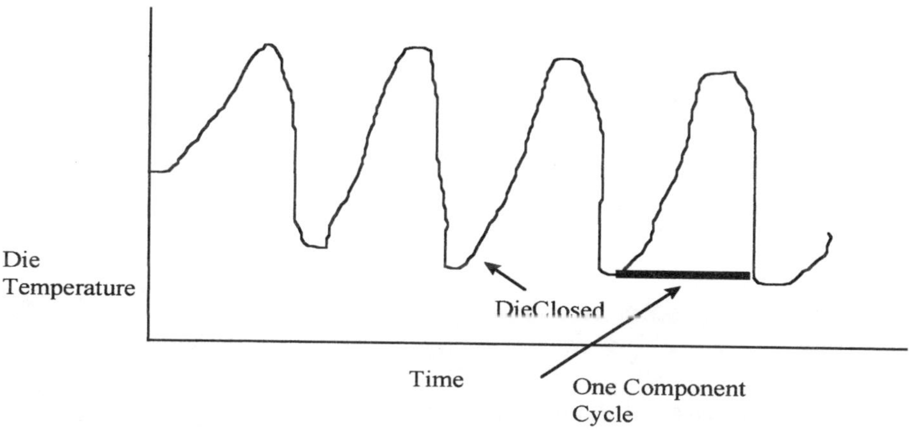

Figure 1

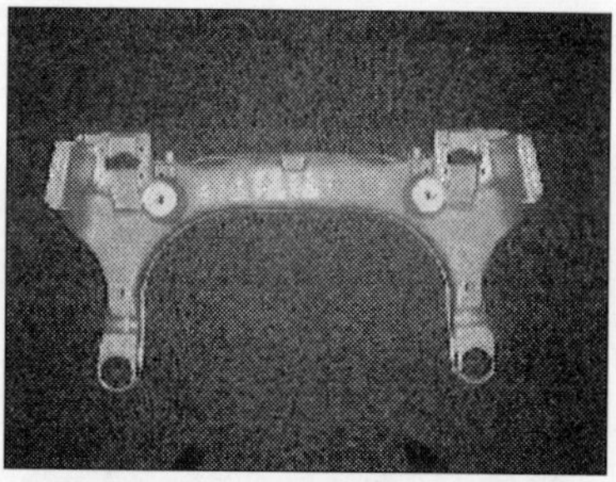

Figure 2

Jaguar XK8 Sub-frame

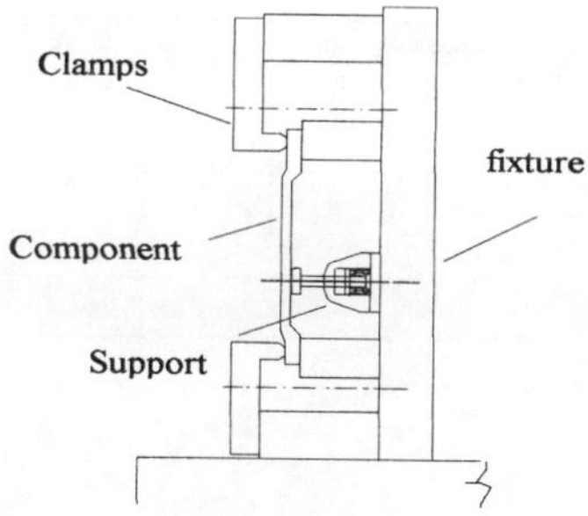

Figure 3

Fixture for X300 Component

A statistical evaluation of friction promoting chemicals

P J LEE and **J M O'SULLIVAN**
Department of Manufacturing Systems Engineering, University of Hertfordshire, Hatfield, UK

This paper is an attempt to evaluate the potential of alternative chemicals in the pretreatment of aluminium alloy substrates prior to such operations as friction welding. The main aim was to determine whether it is possible to use less hazardous chemicals than the solvents used at present.

It was found that under the right conditions methyl containing siloxanes and paraffin oils can promote around a four fold increase in friction when compared to dry / untreated sliding. This is comparable with the results achieved using acetone and trichloroethylene which suggests these less hazardous chemicals would make good alternatives.

1.0. INTRODUCTION.

Solvents such as acetone and trichloroethylene are used to prepare aluminium surfaces prior to such joining operations as friction welding [1]. However, concern has arisen in recent years about the impact such solvents may have on health and the environment [2].

It is known that the cleansing action (removal of dirt, grease and other foreign material from the surface) makes the surface more receptive to adhesion [1]. The extent to which these fluids assist adhesion promotion is however unknown. In fact, very little quantitative data exist on the theory behind lubricated friction let alone an area as specific as this [3].

A category of fluid that has received little coverage is poor boundary lubricants. The best documented of these is probably the methyl containing siloxanes. These have been labelled as positive antagonists since they not only act as poor boundary lubricants but positively encourage friction and wear [4]. The siloxanes are not the only fluids to have been found to behave in this manner. Paraffin oil [5], carbontetrachloride [6] and water [5] have all been said to behave in a similar manner.

Unfortunately, as with the solvents, very little quantitative data exist on the interaction of these fluids with aluminium alloy substrates. What information exists is spread over many years, test techniques and working practices. An aim of this investigation, therefore, was to determine the extent, in terms of both magnitude and consistency of effect, to which the solvents and poor boundary lubricants encourage friction and wear. This is achieved by undertaking tests using each of the individual fluids and comparing them with results from dry / untreated tests. This information will be used to rank the chemicals and to determine the potential of the alternative chemicals in the pretreatment process.

2.0. METHOD.

A number of the test parameters were determined on the basis of the application requirements, previous literature and commercial constraints.

2.1. Fluids tested.

Acetone and trichloroethylene were tested to enable a comparison to be made between the solvents and the poor boundary lubricants. The siloxanes tested were polydimethylsiloxane (PDMS) and polydimethylphenylsiloxane (PDMPS). Although the affect on friction of substituting the methyl end groups for phenyl (which is the distinguishing factor between the two aforementioned siloxanes) has been the subject of much debate these are generally regarded as the two siloxanes that most encourage friction and wear [4]. These fluids are commercially available in a number of viscosities and hence a range were tested.

The paraffin oils tested were pentane (alkane C_5), hexane (alkane C_6) and heptane (alkane C_7). Distilled water was also tested (If found to be suitable in terms of its friction promoting attributes it would be the perfect solution from an economic as well as environmental stand point). Although carbontetrachloride must be regarded the most environmentally unfriendly of the chemicals discussed it was considered appropriate to test to ensure the investigation could be classified as a comprehensive evaluation of the fluids said to increase friction.

2.2. Material to be tested.

Aluminium alloy 6082T6 was considered an appropriate alloy for testing. This belongs to the 6000 series (Al - Mg - Si) which is the most popular extruded aluminium [7]. The cylindrical geometry is most appropriate for the friction welding operation.

The material would be tested in the extruded and the machined condition to enable the influence of the oxide layer and the method of manufacture to be evaluated.

2.3. Normal load and velocity of sliding.

These factors are influential in determining the lubrication regime in which sliding is taking place [8]. For this reason, preliminary dry / untreated tests were undertaken whereby the normal load and sliding velocity was adjusted until the resulting friction fell at the lower limit anticipated for boundary lubrication. The anticipated coefficient of friction range of boundary lubrication is $\mu = 0.1$ to 0.25 [9], therefore the aim was to produce a level of friction on the dry sliding of $\mu = 0.1$. This would enable any increase in friction during the fluid tests to be attributed to the action of the fluid and not the severity of the initial test conditions.

2.4. Method of applying the fluid.

The quickest and simplest method of fluid application was adopted since this is the most probable technique used in industry. This involved wiping the fluid onto the surface to be tested using a lint free cloth.

2.5. Surface finish.

The surface of the extruded substrate is dictated by the extrusion process (Ra = 0.18μm). To ensure a good level of consistency the machined surface was produced on a CNC lathe and could be deemed a good turned surface (Ra = 0.50μm).

2.6. Test method.

A cross cylinder friction test rig was considered appropriate since this permits repeat trials to be undertaken on individual test cylinders. This would then enable the consistency of results to be appraised.

2.7. Sample size.

To balance the desired accuracy with the economics (cost and time) the minimum sample size considered representative of a population was used, namely, thirty [10]. This represented the number of runs that was undertaken under the same test conditions, i.e. using the same fluid on the same surface condition. Ten runs would be undertaken on each of three cylinders. This enabled the consistency both within individual and between cylinders to be evaluated.

2.8. Analysis technique.

The statistical tool "Analysis of Variance" (ANOVA) was considered appropriate since it enables the consistency as well as the magnitude of results to be analysed. ANOVA enables the consistency to be evaluated in three ways [11]:-

- **F Factor.** This determines whether the variation between cylinders is significant or whether it is simply due to the variation inherent in the system.
- **Coefficient of variation (%).** This enables the relative variation in each trial to be determined and enables a comparison between trials to be made.
- **Standard error.** This represents the variation within cylinders.

In all three consistency measures the lower the derived value the greater the consistency.

3.0. RESULTS.

The results from the trials can be seen in Table 1 for the extruded trials and Table 2 for the CNC machined trials. The normal load used was 93 Newtons and the sliding velocity 0.0004 metres per second.

Table 1. The results from the extruded trials.

Fluid type	Mean	F Factor			Standard	Coefficient
	(μ)	Actual	$F_{.05} = 3.35$	$F_{.01} = 5.49$	error	of variation
Dry test	0.11	3.80	✗	✓	0.43	13.68
Water	0.07	5.09	✗	✓	0.15	6.84
5cS PDMS	0.52	3.31	✓	✓	1.19	7.73
1000cS PDMS	0.54	17.14	✗	✗	1.55	9.58
60 000cS PDMS	0.51	1.58	✓	✓	0.79	5.25
DC 510	0.48	7.71	✗	✗	1.17	8.28
DC 550	0.51	2.72	✓	✓	0.85	5.66
DC 710	0.50	0.88	✓	✓	0.72	4.94
Pentane	0.43	22.85	✗	✗	2.64	20.69
Hexane	0.29	3.18	✓	✓	1.80	21.04
Heptane	0.49	14.92	✗	✗	2.24	15.43
Carbontetrachloride	0.57	0.06	✓	✓	2.81	16.75
Trichloroethylene	0.29	2.07	✓	✓	3.58	42.48
Acetone	0.62	1.86	✓	✓	2.64	14.56

The significance levels ($F_{.01}$ and $F_{.05}$) are the two most popular and indicate that 99% and 95% of the population would fall within these limits [11]. A ✓ signifies that the variation is due to variability inherent in the system and a ✗ signifies a real effect / difference between cylinders.

Table 2. The results from the CNC machined trials.

Fluid type	Mean	F Factor			Standard	Coefficient
	(μ)	Actual	$F_{.05} = 3.35$	$F_{.01} = 5.49$	error	of variation
Dry test	0.13	2.24	✓	✓	0.14	3.66
Water	0.12	4.84	✗	✓	0.08	2.24
5cS PDMS	0.50	1.10	✓	✓	0.99	6.69
1000cS PDMS	0.50	0.01	✓	✓	1.89	12.91
60 000cS PDMS	0.47	3.27	✓	✓	1.02	7.32
DC 510	0.42	2.19	✓	✓	0.50	4.06
DC 550	0.45	1.32	✓	✓	1.26	9.45
DC 710	0.47	0.20	✓	✓	1.17	8.49
Pentane	0.40	35.43	✗	✗	1.61	13.57
Hexane	0.19	10.72	✗	✗	0.65	11.65
Heptane	0.49	62.39	✗	✗	1.54	10.62
Carbontetrachloride	0.40	7.25	✗	✗	3.39	28.69
Trichloroethylene	0.49	9.12	✗	✗	3.50	24.27
Acetone	0.33	5.33	✗	✓	1.57	16.04

4.0. DISCUSSION AND CONCLUSION.

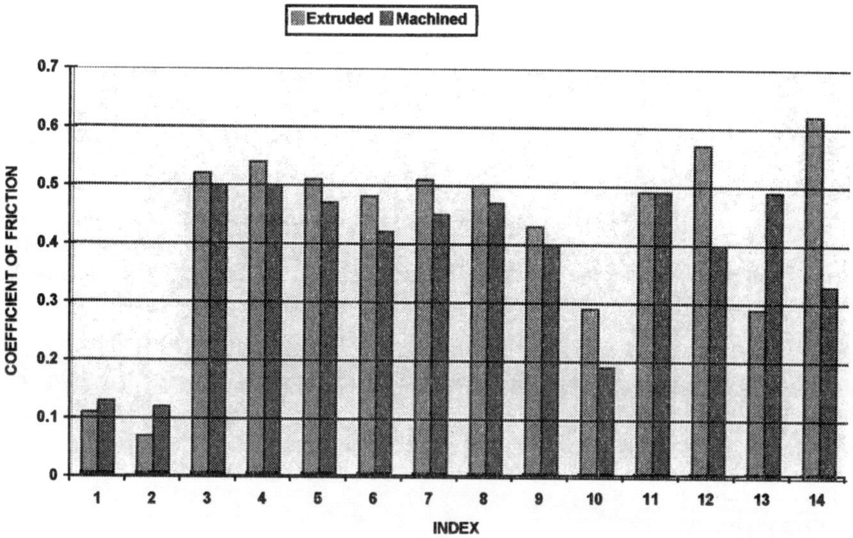

Index:- 1 = Dry test, 2 = Water, 3 = 5cS PDMS, 4 = 1000cS PDMS, 5 = 60000cS PDMS, 6 = DC 510, 7 = DC 550, 8 = DC 710, 9 = Pentane, 10 = Hexane, 11 = Heptane, 12 = Carbontetrachloride, 13 = Trichloroethylene, and 14 = Acetone.

Figure 1. A comparison between the extruded and machined results.

Figure 1 shows that out of all the fluids tested water was the only one which acted as a lubricant, i.e. to have a lower coefficient of friction than the dry sliding trials. Since it is known that water can promote friction on certain other substrate materials this supports suggestions that the type of substrate used is very significant [5]. Acetone promoted the highest friction on the extruded surfaces ($\mu = 0.62$) but was the fourth lowest when a machined surface was used ($\mu = 0.33$). This implies that the relative performance of each fluid in terms of its friction promoting ability is dependent on the substrate condition as well as substrate type. This substrate dependence prevents the substances being ranked in terms of their general friction promoting ability.

Figure 1 also shows that in ten out of fourteen of the cases the extruded condition produces the higher friction. Since the four trials that did not show this trend include the dry and the water trials, and the most inconsistent in the form of the extruded trichloroethylene trial (coefficient of variation = 42.48%) this appears to represent a real effect. In terms of the physical attributes of the surface the extruded substrate would be slightly harder than that of the machined surface, and as discussed in section 2.5., is much smoother. It would be anticipated from a physical perspective, therefore, that the machined surface would produce the higher friction. i.e. the softer the surface the easier it is to damage and the rougher it is the harder it is to slide.

One possible explanation for this trend lies in the method of manufacture. To minimise possible contamination the extrusion process uses no lubricant. Therefore the majority of contamination would be accumulated during transportation, storage and handling between the extrusion process and the test piece manufacture. This would be easily removed by the wiping process. In the machining process coolant is used to prolong tool life and to impart a good surface finish. It may however be the interaction of the coolant with the freshly cleaved surface which causes a stronger surface film to be produced that is harder to remove by the wiping process.

Table 2 shows a distinct relationship between fluid type and consistency (F - Factor). The siloxanes all show a high level of consistency whereas acetone and water were the only other fluids to show consistency at the 1 % level. This implies that the consistency of effect is also influenced by the surface contamination / type and not just the magnitude of effect.

To conclude, this investigation highlights the benefits of the pretreatment process with the majority of fluids inducing at least a four fold increase in friction when compared to dry sliding under the same test conditions. In addition, it has shown that there are a number of alternative fluids that have the potential to be very good friction promoters.

5.0. FURTHER WORK

Further work will be undertaken to prove or otherwise a number of suggestions made within the discussion. For example, an investigation into the influence coolant has on friction will be undertaken by means of more in-depth chemical analysis of the surface.

6.0. REFERENCES.

1. T. Tabata, S. Masaki, and K. Azekura, Bond criterion in cold pressure welding of aluminium, Journal Jpn Soc. Technol. Plast., 1986, Vol. 27, No. 303, pp. 521 - 525.
2. T. Green, Trichloroethylene induced cancer in animals and its relevance to humans. Journal of Occupational Health., 1997, Vol. 39, No. 4, pp. 261 - 273.
3. A. Dorinson, and K. C. Ludema, Mechanics and Chemistry in Lubrication, Elsevier., 1985.
4. A. R. Lansdown, The effects of oxygen availability on the lubricating performance of a dimethyl silicone in the boundary lubrication regime, Wear., 1994, No. 175, pp. 25 - 38.
5. W. B. Hardy, General properties of boundary films, Collected Works, Cambridge University Press., 1936.
6. T. Shirakashi, R. Komanduri, and M. C. Shaw, Conflicting roles of carbontetrachloride as a boundary lubricant, ASME, Journal of Engineering for Industry., 1978, No. 5, pp. 3 - 7.
7. J. V. Rijkom and W. S. Miller, Improved properties of extruded products by heat treatment of AA6082 alloy sections. Proceedings of the 6th International Aluminium Extrusion Technology Seminar, Chicago., 1996, Vol. 1, p. 149.
8. M. D. Hersey, The laws of lubrication of horizontal journal bearings, Journal Washington Academy of Science., 1914, Vol 4, pp. 542 - 552.
9. R. Zechel, Molykote, a publisher of Dow Corning Gmbh Munich., 1991.
10. L. L. Lapin, Statistics for modern business decisions, The Dryden Press, 6th edition., 1993.
11. G. Petersen, Design and analysis of experiments, Marcel Dekker., 1985.

Environmentally clean metal cutting processes – machining on the way to dry cutting

H POPKE, TH EMMER, and **J STEFFENHAGEN**
Otto-von-Guericke University, Magdeburg, Germany

The growing importance of environmental aspects cannot be denied by manufacturing businesses. An important factor to be considered is the use of cutting fluids in machining processes. This paper presents concepts, which need to be considered in companies that extensively use cutting fluids and not only for environmental reasons. Based on these concepts, the improvement and practical realisation of Minimum Cutting Fluid Application and Dry Cutting are shown. In this paper experimental results and effects on tool life and part quality are presented and discussed particulary for drilling, counterboring and reaming processes.

1. AIM AND IMPORTANCE OF ENVIRONMENTALLY CLEAN CUTTING PROCESSES

Considering environmental protection the term “clean production“ has particular significance. More and more research institutions and enterprises are forced to think about ecologically less harmful and environmentally clean manufacturing, facing growing ecological awareness, increasing problems in disposal of waste materials and introduction of very strict environmental regulations. (1), (2)

In Manufacturing Engineering especially the auxiliary supplies used (i.e. cutting fluids, coolants, lubricating oil and grease) have a highly dangerous potential to pollute soil, water, air and to harm people. In machining the emphasis is placed on cutting fluids, which are used in metal cutting processes in enormous amounts to support the process by the simultaneous effects of cooling, lubrication and chip removal by rinsing. Even if these wet cutting processes are used in almost all cases the technology of dry machining has been and still is applied for special tasks in the manufacturing business. Several measures could be taken to qualify, modify, minimise coolant-lubricants and to prolong their lifetime which would lead to positive effects in practical applications.

The task of the research work is to firstly analyse and evaluate those cutting processes, where wet cutting is usually applied (jet application of cutting fluid) and seems to be indispensable. Secondly these processes need to be prepared for a changeover to the new concepts of minimum quantity cutting fluid application (MQCFA) and dry cutting. (3)

According to Figure 1 it is possible to realise a direct change to dry cutting in combination with the use of new cutting tool materials (carbide, CBN) and hard coatings. For cutting processes most applying High Speed Steel tools the MQCFA is more appropriate alternative since a "clean" cutting process can be realised without losses in productivity or quality. The characteristic of these method is, that only the small amount of a highly efficient lubricant needed is fed to the cutting zone and completely used and gives almost dry chips.

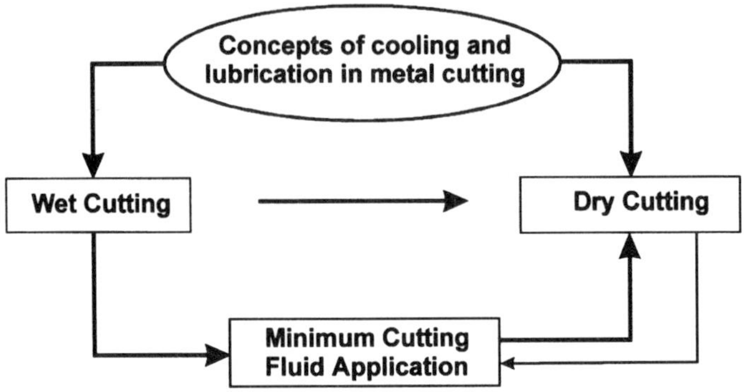

Figure 1: Concepts of cooling and lubrication in metal cutting

In a framework of several research programmes, funded by known institutions, research work is carried out in the field of new concepts of cooling and lubrication in metal cutting. This is done by almost all German universities having a powerful Manufacturing Engineering sector in collaboration with industry in order to prepare ready-to-use solutions for practical application.

2. MINIMUM QUANTITY CUTTING FLUID APPLICATION (MQCFA) AND DRY CUTTING AS OPTIONS TO SUBSTITUTE WET CUTTING (CUTTING FLUID JET APPLICATION)

The Institute of Manufacturing Technology and Quality Management (IFQ) at the Otto-von-Guericke-University Magdeburg for several years has carried out extensive experimental investigations to examine the cutting and tribological phenomena as well as the effects and opportunities of MQCFA and Dry Cutting application for widely used cutting processes.

Despite good initial conditions existing in industry, these methods are used in practice only hesitatingly, even if the results obtained from the laboratory research are positive. It can be said, that the introduction of MQCFA and all the more Dry Cutting for the time being will take place only partially, step by step, since other prerequisites concerning machine tools and cutting tools need to be established first. So for the aim of environmentally clean cutting a complete overall concept is necessary, similar to that developed during several years of research work at the Institute of Manufacturing Technology and Quality Management (IFQ).

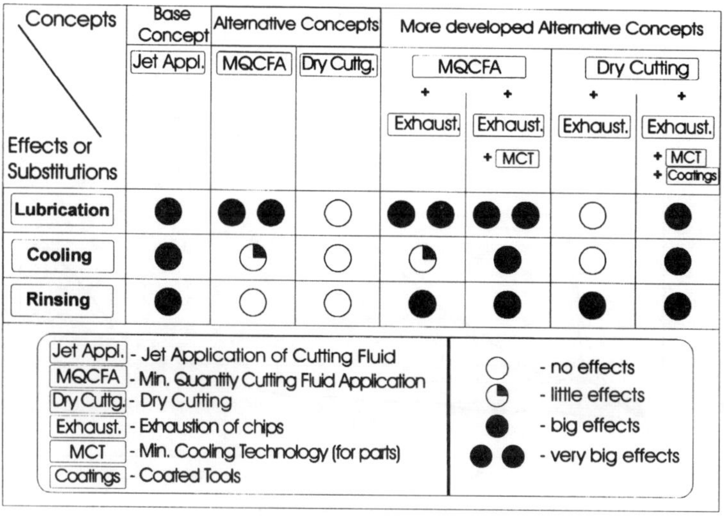

Figure 2: More developed alternative concepts

This concept aims to consider and to substitute **all** effects of the coolant-lubricant in jet application by adequate effects of equal value. Examining the usual cutting fluid jet application, according to Figure 2, the cutting process is supported by lubrication, cooling and chip removal by rinsing. The MQCFA and Dry Cutting as alternatives partly do the same (MQCFA lubricates) or have none of the desired effects (Dry Cutting). These concepts can fulfil the complex requirements in industrial application only by further development of an additional use of minimum cooling technology (to cool down the workpiece) and specially adapted exhaustion or blow-out technologies for chip removal. Also in Figure 2 the contribution of hard coatings to the decrease of thermal load in the cutting zone can be seen due to less friction and good thermal insulation. In the following paragraph the results of MQCFA will be presented for drilling, counterboring and reaming processes.

3. OBTAINED RESULTS IN DRILLING, COUNTERBORING AND REAMING PROCESSES

Extensive systematic experimental investigations concerning the application of MQCFA have been carried out to prove whether the new technology can meet or even surpass the results of the usual method (jet application) in terms of tool wear, cutting forces and torque and quality of produced parts. So Dry Cutting with and without compressed air was included in the comparative tests. To provide overall comparability all experimental conditions for example machine tool, tool, part material, measurement, cutting parameters and other remained unchanged. (4)

The coolant-lubricant was supplied to the cutting edge by impulse spraying when the tool was not in contact with the workpiece. The results gained will be discussed by means of tool life time diagrams. Figure 3 shows the obtained results for the drilling process.

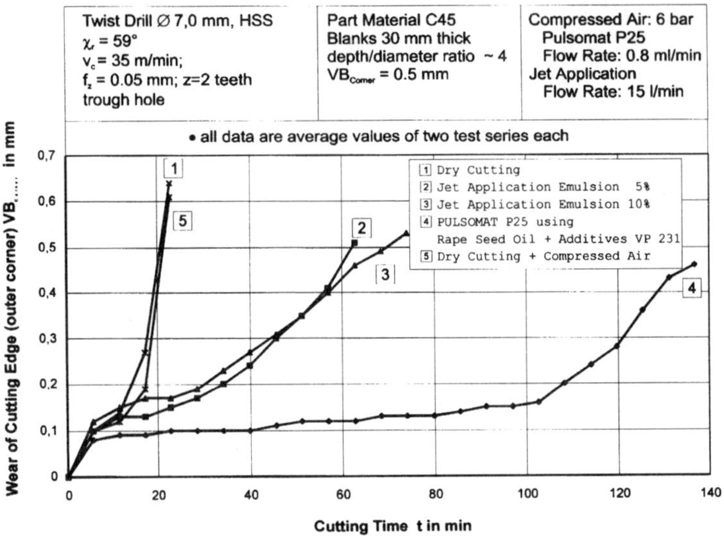

Figure 3: Wear behaviour in drilling

For the selected cutting parameters neither dry cutting nor the use of compressed air led to acceptable results. In jet application of emulsion coolant - within the range from 5% to 10% - the content of oil has no influence on the wear behaviour. Surprising results were obtained from Minimum Quantity Cutting Fluid Application experiments. During a cutting time of 100 minutes the cutting edge (outer corner) wore with a continuous but small rate of wear. So if a wear criteria of VB_{Corner}=0.5mm is chosen, the tool life time is prolonged up to 200% (of the tool life time resulting from emulsion jet application). The rate of flank wear is different according to the different types of cooling and lubrication methods.

For the drilling tool cooled and lubricated by emulsion jet application the maximum wear occurs at the outer corner since it decreases in direction towards the chisel edge. Vice versa is the situation under conditions of MQCFA. The land wear width is approximately constant along the principal cutting edge. Towards the end of tool life time the maximum wear will occur around the chisel edge.

This (atypical) wear behaviour has no influence on the (cutting) torque, but the increased chisel edge wear leads to a considerably higher feed force in comparison to emulsion jet cooling. (shown in Fig. 4) No significant differences between the cooling methods were found comparing the surface quality of the drilled holes.

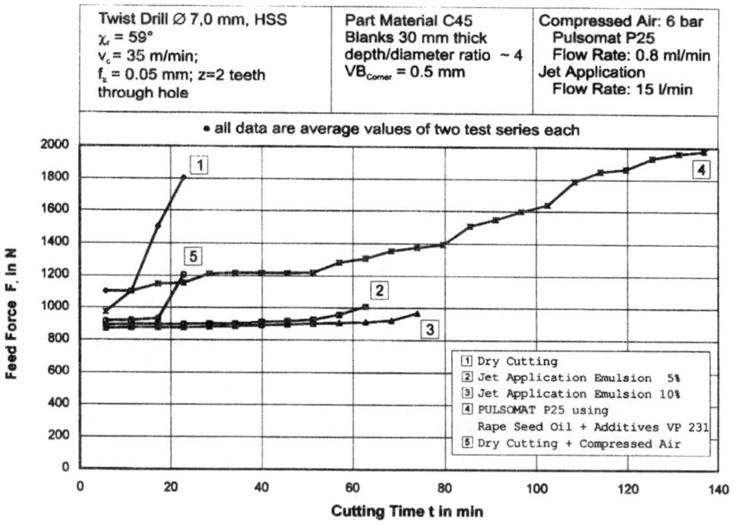

Figure 4: Feed force behaviour in drilling

To complete the investigations for the drilling processes, counterboring and reaming processes have been investigated in order to obtain results for the entire machining (roughing and finishing) of drilled holes. Very good results were obtained using the MQCFA in counterboring processes. (shown in Fig. 5)

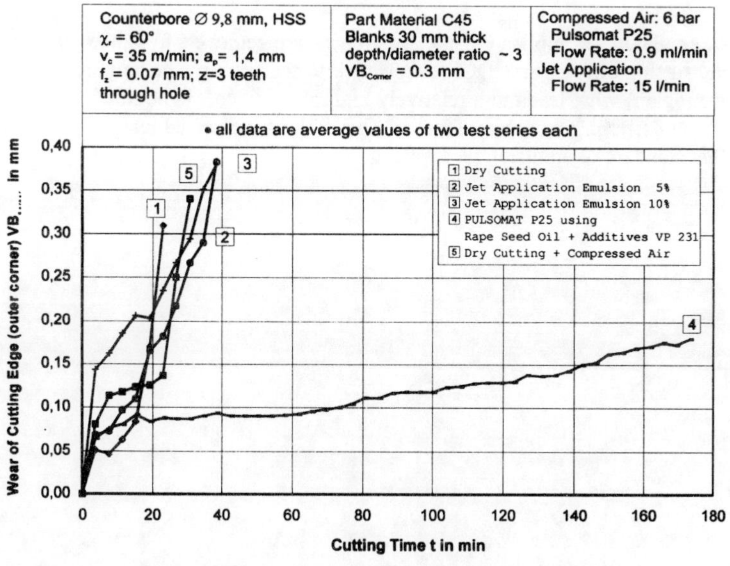

Figure 5: Wear behaviour in counterboring

The long tool life times are obviously due to the outstanding lubrication effects of the lubricant used since there are almost no differences between dry cutting and emulsion jet application.

Investigations of the reaming process have shown very interesting results. The superiority of MQCFA in comparison to emulsion jet cooling is not as obvious as was expected with regard to the results of the other processes. (see Fig. 6) Apart from that the tool life achieved with MQCFA is several times higher than the one achieved under conditions of dry cutting.

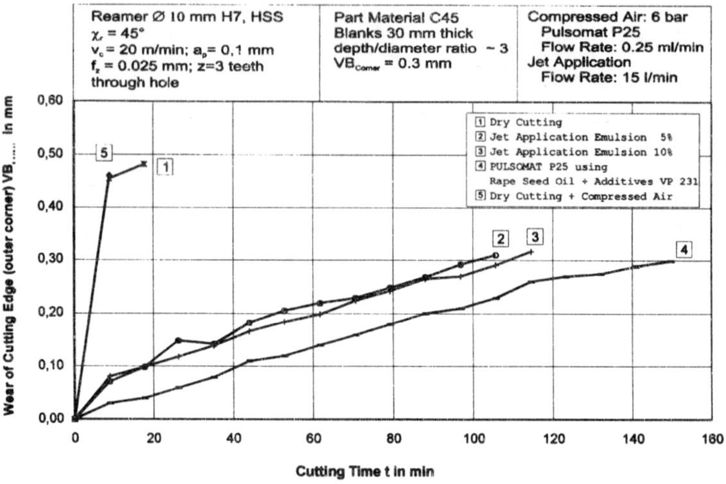

Figure 6: Wear behaviour in reaming

Since in emulsion jet application the little chips are removed from the cutting zone by flushing and rinsing effects, in MQCFA they stick to the cutting face and are drawn into the cutting zone (again) what leads to a relatively higher wear rate - in comparison to the results obtained in counterboring processes. Figure 7 shows the summarised results which are related to the present standard of emulsion jet application.

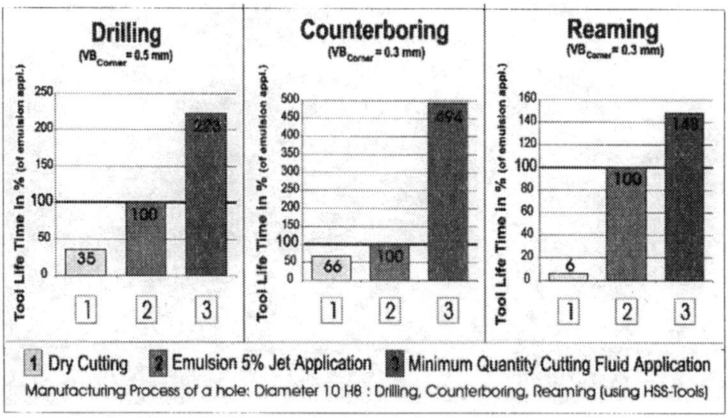

Figure 7: Comparison of wear behaviour in different processes

There are some more results at the IFQ concerning the effects of MQCFA and dry cutting in gear hobbing processes. These results also can be evaluated positively, even if in this cutting process the specifics need to be considered in order to meet the high quality requirements. (5), (6)

The research work concerning the above topics carried out at the IFQ can be summarised: The MQCFA is an ecologically and environmentally clean alternative option in contrast with the conventional use of coolant-lubricants in machining (oil and emulsion jet application), and so furthermore will reveal technical, economical and logistical advantages. For a widespread practical use detailed research work needs to be done in advance and all the conditions of machine tool, part material and tool material need to be considered. The MQCFA can be seen as one step towards the realisation of dry cutting for several machining processes in which High Speed Steel tools are used.

REFERENCES

(1) Team of Writers: Saubere Fertigungstechnologien - Ein Wettbewerbsvorteil von morgen? 13./14.06.1996, RWTH Aachen

(2) Klocke, F.: Trockenbearbeitung - Grundlagen, Grenzen, Perspektiven. VDI-Berichte Nr. 1240, 1996

(3) Popke, H.; Emmer, Th.: Minimal dosiert an die Schneide. Schweizer Maschinenmarkt Nr. 38 / 1994

(4) Popke, H.; Emmer, Th.: Trockenbearbeitung und Minimalschmiertechnik beim Bohren, Senken und Reiben. Maschinenmarkt Würzburg 103 (1997) 18

(5) König, W.; Knöppel, D.: Kühlschmierstofffreie Zahnradfertigung. Bericht zur 35. Arbeitstagung „Zahnrad und Getriebe“ in Aachen 1994

(6) Popke, H.: Hat die Minimalschmiertechnik beim Wälzfräsen eine Chance? Schweizer Maschinenmarkt Nr. 42 / 1995

Standardization of surface finish for adhesion tests

C D WICKRAMATILLAKE and **J M O'SULLIVAN**
Department of Manufacturing Systems Engineering, University of Hertfordshire, Hatfield, UK

ABSTRACT

The significant parameters to be examined in any adhesive system are the absolute strength of the adhesive bond between two surfaces, measured by standardised t-peel tests, lap shear tests and durability tests, usually accelerated, to ensure long term retention of strength.

In testing these parameters conditions of preparation of the surfaces to be joined should be standardised for valid comparisons. The mechanical surface finish of metal surfaces has an effect on the strengths measured. When the surfaces are abraded manually, surface finish can vary widely even when a seemingly constant force and time are applied. The work reported here was to design a simple, pneumatically controlled device to standardise surface finish. Preliminary results of the efficacy of the device and effects on measured adhesive strengths are reported. The tests concentrated on aluminium surfaces bonded with epoxy adhesive.

1.0 INTRODUCTION

Adhesive bonding is becoming very important in the manufacturing industry as it can replace traditional joining techniques such as mechanical fastening and welding (1). Welding of Aluminium material which is becoming far too expensive in volume production, is being replaced by adhesive bonding methods for economic manufacturing, particularly in vehicle, aviation and marine applications. However, aluminium metal surfaces like other metal surfaces provide limitations to adhesive bonding, due to contamination on the surface (2). Some methods for removing this contaminant layer and improving the surface roughness include abrading, sand blasting, anodising and etching (3).

Abrasion of metals surfaces has been used in the past and will be used extensively in high volume production and metal processing industry, to remove the contaminant layer and improve surface characteristics (4). Research and development work carried out around the world looks at methods of improving adhesive bonding, the majority of them using abrasion as a simple process to improve the surface prior to bonding. A common problem in preparing samples of batches for bond strength and durability test, is the existence of increased surface roughness variation caused by the abrasive process.

The objective of this paper is to present the case for achieving standardise surface finish for adhesive bonding. It shows how a pneumatically controlled device was developed to improve the uniformity of surface finish at joints of aluminium specimens, for single lap-shear adhesion tests. Details of experimentation including specimen preparation, surface roughness measurements and comparisons of adhesive bond strength results of machine abraded surfaces with manually abraded surfaces are outlined. Some problems identified during experimentation and some further work to be carried out are also reported.

2.0 EXPERIMENTATION

2.1 Specimen Preparation

All experiments consisted of 6 samples per batch and used 3mm thick 2024T3 Unclad Aluminium material. Sheet material was cut to 1" x 6" long strips, and dimensions conformed to ASTM D1002 standard with a ½" treated area overlap for bonding. All specimens were cleaned initially using a general purpose degreaser and acetone prior to treatment or bonding. The adhesive used to assemble the single lap-shear samples, was a 120°C curing FM73 single part epoxy film adhesive supplied as a 0.25mm thick film. A special clamping system (5) was used to apply a uniform load of 200kPa to all lap-shear bonded joints for curing. This clamping system was essential to reduce additional variation to bond strength caused by the adhesive curing process.

2.2 Abrasive Rig

The rig was constructed as shown in Fig.1, the bottom frame made of steel and majority of the structural components out of aluminium. The abrasive pad fixed to the pneumatic rotor was a 'Scotch-Brite' pad No. 7440 supplied by 3M Abrasive Systems. Specimens were clamped mechanically using compression springs on a bar bolted on to the bottom plate. Individual ramps were controlled via air regulators using electro-mechanical pneumatic controllers. A damping mechanism was used to control the initial load applied by the abrasive head onto the aluminium specimens. The entire device worked to a pre-programmed sequence controlled by a programmable logic controller (PLC). The horizontal bed moved 8 cycles before completing the abrasive sequence for a batch of 6 specimens.

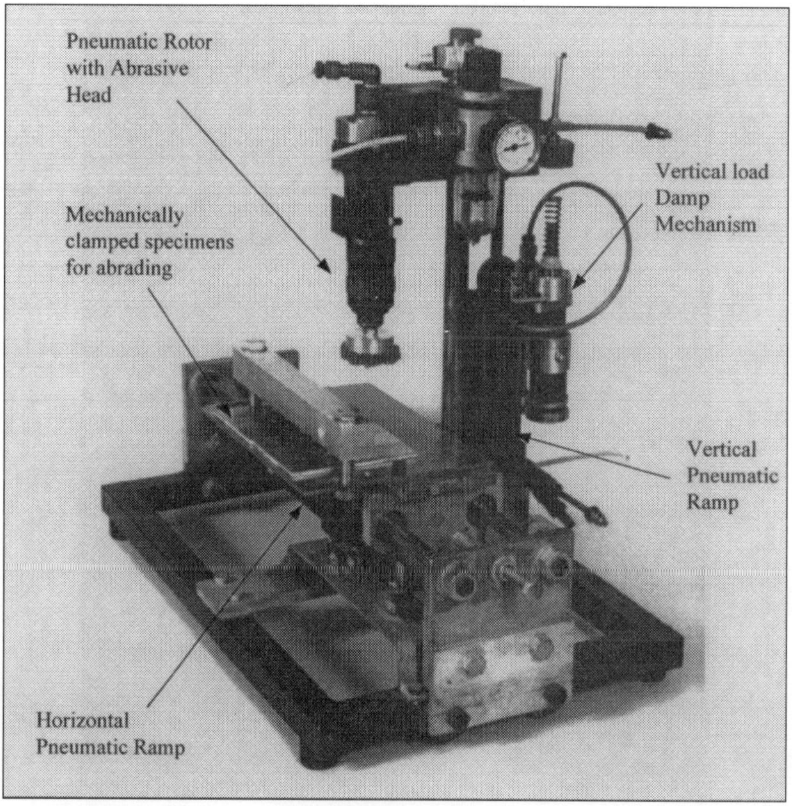

Figure 1 : Pneumatically Controlled Abrasive Rig

3.0 RESULTS AND DISCUSSION

3.1 Surface Roughness

Surface roughness was measured using a 'Surfcom' surface texture measuring instrument which was set for a 3mm horizontal traverse with a speed of 0.06 mm/second at high frequency. Two locations on the treated area were measured and average calculated. Pairs were treated and bonded, each individual pair made up of labelled specimens namely Specimen I and Specimen II. The measured results are shown in the tabulated chart Table 1 and Fig.2. The lowest variation was measured on the untreated (no abrasion) surface. However, compared to manual abrasion, abrasion performed using the device gave improved uniformity in surface roughness with lower variation.

Table 1 : Surface Roughness (Ra) for different Treatments

Sample Number	No Abrasion		Manual Abrasion		Machine Abrasion	
	Specimen I	**Specimen II**	**Specimen I**	**Specimen II**	**Specimen I**	**Specimen II**
A	0.135	0.125	0.540	0.460	0.515	0.380
B	0.115	0.120	0.655	0.655	0.340	0.360
C	0.100	0.105	0.475	0.600	0.430	0.405
D	0.140	0.150	0.560	0.595	0.435	0.350
E	0.235	0.230	0.320	0.395	0.350	0.410
F	0.230	0.250	0.335	0.370	0.475	0.500
Averages	0.159	0.163	0.481	0.513	0.424	0.401
Variation	0.135	0.145	0.335	0.285	0.175	0.150

(Note : Specimen I & II form a pair for bonding)

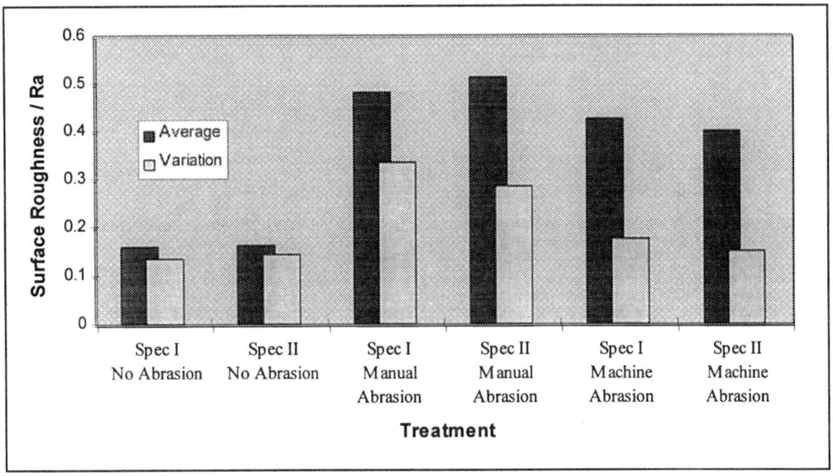

Figure 2 : Surface Roughness vs. Treatment

3.2 Bond Strength Variation

Lap-shear strength tests were undertaken on all samples which had been measured for surface roughness at the bonded areas, using a standard 20kN tensometer. Measured results are shown in Table 2 and Fig.3. Manual abrasion performed by hand using the same abrasive material gave larger variation to that performed using the abrasive rig. The percentage variation reduced from 32.4% to 19.1%, improving the uniformity of surface roughness within the tested batch.

Table 2 : Lap-shear bond strength and variation for different treatments

	Average of six samples (sample numbers A to F)	
Surface Treatment	**Lap-shear Bond Strength / N**	**Bond Strength Variation**
No treatment	9060	17.6 %
Manual Abrasion	7911.67	32.35 %
Machine Abrasion	7915	19.14 %

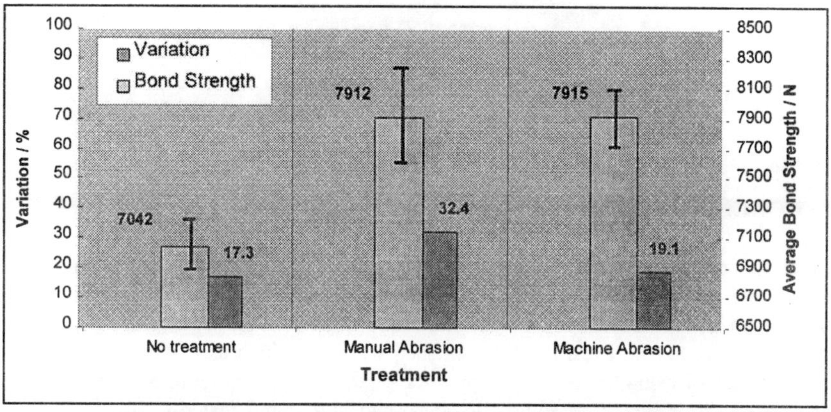

Figure 3 : Average bond strength and % variation vs. Treatment

3.3 Some problems and solutions

Problems identified during experimentation with some possible solutions are outlined below in Table 3.

Table 3 : Abrasive Rig problems and possible solution

Identified Problem	**Possible Solution / Further work**
Excessive abrasive pad wear as shown in Fig 4.	• Increase abrasive pad surface area or pad thickness • Decrease vertical load and increase the number of horizontal cycles, to remove sufficient material
Batch size is limited to six specimens	• Modify length of horizontal stroke to accommodate more area and increase the number of clamped specimens
Mechanical screw type clamp used to hold the specimens	• Modify device to have a pneumatic clamping system to hold specimens, which will reduce time for manual clamping and ease operation
Limited to the use of a single type abrasive pad	• Modify metal head to accommodate abrasive brush, disc and paper, for additional surface roughness experimentation
Cannot use existing device for treatments which uses abrasion in the presence of a liquid (6)	• Design a method to insert chemicals on to the abrasive pad during rotation. • Design a suitable cover to prevent chemical splash to the environment / operator

Figure 4 : Excessive abrasive pad wear

4.0 CONCLUSIONS

Experiments performed denoted how abrasion could vary the conditions of preparation of the surfaces, and provide error data for valid comparisons of bond strength. Results showed the improvement of adhesive bond strength variation obtained using the abrasive rig, compared to manually abraded surfaces. Recent published work (6) which used the abrasive rig noted the importance of using the device for producing large batches of uniformly treated samples which tested the effect of abrading siloxanes onto aluminium material. In conclusion, the design and development of this pneumatically controlled device standardised surface finish for lap-shear adhesion tests. Such a device, considering some of the recommended modifications could be used even on a large scale not only for specimen preparation, but also for large scale research and development work in manufacturing industry.

5.0 REFERENCES

[1] Gledhill R A, Shaw S J and Tod D A, 'Durability of adhesive-bonded joints employing organosilane coupling agents', International Journal of Adhesion and Adhesives, Vol. 10, No.3, (July), 1990, p 192 - 198

[2] Shaw S J, 'Adhesives in Demanding applications', Polymer International, Vol. 41, No.2, 1996, p 193 - 207

[3] Comyn J, 'Surface treatment and analysis for adhesive bonding', International Journal of Adhesion and Adhesives, Vol. 10, No.3, (July), 1990, p 161 - 165

[4] Kalpakjian S, Manufacturing Processes for Engineering Materials, Second Edition, Addison-Wesley Publication, 1991, p 583 - 589

[5] Wickramatillake C D and O'Sullivan J M, 'Preparation of adhesive joints under controlled curing conditions', Proceedings of the 5th International Conference in Structural Adhesives in Engineering, Bristol, IoM, 1998, p 229 - 236

[6] Wickramatillake C D and O'Sullivan J M, 'Modification of Aluminium surfaces using Siloxanes', INALCO'98, Proceedings of the 7th International Conference on Joints in Aluminium, Cambridge, TWI, Vol.1, 1998, p 103 - 110

Deep hole drilling – current and future trends

N WOODFINE
Systems Engineering Faculty, Southampton Institute, UK

Abstract

Drill wander, poor size control and surface finish, flank wear, binding and breakage are all problems that contribute to relatively slow and unreliable production of deep holes when using twist drills. The utilisation of more specialised deep hole drilling methods such as single/double edged gun drilling, BTA (Boring and Trepanning Association) and ejector drilling can improve quality, reduce production times and due to a reduction in operations, be more economical. This investigation considers the current and future trends of deep hole drilling with particular emphasis on the relationship between the cutting tool and support pads.

1. Introduction

The production of holes is arguably the most common machining operation performed in industry, possibly over 75% of all metalworking operations involve holemaking, and therefore any improvements can provide marked economical benefits [1]. The maximum depth to which a twist drill can successfully operate is restricted due to problems such as progressive coolant starvation at the tool tip, chip build-up and wedging of the drill in the hole leading to misalignment and ultimate catastrophic failure [2]. It has been said that tool life will decrease exponentially as the depth of hole increases and the heat in the cutting zone builds up [3]. The application of peck drilling cycles, pressurised through the hole coolant systems improved tool materials, coatings and parabolic-fluted drills (Fig. 1) can greatly improve depth of hole performance up to a length to diameter ratio of approximately 20. Producing deeper holes, in itself, is not a difficult task. What maybe more difficult is maintaining high productivity whilst achieving an accurate diametric tolerance and directional run-out as the depth of a hole increases. A self piloting drill is a most efficient boring method to produce deep holes with a high degree of accuracy and surface finish. A length to diameter ratio of 100 to 1 are common and ratios over 300 to 1 have been successfully drilled [4].

2. Description of deep hole production methods

The self-piloting drill is based on an asymmetrical tool geometry design where the cutting forces are not balanced. The single lipped tool consists of an offset tool point which cuts through the centre of the hole leaving no area of material to be extruded. [5] The main and radial forces that occur due to the cutting action are counter balanced through the hole wall via guide pads located around the tool . The cutting forces are absorbed by pads which support and guide the tool and thus the tip follows the direction of its own axis. The self-piloted drill can therefore cut and then burnish in a single pass as can be seen in Figure 2 [6].

The three commonly used pilotless deep hole drilling systems are gundrilling, British Trepanning Association and ejector drilling. Although BTA and ejector drilling systems are somewhat different to gundrilling, they operate on similar principles to those previously discussed. For gundrilling, a pressurised cutting fluid is pumped to the tool tip through a hole in the centre of the tool and then flushes the chipped swarf back up between the drill flute and the sides of the newly bored hole (Fig. 3). To ensure satisfactory chip removal, the flute of the gundrill is normally apportioned a one third sector of the circle cross section and has a single cutting edge supported by two pads. Modifications to the conventional gundrills have been marketed generally where dedicated machinery cannot be justified. Examples of such tools are "Speed bit", "D" drilling system and "Ventec" where each have increased flute angles to ensure adequate chip removal when using high pressure spray mist coolant systems [7].

The BTA drill has a circular cross section with a single coolant hole running the length of the tool (Fig. 4). The drilling system uses pressurised cutting fluid that is pumped via a pressure head behind the drill bush, down the annular clearance between the drill tube and the newly bored hole. After cooling and lubricating the tool tip, the cutting fluid returns with the chipped swarf, up the inside of drill tube [17 et al]. The ejector drill also has a circular cross section but consists of two concentric tubes. The pressurised coolant flows within the annular space between the two tubes to the drill tip and returns back, with the chipped swarf, through the central tube (Fig. 5). The return flow along the central tube is assisted by a partial vacuum that is created by drawing off approximately one third of the in-flow coolant from the outer tube, through nozzles, straight back into the central tube [17 et al].

3. The Cutting tool

There are three methods of supporting a pilotless deep hole drill cutting tip, a solid tool, a brazed/detachable tip and indexable tips. The use of indexable tips is restricted to holes above the size of 10mm diameter and is therefore is not commonly used for gundrill applications [8][9].

A one piece design, solid carbide drill is said to be stronger and more rigid than conventional brazed gundrills and can be used for drilling holes of under 1.5mm diameter. Augustine [9] explains that the tools have been developed for specialist applications such as medical implants and automotive parts. Increases in cutting speed can normally be achieved using solid carbide tools when drilling softer materials and Noaker [8] considers that the future development of two-fluted solid carbide tools could boost tool life or feed rates by a further 50% due to reduced chip loads. A conventional gundrill tip is normally made from a pre-formed section of tungsten carbide that is brazed onto the shank, which is a hollow and crimped heat treated steel shaft. The driver is also normally brazed at the other end of the shank. The shank and driver can be reused when the end of the working life of the tip has been reached. To ensure straight holes are produced it is important that there is alignment of the tip with the driver and therefore, in some cases, they are ground together. For larger drills, various detachable tips or heads are available which can improve the rationalisation of tooling. As the hole diameter is increased, sufficient space becomes available to utilise indexable tips. Considering gundrills are employed primarily for smaller diameters, the majority of indexable tipped tooling will be found on BTA and ejector drills.

As with any chip removal process, the selection of the cutting tool material is a crucial production and commercial decision. With the focus on faster metal removal rates, improved dimensional accuracy and surface finish, cutting tool materials and coatings have been the subject of intense development for some decades. It is however interesting to note that there appears to be limited information available relating to developments of cutting tool materials for applications of pilotless deep hole drilling. The use of various grades of carbibe, either brazed or indexable tipped, is widespread and currently remains unchallenged. The use of coated carbide tools, although used with indexable tip tooling for larger diameter operations, is not widely adopted for brazed tip tooling such as gundrills. It has been said that coated carbides can be used for gundrills but there are manufacturing problems relating to the some carbide coating processes melting the tip braze [personal discussions]. Polycrystalline Diamond has been used successfully to gundrill cost effectively, highly abrasive non-ferrous materials such as silicon- aluminium alloys, filled rubbers and fibre-reinforced plastics. Saucer [11] states that edge life was 100 times longer than for High Speed Steel, and 10-15 times longer than for carbide. Sharpening of very hard tools such as PCD and CBN can be time consuming and requires diamond wheels to ensure accuracy [12]. This may not be cost effective in smaller machine shops where regular regrinds of tooling are required for small batch quantities of different materials.

The nose grind is a crucial operation for the success of the deep drilling process. The cutting angles influence the shape of the chip, the chip breaking properties, tool life, hole size and shape. The standard grind of 30 - 40 degrees outer angle and 20 degrees inner angle is widely used but special tip geometry's are often developed by trial and error on the shop-floor, the results of which are not always distributed. Standard grinds are available which have been developed based on criteria such as the hole size, material being cut and chip breaking [9] [12]. It appears that there can be difficulties with reproducing optimum tool geometry to precisely controlled parameters. Astakov et al [26] has produced a generalised mathematical expression for the cutting tip geometry of a self-piloting drill with external chip removal. It is said that this combined with a redesigned tool geometry can increase tool life and reduced coolant pressure loss.

Fundamental to the successful production of precision deep holes is the breaking of swarf into chips of a size and shape suitable for uninterrupted flow through the drill flute aided by a pressurised cutting fluid. For smaller diameters where solid carbide cutting tips are used, mechanical chip breaking methods may not be practical and so chipping of swarf is influenced by the adjustments of speeds, feeds, tool geometry and coolant pressure.

4. Support Pad

The support pads or burnishing pads are located around the circumference of the cutting head of the drill. During the cutting process, the forces associated with the cutting action are counter balanced against the pads which are supported against the newly machined bore. The force of the pad against the hole wall produces plastic deformation and burnishes the surface.

The pads are accurately ground onto the circumference of the solid carbide cutting tip to a depth of around 0.050-0.075mm with a back taper (small clearance) of approximately 1/1200 times diameter. Consequently, as the gundrill is reground and becomes shorter, there will be a proportional reduction in diameter. The position of the pads greatly depends on the material being cut and the tolerance on size and finish. Huston [14] describes various diameter relief contours that may be selected for specific conditions and requirements. It appears that the selection of diameter relief can influence boundary lubrication, self stabilisation of the drill, axial alignment, surface texture and hole tolerance. Noaker [8] states as bearing pad contact area increases, the diametrical support and burnishing action of the carbide also increase and therefore yields greater control of finish and size. The extra bearing surface can also benefit applications such as intersecting holes, interrupted cuts, and angular holes. The pads are exposed to extreme pressures, friction and heat generation which can culminate in a build-up of material being welded to the pad face. Irregularities on the support pad will be pressed onto the surface of the workpiece which will affect the hole tolerance, surface finish and it may also erode the carbide tool matrix which can ultimately break down the tool. This problem can occur when machining most metallic materials but in particular those that are soft and exhibit poor machinability properties and when using extra bearing surfaces on tools cutting high-temperature alloys or high-nickel stainless steels [8].

5. Relationship between cutting tool and support pads

The concept of self piloting deep-hole drilling relies on the fact that there should be a balance of the main forces that occur between the cutting tool and the burnishing guide pads [17]. The relationship between the forces generated during the boring process can be related to cutting tool and burnishing pad wear, vibrations, accuracy, power, process planning and computer integrated manufacture [15]. There have been numerous contributions to the understanding of the mechanics of the deep hole drilling process of which most can be listed under the following three categories: the measurement of the torque and complex forces that occur in the cutting and burnishing operation; the consequential affects on the process such as chip breaking and chip removal and finally; the surface texture, lobing, roundness, straightness and microhardness.

Field et al [16] described the surface integrity as having two distinct elements. Firstly the surface topography which considers surface finish, roughness, form and texture and secondly, surface metallurgy which considers the alterations to the surface layers such as plastic deformation and microhardness. The combined cutting and burnishing when deep hole drilling produces a plastically deformed surface layer with a good quality finish. Griffiths [15] [17] has shown that the surface consists of regular a frequency of peaks and valleys that correspond to the feedrate. Ramakrishna Rao et al [18] also found the presence of this pattern after burnishing when investigating surface topology of BTA bored holes. In this work it was found that better dimensional accuracy and surface finish can be achieved at higher cutting speeds and optimum feed rates. It is suggested that higher speeds reduce tearing of the workpiece material and the development of a built up edge. It has also been found [19] that higher speeds may adversely effect roundness error and the straightness error due to increased system instability. A variation in the size of a hole is also said to be effected by feed and speed. Chandrashekhar et al [20] developed a mathematical model to predict roundness error of BTA drilled holes . It was observed that the roundness error remained constant after the length diameter ratio exceeded 4 whilst other contributing factors were feed rate and damping ratio of the spindle/ workpiece.

The causes for the phenomenon of spiralling or multi-lobe formation in BTA drilled holes has been investigated by Gessesse et al [21] and it is said that a number of machining parameters can lead to spiralling. Under certain conditions two common causes were identified, the effect of lateral natural frequency and over-sized edge's circle land. The shaft behaviour whilst BTA deep hole drilling has been studied by Chin et al [22][23] who treated the tool shaft as a pipe and applied Bernouli-Eulerian theory to develop equations to govern its lateral, longitudinal and torsional motion whist conveying pressurised fluid. Wang et al [24] performed machinability trials on nickel-chromium alloyed steel and remarks that tool geometry had a remarkable influence on hole quality. Good stability was said to be partly due to the arrangement of the drilling head with two carbide pads and an additional piloting pad plus soft vibration resisting pads at the back of the drilling head.

Much research has been published regarding the complex force systems applied to pilotless deep hole drilling. Griffiths et al [10][15][17] studied the force system that exists during deep hole drilling operations by equating basic cutting mechanic relationships to experimental data taken from a set of cutting conditions. A force model was produced which distinguished the separate components of the cutting force system, the total force system at each pad and the radial forces at each pad. Latinovic [25] produced an analytical model of the cutting force components. He suggests that optimal radial and angular locations for the staggered cutters ensured that the predetermined resultant cutting-force was transmitted to the bore wall thus preserving the optimal pressure between the guiding pads and the wall. Astakhov et al [26] considers a number of limitations associated with the two previous force models and sets out a proposal which involves the development of the seasoned shear zone model with parallel boundaries.

Crucial to any deep hole drilling operation is the effective breaking of the chips before their removal from the cutting zone. Clearly it is important that no obstruction should restrict the flow of swarf contaminated coolant on its exit path from the tool face. This is particularly relevant when considering hazards such as brazed joints and bushes. Successful chip breaking can contribute to an unimpeded exit of the swarf without producing adverse affects on the surface texture of the hole due to scratching or blockage. Chin et al [4] proposed to monitor the chip state in single-edge deep hole drilling by analysing a computer simulation model and experimental data. He produced two mathematical models and reports that these agreed with the experimental data from the piezoelectric sensor which monitored the formation of the chip. Other reported methods of monitoring chip state include monitoring thrust and torque of tool, noise from the drilling process, measuring frequencies of chips via accelerometer, proximity sensors or ultrasonic sensors.

6. Conclusions

The application of pilotless hole drilling can produce accurate, straight and high quality finished deep (or shallow) holes in a single manufacturing process. The most important measurable output effects of deep hole drilling are surface texture, dimensional tolerance, straightness, roundness, metal removal rate and length/diameter ratio. Table 1 provides a comparison of input variables to the output effects and cross refers this information with current research & development trends in the final column.

Efficient swarf removal is crucial for the success of the process and substantial contributions have been made towards the understanding of the chip condition by studying the mechanics of the process. Some of this work may lead to the development of an effective chip condition monitoring system that would be capable of automatically modifying process variables in real time. Although the geometry of cutting tools and tip coatings have been developed to improve conditions such as metal removal rates there is little evidence of new cutting tool materials or support pad materials being tested in the deep hole drilling environment.

TABLE 1. An analysis of process variables and their effect on deep hole drilling (based on a survey of research and development).

INPUT PROCESS VARIABLES	THE OUTPUT EFFECT [1 = Low impact and 4 = High impact]						R&D
	STex	DTol	Str	Rnd	MRR	L/D	
Improvements to tooling, tool geometry and indexable inserts	3	4	2	4	4	3	4
Chip breaking, chip flow, chip congestion & evacuation	3	4	1	2	4	4	3
Machinability, speeds & feeds	4	3	3	3	3	2	2
Tool stability, vibration, dynamics and whip guides	4	2	3	4	2	4	3
Cutting tool life and wear	4	3	1	3	4	3	3
Drill bushing and pilot holes	3	4	4	3	2	1	1
Support pad wear and stability	4	4	3	3	2	2	2
Support pad and cutting tool materials	4	3	2	3	4	2	1
Force system	4	4	2	3	2	2	4
High pressure coolant, pressure drop, flow rates	4	2	1	2	3	4	3

Acknowledgements

The Author wishes to thank and acknowledge Dr. B. Griffiths of Brunel University, Mr J.Cushing of Hammond & Co Ltd., Mr N. Ryall of Gilbert-Ryall Ltd., Mr L. Edwards of Premier (Deep Hole Drilling) Ltd., Mr P. Dangerfield of Sandvik Coromant UK.

References

1 Nelson, D., (1988) Drilling Technology: The Hole Story, Cut. Tool Eng., V.40, p.20-24.

2 Beard, T., (1995) Deep Holes in a Hurry, Modern Machine Shop, V.68, p.52-58.

3 Bucher, D., (1998) How to Drill Those Deep Holes, Mod. Mach. Shop, V.61(1), p. 52-63.

4 Chin, J., Wu, J., Young, R., (1990) The Computer Simulation and Experimental Analysis of Chip Monitoring for Deep Hole Drilling, A.S.M.E., Prod. Eng. Div,. Vol.44, p.449-464.

5 Azad, S., Chandrashekhar, S., (1985) BTA Deep-hole Machining, Mech.Eng., V.107, p.60-68.

6 Evans, B., Evans. T., Griffiths. B., (1997) The BTA Deep Hole Drilling and Boring System, The 2nd Int Con. on Ind Tooling

7 Pearce, J., (1987) Advances in Modern Machinery Spur Holemaking Tools, Cut. Tool Eng., V.39, p.23.

8 Noaker, P.M., (1996) Gundrills at the Cutting Edge, Man. Eng., V.116(5), p.71-74.

9 Augustine, J., (1995) Gun Drilling Small Holes with Solid Carbide Drills, Mod. Mach. Shop, V.68(1), p.77-82.

10 Griffiths, B.J., Grieve, R.J., (1993) Modelling Complex Force Systems, Part 2: A Decomposition of the Pad Forces in Deep Drilling, J. of Eng. Ind., V.115, p117-183.

11 Saucer, R.H., (1984) Drilling with PDC, Tooling and Production, V.50(9), p.53-55.

12 Stokes, T., (1985) Inside Tips on Gundrill Geometry, Tool.and Prod.,V.50(12), p.77-80.

13 Astakhov, V.P., Galitsky, V.V., Osman, M.O.M., (1995) A Novel Approach to the Design of Self-Piloting Drills with External Chip Removal, Part 1: Geometry of the Cutting Tip and Grinding Process, Journal of Engineering for Industry, V.117, p.453-474.

14 Huston, R.G., (1994) Gundrilling Process Guidelines, Soc.of Man.Eng., p.1-10.

15 Griffiths, B.J., (1993) Modelling Complex Force Systems, Part 1: The Cutting and Pad Forces in Deep Drilling, J. of Eng. for Ind., V.115, p.169-176.

16 Field, M., Kahles, J.F., Cammett. J.T., (1972) A Review of Measuring Methods for Surface Integrity, CIRP Ann., V.21(2), p.219-238

17 Griffiths. B., (1982) Ph.D. Thesis, Brunel University

18 Ramakrishna Rao, P.K., Shunmugam, M.S., (1987) Investigations into Surface Topography, Microhardness and Residual Stress in Boring Trepanning Association Machining, Wear, V.119, p.89-100.

19 Ramakrishna Rao, P.K., Shunmugam, M.S., (1987) Accuracy and Surface Finish in BTA Drilling, Int. J. of Prod. Res., V.25, p.31-44.

20 Chandrashekhar, S., Ahmad, Z., Sankar, T.S., Osman, M.O.M., (1986) The prediction of the Roundness Error in BTA Deep hole Machining based on the Stochastic Description of the Cutting Tool Motion, Int. J. of Prod. Res., V.24, p.879-899.

21 Gessesse, Y.B., Latinovic, V.N., Osman, M.O.M., (1994) On the Problem of Spiralling in BTA Deep-Hole Machining, J. Eng. for Ind., V.116, p.161-165.

22 Chin, J., Lee, L., (1996) The Shaft Behaviour of BTA Deep Hole Drilling Tool, Int. J. Mech. Sci., V.38, p.461-482.

23 Chin, J., Lee, L., (1995) A Study on the Tool Eigenproperties of a BTA Deep Hole Drill-Theory and Experiments, Int. J. Mach. Tool Man., V.35, p.29-49.

24 Wang, Z.M., Ezugwu, E.O., Su., (1997) Advances in Precision Machining of Small Deep Holes, J. of Mat. Proc. Tech., V.68, p.257-261.

25 Latinovic, V.N., Osman, M.O.M., (1989) Optimal Design of BTA Deep-hole Cutting Tools with Staggered Cutters, Int. J. of Prod. Res., V.27, p.153-173.

26 Astakhov, V.P., Osman, M.O.M., (1996) An Analytical Evaluation of the Cutting Forces in Self-piloting Drilling using the Model of Shear Zone wit Parallel Boundaries. Part 1 & 2 Theory, Int. J. of Mach. Tools and Man., V.36, p.1187-1200/ 1335-1345.

27 Nicholson, R., (1979) A Comparison of Neat Oils and Water Mix Fluids for Gun Drilling, Proc. 3rd. Int. Con.on Deep Hole Drilling and Boring, K1-K12.

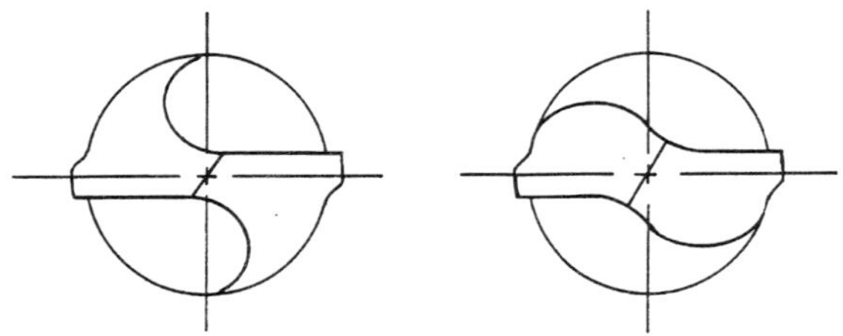

Standard Twist Drill　　　　Parabolic-fluted Drill

Figure 1.

THE PARABOLIC-FLUTED DRILL HAS A GREATER FLUTE AREA AND IS STIFFER THAN A CONVENTIONAL TWIST DRILL

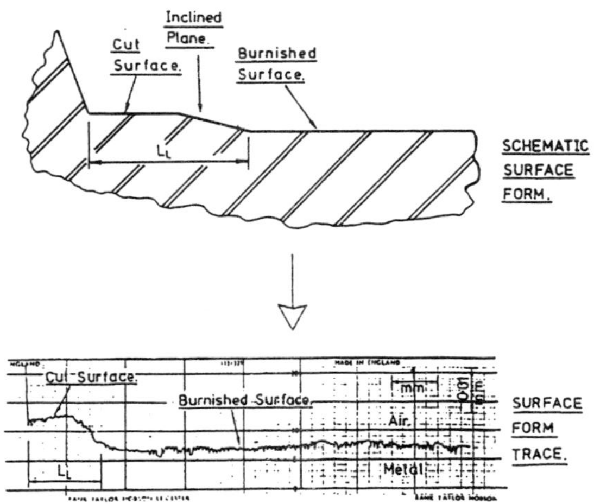

Figure 2. THE 'CUT' AND 'BURNISHED' SURFACES
(Courtesy B Griffiths)

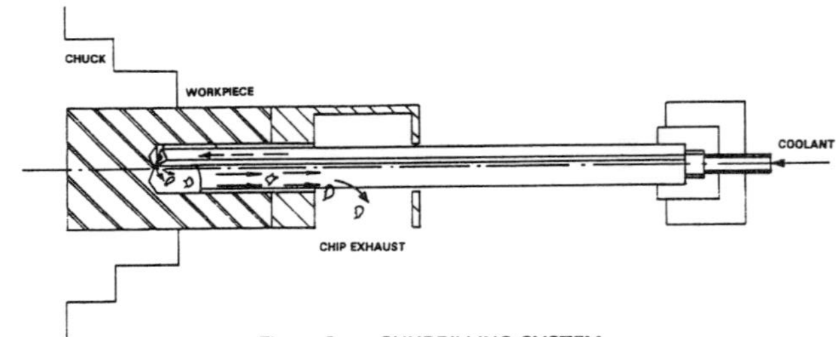

Figure 3. GUNDRILLING SYSTEM

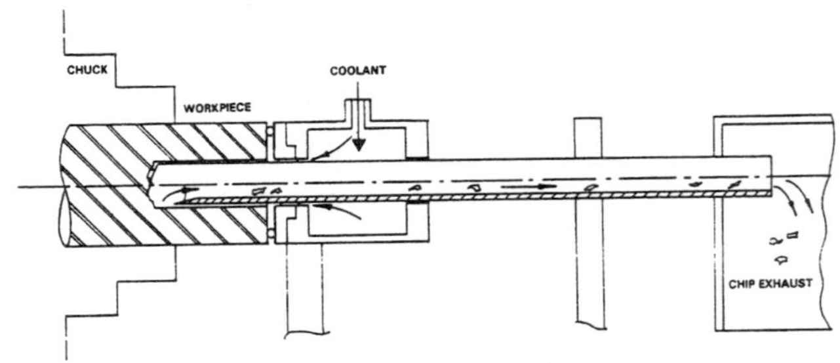

Figure 4. B.T.A DRILLING SYSTEM

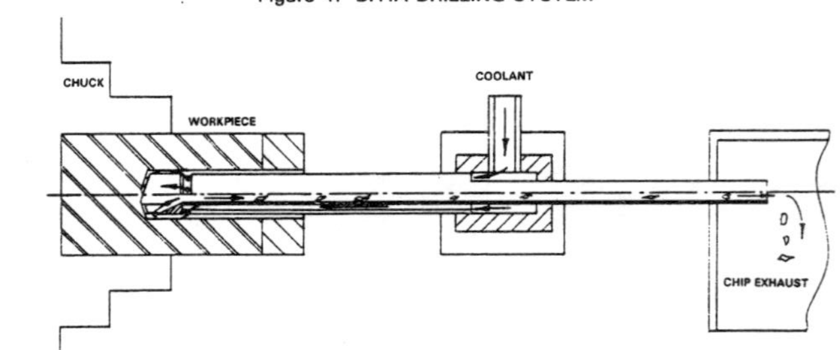

Figure 5. EJECTOR DRILLING SYSTEM

Manufacturing Systems

Modelling and software tool requirements for the design and operation of hybrid flexible assembly and dismantling cells

G BINGER and **V FLEMMING**
Institute fur Produktionstechnic, Technische Universität Dresden, Germany
A BOYLE and **A KALDOS**
School of Engineering, Liverpool John Moores University, UK

Abstract

The paper reports on the development of the necessary system elements required for the creation of a database capable of dealing with assembly and dismantling process planning. Consideration is given to the data transfer and modelling requirements of the product model, the process model, the assembly system model and the human operators needed for the complete specification of the assembly process plan. The role of 3D modelling and simulation as an aid to the process planning function is covered in detail. The application of the technique to the creation of a number of assembly system models is reported.

1 INTRODUCTION

The planning process for the manufacture and assembly functional elements of the manufacturing cycle is influenced by the physical size and geometry of the product, by the specific functional and design features of the product, by the number of customer variants supplied to the market place and by the required batch quantities. Variations in market conditions, fluctuations of demand, economic and quality pressures and the requirements of lean production requires the development of intelligent, accurate and flexible process planning and control systems suitable for use by both specialised planning departments and by production engineers. The functional, ergonomic, capacity and economic solutions required for assembly process planning are closely related to the design, the capacity and the arrangement of the assembly system facilities. Flexible, automatic, hybrid assembly and dismantling systems therefore place considerable but quite specific demands on the quality, the capability, the functionality and the complexity of any computer aided process planning tools used. When comparing the process planning requirements of assembly and dismantling systems with the process planning requirements of other manufacturing activities, the presence of the human operators in the assembly system needs special consideration from the viewpoint of process and system performance. For the human operators, the utilisation of their experience and selective judgement, their manipulative and physical capabilities and their operational flexibility become important factors when deciding on operational sequences and part handling arrangements. All assembly and dismantling processes require a high degree of synchronisation of the assembly system components with respect to a number of factors, including component geometry, robot kinematics, part transfer dynamic behaviour, the functionality of the assembly and dismantling elements and the selected combinations of the assembly or dismantling elements [1].

2 3D MODELLING AND SIMULATION

Visualisation of both the product geometry and the physical arrangement of the assembly and dismantling system under consideration, along with a display of the functional behaviour of the system elements helps to generate acceptable, accurate and efficient process plans. It further helps to improve the efficiency and performance of the plan, it increases the degree of customer acceptance and improves the level of agreement between customer and manufacturer. 3D modelling and a simulation of the assembly and dismantling process are therefore vitally important and have a considerable part to play in the operation selection and validation of a computer aided assembly process planning function. The presentation and animation of 3D models provides particular support for the design and planning process by displaying several important features at different stages of the planning procedure.

Support for the decision making process through 3D presentation and animation is required at the design and planning stage to allow:-

- The suitability of the product for manual or automatic assembly and dismantling to be investigated and evaluated.
- The part assembly or fitting process to be checked for availability, suitability, orientation and accuracy with different assembly sequences.
- Work place design to be undertaken for manual, mechanised and automated assembly and dismantling systems.
- System layout design to be undertaken for the selected assembly system.
- Off-line programming of the material handling system to be completed.
- Presentation of results of various planning stages.

As part of the process plan it is necessary to be able to determine the quality of output of the planning tool through inclusion of the following features:-

- Collision detection between all elements of the model including the individual parts, the subassemblies, products, assembly tools, fixtures, handling devices, assembly and non-assembly machines.
- Work space presentation and out-of-reach-checks for any kinematic elements associated with the operators and/or the handling systems.
- Animation of the kinematics and dynamic movements of the system elements.
- Visualisation of the functional behaviour of both the system and the system elements.

In spite of the recognised benefits of 3D simulation a wide spread general use of simulation tools has not been achieved in practice. The greatest obstacles to their use are the high prices of software and hardware, the considerable expenditure required for modelling and the levels of expertise required. Substantially increasing expenditure is needed for modelling and for training but because of the decreasing prices of hardware, the highest capital investment costs are now associated with the software, particularly the software updates and it is envisaged that user support in system modelling will become even more critical in the future. Currently available 3D simulation systems are generally associated with four different areas of application as follows:-

- CAD systems complete with simulation functionality [2,3,4].
- Robot off-line programming systems.
- Material flow simulation systems.
- Ergonomic simulation and planning systems for material movement.

The functionality of existing 3D simulation packages vary depending on the historical source but increasingly, these systems are incorporating modules which cover the ranges of tasks shown in Figure 1. Users must therefore decide which problems they wish to solve with the particular simulation tool selected. This need arises because of the specific range of tasks made available with the selection and because accurate data interchange between different modules and between different software tools is not easily achieved. For presentation, system layout, planning and testing of assembly and dismantling systems design, 3D modelling is well established [5], but the use of simulation tools is less extensively recognised. In general, the simulation requirement is considered to be economically viable only with relatively simple systems where the model data can easily be imported.

Figure 2 shows the necessary data flow required to create a 3D simulation model of an assembly system. The critical element in the arrangement is the database which is necessary to allow the integration of the module data associated with the various tasks. Unfortunately a fully integrated database which contains workplace details and system element information, in addition to robot data does not currently exist for the simulation of hybrid, flexible assembly and dismantling systems. However the planning and simulation package ERGOplan designed for manual systems and RobCAD or IGRIP intended for robotic systems are typical examples of integrated databases being included in the software for the design, operation and location of system elements. An integrated database also allows accurate data interchange between different modules and between different simulation tools to be more easily achieved.

3 MODELLING AND SIMULATION TASKS

For the product modelling activity, the data exchange between the CAD system and the assembly process planning function is possible via a number of standard CAD interfaces, such as DXF, IGES, STEP, VDAFS or by a CAD direct interface. In the case of assembly and dismantling processes it is necessary to transfer the arrangement of the product elements and the product structure, or the product tree of geometric elements, from the design function to the assembly process planning function. Depending on the complexity of the task to be undertaken different degrees of data abstraction are necessary from the design function. As a general rule, the algorithms used for the reduction of the geometric data do not provide suitable information content or format for assembly and dismantling process planning purposes. For assembly purposes product model information in addition to the geometry requirements has to be evaluated, to give the mass, tolerance, stiffness and ductility of the parts [6]. This is also a prerequisite for the selection of the system elements, including the part feeders and any robots incorporated into the system arrangement.

For the assembly process modelling stage, a control programme is necessary to describe the sequencing and the order of the events in addition to their operating characteristics. For the efficient use of simulation in assembly processes it is necessary for this sequencing information to be automatically or interactively generated during the creation of the process descriptions. Several reports have recently indicated the possibility of generating the assembly sequence in the CAD system [6]. In the future it is envisaged the assembly and dismantling sequences will need to be generated in the CAD system during the design process. Currently it is necessary to import this information as part of the process model [7].

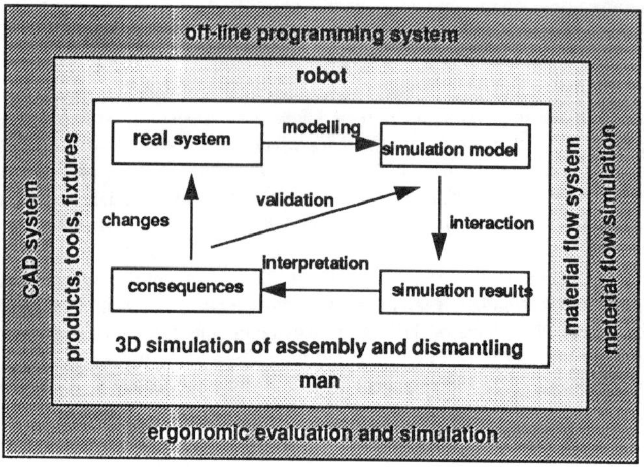

Fig 1. Relationships in 3D simulation systems

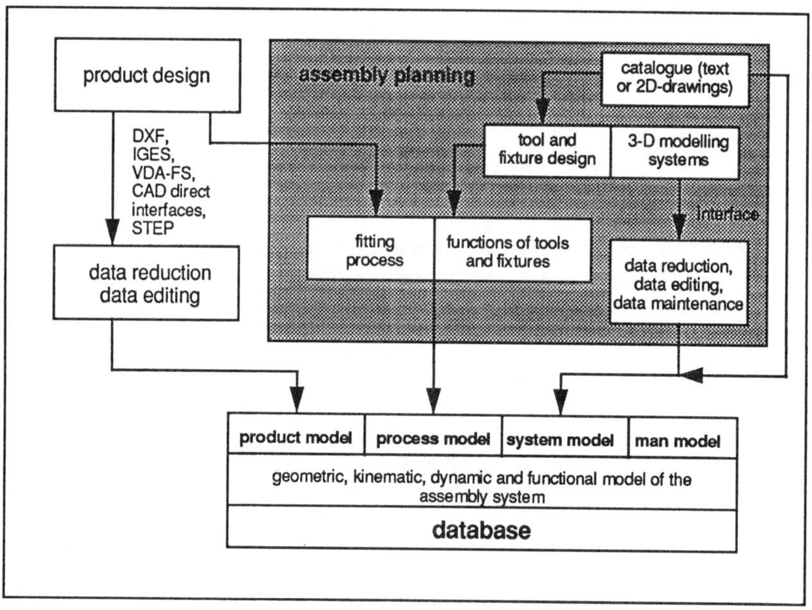

Fig 2. Dataflow in creation of 3D simulation models

The design of the actual assembly and dismantling systems to be used for a particular application is characterised by a mixture of layout planning, facility selection, equipment modification and in-house design. The maintenance of a 3D-equipment database and the user support for the selection of equipment elements presents some major problems. In most cases the assembly equipment to be used in the application is produced by several different suppliers who are not able to supply suitable 3D models of their equipment. It is also necessary to transfer the geometric arrangement, the operating boundary conditions and the functional limitations of the tools and facilities to be used, from the developed 3D model into the in-house design function of the complementary, supporting system elements. As part of the geometric representation the system elements must be specified with kinematic, dynamic and functional properties which have to be implemented as object oriented properties, for instance the angle of approach, the opening width of the gripper, the ability to move the gripper fingers or the ability to hold any part. This applies particularly to gripper fingers, fixtures and other holding or part transfer devices. Only in this way is the simulation useful for supporting the design process.

Modelling any human movement in an assembly system is critical. For a human operator the ergonomic and time evaluations are especially important, in addition to repeatability accuracy, collision detection , the out of reach checks and the area of visibility. Systems offering the simulation of human movements have special interfaces for programming, but as a rule, the effort required to obtain the assembly and handling movements, with their important parameters, direct from the production, process and system model is quite considerable. In comparison to a robot the simulation of a human operator is more important for both the presentation and the verification of the workplace model because the variation in the performance of the human operator is much higher. Static methods can be used to check the workplace but they only consider the average perfomance of the human operators. Often, this is insufficient, because the equipment in addition to the human operator is subject to several demands. Without accurate time evaluation the programme created for any human operator is not useable unlike the robot programmes produced by off-line programming techniques.

Simulation is important when modelling the transition of the individual elements to the complete system. This applies particularly to hybrid systems of assembly and dismantling with non-deterministic time elements or with various divisions of the work content of the human operators. Because the simulation of the whole system generally demands a reduced level of detail compared with the simulation of the individual system elements it is necessary to transform the detailed information about the workplace to an object within a material flow model. To demonstrate the variation in the operation of the whole system at the workplace a closely coupled form of these two simulation methods is necessary. Figure 3 shows the considerations of working principles associated with a simulation study of the complete assembly system.

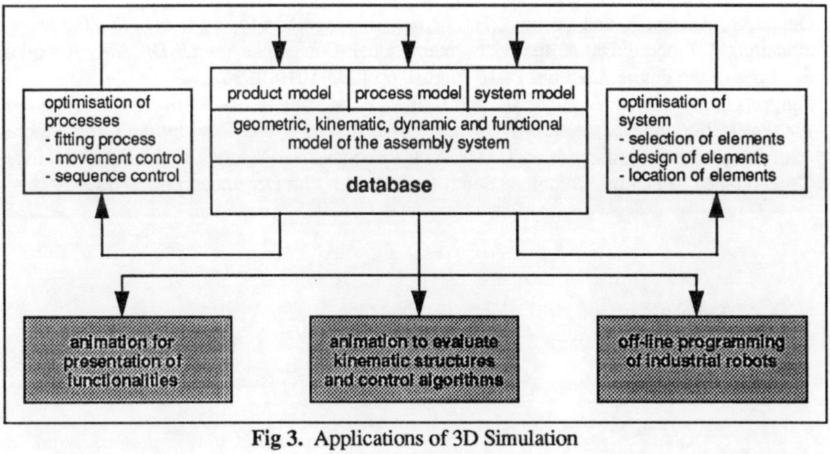

Fig 3. Applications of 3D Simulation

4 CONCLUSIONS

The inclusion of simulation and the planning tools for the assembly and dismantling systems has been used to produce a number of simulation models for a range of applications in both research projects and industrial environment. A suitable database has been established to include the product, the processes and the manual operators. The major problem involved in the operation and development of the database information is in the design data transfer interface to the assembly process planning modules. The use of simulation models is shown to be of extreme importance in modelling industrial applications to determine the workplace conditions and the system performance.

5 ACKNOWLEDGEMENTS

The funding provided by the British Council and the Deutcher Akademischer Austausch-Dienst (DAAD) through the Academic Research Co-operation (ARC) programme in support of this project is greatly appreciated and acknowledged.

6 REFERENCES

1 Sauer, W., Weigert, G., and Goerigk, P. *Real time optimisation of manufacturing processes by synchronised simulation.* Proceedings of the 5th International FAIM Conference. Stuttgart, 28-30 June, pp 271-282, 1995.

2 Kovacs, G.L. and Nacsa, J. *Application of knowledge based control for an integrated robotised CAD/CAM ARC-welding system.* Proceedings of the 12th International Conference on CAD/CAM, Robotics and Factories of the Future. London, 14-16 August, pp 381-386, 1996.

3 Scharf, P. and Simon, M. *New tools for assembly planning: The software system CAPAS supports planning for assembly already whilst product design.* Proceedings of the 5th International FAIM Conference, Stuttgart, 28-30 June, pp 712-723, 1995.

4 Lierath, F., Meyer, R. and Reich, K. *Integrated 3D fixture design by means of feature modelling and rule support.* Proceedings of the 5th International FAIM Conference, Stuttgart, 28-30 June, pp 700-711, 1995.

5 Knackfuss, P., Schmidt, A. and Meier R. *Autonomous robot system for the disassembling of automotive parts.* Proceedings of 12th International Conference on CAD/CAM, Robotics and Factories of the Future. London, 14-16 August, pp 168-173, 1996.

6 Golabi, S., Kazem, A. and Luong, L.H.S. *Contact recognition for assembly planning using solid modelling.* Proceedings of the 12th International Conference on CAD/CAM, Robotics and Factories of the Future. London, 14-16 August, pp 1033-1040, 1996.

7 Binger, G., Flemming, V., Boyle, A. and Kaldos, A. *Modelling and Software Tool Requirements for the Design and Operation of Hybrid, Flexible Assembly and Dismantling Cells.* Proceedings 13th International Conference on CAD/CAM, Robotics and Factories of the Future. Universidad Technologica de Pereira, Columbia, South America, 15-17th December, pp 734-739, 1997.

Requirements of a tool management system for high-variety, high-volume, multi-process manufacture

A BOYLE, C COTTAM, and **A KALDOS**
School of Engineering, Liverpool John Moores University, UK

Abstract

The paper explores the nature of the general tooling problem by identifying direct and indirect tool related cost factors in a system and outlines the requirements needed to develop a tool management system. In addition, the paper reports on a programme of investigation undertaken to analyse and discuss the current tooling system problem at a large manufacturing facility characterised by high volume, high variety process operations. The collection of tooling and processing information, a solution to large duplication of tooling, a general lack of control by centralised management systems and the management of tooling costs via the adoption of a tool management system is proposed.

1 INTRODUCTION

Although manufacturing systems have been developed over many years, there are still a number of technical difficulties existing. The provision and management of tooling is one of the greatest difficulties to be solved in manufacturing systems. Broadly defined, tool management is the capability of having the correct tools available on the appropriate machines at the right time to allow the desired quantity of work-pieces to be manufactured while maintaining an acceptable level of utilisation of the manufacturing resources [1]. Tool management is also defined by its five major functional elements of tool room support, tool allocation, tool distribution, fault detection and tool data flow [2]. The tool management infrastructure is composed of a complex set of interacting factors and many manufacturing companies in Western Europe and the United States have traditionally coped with this situation by breaking the infrastructure into appropriate sets of responsibilities of functions and deploying people to provide the necessary support [3, 4, 5]. The structure and the data organisation of these companies is traditionally based on having specialist groups of staff available with each group having access to its own computer systems.

2 CURRENT TOOL MANAGEMENT SYSTEM

The company is a typical user of a hierarchy type functional system. Many departments within the organisation have links with the tooling function and either require access to tooling information or have a need to create tooling information. Because many of the departments use their own computer software systems each department tends to work independently on tooling projects before combining their work in an attempt to create a final result. The current tool management system interfaces with the tool room, manufacturing, production control, tool maintenance, quality control, industrial engineering and finance. The tool room of the manufacturing company is owned and operated by an independent, externally contracted organisation who is recognised as a leading tool manufacturer and supplier. They have been commissioned to be the main tool supplier, to organise and structure the tool room and to handle the ordering, storage and distribution of the tooling from other suppliers. This arrangement benefits the parent company by providing them with an in-house tool expert who has an excellent technical and operational process knowledge base without the need for employing their own technical staff in such a specialist area.

The primary requirement of the tool supplier is to continually reduce the tooling inventory in the plant. This is being done by establishing re-order points based on tool lead time and tool usage over the previous twelve months period. A second requirement for the tool supplier is to monitor and control the tooling inventory by the introduction and maintenance of a computerised stock control system. All tool distribution, ordering and the collection of tool usage data is done through this system. The system is independent of all other manufacturing company systems and can only be accessed by the tool room personnel employed by the external organisation. Information is supplied to the manufacturing company personnel on request.

Tool usage is tracked by comparing the amount of tooling used with the number of product parts manufactured and assembled. Tool usage cannot be easily tracked to a particular component or to a specific product batch. Scrap parts produced or parts made to stock are not identified or considered in the tool usage calculation. This makes the reasons behind fluctuations in tool usage very difficult to recognise. Consumable tools are distributed to the production machines in sets. A tool set typically contains all the consumable tools necessary to perform a complete tool set-up on the machine for the duration of a work shift. Tool sets are logged into the computer system on issue and are distributed by tool room personnel. Tool sets containing used tooling are picked up by the same personnel and are logged out of the computer system as they are returned. Tool sets can only be traced to an individual machine. The provision does not exist for tracking tooling to a particular station on a machine. If excessive tooling is used the operator will advise the tool room directly or the system will identify the situation at the end of the month or when some other re-order point is affected. The durable and the semi-durable tools are distributed to machines on request of the operator.

Concern action report forms are distributed to manufacturing personnel so that if a tool related problem exists an engineer can be assigned to study and resolve it. Studies can also be performed on any planned tool related changes or on an ongoing operation to provide technical information such as feed rates, recommended tool change intervals and cycle times or an outline of the implementation costs as well as the financial benefits of a recommended change.

The different types of data used by each department and the data flow links between the departments are as shown in figure 1. An examination of the data flow shows the tool development sheets and the tool data sheets used by both the Manufacturing Department and the Production Control Department do not have a formal destination. In addition there is no provision for information feedback to the industrial engineering department. The information flow links are only concerned with the direct information required by the user and the data collection personnel. The tooling data and performance information can not be accessed by the manufacturing departments, but the information is available from the tool room on request.

3 DEVELOPMENT OF NEW APPROACH

From the initial study five areas are considered to be of importance in developing and implementation a new tool management system if tool costs are to be further reduced and controlled. The areas of interest overlap because they all relate to information collection, information flow, departmental communication and the establishment of policies and procedures.

- Tool related costs. To formulate an accurate picture of the most effective management arrangement additional factors to the tool usage costs must be identified. The study has identified nine different types of tool related costs in six categories which are considered to be important for the plant. The actual purchase price of the tooling is the most apparent tool cost. It is directly affected by how many tools are being used and is easily identified. The company does calculate this cost, but does not have the necessary data or information available to offer explanations for why they are spending as much as they are. The company have also identified the cost of keeping a large tool inventory and have addressed the problem by absorbing the first level of the tool supply chain into the company to control the tool room. The over use of tooling decreases the number of times the tool is available for checking and resharpening. This means the tool life can dramatically decrease, thus raising the tooling cost per unit produced. There is no effective way of tracking this information at the company and this cost tends to be overlooked. Tool change schedules tend to be done for operational convenience rather than manufacturing effectiveness which leads to suggestions that tool usage and tool usage cost is not at an optimum, although no data is available to confirm this suspicion. Over used tooling is also expensive because it causes excessive down time of equipment from an increase in the magnitude of the cutting forces generated and the resulting increase in scrap rates. Tool breakage costs relating to poor tool selection resulting from economic purchasing pressures and tools operating at less than expected tool life also considerably increase the tool purchase cost. Poor tool selection also causes excessive down time due to the increased number of tool changes required. Broken tools create scrap which is another cost that must be considered. Set-up time and cost is considerably increased when appropriate tool set-up procedures are not established and followed. This also leads to additional down time due to machine crashes and the scrap produced by incorrectly set tools can be costly. Many tool change schedules are left to the discretion of the machine operators because they have delegated responsibility for the quality of the parts produced. This leads to excessive set-up time, under or over used tooling and increased down time caused by broken tools or continuous excessive strain on equipment due to worn tooling. The recognition of these tool related costs allows a tool management system to be established that focuses on the costs themselves.
- System infrastructure. The impact of infrastructure investments is as critical as tooling choice and in some cases more so. The infrastructure is characterised by its high level of investment, its fixed nature and the functional independence of the departments. It therefore needs to be developed with a clear understanding of the manufacturing and company requirements.
- Tool data. Increased tool data and tool information needs to be collected and analysed in greater depth to determine the tool related costs and why they are being incurred. Many time costs arise and poor quality occurs because of a lack of knowledge about how the process

actually works and possibly more importantly a lack of detail knowledge on how the process should work. The tool information collection system should support and allow both these areas to be addressed.

- Data flow. If the information collected is to serve any useful purpose, it must be made available to all the relevant parties in a format that is applicable to their activity. Standard documents need to be created and a software system must be implemented that is accessible throughout the facility. The independent software systems currently in use are crumbling the system infrastructure.
- Tool usage policies. There must be an end to a reliance on experience. Although experience is an asset to the company, it must be used appropriately. The technical and operational knowledge available must be employed to establish standard tool set-up and tool change procedures. A system must also be developed to ensure the procedures are adhered to, otherwise the data collection and investigative work performed will have a minimum effect on reducing costs.

4 MODELLING THE TOOL MANAGEMENT SYSTEM

A system can be defined as a transformation process that converts a set of inputs to a set of outputs. The system is then displayed as a collection of components which are inter related in an organised way and work together towards a specified logical and purposeful end. The inputs and output of the system are the main interfaces between the system and the external activities. The complexity of a tool management system make it necessary to develop a model from which an understanding of the system requirements, the needs of the system users and the relationships between the users and the outputs can be established. Different modelling approaches and techniques provide different results and they focus on different aspects of a system. Figure 2 shows the structure of the proposed tool management system from the perspective of the system requirements. It is useful to identify the sub-systems that must be considered in the development of the overall tool management system. The relationship given in figure 2 can be used alongside the hierarchical structures of the company to develop the functions required and how they must be utilised in the development, implementation and utilisation of the system. A system model can be developed to show how the system elements operate and the degree of integration achieved. The model of the proposed tool management system requires the consideration of three areas.

- Auditing sub-system. This is responsible for monitoring the system performance. The proposed tool management system element that assists in this function is the fault detection system. Fault detection is achieved by mechanical means or manual means such as visual inspection. It is a way of identifying potential tool problems.
- Management sub-system. This sets the achievements for the decision making and exercises control over the operational subsystem. It includes the tool room support, policies and procedures. The tool room is the heart of tool procurement, tool control and tool distribution. If the tool room is not functioning effectively and tools are not being used to their full capability then the tool management system will fail. Policies and procedures help to ensure that the developed best practice tool usage methods are put in to operation
- Operation sub-system. This contains components which carry out specific actions of the system to convert inputs to outputs. This includes the data collection system and the data flow system. The data collection system makes tool data and information available for

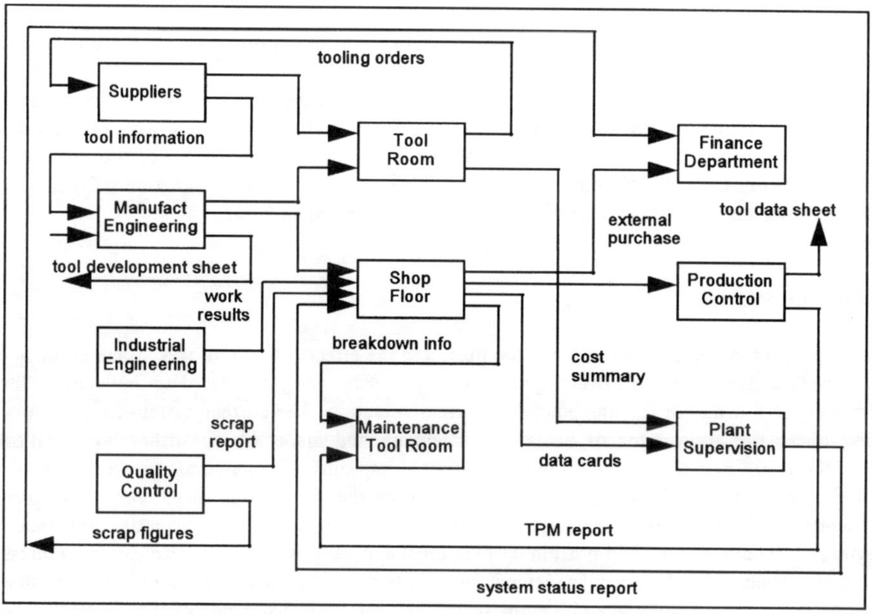

Fig 1. Current tool management system data flow

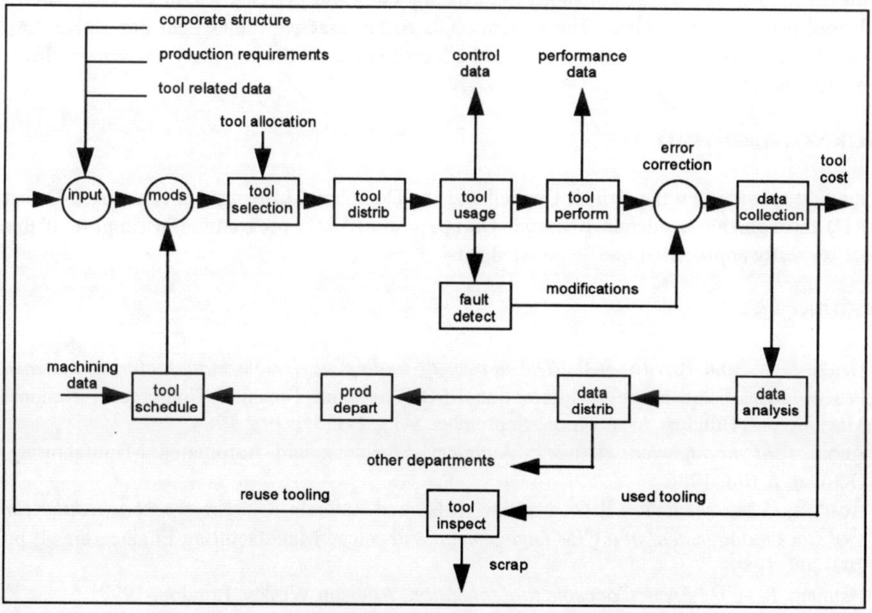

Fig 2. Proposed tool management system

scrutiny. The data flow system distributes the data and information to the tool management support functions to analyse function specific information for the purpose of identifying potential areas of concern.

The model offers an explanation of what systems and subsystems are involved in the tool management system and their sequential relationships with each other. The drawback with the model is that it fails to show what functions are involved or the relationships of the functions, what activities must occur within the systems and subsystems if they are to function correctly. A further drawback is that it fails to outline the design and implement strategy.

5 CONCLUSIONS

In investigating the current tool usage in the plant and the effects of current tool usage strategies it is revealed there are no formal procedures for data analysis, for data collection, for determining tool usage or for implementing corrective action. The current system tends to give tooling issues the appearance of being overwhelming and insolvable. Further investigation reveals these are not the actual problem but are symptoms of an overall greater problem. Tooling issues are currently being approached from the bottom up in a very functional environment. Although this approach offers some benefits, it has some substantial drawbacks. One of the main advantages of a bottom up functional approach is that a department can address a problem immediately. Upper management support is not required nor is the reliance on other departments. Also a problem can be approached with the specialist needs of the department in mind. The current bottom-up approach supports department independence and does not provide an infrastructure that combines the existing skills and systems within the company to attack tool management issues. The system tends to be re-active rather than pro-active. An issue is not addressed until it has a recognised adverse effect on the company. Normally this is of a budgetary nature.

6 ACKNOWLEDGEMENTS

The funding provided by the British Council and the Deutcher Akademischer Austausch-Dienst (DAAD) through the Academic Research Co-operation (ARC) programme in support of this project is greatly appreciated and acknowledged.

REFERENCES

1 Hankins, S.L.and Rovito, V.P., *The impact of tooling in flexible manufacturing systems.* Proceedings of 2nd Biannual International Machine Tool Technical Conference, National Machine Toll Builders Association, September 5-13, 1-175/1-198, 1984.

2 Anon. *Tool management at work.* American Machinist and Automated Manufacturing, October, p 106, 1986.

3 Acaccia, G.M., Michelini, R.C., Molfino, R.M. and Raffaelli, G., *An expert scheduler for tool stock management in a CIM environment.* Advanced Manufacturing Engineering, 1, pp 203-209, 1989.

4 Hannam, R., *CIM From Concepts to Realisation.* Addison Wesley, London, 1997.

5 Torvinen, S.J., Salminen, K. and Vasek, L., *Integration of a CIM tool management system into an intelligent feature-based process planning system.* Computers in Industry, 17, pp 207-216, 1991.

Depth estimation of an object using stereoscopic vision

C D'SOUZA, K SIVAYOGANATHAN, D AL-DABASS, and **V BALENDRAN**
Department of Mechanical and Manufacturing Engineering, Nottingham Trent University, UK

In robotic applications, the location and orientation of an object is often required for the robot to approach and pick the object. The depth or distance of an object from a sensor can usually be obtained by range sensing or using stereoscopic vision. In the work reported here, stereo vision is used. Two grey-level images of the same object are taken slightly apart from each other using two cameras. A calibrated stereo rig is used for this purpose. Using image processing, features such as lines, corners and other interest operators are extracted. Similar features in the two images are then matched. The depth is computed using camera geometry. Corners are extracted as features for matching. This paper describes the various corner extraction and matching schemes available to estimate the depth of an object using a calibrated setup. Results obtained in experiments using stereo vision using a median filter for corner extraction are presented and discussed.

1. INTRODUCTION

Viewing a scene from two (or more) different positions simultaneously allows us to make inferences about 3-D structure, provided that we can match up corresponding points in the images. This technique is called stereoscopic vision, or stereo vision for short. It is thought that the visual systems of humans and some other animals makes use of this, and it is very important in attempts to develop practical computer vision systems. Stereo vision which utilises two cameras to view an object from slightly shifted positions (refer figure 1) allows one to gauge the depth of the object with respect to the camera in a non-contact manner by making use of the disparity between the two images. This is a passive method as compared to range sensing. It is useful for guiding robots to approach objects to perform assembly tasks.

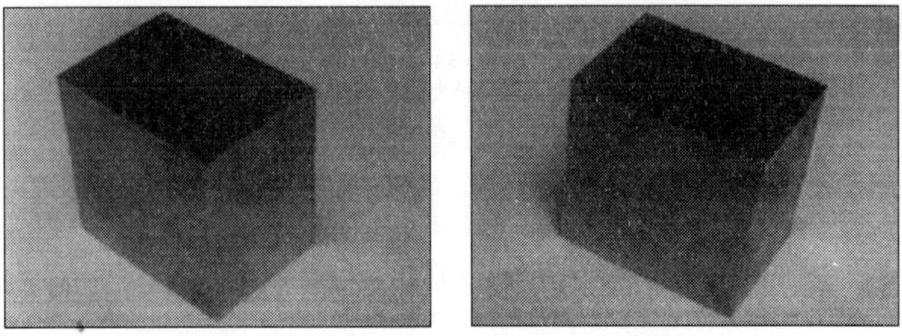

Figure 1 Views from two different cameras in a stereoscopic set-up

2. PERSPECTIVE PROJECTION

Assuming the pinhole camera model, the projection of a point $P(X,Y,Z)$ in world coordinates onto an image plane is given by

$$x = \frac{Xf}{Z} \;; \qquad y = \frac{Yf}{Z}$$

where f is the focal length, and x and y specify the position of the point in the image plane. (refer figure 2).

Finding the value of f in pixel units is the basic camera calibration and is done by making measurements of the image of an object of known size and distance. X, Y and Z are measured in the same units (ie inches/cms) while x and y are measured in pixels.

2.1 Stereo Geometry for Parallel Cameras

If two cameras are parallel and placed side by side (figure 3), the image of a point will have the same y coordinate but two different x-coordinates xL and xR. It is convenient to describe positions relative to an imaginary central camera (cyclopean coordinate system). D is the separation between the camera centres and is also called the baseline.[1]

Now if a point at (X,Y,Z) in cyclopean coordinates produces images at (xL,y) and (xR,y) in the two cameras, its position in space can be found from the image positions using the formulae:

$$X = \frac{D(xL+xR)}{2(xL-xR)} \;; \qquad Y = \frac{Dy}{xL-xR} \;; \qquad Z = \frac{Dy}{xL-xR}$$

The quantity xL-xR which appears in all three formulae is called the stereo disparity (or just the disparity) of the point. The equations are derived using the fact that the perspective

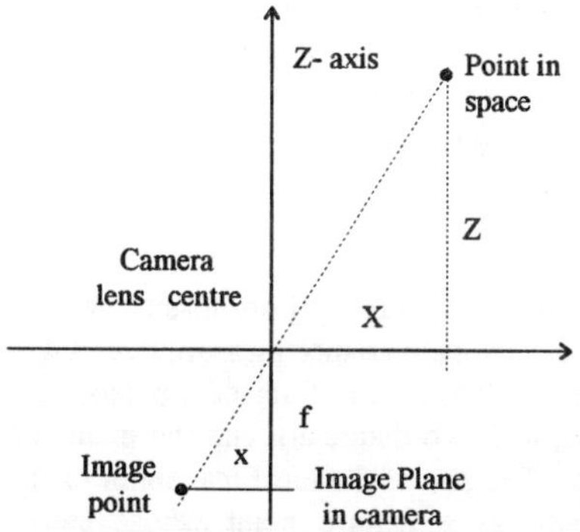

Figure 2 Perspective projection

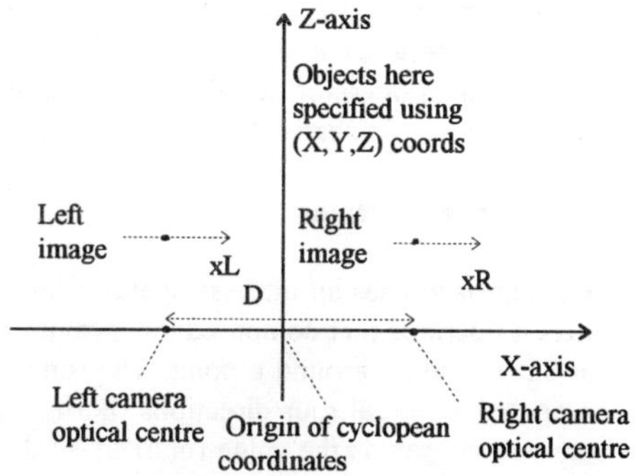

Figure 3 Stereo geometry for parallel cameras

projection equations in x for the left and right cameras are $xL=(X+D/2)f/Z$ and $xR=(X-D/2)f/Z$. A small disparity yields a large depth, and vice versa.

2.2 Stereo Geometry for Converging Cameras

It is not usually convenient to set up cameras with their axes parallel, because this limits the region of space in which objects are visible in both images. It is more normal to aim the cameras so that their axes are angled inwards, and converge on the objects of interest. The angle by which the cameras converge is termed as the vergence angle. The point in space where the optical axes intersect is called the fixation point(Z_0) for the cameras and it has zero disparity.

This makes the geometry more complicated. Provided the amount of convergence is small, though, the vertical disparity can be ignored, and there is a reasonable approximation which can be used to calculate the depth [2]. The disparities can be adjusted by adding the quantity fD/Z_0, which is the disparity an object at the fixation point would have if the axes were parallel. The new equations become:

$$X=\frac{D(xL+xR)}{2p}\ ; \qquad Y=\frac{Dy}{p}\ ; \qquad Z=\frac{Df}{p}\ ; \qquad \text{where } p=xl-xR+\frac{fD}{Z_0}$$

Z_0 is found during calibration.

3. FEATURE EXTRACTION

To recover the depth, features from two images must be matched. These features can be corners, edges or other interest operators, which are discussed below. Hence the first stage of stereo vision involves preprocessing to extract these features. The second stage involves actual correspondence and in the third stage, the use of equations described above is made in order to calculate the depth.

There are several methods for grey level corner detection some of which have been described in more detail below. Kitchen and Rosenfeld [3] have implemented four types of operators based on the gradient magnitude of gradient direction, change of direction along the edge, angle between most similar neighbours and the turning of the fitted surface. Beaudet[4] has defined an operator called DET which responds at corners and saddle points of a surface. Near a corner of a shape DET responds (with opposite signs) on both sides of the edge.

3.1 Moravec detector

One approach uses an interest operator located in an image. Moravec[5] proposed one such interest operator that computed the local maxima of a directional variance measure over a 4x4 (or 8x8) window around a point. The sums of squares of differences of adjacent pixels were computed along all four directions (horizontal,vertical, and two diagonal), and the minimum sum was chosen as the value returned by the operator. The site of the local maximum of the values returned by the interest operator was chosen as a feature point whose stereo counterpart was to be found.

3.2 Median filter

If a corner point is observed in a window, it is found that there is difference between the value of the central pixel and the median value of the neighbourhood[1]. This suggests the following algorithm; pass a median filter over an image and calculate the difference in magnitude of the median and centre pixels. If this difference exceeds some threshold, regard the pixel as a likely corner[6]. This algorithm is disturbed by noise and small features (features thinner than the window) but can be used in conjunction with information from the Hough transform to extract line ends.

3.3 MIC (Minimum Intensity Change) algorithm

This algorithm[7] is based on the variation of image intensity along arbitrary lines passing through the point of investigation within the neighbourhood of the point. A corner is detected if the variation of the image intensity along such lines is high for all line orientations. This algorithm employs linear interpolation to compute the directional first-order derivatives. The algorithm is very fast as it effectively rejects points with small intensity variations. The points whose minimal intensity change over all directions is high are declared corners.

Apart from the above, two other corner detectors which are extremely common are the Harris[8] and SUSAN[9] corner detectors.

4. THE CORRESPONDENCE PROBLEM

The problem of pairing up the features is known as the correspondence problem. If we apply no constraints, any feature in the left image could match any feature in the right image.

One of the most basic constraint, when looking for matching features in a single pixel row of the image which has been assumed is that a feature with a particular y value can only match a feature in the other image with the same y value. This is known as the epipolar constraint.

The other two constraints are that similar features match each other and that features close together in the image should have similar disparities.

There are many ways of measuring feature similarity. For example, some systems make use of the orientation of the edge. Another which adopts a measure based on the grey-levels in the image is: the differences in grey-levels in a patch of pixels round the feature(i.e. an edge) are added up on the assumption that matching regions should have a similar local structure and overall grey-level. Two patches are considered, one on each side of the feature, and a match will be accepted if either the grey-level differences to the left of the feature are small, or the differences to the right of the feature are small.

Marr and Poggio [10] proposed a theory which was later implemented in a better way by Grimson[11]. In this approach the image is first preprocessed using several Laplacian of Gaussian (LoG) filters and the zero crossings are found. The overall matching strategy uses a coarse-to-fine iterative approach with disparities found at coarser resolutions used to guide match-point search at finer resolutions. In one approach [12] to point pattern matching a merit score is assigned to each pair (P_i, Q_j), according to how closely other pairs (P_h,Q_k) match when P_i is mapped into Q_j. The scores can then be recomputed, giving weights to the other point pairs (P_h,Q_k) based on their own scores; and this process can be iterated. When this is done, the scores of pairs that correspond remain relatively high, while those of other pairs become low. In uncalibrated setups, the camera geometry is unknown and cameras can be placed in any orientation and need not be parallel. A few points are first detected from both views and matched and from this the epipolar geometry is computed. This is done by finding the fundamental matrix[13]. This requires fast computation but near real-time rates are being achieved.

5. EXPERIMENTS AND RESULTS

For the work reported, the features used were corners. The use of the median filter was made for detecting corners in grey level images. Filters of sizes 3x3, 5x5 and 7x7 were used. It was observed that a larger filter could detect corners where two lines (edges) met obliquely in the image, compared to a smaller filter[5]. Also, rather than a single point, the filter response gave a group of closely lying points which then had to be averaged to find a single corner point. To avoid spurious points from being detected, with the help of a mouse, a small area was defined where the corner was to be found.

For matching the corners, first a band for the y-coordinate was kept to detect all features lying about a horizontal line. Then a band for the x-axis was kept after experimentation and all points appearing in sequential order in the horizontal line were matched.

Calculated depth (cm)	Measured depth (cm)	Absolute Error found	Percentage error
114	114.3	0.3	0.26%
140.7	140.9	0.3	0.21%
169.7	171.5	1.8	1.05%
197.6	201.9	4.3	2.1%
220.2	227.3	7.1	3.1%

After this, the equations described earlier were applied to calculate the depth of a single corner point at different distances. The constants f and Z_0 were found from the calibration procedure. The fixation point was found by centering the two cameras by focusing on a point which was orthogonal to the baseline. Different vergence angles were also tried. The depth was first calculated by the using the algorithm and then verified manually. The results obtained have been tabulated above.

6. CONCLUSIONS

In this paper, the theory of stereo vision, various corner detection methods and matching schemes were described. Results obtained using a calibrated setup were presented. The error obtained is due to several factors such as accuracy of the corner detector, estimation of Z_0, and the approximation in the formulae. However this error can be minimized by software using error compensation techniques.

Acknowledgements
The authors wish to thank Rolls Royce and Associates for providing the manipulator for this project.

REFERENCES

(1) R.D. Boyle, R.C. Thomas; *Computer Vision- A first course*, Blackwell Scientific Publications, 1988
(2) David Young; *Sussex Computer Vision Teach Files*, Sussex University, WWW address http://www.cogs.susx.ac.uk/users/davidy/teachvision/vision0.html
(3) Les Kitchen, Azriel Rosenfeld; Grey-level Corner Detection, *Pattern Recognition Letters*, vol 1, pp 95-102, 1982
(4) Paul R. Beaudet ; Rotationally Invariant Image Operators, *International Joint Conference on Pattern Recognition*, pp 579-583, 1978
(5) Hans P. Moravec; Towards Automatic Visual Obstacle Avoidance, *Proceedings of the 4th International Joint Conference on Artificial Intelligence*, pp 584, 1977
(6) K. Paler, J. Foglein, J. Illingworth, J. Kittler; Local ordered grey levels as an aid to corner detection, *Pattern Recognition*, vol 17(5), pp 535-543, 1984
(7) Miroslav Trajkovic, Mark Hedley; Fast corner detection, *Image and Vision Computing*, vol 16, pp 75-87, 1998
(8) C. Harris, M. Stephens, A combined corner and edge detector, *Proc. 4th Alvey Vision Conf.*, 1988, pp.147-151
(9) S. Smith, M.Brady, SUSAN – a new approach to low level image processing, *DRA Technical Report TR95SMS1*, 1994
(10) Umesh R. Dhond; Structure from Stereo- A Review, *IEEE Transactions on Systems, Man and Cybernetics*, vol 19(6), pp1489-1510, 1988
(11) W.E.L.Grimson; Computational Experiments with a Feature Based Stereo Algorithm, *IEEE Transactions on Pattern Analysis and Machine Intelligence*, vol 7(1), pp 17-34, 1985
(12) Sanjay Ranade, Azriel Rosenfeld; Point Pattern Matching by Relaxation, *Pattern Recognition*, vol 12, pp 269-275, 1980
(13) Faugeras, O; *Three-Dimensional Computer Vision*, MIT Press, 1993

Development of an interactive process parameter selection model for high-speed machining

A KALDOS and **A BOYLE**
School of Engineering, Liverpool John Moores University, UK
D FICHTNER
Manufacturing Automation, Technische Universität, Dresden, Germay
I F DAGILOKE
Sencam (UK) Limited, Manchester, UK

Abstract

High speed milling (HSM) is one of the emerging cutting processes having tremendous potential for increased metal removal rates and for producing improved surface finish, burr free edges, dimensional accuracy and a virtually stress free component after machining. The paper presents an interactive process parameter selection model to predict the effects of the input cutting parameters on the output process variables. This interactive system can be used to select the most suitable cutting parameters in various high speed machining applications when cutting a number of different materials. The system provides the user with an open model structure to incorporate company specific features.

1 INTRODUCTION

The implementation of high speed machining presents a challenge since various factors limit the extent to which cutting speeds may be advantageously increased. These factors include the cutting temperature, the cutting tool material, the machine tool design, the geometry of the cutting tool, the cutting power and after machining the surface integrity and metallurgical condition of the workpiece. Since the actual cutting speed achieved depends on the workpiece material, the type of cutting operation being employed and the cutting tool material used a full understanding of the technology requires a structured process model to describe the high speed milling process. A process model which allows for the prediction of the process output parameters, complete with supporting experimental data, is essential in determining the factors which influence productivity at high cutting speed. The aim of the paper is to verify a modular structured process model that can be used to predict the effects of the independent cutting parameters on the milling process. An indication of the structure of the model developed for the computer aided cutting process evaluation and parameter selection is also presented.

The strategy adopted is to provide a number databases for the input data, which enables the input cutting data to be retrieved during simulation and the calculated results to be displayed both numerically and graphically. Alternatively, the results can be stored in the database for recall at a future date. The computer model is designed with a database which has separate data modules for work materials, machine tool configurations, cutting conditions, cutting parameters and workpiece geometry. The data modules allow for the inclusion of interactive data input for database extension or for upgrading the stored values.

Cutting temperatures, cutting forces and the material removal rates are established with particular consideration being given to workpiece and cutting tool properties. The developed computer model allows for the process parameter selection for the high speed milling process relevant to the aerospace industry but due to its generic structure it can be used in the tool and die industry as well as many other industries. The simulation model provides a method of calculating cutting parameters in an interactive way and gives a numerical indication of some process performance indicators including cutting temperature, cutting forces and material removal rate. The developed and verified computer model provides a foundation upon which optimisation strategies for high speed machining may be developed by integrating process parameter selection and manufacturing constraints at the process planning stage. The numerical results and their graphical representation make it possible to apply the model to complex machining analyses. The results are useful in determining a suitable chip management system for a high speed milling operation. The developed system provides an excellent aid for the determination of machine tool and main spindle specification for a particular operation.

2 NUMERICAL MODELLING STRUCTURE

The strategy used is to create and implement an extensive and comprehensive theoretical model, based mathematically on cutting process parameters. Application of such a model makes it possible to show the dependency of the cutting parameters and their effects on the simulated results. By solving the equations simultaneously as a set, it is possible to see how the choice of input values by the user influences the high speed machining process. The simulation model is supported by an open integrated database for the data acquired for workpiece and tooling, mechanical and thermal properties, cutting conditions, workpiece geometry, tooling geometry, machine tool parameters, generated numerical and graphical results [1][2]. The computer model consists of three major modules namely the integrated input database module, the numerical simulation module and the output module. The input module has an integrated database consisting of the following parameter data sets :-

- Workpiece materials

  A range of workpiece materials can be accommodated by the computer model requiring the following material properties:-

  - workpiece density
  - thermal conductivity
  - specific heat capacity
  - initial workpiece temperature
  - Brinell hardness value.

- Tool materials

  A range of special high speed machining tool materials can be accommodated by the model for which the following parameter values are required for the database :-

- tool is coated or uncoated
- tool material density
- thermal conductivity
- specific heat capacity
- initial tool temperature

- Tool geometry
  The following tool geometry values are entered into the tool geometry data set :-
  - tool diameter
  - number of teeth or flutes
  - tool shank diameter
  - rake angle
  - flute length.
- Workpiece geometry
  The geometry of the workpiece is assumed to be predominantly prismatic for which the following values are required :-
  - blank depth
  - blank length
  - blank width
  - geometrical features.
- Cutting parameters
  The following cutting parameters can be selected and entered by the user to suit the machining conditions of the component :-
  - feed value
  - feed rate
  - depth of cut
  - width of cut
  - cutting speed.
- Machine tool parameters
  The choice of machine tool for the machining operation is allowed for in the model by the provision of the following machine tool limits :-
  - range of operation for maximum feed travel in x, y and z axes
  - range of operation for feed rate in x, y and z axes
  - spindle speed
  - spindle power
  - tool length range
  - tool diameter range.

Having received the input information, the coded program is designed to transform the geometric, mechanical and thermal properties information of the workpiece, the cutting tools, the machine tool and the cutting conditions into a suitable form for simulation. Alternatively, a new set of input data can be entered into the model. The workpiece, cutting tools, machine tool and cutting conditions data file can be edited making it possible to update and expand the databases. The input module can be edited prior to the cutting simulation. The numerical simulation module has an inbuilt numerical data controller with four sub-modules [2], which determines the cutting parameters, the cutting forces, the cutting energy and the cutting temperatures respectively. Once the input data is entered, the numerical simulation takes approximately 20-25 seconds to execute all its operations. The output module generates both the numerical output data and a graphical representation of the cutting process.

3 EXPERIMENTAL WORK

Since one of the objects of the study is to verify the model against experimental results with particular reference to the milling of aluminium alloys suitable for aerospace components at high cutting speeds, aerospace aluminium alloy is employed in a set of experiments. The workpiece material is received in the form of rectangular bar with a cross section measuring 45 mm by 300 mm and a length of 300 mm. The machine tool used is a Marwin three axis, vertical spindle, milling machine. The machine tool is equipped with a 30 000 min^{-1} high speed spindle with ceramic ball bearings and an operating power of 25 kW. It has a maximum feed rate in the X, Y and Z directions of 20 m min^{-1} and the execution time of the control system is 20 ms.

The cutting tool material selected for the experiments is tungsten carbide brazed on to a high speed steel body. This material can be easily machined in its annealed state to the required shape, then heat treated to produce the desired mechanical properties for metal cutting. The cutters chosen are 50 mm and 25 mm diameter with two cutting points. The tool geometry is unique since it is designed and made for high speed milling of aluminium alloys. A three component Kistler dynamometer is used to measure the cutting forces in the X and Y directions in the horizontal plane. The processing of the signals is performed by a 12-bit data acquisition card Lab PC Plus complete with Daqware software supplied by National Instruments UK.

4 MODEL VERIFICATION

The chip formation process is most frequently explained by means of plasticity theory supplemented by the analysis of fracture. Plastic flow theory implies the strain and strain rate are very high for the chip to form because of the geometry of the process. The effect of rake angle on cutting forces is of considerable interest since if the forces reduce with large positive rake angles the power requirement will be less. The aim is to establish the influence of the rake angle on cutting force and to see how it affects the power requirements in high speed machining. Using the input data from Brown [3] with a range of rake angles from 6 degrees positive to 20 degrees negative the predicted cutting forces from the model are as given in Figure 1 along with the experimental results for an aluminium 2017T4 workpiece material being machined at 1 000 m min^{-1} at a depth of cut of 2.54 mm. The width of cut is 5 mm, the cutting tool diameter is 40 mm with 2 cutting points. The results obtained suggest that higher negative rake angles do produce higher values for both the principal cutting force and the feed force. Although the increase in the principal cutting force is not as dramatic as the increase in the feed force. The reduction in the cutting forces as the rake angle is increased positively is attributed to an increase in shear angle with the implication being a lower power is required in the chip formation process. The predicted cutting forces plotted in Figure 4. compare well with the experimental work of Brown [4] generally being 15% greater for the principal cutting force and 9% greater for the feed force. Similar decreases in force with increasingly positive rake angles are reported by von Turkovich [5] and Fenton, Oxley [6] and Schulz[7].

Figure 2. shows the relationship between the measured principal cutting force, the simulated or theoretical cutting force and the cutting speed. As can be seen, the cutting force increases with cutting speed in both the lower speed range up to 2400 m min^{-1} and in the higher speed range from 2800 to 5000 m min^{-1}. The experimental cutting force versus cutting speed correlates well with the simulation results.

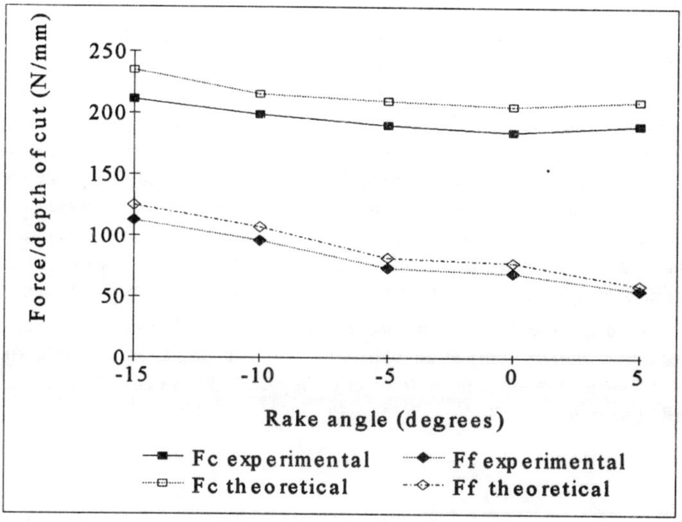

Fig. 1. Rake angle and cutting forces

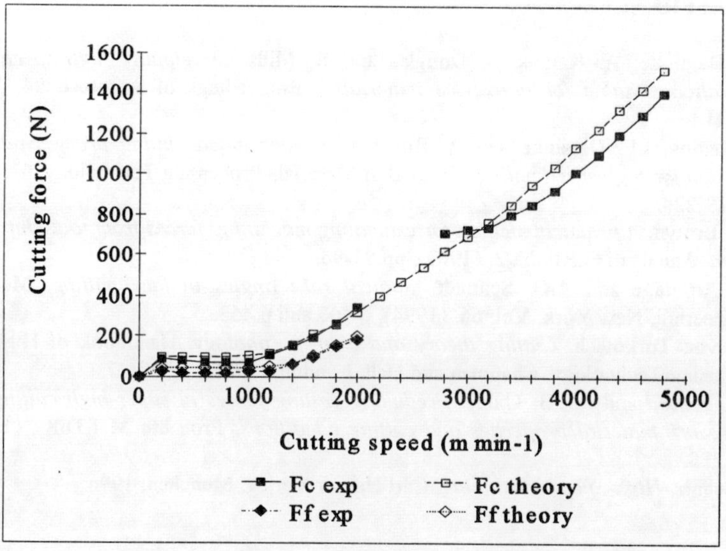

Fig. 2. Cutting forces versus cutting speed

5 CONCLUSIONS

The developed computer simulation model provides a sound foundation for process parameter selection and process optimisation for high speed milling primarily for the aerospace industry. It is envisaged that the model can also be used for predicting cutting parameters when cutting other materials such as hardened steels, which is of vital importance for the tool and die industry. The results obtained from experiments correlate well with the process parameters predicted by the computer model. In the study it is established that in high speed milling the cutting forces increase as the cutting speed increases. This is duly demonstrated experimentally and by means of the developed computer simulation model. This increase can be attributed to the effect of the momentum force which is in agreement with the findings of a number of published works. During the study it became clear that as the cutting speed increased the removal rate increased in the region up to 4700 m min^{-1}. As a result of computer model verification sufficient evidence has been gathered to justify the application of neural networks in system modelling and the initial neural system model is being trained at present.

6 ACKNOWLEDGEMENTS

The funding provided by the British Council and the Deutcher Akademischer Austausch-Dienst (DAAD) through the Academic Research Co-operation (ARC) programme in support of this project is greatly appreciated and acknowledged.

7 REFERENCES

1 I.F. Dagiloke, A. Kaldos, S. Douglas and B. Mills. *Developing high speed cutting simulation suitable for aerospace materials.* Proceedings of Aerotech 94, IMechE, (1994)

2 A. Kaldos, I.F. Dagiloke and A. Boyle. *Computer aided cutting process parameter selection for high speed milling.* Journal of Materials Processing Technology, 61, (1996), pp 219-224.

3 C.A. Brown. *A practical method for estimating machining forces from tool-chip contact length.* Annals of CIRP, 32/2, (1983), pp 91-96.

4 J.B. Armitage and A.O. Schmidt. *Radical rake angles in face milling.* Mechanical Engineering, New York, Vol. 66, (1994), p 403 and p 453.

5 B.F. von Turkovich. *Cutting theory and chip morphology*, Handbook of High Speed Machining Technology, Chapman and Hall, London, (1985), pp 27-47.

6 R.G. Fenton and P.L.B. Oxley. *Predicting cutting forces at super-high cutting speeds from work material properties and cutting conditions.* Proc 8th M.T.D.R., (1976), pp 247-258.

7 H. Schulz. *High-Speed Machining.* Carl Hanser Verlag, Munchen, 1996

An integrated approach to assembly line design

J D LANHAM
Faculty of Engineering, University of the West of England, Bristol, UK

1. ABSTRACT

Currently the approach to assembly system design often consists of a series of loosely linked tasks that use a range of analysis methods and tools. The task is typically broken down into a set of sub problems that focus on separate facets, such as assembly sequence selection, assemblablity review and line layout, often without providing a wider global perspective. The approach presented in this paper provides a new approach for representing assembly sequences in such a way as to facilitate assembly sequence selection, design for assembly review and assembly line balancing in an holistic manner allowing engineers to assess the overall impact of alternative product designs.

2. THE NEED FOR INTEGRATION IN ASSEMBLY LINE DESIGN

Assembly is a task germane to virtually all manufactured products and research undertaken by the DTI in 1987(1) revealed that assembly costs contribute about 40% of the total product cost. Thus any improvement in the efficiency of the assembly process can bring useful cost benefits. At present the methods and tools currently available address the problem of assembly line design by taking a piece-wise approach that decomposes the problem as a series of loosely coupled activities, typical methods are summarised in the Table 1. Each of these methods focuses on a specific facet of the problem and aims to determine an optimal solution for each facet, for example minimising the number of parts in the product in isolation from other activities in the process. Current DFA based tools are in effect open loop systems that only assess the product from one standpoint, that of assembly costs. The emphasis of a design for assembly study is largely on reducing costs by reducing parts count. Whilst it is undeniable that this type of method has been shown to reduce costs, in some cases it may be cheaper to have a part design that is less than optimum from an assembly standpoint but is in fact a less expensive piece part. Although there is a greater assembly cost, the overall cost may in fact be lower as the lower manufacturing costs may outweigh the increased costs incurred in assembly. This point is illustrated by the experience of British Aerospace with the assembly of a part on a A300 series airliner (2).This example highlights the need to take a wider perspective than is often taken at current and highlights the inability of current methods to support this wider perspective. These local optimisation methods though take no account of the overall cost performance which in reality is the more significant factor. Furthermore it is often difficult to quickly and easily assess the impact of design options/alternatives on the assembly system as a whole due to the different tools and methods used at each stage and the incompatibility of the representations used.

	Assembly Sequence	Product Design	Line Design
Typical Tools / Methods	Bourjalt (3),de Fazio and Whitney (4),Homen de Mello and Sanderson (5)	DFA Boothroyd & Dewhurst (6), LUCAS & Swift (7)	Kilbridge and Wester (8),Ma (9)
Representation form used	Liasion Graph, Diamond Graph, AND/OR Graph	Assembly drawings	Precedence network
Evaluation metric	Shortest assembly sequences, multiple assembly sequences	Reduce costs by reducing parts count	Line Balance, station/station variance

Table 1 : Current tools and representations.

As a consequence of these limitations a new approach has been devised that creates a single representation of the product that supports assembly sequence evaluation, design for assembly review and line design.

3. THE ASSEMBLY STATE VECTOR REPRESENTATION METHOD

If we consider a product with n component parts. It is possible to represent this as an n dimensional vector where each element corresponds to a component in the product. The vector (1,1,1, ... 1) corresponding to the completed assembly and the complementary (0,0,0, ... 0) vector corresponding to the completely dismantled product. If a weight is assigned to each element in the vector, calculated as 2^{j-1} where j represents the part number, or position in the vector, we can express the assembly state as a number that is the weighted product of the part numbers present. This representation, termed the Assembly State Vector, ASV, allows multiple sequences to be identified from any given assembly state, yet the numbering method implicitly contains sequencing information which enables an overall assembly sequence to be determined that satisfies the global product sequence constraints. Thus the engineer can focus on the localised precedence requirements, or partial assembly sequences, and the coding scheme ensures the overall precedence constraints are satisfied.

For example given a five part product the complete assembly can be represented as a binary vector (1,1,1,1,1) which as the weighted vector becomes.

$$1 \times 2^{(5-1)} + 1 \times 2^{(4-1)} + 1 \times 2^{(3-1)} + 1 \times 2^{(2-1)} + 1 \times 2^{(1-1)} \quad \text{or} \quad 16 + 8 + 4 + 2 + 1 = 31$$

Correspondingly the weighted vector for each part is given by the vector in which each part is the only part present.

Thus for part n = 5 the weighted vector is given by :

$$1 \times 2^{(5-1)} + 0 \times 2^{(4-1)} + 0 \times 2^{(3-1)} + 0 \times 2^{(2-1)} + 0 \times 2^{(1-1)} = 16$$

Thus it is possible to represent the state of an assembly by a unique vector calculated from the sum of the vectors of the parts present. This representation of a product and possible assembly sequences can be represented in a graphical form that shows all feasible assembly states for the product. An example is given for the simple pen as shown in Figure 1. The assumption , as used by many others, is that only one part is added at each step, or two previously created sub assemblies are joined. Thus the product is built up by adding parts one at a time.

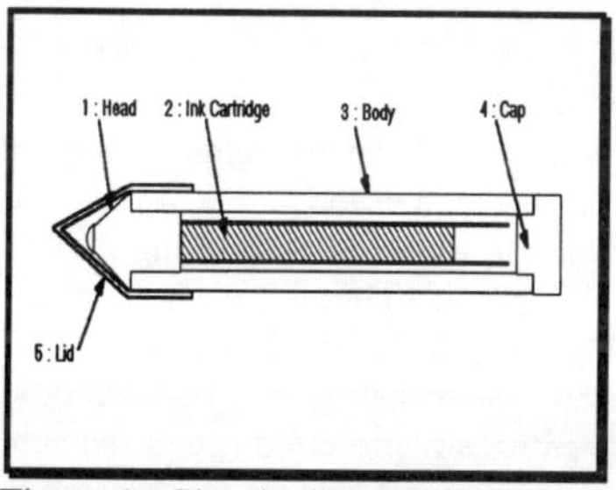

Figure 1 : Simple disposable plastic pen

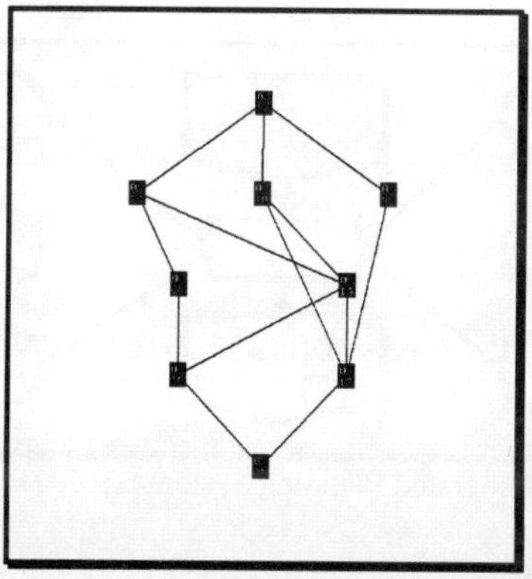

Figure 2 : Directed Graph of assembly sequences for pen

In this form the top node corresponds to the unassembled product and the bottom node the fully assembled state, the node numbers are the ASV values for that node and indicate which parts are present in that sub state. The interim nodes correspond to the range of feasible sub-assemblies and the arcs linking the nodes correspond to the parts, or sub-assemblies, that can be added at each node. An assembly sequence is generated by traversing the graph from the top unassembled state to the bottom fully assembled state. This representation allows for any assembly state to be reached in a number of different ways reflecting the range of possible feasible sequences that the engineer may wish to consider.

4. AN INTEGRATED APPROACH TO ASSEMBLY LINE DESIGN

Using the ASV method an integrated system for the design of assembly lines has been developed. As part of the development of this method a system has been coded in the C language to run on a PC. The overall structure of this system is shown in Figure 3. This data is stored as a dynamic linked list and is accessed by a number of modules which undertake the specific tasks required. The individual modules will be briefly described to give an overview of the whole system.

User Input Interface : The engineer enters details of the steps in the assembly sequences. It is this list structure that is used to support the subsequent analysis. Details entered include handling and mating/joining times, DFA part classification and details of the part cost. Global information such as production volumes required, cost of labour, duration of working period and capital costs of workstations can also be entered at this point. Through an interactive dialogue the user is prompted for details of feasible assembly sequences. Details are entered of the part numbers at each step and these are converted to ASV values before being stored in a list along with the related data. The user can additionally identify any preferred / required sub assembly combinations, i.e. for the purposes of testing or adjustment during the assembly sequence.

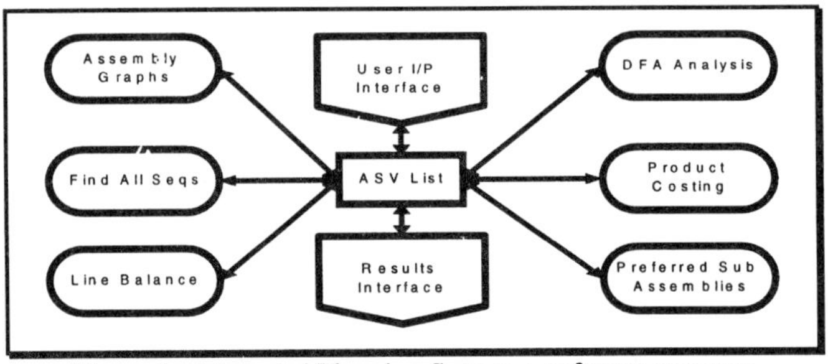

Figure 3 : Integrated Planning System : software structure

All Sequences : An assembly sequence can be generated by undertaking a "walk" through the graph as shown in Figure 2. This is effected by traversing the ASV list from the ASV state corresponding to the fully assembled product and recursively identifying from the current state which sub assembly should be taken as the next step until the state reached corresponds to a single piece part. Some states, or part combinations, may in fact be reached by several paths, or assembly sequences. For example a sub assembly of parts A, B, C & D may be made in two ways. One sequence A & B, then add C, then add D and a second sequence D & C, then add B, then add A. From a line balancing standpoint it is desirable to consider both sequences as one or the other may fit better within the cycle time for the product. Thus the sequence searching algorithm has to deal with multiple instances of some states and allow for this in generating sequences. The situation was further complicated by the inclusion of sub assemblies that are themselves made by joining two sub assemblies together. So for the four part sub assembly A,B,C & D the product could also possibly be made as A & B, C & D then the sub assembly A & B added to sub assembly C & D. This creates additional complexity as there are two possible sub sequences for every sub assembly mating of this type.

Required Sub - Assemblies : As discussed it may be appropriate to identify specific sub states of the assembly that should be included in the sequences to facilitate testing, inspection or adjustment during the assembly. The system prompts the user for details of the sub states required and using the ASV values checks that only those sequences that incorporate the required state are processed by the line balancing algorithm.

Line Balance : An algorithm has been developed that again uses the same basic ASV list to determine the allocation of tasks to workstations. Using the sequences identified the task list is scanned and tasks allocated in precedence order. If it is not possible to allocate a task due to insufficient remaining time at the station the ASV coding is used to indicate the next feasible task that does not violate precedence constraints. Furthermore the algorithm has been developed to allow for task times greater than the required cycle time. In this case parallel stations are set up with duplicate tasks on each station. Subsequent tasks are then duplicated across the parallel stations until all the slack time on the parallel stations is used. Thereafter assuming no tasks have times greater than the cycle time the algorithm reverts to allocating tasks to single workstations. Furthermore the system allows for different cycle times to be entered and the analysis rerun using varying cycle times to determine if a different sequence is more appropriate.

DFA Analysis : DFA methods focus on identifying those parts which are potentially redundant in a product and classify parts as either essential or non-essential. As discussed earlier one of the limitations of current DFA tools is that they provide no feedback to the designer as to the overall impact of design decisions, or the relative overall benefits of a number of design alternatives. The system developed allows the designer to modify details of part costs, operation times and part status to determine the relative merits of a number of design alternatives. The essential /no-essential status of a part can easily be altered and the redundant part is then deleted from the ASV list in all sub states where it was previously and the analysis previously described repeated.

Assembly Graphs : Using the data held the system has option to display the possible assembly sequence as a directed graph as shown in Figure 2, or the more familiar precedence network which shows which part is added at each step.

Product Costing & Results Interface : After undertaking the analysis the system produces a range measures including unit cost, line performance, number of workstations, workstation to workstation variation. These can then be analysed by the user to select the assembly sequence that overall gives the desired results.

5. SUMMARY AND FUTURE WORK

The work described in this paper has presented a method that allows an holistic perspective to be taken to the problem of assembly line design. The aim has to integrate the product design evaluation task as undertaken with DFA analysis with the issue of line design. The ASV method described supports the evaluation from both perspectives. Changes to the product design can easily be propagated through the system and their effect on the whole assembly system evaluated. Thus it is very easy for the designer to experiment with a range of design scenarios and quickly gauge the relative merits of the options being considered.

The approach developed has been coded in C and currently provides a functional, if somewhat basic, system that can be used to evaluate a range of design options. There are two main issues for the future relating to this work.

1) Currently the system has only been evaluated using synthetic data generated to evaluate the integrity of the system. An industrial based case study is currently being planned to evaluate the system using an actual product.

2) Work is required to develop an interface to a CAD system to automate the generation of assembly task information. This is currently the processing bottleneck and is seen as the main area for future work.

---

1. REFERENCES

1 FAMOS-Eureka Project EU72 : Department of Trade and Industry : 1987

2 AirBus News : Anon : When two into three really does add up : AirBus News : January 1992.

3 Bourjault A. : Methodology of assembly automation : a new approach : Procs. 2nd Int. Conf. on Robotics and Factories of the Future, 1987.

4 de Fazio T. & Whitney D. : Simplified generation of all mechanical assembly sequences : IEEE Jnl. of Robotics and Automation : Vol. RA-3 : No 6 : 1987

5 Homen de Mello L.S and Sanderson A.C. : AND/OR Graph Representation of Assembly Plans : IEEE Trans. on Robotics and Automation : P 188 - 199 : 6 : 2 : April 1990

6 Boothroyd and Dewhurst : Design for Assembly : A Designers Handbook : Dept. of Mechanical Engineering, MIT : 1983.

7 Lucas Engineering Systems : Design for Assembly Practitioners Manual Ver. 7.1: Sect. 8.1 : 1990.

8 Kilbridge M. & Wester L. : A Heuristic Method of Assembly Line Balancing : Jnl. of Industrial Engineering : Vol. 12, No. 4 : p 292 - 298 : 1961.

9 Ma X. : Finding the best possible solutions to simple assembly line balancing problems : Procs. of the Institution of Mechanical Engineers : Vol. 211 : Part B : p 53 - 61 : 1997.

An overview of the use of internet telemetry systems in manufacturing

P G LEE and **C E R WAINWRIGHT**
The CIM Institute, Cranfield University, UK

ABSTRACT

The information and access to knowledge readily available on the World Wide Web (WWW) is well known for its worth to both individuals and organisations alike. In recent years, however, the very structure of the Internet that facilitates the WWW is allowing organisations to increase profitability through a technology known as the Virtual Private Network (VPN). This paper outlines the underlying technologies and suggests the possibilities of using the Internet to transmit telemetry data.

1.0 INTRODUCTION

There has been much hype written (and spoken) about the Internet, what it is, what it can do and whether it should be regulated or not, but essentially *apart* from the information available on it, the Internet can offer organisations very attractive benefits. These benefits may come about by utilising the structure of the Internet instead of the manufacturer's own Wide Area Network (WAN).

If Internet technology and telemetry systems can be combined, a very powerful and cost effective means of distributing control of manufacturing processes may ensue. By utilising Internet Telemetry Systems (ITS) it is feasible for a member of staff to ring a local number anywhere in the world and affect a change in the production process.

The development of ITS is still in its infancy and the many applications of the technology are yet to be realised. This paper is written at an embryonic stage of research and it aims, therefore, to briefly outline the technologies involved.

2.0 EXPLANATION OF THE TECHNOLOGIES INVOLVED

Several technologies combine to provide ITS, but all of these are in existence already and have been combined successfully to provide ITS for other areas other than manufacturing processes.

2.1 What is the Internet?

The Internet is a series of rooms containing networked computers, banks of modems and routers, across the entire world connected together by dedicated telephone lines. The telephone lines are not necessarily land-lines, but represent all current telephony technologies.

Currently, the terms Internet and World-Wide Web (Web) are used interchangeably. However, they are not the same thing. The Internet is the global network infrastructure that allows the Web to operate. The Web is a collection of documents linked together using a specific Internet Protocol known as Hypertext Transfer Protocol (HTTP). The Web could not exist without the Internet, but the Internet is totally independent of the Web, [1].

2.1.1 How the Internet works.

The history of the Internet is not so important, since most users are aware of its US Military and Academic origins. How it works is less well known.

The Internet uses a network protocol called TCP/IP. The term actually refers to two separate parts: the Transmission Control Protocol (TCP) and the Internet Protocol (IP). Every computer that is connected to the Internet understands these two protocols and uses them to send and receive data from any computer on the network using a *packet-switched network*, [2].

First, TCP breaks down each piece of data into smaller, individual pieces called *packets*, each of which is wrapped in an electronic envelope with Internet addresses for both the sender and the recipient. The IP protocol then figures out how the data will go from point A to point B by passing through a series of *routers* just as postal mail passes through several sorting offices on its way to the addressee. Each router examines the destination addresses of the packets it receives and then passes the packets on to the next router as they make their way to their final destination. Packets may travel completely different routes on their way to a destination computer. At the destination the packets are checked and finally reassembled, Figure 1.

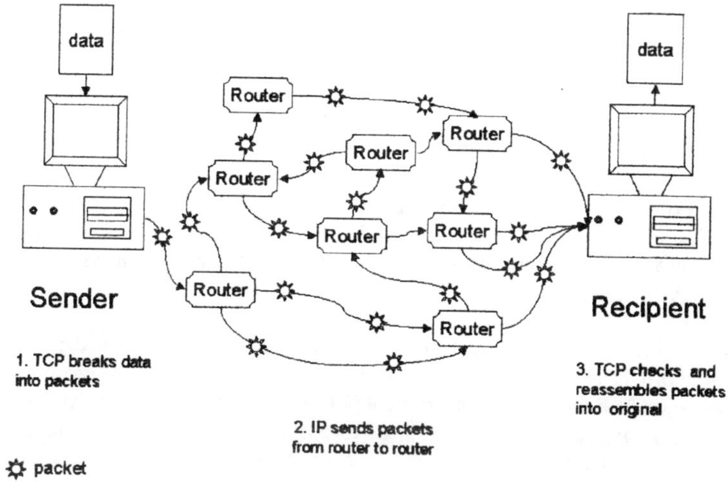

Figure 1: Transmission of data over the Internet.

2.2 Telemetry systems.

"Telemetry is the technology which enables a user to collect data from several measurement points at inaccessible or inconvenient locations, transmit that data to a convenient location, and present the several individual measurements in usable form." [3]

There are three main types of telemetry system, all with similar system components, the difference lies in the type of multiplexing they employ. The three main types of multiplexing are Frequency Modulation (FM), Pulse Amplitude Modulation (PAM) and Pulse Code Modulation (PCM). PCM whilst it is less efficient in translating input data bandwith into multiplex bandwith its output is digital, which lends itself to inputting into a computer.

The advent of computing and rapid increase in processor power against relative cost has allowed telemetry systems to become widespread and to monitor and control increasingly larger applications. The water industry traditionally a large investor in telemetry systems typifies the extent of current telemetry systems. One of the larger regional water companies in the UK has over 4700 outstations where anything from 16 to over 130 different measurements from each station will be collected and transmitted for processing and control purposes. [4]

In manufacturing telemetry systems can be used to monitor any physical conditions such as temperature, pressure etc, and the use of telemetry systems are therefore more widespread in the continuous or process manufacturing industries like petrochemicals. However there is a case for using telemetry systems when combined with Internet technologies in other types of manufacturing industry. In theory, telemetry data could be transmitted over the Internet to the central site, and a VPN may facilitate this.

2.3 Virtual Private Network (VPN)

A Virtual Private Network uses public and private network infrastructure to connect nodes using the Internet as the medium for transporting data. These systems use encryption and other security mechanisms to ensure that only authorised users can access the network and that the data cannot be intercepted.

The communications protocols that allow VPNs have been established for several years having been originally included in Microsoft NT 4.0™ . The protocol Microsoft developed is known as Point To Point Tunnelling Protocol (PTTP). Rival protocols existed (and continue to exist), most notably Level Two Forwarding Protocol (LTFP) from Cisco Systems. By the end of 1997 Microsoft and Cisco had agreed to merge the best parts of their respective protocols to produce Layer 2 Tunnelling Protocol (L2TP). [5]

The principle that the protocols work on is the establishment of a secure tunnel between two computers that the data passes along. The data cannot be intercepted and enters and leaves the computers through the tunnel. To ensure absolute security, encryption by either software or hardware may be used together with a firewall.

The simplest way of using tunnels is the use of tunnelling software on both the client and the tunnel server which is usually located at the main corporate site. It is possible for the tunnel server to exist at the Internet Service Provider (ISP)'s site which the main corporate site uses. The client software starts the tunnel whilst the tunnel server ends the tunnel which is established on the basis of user identification and password protection. There is absolutely no need for the ISP to support tunnelling in any way and the transfer of data proceeds without any interference from the ISP. The Internet is therefore being used as a private network and this instance could offer a link between the telemetry system to the collection point and allow access to the telemetry data remotely using the Web.

2.3.1 Benefits of VPNs

A traditional WAN requires leased communication lines, banks of modems, routers, network cards, software and many support personnel. The VPN utilising the public Internet obviates the need for much of the corporate investment in WAN since the hardware, communication lines, etc do not belong to the organisation. Listed below are the benefits of VPNs:

- Low capital investment required making VPNs ideal for expansion or start-up situations.
- Low maintenance costs since the Internet is maintained by both public and private bodies.
- Access to the VPN is made by making a local phone call, further reducing operating costs.
- Uses Web Browsers which are free and simple to operate (reduces training costs).
- The tunnel improves Internet performance according to (VPN solution providers).
- Investment in upgrades to the Internet infrastructure benefit the VPN at no direct cost.

These are just a few of the benefits of VPNs which can be summarised as financial and operational.

3.0 INTERNET TELEMETRY USING VPNs

The Internet in an Internet Telemetry System is the means of transferring the data between sensor and collection point (using the VPN), and the means of viewing and controlling the system is facilitated by the Web.

This presents several problems associated with the Internet that may have an adverse effect on a telemetry system. The first problem is speed. Anyone who has ever used the Internet will attest to the lack of speed when a large percentage of the purported 50 million users are on-line. According to vendors of VPN solutions VPN's tunnelling improves performance significantly. However, the lack of guaranteed performance would suggest that connecting a "mission critical" telemetry system to the Internet would be ill advised.

The second problem is the low refresh rate of browsers. In a control system relying on telemetry data for input the speed of display of the information is crucial. This is easy to rectify since Java™ and Microsoft ActiveX Controls™ can display real-time information.

The final problem is the loss of ownership and therefore control over the network. If the ISP encounters problems either financial or operational the VPN user may suffer. Also as this is a relatively new technology, ISPs are still considering charging schemes.

3.1 Implementation of Internet telemetry.

Since PCM telemetry systems transmit their data in binary form it is feasible that the actual transmission from the telemetry site to the collection site could be accomplished over the Internet by using VPN technology. The database that the data is collected in could be displayed as a Web page using the various programmes that allow software sockets in Web pages as previously mentioned. Remote users could in effect use a VPN to connect to the Web page and therefore affect the control changes necessary. With the advent of Handheld Personal Computers (HPC) it is also possible to connect anywhere in the world to the control site Web page.

3.2 Example of an Internet telemetry system.

Clover Systems Inc. (CSI), of New Jersey, USA, have already installed working systems. CSI offers a vendor managed inventory system (VMI) entirely based around an Internet telemetry system and have a working demonstration available on their web-site. [6]

The WEB-VMI™ system assigns the responsibility for running and maintaining a manufacturer's inventory (in this instance, chemicals) through a customer care and web logistics centre. The whole system is installed on a secure server at CSI and uses the Internet to allow 24 hour access to reports, re-order and critical system events in real-time. The convenience of the system is realised through the ability to access the secure web-site using a normal Internet browser and a local telephone call anywhere in the world.

The screen shot below is from the CSI demonstration site [6] and shows a fictitious chemical company's US sites broken down into states and then into cities and into individual sites within the selected city.

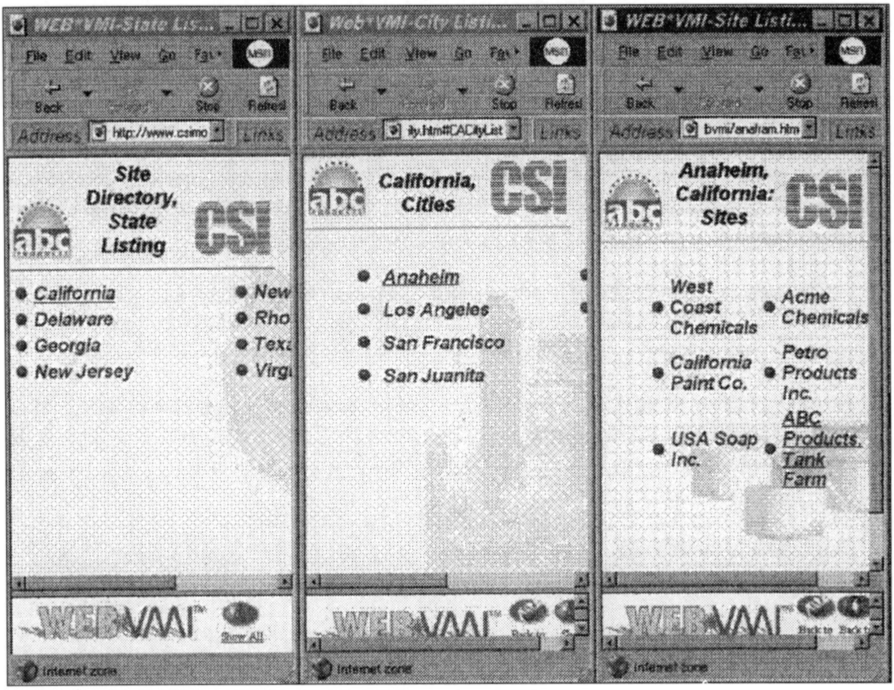

Figure 2: The WebVMI demonstration.

It is easy to imagine the benefits of such a system being realised in many and diverse manufacturing situations. Applications in marketing and sales management would enable the profitable fulfilment of orders through the access to up-to-the minute data on stock positions etc.

The following screen shows telemetry data arriving at the VMI web-site from telemetry devices reporting chemical storage tank levels. The telemetry device transmits the data to the CSI's central monitoring site where it is converted into a secure web page showing a range of reports including inventory management and consumption. The alarm condition of two of the tanks is clearly visible and would be communicated directly to the chemical company by fax, electronic mail, pager or through the WEB-VMI Internet page.

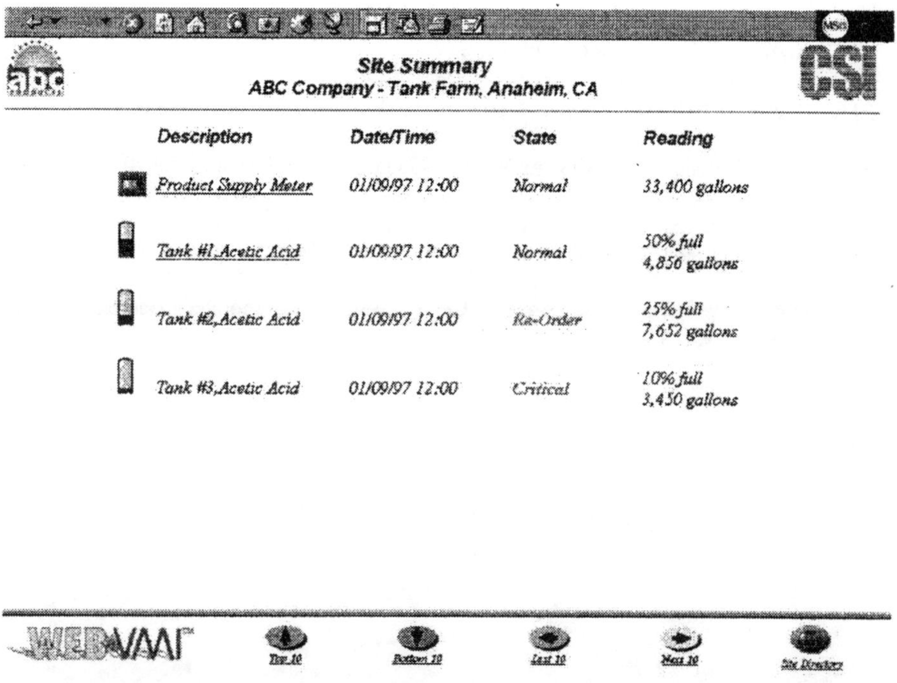

Figure 3: WebVMI Internet telemetry for chemical storage.

4.0 CONCLUSION

It is apparent, that if it is possible to route telemetry data via the Internet that great cost savings can be realised. The question is: "Whether the enabling technologies are mature enough (in IT terms!) or reliable enough ?" and this requires further investigation. Global manufacturing strategy and planning will benefit immensely from the readily available information. The low capital investment required to implement an Internet telemetry system will also feature prominently in the assessment of the viability of these systems.

New technology like the HPC adds the further possibility of the remote access into a telemetry and control system and this will also require investigation. However, the display of real-time data using the Web is already in existence, for instance, BT's SuperRoute 66 real-time Ipswich bus timetable, and this use will be the growth area of the Internet in the very near future.

REFERENCES

[1] Stein, Lincoln D., How To Set Up And Maintain A World-Wide Web-Site – The Guide For Information Providers, Addison Wesley Publishing Company, 1995

[2] Comer, Douglas E., Internetworking With TCP/IP Vol. 1: Principles, Protocols And Architectures, Prentice Hall 1995

[3] Strock, O.J., Introduction To Telemetry, Instrument Society of America, 1987

[4] Buswell, Philip, Improving As We Go Along, Control Systems, Vol.10, No.4, 1996, pp80-81

[5] Jennings, Roger, Using Windows NT Server 4.0 TM, Que Corporation, 1996

[6] http://www.csimonitoring.com/ Clover Systems Incorporated, Berkeley Heights, New Jersey, USA,

Applications of the electronic nose in manufacturing systems

P LOWERY, P CORCORAN, H STRATTON, and **M ELSHAW**
Sensing and Control Research Group, University of Derby, UK

Abstract

The concept of an electronic nose has been in existence for over 10 years. The basic components of the instrument are a multi-sensor array, signal conditioning and pattern recognition engine. Various instruments and configurations have been applied in a range of tasks including odour classification, description and detection. However, instrument manufacturers have failed to develop analytical devices which exhibit repeatable performance when applied in challenging measurement environments. This failure is a consequence of the inability of manufacturers to overcome the limitations of existing sensor technology in terms of sensor repeatability, drift and reproducibility.

The utilisation of intelligent information processing techniques has been shown to enhance instrument performance and has facilitated the development of application specific instruments for quality monitoring in manufacturing systems. These systems can be used to detect process faults or contamination and hence offer the potential benefits of rapid on-line detection and analysis of quality problems. Hence there are potential benefits in terms of avoiding process wastage and reducing the risk of product recall. The qualitative and quantitative information derived from these systems can also be used to select effective remediation strategies.

This paper identifies the application of electronic noses in a range of manufacturing systems and describes how the development of embedded intelligent information processing techniques can enhance system performance.

The Development of the Electronic Nose

The mammalian sense of smell, which is one of the least copied mammalian senses, is a primary instrument used by industry for quality control, process monitoring and environmental monitoring in the foodstuffs and brewing industry.

The concept of an artificial odour detection instrument was first proposed by Wilkins and Hartman in 1964 (1) with refinements identified by Buck *et al.* in 1965 (2) and Dravieks and Trotter in 1965 (3). With advances in technology over the following two decades, the concept of the electronic odour sensing instrument, as an intelligent system for classification of odours, emerged. (4) It was not until the late 1980's, that electronic odour detection machines were termed as *Electronic Nose's* (5). The past ten years has seen considerable development of commercially manufactured electronic nose instruments, for the automated detection and classification of odours. All systems are based upon the three essential components of an electronic nose instrument, the multisensor array, signal conditioning circuitry and microcomputer pattern recognition engines (see Figure 1).

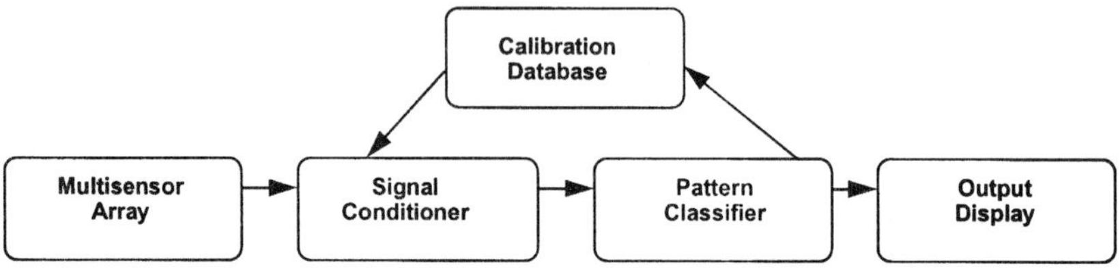

Figure 1 : Electronic nose architecture.

However, commercially available instruments seem to exhibit limitations in terms of repeatable performance when used in open measurement environments. This is due to limitations in sensor repeatability, reproducibility and drift. The major constraint that has limited the employment and mass production of electronic noses is the drawback associated with the gas sensor technology. The key problem being addressed by researchers is the poor repeatability that sensors show over time due to drift caused by ageing, humidity and temperature effects. The outcome of this is that odours are classified incorrectly as the measured responses migrate from the calibrand fingerprints stored in a calibration database for identification purposes. This problem therefore leads to the need for adaptive learning or frequent recalibration and reconfiguration of the instrument. Furthermore, systems are required to accurately control temperature and humidity or the effects of these factors need to be taken into account when processing the data. To enhance the performance of such systems, novel processing tools are required, such as thermal modulation techniques (6) along with customised odour sampling systems. The utilisation of intelligent pattern recognition techniques (7) (e.g. artificial neural networks, fuzzy logic, rule based systems) has also been shown to enhance the performance of electronic nose instruments in real applications.

With this enhanced performance, specific instruments have been developed for quality monitoring of a wide range of products, including foodstuffs (fish, cheeses meat etc), beverages (beers, wines, coffees, whiskies etc) and cosmetics (perfumes, soap deodorants etc). The monitoring and controlling of the environment (air pollution, emission control) is also possible, improving health and safety.

Electronic Noses in Manufacturing

The utilisation of electronic noses for quality control and process monitoring, and their mass production for sale appears to offer manufacturers a major opportunity. The quality of raw materials can be established by such artificial odour sensing devices, and therefore ensure that manufacturers are not producing goods with substandard raw materials and being exploited by their supplies. For example, they are able to ascertain the freshness of meat and fish, or whether pork will give off a disagreeable odour known as 'boar taint' when cooked, or they can be used to detect malodours caused by grain containing a large amount of insects and micro-organisms (8),(9),(10). Electronic noses have also been used determine the quality of perfumes, cola, drugs and paper, or to establish if underarm deodorants keep their scent for a sufficient period and whether packaging will taint products. Moreover, they could assist in the research and development of new products to establish whether mayonnaise is rancid, and if tobacco, tea and coffee have been blended correctly. (11),(12),(13)

Electronic noses can be employed throughout the production cycle. For example, odour samples could be tested by electronic noses to enable process problems to be detected at the earliest possible opportunity. As noted by Pendick (1997) (14) and Hodgins and Simmons (1995) (16) it is possible to establish whether diacetyl and amylacetates occurs in wine or if beer has developed a buttery off taste. In the case of beer this would save the expensive loss of the complete batch by simply adjusting the timing of the different stages in the production process. Furthermore, electronic noses can be employed to assist in determining where production problems exist and identify remediation strategies. Additionally, based on the observations of Keller et al. (1996), through the reduction in need for expert organaleptic panels whose performance is dependent on physiological and sociological factors, greater repeatability may be obtained in the testing of foodstuffs and beverages.

Another application of artificial noses that would be of interest to manufacturers is the control of odour dissemination into the atmosphere in the workplace. Various studies have shown that certain scents can make employees feel more positive, work harder and concentrate for longer periods (15). Finally, as noted by Keller et al. (1994b) (16) manufacturers could be assisted in the measurement of emissions in line with environmental legislation, and in helping to identify and locate leaks that may occur in a plant.

Despite the opportunity for the employment and mass production of electronic noses there has been limited interest from manufacturers, as outlined by Pearce (1997)(17), Holmberg et al. (1996) (18), Craven et al. (1996) (19) and Gardner and Barlett (1994) (20). A key problem associated with electronic noses is their large size, weight and power consumption when most manufacturers would prefer light and mobile devices. It has taken a great deal of time for commercial systems to become available following the initial research and the instrument cost of between £20,000 and £50,000 has deterred many small and medium sized manufactures from adopting the technology for process and quality control. Furthermore, the early claims made that artificial odour sensing devices would have the same abilities as biological ones have proved to be overstated, with the majority of existing devices being application specific. In fact, there has been a general shift away from developing generic electronic noses based on the architecture of the biological olfactory system to custom devices utilising those sensors and pattern recognition techniques that are readily available. Despite over 15 years of research the approach is basically the same as that put forward by Persaud and Dodd (1982) (21) which was based upon a simple analogy of the olfactory system.

Odour Sensing Tasks in Manufacturing

As multi-sensor systems become more common in engineering applications, it has become apparent that appropriate information processing tools and techniques need to be identified. Most complex systems will consist of a wide variety of distributed sensor systems that generate a great deal of multi-variable information which can prove difficult to interpret. It is not difficult to realise that important information must be encoded within these complex multivariate sensor outputs and the challenging task is to devolve meaningful relationships associated with the system as a whole, or sub systems within it. Moreover, these complex transformations provide the key element in the overall system performance and the accurate description of complex odours. In reality, as previously stated, these transformations are generally non-linear in nature and standard information processing techniques are of limited use. Traditionally electronic noses have been based on steady state responses from multi-sensor arrays in controlled environments which related to a high degree of success. However, to translate this success to real world applications a new set of information processing tools have to be considered to overcome sensor drift, non-linearity and interfering measurands. The task of extracting meaningful information from noisy signals follow a number of important processing stages, namely signal conditioning, pre-processing, feature extraction and finally some form of pattern classification / description.

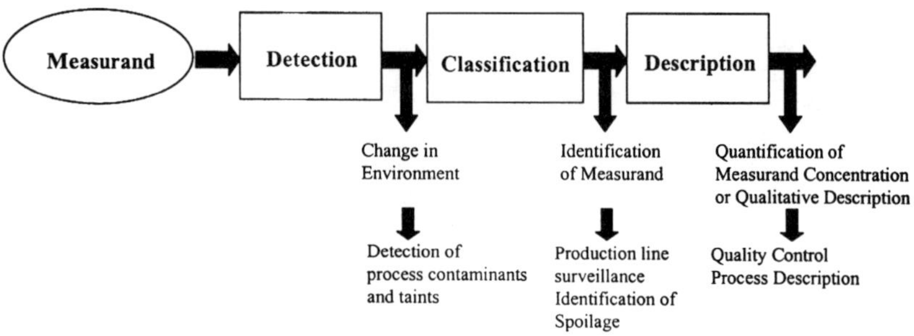

Figure 2 : Electronic nose task requirements

There are a number of tasks that can be addressed by electronic nose instruments, each having their own specific requirements. The hierarchy is shown in Figure 2. The first task is that of detection (which may not be required for some applications such as bench tests) which is followed by classification and then description. The detection task is most applicable in open environments where, due to the uncertainty associated with the multisensor array, this task often necessitates the use of multisensor detection theory analogous to that used in radar systems, although more simple multivariable thresholding techniques can be applied. This technique is applicable to the on-line detection of process contaminants and taints.

Following detection, the system may be required to classify the detected odour. Given a known set of species and reference data describing each species, the task is to identify individual samples as belonging to one of the classes. In recent years much emphasis has been placed on the use of novel neural network pattern classification algorithms due to their ability in accommodating non linear problems. Moreover, many groups have succeeded in embedding intelligent pattern classification tools into real systems (10),(11),(16). In most applications, the classifier output is able to discover non-linear relationships between inputs and outputs and this feature enables these techniques to be used in a diverse number of applications including production line surveillance (identification of contaminants) and identification of spoilage.

The most challenging task is description, which can be divided into two tasks. The first is quantitative description which would typically involve determining the concentration of the various classes of measurand present. Since the sensitivity characteristics offered by the sensors are non-linear, with this non-linearity exacerbated by the combinatorial effect of multiple odours, the task of determining odour concentration has resulted in the need for complicated pre-processing and pattern recognition approaches. The second task in this category is qualitative description, where the system is required to extract meaningful features and concepts that are representative of the environment under test. This will only be possible if the quantitative output from the multisensor array can be mapped to qualitative odour descriptors (e.g. fruity, floral, pungent) or measures of certainty associated with qualitative process descriptors (e.g. risk of malfunction).

System Development

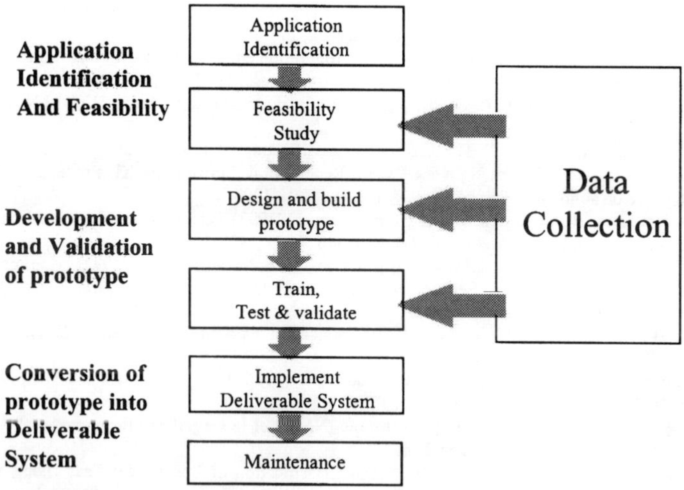

Figure 3 : System development life cycle

The development cycle of the electronic nose is shown in Figure 3. The first stage of system development is to identify an application and investigate the applicability of the electronic nose through a feasibility study. This requires data to be collected that is representative of the problem to be solved. For example, for an application involving the detection of a process contaminant. data would be required that represented the *change of condition* and *no change of condition* in the presence of a range of environmental interferents. The second stage involves the development and validation of a prototype device. The device needs to be designed and built and then optimised through a cycle of training, testing and cross-validating the system pattern recognition engine. The conversion of the prototype into a deliverable system requires the implementation of the data acquisition, pre-processing and information processing algorithms in an embedded processor. For on-line adaptive learning, detailed consideration should be given to the updating the calibration database to accommodate the effects of sensor array drift.

Conclusions

Electronic noses have yet to be established in viable application domains. The greatest opportunities for taking bench instruments and converting them to robust commercial systems may lie in the field of manufacturing. Applying electronic noses to the tasks such as detection of contaminants, production line surveillance and process description may lead to rapid and cost-effective solutions for the improvement of manufacturing processes.

References

1 T.M. Buck, F.G. Allen and M. Dalton, " Detection of chemical species by surface effects on metals and semiconductors, in Surface Effects in Detection " (eds. T. Bergmanand A. Travnieks), Spartan Books Inc., USA, 1965.

2 A. Dravnieks and P.J. Trotter," Polar Vapour Detection based on thermal modulation ofcontact potentials," J.Sci. Instruments, 42 (1965) 624.

3 K. Persaud and G.H. Dodd " Analysis of Discriminant mechanisms of the mammalian olfactory system using a model nose, Nature, 299 (1982) 352-355.

4 J.W. Gardner,"Pattern recognition in the Warwick Electronic Nose" 8th International Congress of European Chemosensory Research Organisation, University of Warwick, July 1987

5 P. Corcoran, P. Lowery & J. Anglesea " Optimal Configuration of a Thermally Cycled Sensor Array with Neural Network Pattern Recognition" Sensors & Actuators B (in press) 1998

6 Corcoran P and Lowery P, "Neural processing in an electronic odour sensing system", Vol 15 No 4 pp 15 - 18. MCB University Press.

7 Lowery, P. and Corcoran, P. (1997) The Enhanced Classification of Loose Leaf Teas Using a Thermally Cycled Gas Sensor Array and Neutral Network Pattern Recognition. Eurosensors XI: Proceedings of the 11th European Conference on solid-state Transducers. Vol. 3, pp. 1301-1304.

8 Winquist, F., Hornsten, E., Sundgren, H., and Lundstrom, I. (1993) Performance of an Electronic Nose for Quality Estimation of Ground Meat. Meas. Sci. Technol. Vol. 4, pp. 1493-1500.

9 Olafsson, TR., Martinsdottir, E., Olafsdottir, Sigfussion, P, and Gardner, W. (1992) Monitoring of Fish Freshness Using Ton Oxide Sensors. Sensors and Sensory Systems for an Electronic Nose. pp. 257-272. Gardner, J. and Bartlett, P. (Ed) Nato ASI Series.

10 Bourrounet, B., Talou, T. and Gaset, A. (1995) Application of a Multi-Gas-Sensor Device in the meat Industry for Boar-Taint Detection. Sensors and Actuators B. Vol. 26-27, pp. 250-255.

11 Keller, P., Kangas, L., Liden, L., Hashem, S. and Kouzes, R. (1996) Electronic Noses and Their Applications. IEEE Northcon/Technical Applications Conference (TAC '95) Portland.

12 Hodgins, D. (1995) The Development of an Electronic 'Nose' for Industrial and Environmental Applications. Sensors and Actuators B. Vol. 26-27, pp. 255-258.

13 Lowery, P. and Corcoran, P. (1997) The Enhanced Classification of Loose Leaf Teas Using a Thermally Cycled Gas Sensor Array and Neutral Network Pattern Recognition. Eurosensors XI: Proceedings of the 11th European Conference on Solid-State Transducers. Vol. 3, pp. 1301-1304.

14 Pendick, D. (1997) Best Nose in the Business. New Scientist. Vol. 22 February, pp. 24-29.

15 Munson, M., Walsh, T. and Yeykal, T. (1995) Fragrant Fuel (Effect of Scent on Solving Puzzles). Prevention. Vol. 47, pp. 38.

16 Keller, P, Kouzes, R and Kangas, L. (1994b) Neural Network Based Sensor System for Manufacturing Applications. Advanced Information and Technology Conference.

17 Pearce, T. (1997) Computational Parallels Between the Biological Olfactory Pathway and its Analogue "The Electronic Nose": Part I. Biological Olfaction. Biosystems. Vol. 4, No. 1, pp. 69-90.

18 Holmberg, M., Winquest, F., Lindstrom, I., Davide, F., DiNatale, C. and D'Amico, A. (1996) Drift Counteraction for an Electronic Nose. Sensor and Actuators B. Vol. 35-36, pp. 528-535.

19 Cravern, M., Gardner, J. and Bartlett, P. (1996) Electronic Noses – Development and Future Prospects. Trends in Analytical Chemistry. Vol. 15, No. 9, pp. 486-493

20 Gardner, J. and Bartlett, P. (1994) A Brief History of Electronic Noses. Sensors and Actuators B. Vol. 18, No. 1-3, pp. 211-220.

21 Persaud, K. and Dodd, G. (1982) Analysis of Discrimination Mechnaisms in the Mammalian Olfactory System Using a Model Nose. Nature. Vol. 299, pp. 352-355.

A framework for the implementation of computer integrated manufacturing in small and medium enterprises

H B MARRI, A GUNASEKARAN, and **R J GRIEVE**
Department of Manufacturing and Engineering Systems, Brunel University, Uxbridge, UK

Abstract

Small and Medium Enterprises (SMEs) play an increasingly important role in all aspects of competitiveness both in terms of products and production techniques. Also these include management methods, the organisation of the firm and human resource training. During the last two decades, several factors have forced global manufacturers to make dramatic changes in their products, markets, and manufacturing strategies. One of the ways by which SMEs can achieve a competitive advantage in manufacturing is through the implementation of Computer Integrated Manufacturing (CIM). Realising the importance of SMEs in the national economies an attempt has been made to review the application of CIM in SMEs. The characteristics of CIM have been studied with reference to the application and infrastructure in SMEs. Also, a framework has been offered for the implementation of CIM in SMEs.

1. INTRODUCTION

SMEs across the world are initiating the CIM. They face growing international competitive pressure combined with the need to interface and integrate their computer aided design (CAD) systems, numerically controlled (NC) machine tools, robots, and their control systems (1). In a global competition the harnessing of technological resources has become a major strategic challenge for SMEs. The globalization of markets, growing interpenetration of economics and increased interdependence between players change the conditions of competition and oblige enterprises to alter the foundations of their competitiveness and economic performance. Many SMEs have to respond to these challenges and supply competitive goods and services by adapting the mix and quality of their factors of production to the nature and acceleration of the change now in progress. In other words, improving their competitiveness is absolutely imperative to SMEs, as it is to large enterprises.

CIM is a multi-disciplinary subject and is a complex, multi-layered system designed for the purpose of minimising waste and creating wealth. In this context "waste" means various non-value adding resources, such as human labour, finances, time, machinery and energy. According to Lefebvre et al. (2), CIM concerns with providing computer assistance, control and high level integrated automation at all levels of the manufacturing industries, by linking islands of automation into a distributed processing system. These isolated automated production islands include NC, Distributed Numerical Control (DNC), Computerised Numerical Control (CNC), Manufacturing Resource Planning (MRP), CAD, Computer Aided Process Planning (CAPP), Computer Aided Manufacturing (CAM), automated storage, automated material handling systems, and robotics.

Studies by Cromie (3) and Smallbone (4) have identified that the lack of capital as the common problem for the start-up failure of both young firms and SMEs. A study conducted by Kellock (5) found that 23% of the SMEs examined suffered from insufficient working capital. The associated report also indicates that up to 60% of all SMEs in the United Kingdom were financed by overdraft, suggesting a general lack of long-term investment in this sector. Thomson and Thomson (6) and Balic et al. (7) propose that the implementation of an effective policy is an essential element in the successful management of SMEs. An effective policy as one that provides a common set of rules to govern the decision-making process within the organization and is also communicated to all the organization's employees.

2. SMALL AND MEDIUM ENTERPRISES

A unique definition of SMEs is not possible, the concept varies from country to country and from sector to sector. However, in terms of the community's structural funds and leading instruments, it has always been accepted that the SME should not have a workforce exceeding 500, net fixed assets exceeding not more than a third of its capital held by a large firm. In particular, it has been recognised for some 20 years that their dynamism, related in part to the technological and economic changes which have occurred over this period, has made an important contribution to the creation of new jobs, the economic revival of certain regions and also to the technological progress. It cannot, however, be taken for granted that this dynamism will be maintained in the future in view of trends such as the globalization of markets and technology, the intensification of competition and the acceleration of technological development and change. These factors explain the interest shown by the governments of all countries in the competitiveness of their SMEs which, along and inter-related with that of large firms, constitutes the basis of national competitiveness (8). Following are some of the characteristics of SMEs:

- In SMEs, shop floor tasks which are closely related to production have mostly involved a high degree of human decision making and execution.
- Personal relationships in SMEs are of highly important.

- SMEs have a higher inherent innovatory potential than large enterprises.
- Ability to react quickly to keep abreast of fast changing market requirements.
- Lack of bureaucracy. Dynamic, entrepreneurial managers react quickly to take advantage of new opportunities and are willing to accept risk.
- Efficient and informal internal communication networks. Afford a fast response to internal problem solving: provides ability to reorganize rapidly to adapt to change in the external environment.
- Large number of SMEs have some sort of external linkage which is of importance to the development of their business.
- Small enterprises tend to concentrate on traditional industries where entry barriers are low, minimum production scales are low, and labour intensity relatively high.

3. A FRAMEWORK FOR THE IMPLEMENTATION OF CIM IN SMEs

Implementing a new technology is a greater burden for SMEs than it is for large companies, which have better resources through a large number of employees and a broader basis of know-how. Solutions for shop floor control have been worked out already by research institutes. One special problem is given by the necessity to adapt these solutions to the needs of SMEs. A conceptual model for the implementation of CIM in SMEs is shown in Figure 1. The issues of implementation of CIM in SMEs are discussed under four perspectives such as strategic, tactical, operational, and the organizational perspective. The details of the proposed model is presented hereunder.

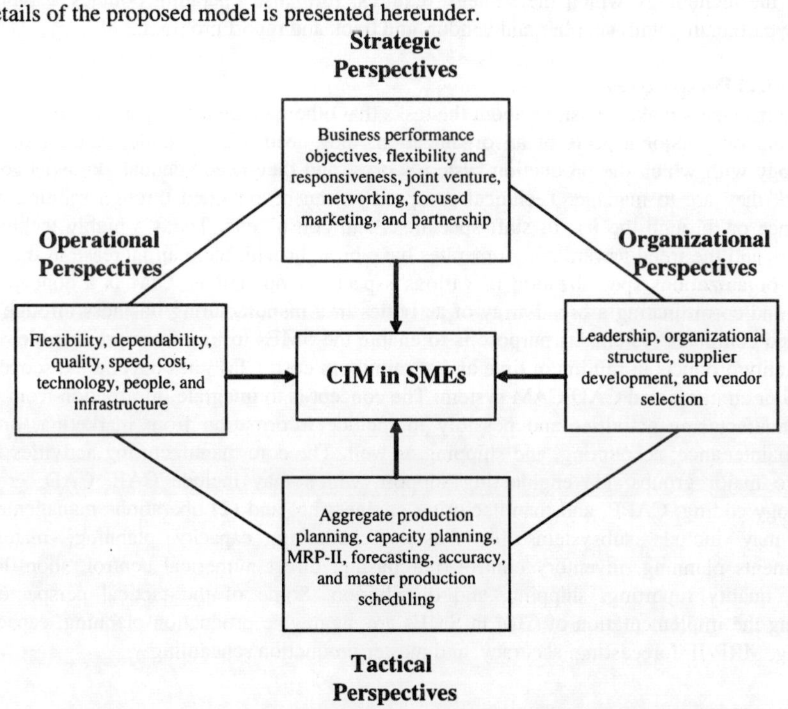

Figure 1: A Conceptual Model for the Implementation of CIM in SMEs

3.1 Strategic Perspectives

Top managers are responsible for making the vital decisions that set the company's overall goals and keep all parts of the company pulling together towards these goals. Decisions that have a long-range impact on the general direction and basic character of a company are called strategic decisions. Through strategic planning, managers evaluate the company's relationship with its external environment and establish the basic directions for the company. Each business unit or a single business company has a business strategy that guides business practices and directs how it will deal with its customers, competitors, and conditions. The business-level strategy can be based on corporate strategy to plan how the business can provide, what the customers will perceive as the best way to satisfy their needs. If a company is to survive and prosper it must consistently provide customer satisfaction. There are certain fundamental strategies that can be employed to improve productivity in manufacturing operations. Since these strategies are often implemented in SMEs generally by means of automation and particularly by CIM technology. The characteristics of a company in terms of business performance objectives, flexibility and responsiveness, joint venture, networking, focused marketing, and partnership should be considered while designing and implementing CIM in SMEs.

The implementation of CIM requires a clear, precise corporate strategy, the success of which will depend upon careful planning of several logical steps namely: prime the corporate culture for change, clearly define expectation, appoint a champion for CIM design and implementation, establish a project team, perform a comprehensive environmental analysis, identify the technology which the strategy requires, formalise operating policies, establish working partnership with supplier and vendor, and track and report progress.

3.2 Tactical Perspectives

Since managers make decisions about the tasks that other people are to perform, they must understand two major aspects of an organization: they need a basic understanding of the technology with which the production system works, and they need adequate knowledge of the work they are to manage. Technical competence can be obtained through training and experience or through the use of staff specialists and consultants. Today's highly technical processes and the trend toward conglomerates have brought with them an increase in the use of staff organizations specialization in various aspects of operations. CIM is a concept of linking and co-ordinating a broad array of activities in a manufacturing business through an integrated computer system. Its purpose is to enable the SMEs to transform product ideas to high-quality products in minimum time and at minimum cost. CIM goes beyond the scope of an FMS or an integrated CAD/CAM system. The concept is to integrate information from the core manufacturing activities and possibly to include information from marketing, order entry, maintenance, accounting, and shipping as well. The core manufacturing activities fall into two major groups: (1) engineering support, which may include CAE, CAD, group technology coding, CAPP, and manufacturing engineering; and (2) operations management, which may include subsystems for master scheduling, capacity planning, material requirements planning, inventory control, purchasing, direct numerical control, shop-floor control, quality reporting, shipping, and distribution. Some of the tactical perspectives regarding the implementation of CIM in SMEs are: aggregate production planning, capacity planning, MRP-II, forecasting, accuracy, and master production scheduling.

3.3 Operational Perspectives

In a company's organisation chart, operations often enjoys parity with the other major business functions: marketing, sales, product engineering, finance, control (accounting) and human resources (personnel, labour relation). Sometimes, the operations function is organised as a single entity which stretches out across the entire company, but more often it is embedded in the distinct, typically product defined divisions into which most major companies are organised. In a manufacturing company, line management frequently extends to the stockroom, material handling, the tool room, maintenance, the warehouses, and distribution, as well as the so-called "factory floor". Support services for line management's operations can be numerous. Within a manufacturing environment, support services carry titles such as quality control, production planning and scheduling, purchasing, inventory control, production control, industrial engineering, manufacturing engineering, and field service.

Some of the operational perspectives regarding the implementation of CIM in SMEs includes: flexibility, dependability, quality, speed, cost, technology, people, and infrastructure. For instance, the implementation of CIM in SMEs will improve the flexibility by reducing change over time and GT cells. Similarly, the reduction in overall lead time of the product may improve delivery schedule performance and responsiveness. Also, operations of CIM requires different skills and relations employees to function as a team work.

3.4 Organizational Perspectives

SMEs may find it worthwhile to set up a specialist team to play a consultancy role in the organization in planning and implementing CIM projects. This can be valuable for SMEs located on different sites where the specialist team can be useful in technology transfer. The sensitive aspect of personnel management in the CIM context is its effect on numbers. Labour reduction is almost always a factor. Linked with this is the change in job description. Many of the totally unskilled jobs, such as machine loading, will be eliminated but they will be replaced by job of greater responsibility because the machine loading operator in the CIM context is likely to look after a number of machines, use a terminal to receive and process data, and handle minor maintenance problems that occur. Some of the organizational perspectives are leadership, organizational structure, supplier development, and vendor selection etc.

Organizational structure can be seen as the assignment of people with different task and responsibilities. Technological measures such as automation and integration are of no value if applied to an immature, complicated, and chaotic organization. The implementation of CIM in SMEs will change the pattern of organizational structure, supplier development, and vendor selection.

4. CONCLUSIONS

An attempt has been made in this paper to present a model for the implementation of CIM in SMEs. The experiences of SMEs with CIM along with a framework for the implementation of CIM in SMEs are described. Finally, future research directions are presented.

There is a reason to believe that the adoption of new efficient manufacturing techniques may be problematic in SMEs. It has been claimed that SMEs frequently lack expertise, time, money, and support to upgrade their current manufacturing operations, introduce new technologies and methods, implement better quality control, and improve work force training. While CIM technologies have the potential to provide significant productivity and quality improvements, they also require relatively steep capital investment, including significant start-up and learning costs. Further, these technologies may require modification and adoption before they can be utilised productively. Further research is required in the area of the human factors is designing a CIM and there is a need to establish the level of skill and training required in order to implement CIM in SMEs. The reasons for the implementation of CIM in SMEs failure are due to the: lack of developing an effective support system, lack of planning for a higher level of system integration, lack of experience with modern technologies, inadequate understanding of new technologies, and lack of top-management knowledge and support of CIM. It is suggested that further research is needed for total system integration rather than stand-alone technologies for the implementation of CIM in SMEs.

REFERENCES

(1) CASA/SME, 1985, (Computer and Automated Systems Association of the Society of Manufacturing Engineering), A program guide for CIM implementation, *CASA/SME,* Dearborn, Michigan 48121.

(2) Lefebvre, L.A., Lefebvre, E., and Harvey, J., 1996, Intangible assets as determinants of advanced manufacturing technology adoption in SME's: toward an evolutionary model, *Int. journal of IEEE Transactions on Engineering Management,* vol. 43, no. 3, pp. 307-322.

(3) Cromie, S., 1991, The problems experienced by young firms, *International Small Business Journal,* vol. 9, no. 3, pp. 43-61.

(4) Smallbone, D., 1990, Success and failure in small business start-ups, *International Small Business Journal,* vol. 8, no. 2, pp. 34-47.

(5) Kellock, 1993, The barriers to growth for Britain's small medium sized companies, *Cranfield Institute of Technology.*

(6) Thomson, S.C., and Thomson, M.C., 1990, Policy: how it benefits small businesses, *Practising Manager.*

(7) Balic, J., Zivec, Z., and Cus, F., 1995, Model of a universal manufacturing interface in CIM for small and medium-sized companies, *Journal of Materials Processing Technology,* vol. 52, pp. 102-114.

(8) OECD, 1993, Small and medium-sized enterprises: technology and competitiveness, *OECD Publications Service, France.*

The installation and validation of machine vision systems within an aseptic suite environment

M McGOWAN
Ethicon Limited, Livingston, UK
D K HARRISON, G CULLEN, and **D S ROSS**
Department of Engineering, Glasgow Caledonian University, UK

Abstract

The installation of machine vision systems within manufacturing companies as a form of inspection is becoming more common. This paper deals with the issues concerning installation and validation within a cleanroom environment. Once installed systems have to be validated under the auspices of the MDA, FDA, etc. requirements, so that the medical products produced can be marketed world-wide.

Operating within a cleanroom environment results in stringent criteria requiring to be met, which are not normally required within most manufacturing sites. These criteria are addressed along with the issue of validating the machine vision system.

Introduction

Machine vision systems are now used in all areas of manufacturing. They are excellent when inspecting parts where the areas of interest have physical characteristics that are quantifiable.

Once a Machine Vision system has been selected [1] and purchased the installation and validation of the system become the critical issues; with the validation process especially critical in the medical and pharmaceutical industries where there is legislation requiring the validation process before products may be sold. This legislation comes from governmental organisations e.g. Medical Device Agency (MDA), UK, and the Food and Drug Agency (FDA), USA.

The development of the machine vision system software is assumed to form part of the installation process. This is very much problem dependent and is not discussed in this paper.

When installing any type of equipment within a cleanroom environment the selection process is critical. The preferred selection is equipment specifically designed for operation within a cleanroom environment. However this is not always possible, as is the case with most machine vision systems and instead the installation must satisfy the criteria laid down by the classification of cleanroom. Satisfying this criteria is the first priority of the installation, before even consistency of captured image; which in most cases is the starting point of a machine vision solution.

Machine Vision Background

A machine vision system typically comprises of a Charge Couple Device (CCD) video camera, a frame grabber and a microprocessor. The camera acquires the image, the frame grabber then converts it to a format that can be used by the microprocessor and the microprocessor then carries out the digital signal processing on the image, see Figure 1.

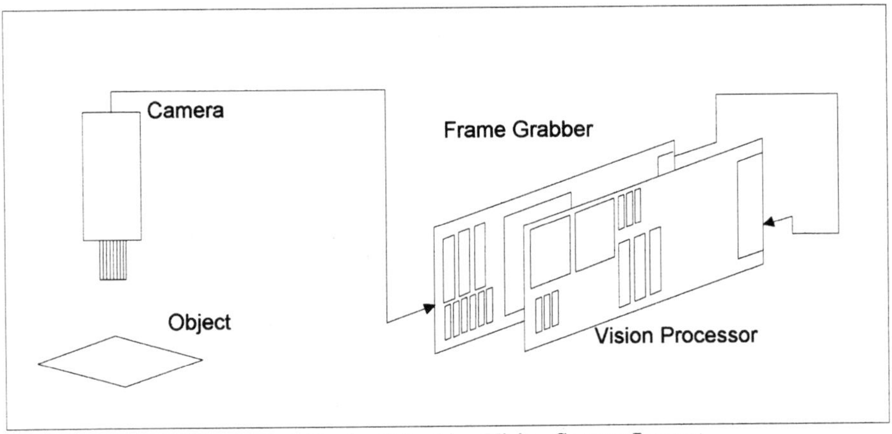

Figure 1: A Typical Machine Vision System Layout

The most common vision systems currently on the market are 256 (2^8) level greyscale systems. These systems convert the captured image into a format where each pixel is represented by one byte giving the 256 greyscale levels. Typically these systems can range in price from several thousand to tens of thousands of pounds depending on level of technology required.

Once an image is captured, tools are required to extract the required information. These tools are essentially DSP algorithms which extract specific information from the captured image. Common tools are:

- Blob Identifies masses with related pixel values
- Edge Detects edges, usually independent of rotation
- Correlation Compares an acquired image against a stored image and returns information on the similarities

- OCR/OCV[1] Text reading and checking tools

As well as these tools, filtering is often available to enhance specific characteristics of the image. This combined with measurement tools that allow distances in the image to be related to real world units, e.g. pixels to millimetres, allows for a very powerful automated inspection system.

Information on machine vision manufacturers and integrators can be obtained from the relevant national bodies [2].

Cleanroom Environments

This paper is primarily focusing on cleanrooms classified by British Standards [3] to be either class E or F (class 100 using US classifications). This standard of cleanroom provides an environment that is bacteria and particulate free, therefore this class of cleanroom is very common amongst medical and pharmaceutical manufacturers. To attain this standard a cleanroom will the require the use of unidirectional airflow and High Efficiency Particle Air (HEPA) filters to remove air borne particles.

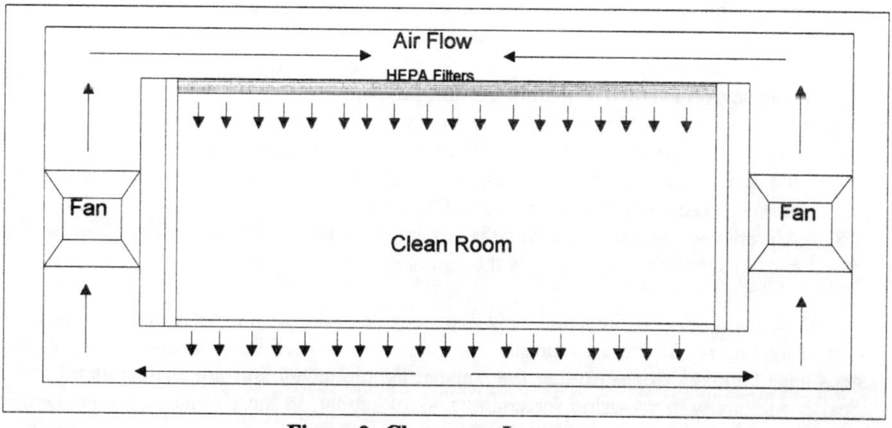

Figure 2: Cleanroom Layout

The HEPA filters clean the air before it flows back into the clean room. The air flows out through grills in the floor or lower parts of the wall. This exhaust air is mixed with fresh air and is supplied to the cleanroom through the HEPA filters in the ceiling. Typically volumes of air movement can be 10 to 100 times greater than a normally ventilated room.

Clothing within cleanrooms is very specialised and can be very restrictive on the time and type of work that can be undertaken whilst the cleanroom is in operation. Typically multiple layers of clothing will be required that minimise personal particle shedding. All clothing must

[1] Optical Character Recognition/Optical Character Verification

be made from specialised fibres and will cover virtually the whole body, some form of hood will be worn to cover the head, and gloves will be required.

Installation Within a Cleanroom

As mentioned previously, commonly when installing machine vision systems the most important criteria for success is the consistency of captured image; however when operating within a cleanroom this takes second place to ensuring that the operation of the machine vision system does not cause the cleanroom to deviate from the specification for its classification.

The biggest single problem which is encountered within cleanrooms is lighting. In many cases specific lighting solutions cannot be used within a cleanroom environment, e.g. using a tent or shroud to supply diffuse light. In this case it is the physical characteristics of the lighting rig that cause the problem. Before discussing this further a common by-product of generating light will be discussed.

When generating light, very often significant levels of heat are also generated. This heat is dissipated into the air where it can cause constant localised turbulence. In cleanrooms there is always turbulence disturbing the unidirectional airflow. However, in most cases this quickly dissipates and the unidirectional air flow is quickly resumed. However, when generating light the heat causes constant turbulence which can cause severe problems.

There are several alternative solutions to this problem.

1. Use a light source that generates little heat e.g. fluorescent tubes
2. If a light source generator is being used, ensure the fan operates in the direction of the airflow (check whether vents require filtering)
3 A preferable option to 2 is for the generator to be placed outside the cleanroom and use fibreoptic cabling to supply the light within the cleanroom.

In some cases the environmental conditions may be such that it is not possible to install any kind of lighting rig. In these circumstances where a specialised lighting set-up is not essential steps must be taken to ensure that the ambient lighting conditions are as near as possible to ideal conditions with provision for consistency of supply. In most cleanroom manufacturing sites the room lighting is supplied by fluorescent tubes. Where they are not already, the ballast's should be changed to high frequency types. This will alleviate a common lighting problem where the vision system detects the variations in ambient lighting caused by supplies being at mains frequency.

This will alleviate one type of problem, however, the light will still be effectively point sources which can cause severe problems especially where metallic or shiny surfaces are involved. The use of clean room specification diffusers can reduce the problem of glare with highly reflective materials. These can be placed near the lighting or near the cameras, experimentation will identify which is most beneficial, though if they are placed near the cameras, fewer diffusers will be required which would be beneficial from a cost perspective.

In many cases machines are guarded for safety reasons. These guards are commonly manufactured from clear perspex. If these guards are tinted or opaque there can be additional lighting benefits. With these alterations there can be a substantial decrease in the amount of light hitting the object at an acute angle. Light striking an object at an acute angle can be very unpredictable, its effect being very much dependent on the part it strikes; however again with reflective parts there is always the potential for severe glare problems and any steps to minimise this risk are beneficial. The diagram below shows an example of a set-up that has proven to be very effective in supplying consistency of captured lighting despite using only ambient lighting

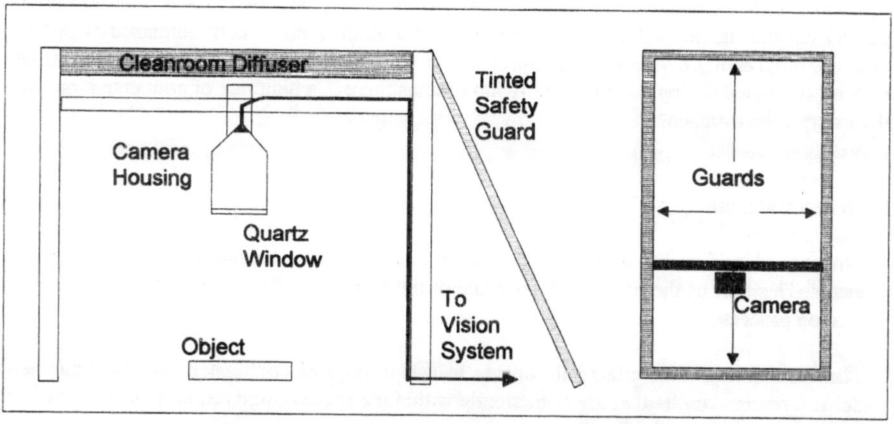

Figure 3: Vision System Cleanroom Set-up

The cameras are the next obvious area of discussion, and from the diagram it can be seen that the cameras are actually contained within metal housings which have a quartz window. There are two main reasons for this; firstly, the housings are more aerodynamic than a camera alone, thus a camera within this type of housing will cause less turbulence; and secondly, the casing protects the camera from the chemicals that are used within the cleanroom e.g. bactericides, detergents and antistatic agents.

When developing the support rig that will hold the cameras and any lighting that is being used some important issues should be addressed. Where possible stainless steel should be used for the framework. Also no crevices or areas that particles can build up in should be present, even if they are allowed within the cleanroom the work that would be involved in cleaning would be a severe hindrance. On the same theme all cables used should be smooth, ribbed cables or sheaths will cause the same problem by being natural areas for particle build up and being difficult to clean.

Where possible perform a smoke test on the rig before installation in the cleanroom, this will indicate any areas of concern there may be as far as turbulence is concerned. If a lighting rig is being used this should be operational during the smoke test so that any heat generated turbulence will be noticeable.

Scheduling

Installation within cleanrooms is a difficult process and scheduling is the key to reducing the complexity of the installation. It is impractical, though not impossible, to carry out an installation whilst the cleanroom is operational. The clothing and need to keep particle counts low mean that the job is at best very slow at worst impossible.

A better solution is to schedule a period where the cleanroom can effectively be dirty (reduction in cleanroom specification), weekends and national holidays being the obvious times. With a reduction in cleanroom specification, the clothing requirements will be less restricting manoeuvrability. Also in many cases drilling may be required and this can only be carried out during this time. A major issue if the system is not totally contained within the cleanroom is cabling between the cleanroom and the exterior environment and again this can only be addressed during such a shutdown period as there is a high risk of contamination from the exterior environment.

System Validation

As mentioned earlier for medical and pharmaceutical industries the validation process is essential. This part of the paper outlines some of the areas that should be addressed during the validation process.

The validation process effectively proves to a high level of confidence that the equipment under test returns results that are consistently within the specification required, and where it is necessary the system will fail-safe.

The lighting of machine vision systems is, as mentioned, an important success criteria. However, in manufacturing environments even with specific lighting solutions there will be variations in the lighting conditions. The validation procedure should take account of this and test for all possible lighting conditions.

Lighting variations are likely to occur due to; movement of people within the suite; reduction in power or failure of lighting; new lighting being fitted; direct sunlight if exterior windows are in view, etc. It is possible to mimic these conditions and to test for reaction to extreme conditions. By varying the rooms lighting conditions and by placing masks to block light, the conditions can be varied from normal ambient to total darkness. To increase lighting levels a portable fluorescent tube can be used and the level increased by placing the tube closer to the system. For consistency of testing a light meter should be used to measure the light level at the object being inspected.

During actual validation tests that have been undertaken these two changes in lighting were used to simulate the expected variations in lighting that would occur in the cleanroom environment. The tests were carried out with small incremental changes in the level of lighting taking place. The changes were continued until the system failed. It is important that when the system fails it fails safe and the tests should be thorough to check that this is the case with lighting extremes; poor tolerance specification of the vision system can lead to the system effectively 'guessing' when not enough information is present.

When validating the system it is important that the parts that are being inspected by the machine vision system are considered. The validation should include inspection of parts with all variations in tolerance being present in the sample population used for the validation procedure. As well as changes in physical dimensions this should cover any artwork or text that is present on the pack. This can be of great importance as many vision systems are used to inspect text to ensure that labelling is correct for the product, a very serious issue for medical and pharmaceutical companies.

Software validation [4] is becoming increasingly important as part of the overall validation process where computerised systems are becoming even more prevalent within manufacturing processes. Many of the areas that will be covered as part of standard validation will be relevant to software validation. However, there are many areas required for a software validation that are not covered and where appropriate these should be investigated and implemented.

The validation work should be thoroughly documented with all test data being tabulated, because it is this information that one of the mentioned bodies will possibly review as part of their external auditing schemes. Failure to comply to the requirements can lead to being barred from markets.

Conclusions

The installation of vision systems within a cleanroom environment has been presented within this paper as being a difficult but not impossible task. The major source of problems is lighting. When using vision systems it is rare to use only ambient lighting, however, within a cleanroom it is possible to use ambient lighting as it will be more consistent within the cleanroom environment. This is caused by the steps mentioned within the paper, but also due to the minimal effects of sunlight and the fact that cleanrooms generally operate with as few people present as possible. The set-up shown earlier can also cope with people moving close by as its main source of ambient lighting is from above.

The validation process is mentioned as being critical for medical and pharmaceutical companies, however, though not officially required in many other industries it provides an excellent structure for testing new processes and equipment. The information given for validation is also not exhaustive and the level of testing required is dependent on the problem.

References

[1] M McGowan ,D K Harrison, G Cullen, D S Ross; Machine Vision Selection Criteria an End Users Viewpoint, *Proceedings of the Thirteenth National Conference on Manufacturing*, pp 331-336, 1997.

[2] UK Industrial Vision Association Ltd: PO Box 25 Royston, Herts, SG8 6TL
Tel: 01763 261419 Fax: 01763 261961 Web: www.ukiva.org

[3] British Standards 5295:1989: British Standards Institution, Linford Wood, Milton Keynes, M14 6LE, England

[4] WWW.FDA.GOV/cdrh/comp/swareval.htm

Effect of object location on measurement accuracy of a vision system

A ROY, G BURNS, and **D K HARRISON**
Department of Engineering, Glasgow Caledonian University, UK

ABSTRACT

Inspection is a very important activity in a manufacturing process. One of the main obstacles in industrial automation is the difficulty in automating the inspection process. Vision systems are used in many inspection activities including non-contact measurement This has the potential of producing enormous benefits for the industry but the vision applications did not expand at the rate predicted by some researchers. This is because the commercially available systems have limited ability and are non-flexible in nature.

Against this background a knowledge based vision system has been developed which takes a novel approach to help producing multi-dimensional measurements. For any system consistency in measurement is very important. This paper discusses the effect on the measurement accuracy of the knowledge based vision system due to the variation of object location while the object distance remains unchanged.

1 INTRODUCTION

Any inspection process which involves human inspectors checking out components by visual inspection can be very inefficient and inconsistencies are likely to be present. An ideal inspection system should have both speed and accuracy. Unfortunately, they do not always go hand in hand. The only meaningful and economic way of achieving this is by replacing the human inspection process by an automated inspection system wherever possible.

Vision systems can play a dominant role in automating the inspection process in industry. A lot of the Vision systems currently in use are dedicated systems and the work is carried out by private companies who are secretive about their applications. Hence, a significant proportion of the work is not reported in open literature.

Braggins [1] carried out a study of the European market in 1986. The European market for commercially viable vision products was estimated to be $47.7 million and he predicted that the market will rise at a compound growth rate of 52%. If this growth rate is assumed, the current European market can be estimated to be in the region of $4.1 billion. The American market is estimated to be around four times larger. This would tend to suggest that the vision systems play an extremely important role in the current industrial scene

A recent literature survey carried out to identify the research and development work currently being carried out to promote the application of vision system technologies in inspection, shows that these are being applied in broadly the following categorised:

- Dimensional measurements/inspection
- Part/object identification
- Inspection in Electronic and related industry
- Surface inspection
- Other applications involving inspection

In spite of the developments of CCD cameras and a substantial research activity to apply vision systems in inspection, it still remains a difficult task. An intelligent approach is required to process the information provided by the vision system. This prompted a lot of researchers such as Novini [7], Perkins [8] etc. to use AI-technology to assist in vision system applications. But to date very few systems are available which are sufficiently flexible to handle multi-dimensional measurements. However, a knowledge based vision system has been developed at this university which is capable of carrying out automatic analysis of captured images to provide complete inspection data. In any inspection activity consistency of measurements very important. This paper discusses the effect of positional data of object to be measured, on the accuracy of dimensional measurements.

2 SYSTEM OVERVIEW

2.1 Equipment - Microsight 1

The Vision system used for this research is Digithurst's Microsight 1, which captures and digitises images, transferring them as 8-bit data for storage within an area of the computers own memory. The System can be switched to resolution of 256x256 pixels or 512x512 pixels, each pixel having a grey scale value of between 0 - 255.

The main components of Microsight Scanning system are :

Microeye Scanning Card suitable for plugging into a PC

One volt peak to peak interlace video Camera

Software containing ICS (Image Capture System) routines which allow storage of full grey scale images on disc

2.2 Captured Image Quality

The image captured by using the Digithurst vision system Microsight 1, was to be used by the image analysis software for automatic identification and non-contact measurement of the component. Hence, it is extremely important to capture a good 'quality' image. For this research a good quality image is an image having a sharp contrast between the background and the component surface. The holes, cut-out etc. should be shown in a different contrast - similar to the background, so that they can be identified. The method used by the image analysis software to detect the edges from the digitised image is such that any sharp change in grey scale value on component surface can be mistaken for an 'edge'. So the image quality had to be ensured by employing specially designed lighting rig with back-lighting facility.

2.3 Image Processing and Analysis

Any vision system capable of carrying out non-contact measurement must be able to analyse the image that has been captured by the camera. In many vision system applications the raw image data is pre-processed to smooth out the image by eliminating most of the noise in the system. In this system, steps have been taken to ensure that sharp images are captured and hence no pre-processing is required. However, each image still needs to processed and analysed to establish the location of various edges of the component in the field of vision and then all the information stored within the image is to be converted to meaningful data for dimensional analysis. This is invariably the most important part of image analysis and this is where the main difficulties lie.

The edge detection is usually a complex activity and various edge operators such as Robert's or Sobel's are often used. But in this case a much simpler technique has been successfully employed to detect the edges, which uses simple masks with a high-low or low-high combination of adjacent pixels in the image file.

Understanding the relationships between the edge points and extracting the dimensional information contained within the image, however, seem to be the most difficult task. A multiple xy-scan software was developed based on a knowledge based approach which was capable of automatic identification of a component and carrying out a complete dimensional analysis. However, since the software relies on finding information about the component in the knowledge base, it can only analyse those parts which already exist in the database. The component dimensions were accurately measured, by a Co-ordinate Measuring Machine (CMM) and these were taken to be the reference or true dimensions when the errors for the critical dimensions were calculated.

3 EVALUATION OF THE KNOWLEDGE BASED SYSTEM

In an industrial environment, the object can be expected to be located either:

- at the same location in the field of vision or
- at many different locations within the field of vision of the camera.

To simulate this, many images of the same part were analysed using auto-analysis software. It may be unsuitable to carry out the non-contact measurements under controlled lighting conditions. For this reason, the objects were exposed to normal diffused fluorescent lighting as much as possible, although the back-lighting was used to ensure that the edges were clearly defined. The object distance was kept constant during all these tests.

The results obtained by keeping the object location fixed produced the same measurements, as expected, since the software is virtually analysing the same image every time. The results obtained by varying the object locations are summarised in the tables below :

Table 3.1 Shows Vision System dimensions against CMM measurements by repeated analysis of rectangular parts at different locations within the field of vision

Location Number	Length	Breadth	Hole 1	Hole 2	Hole 3	Hole 4	Hole 5
1	112.644	91.896	30.326	20.1261	20.4245	19.5866	19.5866
2	112.874	91.909	30.6358	20.8587	19.7877	19.5437	19.5437
3	112.984	91.569	30.2571	20.5828	19.957	19.6957	19.6957
4	111.333	90.733	30.1051	20.4009	20.2485	19.1678	19.4436
5	112.496	90.67	29.6481	19.9849	20.4381	19.157	19.157
6	112.832	91.808	30.2789	20.7	20.1058	19.449	19.449
7	112.874	91.909	30.6358	20.8587	19.7877	19.5437	19.5437
8	112.617	90.887	29.338	20.2377	20.5893	19.1854	19.6299
10	112.023	91.582	30.2789	20.7	20.1058	19.449	19.449
11	111.923	91.733	29.9627	19.4452	19.5761	19.3763	19.3763
12	112.146	91.211	30.2423	20.6927	19.862	19.5982	19.5982
Mean(mm)	112.4315	91.44609	30.2423	20.41706	20.08023	19.43204	19.49752
CMM(mm)	111.982	90.876	30.048	20.0427	20.04	19.848	19.839
Error(mm)	-0.4495	0.579	0.1943	0.3743	0.0402	-0.356	-0.3415
% Error	0.04	0.06	0.07	1.86	0.02	-1.79	-1.72

Table 3.2. Shows Vision System dimensions against CMM measurements by repeated analysis of circular parts at different locations within the field of vision

Location Number	O/Dia	I/Dia	Hole 1	Hole 2	Hole 3	Hole 4
1	111.095	51.003	12.3102	12.2404	12.2404	12.5403
2	111.153	51.022	12.6841	12.4359	12.4359	12.6214
3	111.231	51.208	12.8392	12.7606	12.7606	12.6108
4	110.782	51.302	12.696	12.4713	12.4713	12.6771
5	110.774	51.09	12.6233	12.3864	12.3864	12.1253
6	111.137	51.385	12.144	12.3648	12.3648	12.3133
7	111.279	50.876	12.1349	12.5078	12.5078	12.681
8	111.031	51.137	12.7083	12.5923	12.5923	12.4634
9	111.106	50.78	12.8133	12.1869	12.1869	12.4816
10	110.925	51.471	12.3671	12.1964	12.1964	12.3341
11	111.198	51.387	12.2779	12.23393	12.2393	12.4541
12	110.866	51.483	12.2987	12.2599	12.2599	12.4751
Mean(mm)	111.0481	51.17867	2.49142	12.38639	12.38683	12.48146
CMM (mm)	109.926	51.453	12.708	12.696	12.674	12.677
Error(mm)	1.1221	-0.2744	-0.2166	-0.3096	-0.2872	-0.1955
% Error	1.02	0.05	-1.7	-1.85	-2.26	-1.54

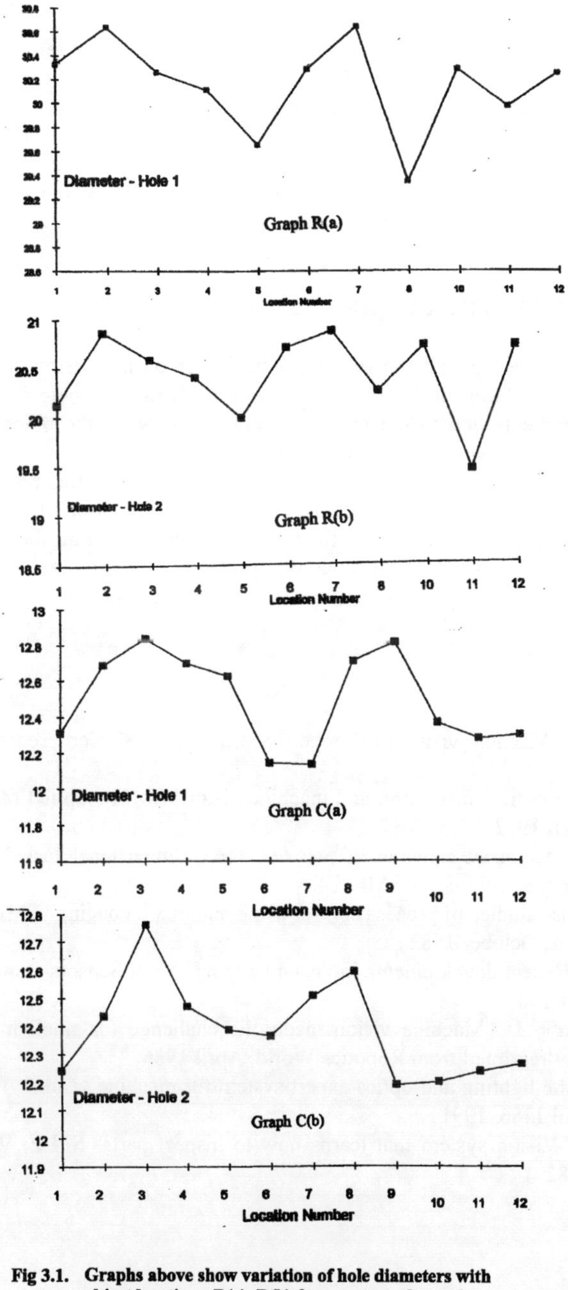

Fig 3.1. Graphs above show variation of hole diameters with object location - R(a), R(b) for a rectangular and C(a), C(b) for a circular component.

4 DISCUSSIONS

The above tests show that the results obtained by the vision system are repeatable provided the object is located at the same place in the 2D-space. However, there appears to be some small variations in the results, as the object is randomly located in the field of vision of the camera. This is demonstrated by the graphs shown in Fig 3.1 and Fig.3.2 above. These variations are likely to be due to the angle of vision of the camera from the centre of the object. However, the Tables 3.1 and 3.2 show that the mean error levels remained below 0.4 mm for rectangular components and below 0.3 mm for circular components.

5 SUMMARY AND FURTHER WORK

These investigations have shown that a knowledge based approach to vision system analysis can make automation of multi-dimensional analysis possible. However, there appears to be slight variation in measurements due to the object location, even when the object distance is not altered. This was not anticipated prior to this investigation.

Further investigation is to be carried out to establish the root cause of this problem. One of the possibilities may be due to the distortion of the image causing variation of the pixel size as the object moves away from the centre of the field of vision, thus affecting the scale factor. The overall accuracy of such a system still remains a problem, this must be resolved so that the vision system can be widely used for dimensional measurements.

REFERENCES

[1] Braggins D W "Machine vision in Europe" SME, Vision '87 Conference Proceedings, USA

[2] Prior T "Electro-optical inspection and machine vision" SME, Applied Machine Vision Conference, April 1982

[3] Bothel M "Linear array cameras in two and three dimensional tasks" Marc Bothel Everett, Washington - printed by SME, USA

[4] Beavers JC "Case studies of process control using automatic guaging" SME, QualTest-1, Conference, USA, October 1982

[5] Trepagnier P "Recent developments in visual gauging" SME Sensors Conference, SME, USA, November 1985

[6] Garakani A, Cobb J "Machine vision meets the challenge for small measures"SME, Michigan, USA - reprinted from Robotics World, April 1986

[7] Novini A R "The lighting and optics expert system for machine vision" Proceedings of SPIE (USA), Vol 1386, 1991

[8] Perkins W A "Vision system that learns how to inspect parts" SPEE, Vol 336 Robot Vision, May 1982

Multi-scopic neuro-vision for two- and three-dimensional object recognition/classification

T SHAW, S BOUSSAKTA, F NABHANI, and **A WATSON**
School of Science and Technology, University of Teeside, Middlesborough, UK

ABSTRACT

Machine vision systems are being increasingly used for sophisticated applications such as classification and process control. Though there is significant potential for the increased deployment of industrial vision systems, a number of important problems have to be addressed to sustain growth in the area of industrial machine vision. Artificial neural networks coupled with machine vision systems offers a new methodology for solving difficult computational problems in many areas of science and engineering. As a consequence, the research work presented in this paper investigates several novel uses of machine vision and artificial neural networks in the processing of single camera multi-positional images for two-dimensional and three-dimensional object recognition and classification. Many industrial applications of machine vision allow objects to be identified and classified by their boundary contour or silhouette. Boundary contour information was chosen as an effective method of representing the industrial component, a composite signature being generated using vectors obtained from the generation of multi-centroidal positions and the boundary pixels.

1 INTRODUCTION

Over recent years the use of computer vision systems in industry has become widespread in many countries, the main users being in the fields of manufacturing and industrial inspection (1). The application of image analysis, together with the integration of neural networks in industrial inspection has opened up a new dimension in the area of component recognition and classification.

Many industrial applications permit components to be described and recognized by their boundaries or silhouettes. Components which are similar have similar boundaries, and the use of boundary information is beneficial in that it reduces the dimensionality of the input data quite considerably. However, the boundary information can often be corrupt due to the resolution of the digitization as well as noise.

Neural networks offer a very different methodology for performing recognition and classification processing. Neural networks being mathematical models of 'simulated' brain activities operate in a similar fashion to human processing systems. The neural network technology utilized is by nature a "fuzzy" system which is not designed to provide precise answers and solutions to every problem. It is a tool which attempts to classify patterns according to other patterns it has "learned" and to give the most reasonable answer based upon the variety of learned patterns.

Three-dimensional component recognition has long been considered a difficult problem (2). The task of three-dimensional component recognition and classification has been chosen for investigation, since it is 'mathematically and computationally complex' and extremely relevant to the needs of the machine vision industry. Neural networks offer the opportunity of removing some of the elements of this 'mathematically and computationally complex' problem and thus open up the area of inspection and classification of three-dimensional components.

While many applications could be solved by performing only two-dimensional component recognition, the more complex applications require the acquisition of three-dimensional information along with three-dimensional recognition. The object of this research is to combine imaging equipment and a software simulated Artificial Neural Network (ANN) in the development of a robust, three-dimensional component recognition system capable of recognizing and classifying manufactured components, regardless of their placement and orientation within the field of view of the camera.

2 SHAPE REPRESENTATION AND DATA CAPTURE

The experimental set-up used for the case studies presented in this paper follow a fairly conventional approach, as shown in Figure 1 and Figure 2. The hardware was comprised of a host PC -Western Systems 486 DX 50 MHz unit, with maths co-processor and colour monitor, Colour CCD - Sony DXC-151P, Image Processing Unit - IV120C (Megascan by J-L Automation, this package provides a large number of ready to use subroutines for manipulation of the captured images) and a Video Monitor. The UMI RTX robot was used to position the CCD camera in the various target locations. The PC houses and controls the IPU, and the IPU (IV120C) utilizes the Texas Instruments TMS320C25 digital signal processor.

The image memory of the IV120C comprises a 512 x 512 x 8 bits Video Ram (VRAM) and a 1024 x 1024 x 16 bits Image Memory (DRAM). Library routines and software interfaces allow the loading of data and variables from the host PC environment to the TMS320C25 digital signal processor.

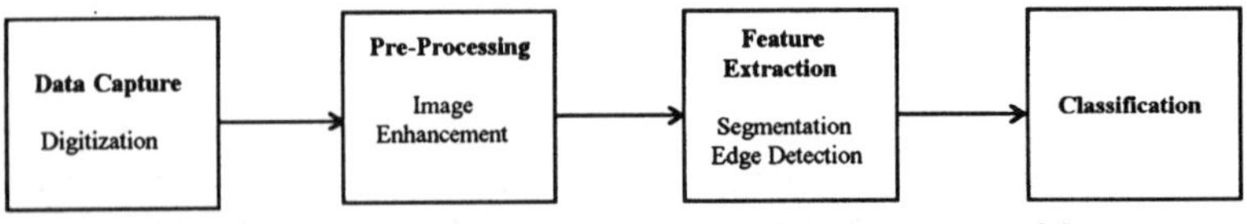

Figure 1 Block diagram of a conventional visual pattern recognition system

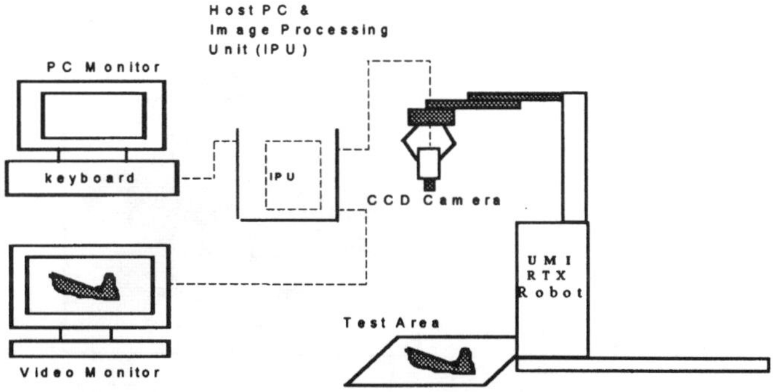

Figure 2 Schematic representation of the test equipment

3 CASE STUDIES

3.1 Automobile filter

The two case studies presented in this paper are (1) an automobile filter and (2) an automobile carburettor. In case study 1 the automobile filter is used to demonstrate the systems ability to recognize and classify components irrespective of translational or rotational orientation. Although the filter is a three-dimensional component it is possible to represent it by a closed two-dimensional boundary.

The raw image data was captured by a CCD camera and digitized into a 256 x 256 x 6 bit grey image and processed by an IV120C image analysis card. The filter was positioned on a light table so as to produce high contrast images, and a Laplacian edge detection routine was used to segment the images. Following this the position of the centroid was determined and a contour sequence was produced in an ordered manner representing the Euclidean distance between the centroid and the boundary pixels of the digitized shape. This centroidal profile is a 1D representation of the component boundary. The length of the sequence is determined by the number of pixels on the perimeter, Freeman chain coding was used to trace around the component boundary.

A single start point (SSP) algorithm based on the longest Euclidean distance was used in this research, however, during initial experiments whilst analysing symmetrical type components it was discovered that it may not be possible to use a SSP because some components may have several longest Euclidean distances. Typically a pair of electrical pliers could be considered symmetrical and would possess two start points, furthermore, the pliers may be open, partly open or closed and therefore the shape boundaries would also vary.

Results of case study 1

Figures 3 and 4 below show an outline, orientation and signature of the filter used for training within the neural network. Several parts of the filter are capable of being rotated / removed, however, in order to prove the systems capabilities it was decided to maintain the filter in a fixed position, but allow the filter to be translated and rotated as a complete unit within the field of view.

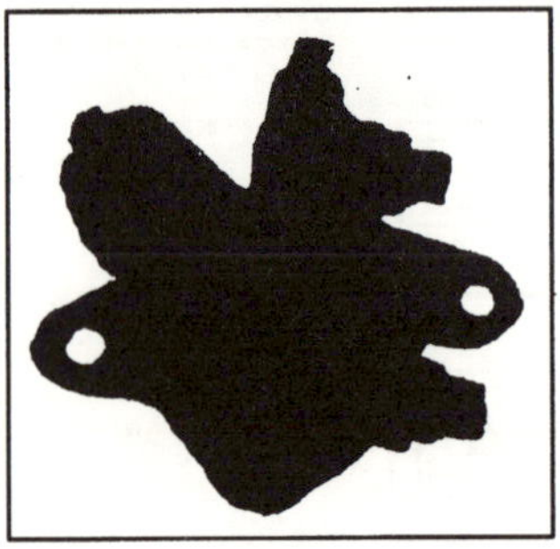

Figure 3 The boundary outline and orientation of the training filter

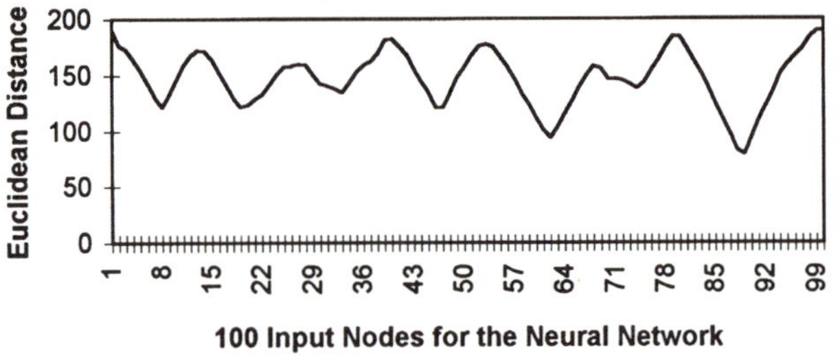

Figure 4 The signature of the training filter

Many positions within the field of view were tested, but for brevity initial trials were based on a sample size of ten positions / orientations, including the training set of one as shown in Figures 3 & 4. The following system variables were used:
Input Nodes - 100; Hidden Nodes - Various; Output Nodes - 1; Output Threshold- 0.99; Learning Rate - 0.6; Momentum - 0.6; Learning Threshold - 0.0001.
Figure 5 below shows the training signature with two of the test signatures superimposed onto it.

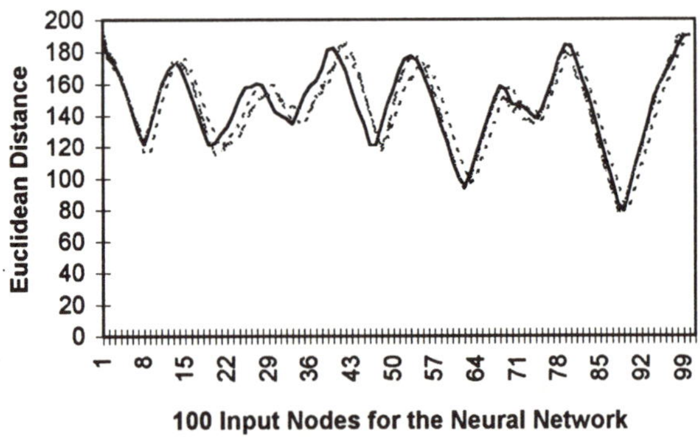

Figure 5 The signature of the training filter (solid line) and two test signatures

Table 1 shows a summary of the outcomes for 21 hidden nodes.

Table 1 A summary of the outcomes using 21 hidden nodes

Strategy: Single Start Point **Number of Hidden Nodes = 21** **Number of Training Images = 1** **Number of Test Images = 9**				
Filter	**Status**	**Boundary Pixel**	**Boundary Origin**	**Classification**
1	**Train**	**1404**	**x = 258 y = 75**	**N/A**
2	**Test**	**1332**	**x = 180 y = 126**	**0.993618**
3	**Test**	**1372**	**x = 174 y = 106**	**0.992550**
4	**Test**	**1358**	**x = 197 y = 79**	**0.993297**
5	**Test**	**1347**	**x = 182 y = 27**	**0.993370**
6	**Test**	**1319**	**x = 254 y = 163**	**0.992997**
7	**Test**	**1371**	**x = 319 y = 12**	**0.994779**
8	**Test**	**1316**	**x = 240 y = 140**	**0.992870**
9	**Test**	**1334**	**x = 351 y = 80**	**0.992439**
10	**Test**	**1350**	**x = 293 y = 188**	**0.994112**

As can be seen from Table 1 the system proved very successful in its classification of non-symmetrical components, however, to deal with symmetrical components a Multiple Start Point (MSP) algorithm will be needed. McNeil (3) developed a MSP algorithm to deal with simple two-dimensional planar shapes. The electrical pliers shown in Figure 6 and their respective signatures in Figure 7 below clearly demonstrate that a MSP system is required for symmetrical components.

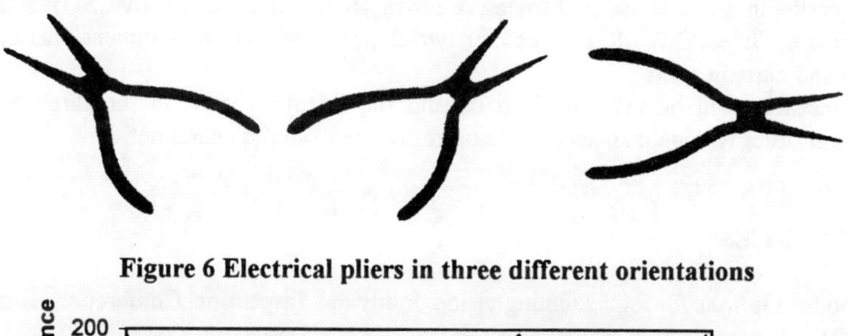

Figure 6 Electrical pliers in three different orientations

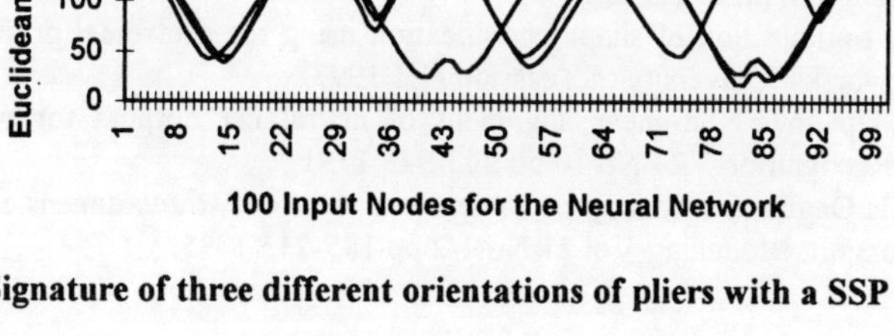

Figure 7 Signature of three different orientations of pliers with a SSP

3.2 Automobile Carburettor

An automobile carburettor was chosen for case study 2. The RTX robot was used to position the single CCD camera in five positions within the field of view, position 1 was chosen normal to the surface and positions 4 to 5 were at 90^0 to each other with the camera at an angle of 70^0 to the viewing surface. The five positions chosen together with the inclined camera angle provided an approach to dealing with components (carburettor) with attachments which may be partially occluded. A considerable increase in flexibility of the classifier is achieved by adding the requirement of invariance to partial occlusion (4). Unlike non-occluded object classifiers that operate on a global feature vector which characterizes the entire shape, partial shape classifiers are forced to operate on a set of local features which characterizes parts of the shape.

Several techniques are available for the acquisition of three-dimensional information. Some of the most popular and conceptually straightforward techniques have been those involving the stereoscopic disparity principle (5). The approach used in this research requires the camera to be positioned in five different locations providing almost full coverage of the component. At each camera location raw image data was captured by the CCD camera, following this a composite image was generated and an Euclidean vector sequence was subsequently produced. The number of elements in the contour sequence corresponding to the input vector for the ANN was set at 250 due mainly to memory restrictions. The initial data was used as a training set and procedure was repeated with various items of the carburettor removed or occluded to test the system. At present the testing is still ongoing and results should be available in the near future.

5 CONCLUSION

The main objective of this research paper is to outline a method of representing two-dimensional and three-dimensional components which can be translated into a form which can presented to an ANN for recognition and classification. The system should provide several novel applications of neural networks in processing of Moving Camera, Stationary Object (MCSO) and Moving Camera, Moving Object (MCMO) images for two-dimensional and three-dimensional component recognition and classification.

Further research will be performed to expand the feature vector, in the three-dimensional component, in order to obtain an acceptable performance from the neural network.

6 REFERENCES

(1) P Woods. Options for implementing vision. Industrial Inspection Conference. Birmingham. pp 3-13 1991.

(2) Chun-Shin Lin and Chamnong Jungthirapamich. Invariants of three-dimensional contours. Pattern Recognition V23 No 8 pp 833-842 1990.

(3) A R McNeil. The optimisation of shape classification using the centroidal profile sequence and artificial neural networrks. University of Teesside Phd 1997.

(4) L Gupta and AM Upadhye. Non-linear alignment of neural net outputs for partial shape classification. Pattern Recognition V24 No 10 pp 943-948 1991.

(5) MB Lynch and CH Dagli. Stereoscopic neural-vision for three-dimensional object recognition. Mathl. Comput. Modelling Vol 21 No 1/2 pp 185-215 1995.

The correlation between surface leakage current and contamination level of printed circuit boards

A SUKSRI and **J C BEAMISH**
Department of Manufacturing System Engineering, University of Hertfordshire, Hatfield, UK

ABSTRACT

Surface Insulation Resistance (SIR) testing has been a traditional method of measuring the cleanliness of a printed circuit board (PCB). This practical test is to ensure the circuit reliability and cleanliness of the PCBs. Published work has established that the SIR test is a very misunderstood test method as surface insulation resistance is not being accurately measured. Various effects such as current leakage, which may arise from the air, PCB substrate, PCB surface and contaminants have been said to influence the level of accuracy of SIR measurement. The test continues to be used as a tool to measure contamination level.

Previous work has suggested that the application of a DC bias voltage to the SIR measurement produces a voltage dependent resistance (VDR) effect as well as a surface leakage current (SLC) on PCBs. However, with AC voltage, results revealed that the application of AC biasing voltage produces more acceptable SIR reading. It is apparent that much of the previously published experimental work on SIR measurement is not well characterised. The research being undertaken by University of Hertfordshire is to establish a methodology which can quantify the level of contribution of the aforementioned effects in order to identify the leakage current as well as modes of conduction of contaminants on PCBs.

1 INTRODUCTION

Insulation resistance of PCBs is considered to be a very important factor when a printed circuit board is operated in any normal/hostile environment. In order to maintain a high resistance, the PCBs must be free of ionic or non-ionic contamination. Unfortunately, contamination may be introduced during manufacturing processing, such as etching and drilling as well as the assembly processes such as soldering.

As current technology is developing, the line spacing of the conductors on PCBs is getting much closer. Consequently, the presence of contamination can influence the performance of the circuit due to the possibility of leakage current. This leakage current can seriously impair product performance and reliability as discussed by Chan *et al* [1]. It is imperative that an electronic assembly is capable of operating reliably in a condition of which it was designed for.

2 CURRENT PRACTICAL TEST ON PCB CLEANLINESS

Two practical tests that have been widely use to qualify PCB cleanliness are Solvent Extract Conductivity Test (SEC) and Surface Insulation Resistance (SIR). The principle of the SEC test is to immerse the test PCB into a standardised isopropanol-water solution. Any ionic contamination is dissolved in the solution. The increase in electrical conductivity of the solution is translated to microgram (μg) of equivalent sodium chloride per unit area, $\mu g.\ cm^{-2}$ equivalent NaCl. SEC is a very quick method of measuring the level of contamination. Unfortunately, as it was implied earlier, this method can only detect ionisable residues that can be extracted by the test solution. Anything not dissolved is not detected and its effect on performance cannot be assessed. Further, estimated cleanliness is an average value; local concentrations on the board are not highlighted even though they may affect performance. Finally, no account is taken of the risk posed by residues that are detected.

The second traditional test method is the Surface Insulation Resistance (SIR). Historically, SIR began as a material property to be measured on the base material, regardless of the test conditions imposed on the test specimens[2]. During the test, the interdigitated comb pattern coupons are placed under a bias voltage and periodically the bias voltage is removed and the resistance measured with a test voltage. Biasing voltage are applied to a specimen at about 45-50 volts DC and a testing voltage of -100 volts DC as specified in IPC-TM-650[3]. The test is carried out while the specimens are in an environmental chamber at 85±2°C and 85±2 %RH. A standard SIR test normally lasts from three to 56 days, typically seven days. The reading is obtained at 24, 96 and 168 hours in a 7-day SIR test or on a daily basis. Work has been carried out which correlates the SEC and SIR measurement. A relationship between SEC and SIR has been investigated by Chan *et al*[4]. The research concluded that, increasing the ionic contamination by a factor of 2.5, decreases the SIR by about a decade.

3 FACTORS AFFECTING SIR MEASUREMENT

SIR have been used as a tool to qualify contamination, flux, solder mask, as well as electrical performance of circuit boards. The objective of SIR measurement is to measure the current due, purely to ionic migration. The ionic migration is a result of contamination. However, there are other factors that affect the measurement of SIR including; effect of leakage current, electrochemical cells, bias voltage, as well as emf induced from other sources. These effects are summarised and explained below.

3.1 Effect of leakage current

According to Ellis[5], there are four possible routes of leakage current between conductors on PCBs that contribute to the lowering of SIR readings. These are;

- Leakage current through the air
- Leakage current through contamination on the conductor surface
- Leakage current through the substrate surface
- Leakage current through the body of the substrate

At a high level of temperature and humidity, all of the above factors must be considered. Air at high levels of humidity, near the conductor, may not be the same as in bulk of the atmosphere. It is also worth recognising that surface topography of PCBs under high magnification possess a sponge-like surface, containing microporosity and interconnecting channels. Leakage current may be induced through contamination, which can behave as ionic, or ionic-like electrical leaking species leading to a lower SIR reading. Leakage current may also occur through the body of the PCB substrate when the laminate is not properly cured during the manufacturing process.

3.2 Effect of electrochemical cells

It is important to consider that it is not necessary for any external DC bias voltage gradient to be applied for leakage current to occur in the presence of contamination and moisture. In general, there is an electrical leakage current flowing between two dissimilar metal conductors[6], or a concentration gradient between similar metal conductors, on the printed circuit board. Due to the difference in standard electrode potentials, under a condition of high humidity, high temperature with ionic contaminants, a complete circuit will be formed and leakage current will start to flow.

3.3 Effect of bias voltage

Previous work conducted by Ellis[7] has revealed the phenomena called "VDR effect" (Voltage-Dependent Resistance). This effect is such that the insulation resistance values are higher than expected under the operating conditions for printed circuits. The VDR effect is directly related to the variation of DC bias voltage and the test voltage used. If a test voltage is applied while the ions (within the molecular structure of PCB substrates) are in disequilibrium, one of three phenomena can occur. Firstly, if a test voltage is applied in the opposite polarity to the bias voltage, its field will tend to cause the ions to recombine, the resistance obtained will be lower than expected. Second, if a test voltage is of the same polarity but lower or equal in value compared with the bias voltage, little change in the ionic structure will occur, so that the measured resistance will be higher than expected. Third, if a test voltage is higher than the bias voltage and of the same polarity, the ions will tend to separate further, requiring a supply of additional energy, the resultant measurement of resistance will still be higher than expected. This VDR effect can strongly mislead the result of the SIR readings.

DC continuous voltage bias causes mobile ions to migrate towards the electrodes and are thus being depleted from the insulating surface. For this reason, an application of AC bias voltage was used rather than reversing the polarity of DC biasing voltage. AC leakage current is measured as an AC voltage is applied. AC-SIR measurement was investigated by Chan[8]. Under AC voltage, the mobile ions merely oscillate back and forth resulting in no net flow. Thus, the ions on the insulating surface will not be depleted. Conversely, the AC voltage in any large magnitude may cause voltage-acceleration of some real-life degradation failure. The application of AC bias voltage to the test sample is questionable due to the potential change of impedance of the test coupon at different frequencies and levels of contamination.

3.4 Effect of emf induced from other sources

Errors in SIR measurement may also be introduced by other sources such as radio wave interference, thermal emf, as well as triboelectric currents effects. The use of guard electrodes to shield the conductors on the specimen enable the conductors to behave as antennae. This will result in an emf being induced through the pick up of interference which will produce misleading results to the SIR reading.

Thermal emfs generated by conductors will be an additional source of measurement errors. Thermoelectric potential will be generated as thermal emf varies with temperature increments. Other errors source that contribute to the SIR reading is the triboelectric currents effect. The effect of triboelectric currents are generated by charges created at the interface between a conductor and an insulator due to friction. A typical example would be currents generated by insulators and conductors rubbing together in a coaxial cable during the SIR testing.

4 IDEAL EQUIVALENT CIRCUIT

From the aforementioned effects that contribute to the inconsistency of SIR measurements, an ideal equivalent circuit of sample PCBs can be modeled as an impedance containing a network of resistances and capacitors. The diagram shown below in figure 1, is adapted from Ellis[9]. With parallel metal conductors on dielectric surface, the conductors contribute an electromotive force (Emf) to the test pattern. During power on, these conductors will exhibit overshooting of voltage due to charge/discharge of capacitance. The current flowing in each branch is unlikely to comply with Ohm's law due to the aforementioned parameters.

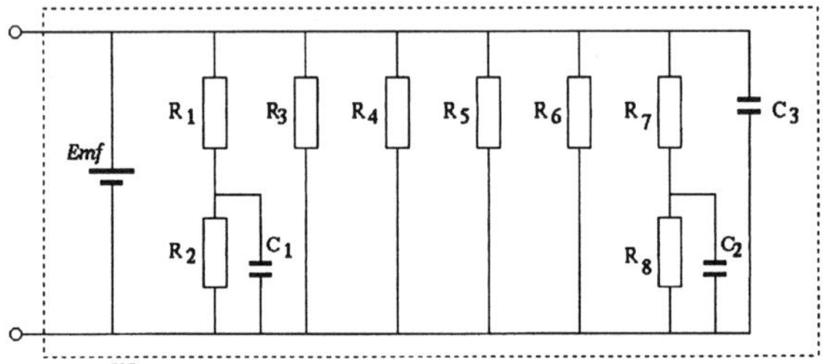

Where:

R1-Ionic charge resistance ca. 1 MΩ
R2-Ionic discharge ca. 100 TΩ
R3-Air resistance ca. 1 TΩ
R4-Contamination resistance ca. 1 TΩ
R5-Surface resistance ca. 1 TΩ
Emf- Electromotive force
R6-Substrate resistance ca. 1 TΩ
R7-Migration charge ca. <1 MΩ
R8-Migration discharge ca. 1 TΩ
C1-Ionic capacity ca. 1 F
C2-Migration capacity ca. 0-infinity
C3-Pattern capacity ca. 100 pF

Figure 1. Equivalent circuit of interdigitated comb conductors

5 RESEARCH METHODOLOGY ON SIR AND CONTAMINATION LEVEL

The literature identified a number of parameters that contribute to the inconsistency of SIR testing. A research framework will be developed to quantify the contribution of individual effects that have been said to influence SIR Measurements. Among the test parameters for investigation are:

-The nature of the PCB substrate
-Level of contamination
-Biasing voltage, DC and AC, forward and reverse (polarity) bias application
-Leakage current on different media i.e. air, surface, contamination and substrate

Test specimens chosen for preliminary work are, standard FR-4 (epoxy resin) interdigitated comb pattern with spacing between conductors of 0.008/0.012 and 0.012/0.013 in., bare gold and tin-lead hot air soldering reflow (HASL). Contamination will be introduced on to the surface of PCBs. During the test, the clean, controlled samples will be placed under the same environmental chamber as the contaminated samples. The leakage current monitoring will register the trends in behaviour of the contaminated as well as the controlled samples. The values will then be used to identify the relationship between leakage current and contamination levels.

6 CONCLUSION

SIR offers significant benefits on cleanliness and electrical reliability of PCBs. However, the deficiencies of the SIR have been identified, of which, the effects that lead to the inconsistency of measurement needs to be established. Furthermore, with the development of today's electronic manufacturing, electronic assembly must be capable of operating reliably not only in mission-oriented, but also in commercial, applications. This paper outlines a procedure for quantifying the factors that contribute to the SIR readings.

7 FURTHER WORK

From experimental work, results obtained will attempt to:

-Identify the relationship between leakage current and the level of contamination
-Identify the mode of conduction of the leakage current
-Model the characteristic impedance of a comb pattern

Ultimately, the objectives of this methodology are firstly, to quantify the effect mentioned and second, to establish the relationship of leakage current due to the contamination of PCBs so that the SIR test method will be used as a qualifying tool that will ensure PCBs cleanliness and electrical reliability.

8 ACKNOWLEDGEMENTS

The authors wish to express their gratitude to John Burbridge of ALPHA METALS for the help on qualifying for cleaning process as well as ionograph test on specimens.

9 REFERENCES

[1] A. S. L. Chan, D. Culver, and T. A. Shankoff, "Printed Wiring Board Inner Layer Contamination Study," *Circuit World*, vol. 17, pp. 10-14, 1990.

[2] IPC-9201 SIR Handbook: IPC, 2215 Sander Road, Northbrook, Illinois, 60062-6135, USA, 1996.

[3] IPC-TM-650 Test Method Manual, Method 2.6.3.3, IPC, 2215 Sander Road, Northbrook, Illinois, 60062-6135, USA, pp. 1-4, 1995.

[4] A. S. L. Chan and T. A. Shankoff, "A Correlation between Surface Insulation Resistance and Solvent Extract Conductivity Cleanliness Tests," *Circuit World*, vol. 14, pp. 23-26, 1988.

[5] B. N. Ellis, "The Determination of Flux Residue Safety: The State of The Art," *Circuit World*, vol. 20, pp. 26-32, 1994.

[6] Q. N. Xiao, F. Grunwald, and K. Carlson, "Surface Leakage Current (SLC) Methodology for Electrochemical Migration Evaluation," *Circuit World*, vol. 23, pp. 6-10, 1997.

[7] B. N. Ellis, "Low Voltage Insulation Resistance Measurement of Printed Circuits and Its Implication," *Circuit World*, vol. 14, pp. 21-25, 1988.

[8] H. A. Chan, "Surface Insulation Resistance Methodology for Today's Manufacturing Technology," *IEEE Trans. on Components, packaging and Manufacturing Technology-Part C*, vol. 19, pp. 300-306, 1996.

[9] B. N. Ellis, "On Insulation Resistance," *Circuit World*, vol. 21, pp. 5-11, 1995.

Wireless communication networks for monitoring goods transportation among remote manufacturing factories

G Y TIAN, Z X ZHAO, and **R W BAINES**
School of Engineering, University of Derby, UK

ABSTRACT

This paper presents a wireless communication system for monitoring goods in an automotive corporation. The system includes a portable transmitter with monitoring sensors and a remote receiver linked by cellular radio networks of mobile telecommunications. A mobile telemetric system for weight measurement and warehouse management was implemented and used in an automobile manufacturing enterprise that has different car parts factories and assemble plant in different geographical areas. To monitor the material and parts, portable transmitters, integrated by modified mobile phones and microcontrollers with the sensor interface, signal processing and data encoding, are installed on lorries and cranes and other transport vehicles. The microcontrollers are embedded in both the transmitter and the receiving units. The real-time data for such entities as weight, delivery destination and the delay of goods could be made available in the mobile telecommunications networks. The communication protocol, methodology of network management, mobile integration, microcontroller control of signal processing and data coding, and sensor interface are also discussed. The information system for a manufacturing scheduling is investigated and the advantages of the system are highlighted.

1 INTRODUCTION

In recent years it has been a proliferation of portable computers and a trend toward the production of smaller and more powerful units. A serious drawback of current portable computers is that users have to use awkward file transfer mechanisms to upload and download files they may need while away from their office. There is strong desire, for having the same file accessing environment that enables users to access the same data both in the workplace and away from it.

While modern portable computers are potentially capable of providing this functionality, they are hindered by the lack of continuous network connectivity.

A step forward is to add already available high-bandwidth mobile networking capabilities to portable computers. The technology to support this idea already exists. For example, beyond the near-ubiquitous LAN in business environments, high-speed wireless interfaces are emerging with a data processing speeds of up to 16 Mbps, using radio frequency (RF) or infrared (IR) technology [1], [2]. This provides the necessary tools for high-bandwidth access to computation and information resources.

In this paper, the concept of providing continuous network access to mobile computers is explored and a set of IP-based protocols presented that achieve the goal [3]. As an example of using sensing and electronic technologies for obtaining information from computer network, an example of weight measurement is provided. The modification mobile phone system for wireless network node is also discussed. The model of wireless communication networks for monitoring goods among remote manufacturing factories is designed and implemented in the research.

2 SENSING TECHNOLOGIES

A product is assembled using different parts that are manufactured in different remote factories. The transportation of the parts is an important issues of manufacturing management. In various manufacturing factories, a wireless communication network and mobile phones can be used for monitoring goods among remote manufacturing factories. A new sensor for goods weight measuring and the interface between sensor and mobile phone as a wireless node are implemented.

A system of monitoring goods transportation should not only access the real-time information of goods flow, but also improve the reliability and safety. For example, vehicle overloading is a problem of road safety. According to one study it has been the case of 32% of serious accidents [4]. Overloaded heavy vehicles (trucks) also contribute to the severe degradation of road surfaces. An on-board load indicator is proposed. There are various ways to determine the load including tyre pressure measurement, head counting and measuring the strain of the axles or chassis. Tyre pressure has to be measured while the wheels are turning and it is affected by temperature. Strain measurement is a better option but the effect of varying load distribution on the axles has to be considered. Resistive strain gauges to measure the deformation of vehicle structure are a way forward.

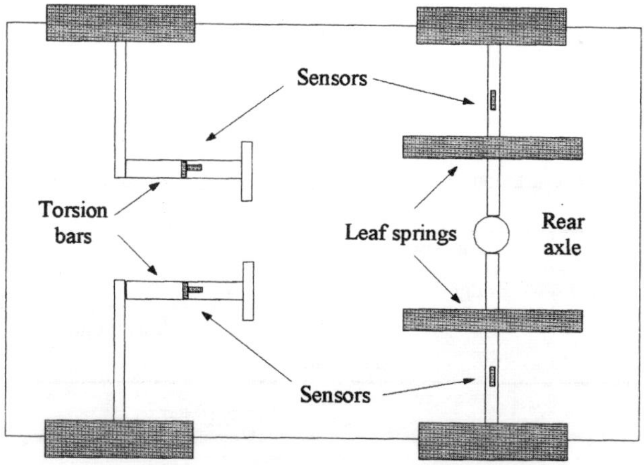

Figure 1 The positions of the sensors on a vehicle

As shown in Figure 1, four array resistive strain sensors on a vehicle for weight measurement are symmetrically distributed on the rear axle and front torsion bar which can eliminate the influence of varying load distribution. Based on the measured deformation of axles, the weight of goods on the vehicle can be calculated by combination the output signals from the rear axle and the front torsion bar. To simplify the interface of sensors and microprocessors, frequency output sensors [5] are used. The strain sensor with frequency output and its interface are shown in Figure 2, where a Wien bridge is employed to the sensor stability using differential resistive strains in the resonant bridge.

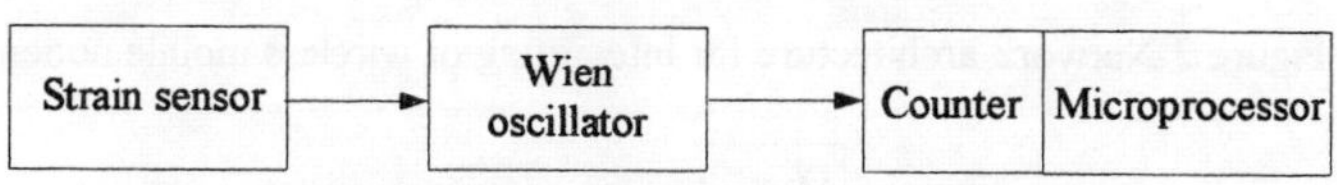

Figure 2 Strain sensor with frequency output and interface

3 NETWORK ARCHITECTURE

In order to show the routing between mobile nodes and fixed stations, the interconnection of wireless subnetworks with the wired factory network is presented. A network architecture of integrating the wireless mobile nodes is illustrated in Figure 3.

3.1 Wireless cell structure

Within a given area or "cell", mobile stations operate in the same physical and logical channel and form a wireless LAN segment. Mobile stations cannot reach each other directly, but have access to only one central station (the hub) which is also the access point (gateway) to the wired network. Mobile stations can roam from cell to cell by registering with another access point named by the handover.

The wireless cell as shown in Figure 4 is modified from a mobile phone. The key board of a mobile has double functions for dialling phone numbers and input appropriate data such as good materials, date and transporting places. The goods weight and statistical data are controlled by an MC 68040 microprocessor. The serial port is used for dual communication to transmit and receive data. A manual switch is used for controlling oral communication and data communication.

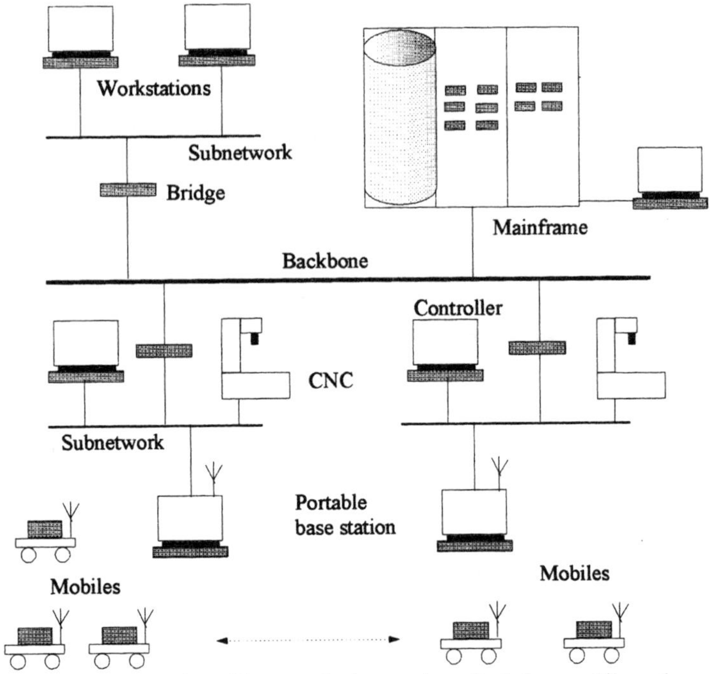

Figure 3 Network architecture for integrating of wireless mobile nodes

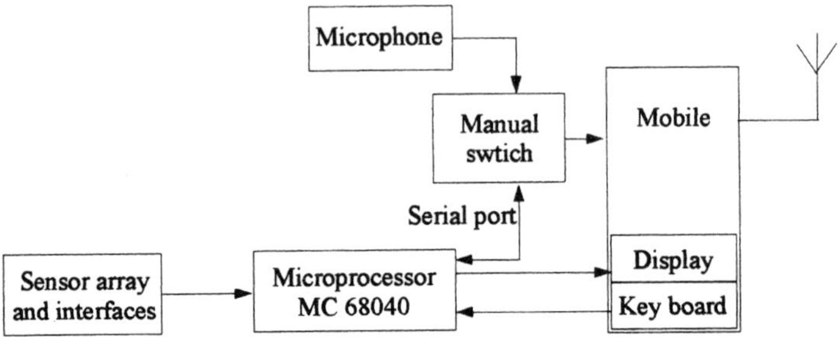

Figure 4 Wireless node modified from mobile phones

3.2 Network architecture

The architecture, as shown in Figure 3, is established based on a factory floor divided into fabrication cells. Each fabrication cell has its own subnetwork with one or more wireless cells. A mobile can roam between fabrication cells, but logically belongs to one particular fabrication cell. Existing distributed applications must continue to work without interruption when a mobile host moves between adjacent cells. Furthermore, a mobile node should be isolated, by the network from other systems that attempt to communicate with it.

3.3 Mobile IP

Several communication protocols can be adopted such as MAP/MMS [6]. For compatibility with the World Wide Web (WWW) which is based on Internet protocol (IP), IP is recommended in the research. This requires the IP enhancements to be transparent for (i) the mobile users, (ii) fixed users, and (iii) existing routers in the network. The mobile-IP proposal achieves these requirements by providing mobile hosts with a permanent "home address", regardless of its current point of attachment to the Internet, and a " care-of address", which provides information about the current point of attachment. Based on Internet Protocol, the mobile IP is easily implemented. However, since the mobile IP is still not well understood for integrated mobile data communication, it is difficult to integrate the system with WAN (it can be used as wireless LANs like Ethernet).

4 CONCLUSIONS

This paper has considered a problem related to the distribution of a control application in a mobile automated system. As an application example, a wireless communication network for monitoring goods among remote manufacturing factories was developed for specific wireless communication. A new sensor technology of frequency output has successfully been used for real-time measurement. The example of integrated system has demonstrated the possibility of further applications.

Future work consists of implementing a field test and the evaluation of other wireless standards in the framework of industrial communication network.

REFERENCES

1. Davis D. S. 1991 "Wireless LANs broadcast their benefits over cable", Electronic business, pp 58-62.
2. Reichert F. 1994 " The walkstation project on mobile computing ", IEEE/ICCC conference wireless networks, pp 974-78.
3. Kuwano H. et al 1997 " New multimedia communication services using sensing systems", NTT review, Vol. 9, No.1, pp 100-107.
4. Lacquet B. M., Swart P. L. And Kotze A. P. 1996 " A low-cost on-board vehicle load monitor ", Meas. Sci. Technol. Vol. 7, pp 1761-65.
5. Tian G.Y. 1998 "The design of frequency output sensors and their information control methodology", PhD thesis, University of Derby (to be submitted).

Morel P. And Decotignie J.-D. 1996 "Integration of wireless mobile nodes in MAP/MMS", Control engineering practice, Vol. 4, No.6, pp 821-829.

Development of an adaptive controlling equipment to guide operator and manufacturing machinery to an individually optimized production process

M USENER, I BANIK, A DAVIES, and **E G JOHN**
Cardiff School of Engineering, Cardiff University, UK

ABSTRACT

This paper details the problems and solutions involved in how to improve the manufacturing efficiency of a thermoset moulding department. It outlines necessary work required to define problems, to analyse existing work patterns and design the appropriate monitoring and controlling equipment. The improvement of the shop efficiency is based on maximised labour utilisation with adaptive machinery control. This equipment is guiding one machine operator to control two independent thermoset moulding machines. The monitoring and analysing routines allow the system to guide each operator individually to his personally best production efficiency in respect of the manufacturing process, production and quality targets.

1. INTRODUCTION

The basics of thermoset plastics have been understood since the end of the 19th century with the discovery of formaldehyde polymers and the first commercial exploitation under the name of Bakelite. At that time thermosets were mainly used for electrical insulation. Until 1930 with the arrival of the first thermoplastic materials and the development of associated processing techniques, thermoset plastic was the most important artificial material. Due to the nature of thermoset materials and its processing requirements the demand increased steadily while the production of thermoplastic products virtually exploded. Today, thermoset material can still be found in various products. Electrical products like wall plugs, sockets, heavy duty electrical components, fuse boxes, products under high mechanical, chemical and thermal stress and a huge variety of catering equipment from school plates to in-flight meal trays are made of thermoset materials.

The project, which formed the basis of this paper, was undertaken in a medium sized company in South Wales. The aim was to reduce the cost of the thermoset moulding products, to increase efficiency and modernise the working environment. The manufacturing department consisted of 16 individual compression-moulding machines. The machine park was about 25 years old, automation was reduced to temperature control and press timing. A typical production run starts at a pelletising machine, where raw powder is pressed into pellets of specific sizes and weights. In the following production step these pellets are preheated and placed into a hot moulding press to be pressed and cured to the final product. After this main forming step the products can be decorated and finished with a second and third curing operation together with a decorative foil or a different coloured material and a glazing powder. These final steps take place in the same cavity and require a slightly different process control. After the product has been finally cured, it is removed from the press deburred and polished.
The company product range consists of different products for airline in flight catering, general catering equipment, disabled crockery and promotion products. Direct labour involvement in the making of the products is very high and due to the fact that improvements to the manufacturing equipment have been reduced to necessary maintenance work, the whole project had to focus on these two major production factors. The main competitors are to be found in the Far East, where finished product prices are constantly 30% lower than the manufacturing prices of the company. Given of these facts it is obvious why the re-engineering of the production facility was urgently needed.

2. ANALYSIS OF THE MANUFACTURING PROCESS

As a result of non existent or inaccurate data the project started with a detailed study of the current manufacturing process. This study forms with its detailed analysis the basis of the project. After starting the study it was found very quickly that a costing system with the ability to cost in detail the smallest manufacturing steps had to be developed. With this system it is possible to define the real manufacturing price for a product, a production run or a new manufacturing scenario. The system itself started very simple to define the basic manufacturing steps and ended in a programmed excel spreadsheet system with the ability to store data in a database for future reference and advanced cost analysis. It is now possible to compare the impact of different manufacturing scenarios or set-ups in terms of final production costs, cost variation and cost distribution among different production departments, different machines, operators or materials. With this system it was now possible to benchmark changes and upgrades and to assist management with realistic data on costs of future production runs.
The new system itself divides the thermoset manufacturing department into 4 subdivisions: Pelletising, Moulding, Finishing and Final Inspection and Packing. The analysis itself is split in 4 sections: Data-Collection, Data-Analysis, Detailed-Results and Result-Presentation as a final summary. With the use of Microsoft Excel and some programming routines the program is easy to use. Data entry in the first section comprises of two parts. Part one is for each product and collects data directly from the shop floor or from an archive into a systematic grid. Part two contains fixed or uniform data about the company itself like standard manufacturing materials, overhead costs/factors, maintenance costs etc. This part hardly changes at all, but gives the possibility to experiment with the impact of fundamental changes like: "what happens if ...". After the first part of the Data-Collection is finished and entered, Excel performs the analysis automatically.

3. RESULTS OF THE ANALYSIS

The collection of vital data for the analysis proved to be the most challenging task of the entire project. Due to the nature of collecting data and asking questions, the first improvements the operations procedures could be made. At the same time it was found that several production steps had never been analysed and documented. Finally, with the requirement of accuracy and completeness, a first comprehensive picture of the department could be drawn.

One of the areas for improvement was found, because of contradicting production data. It was not always possible to find production processes with identical processing times in the moulding shop. For instance the individual moulding time for a simple white plate had processing times from 43seconds to 63 seconds. A careful analysis revealed, that these differences were caused by usage of different moulding presses and different settings of the machine controllers. An efficient manufacturing shop has the possibility to slow processes down if required, but in the instances recorded this was not the case. The final reason was the difficulty to set the antique machine-controllers. Only one of three supervisors had the ability to tune the process control to an optimum, but even he could not guarantee identical settings from production run to production run. As a result of this it was agreed that these machine controllers had to be replaced / modernised.

The analysis itself revealed that production is dominated cost-wise by two major factors. According to the type of production the distribution of the cost factors vary. For example a product, which remains for three cycles in the press creates a significantly higher labour cost factor than a single cycle operation. But vice versa the single cycle product is dominated by the material costs, due to a different cost distribution. This is shown in figures 1 and 2.

Dinner Plate white B1000, single cycle operation

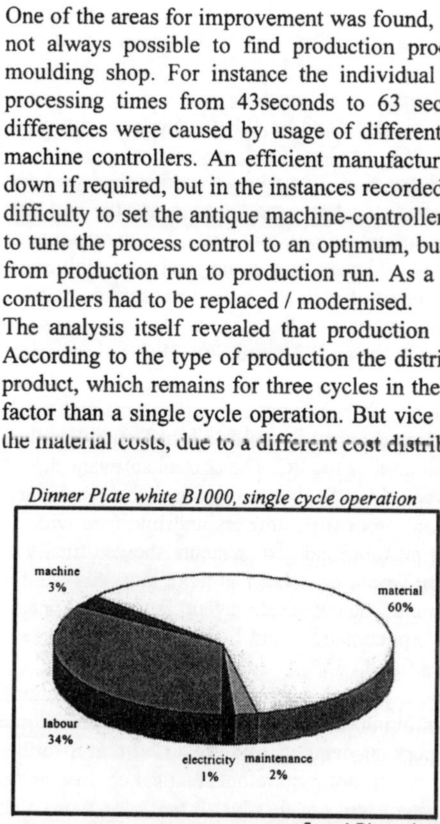

figure 1:Dinner plate

Swiss Air small plate, three cycle operation

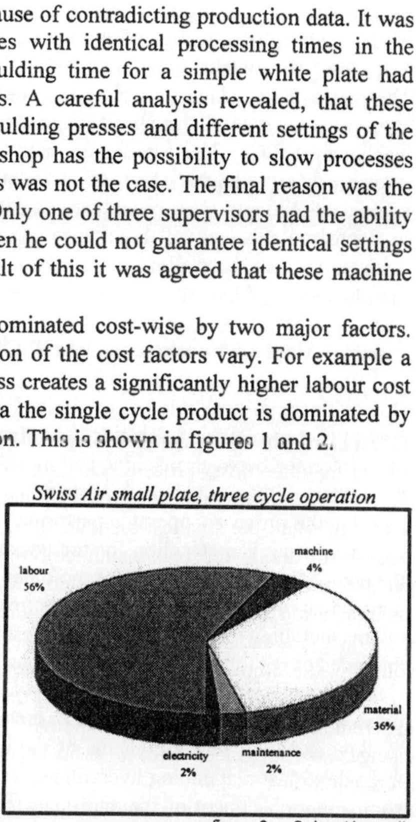

figure 2: Swiss Air small plate

This pattern of cost distribution was found for various different products. The content of the material costs varied from product to product but remained always one of the highest factors even in a far more detailed breakdown of the product costs. Due to the nature of the project the search for more competitive raw material prices was left to the material purchasing department. Hence, the reduction of labour time or the maximisation of labour time utilisation is the major target of this project. In order to get the labour time utilisation to a more acceptable level, the upgrade of the machine controllers was vital and had to be done urgently.

4. MACHINE CONTROLLER

A thermoset moulding press controller is a very simple device compared to a thermoplastic moulding one. The functions controlled are: temperature of the mould, oil pressure, oil temperature, top-ram, side rams, bottom-ejector and necessary guarding to protect the operator. The system found on the machines are a combination of several independent controllers with the major disadvantage that relevant process parameters are scattered over the machine. Changes of one parameter which influence other parameters are not linked and the performance of the machine is a compromise of possible tolerances within the system. A modern system is in charge of all parameters and can therefore operate the machinery always at its best performance.

The new system which was planned should incorporate all machine parameters, be able to store previous manufacturing set-ups, be easy to program and be able to interface with other machines or PCs. The interfacing with other PCs was required, because of the aim to implement an expert-system to give supervisors and operators a comprehensive and competent guide during a production run. It was found that an expert-system was the best way to overcome the lack of experience of the operators and supervisors. A combination of modern machine controller with an incorporated or closely linked analysing and optimising facility could stabilise the quality- and quantity-output of each machine. The requirements for such a system showed, that an upgrade of the old cam-controlled machine controller was not possible. Alternatives were a modern PLC-controller or an PC-based Soft-PLC with a variety of different input and output cards.

Due to financial constraints a minimised package with minor press modifications and a self made PC-controller was selected. The final system should be a self designed relay-board as ideal interface between machine and input/output card of the PC. The control software should be written with a standard programming package with facilities to control the machinery, monitor the process / operator performance, store process parameters and interface with an expert system. The selection for the package of custom made components showed finally at the end of the whole project, that no other system would have been as flexible as this one for the amount of money spent. During the realisation of the controller several faulty components of the moulding press were detected, electrical, pneumatical and hydraulical systems were upgraded or repaired and relevant interface relays fitted.

Once the system was implemented, it was found that a major problem of the final moulding press was the inertia of the system. Fast electrical impulses of the PC could not be realised due to slow and long reaction-chains of the machinery. A detailed study revealed that a further upgrade to fast responding hydraulical systems, which not require pneumatical control, and a complete replacement of the entire electrical system was not feasible. It was also found that fast reacting systems are not required for systems with these relatively long (out of the perspective of a PC) processing times like in a thermoset moulding process. In actual facts the PC was far too sensitive to the machine output signals. These fact were finally incorporated into the controlling software and from that point on the system was functioning very well.

The realisation of the new controller was compared to price quotes of £20000 for a less flexible system the best option. The hardware components including press modifications, custom made PCB, PC and I/O-card were about £3000. Software was developed in house and together with the labour costs for hardware installation and testing a estimated £4500 had to be spent. A total amount of £7500 for the first fully automated system customised to the requirements of that shop floor with the option to recover most of the costs with upgrades of the remaining 15 presses was a cheap deal.

4. LABOUR TIME UTILISATION

As shown in the cost analysis the most important cost factor is labour costs. Before any reduction plan can be detailed a precise cost splitting has to be done. It was found that the moulding department with up to 70% of the total labour time is the biggest single sub-division. At that point an in-depth study of the work in the moulding shop started. Several products, operators and machines were timed, videoed and analysed. Finally, it was revealed that up to 54% of the 70% total departmental labour time was wasted due to the nature of the moulding process. In a three cycle process the operator is waiting up to 54% of the process time (54% consists machine cycle 1, machine cycle 2 and the machine glazing cycle). This is shown in figures 3 and 4.

Labour time distribution within the department

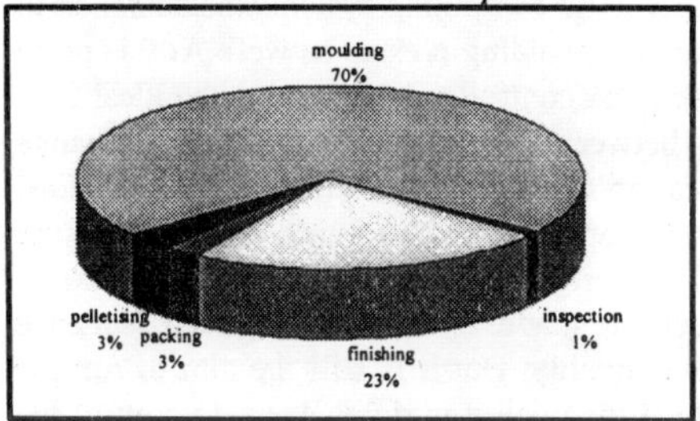

figure 3: Time distribution, department

Work activities during the moulding process

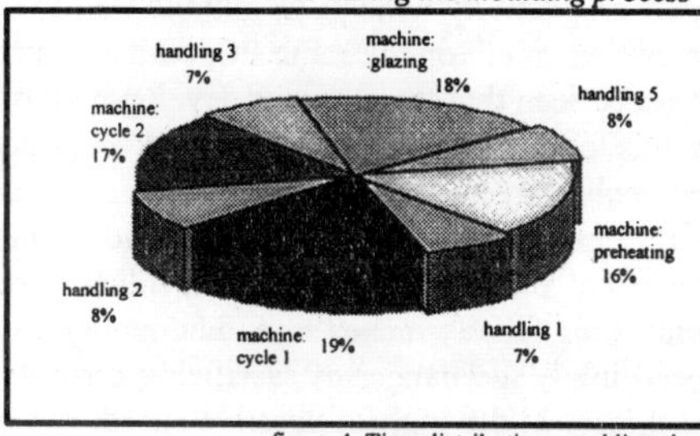

figure 4: Time distribution, moulding shop

Therefore, it was the aim to optimise time utilisation. Possibilities to optimise the labour performance was analysed and several options occurred. One option was to reduce the time between the waiting sessions, or in other words: make the operator work faster. The total saving could be 21% of operator time in case of total automation. A comparison between 54% waiting time and 21% necessary handling time shows that this option was not feasible. Then, the changing of process parameters showed virtually no impact on the required waiting times (without a loss of product quality). Hence, the process dictates fixed curing times of the product and a solution had to be found somewhere else.

The solution was found by looking at the manufacturing shop itself. It was very common to see operators who were working on the presses were moving boxes, organising new raw material, new foils or glaze powder, stacking finished products or chatting to other colleagues. These men had spare time at the same time. The conclusion was to co-ordinate these waiting cycles in a way that while one machine was working a second one could be prepared for the next cycle and once this machine was started, the other machine had to be prepared for the next run. In other words, one operator should control two machines.

First tests by the authors showed the possibility of the 2-machine-1-man-project, but important constraints occurred. Mechanical assistance to the handling of the press like automatic gates or simplified machine control were needed as well as shorter walking distances between the presses, less finishing work to be done on the machine (no spare time left) and a superior controlling mechanism to keep both machines in a fixed timing sequence between themselves. This last point proved to be the key factor to operate both machines in sequence. The reason can be found by looking at a typical production scenario: machine one **(A)** runs a single cycle process with an average processing time of 65seconds. Machine two **(B)** runs a complicated process with three cycles at a total time of 250seconds. This scenario

means that the operator himself has to keep each individual process in mind and to follow the operation step by step. But to make things difficult he can not just remember a fixed sequence, because press one will overtake press two. Within one complete operation to produce one **B**-product you are able to make 3.8 **A**-products. This means only after every fifth complete sequence of machine number two (B) the production steps will be repeated in the same order. Additionally, a single cycle sequence like on machine **A** contains up to 12 individual steps. A triple cycle like on machine **B** easily requires 40 steps. Therefore, one full sequence will include in our scenario 416 consecutive steps to be performed on time and in the right order. An added difficulty is, that many of these steps are similar or identical. Therefore, a new additional system was required to guide the operator. Ideally this guide should be in direct connection with the machine and the operator so that mistakes are recorded and corrective action takes place with the following steps. As a solution a guiding system with different lights and activity-done-push-buttons was developed. The computing system was designed to be implemented on the same PC, which controls both moulding presses as well. At this point it can be seen that the previous development of the press controller was forming the ideal base to this operator guiding project. As interfaces between the three systems data exchange protocols were created and defined as communication channels between the three programs. Additionally, the system is designed to monitor the operator and to guide him through the operation in respect to his individual working performance/speed. Only this flexible controlling facility makes sure, that quality products are made. Systems with a fixed sequence speed imply the danger of sacrificing quantity over quality. But it is still the aim to run the machinery at the highest possible speed, but only if the quality is right. Areas to control the speed of the sequence can be found among many fixed parameters of manufacturing process. These areas were created as flexible buffers in the system to buffer certain difficult steps of the process.
As soon as the development of the hardware and the basic layout of machine and software were finished the company went into receivership and the project came to a standstill. But the with cost analysis system on hand it can be proven, that an average cost saving of 25-35% can be achieved, if the system were fully developed, tested and maintained. Such a developed system would finally look like that shown in figure 5.

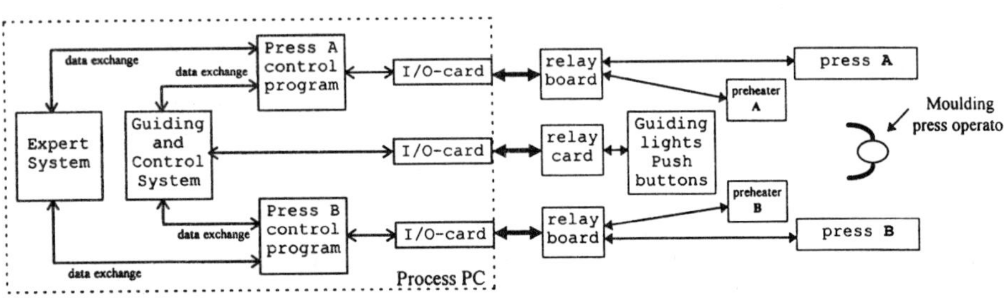

Figure 5: Thermoset Moulding Control-System

5. CONCLUSION

The developments of specific and for the operation modified control system can transform an antiquated manufacturing department into a competitive modern production facility. As shown in the report above cost savings between 10% and 35% can be obtained for a reasonable price.

Production and Inventory Control

Implementing MRPII in bespoke manufacture – a case study

R ASTLE
Pashley Cycles, Stratford-upon-Avon, UK

1 ABSTRACT

MRPII systems have generally been applied to medium and large–scale batch manufacturing operations. Reductions in the cost of computer hardware and software systems have resulted in increasingly viable opportunities for smaller companies to apply MRPII.

This has provided W.R Pashley Ltd, a medium size manufacturer of bicycles with the opportunity to develop new manufacturing planning and control systems for its small batch production facilities. This system is controlled with the Alliance\MFG™ MRPII software, which is particularly suited for use in small/medium enterprise (SME) companies. This paper examines the implementation procedure within SME companies and the problems experienced at W.R Pashley Ltd.

2. W.R PASHLEY LTD OVERVIEW

W.R. Pashley Ltd is a Midlands based manufacturing company that until recently had specialised in the design, manufacture and retail of industrial work bikes, an example of this type of product being the familiar Royal Mail post office bicycle.

Three years ago the company began a programme of restructuring led by a management buy out of the existing family owners. The new management team saw the company's future as a key player in the developing niche market. New and innovative products began to be designed and manufactured that capitalised on the company's 60-year heritage and success. The company pushed into new markets whilst consolidating its existing product line, enabling the company to progressively increase its annual turnover and profit levels.

The growth of the company's depth and scope of business whilst having a beneficial effect on the company's long term viability in a difficult market sector was also having a negative effect on the companies internal systems.

It was with these problems in mind that the management took the decision to develop the companies IT systems infrastructure.

2.1 Teaching Company Scheme

W.R. Pashley Ltd, realising that it had deficient human resource to tackle any major systems project began to employ skilled professionals to manage many of the changes. The company had lacked expertise in manufacturing, information systems and operations and thus looked to recruit in these areas.

The government initiative entitled the Teaching Company Scheme allowed W.R. Pashley Ltd to attract and employ a graduate to act as an instrument of change within the company. The scheme runs as a government funded joint venture between the company and an academic institution, with both parties benefiting from the inter-disciplinary contact. The employee enjoys the benefit of academic support through an academic institution and financial support for training, equipment and travel.

3. THE MRP IMPLEMENTATION PROCESS

3.1 Systems development group

In order to set out the scope and detail of the implementation project a management systems development group was instigated. The group would act as an effective solutions development body and as a means of communication between departments. Group sessions focused upon the operational problems experienced by the company and the systems that would have to be developed.

Operational system improvements that were sought at this point were:

- Reduction in stock levels.
- Reduction in stock-outs.
- Improved stock visibility.
- Improved cost analysis function.
- Lead time reduction on finished product.
- Improved sales related information.

It was important that the solutions developed would be viable within the company's existing structure, maintaining a realistic view of the company's business capabilities and the restrictions on time, manpower and monetary resource.

3.2 Company operation and solution selection

Before any decision on the solution that would be followed the commitment to change within the company, especially at management level, was investigated in order to assess the possible barriers that the implementation programme may face. The company's present operational state was also analysed, in order to determine the scope of the project and the specific objectives.

When analysed it was apparent that the company had fallen behind in manufacturing and operational technology. An on-going lack of investment by the previous family owners in manufacturing plant, information technology and staff training now impeded the company's growth and development plans.

The number of product variants produced now numbered in the hundreds and as the volume of sales increased the size of inventory tripled. The purchasing and stock control functions had become increasingly difficult areas to manage, with excessive safety stock levels compounded by low overall stock turns and high component obsolescence.

The project team thus thoroughly investigated all the problems that the company might encounter during the implementation programme whilst looking into all the systems solutions options open to the company.

3.3 Solutions developments

The development of low cost manufacturing software solutions specifically tailored to the SME market now allows systems possibilities that would have been out of the scope of a company such as W.R Pashley Ltd only a few years previous. Sophisticated application packages are now freely available at low initial purchase cost and without the high associated capital expenditure of intensive training, IT consultancy and software/hardware maintenance.

Whilst the reduction in risk and expenditure associated with an IT systems implementation project through the use of generic Windows based applications can create numerous benefits for the small/medium size company the variety of applications and implementation solutions can also create problems.

Difficulties arise from the selection of an over complex systems choice, as small companies pursue over ambitious without the necessary skill, experience and resource to carry them out. A poor match between the company and the solution chosen will result in lengthened project time scale, increased project expenditure and low benefit returns at project completion.

3.4 MRPII benefits

To accurately assess the possible benefits of an MRPII programme the project group looked into the percentage improvements that could be expected from a model class A MRPII implementation. These figures were assessed by the group and brought into the real world with the company's own experience and ambitions.

The expected benefits proposed by the group of a MRPII implementation programme were:

Reduction in cost of inventory.	20-50%
Reduction in purchased material cost	5-10%
Reduction in scrap/ rework/ obsolescence	50-75%
Increase in productivity	10-25%
Increase in sales units	10-20%

With the proposal of these benefits the group made the decision to push ahead with an MRPII programme, with management backing the full scope of developments and progress that would have to be made.

4.0 SYSTEMS SELECTION

The Alliance Manufacturing Software companies MRPII application, Alliance MFG™ for Windows was the chosen MRPII software. The package was viewed as offering an ideal combination of functionality and systems support, at a price that was right for the company. The application comprises bill of materials, sales orders, forecasting, materials requirements planning, capacity requirements planning, purchase orders, works orders, inventory control and shop floor control in a simple, easy to use well integrated package.

The package also allows seamless integration with CAD packages within the bill of materials module and with accountancy applications for invoice passing between sales order and purchase order modules.

5.0 MRPII IMPLEMENTATION

5.1 The MRPII implementation plan

The MRPII implementation plan was a development of standard MRPII implementation practice with special consideration being made to the limits on human resource and capital expenditure. The plan addressed issues such as employee skill levels, organisational structure, stock operations and movements and sales co-ordination amongst others.

During the development of the implementation plan a test cell was set up for the fabrication of low volume, high cost product. This test cell would allow the project group to test sales, production, purchasing and stocking procedures within a controlled non-critical environment. The cell was designed to enable stock movements, manufacture times and process flow to be accurately monitored and give an accurate picture of the development of MRPII control in the mainstream manufacturing operation.

5.2 MRPII implementation difficulties

The main implementation difficulties that were anticipated were the lack of accurate, developed production planning and control information and the difficulties of migrating from the existing IT systems, both in technical and human resource terms.

Applying the MRP and CRP elements of the Alliance package would require improved production data before implementation. Whilst it would be time consuming to carry out the data collection and information processing the long term benefits to the company, such as improved stock control, production planning and forecasting justify the effort involved. In order to gain the required data accuracy the information held on the legacy information system would be audited and restructured whilst data transfer was taking place.

The MRPII implementation will be fully tested before being rolled out into mainstream working practice and brought into full scale practice only after changes have been made to operations practice.

6.0 CONCLUSION

The MRPII implementation project is now poised to enter the critical stage of data transfer from the legacy IT system and the development of operational practice in line with the implementation plan beginning to occur.

The pilot cell stage of the project will determine whether rigid rules associated with MRPII control can be accept the variable nature of the cycle market where production scheduling is not of a determinable nature.

REFERENCES

Anderson, J. C., Schroeder, R. G., Getting Results from your MRP System, Business Horizons, May-June 1984.

Burns, O. M., Turnipspeed, D., Critical Success Factors in Manufacturing Resource Planning Implementation, International Journal of Operations & Production Management, Vol 11, No 4 1991

Luber, A, D., Solving Business Problems with MRPII, Digital Press 1995.

Tincher, M. J., Sheldon, D, H., The Road to Class A Manufacturing Resource Planning, Buker 1995.

Efficient planning and control in small manufacturing firms using white board systems

S BERRY and **W MURPHY**
School of Mathematics and Computing, University of Derby, UK

Abstract
This details an investigation into the ways in which Small Manufacturing Firms can carry out their planning using a Minimal Information set.
Small manufacturing firms tend to plan either using regression model of the number of jobs in the system, or planning board models.
The benefits of these alternative approaches have been assessed through simulations of models for small manufacturing firms which had been constructed from survey data.
The investigation shows that for a typical small manufacturing firm a "White Board" system needs only the orders arrival time and the average job duration to provide an efficient and effective planning and control system.

1. Introduction

This paper demonstrates the use of Simulation Modelling to gain an insight into the production planning and control tasks encountered by small manufacturing firms. A set of 10 case studies provided information from which a model was derived of a typical production system in a small manufacturing firm. Typically a firm needs to be able to estimate the "job completion time" or queuing time in the system at the instant it receives an order in order to plan and control their production systems (Chang, 1997). This forecast informs the firm of the need for:

a) overtime or subcontract working so that the job can be delivered on time to the customer
b) or whether they should extend their delivery lead times to ensure on time delivery.

Two commonly used approaches to production planning and control in small manufacturing firms are considered; that is "the number of jobs present" and "a planning white board". It extends the investigation to the potential of "White Board" planning systems as an accurate "minimal information set" planning and control device.

The simulation analysis shows that the prediction of job completion times can be achieved through the use of a "simple" piecewise-linear model that relates "time in the system" to the status of the system at the time of the jobs arrival; that is the forecasting process is non-trivial. However, the alternative planning white-board approach can produce realistic forecasts using a simpler set of data.

2. Regression Models.

Consider a typical small firm where the production process, and therefore the model, consisting of three work stations in series operating as a flow shop.

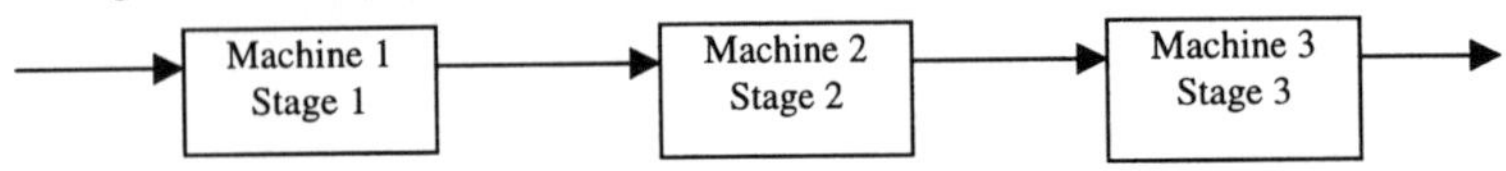

Let b_i be a[0,1] variable where $b_i = 1$ indicates that stage i is busy at a given time, B is the vector showing the system status for example B=[0,1,0] indicates that machines 1 and 3 are free whilst machine 2 is busy, and T_i the average processing time at stage i.,
For simplicity it is assumed that the average processing time at each stage is T, that is $T_i = T$ all i.

Here a good estimate for completion time must depend on the "status of the system" at the time of the job's arrival.
For example consider the simplest case when there is a single job in the system and this job is at the final stage, the system status is therefore given by B = [0,0,1].
In this case there is no delay for the new job and the completion time can be estimated by the total average processing time,

$$\text{Estimated Time} \qquad E = 3T$$

In the alternative case where the single job is at the first stage, here B = [1,0,0].
The new job will be delayed by the existing job at the first stage and the estimated time is given by

$$E = 3T + \text{Delay},$$

and the delay (D) is a function of the remaining processing time for the job at stage 1.
In the general case when there are n_j jobs at stage j it follows that the cumulative jobs to this stage C_j is given by

$$C_j = \Sigma^j n_i$$

the delay is now given by

$$D = T.\ \text{Max}\{C_1, (C_2 - 1), (C_3 - 2)\}$$

and the completion time can be estimated from

$$E = 3T + D.$$

Notice that when there are several machines in series, then D will be given by

$$T.\text{Max}((C_j - (j-1)) \text{ all } j$$

Hence the total processing time can be estimated using a "non linear model" which may be approximately linear, for example if the same machine always dominates the process then there will be a linear model available by planning around this machine. In general an estimate can be obtained from

$$E = A + B\, f(C_1, C_2, \ldots\ldots, C_n)$$

Or

$$E = A + B\, g(w_1, \ldots, w_n, a_1, \ldots\ldots, a_m) \qquad \text{A and B constants.}$$

Where a_j is the inter arrival time for the jth item, that is the time between the arrival of the $(j-1)^{th}$ and the jth items, and w_j indicates the work in progress at the jth machine. However case study research has shown that a small firm is not able, or willing, to gather, or maintain, a volume of quality information about the status of the production system. Therefore the aim is the determination of the minimal information planning system for a small manufacturing firm.

3 Simulation

In each simulation, information about the state of the workshop was gathered at the arrival of each job leading to a series of "snapshots" of the workshop relating to the following parameters:

a) Number of jobs queuing at stage one
b) Number queuing at stage one at the arrival of $job_{(n-1)}$
c) Number of jobs at stage one
d) Number of jobs in the system
e) Number of jobs queuing in the system
f) Time since the arrival of the $job_{(n-1)}$
g) Time since the arrival of $job_{(n-2)}$
h) A weighted moving average for job completion times

This approach was adopted by Chang (1997) in the analysis of a more general job shop. Chang assumed that information relating to the parameters would be available to the production planner or could be calculated by the planner.
The simulation calculated the job completion times and this information together with the "snapshot" data was analysed to determine the minimal quantity of information needed to be able to produce an accurate estimate a job's completion time; this estimate to be available at that instant when the job enters the workshop. In this study regression techniques were applied to the data generated by each simulation in order to determine the minimal information set that would provide an accurate estimate for job completion times. Models to enable the estimation of job completion times were then constructed to investigate the efficiency of the process.

3.1 Effectiveness of the parameters

Two variations were simulated to represent the extreme locations for the dominant machine

(i) $T_1 > T_2$ and $T_1 > T_3$ machine one dominant [DNN]
(ii) $T_3 > T_1$ and $T_3 > T_2$ machine three dominant [NND]

The effectiveness of the eight parameters were assessed using the two variants. The results below indicate the relevance of the two parameters most often used by the firms: namely, "number of jobs at stage one" (the minimal information), and "the number of jobs in the system (in effect the order book).

Variation 1 the [DNN] model.
Scatter plots showed that in this case the "number of jobs at stage one" provides an efficient forecast for job completion time, see graph 1.
The scatter plot of "time in system" against "number in system" indicated the need for a more complex piecewise linear model, see graph 2.
Notice that

- there is a large spread at each data point.
- The greatest variation occurs when there are 3 jobs in the system, the point of greatest uncertainty.

Therefore the parameter "number of jobs in the system" does not, by itself, produce as good an estimate for job completion times as the parameter "number of jobs at stage one" where machine 1 was dominant.

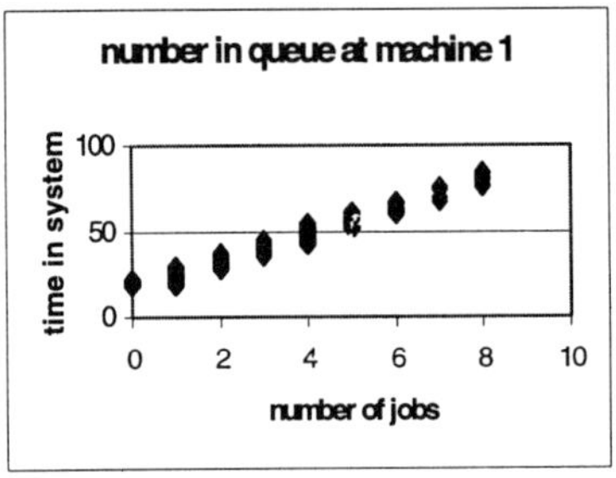

Graph 1

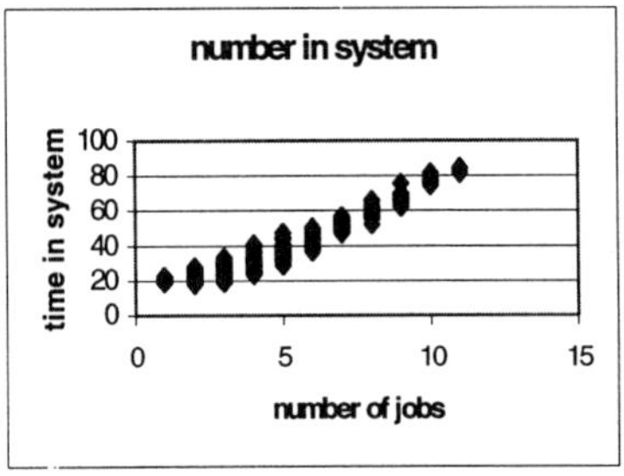

Graph 2

Variation 2 the NND model.
Scatter plots showed that in this case the number at stage one does not provides an acceptable forecast for job completion time, see graph3.
The scatter plot of "time in system" against "number in system" indicated the need for a more complex , piecewise linear model, see graph 4.
However the regression analysis of the full data set showed that the combination of parameters "number of jobs in the system" (x_1) and the "number of jobs at stage one" (x_2), could provide a satisfactory model to enabling the estimation of the job completion time (y) for both variants, [DNN] and [NND].

The model had the form $y = a_1x_1 + a_2x_2 + c$ i.e. a more complex linear relationship, and hence requiring more and better quality data to enable effective control.

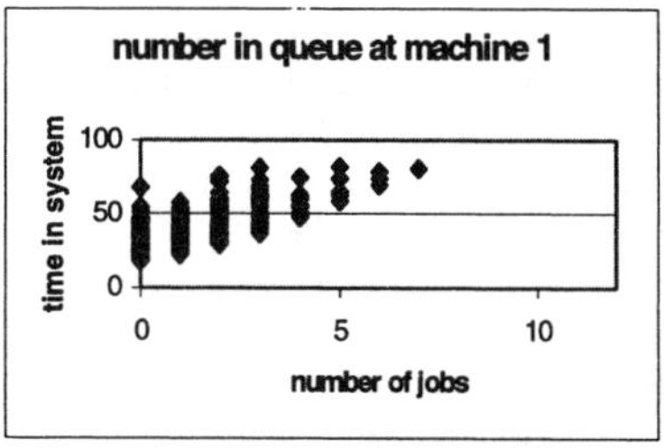

Graph 3

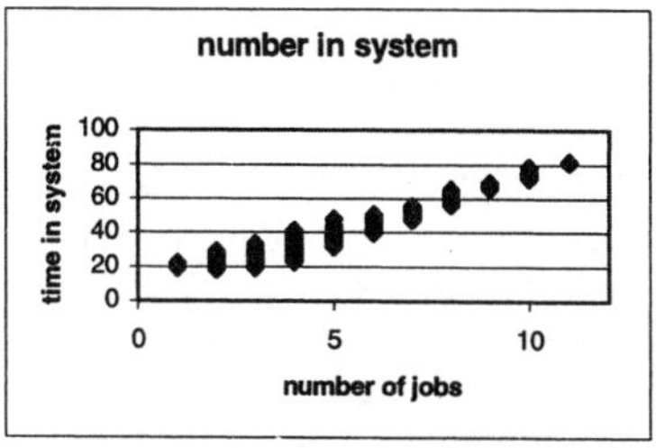

Graph 4

The following diagram shows the results from the use of a piecewise linear model. The results from these two variations showed that

- not all small manufacturing firms will experience similar planning and control problems,
- the location of the dominant machine is of great significance in the analysis of planning and control problems,
- not all configurations will have the same information needs.

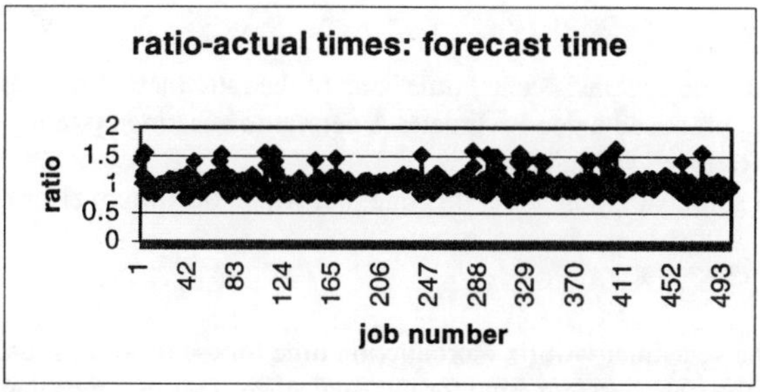

Graph 5

4 "White Board" Planning Systems

The term "White Board" is used to describe a planning and control system commonly used in small manufacturing firms. The White Board being used to facilitate planning and then control the production processes.

To illustrate this system consider a *double glazing window* manufacturer, at that time when an order has been received the window *fitting* date is fixed onto the first available, and suitable, slot on a wall scheduler and production processes then allocated so that this date will be achieved.

To assess the worth of this method for small manufacturing firms each of the variants, [DNN] and [NND], was simulated and for each job the estimated job duration compared with the actual job duration. The simulations used the simplest available data set, where the times displayed on the white board were not updated following the completion of a job. This data was used so that the simulation could imitate the lack of information gathering resources in a small firm and hence evaluate this "minimal data collection method" for use within small firms.

Variant 1 [DNN]

At the time of arrival the scheduler will fix a production time for the first stage using the "white board", and not considering the real status of the first stage (the true available time), the estimate for the job completion time is now given by

"White board" ending time + Σ average times on the following machines

the simulation indicated that this was an efficient way of estimating the job completion time.

The graphs of "forecast time" against "actual time" and of the ratio (actual/forecast) showed that this method produced a good estimate. A perfect forecasting system would have produced Ratio = 1, all jobs.

The actual plot showed some variation about this line, but with a maximum error of approximately 20%.

Variant 2 [NND]

At the time of arrival the scheduler will fix a production time for the final stage using the "white board" here the board assigns time for the final stage. This assumes that because the final stage is dominant the job will have been processed on the preceding stages by this time. The job completion time is now given by

"White board" starting time + average time on the last stage

Again the simple time estimates, not updated, are used to imitate a small manufacturing firm.

The following graph indicates the results from the simulation of a non-updated planning board,

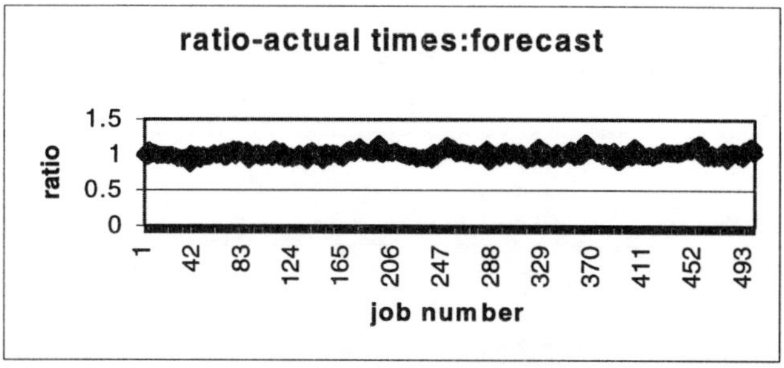

Graph 6

5. Conclusions

A non updated planning board system can produce a more efficient and effective planning and control system for a small manufacturing firm than the use of the number of jobs in the system, a method adopted by many small firms.

This method works well because data shown on the "White Board" implies the "collection" of quality data about the number of jobs in the system and the position of the jobs in the system. A comprehensive but minimal information set of data can be collected in a very simple way.

The work has been extended to show how this approach can be used in more complex environments.

References

Chang, F.R.C, 1997, Heuristics for dynamic job shop scheduling with real time updated queuing time estimates, Int. J. Prod. Res., 3 5(3), 651-665.

Kaplan, A.C. and Unal, A.T 1993, A probabilistic cost-based due date assignment model for job shops, Int. J. Prod. Res., 3 1(12), 2817-2834.

Merit D, 1991, Scheduling is in the cards, American Printer vol 206 no4, pp60.

An investigation of the performance of MRP planning in an uncertain manufacturing environment

M D BYRNE and **H MAPFAIRA**
Department of Manufacturing Engineering, University of Nottingham, UK

The logic of MRP planning systems requires that lead times are predetermined and are assumed fixed. This involves an implicit assumption that there are no capacity constraints and any amount of production is presumed possible on each facility in each time period.

The simulation study presented in this paper addresses the more realistic situation of variable lead times and constrained capacity. The results confirmed that under these circumstances system performance deteriorates as capacity loading is increased.

1. INTRODUCTION

MRP planning is the most common procedure for order release planning in a manufacturing environment. The MRP system was originated by Orlicky (1) in the early 1960s in the USA as a computerised approach for the planning of materials acquisition and production. However, its manual practice dates back to the 1940s. The system is built around the principle that large inventories are necessary to support production, the purpose of the inventory being to smooth out the adverse effects of lead time variability. It also assumes that resource capacity is infinite and that lead time is fixed and can be predetermined. Real manufacturing environments however are constantly changing, and nowadays lean production practices are vital for competitiveness. This requires accurate lead times in order to reduce inventories and increase responsiveness. A number of experimental studies on lead time uncertainties and lead time determination (due date assignment) have been undertaken by various researchers. Many of these studies have attempted to assess the degree of instability or nervousness, and the associated cost of the MRP planning system under lead time uncertainties. The aim has been to understand the planning environment better with a view to predicting system performance.

In this paper it is assumed that lead times are dynamic and variable. The objective is to test the performance, via simulation, of MRP planning in a manufacturing environment subject to uncertainty.

2. PAST RESEARCH ON UNCERTAINTY IN MRP SYSTEMS

Much of the research into MPR uncertainty and due date assignment has shown that shop loading has an effect on shop performance. Paradoxically, many lot sizing procedures in MRP systems are insensitive to capacity availability. As pointed out by Bellington et al (2) capacity planning modules associated with MRP systems do not deal with how to plan capacity, but rather how to get the information necessary to plan capacity; it remains the job of the scheduler to "fix" the MPS once capacity unavailability is detected.

Problems created by uncertainties in MRP systems have been well recognised. Murthy and Ma (3) categorised uncertainty in MRP systems into two groups; environmental and system uncertainty. Environmental uncertainty is made up of demand and supply uncertainty. Demand uncertainty consists of uncertainty relating to customer orders and forecasting errors. Supply uncertainty concerns the unreliability on suppliers in meeting supply schedules in their right quantity and right time. On the other hand system uncertainty relates to the uncertainty within the production system itself. It includes variation in product quality, variation in product structure, variation in production lead times, etc. Various researchers and practitioners in dealing with these problems have suggested a number of approaches. Plossl (1) suggested the use of safety stocks, safety lead times, and lot sizing for buffer protection. His recommendation is based on the original MRP logic, built around the principle that large inventories are necessary to support production. Whybark and Williams (4) also suggested that safety lead times should be used to deal with lead time uncertainty in single level systems. In contrast Grasso and Taylor (5) concluded in their study of uncertainty in purchasing lead times in multi-level systems that safety stocks should be used instead of safety lead times to buffer lead time uncertainties. Murthy and Ma (3) as well as using safety stocks and lead times, suggested the use of hedging and over-planning, and yield factor for buffer protection.

With the current customer-driven market, safety stock is a luxury which can rarely be afforded. Methods have therefore to be found of accurately determining lead times. Harl and Ritzman (6) developed a heuristic algorithm which alters (decrease or increases) lot sizes to match available capacity. As pointed out by Seidmann and Smith (7) most of the problems associated with order release planning are due to random assignment of due dates to jobs. They suggested an analytical formulation which considers shop dynamics and assures the optimality of the due date assignment problem. In the same vein Hoyt (8) suggested a straightforward method of determining accurate lead times. He suggested that lead times should be determined dynamically based on the most recent production period's queue divided by its average output. The method however suffers from a common criticism of MRP planning systems: the use of historical data for lead time determination. A recent development in order release planning is by Watson et al (9). They developed a discrete simulation based model for order release planning which generates order release plans via a backward explosion logic. The model calculates lead times based on anticipated queues in the system each time the model is executed. They concluded that the simulation model performs better than the standard MRP system due to its ability to accurately capture real-world capacity constraints and operation rules. However with the excessive execution time of simulation models its use as a day to day planning tool is limited.

3. PROBLEM DESCRIPTION

This section describes the illustrative example used in the investigation.

3.1 Product Structure

The product structure used is shown in Figure 1. One each of components D and E are required to make to make one of subassembly B, and similarly one each of B and C are required to make one of A.

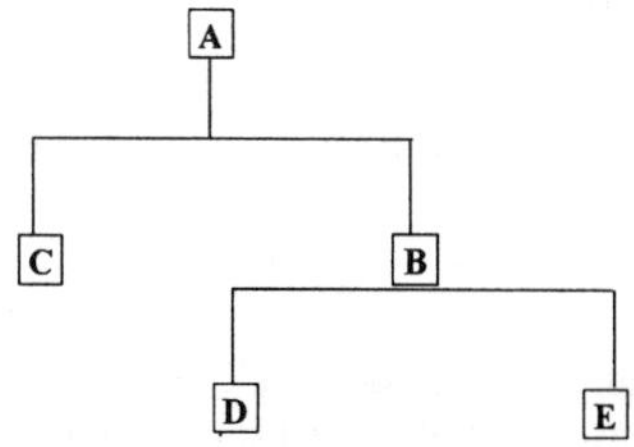

Figure 1 Product Structure

3.2 Routing Information

The routing information for the product is shown in Table 1. The routing sheet specifies the route a job is to take, the particular operations involved, the standard set-up time, processing time per unit, and the transport time. Each work centre consists of one or more operations and the times shown are totals for each work centre. After completion of operations at one work centre the items are send to the next work centre according to the sequence (Op. No.).

	Work Centre 1				Work Centre 2				Work Centre 3				Work Centre 4			
Component	Op. No.	PT	ST	TT	Op. No.	PT	ST	TT	Op. No.	PT	ST	TT	Op. No.	PT	ST	TT
A		/	/	/		/	/	/	1	2	2.4	2.8	2	2.2	2.5	2.7
B		/	/	/	1	1.8	2.1	2.3	2	2	1.8	2.5	3	1.8	2.1	2.4
C	1	2.2	2.6	2.8	2	2	2.2	2.8		/	/	/		/	/	/
D	2	1.8	2.4	2.6	3	2	2.5	2.6	1	2	2.2	2.4		/	/	/
E	1	2.4	3.2	2.2	2	2.0	2	2.4		/	/	/		/	/	/

Op. No. - Operation Number (processing sequence)
PT - Processing Time
ST - Set-up Time
TT - Transport Time

Table 1 Routing information for product A and its components

3.3 Operational and Planned lead times

From the routing information, the operational lead times can easily be determined and this is shown in Table 2. The operational lead times represent the work centre loading of each component at the various work centres it visits. Addition of the operation lead times for each component gives the planned component lead time.

	Operation Lead Times - Hours				Planned Lead Times	
Component	WC1	WC2	WC3	WC4	Hours	Days
A			7.2	7.4	14.6	2
B		6.2	6.3	7.6	20.1	3
C	7.6	6.7	6.4		20.7	3
D	6.8	7.1			13.9	2
E	7.8	6.6	6.6		21	3

Table 2. Operation lead times and Planned lead time

3.4 Order release plan

Using the obtained planned lead times an MRP calculation is performed to determine order release plans. It provides the basis on which the experimental production system is modelled. The planning horizon for the Master Production Schedule (MPS) was set such that a substantial number of products are scheduled. A planning horizon of 30 days was adopted with time buckets of one day. The MPS was designed to be repetitive and continuous. To enable realistic mimicking of an actual system the MPS was designed with a warm-up period of 20 days. The warm-up period insures that the system had reached steady state before results were collected. The order release plan for these experiments consists of orders to meet the MPS requirements for the current planning horizon period, the part orders to meet the next planning period (rolling order release), and orders to satisfy the warm-up period requirements.

3.5 Model Development

A simulation model of the system was developed using the ARENA simulation software package. To represent uncertainty in the system the lead times were taken to be stochastic. A normal distribution was chosen to be suitable since the lead times as used are the sum of various work centre process times. The arrival pattern of orders was considered deterministic and was scheduled in accordance with the order release plan.

3.6 Simulation Logic

The system consists of five stations each with its own queue. Orders arrive according to the order release plan and join their respective queues. When the station is available the order seizes a machine and is delayed for the duration of its lead time. When processing is finished the component is released to the following queue or out of the system for the finished product. The time the component spends in the system is recorded.

4. EXPERIMENTAL METHODOLOGY

Three simulation experiments were designed for this study. The aim was to study the performance of MRP system in an uncertain environment, and with varying capacity loading. Experiments were therefore designed with different levels of variability and capacity loading.

The three levels of uncertainty used were designated: no variability (deterministic lead times), low variability, and high variability.

Two sets of experiments were performed with different levels of capacity loading. The first was a benchmark allowing different amounts of excess capacity at all work stations. In the second experiment loading was increased gradually (in steps) by up to 50% to create capacity bottlenecks at some work stations.

A full factorial experimental design was executed, combining each level of variability with each level of capacity loading.

4.1 Performance Measures

The following performance measures were used to assess the performance of the MRP planning system in each set of conditions:

- Average machine utilisation
- Average queue length
- Average queue delay
- Average component waiting time (waiting for other components for assembly)
- Average component flow time
- Average order lateness
- Average number of late orders

5 RESULTS

The major findings of the simulation are summarised below. Full details may be found in (10).

For all the performance measures used the system's performance deteriorated when capacity loading was increased. This deterioration in performance was however only found to be significant where resource capacity was constrained. For components using only resources where capacity was abundant, the deterioration was for all practical purposes negligible. This appears to be the basis on which earlier conclusions that lead times are fixed and can be predetermined were made. However when capacity is limited, as is the more usual case, lead times were found to increase substantially with increased capacity loading. Such increases were mainly due to increases in queue and waiting times as capacity loading was increased.

The investigation also confirmed the following factors about MRP planning systems:

1. The lead time performance of components depends on their position in the BOM structure. For "simple" components (i.e. those manufactured directly from raw materials), performance is only influenced by the processing stations responsible for their processing. For subassembly and assembly components, as well as being affected by their own processing stations, performance is also influenced by the performance of their feeder components. This phenomenon was recognised as the effect of "dependent events" as discussed in the OPT/TOC literature.
2. Queue time accounts for the greatest share of lead times for "simple" components. Conversely, for subassembly and assembly components waiting time accounts for the greatest part of lead time.
3. The performance of batch production systems is seriously affected when the system is overloaded. Even under optimum batch sizes, a certain level of WIP always exists. Under overload the system deteriorates, producing less than it would under normal loading.
4. The degree of uncertainty was found to have an influence on the performance of the system. The system with low variability was found to outperform both the deterministic and high

variability systems. Although surprising at first glance, this finding is explained as being a consequence of positive knock on effects from the normal variation of lead times in this particular system. With higher variability the negative effects become dominant.

6. CONCLUSIONS

This study set out to investigate the performance of MRP in a manufacturing system subject to variability. A simulation model of the system was developed and its performance tested under different levels of variability and capacity loading. Several performance measures were used to quantify the performance of the model, and for all these measures the performance of the system deteriorated as capacity loading was increased. This deterioration was found to be serious when resources become capacity constrained. The level of uncertainty was also found to have a significant effect on the performance of the system. The system with a low lead time variability was found to out-perform both deterministic and high variability systems.

A good understanding of the causes and nature of lead time variability is particularly important to practitioners since it is the basis for quoting reliable delivery dates, a qualifying criterion for winning orders in the current manufacturing environment.

REFERENCES

1. Plossl G, 1994, "Orlicky's material requirements planning", 2nd ed., McGraw-Hill, New York
2. Bellington P J, McClain J O & Thomas L J, 1983, "Mathematical programming approaches to capacity constrained MRP systems: Review, formulation and problem reduction", *Management Science*, Vol 29. No 10, Pp 1126-1141
3. Murthy D N P & Ma L, 1991, "MRP with Uncertainty: a Review and Some Extensions", *International Journal of Production Economics*, Vol 25, Pp 51-64
4. Whybark D C & Williams J G, 1976, "Material requirements planning under uncertainty", *Decision Sciences*, Vol 7, Pp 595-606
5. Grasso E T & Taylor B W, 1984, "A simulation based experimental investigation of supply/timing uncertainty in MRP systems", *International Journal of Production Research*, Vol 22, Pp 485-492
6. Harl J E & Ritzman L P, 1985, "A heuristic algorithm for capacity sensitive requirements planning", *Journal of Operations Management*, Vol 5, No 3, Pp 309 - 326
7. Seidman A & Smith M L, 1981, "Due date assignment for production systems", *Management Science*, Vol 27, No 5, Pp 575-581
8. Hoyt J, 1983, "Determining dynamic lead times for manufactured parts in a job shop", in Computers in Manufacturing Execution and Control Systems, New Jersey, USA: Auerbach Publishers
9. Watson E F, Medeiros D J & Sadowski R P, 1995, "Order release planning using variable lead times based on backward simulation model", *International Journal of Production Research*, Vol 33, No10, Pp 2867-2888
10. Mapfaira H, 1997, "An investigation into the performance of MRP planning systems under varying capacity loading in an uncertain manufacturing environment", Unpublished MSc dissertation, University of Nottingham

Improving competitiveness in contract electronics manufacturing through better shop floor materials management

R HEATON
Celestica Limited, Stoke-on-Trent, UK (also at Loughborough University, UK)
E KYRANTONIS
Celestica Limited, Stoke-on-Trent, UK
P G LEANEY
Department of Manufacturing Engineering, Loughborough University, UK

The contract electronics manufacturing (CEM) market has a very high growth rate. Competition is fierce and is based on reduced cost, improved responsiveness and added value. In printed circuit board assembly, the majority of product cost is materials, so there is the opportunity to gain competitive advantage by improving the way in which materials are managed and reduce the overheads attached to materials.

This paper reports on efforts at the Kidsgrove plant of Celestica Ltd., a CEM, to improve the in-plant materials management systems. The strategy is to improve the physical flow of materials and manage its information in real-time to provide dynamic and efficient operational control. The approach includes process flowcharting of original and proposed materials processes, analysis of how other companies manage materials and identify best practices. More specifically, implementation of shop floor material control software, barcode technology and associated working practices. The expected tangible benefits, for the Kidsgrove plant as a whole, amount to £290,000 savings each year. Implementation of this shop floor material control system began in Q1, 1998. Its progress is reviewed, noting some valuable lessons and defining the way forward to further "roll out".

1. INTRODUCTION

Celestica, based in Toronto, Canada, provide contract electronic manufacturing (CEM) services to a number of large original equipment manufacturers (OEMs). There are 3 major players in the CEM market, of which Celestica is one. The Kidsgrove plant manufactures printed circuit board assemblies using surface mount technology. It is the main European production site in the group with 2100 employees and revenue of $575 million. In 1994, the site entered the CEM market as Design to Distribution Ltd. as a wholly owned subsidiary of ICL. It was bought by Celestica Inc. in March 1997. In 1994, the world-wide electronics market amounted to $420bn with an average annual growth rate of 7%. Though the CEM market accounted for $35bn (8.3%) of this, the projected annual growth was 22%. Some captive manufacturers are reducing or merely maintaining their capacity in the face of cost effective CEMs. The competition for this business is fierce as increasing capacity is being strategically outsourced. Customers demand flexibility and responsiveness which require reduced product introduction lead times (1) and effective shop floor operations. To be competitive, Celestica Ltd. has set-up teams as customer focused business streams, which have one or more dedicated production lines, and are implementing appropriate shop floor I.T. systems in order to meet these requirements.

The aim of this paper is to look at improving materials management (2) at the Kidsgrove plant, as part of Celestica's strategy to provide increasing shop floor flexibility and responsiveness. (3) This strategy includes the implementation of TIMMS shop floor data management software as well as its support and development in conjunction with Sony. TIMMS stands for Totally Integrated Manufacturing Management System, (4). It is a software package developed by the Sony Corporation over 10 years ago. It is aimed at the electronics industry, providing an integrated solution for managing production in real-time. It currently consists of 3 main modules; (a) the Advanced Intelligent Programming module balances a complete auto-mount production line, containing several different makes and types of machine, and maximises efficiency; (b) the Real-Time machine Monitoring module collects messages from the machinery and interprets this to provide management information; (c) the Parts Control System (PCS) enables management of materials from stores to production, using barcoding technology. The team implementing TIMMS within Celestica are very closely involved with Sony in developing this software to meet the needs of the wider electronics industry.

This paper relates to the PCS module. The objectives are to investigate the original materials processes at the Kidsgrove plant, carry out a critical analysis and address how improvements can be made by effective implementation of software and working practices combined. Then look at the progress to date and the way forward.

2. ORIGINAL MATERIALS PROCESS

Figure 1 shows a flowchart of the original situation. Suppliers deliver materials to goods in. Then it passes through inspection to main stores, where it is recorded using a card system. The main stores are remote, 650 metres from the production plant. Each business stream has strategically positioned "point of use" stores on the shop floor, containing 24 hours of material for all major boards, providing flexibility to swap between boards quickly. Material infeed workers check the raw material levels on the "point of use" stores every shift and manually raise pick lists on the MRP (materials requirement planning) system. The stores people pick these lists, recording any shortages, and deliver them to the shop floor.

Small and inexpensive components, such as resistors and capacitors, are supplied mounted on reeled tapes. Surface mount machines mount these components on to the surface of the bare boards to make up printed circuit board assemblies. Large, more expensive chips are supplied on plastic trays, called flat packs. A surface mount machine signals that a reel has run out. They unload the spent reel and fetch a replenishment reel from the "point of use" store. Then the operator does a visual "buddy" check with another operator or supervisor. They load the fresh reel on to the machine, manually recording the reel change. Replenishment of flat packs is similar, except these are topped up on the machines.

Attrition is material that has failed to go through the process correctly, eg. rejected by machines or wasted by operators. Low-value attrition is scrapped, since this can not be re-used. Quantities are manually scrapped off based on the machine spoilage print outs. High-value parts are better controlled, because these parts can be re-used and scrap is recorded manually. To improve data integrity, cyclic stock checks are carried out. The WIP (work-in-progress) tracking system provides the MRP system with usage information by transmitting the number of boards produced through each stage. If the bills of materials used by MRP, engineering systems and production are not the same then this will have an adverse effect on data integrity. Traceability is by a manual record of reel changes, containing part number and time. If wrong or faulty material is placed on to boards, this is roughly linked to built boards, using the WIP tracking system.

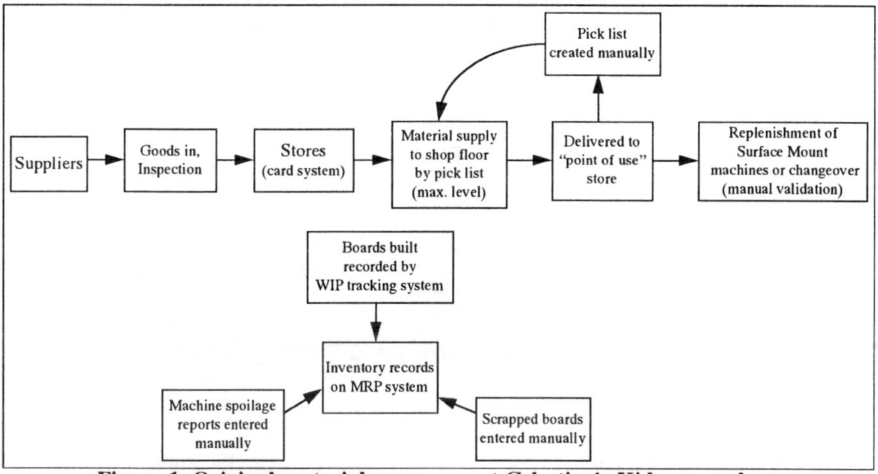

Figure 1 Original materials processes at Celestica's Kidsgrove plant

3. CRITICAL ANALYSIS

3.1 Kidsgrove plant

The data integrity of the inventory records has a big effect on costs. In the original materials process, at several stages it can be affected by human error. In addition, around 80% of low value scrap is not recorded, which will also affect the inventory data integrity. Before a stock check is done the data integrity may have gone so far out as to incur excess cost such as a rush order to correct a shortage or, in the worst case, cause a line stop.

Visual validation of parts is open to error, since there are many similar parts with similar part numbers. The effect of loading a wrong reel on to the machine can be devastating, meaning the reworking of many boards, lost production time and late delivery to the customer. Traceability is limited, since only the part number and time of change is recorded. This means that more boards than necessary need to be inspected for suspected faults. There are several actions involved in replenishing surface mount machines, which take time and since the operator is reacting, there is always the risk of down time and hence lost production.

The "point of use" stores contain enough material to sustain at least 24 hours of production, in order to avoid line stops by triggering "red alerts" for shortages. To optimise cashflow, stock must be kept to a minimum and on the move at all times. There is the opportunity to reduce stock levels by improving inventory data integrity and optimise levels on the shop floor by using scheduling information. This would reduce the risk of parts being damaged on the shop floor, e.g. through handling.

3.2 Company visits

Part of the methodology of this research is to visit similar companies and see how they manage materials, (5). A framework (Figure 2) was developed from the common basic elements of all 5 companies visited to date. Then flowcharts were constructed around this template, so as to make comparisons about how the companies manage their materials. Each company has different solutions, dependant on their situation; e.g. type of manufacturing. On the whole, they are behind Celestica in systems and technology, but are rapidly catching up.

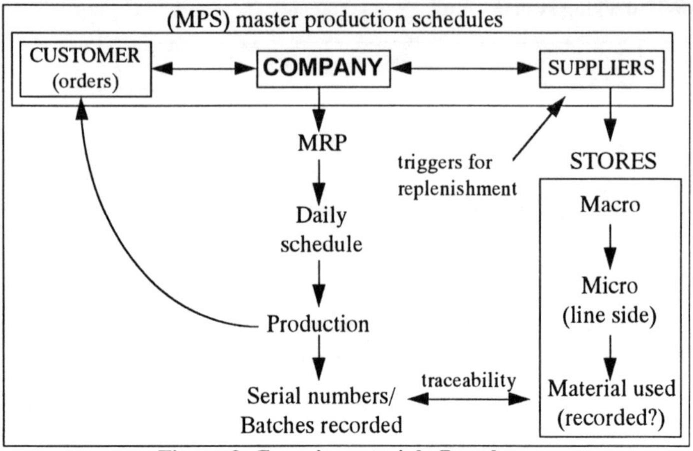

Figure 2 Generic materials flowchart

The following is a summary of one of the company's visited. In the master schedule, they include forecast error and simulate abnormal orders. Materials are ordered from suppliers by phone, fax, EDI and physical communication, to satisfy the MRP requirement. Stores are split; small parts on-site, large parts off-site. These are supplied to the line side, where portable barcode scanners record the required replenishment. Production assemblies are controlled by magnetic tags, kanban racks or dual card kanban. Hand held barcode scanners are used to track the serial numbers of board assemblies and complete systems.

4. PROPOSED IMPROVEMENTS

The Kidsgrove plant is currently implementing barcoding technology and software to improve material control from goods in through to production. Stores will be controlled using portable radio frequency barcode software. The PCS module using hand held barcode scanners, will provide a shop floor material control system based around the auto mount machines and their "point of use" stores, see Figure 3. A major benefit of implementing barcode technology is improving the data integrity of the inventory records, (6).

Using PCS, all materials transactions and movements are accountable to individuals using barcoded user I.D.'s. Barcodes on reels and flat packs will be scanned to "book in" parts to the shop floor and provide a handshake with main stores. The material database and schedule will be used to provide automated pick lists and identify shortages. Schedules will be used to predict the time of changeovers and optimise kitting.

PCS automatically records boards built, machine pick up errors, part exhausts and operators can manually assign spoilage, which encompasses all consumption. Scrap quantities will be transmitted to the MRP system. The software will be used to monitor attrition and spoilage as it happens, which will improve data integrity and reduce attrition through problem solving and process improvement.

PCS displays up-and-coming reel changes for each surface mount machine, so reel changes can be prepared for and down time minimised. When parts are loaded on to and removed from machines, including attrition, they will be validated by part number and a traceability record kept of each individual reel or pack. In addition, a fixed head barcode scanner will be used to record each board that passes through the machine. Linking this information will provide complete traceability of parts on to boards, which will reduce the effort required to tackle suspected components faults. Of added value will be the ability to link test and performance data to batches of parts from suppliers, enabling us to try different suppliers or versions of a particular component.

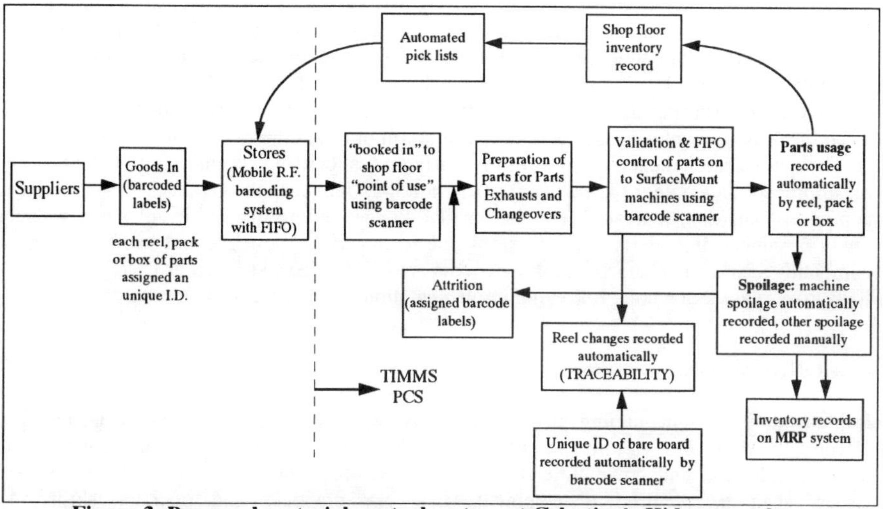

Figure 3 Proposed material control system at Celestica's Kidsgrove plant

Table 1 Summary of original versus proposed processes

Area	Original	Proposed
1. Supply to production and stock levels	Pick list manually created. No "handshake" between stores and shop floor.	Automated pick lists, optimised kitting and a "handshake" with the shop floor.
2. Inventory data integrity	Affected by human error and unrecorded data.	Control and record usage of reels or packs by barcode.
3. Attrition management	Machine spoilage reports printed out and scrap manually entered. Cyclic stock checks.	All attrition accounted for; machine reports automatically recorded and operators manually enter scrap.
4. Parts run out on machines (exhaust)	Operator reacts to exhausts.	Prediction of exhausts enables preparation of replenishments to minimise down time.
5. Validation of parts loaded on machine	Visual checking of part number and carriage location plus a "buddy check".	Automatic checking of new reel or flat pack and feeder using barcodes.
6. Traceability of parts to built boards	Manual record of material replenishments, by part number and time, linked via WIP tracking system.	Automatic storage of all material replenishments and built boards by unique barcode.

5. PROGRESS TO DATE

A full trial of the materials process using barcoding technology, as described above, started Q1, 1998. PCS is on trial on 1 of the 10 production lines. 40 users have been trained in the business stream on all shifts. The biggest challenge of implementing PCS is that it must be run 24 hours a day, 7 days a week, so training and support has to cover all shifts equally. Initial training was off-line and set-up to be a simulation of the shop floor. The most learning has been done on the shop floor, where users have had to get used to the software and integrate it with working practices. The aim is now to get the production team to a state where they are starting to see some benefits from the system and can run it on their own. Then the development team can concentrate on "roll out" to other production lines.

Though the implementation is still in its early stages, evidence is emerging of the benefits of improved data integrity, reduced stock holding, reduced obsolescence costs, reduced component attrition, increased machine availability, complete traceability of parts, reduced rework, reduced line alerts and line stops due to materials shortages and overall enhanced competitiveness. The planned tangible benefits amount to £290,000 savings each year, for the Kidsgrove plant as a whole, following implementation.

6. CONCLUSIONS

Implementation is an ongoing process and it is very important that software is managed effectively. Monitoring of improvements will be done through hard measurables, such as data integrity, attrition and down time due to machine replenishment. There will also be soft issues, such as required developments in work practices and operator skills. As implementation progresses the measured benefits will justify continued capital expenditure.

A big challenge is to get the users in to a real-time mentality. For PCS (and TIMMS as a whole) to be effective, users will have the skills and knowledge to act appropriately in the dynamic manufacturing environment on the shop floor.

The original materials process has been characterised and opportunities for improvement identified, making best use of PCS. The proposal has been discussed and flowcharted, combining working practices and software functionality. The results of the software trial to date have been encouraging and we will progress improvements in materials management by "roll out" to other production lines, development of the software and how it is utilised on the shop floor.

REFERENCES

(1) Obank, A., Leaney, P.G., and Roberts, S., "Data Management within a manufacturing organisation, Integrated Manufacturing Systems Journal, Vol. 6, No. 3, 1995, Pub. MCB, ISSN 0957-6061, pp. 37-43

(2) Baker, M., Prabhu, V., "Materials Management", McGraw-Hill, ISBN 0-07-084935-8.

(3) Bauer, A., Bowden, R., Browne, J., Duggan, J., Lyons, G., "Shop floor control systems: from design to implementation", Chapman & Hall, 1991.

(4) Anon, "The eyes and ears in production", TIMMS brochure version 1, Sony Corporation, October 1997.

(5) P.G.Leaney., "Assessing best practice in contract electronics manufacture in the supply-chain", IPPS Group report no 97/4, Loughborough University Department of Manufacturing Engineering, May 1997.

(6) Marcel, B. "A black and white case for more barcoding", Works Management, September 1996.

Finite capacity scheduling and MRPII hybrid production control systems – conditions for successful implementation

D J PETTY, M D STIRLING, and **L C TRAVIS**
TIME Research Institute, University of Salford, UK
R J BENNETT
Federal Mogul Technology Limited, Rugby, UK

Manufacturing Resource Planning (MRPII) is the most widely employed methodology for production planning and control. Many authors however, have stated that the fundamental basis of MRPII is flawed: that is to say, scheduling is based on fixed manufacturing lead times/lot sizes with infinite capacity. There has also been great interest in alternative methodologies such as Optimised Production Technology (OPT) which plans to finite capacity. As a result, there are now several software houses who offer Finite Capacity Scheduling (FCS) tools designed to operate alongside MRPII packages. These provide detailed Finite Capacity Planning while retaining the database infra-structure of MRPII. Such hybrids claim to give the benefits of FCS to established MRPII users without the need for re-implementation. This paper describes the operation of such hybrids. Based on a case study drawn from the automotive components industry, it discusses the pitfalls that might be experienced without careful consideration of planning methods in relation to Bill of Material (BOM) structures. In particular, it shows that for a hybrid to be successful, it is necessary to consider the system as a whole, rather than considering FCS as a simple addition to MRPII.

1. Introduction

1.1 Background

FCS systems have become increasingly popular over recent years. This is due in part to the fact that powerful personal computers, capable of running FCS, are now available for less than £1000. A more significant factor is that many companies have relatively mature MRPII systems in operation. Many FCS systems are now designed to augment MRPII systems, using the same database. The success of FCS is mixed however, [1] and appropriate design of such systems is crucial for success. This paper will discuss the operation of FCS/MRPII hybrid systems and the potential pitfalls that might arise if the complexity of the manufacturing system is not adequately considered.

1.2 Early Development of MRPII and FCS

The development of both MRPII and FCS is closely related to the availability of Information Technology (IT). While the basic concepts of these two approaches can be applied manually in simple cases, in practice a computer is required in realistic situations. The development of MRPII took place largely in the USA in the 1960s and 70s. Several authors [2, 3, 4] were prominent in this process. By the late 80s, MRPII was widely applied and many commercial packages were available.

The American Society of Production and Inventory Control (APICS) discussed an MRPII Crusade [5]. By contrast, the development of FCS was more muted and more focused in Europe. The reasons for this were rooted in the aftermath of World War II [6]. The USA had enormous, undamaged production capabilities. Factories in much of Europe however had been devastated. The effect of using early FCS was to optimise utilisation. MRP on the other hand, minimises inventory by only planning orders when the calculated stock balance falls below zero. FCS developed from early Operations Research (OR) techniques for sequencing and scheduling. In practice these techniques were limited when confronted with the complexity of manufacturing companies. Overall, FCS was rarely applied.

1.3 Just in Time and Optimised Production Technology

By the late 1980s, some authors were beginning to challenge MRPII orthodoxy. Schonberger [7] argued that MRPII was cumbersome and insufficiently proactive, simply accepting inherent limitations in the manufacturing systems. He promoted the Just in Time (JIT) philosophy which emphasised simple, Kanban based, control systems. Another, more fundamental, challenge to MRPII was Optimised Production Technology (OPT) [8]. The view was advanced that MRPII was inherently flawed by its infinite capacity, fixed lead time, backward scheduling technique. OPT was based on finite capacity logic with an emphasis on the management of bottlenecks. The OPT scheduling package is complex, but at a fundamental level however, the differences between forward/finite and backward/infinite capacity scheduling methods are simple:-

Figure 1: Backward Vs Forward Scheduling

- **Backward Scheduling.** The starting point for backward scheduling is the due date for the job. The start date is calculated simply by subtracting a fixed lead time from the due date. Note that using this method, it is possible for jobs to be scheduled before the current date (as in the example above).

- **Forward Scheduling.** In this case, jobs are loaded forward, with a theoretical earliest start date of the present. Typically, a fixed lead time will not be employed, and the job will be forward loaded with other jobs to finite capacity. Note that the job will not automatically be planned to finish on the due date: in general dues date are only used to determine the order in which jobs will be loaded.

2. Application of FCS in an MRPII Environment

2.1 Hybrid Control Systems

There is a growing consensus [4] that companies are best served by employing hybrid systems. That is to say, draw on the best features of the different approaches available and derive an optimal solution for the company. Certainly, MRPII has evolved beyond the planning systems of the 1970s. MRPII packages are now highly integrated databases that provide a structure for the overall management of a manufacturing company. Indeed, in most modern MRPII packages, much of the software is devoted to support of commercial activities such as Sales Order and Purchase Order Processing.

2.2 Short Interval Scheduling Systems

One area that is now being explored by that many companies is the combined use of MRPII and FCS [9]. In particular, several commercial FCS packages have been designed to work alongside an existing MRPII database. Such systems are often referred to as Short Interval Scheduling (SIS) systems. Their method of operation can be summarised in Figure 2:-

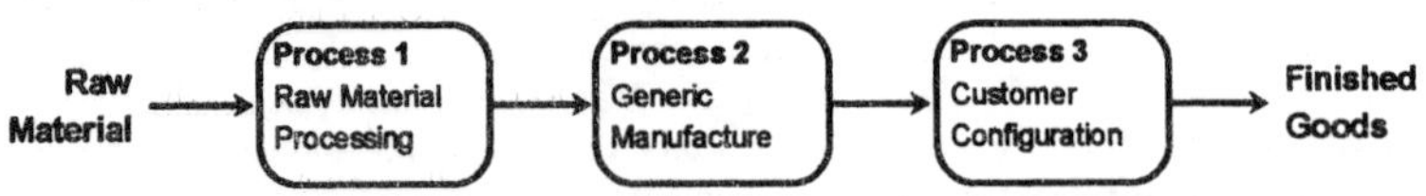

Figure 2: Short Interval Scheduling Systems

The basic idea is that a well established MRPII system is used to plan production in the usual manner. MRP is used to define the start and finish date of works orders (WOs) using the standard infinite, backwards loading mechanism. Normally, those planned orders due for release would be printed. If an SIS systems were employed however, WOs would be extracted from the MRPII via a suitable interface and imported into an SIS system. Here, the orders would be subject to finite planning and the start and finish date of the orders recalculated. Then the order would be sent back to the MRPII systems and then dealt with in the normal way. It is claimed by the suppliers of SIS packages that this approach combines the advantages of FCS with the well established commercial database of an MRPII system.

3. Practical Application of SIS

3.1 The Case Study Company

The case study company is a small manufacturing firm supplying product to automotive industry. The company had a implemented an MRPII package during the early 1990s. They believed however, that the fundamental scheduling limitations of MRPII was a problem for the business. They decided therefore to purchase an SIS package. Following a selection process, the serious work of implementation began during the summer of 1997.

3.2. Overview of Company Operations

The manufacturing process can be represented as a three stage process as shown in Figure 3 below.

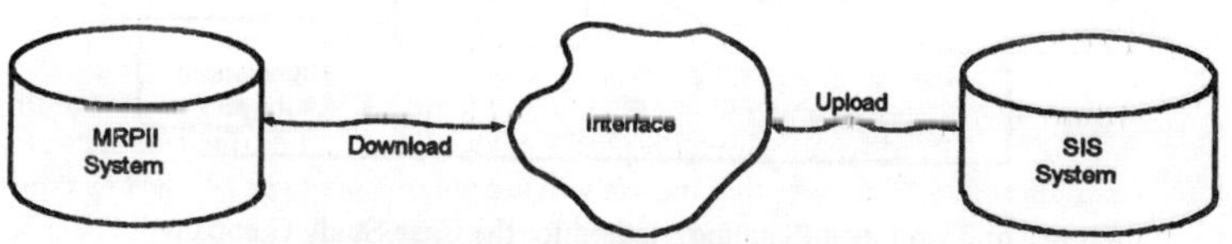

Figure 3. Production Process for the Case Study Company

Raw material undergoes initial processing to convert it into a form that can be used by later operations. It is then converted into a semi-finished product. Item varieties at both process 1 and 2 are low. Like many companies however, at the point of final configuration, there is a very large increase in variety.

3.3 Planning System Design

The use of SIS systems is often presented by a straightforward process. In practical cases however, such as in the case company, the situation is more complex. There were two key issues that needed to be addressed as part of the overall systems design:-

Bill of Material (BOM) Structure. The BOM for the product is shown right. This structure was chosen because of the high degree of commonality of the basic product within different saleable items. While this is logical from a conceptual point of view, this does create problems for the application of the SIS system. Consider a case where the company has three customer orders for three different saleable items (1, 2 and 3) with slightly different due dates. If these three items were configured from the same basic product, this would lead to a dilemma. The question is, whether three WOs for the common basic product should be created (see Figure 5a) or only one (see Figure 5b). This is not a straightforward issue: if the former is chosen this will allow two set-ups in the factory to be saved. It will lead to excess material being produced which will not be consumed until product of the last saleable item has started.

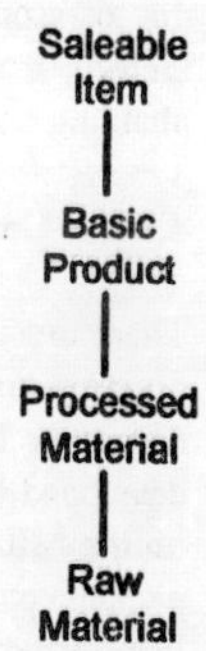

Figure 4. BOM Structure

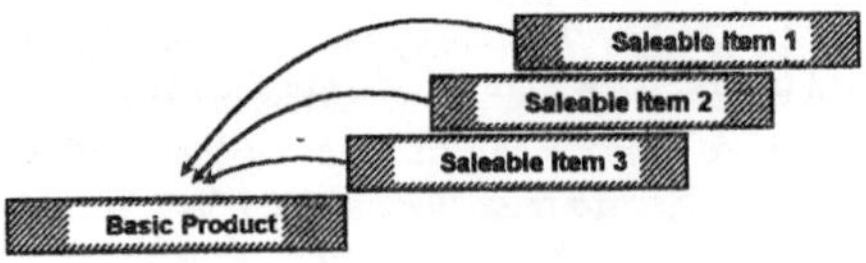

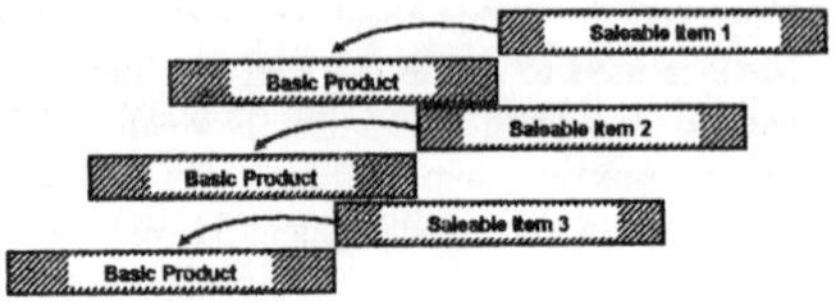

Figure 5a. Aggregating Method

Figure 5b. Lot for Lot Method

It was decided that the first method needed to be employed, to prevent excessive set-ups in the factory.

Planning Methods. As a consequence of the above, it is essential to carefully establish the sequence by which planning takes place. The key issue is that the SIS system simply schedules WOs to finite capacity. It does not recognise that in fact, WOs in this case are connected; saleable items cannot be physically produced if processed material is not available. Thus it is not possible to simply export all WOs into the SIS package and subject these to finite capacity planning. If this approach were adopted, there could be no guarantee that orders for different levels in the BOM were synchronised. Thus the following method had to be devised to ensure consistency between different levels in the BOM.

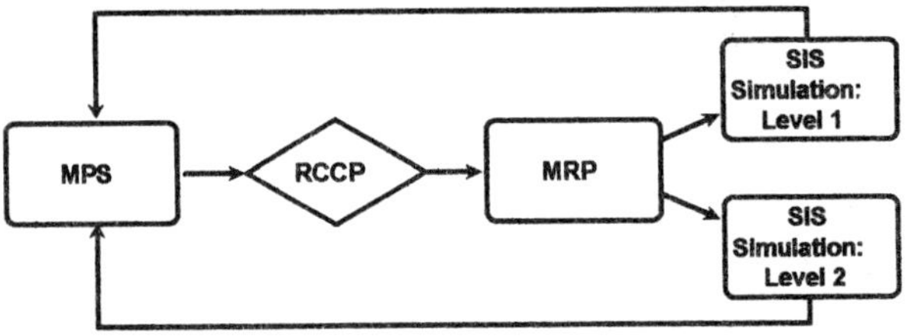

Figure 6: Two Level Planning Method for the Case Study Company

4. Discussion

4.1 Application Within the Company

The methods described above addressed the problems created by the BOM structure in the company. The planning method devised allowed the SIS to provide Finite Capacity Planning while still retaining the lot sizing and demand aggregating logic of MRP essential to maintaining efficiencies within the factory. It must be recognised however, that in order to achieve this, there is a penalty in increased planning complexity.

4.2 General Application of SIS

There are undoubtedly significant benefits from the application of a SIS in certain circumstances. In practice however, SIS operates best for relatively simple products. In more complex cases, it is necessary to ensure consistency between different levels in a BOM using techniques such as that developed by the case study company. An alternative approach to the SIS is as a technique for fine tuning MRP generated schedules. This approach is summarised in Figure 7. In this case, the MRPII package operates in a more orthodox manner. A Master Production Schedule (MPS) drives MRP after being validated by infinite capacity Rough Cut Capacity Planning (RCCP).

This approach is a simple application of MRP. The WOs generated by MRP can then be downloaded into the SIS system and subjected to Finite Capacity Planning. In this way, capacity problems can be detected far more accurately than RCCP.

In effect, this method can be characterised as a hierarchical approach: the different levels in the BOM being considered separately. Where problems were detected, these would be solved by adding or redistributing capacity locally. As a last resort, it would be necessary to re-schedule at the MPS level and re-run MRP. Note that to isolate the levels, WOs are not uploaded into the MRPII system under any circumstances. This approach can be considered as replacing the normal detailed Capacity Requirements Planning (CRP) logic of classical MRPII with FCS.

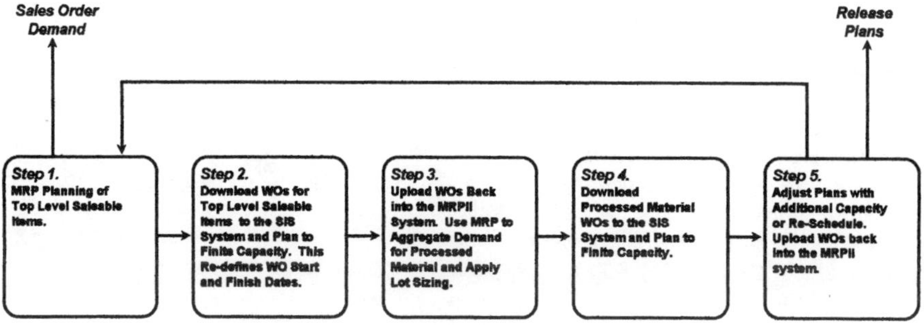

Figure 7. Simplified Application of SIS.

5. Conclusions

SIS offers the opportunity for more precise planning than could be achieved with MRPII alone. Problems can arise however, where FCS is required at more than a single level in the BOM. In such a case, if a fully integrated planning system is required, the planning systems must be carefully designed to ensure different BOM levels are synchronised. An alternative is to apply SIS in a more simplified form where it is applied as a simulation tool used to verify a manufacturing plan.

6. Acknowledgements

The authors would like to thank Federal Mogul for their permission to publish this paper. Dr Petty's post of Senior Lecturer in Manufacturing at the University of Salford is funded by Federal Mogul.

7. References

1. Burgoine J, *Beyond MRPII*, Production Engineer, Jan 1988.
2. Orlicky J A, *Material Requirements Planning*, McGraw Hill (2nd Ed), 1975.
3. Plossl G W and Wight O W, *Production and Inventory Control*, Prentice-Hall, 1967.
4. Wallace T F, *MRPII: Making it Happen - The Implementers Guide to Success with Manufacturing Resource Planning*, Oliver Wight Publications (2nd Ed), 1990.
5. Browne J, Harhan J and Shivnan J, *Production Management Systems - A CIM Perspective*, Addison-Wesley, 1989.
6. Wight A B, *Manufacturing Control Systems: American Vs European*, BPICS Control, Feb/Mar 1983.
7. Schonberger R J, *Japanese Manufacturing Techniques - Nine Hidden Lessons in Simplicity*, Free Press, 1980.
8. Fox R E, *OPT vs MRPII - Thoughtware Vs Software Part I*, Production and Inventory Control Magazine.
9. Roy R and Miekle S T, *The Role of Discrete Event Simulation Techniques in Finite Capacity Scheduling*, Journal of the Operations Research Society, No 46, 1995.

Proposed methodology for developing generic solutions to production and operations problems utilizing spreadsheets

P J VIVIAN
University of Central Lancashire, Preston, UK

This research, which is in it's infancy, is based on the premise that if the solution can be modelled on paper, then a good starting point is to mimic that solution on the spreadsheet. The current methodology is an iterative approach limited only by the users ability to recognise the generic nature of the problem, and, expertise with the particular spreadsheet. (In this case, Excel5) The methodology is illustrated using the development of the humble Gantt chart into a generic solution for sequencing processes/activities within multiple resource constraints.

1. USE THE "PAPER" Model

Experience has shown that a good starting point is to have a paper model that is already understood and used. The users may not intuitively understand the fundamental nature of the problem, but they are familiar with "their" paper model. Since this model does work, to a greater or lessor extent, it can provide an interesting basis for development.

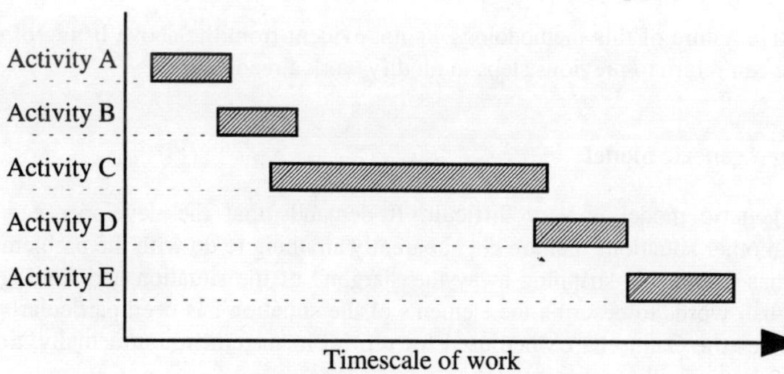

Fig 1: The ubiquitous Gantt chart

The Gantt chart is a well known paper model in which activities are represented by horizontal "bars" drawn on a chart against a time scale. GANTT.XLS is a solution for creating Gantt charts. The starting point of its development was that the bars would be created by the user and any information required would be elicited from the resultant chart.

Perhaps the most important benefit of this approach is that it forces the developer to be creative with the spreadsheet, rather than attempting to force the problem into a familiar spreadsheet structure, which usually turns out to be a simple list.

2. SEVEN STEPS

The following steps are suggested as way of creating an effective spreadsheet solution:

Step 1. Understand the "problem" - Have a paper model.
Step 2. Develop the generic nature of the problem.
Step 3. Design the spreadsheet model.
Design is done on paper!
Mimic the paper model.
Define the model's Objects, Properties and Methods

Now you can start using Excel!...

Step 4. Create the model's Objects and Properties.
Define range names.
Develop and test the (custom) functions.
Step 5. Structure the methods.
Define level of automation.
Create custom menubars and toolbars.
Define sub-routines.
Step 6. Define order of implementation of methods.
For each sub-routine....
Step 7. Implement the methods.

At some point you have to accept the model as it is!

Step 8. Use the solution.

The iterative nature of this methodology is not evident from the above listing of steps. At any step, one can return to previous steps to modify work already done.

2.1 Creating a generic model

Creating a generic model is very difficult. It demands that the developer can identify similarities in other situations that are not apparently anything to do with the problem at hand. The author has found that stripping away the "jargon" of the situation and finding normal, generic, English words to describe the elements of the situation has been particularly helpful. Fig. 1 illustrates the Gantt chart when used for project management, but actually, time based bar charts are used in a whole variety of situations. Any chart which has bars representing something happening over time, is a Gantt chart.

The bar represents a "process", the vertical axis shows the "entity" being utilised by that process and the horizontal axis represents time in discrete intervals, i.e. seconds, minutes, hours, days, etc. That axis can change its units. Near time can be in smaller units (e.g. days) than far time (e.g. weeks, months). In fact, the actually units of time are irrelevant to the model, though not to the user. The generalisation of this model can proceed to include how the chart is used and the nature of the information derived from the chart. Interesting aspects that arose from this analysis were the use of "multiple" charts, in which the process is defined by more than one bar, and the concept of "ambiguity" which is encountered when attempting to simplify the chart, and hence its representation of reality.

The design of the spreadsheet model means considering how to replicate the paper model in the medium of a spreadsheet, and then considering how this medium can enhance the model. It is at this stage that one must resist the urge to rush into implementing the model. The simpler, and more generic, the model, the easier will be the later implementation. This faith, however, tends to come as the result of bitter experience!

3. ENJOY THE DEVELOPMENT

Once satisfied with the model, the developer can use the facilities within Excel to make the development of the model more rewarding. Excel 5 and VBA is based on the concepts of Object, Property and Method. These concepts should be extended to the creation of the spreadsheet model. The objects and properties of the model are defined using range names. Great care should be exercised here to minimise the number of names and ensure that the ranges so defined are appropriate to their anticipated use. Appreciation of this comes with familiarity with VBA. The spreadsheet model needs to be tested using appropriate data, and this particularly applies to any new custom functions that have been designed. In effect, the paper model is developed into an electronic model, on screen.

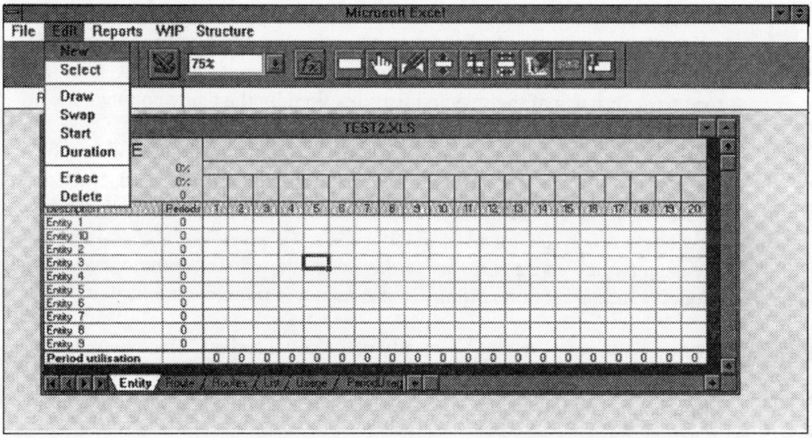

Fig. 2: The development of GANTT.XLS

4. METHODS, MENUS & TOOLBARS

A model is also defined by its Methods. The developer has to decide which methods to automate and which to leave to the user. A big advantage of a spreadsheet over conventional programming is that every method does not have to be programmed. The user can be left to do certain things manually. The Pareto principle applies here! 20% of the programming will achieve 80%, or more, of the methods. Even better, the basic spreadsheet functionality can be left available to the user, to a greater or lessor extent, so that the user can do anything (s)he wants within the overall constraints of the solution.

Fig. 2 shows the GANTT.XLS screen in an early stage of development. In fact the only menus working are the File and Window menus. Spreadsheet applications can be classed as "Self" or "Others". A "Self" solution is a workbook that looks after itself, whereas an "Other" solution is a workbook that needs, and manipulates, other workbooks. Having decided which solution is applicable, the developer can use a basic template to immediately create an automated solution that works. It doesn't do anything significant, but it is operational.

The menus and toolbars can be created and the basic command macros defined. This encourages a structured approach to the design of the solution. It is also helpful, at this stage, to consider if any common sub-routines are possible. The ability to do this comes with experience. However, once detail design and coding is underway, one soon starts to recognise similarities and modify existing sub-routines for wider use.

4.1 The Gantt Chart Toolbar

With GANTT.XLS, the following "Edit" commands were identified. One should note how the commands relate to the physical, i.e. paper, nature of the model.

New: The spreadsheet draws a bar in the selected range of cells, provided that the range is valid. i.e. in the chart and empty. The new process is automatically selected.

Select: This became an evident requirement when it was realised that to manipulate a bar, one had to note its existence. Physically, we do this visually. We see the bar. However, the spreadsheet has to take note of the characteristics of the selected bar. Inspired by the Copy command.

Draw: .The spreadsheet draws a bar with the same characteristics as the selected bar, but in another row, i.e. for a new entity. Very useful for multiple sheets. Not necessary to redefine the process characteristics. The new bar can only be drawn in empty cells so "clashes" are impossible. In the majority of scenarios, this is desirable, though for some, since this model is intolerant of ambiguity, it can lead to a more complex chart

Swap: This moves the bar vertically up or down a chart. Since there is no horizontal movement, other charts are not affected. Like the draw command, only the destination row needs identifying. The bar, in the new position, remains selected.

Start: This moves the bar horizontally. Now in this case, all other bars related to the same process have to be considered and moved. As before, all destination cells have to be empty. The original bar, in the new position, remains selected.

Duration: This alters the length of the bar. In this case, all other bars related to the same process have to be considered and altered. As before, all destination cells have to be empty. The original bar, in the new position, remains selected.

Erase: This "rubs out", i.e. removes, the selected bar. This does not necessarily delete the process, because other bars may exist. No bar remains selected.

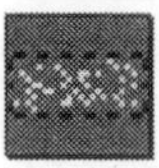

Delete: This removes all of the bars relating to the selected process. No bar remains selected.

4.2 Implementation strategy

In this case, the strategy for implementation was to first implement the edit commands, in the above order. These enable the user to draw and alter Gantt charts. Commands to create new charts from templates are universal in nature and already exist in the workbooks template, in this case "OTHERS.XLS". The Reports menu has commands that produce the reports that can be elicited from a Gantt chart. Firstly there are the "Entity Lists", with data such as process start, duration and finish times. Then there are the "Process Routes", which show how an entity is utilised by various processes over time. Finally, statistical information on utilisation, by entity and over time, can be displayed in graphical form. The relevance of this data is dependant on the scenario being modelled.

5. SPECIFIC APPLICATIONS

It is at this point that the jargon can be re-introduced, to make the model scenario specific. This can be done at the superficial level by just changing the text on the spreadsheet(s) involved. Further modifications can be achieved by altering the VBA code and adding further, specialised, commands.

TIMETABL.XLS is an example of such a specialised Gantt chart application. This is a multiple Gantt chart consisting of 4 sheets: Student groups (46), Lecturers (37), Rooms(32) and Modules(88). The bars relate to the classes and, despite their size, the sheets are very easy to manipulate. In this case, the primary report is the entity's process route. In other words, the timetables for students, lecturers and rooms. Module timetables are also possible but rarely necessary. Other statistical information is readily available

One modification from GANTT.XLS was to produce graphical representations of the individual timetables. This was simply achieved by copying the entity's row on the Gantt chart and pasting the copy into another specially designed template sheet. TIMETABL.XLS looks like a specialised Time Tabling spreadsheet whereas it is, in practice, a slightly modified generic GANTT.XLS. It is interesting to note that GANTT.XLS could do the same thing, but that the terminology would appear alien to the academic administrator and students alike.

To be completely honest, the original version of TIMETABL.XLS was written before GANTT.XLS. It was during its development that the generic nature of the problem became evident, so GANTT.XLS is the product of the second run through of the methodology. This illustrates the last point of the methodology. There are always ways of improving a spreadsheet model, and the best test is usage. The advantage of the spreadsheet is that, at all stages, it can be used to produce practical results. The raw data is visible and always available. Once in use, the improvements can be noted for implementation at a later date. If the model's objects have been appropriately defined, then it is possible for the user to modify the model without detriment to it's automated operation. In many applications, it is desirable to allow the user some scope for further, personal, modification.

6. CONCLUSIONS

This research into the use of spreadsheets, is in its infancy. There is no anticipated outcome at the moment, but four inter-related strands are evident.

If the University is anything to go by, most of the functionality of Excel5 is unused, with the most common usage being the creation and updating of "Lists". Experience with students, full-time and part-time, is similar. A lack of appreciation and modelling skills is one reason, but perhaps the other is that most of the functionality is inappropriate.

Modern "How To" books show how to use Excel's commands but do not necessarily help the user to resolve their own problems. (1) (2). The development of a methodology, and the teaching of it, is providing opportunities to develop ways of improving the appreciation of what is possible, and developing skills in object based modelling and Windows software design. It is also introducing students to a structured approach to programming in particular and problem solving in general.

This methodology is not disciplined enough for large scale projects, but is ideal for the non-professional programmer to develop applications and learn what (s)he wants. It is a good prototyping medium, and, the sometimes relatively large, investment in data generation is not lost.

Finally, the creation of generic solutions to operational problems has interesting development potential (3). Most people, after a little training, could use a Word processor to write a letter, and use a spreadsheet to add up a column of numbers. Could a similar situation be developed for operations management? However, these people have had an life-long education to develop an appreciation of the written language and of arithmetic. They are familiar with the basic model. Does such a pervasive model exist for Production and Operations Management?

REFERENCES

(1) William J. Orvis."Visual Basic for Applications by Example." Que Corporation. (1994). ISBN: 1-56529-553-6

(2) Lee Hudspeth, Timothy-James Lee. "The Underground Guide to Excel 5.0 for Windows." Addison-Wesley (1994). ISBN: 0-201-40651-9

(3) Vivian, Peter J., "Spreadsheets: A Source of 'New' Solutions for 'Old' Problems. Proceedings of the 14th Conference of the Irish Manufacturing Committee. (Sept 1997) pp 813-822

Process Planning and Scheduling

The application of an embedded GA to the improvement of order scheduling in an injection moulding environment

I BANIK, A DAVIES, E G JOHN, and **M USENER**
Cardiff School of Engineering, Cardiff University, UK

ABSTRACT

This paper details the problems of scheduling in a plastics moulding environment, and outlines a methodology to reduce the throughput time of customer orders by the employment of a commercially available Genetic Algorithm (GA), Evolver. The difficulties of data acquisition and the criteria for processing and scheduling are discussed and a functionally tested system described. The paper includes a detailed description of the package and the design of the graphical user interface (GUI).

1. INTRODUCTION

In recent years, many of the traditional problems, associated with manufacturing and its management, have been re-examined using Genetic algorithms (GAs). As such GAs, which are based on the Darwinian concept of the "survival of the fittest", are often applied as solutions to iterative optimisation problems [1]. Production scheduling is one such problem area and in a review of fifty papers [2], it was shown that the majority of work carried out in this field focuses on either the single machine earliest due date problem, or the job shop dispatching problem. There is also evidence to show that research effort has been directed at other aspects of scheduling, such as the machine breakdown problem, although most of these studies have been undertaken using simulated data. This paper reports on the throughput scheduling problem associated with a parallel machine environment of an injection moulding facility. The study therefore concentrates on the application of GAs to schedule production in a real world environment.

2. DEFINITION OF THE PROBLEM

Injection moulding is a single stage process. In a typical industrial environment the items produced within the facility can often be completed on one of a selection of possible machines. Such a condition implies that, in this situation, scheduling production is a parallel machine problem, which poses the question to which of the available machines are the products in the current demand cycle to be allocated.

In this study the actual injection moulding machines vary in age, reliability and rate of production. The facility consists of 23 injection moulding machines, which are arranged in eight different work centres, five of which contain four machines and the remaining three each contain a single machine. Some of the machine cycles can be operated on a fully automatic basis, whilst others require varying demands of operator time. This demand of time varies from product to product and may also vary from machine to machine. Factors which affect the demand of operator time are mainly related to the amount of external work (e.g. de-flashing, packing, assembly) required. The actual layout of the machines, the number of available operators (nominally eight) and the current product mix being manufactured all have a bearing on the possible solution space of assignments. To further compound the problem, customer importance, order value and variability, and changes in delivery requirements also have a significant effect on the scheduling difficulties.

The foregoing would clearly suggest that, in such a situation, production scheduling requires the solution of a dynamic multivariate problem. Exhaustive enumeration of the possible solutions to such problems is not an option, consequently either heuristic or other solution strategies, such as the application of genetic algorithms, have to be explored.

3. FORMULATION OF THE SOLUTION

The problem solution was developed using a three stage strategy:-

Stage 1: Create a model of the production variables via a spreadsheet approach.
Stage 2: Generate a near optimal machine loading assignment using GAs.
Stage 3: Modify the generated loading assignment to accommodate operator availability.

3.1 Stage 1 - Creation of the Spreadsheet Model

A standard product data table, incorporating product code and process time information from existing data files within the company, was created, in Microsoft Excel for Windows 95, Version 7.0, (see Table 1). Using the functions, formulas, links etc., within this version of Excel, it was then possible to create a representative model of the production situation and incorporate the associated variables. The model incorporates and represents the following features of the situation.

1. All Machine/Product availability and assignment possibilities.
2. Non availability of machines (Breakdown and pre-assignments)
3. Preferred machine priorities.
4. Specified working hours.
5. Customer priority conditions.
6. Assignable objective function (maximise order throughput, due date delivery criteria, minimise total throughput time, or any weighted combination).
7. Initial conditions of the facility set.

Table 1 Section of the Product Data Table

		MACHINE TYPE		S1			S2			S3			S4			S5			B2		
ID	CODE	DESCRIPTION	TIME	Pieces/Hr	Operators	Preference	Pieces/Hr	Operators	Preference	Pieces/Hr	Operators	Preference	Pieces/Hr	Operators	Preference	Pieces/Hr	Operators	Preference	Pieces/Hr	Operators	Preference
1	0007.N	FILTER,PART No.43611503M	90	90	1	1	90	1	2	0	0	0	0	0	0	0	0	0	0	0	0
2	0008.N	FILTER,PART No.43613501M	70	0	0	0	0	0	0	0	0	0	0	0	0	0	0	0	0	0	0
3	005C	VEG RACK X3 NETTO GREEN	60	60	1	1	0	0	0	0	0	0	0	0	0	0	0	0	0	0	0
4	007E	BUTTER DISH WHT PP 10.5X6.5X4	1500	1500	1	1	0	0	0	0	0	0	0	0	0	0	0	0	0	0	0
5	09G	8LTR & 5010 BOWL WW WHITE/CLR	106	0	0	0	0	0	0	0	0	0	0	0	0	0	0	0	0	0	0
6	101101B	500ML BASE CLEAR (RANDOM) SAIN	360	360	0.333	1	360	0.333	1	360	0.333	1	360	0.333	1	0	0	0	0	0	0
7	101101D	500ML BASE COOLBAGS & BOXES	360	360	0.333	1	360	0.333	1	360	0.333	1	360	0.333	1	0	0	0	0	0	0
8	101101E	500ML CL/CL COOLBAGS & BOXES	250	250	0.333	1	250	0.333	1	250	0.333	1	250	0.333	1	0	0	0	0	0	0
9	101101GBR	500ML BASE CLEAR (RANDOM)	360	360	0.333	1	360	0.333	1	360	0.333	1	360	0.333	1	0	0	0	0	0	0
10	101101GLC	500ML LID CLEAR	288	288	0.333	1	288	0.333	1	288	0.333	1	288	0.333	1	0	0	0	0	0	0

3.2 Stage 2 - Generation of the Machine Loading Assignment.

The model outlined in section 3.1 is linked to appropriate GA software. The software used in this study is a commercially available product called Evolver. [3] which is fully compatible with Microsoft Excel. Evolver uses a proprietary set of genetic algorithms to search for optimum solutions to a given problem providing the problem is capable of being modelled within the constraints of the software environment.

In its operation mode Evolver permits the user to define and specify:-

1. The choice of algorithmic method (recipe, order, group etc).
2. The GA chromosome (by variable definition).
3. The mutation and crossover rates.
4. The initial population size.
5. The objective function.
6. The terminal sampling stopping criteria.

In this particular study, it was decided that the order solving method would be appropriate. This algorithmic method permutates the contents of an array in order to create new chromosomes. The variables which form the chromosome string, correspond to the cells 'TaskID' of each product to be scheduled (see Table 2, first column). The initial parameter values, were then set. (i.e. Population size 50 chromosomes, crossover rate = 0.5, mutation rate = 0.001 and number of iterations 2,000.) These values ensure a compromise between search effectiveness and computational time required.

Different objective functions were then tested. Of these, the most effective was found to be the maximisation of orders delivered by due date. Cost penalties, according to the priority of the customer order, were also assigned to any overdue orders (see Table 2, end column). With such a approach it is then possible to minimise the total penalty value.

The output from this exercise shows when, and to which machine, each order should be assigned to meet the conditions of the specified objective function. Under test conditions the software was able to create near optimal production schedules with limited amounts of input data.

Table 2 Section of the Spreadsheet Model

TaskID	Code	M/C	Time(mins)	Tasks Used	Operators	Available Time	Priority
29	C248K	B10	19.05	1	0.5	-261.33	25
230	6100.W	B10	1.94	1	0.33	122.67	200
349	A192B	B13	5	1	1	122.67	100
127	5504.GT	B15	2	1	1	-277.33	25
153	407012GUZ	S2	11.11	1	0.5	-69.33	200
117	8048B	D2	18.29	1	0.5	-149.33	200
401	CISA562NB	B14	18.52	1	1	122.67	25
381	242216GBW	D2	10.42	1	1	122.67	200

3.3 Stage 3 - Modification of Machine Loading Assignment

Having created, what could be argued to be an 'idealised' schedule, the next stage of the strategy was to adjust the created solution in order to accommodate the number of available operators. The number of available operators was not included in the initial 'idealised' schedule solution because the schedule was designed to be planned on a weekly basis, whilst the operator availability varied twice daily, according to the shift pattern. If the available operators were less than that required in the 'idealised' schedule, a strategy of assigning high priority to those orders having the earliest due date was adopted. In the event of a tie, priority was then given to the order with the shortest operation time (SPT rule).

Under certain solution conditions of machine assignment and product mix it is possible for one operator to control several machines within a given work centre. If such an option is part of the solution the programme undertakes a further adjustment of the schedule to accommodate this condition.

Once the foregoing stages have been completed the programme creates a graphical representation of the machine/order assignment (see Figure 1, the shaded areas representing machine idle time). To illustrate the effect of operator availability on a given solution, a further run was undertaken. (Figurel - 10 operators, Figure 2 - 6 operators). From this it is evident that amount of machine idle time is noticeably greater in the second case.

4. GRAPHICAL USER INTERFACE (GUI)

In order to enable production scheduling staff to use the programme on a real time basis, Visual Basic for Applications (VBA) was used to create a user friendly interface. VBA permits the creation of dialogue boxes, buttons, message texts etc. which enables the user to interact with the programme. The interface and its procedures was tested on prospective users during the programme design stage.

Figure 1 Example Graphical Output Produced with 10 Operators Available

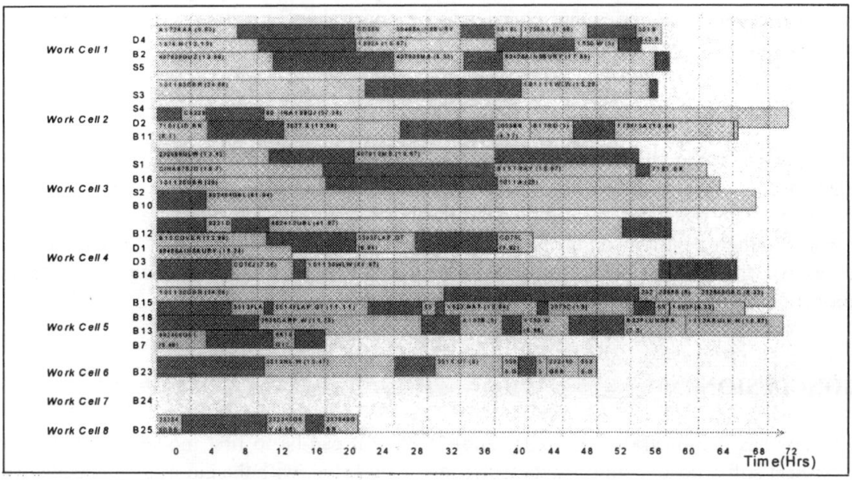

Figure 2 Example Graphical Output Produced with 6 Operators Available

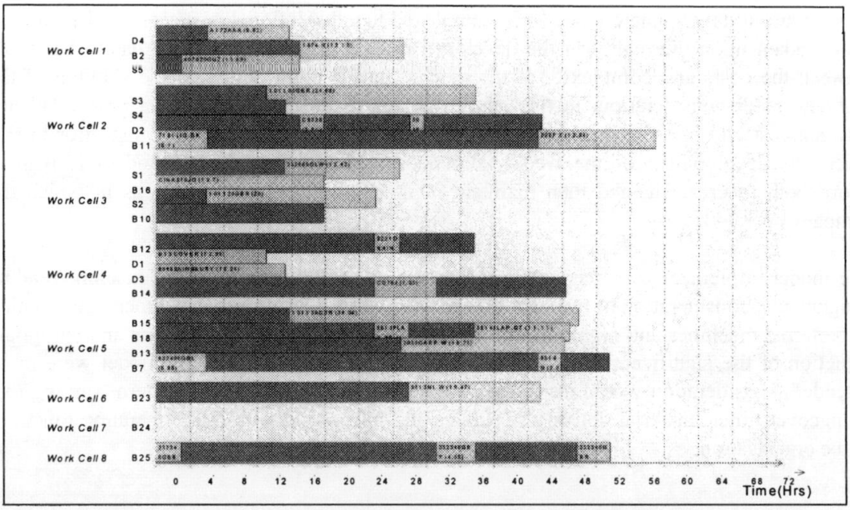

4.1 Data Input

The data input required to create a live schedule consists of detail on the current works orders (i.e. product code, machine, quantity), together with details of all other existing (live) orders (i.e. product code, quantity, due date, urgency of works order). The data input interface was designed to enable the user to easily and accurately input the required information. The data input problem was also minimised by adopting a net change, rather than a total regeneration approach.

Provision was made to enable product change and new product information to be incorporated in the data table. This was achieved by the development of an enquiry and update feature, which permitted data queries and changes to be undertaken by authorised employees via a restricted access system.

4.2 Solution Output

As previously stated, the programme creates a graphical output (Figures 1 and 2). The horizontal Gantt chart shows when and where each of the current works orders are to be completed. The machines are labelled on the left hand side of the chart and are arranged according to their positions on the shop floor. Each bar either indicates work to be done or, in the case of the shaded bars, machine idle time. The work bars are labelled with the appropriate product code and expected time of manufacture.

5 CONCLUSION

This study has demonstrated how easy and effective it can be to first model, using standard spreadsheet software and simple data tables, and then solve, with the aid of an embedded GA software package, a parallel machine scheduling problem in an injection moulding facility.

As with any situation where modelling is employed to aid solution, the success of the exercise is dependant upon the reality of the developed model. However a balance has to be struck between the costs and complexity of the model, and its applicability to the solution of the problem under consideration. In this study the authors are confident that a proper balance between cost and benefit has been achieved. This is evidenced by the fact that prior to this study, the development of the weekly production schedule required considerable manual effort, both to create it, and then maintain it, under the dynamic conditions in which the company operated.

The model implemented in this study, incorporates the options of varying machine/product assignments, consideration of machine breakdown, biasing for priority customers, designation of preferred machines and regard of existing production loads and is therefore an appropriate depiction of the significant factors of the operating environment. If the model were to be extended, consideration would also then be given to variability of shift pattern, varying tool changeover times, material delivery schedules and integration with other operating functions of the organisation.

REFERENCES

[1] School of Electronic, Communication and Electrical Engineering, University of Plymouth, *3 Day Modular Course on Genetic Algorithms, Neural Networks and Fuzzy Logic*, 1997.

[2] Ali, I., *Manufacturing Shop Scheduling Using Genetic Algorithms,* Phd, UWC, 1997.

[3] Axcélis, Inc., *Evolver User's Guide*, 1990-1996.

A case for improving the industrial measurement of schedule adherence

H HARVEY
CAMM, The Industry Centre, Sunderland, UK

Abstract

This paper reports on some recent research which has shown that industrially practised measures of schedule adherence can vary widely in the degree and form of item and time aggregation. Very little debate has taken place on normative forms of schedule adherence that would give effective inter-company comparison or accommodate internal changes within a company over time. It is also argued that schedule adherence should be a measure which is of assistance to operational decisions. This will benefit the interpretation of more aggregated forms. As a preliminary step to the improvement of this measure, it is suggested that the activities of schedule construction and schedule execution should be measured separately.

1 INTRODUCTION

The source of this paper was a structured empirical investigation into scheduling and control practice of Make-To-Order, (MTO), small to medium enterprises, manufacturing discrete items (1).

A set of 12 case studies was employed, divided into 4 groups, varying from 5 down to 2 companies. Each company was investigated through the impact of choices on a single product. Each group consisted of members who put the same number of part types together in Final Assembly for a range of demand in annualised averages of units of the particular end item. See Figure: 1 for the distribution of the case studies in the chosen domain. The product was not selected on any criteria of optimisation in respect of its fit with the company's systems.

The objective of the research was to observe how a product 'performed' under what was hypothesised to be a changing set of choices in system parameters as the demand increased, e.g., in layout, product decomposition, resource sharing, etc. Trends between system parameters were open to confirmation or negation in the other groups.

Since case study methodology is largely constrained to elicit data which is already collected, it was decided to ask each company how they measured the performance of their scheduling and control activities. Schedule adherence was the most common form of measure, although it was not always reviewed in isolation as will be discussed later. The replies showed a wide variation in the form of time and manufactured item aggregation. The implications of this variation and the measurement of schedule adherence in general is the subject of the rest of this paper.

2 ANALYSIS OF THE RESULTS ON SCHEDULE ADHERENCE

The values of schedule adherence and the aggregation in terms of manufactured items and time are given in Table: 1. This table demonstrates the differences in the levels of aggregation of manufactured items, from individual works orders to completion of many customer orders for a given transport load. There is also range of aggregation in respect the measuring and reporting periods, e.g. from being measured hourly to weekly and reporting the weekly or monthly average. It was common for the reporting period to be coincident with the period of the schedule which made it difficult to interpret what constitutes the average value and would contribute to the difficulty of interpreting the results for operational decisions. In a number of cases, the time series of data collected showed that despite the average values given, the range of values about that average could be large: e.g. 30% or greater. Other data sets collected, for example on demand, showed up the significance of the difference in reporting periods and item aggregation. A number of other researchers have reported on the difficulties associated with the decomposition of hierarchical scheduling systems (2,3), which could be considered to have some similar attributes.

The consequences of these variations in the form of measurement for direct inter-company comparison or survey based research are likely to be serious. A number of survey questionnaires (4, 5, 6) have been found to have no questions that would permit the normalisation of the solicited values of schedule adherence. However, it is equally important how the measurements of schedule adherence can be used at individual company. Discussion of the benefits of employing an identical measure of schedule adherence of the current general form but non-comparable units is found in Section 3.

Despite the variation in the units of measure for schedule adherence, a number of similarities were observed between the 12 case studies. It is recognised that these results are not validated by the case study structure or are in numbers that can be statistically generalised. Nevertheless, this data set may provide some indication for direction of future work and provides a useful point from which to proceed in an area which is under-reported.

The first similarity to be noted was that all the values given for schedule adherence were for the manufacturing unit as a whole. In most cases it would have been difficult, if not impossible, to break down the values for individual products. In the case of those sites which had multiple Final Assembly lines, the value given was for all those lines. It is precisely the values at the level of sub-units of manufacture that are likely to assist in operational decisions. Those sites that have systems that sum compliance with due date by work centre independently of product flows, are likely to have particular difficulty in breaking down these values in a meaningful way. The values can only remain the sums of work centre 'islands'.

A second similarity between the sites was that no account was taken of recovery actions to try and return to schedule. The differences in tactics that individual sites adopted in this respect was quite complicated and related to a system's capability to operate with an order backlog. It is possible to review the schedule adherence figures in conjunction with measures such as direct labour costs as a percentage of sales or labour efficiencies and this was the response in two cases. However, it is suspected that these reviews remain at a fairly high level and would be difficult to disaggregate to an appropriate operational level. Furthermore, these other measures are also process centred and don't reflect the large amount of time which is not associated with material/product transformation.

A final similarity between the sites was the lack of qualification given to the magnitude of lateness (or earliness) and in a number of cases the quantity of items that were late. For those systems that report and operate with a backlog, anything remaining in the backlog pool will be reported as late in terms of the unit of reporting period. This was commonly one week and would be too coarse a measurement for the purposes of seeing the benefits of improvements in many situations. From Table 1, it can be seen that the product aggregation can be in terms of item lines of a sales order or the proportion of customer orders that are late. These measures fail to indicate the size of the orders and would only be of benefit if the item magnitudes or order magnitudes were similar for different orders.

3. THE VALUE OF SCHEDULE ADHERENCE MEASURES TO OPERATIONAL DECISIONS

A further and fundamental difficulty was found in the practice of schedule adherence measurement at the case study sites. In general, it was not possible to recognise whether or not a good schedule was being poorly executed or a poor schedule is being recovered by agile execution. The more probable practice is a compromise between these two extremes in which both constructor and executor of the schedule try to accommodate one another in order to produce a value of schedule adherence that is acceptable.

It would be possible to view such data of schedule adherence in conjunction with Return On Investment but this is likely to be at too high a level to assist operational decisions despite being able to indicate the need for operational improvements.

One of the results to come from the research already cited (1) was that much of the anomalous behaviour of the commercially available scheduling software, e.g. (7), could come from the under-reporting of the impact of informal scheduling. Informal scheduling here is not referring to 'rule' sets of individual schedulers though this too can be very significant. It is referring to the scheduling activities of first line supervision which may or may not be formally assigned. It includes such activities as the allocation of staff and the re-batching and re-sequencing of materials. These activities can have a significant effect on the outcome of the schedule.

It might be argued that provided conditions remained the same, i.e. the same resources were employed and the same product mix was being produced in an adequately randomised permutation in relation to the periodicity of the schedule, then a change in schedule adherence as presently measured could be used as a limit switch for action. It may not be obvious where to take the action. There may be quite a large number of factories that do not have this kind of stability. Quite large changes have been declared in schedule adherence on the introduction of factory reorganisation, (8). However, it is often difficult to assess real magnitude of these changes without looking outside the newly created boundaries and conditions within them. In spite of these caveats, there will be some cases where a self-referenced measurement can identify changes in performance.

4 A PROPOSAL FOR AN IMPROVED METRIC OF SCHEDULE ADHERENCE

An improved metric of schedule adherence is being proposed in order to facilitate inter-company comparison, assess the benefits of internal change and provide data which can assist operational decisions. However well it may be constructed, it is not a measure that will be able to be viewed in isolation. Part of its development will require the development of other measures, such as utilisation/capacity. Schedule adherence cannot be seen in isolation of the 'costs' to achieve it, such as the stock levels.

The first step proposed in the development of this measure is to divide the subject of measurement into two parts. The first measure will look at how good the schedule is as a schedule of a particular set of resources. The second measure will look at how well that schedule is executed.

As an example of the first measure, the scheduler could be given the objective of trying to maximise the ratio of the sum of the set up and process times of the constituent items to the total duration in manufacture of those same items from the first operation on the first item to the end of the last operation on the last item, i.e. minimise queue and move times. Undoubtedly, this would show up some of the limitations of the current industrial software or activity that comes under the title of scheduling. However, it should help focus the object of collaboration between the scheduler and the designer of the manufacturing system - a collaboration that would appear too infrequent. Such a ratio would need to be viewed in conjunction with utilisation of the resources since the potential for reducing queue and move times is highly dependent of this parameter. There would be advantages to implementing this measure over well defined boundaries of manufacture with respect to item completions and end product identity. If some items are being made to stock, e.g., components or sub-assemblies, it would not be comparable if these phases of manufacture were simply added, assuming zero time between phases to those products where no stock was held within the manufacturing process.

The objective of the second measure is to assess how well the schedule is executed, i.e. this is being measured as a distinct activity from schedule construction. An example of this measure would be a numerator of the time taken for the execution for the above schedule plus the time employed in recovery (or a converted cost equivalent) divided by the denominator of duration of manufacture of the items as employed in the previous ratio. At some sites it would be necessary to include the extant backlog in the numerator. If a problem is found to exist within the execution of the schedule, then the effort can be focused on decision support for staff allocation or for local planning of material flow. Such solutions will be site dependent and may require a more sophisticated set of measures to identify the source of the difficulties.

It may be the case that a more elaborate set of measures and/or boundaries of application is required than the current proposal to separate the measurement of construction and execution. However, this division is seen as an important first step.

5 CONCLUSION

It is not going to be easy to arrive at a formulation of scheduling performance measures that could be applied over a significantly large range of manufacturing enterprises. However, the reward of successful formulations should not only benefit the individual companies but also begin to establish a framework of measurements that will give robust and comparable empirical evidence. This is seen as essential for the development of manufacturing system theory.

REFERENCES

1. Harvey, H., 1997, "Scheduling and control activities in industrial, manufacturing system frameworks", PhD Thesis.
2. Bakke, N.A., 1993, "The challenge of capacity planning", International Journal of Production Economics, Vol. 30-31, pp 243 - 264.
3. McKay, K.N., 1995, "A review of hierarchical production planning and its applicability for modern manufacturing", Production Planning and Control, Vol. 6, No. 5, pp 384 - 394.
4. Kearney, A.T. (Consultants)/Institute of Materials Management, "Leaders and laggers in logistics", 1990.
5. FORCAST Feasability Study, EUREKA/DTI project, Final Report, 1993, DTI, London
6. Whybark, D.C., 1993, "A worldwide survey on manufacturing practices", Discussion paper, Indiana Centre for Global Business, Indiana University, USA.
7. Burns, O.M. 1991, "Critical success factors in Manufacturing Resource Planning implementation", International Journal of Operations and Production Management, Vol. 11, No. 4, pp 5 - 19.

8. Harvey, H., 1993, "Report on the impact of Lucas methodology for factory re-design on manufacturing planning and control systems" Sub-project 1, FORCAST/EUREKA DTI project, DTI, London.

Table: 1: Data on item and time aggregation in schedule adherence measures

COMPANY (see Note: 1)	SCHEDULE ADHERENCE	PRODUCT AGGREGATION	TIME AGGREGATION
LYCOL	Not measured	-	-
FILTAIR	85 to 95%	Item lines late of all sales orders	Weekly average of day's output, early or on time
IOSA	see Note: 2	"	Weekly average of daily output of 3 shifts
HAZLITE	Not measured	-	-
KITCAB	Not measured	-	-
'α'CTRL	85%	Item lines late per sales order	Weekly average of daily output, single shift
PYROBE	97%	Order sets on scheduled vehicles	Weekly average of each half shift
CIRTRANS	82%	Committed customer orders	Weekly average of hourly outputs
BEEDRY	75%	End item completions	Weekly average of daily results
N.E.RATIO	84%	Committed customer orders	Monthly average of weekly completions
M.V.GEAR	46%	Works orders by cell, overdues + dues	Weekly average of weekly completions
JACCO	87%	Individual end item completions	Weekly average of daily results

NOTE: 1: The names of the companies given in this column are fictitious in accordance with the aggreement on confidentiality.
NOTE: 2: This company measured the number of order lines that were late each week. Mean of late items was 3.7 times target value of 15. Standard deviation was 18.3. The target was going to be increased to 30.

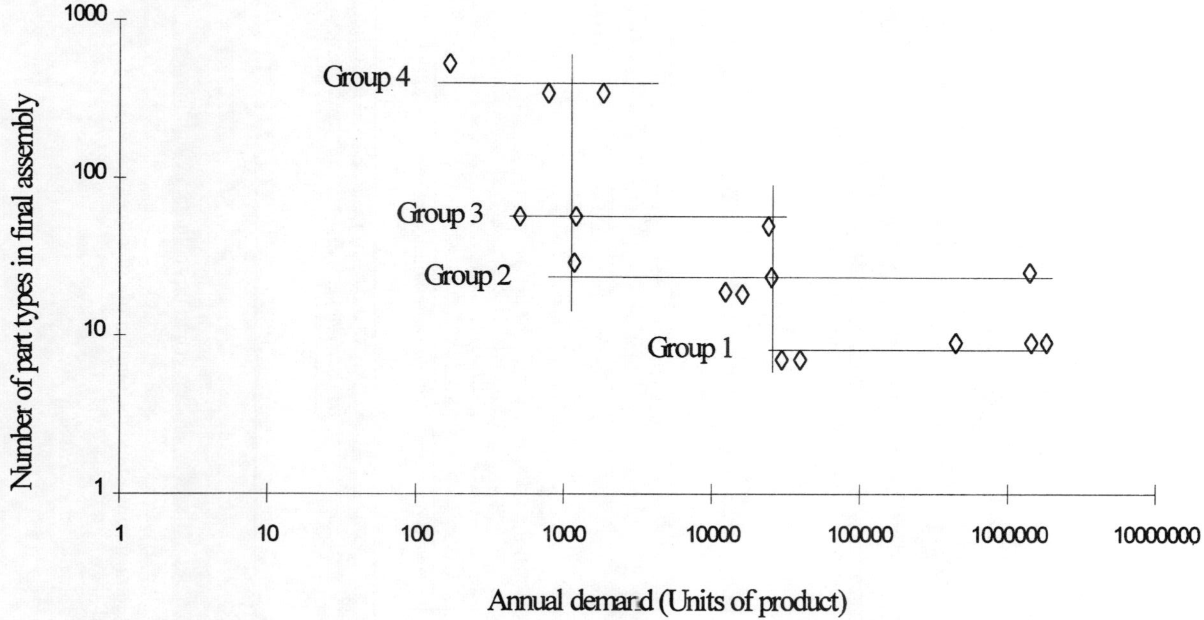

Figure: 1: Distribution of case study groups in the domain

Artificial neural networks for real-time scheduling

H M NURELDIN, R F O'CONNOR, and **A W DUFFILL**
School of Manufacturing and Mechanical Engineering, University of Birmingham, Edgbaston, UK

ABSTRACT: In a real manufacturing environment, reactive scheduling takes place in real-time, not a predictable smooth one. Reactive scheduling is a complex process and it is possible to produce a large number of schedules for the same job. As a result many different models and practices have been introduced, by researchers, to generate the best or optimum schedules, with respect to a specific production situation. This paper presents artificial neural networks (ANN) as an alternative tool to provide a solution to the real-time scheduling issue. The initial work attempts to utilise the capability and performance of an ANN with the goal of developing a real-time reactive schedule and control system.

1 INTRODUCTION

Manufacturing industries now operate in world markets where global competition is fierce. Manufacturers need to introduce the right product to the market in advance of the competitor. To meet this competitive challenge, it is essential that manufacturing companies effectively utilise resources when producing competitive products. To this end the implementation of advanced technology resources, such as computer numerically controlled (CNC) machines, has been a basic requirement for improving manufacturing operations. However it is also vitally important that these advanced technology activities are effectively planned and controlled; after all the potential gains from the advanced manufacturing resources cannot be fully realised if the methods and tools which serve these resources are sub-optimal.

A key element of the planning and control functions of manufacturing systems is how materials (i.e. bought-out items, raw materials, components, sub-assemblies etc.) are released onto the shop floor. It is now essential that materials should be delivered to the point of use 'at the right time, in the right quantity'. This follows the aims of lean manufacture where methods of production and control are continually improved and reductions in inventory levels continually sought. To facilitate these aims a plan of how materials are supplied and, at the same time, to make manufacturing operations effective must be produced. This plan, usually referred to as a schedule, represents what has to be made and when and how the different materials must be supplied to ensure timely production outputs.

The nature of many manufacturing systems is such that several different products must be processed using a defined set of resources. This is particularly the case for companies who get steady demand for a range of products. This type of company will typically operate a batch production environment where, at any instant in time, there will be several products being produced in terms of timing and quantities. The task of scheduling the numerous product orders through the limited, and often variable, production resources and the continual need for re-scheduling is a complex process. This area of management methodology has afforded much research effort. Many different rules, algorithms and heuristics have been developed in an attempt to produce suitable analytical tools that seek to optimise a particular schedule within specific constraints.

2 REAL-TIME SCHEDULING

2.1 The Need For Real-Time Scheduling

Scheduling usually represents a static solution to manufacturing process at hand. Roaddammer (1) said that the scheduling theory is only applicable to static systems, which are, in practice, non-existent. Usually a schedule is prepared, checked and then applied to a process. Unfortunately the manufacturing environment is dynamic and, more often than not, un-predictable. A machine might break down, a tool might break, supplies might be late, the product quality is incorrect (i.e. causing scrap and re-work), operator(s) may be un-available and the continual arrival of new orders usually adds complexity to any scheduling problem. Hence any scheduling solution prepared earlier could quickly become useless. This resulted in a search for a better solution, which is dynamic and offers real-time scheduling. This dynamic scheduling or continual re-scheduling takes place in real-time, not in a predictable smooth environment. The characteristics of manufacturing activity have changed dramatically and this, in turn, has resulted in the change in computer based scheduling requirements from the predictable ones (e.g. materials requirement planning) towards dynamic and real-time ones.

Schedule changes, on the shop floor, in response to changes of priorities in the demands of the incoming work and changes in production resources, such as machine breakdowns etc., happen without stopping of the manufacturing processes. Therefore a schedule capable of changing its internal representation and re-scheduling, without hindering the on-going process is required. But if one could monitor key influential factors that affect a schedule, relate to certain situations, and in turn teach and learn what action to take, then the basis for a real-time scheduling will be there. An important issue relating to ANN relates to the applicability of this 'new' technology to such a problem area.

This work, to date, forms a preliminary investigation into the applicability of using an ANN to form the prime decision-maker within the formulation of a schedule and the following re-schedules. A simple manufacturing system has been considered and an ANN approach has been developed to manage this real-time scenario.

2.2 Artificial Neural Networks (ANN)

The concept of ANN is a branch of computing, which emerged when McCulloch and Pitts proposed the neuronic equations by threshold functions in 1943 (2). Various ANN models have been proposed since that date, but they did not become popular until the late 1980s. By then the accelerated pace of high performance hardware development and the emergence of parallel processing architecture made ANN easier to calculate.

The design of an ANN, also referred to as connectionism or parallel distributed processing, takes up the brain metaphor of processing information. It is based on the structure of the human brain where intelligent behavior appears to come from the interactions of the huge number of brain neurons. When comparing the brain to a computer it has been found that the computer is limited by its processing power and speed. The ANN information processing simulates the brain through the interactions involving the large number of simulated neurons. Its structure is simple an ANN is built from an array of elements, known as the neurons, spreading over a number of layers and interconnecting to each other with positive and negative weights.

The ANN concept differs from the classic computer solutions through the absence of programmers. However it still allows the user to analyse problems, produce a structure and stepped procedures and to present a solution. In using an ANN all that is required is examples of similar problems with the desired set of results. The ANN is initially trained using these desired set of results. After training the ANN will be ready for use and should be able to produce good results to new similar problems.

The main strength of ANN is its excellent modeling of complex problems, thus offering a powerful predictive capability to engineers. Its massive parallelism makes its processing speed very fast thus making it suitable for real-time reactive applications. Its ability to learn through the self-acquisition of knowledge and robustness and fault tolerance gives it advantages over traditional data processing practices. The ANN properties make it a promising system to develop solutions to problems that are intractable, have unknown algorithms and to which hard and fast rules cannot easily be applied.

2.3 Research into Production Scheduling

Many artificial intelligence (AI) researchers have used different methods to solve the scheduling problem. All these methods were limited as they referred to specific conditions, such as a certain number of machines, that have no breakdowns and contain products which meet exact delivery dates. While in the real world there are more and varied machines, breakdowns do occur and delivery dates and quantities can vary. Hence in the real situation traditional techniques, such as the enumeration or branch and bound methods, become impractical as the situation becomes more and more complicated. Yet all these traditional, or basic methods, form the basis of current scheduling research, especially AI.

Hadavi, Hsu, Chen and Lee designed a real-time distributed scheduling system which they called ReDS (3). ReDS uses an algorithm called FORCE which tries to find near optimal release dates for the incoming orders, matched these with the manufacturing resources and coupling this with predictive and reactive scheduling. ReDS integrates the techniques of artificial intelligence and operational research in its design to meet the objectives of the management, such as reducing work in progress and finished goods inventories, meeting due dates and maximising machine utilisation.

Another AI scheduling system called ADDYMS was developed by Butler and Ohtsubo (4) is an opportunistic scheduler, it takes advantage of the state of the manufacturing environment at the time of scheduling, and it can handle the sharing of local resources among different sites very efficiently. ADDYMS uses a database of heuristics to produce the dynamic schedules. Unfortunately, it does not possess a centralised scheduler and hence the constraint handling and schedule quality have not been investigated.

2.4 On-Line Scheduling

Conventional on-line scheduling represents the whole schedule as a single entity. Past works (5) have shown that this makes the schedules very fragile. Therefore changes to the underlying system layout will lead to the whole schedule becoming obsolete.

Changes to a single cell of the system require the whole system to be re-scheduled. In addition to the difficulty in modifying such a reactive schedule, they are usually constructed with a database, containing all possible system conditions. The creation of such a database will be a lengthy onerous task. Furthermore time to access and respond to a rapid change to the production environment would not be instantaneous. This does not look feasible for time sensitive operations. What is required is a system that is predictive and has the ability to react to events quickly and accurately.

2.5 Why to use ANN for Scheduling?

Producing optimised and 'best' schedules involves operations of varying sizes of complexity. The scheduling forces the user to face different sources for its computations that are, in-turn, difficult to track (6). Traditional methods, involving heuristics (7), branch and bound (8), enumeration and sampling, have been proposed to change the scheduling problem from being computational intractable into computational tractable. Moving from one stage to another does not come with zero cost. The solution quality will no longer be as good as it should be, as some rules are typically relaxed and some constraints could be adjusted. This could sometimes result in a schedule that might be un-realistic.

A scheduling system is required that can continuously monitor the output and any changes that might occur. This type of system will react to the constantly monitoring and intervene if anything changes that will reduce the credibility of the schedule. ANN is the tool which can be used to build such a system because of the following benefits that ANN offers:

- The ability to tackle problems, such as non-polynomial complete that frequently occur in scheduling, whose solutions are difficult to define.
- It has the ability to work with fuzzy data as well as with incomplete data. It is able to deal with situations that have not been encountered before. Both situations being normal in real time scheduling.
- It is robust for the failure of one component does not affect the whole system. An ANN scheduling system would have the property of distribution, in contrast to centralised systems where failure of one component results in the system failure.
- Traditional systems have to be continually updated and maintained. In contrast the ANN has the ability to learn by experience and so the performance of a scheduling system will continually self improve.
- Real time scheduling systems should react instantaneously to be effective. Traditional systems use serial computation. A important quality of ANN is its fast processing speed which is achieved by parallel processing architecture.

These qualities of ANN renders it a suitable system to be used in building a real-time scheduling system. It will overcome many of the problems faced by the traditional models.

3 EXPERIMENTAL RESULTS

3.1 Overview of the ANN Model.

The ANN model of the real-time scheduling problem was developed using back propagation (9); which is the most popular rule learning algorithm. The ANN model took ranges of the input variables and produced, as outputs, suitable reactive actions. The model input parameters were:

- Setup time
- Duration of all the operations.
- Amount of work completed of the total product.
- Utilisation of each machine.
- Tardiness penalty, calculated using current date, delivery date and customer priority and value of the business.
- The existence or absence of similar machines.

The output value was then interpreted to a suitable reactive action to keep the schedule up to date.

3.2 Applying the ANN Model.

Initially, the proposed real-time reactive scheduler was applied to a simple system in a table manufacturing company to evaluate its performance. The system was trained with a relatively limited number of input constraints such as:

- Delivery dates had to be met
- The orders arrive at expected dates, with some exceptions of when a customer has an urgent order which had to be finished very early
- Queues needed be kept to a minimum
- Sequence constraints were left un-changed.
- Machines break down and there were other unexpected events

The ANN model was trained using samples that span the allowable ranges of the input variables. Fig. 1 presents the result of the training of the ANN model. One line represents the expected output, while the other represents the actual output from the training. The training results are quite encouraging when compared with the actions of a human scheduler in similar circumstances.

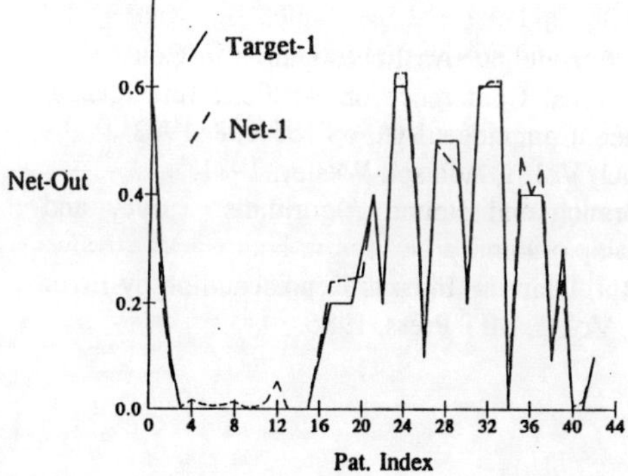

Fig. 1 ANN Model Training Set

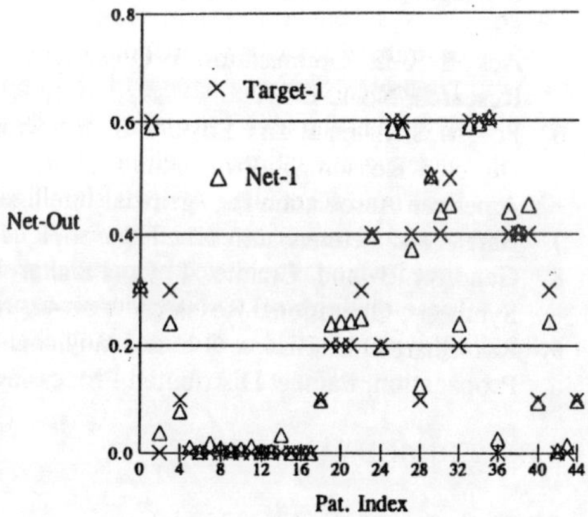

Fig. 2 ANN Model Test Set

The research then showed that once the ANN was trained, the system could then be used, in real-time, with new situations. The output was then compared to the action that could have been taken by a scheduling expert in similar situations. The results shown in Fig.2 represents the test set. The Δ represents the actual output and X represents the action taken by an expert

4 CONCLUSION

The research into an ANN model for real time scheduling has been shown that it can satisfy the project aims of having continually improving production control systems and the reduction in inventory levels. Future work will extend this preliminary work, so far completed, with the use of more constraints, larger domains and the possibility of using on-line data collection systems, e.g. supervisory control and data acquisition (SCADA). This will feed the shop floor status, in real-time, to the ANN scheduling model.

The experimental work has shown promising results. The re-schedule acted as expected without the need of the presence of an expert in scheduling. Further work will extend on the primary work carried out, through the use of more constraints, larger domains and use of SCADA systems to feed the shop floor status in real-time to the ANN scheduler.

REFERENCES

1. Roadammer F A and Whilte K P, A Recent Survey of Production Scheduling, IEEE Trans. On System, Man. and Cybernetics, Vol. 18, No 6, pp 841-851, Nov/Dec 1988.
2. McCulloch W S and Pitts W A, Logical Calculas of the Ideas Immanent in Nervous Activity, Bulletin of Mathematical Biophysics, Vol. 5, pp 115-133, 1943.
3. Hadavi K, Hsu W, Chen T and Lee C, An Architecture for Real Time Distributed Scheduling , Artificial Intelligence Applications in Manufacturing, MIT Press, 1992.
4. Butler J and Ohtsubo H, ADDYMS: Architecture for Distributed Dynamic Manufacturing Scheduling, Artificial Intelligence Applications in Manufacturing, MIT Press, pp 199-213, 1992.
5. Ackoff R L, Optimisation + Objectivity = Opt Out, European Journal of Operational Research, No 1, 1977.
6. Fox M S, Allen B and Strohm G, Job-shop Scheduling: An Iinvestigation in Constraint-Directed Reasoning, Proceedings of the National Conference on Artificial Intelligence, American Association for Artificial Intelligence, Cambridge, MA, pp 155-158, 1982.
7. Barr A and Feugenbaum, The Handbook of AI, Vol. 1, Addison-Wesley, 1981.
8. Gendron B and Crainic T G, Parallel Branch and Bound Algorithms: Survey and Synthesis, Operational Research, Vol. 42, No 6, 1994.
9. Rumelhart D E, Hinton G E and Williams R J, Learning Internal Representation by Error Propagation, Parallel Distributed Processing, Vol. 1, MIT Press, 1986.

Scheduling within a specialist make-to-order manufacturing company

L PACEY, A C LYONS, and **A BOYLE**
School of Engineering, Liverpool John Moores University, UK

Abstract

The paper reports on the methodology used to define and resolve issues relating to the selection of production parameters and the identification of the functionality requirements in scheduling from order receipt through to product delivery, by a specialist make-to-order capital goods manufacturing organisation. The development of a decision support tool to re-schedule and re-optimise a manufacturing system subject to multiple work orders, a large number of engineering design and performance changes, changing work structures and a fluctuating resource provision is considered and examined.

1 INTRODUCTION

In response to the continuing pressure from customer demand, for increased product variety, greater customisation, lower cost, higher quality and shorter lead-times, manufacturing organisations are developing more dynamic approaches to managing and controlling the manufacturing cycle, from the receipt of the order to the delivery of the product, than has been the case in the past. A considerable number of current manufacturing companies are in a transition state as they move from a more traditionally based make-to-stock manufacturing culture to the present and future requirements of a make-to-order manufacturing arrangement. The requirement of a make-to-order system has fundamentally different strategic needs to a make-to-stock system [1]. This mis-match in the strategic needs of the two types of manufacturing system is often seen to be hindering the development of make-to-order manufacturing systems as the weaknesses of make-to-stock arrangements increasingly struggle to meet the functional requirements of a make-to-order market place [2]. This is seen by the substantial body of work, particularly in the scheduling field, that is addressing the problem of optimising the scheduling of a manufacturing environment having relatively fixed parameters.

The fixed parameters of the system are typically seen as the work structures, the part routes, the production process and the product variety. This is a situation a make-to-order manufacturing environment typically does not experience. The nature of any make-to-order manufacturing environment is product change. Product change is a fundamental requirement which is critical to meeting the order requirements of the customer, but means the manufacturing environment is very dynamic and therefore difficult to both manage and control [3]. The fluctuating demands on the system indicates a new approach is required for addressing issues involved in scheduling the design and the manufacture process.

2 OPERATIONAL CHARACTERISTICS OF THE MAKE-TO-ORDER SYSTEM

For the capital goods manufacturing organisation concerned changes to the content of the product specification of the customer occurs both frequently and throughout the lead-time of the manufacturing cycle. Although a product specification is normally agreed between the customer and the manufacturer as part of the tendering, quotation and ordering functions, changes to the product specification are both expected and extensively accommodated as part of the normal operating circumstance of a make-to-order manufacturing system. In addition many product changes are positively encouraged because of the potential uplift effect on the profit margin as late but expensive modifications are incorporated in the final product requirements. Any product changes impact on the product design, on the work structure, on the part routes and on the manufacturing content of the order. This inevitably modifies the mix between the bespoke elements and the standard elements of the manufactured output. However the increasing use of modular design strategies to accommodate product change is also increasing the commonality of product elements in design and hence is tending to increase the predictability of the process utilisation.

Initially the design, purchasing and manufacturing content of an order is based on estimated activities. The organisation, management and production scheduling of the bespoke elements of the order must inevitably rely on the best possible estimate of the anticipated design function activities, of the manufacturing complexity and content and of the variability of the process demands. Uncertain process and resource demand gives rise to changes in the time and the volume of work an activity requires for its completion. As the order progresses through the manufacturing cycle, the estimated number of hours required to complete each of the remaining tasks requires continuous changes and updates.

The duration of any activity is a function of the work content or processing requirements, the number and type of the resource allocated to the activity and the capacity of each resource. Any changes in the work content and in the resource allocation are thus reflected in the duration of the activity, which must then impact on all the other downstream activities in the work structure and on the work synchronisation. The product complexity is also a factor requiring consideration as it reflects the required interdependence of the design and manufacture activities and the necessary logical precedence of the activities. A change to any one element of an order therefore has an impact on many other elements downstream in terms of both work synchronisation and resource demand. In addition changes to any one order has an impact on each of the other orders in the system because the modifications have the potential to free resources or demand greater resources to allow progress to be maintained against the established schedule. Product assembly, commissioning and testing is the final function in the manufacturing cycle and gives rise to the most obvious demands for re-design, re-purchasing, re-manufacture and changes in resource requirements. It is clear product changes emanate from day one during the design, manufacture and assembly stages involving order receipt, customer specification, design solutions and manufacturing. The required product changes have to be addressed sufficiently quickly to enable effective management and manufacturing control to be maintained. Rescheduling and work progress review must be continuously undertaken and re-evaluated to respond to the demand changes in the production process. In addition production control is required to rapidly direct and drive the resources in the production process. This allows the synchronisation of activities to be maintained and hence the requirements of the critical path to be achieved. In many manufacturing organisations the work re-scheduling and the production control functions are supported by manual, document based systems that are limited in scope, in the depth of analysis achievable by the a group of manual operators and in the effort required. Control output is also constrained by the manual capability of such systems.

3 REQUIRED SCHEDULING CHARACTERISTICS

The complex interdependence of the order generation, the product design and development, the organisation of the production resources and the design, purchasing, manufacture, assemble, test and commissioning functions within a make-to-order manufacturing organisation means a scheduling support tool for controlling the production resources cannot be developed independently of the greater overall manufacturing system and company organisation. The initial starting point in the analysis of the existing manufacturing system of the company is the generation of an IDEF0 model of the complete order receipt to delivery system [4]. This enables the identification of the current information flow and classification structures used in the business system to be established. The IDEF0 definition of the present planning / scheduling system also allows function identification and analysis to be undertaken of the attributes and capabilities of the current system [5].

In addition, identification of a best possible manufacturing flow, using a precedence model, enables the complexity and sensitivity of the product range, to change, to be established. This is then combined with the results obtained from the generation of a problem cause list and a problem requirements list to provide a summary of the generic, specific and operational characteristics of the system. It also allows prioritisation of the scheduling requirements to be undertaken and provides valuable information to establish the necessary implementation approach and the implementation methodology.

For a complex product in a make-to-order manufacturing system the re-scheduling function is both an important and a significant challenge. The very high number of changes experienced throughout the order life cycle, the complex interdependency of activities in the work structure, the impact that change has on many other functional elements downstream in the system and the continued synchronisation of the work itself, places great demands on the scope, functionality and flexibility of any scheduling system. The scheduling system must be capable of dealing with the resource demand, the resource allocation, uncertain activity processing time and more significantly the re-processing requirements of re-design, re-purchasing and re-manufacture. The response time of the system to a re-scheduling request, its ability to explore "what if" type scenarios and the capability to re-synchronise efficiently both the work requirements and the resource requirements in order to deliver the end product without impacting on either the critical path or the promised delivery date are further important considerations. The synchronisation of the activities in the work structure and the part routes is often seen to be the most effective method of maintaining the minimum critical path. In addition, given the complexity of the re-scheduling problem the scheduling horizon is often extremely limited. This time window, from identification of the required changes to the time of actual impact on the schedule, defines the time available to combat any possible slippage to the schedule. The larger the time window, the greater is the opportunity to instigate proactive action for the remainder of the work schedule.

4 THE CURRENT SCHEDULING SYSTEM

In the company, the product design, purchasing, manufacture and assembly functions are identified and structured by an alphanumeric design code. This code provides the only classification tool used to communicate the structure and contents of any particular product. It is generic and applicable to all product types produced by the company. It also provides the necessary forward control mechanism to structure both the Bills of Material and the financial budget. In addition it provides feedback control to allow work progress information to be collected and anticipated work progress reports to be generated.

The current company procedures, as shown in figure 1, are based on employing a number of disparate decision support tools to control the scheduling activity. The information system relating to both actual work progress and anticipated work progress is a manual document based arrangement in which the scheduler forms the links manually between the different decision support tools. In addition, the scheduler generates the necessary schedule input data and analyses the schedule output information. Such an arrangement automatically constrains the depth of detail and the scope of the analysis obtainable and given the normal operational time constraint pressures it makes re-scheduling a particularly demanding challenge. The complexity of the interactions within a single order and the impact they have on other orders can not be evaluated in detail, thus making the exploration of "what if" scenarios very limited. The size of the scheduling window is therefore reduced but the system is very flexible and can adapt to suit the requirements of any order.

5 THE SCHEDULING SUPPORT TOOL

An integrated finite capacity-scheduling tool is under development and is being employed to replace the current scheduling system. The new approach as shown in figure 2, is based on the integration of all the necessary system elements into one analysis environment. In addition there is an increase in the number of elements taken onto account in terms of its design, purchasing and manufacturing content and more resource elements are now accounted for by the decision support tool. The decision support tool has the flexibility to cater for changing work structures, different part routes, modifications to activity processing requirements, variation in the resource allocation and capacity and the reorganisation of shift patterns. These changes can be accommodated through out the manufacturing cycle. It also possess the functionally for "what if" scenarios to be evaluated and comparisons to be made against previous options.

Manufacturing and scheduling control documents in the form of route sheets, work to lists, Gantt charts, load charts and late jobs reports are an integrated and automatic output from the system. The new scheduling support tool now enables the scheduling function to drive, synchronise and re-sequence the available resources. It utilises the current alphanumeric design code to collect information automatically on the content of the orders, on the current

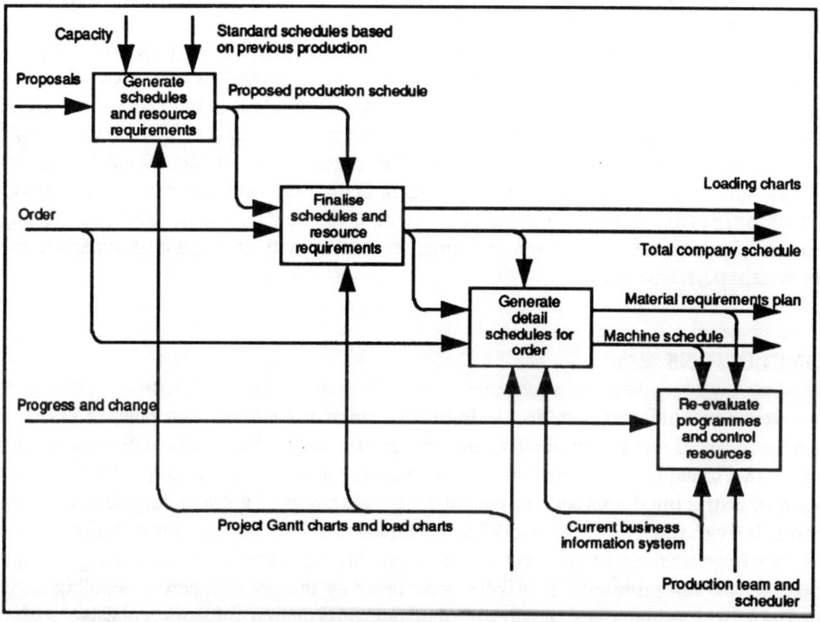

Fig 1. Arrangement of original scheduling system

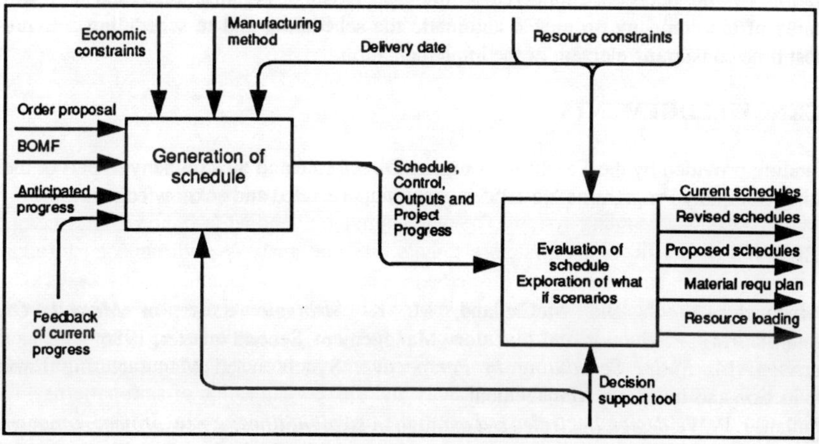

Fig 2. Proposed new scheduling system

and anticipated progress of the orders. Total re-scheduling is now achievable in a response time of seconds and a full update on current work progress and anticipated work progress is available at the end of a 2 minutes operation.

Implementation of this new decision support tool is being undertaken incrementally. This has the advantage of allowing system corrections to be implemented, real time development to take place and a suitable evaluation period to be agreed. In addition the end users have been able to become accustomed and familiar with how the decision support tool operates and the content and format of its output reports. The introduction of the control outputs to synchronise and re-sequence the resources is again being achieved incrementally to provide a suitable transition from the older established system to the new approach. Operation testing of the new decision support tool is using real-time, real orders with an ongoing comparison to the current system performance.

7 CONCLUSIONS

Make-to-order manufacture is subject to frequent change and requires flexibility to enable it to respond both quickly and accurately to the change demands. Very little published work has addressed the scheduling of such a flexible manufacturing environment. This makes the definition of a structured approach to enable the development of a scheduling decision support tool particularly challenging. This work has resulted in a technically complex product, which is subject to a large number, to frequent and to many different types of change being scheduled, re-scheduled and re-optimised. This is being achieved by the use of a new scheduling decision support tool, which integrates previously disparate information into one cohesive analytical environment with the ability for the user to explore extensively "what if" scenarios. The integration of the decision support tool into the existing organisational and information structures of the company, to enable automatic the scheduling and re-scheduling activities is the most time consuming element of the implementation.

8 ACKNOWLEDGEMENTS

The funding provided by the Teaching Company Directorate and the company as part of the Teaching Company Programme No 1955 is greatly appreciated and acknowledged.

REFERENCES

1 Marucheck, A. S. and McClelland, M. K. *Strategic issues in Make-To-Order manufacturing*. Production and Inventory Management, Second quarter, 1986

2 Harrison, M. *Finite Scheduling in Perspective*. Synchronised Manufacturing Limited. Production and Inventory Management

3 Scanlan, J. P. *We design in circles but manage in straight lines; a new project management paradigm for concurrent engineering*. Advances in Manufacturing Technology XI. Proceedings of 13th National Conference on Manufacturing Research, pp 502-507, 1997.

4 Wu, B. *Manufacturing Systems Design and Analysis*. Chapman and Hall, London, 2nd Edition, 1994.

5 Mayer, R.J., *Information integration of concurrent engineering*. IDEF Object-Oriented Design Method Report A1-TR-1992-0056, Wright-Patterson Air Force Base, Ohio, USA, 1992.

The integration of object-oriented and logic programming to develop automatic process planning

J PANG, C F McLEOD, and **D K HARRISON**
Department of Engineering, Glasgow Caledonian University, UK

ABSTRACT

Logic programming and the object-oriented approach have made significant contributions to the field of artificial intelligence, particularly with respect to knowledge-based system development tools. The relative strengths of these two technologies are complementary. The ability to combine object-oriented and logic programming paradigms produces an extremely flexible and powerful environment for developing knowledge-based applications.

One of the important activities in Computer Integrated Manufacturing (CIM) is Computer Aided Process Planning (CAPP). The nature of process planning is knowledge-intensive. The effective representation and manipulation of process planning knowledge is one of the key issues in CAPP systems. This paper explores the benefits of integrating logic programming and the object-oriented approach to develop a generative CAPP system, utilising each technique to obtain maximum benefit. Three generic orthogonal OMT models - object model, dynamic model and functional model have been constructed and the prototype CAPP system is implemented with an object-oriented paradigm within the logic programming environment by using Prolog^{++} software.

1 INTRODUCTION

A process planner works in a technical environment, he or she needs to recognise the initial state raw material blank, goal state design specifications of a part, determine required operations and operations sequence and select resources used in performing these operations. To cope with these, process planners are required to naturally match and associate the knowledge, data and algorithm patterns of process planning through their instinctive logic. Computer aided process planning is an attempt to simulate this kind of human intelligence on computers. It has attracted great research interest for the last two decades and evolved from studies on variant systems to semi-generative, generative, and lately, systems which are intended to be integrated seamlessly with other factory systems [1]. However, very few generative CAPP systems are seen in the real industrial environment since these systems require establishing a large amount of logic and rules, developing complicated knowledge bases, applying sophisticated search algorithms and managing empirical knowledge [2]. With the rapid development of computer techniques, especially artificial intelligence (AI) methods, the research aspects of development of CAPP systems have been changed dramatically compared with the initial research activities. The ability to effectively capture, represent, and utilise knowledge has had an important effect on automatic process planning.

Object-oriented methodology is a new way of thinking and representing real world problems in terms of objects. It provides us with a vehicle to organise objects into inheritance class hierarchies. The technique has been developed from early methods of programming technique to later methods of systems modelling. Object-Oriented Modelling Technique (OMT) defined by Rumbaugh etc. [3] thinks about problems using models organised around real-world concepts. The fundamental construct is the object, which combines both data structure and behaviour in a single entity. Compared with other object-oriented approaches, OMT contains most of the knowledge presented in either object-oriented or traditional system analysis and design approaches, it produces better abstract objects and is one of the most complete and powerful systematic methodologies for constructing a new system model. OMT is based on building three interrelated models to describe a system: object, dynamic, and functional. The object model captures the objects in the system and their relationships to one another. The dynamic model describes the control aspects of the system as it changes over time and the interaction between objects. The functional model specifies the transformations of object values and constraints on these transformations.

Logic programming can implement logic in a computer language for problem solving since logic has traditionally provided a firm conceptual framework for representing knowledge. The logic is made up of logical statements from which deductive mechanism can be drawn automatically. A logic programming language can represent these aspects by having a syntax similar to the logical statements and execution process simulating the deductive mechanism [4]. Prolog is the most popular and widely accepted implementation of this kind of language. It not only provides quite a few powerful features like pattern matching, depth first search, and automatic backtracking, but also has a declarative logic-based syntax and an in-built inference engine. As a result, programs that perform logical reasoning are much easier to write in Prolog than other languages. The program developer is allowed to express problems in a completely 'declarative' way and describe problems as logical rules and facts, without giving instructions for how the problem is to be solved. An execution mechanism embodied in implementations of Prolog can then be used to search efficiently and systematically for a solution to the problem [5].

Both OMT and logic programming are powerful AI methods. It is very helpful and effective if we combine these two schemes to take advantage of the best features of each for representing process planning knowledge. In this combination, OMT can benefit from logic programming in the following ways. First, logic programming can be easily used to construct objects with reasoning capabilities. Second, a logic programming language like Prolog is a type free language. This allows greater freedom for the programmer than a typed language, leading to an increase both in productivity and creativity [6]. Third, all data structures in Prolog are dynamic. The creation and garbage collection of data structures are automatic. A Prolog-based OMT has the ability to allow truly dynamic class hierarchies. One can not only add classes to a class hierarchy dynamically, but also can dynamically augment classes that have already been established. In addition, logic programming can be enhanced by OMT since it is based on modelling objects from the real world and then uses the model to build a language-independent design organised around those objects. OMT promotes better understanding of requirements, cleaner designs, and more maintainable systems [3]. Logic programming languages like Prolog impose very few restrictions on the structure and organisation. This can lead to problems for organising and managing their code when constructing large applications. The imposition of OMT class hierarchies will bring about more efficient and easily manageable logic programming structures. This paper describes how to build up three logic programming-based OMT models for a CAPP system by combining the two technologies.

2 DEVELOPMENT OF LOGIC PROGRAMMING-BASED OMT MODELS FOR A CAPP SYSTEM

This section demonstrates how the integration of OMT and logic programming was successfully applied to CAPP system design. Three logic programming-based OMT models: object, dynamic and functional are proposed on the basis of the analysis of information required and processed in process planning. The object-oriented logic programming language Prolog^{++} was used to construct the system. Prolog^{++} inherits the problem solving and database characteristics of Prolog, and also inherits the sound methodology of the object-oriented programming approach. It is very suited to be used for implementing the integrative strategy to the CAPP system.

2.1 Object Model

The object model describes objects in the CAPP system along with their relationships to one another. It consists of three modules: product information module, process planning module, and manufacturing resource module. Each module is a logical grouping construct which captures one perspective of the system. An object diagram is depicted to show the object classes and their relationships in the model (see Fig. 1). The essential task in the framework design is to develop an intelligent and integrated process planning module which serves as a linkage between the other two modules.

2.1.1 Product Information Module

The product information module describes the design requirements of a part. It is a basis for the CAPP system. The module consists of part class and its deeply nested classification subclasses about part shape and feature. The part class has attributes and input operations for general part information such as name, description, batch size, material data etc. The subclasses are composed as a multiple inheritance hierarchy to describe a complicated part shape with a variety of features. Polymorphism is used for generalising the name of input data across many different types of subclass, each having their own attributes and operation definition of that task. The superclass part has an operation which can not only search its subclasses to build up a classification tree for part specifications according to the inheritance hierarchy, but also allow sub-tree to be selected and traversed automatically. Hence, a user can conveniently select part shape and features on it, input relevant data and store them as attribute instances according to the classification tree.

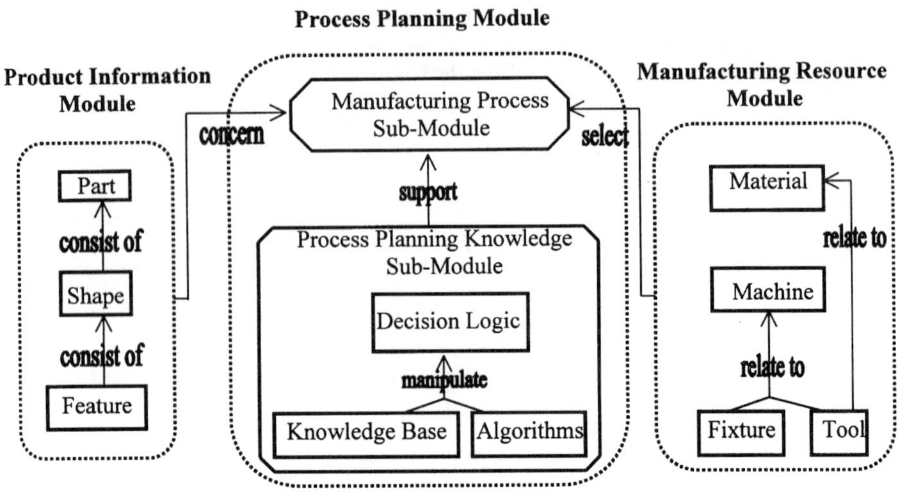

Figure 1: Object model for the CAPP system

2.1.2 Manufacturing Resource Module

In this module, we apply the integration of object-oriented and logic programming to the area of resource management. The approach inherits all of the database features of Prolog, which can be thought of as a deductive, relational database manager. It augments this with the ability to structure the databases as instance hierarchies, using inheritance to avoid duplication and redundant values. Four class hierarchies about material, machine, tool and fixture were built in the manufacturing resource module. They are represented as hierarchies with instances at the bottom level, utilising default values and inheritance to classify them. An important feature exhibited by this method is ability to store database records locally, rather than having a global repository. An instance hierarchy can have several local databases distributed throughout it according to classification. Each hierarchy has operations to manage data, which can view data in the hierarchy, add new data, change or delete old data. There are some associations between these class hierarchies. For example, the tool class is related to both the material class and the machine class. This means that the data for tool depends on the data for both the material to be cut and the machines to be employed.

2.1.3 Process Planning Module

The process planning module contains all technical process information on manufacturing a workpiece from stock to finished part. It is the essence of the object model and consists of two interrelated sub-modules: manufacturing process sub-module and process planning knowledge sub-module. The latter sub-module gives necessary support to conduct general planning activities for production in the former.

The manufacturing process sub-module describes all necessary work steps in planning process. It includes eleven object classes, which are: (1) part interrogation class, (2) process selection class, (3) stock determination class, (4) machine selection class, (5) set-up selection class, (6) fixture selection class, (7) operation sequencing class, (8) tool selection class, (9) cutting conditions decision class, and (10) time and cost calculation class, (11) process planning output class. The sub-module can communicate with the product information model and the manufacturing resource model and perform the planning process by using relevant knowledge in process planning knowledge sub-module.

The process planning knowledge sub-module is the core to automate the CAPP system, which includes expert knowledge as well as scientific laws to assist process planning functions. In this sub-module, manufacturing domain-specific knowledge is represented as a set of logical statements in terms of rules and facts. These statements can then be manipulated by repeatedly using a particular rule of logic, in accordance with a defined strategy, to accept user input, access the specified statements and draw appropriate new statements. The original set of the rules and facts are classified and organised as the knowledge base class hierarchy, the manipulating of these to obtain new information or knowledge is defined as the decision logic class. Some necessary calculation and optimisation operations are contained in the algorithm class, which also can be used in process planning by the decision logic class.

There are various kinds of rules built into the knowledge base class which are classified as relevant subclasses according to the work steps required in the manufacturing process sub-module. They include: rules to create machining processes subclass; material selection rules subclass; rules to set-up machining processes subclass; machine and tool selection rules subclass; operation sequencing rules subclass; fixture rules subclass; cutting conditions selection rules subclass.

Decision logic class is the core to represent process planning knowledge and is used to manage, control and co-ordinate information in the CAPP system. It can manipulate knowledge from the knowledge base class, algorithms from algorithm class and attribute instances from all object class to obtain new data or knowledge. The decision logic utilises three main types of operators: domain, search and control. The domain operator recognises (by matching) which rules in the knowledge base class can be satisfied and what algorithm from algorithm class is required to particular process planning problems or to part of the problem solving procedure. The search operator is mainly concerned with the search strategy that can be depth-first for blind search. The control operator is used to determine which domain operator or search operator is used and when to carry out certain process planning tasks during the problem solving process. The three operators are often combined together for reasoning a particular process planning domain problem.

2.2 Dynamic Model

The dynamic model specifies allowable sequences of changes to objects from the object model. It deals with flow of control, interactions, and sequencing of operation in the system. To build the model, a scenario of planning procedure needs to be identified. In the CAPP system, the manufacturing process sub-module has significant dynamic behaviour and contains all work steps to produce process planning, so it is a key object under the scenario. A state transition diagram is illustrated as Figure 2, it describes all of the behaviour of the sub-module. States are equivalence classes of attribute and link values for the object. Events can be represented as operations on the object model. In Figure 2, an event trace for the scenario has been constructed and the events are consistently and completely matched among object classes in the manufacturing process sub-module. The state diagram describes the states in the sub-module and the ways the sub-module transits from one state to another.

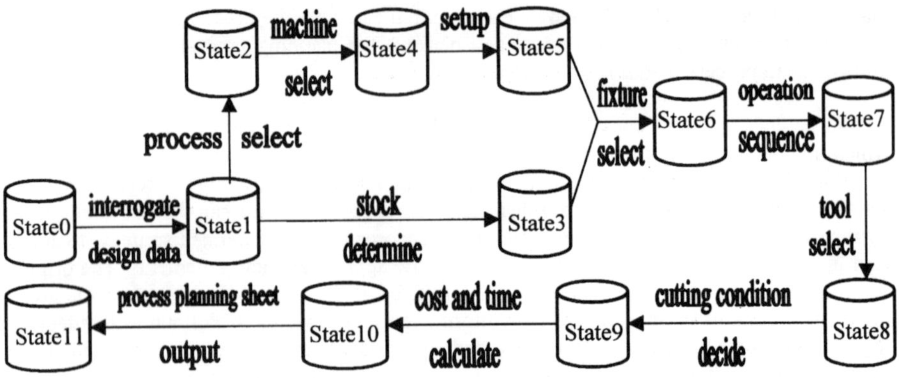

Figure 2: Dynamic model for object manufacturing process sub-module

2.3 Functional Model

The functional model specifies the operations in the CAPP system. In particular, it describes how the execution of an operation affects the values of the objects in the system. We have integrated logic programming into the OMT model, so the functional model can not only show the data value transformation but also knowledge and decision logic transformation within the system. Three major functional models are stock determination, process selection and machine selection. These models show how output values in computation and/or reasoning are derived from input values. The functional models consist of information (data and knowledge) flow diagrams which show the flow of information from external inputs, through operations and internal information stores, to external outputs. For illustrative purpose, the graphical model for process selection is shown as Figure 3. In process selection, process selection logic in decision logic class is core of the operation. First the decision logic loads desired features from product information model and process selection knowledge from the knowledge base class into the working memory. Then it associates the desired features with the corresponding process and automatically compares them with the process selection knowledge to issue the required process list as output. Finally it stores the output value in the process selection class as attribute instances.

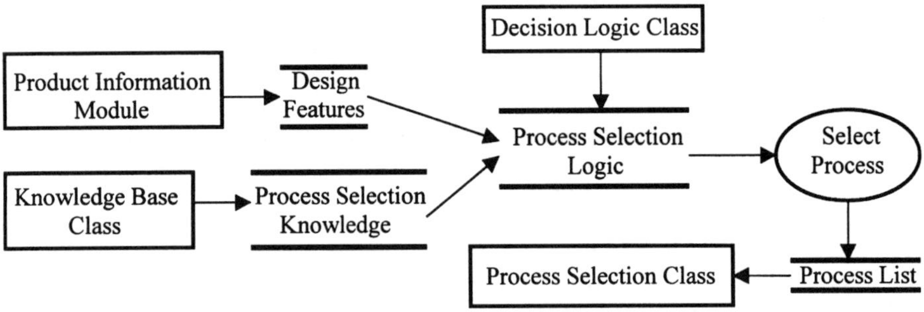

Figure 3: Functional model for object process selection

3 CONCLUSION

In this work, we exploit the potential for developing a CAPP system by integration of object-oriented and logic programming paradigms. The efficient mixture of the two technologies has been used to solve in an elegant way some difficult problems in CAPP which could hardly be managed in a 'pure' object-oriented or logic programming environment. In the logic programming-based OMT models, we make a clear separation of knowledge base from the reasoning control engine (decision logic). It is also necessary to separate dynamic data structure (including part design input, manufacturing process temporary output data or decisions made, intermediate states and final process planning output), static resource module, algorithm class and knowledge base class from each other. The core of the models, decision logic, employs search techniques and conflict resolution methods to communicate and process information between the knowledge, data and algorithms. The inputting and outputting of information within these models is very free and easily collected together, separated and stored. So the architecture has considerable capabilities in recognising and matching patterns and relationships in even large amount of changing and complex information. It is well suited for automated process planning. Although the system is currently restricted to rotational parts, the techniques developed can be expanded to address a wider range of engineering components. The cleaner system design in a precise notation facilitates integration, maintenance, and enhancement of the entire software life cycle. The development of the CAPP system based on such logic programming-based OMT models gives a foundation toward the integration between various components of CIM.

4 REFERENCES

1. H. C. Leung, "Annotated bibliography on computer-aided process planning", *International Journal of Advanced Manufacturing technology*, Vol. 12, No. 5, pp 309-329, 1996
2. W. Eversheim and J. Schneewind, "Computer-aided process planning - state of the art and future development", *Robotics & Computer-Integrated Manufacturing*, Vol. 10, No. 1/2, pp 65-70, 1993
3. J. Rumbaugh, M. Blaha, W. Premerlani, F. Eddy and W. Lorensen, "Object-oriented modeling and design", *Prentice-Hall*, 1991
4. K. Darby-Dowman and J. Little, "Critical factors in the evolution of logic programming and Prolog", *European Journal of Information Systems*, Vol. 6, pp 67-75, 1997
5. M. Spivey, "An introduction to logic programming through Prolog", *Prentice-Hall*, 1996
6. R. Bose, "Strategy for integrating object-oriented and logic programming", *Knowledge-based Systems*, Vol. 7, No. 2, pp 66-74, 1994

An intelligent production planning and scheduling using genetic algorithm

J G QI, G R BURNS, and **D K HARRISON**
Department of Engineering, Glasgow Caledonian University, UK

ABSTRACT

For solution of optimization problems, genetic algorithms (GAs) have been investigated and shown to be effective and flexible in exploring a large and complex space. Also, they do not require derivative information or other auxiliary knowledge, and provide a number of potential solutions to a given problem. The choice of final solution is left to the user.

This paper describes a hybrid GA-heuristic search strategy that can find the best and best average cost of optimised scheduler and optimised machine output. The proposed approach employs a genetic algorithm to optimise the parameters of a simple heuristic problem. A production planning and scheduling example is given in order to demonstrate the effectiveness and extension of approach.

1. INTRODUCTION

Planning and scheduling have made significant contributions to the field of manufacturing, and at the same time play a crucial role in customer satisfaction (1). Integrating planning and scheduling by using genetic algorithm will produce a range of benefits such as increased throughput rate, reduced work-in-process, maximized machine utilization, meet order due dates, satisfy production constraints, and further achieve the business objectives. Planing can be generally described as selecting and sequencing activities such that they achieve one or more goals and satisfy a set of domain constraints. While scheduling is an optimization task that assigns resources and times for each activity so that the assignments satisfy the temporal restrictions of activities and the capacity limitations of a set of shared resources. In practice, planning and scheduling are usually interleaved, and techniques developed in one area are often applicable to the other (2).

Planning and scheduling are known to be hard, and belongs to the class of NP-complete problems (3). The traditional approach for resource planning and scheduling problems is through using dispatching rules. Panwalkar and Iskander (1977) (4) presents a list of over 100 rules. A drawback with such dispatching rules is that there is no single rule, including the commonly used rules such as 'shortest processing time' (SPT) and 'earliest due date' (EDD), that outperform or dominate other rules under varying conditions. Recent advances in computing technology have led to the development of a hybrid genetic algorithm approach which overcomes the drawback of heuristic in solving complex resource planning and scheduling problems (5).

The remainder of this paper is organized as follows: Section 2 briefly reviews genetic algorithms. Section 3 gives a description of the proposed genetic algorithm for resource planning and scheduling. Simulation results from a case study are provided in Section 4. Finally, Section 5 represents some conclusions of the paper.

2. REVIEW OF GENETIC ALGORITHM

A genetic algorithm is a computational model that emulates biological evolutionary theories to solve optimization problems. A GA comprises a set of individual elements (the population) and a set of biologically inspired operators defined over the population itself. According to evolutionary theories, only the most suited elements in a population are likely to survive and generate offspring, thus transmitting their biological heredity to new generations (6)(7).

For solution to an optimization problem, five components are required:

- *Encoding.* This is a way of encoding the decision variables of the optimization problem in a string of binary digits (1's and 0's) or other representations called a *chromosome*.
- *Evaluation or Objective Function.* This function is used to evaluate the given decision variables and return a value. The value of a chromosome's objective function is a measure of fitness of that chromosome.
- *Initialization of the Population.* A method of initializing the population of chromosome is required. In general, the population of chromosomes is initialized randomly.
- *Genetic Operators.* Genetic algorithms employ parent selection techniques that selects strings to create a new generation, where the fittest members reproduce most often. After the parent selection, the process of crossover, mutation and other operations are applied.
- *Optimal Parameters.* A set of parameters is predefined to guide the genetic algorithm, such as the length of each decision variable, the number of chromosomes, the crossover rate, the mutation rate and stopping criterion.

3. THE PROPOSED APPROACH

3.1 Chromosome definition

3.1.1 Multiple sequence chromosome

In this representation, a large number of orders are broken down into smaller ones, The genes corresponding to the number of smaller jobs units are sequenced according to order numbers. This is called a sequence chromosome definition. Multiple sequence chromosomes allow the addition of further sequence chromosomes with a defined numbers of genes each related to the original orders sequence chromosome. The genes of the additional sequence chromosome describe the sequence by which machines should be considered when scheduling orders in the original orders sequence. A multiple sequence chromosome example is shown in Figure 1, which comprises 2 machines and 5 reduced smaller orders units.

3.1.2 Non-sequence chromosome

Non-sequence chromosome can be described as genes with the quantities of each operation on the various machines. It is important to note that the operations that cannot be handled by certain machines have no corresponding genes. The constraint is applied to remove some unnecessary genes. A non-sequence chromosome is given in Figure 2, which comprises machine A and B, 900 Order1 and 500 Order2. The 900 Order1 is replaced by 3 orders of 300 units each. The 500 Order1 is replaced by 2 orders of 250 units each.

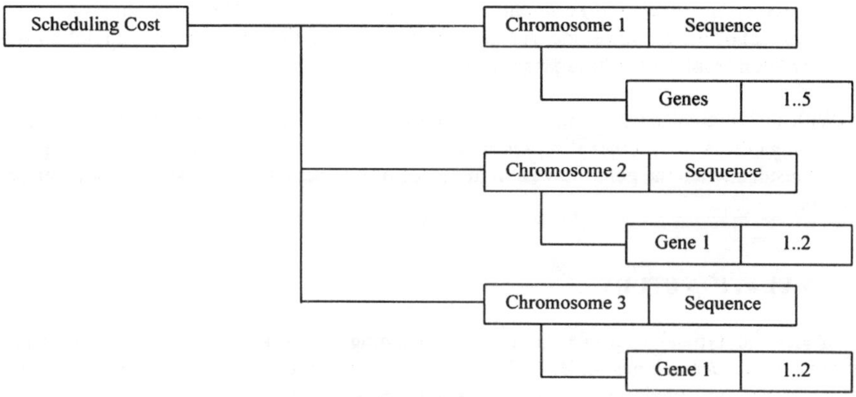

Figure 1. A multiple sequence chromosome

3.2 Advanced optimal operators

3.2.1 Crossover

Each chromosome consists of a number of genes. Each of these genes in turn comprises 16 bits. Thus each chromosome effectively consists of a string of bits. For sequence and gene bound chromosomes, the crossover operator randomly selects two-bit position along the chromosome of the first parent and then exchanges bit information with the second parent in order to make an offspring. For non-sequence chromosomes a random number of bit positions are exchanged between the two parents.

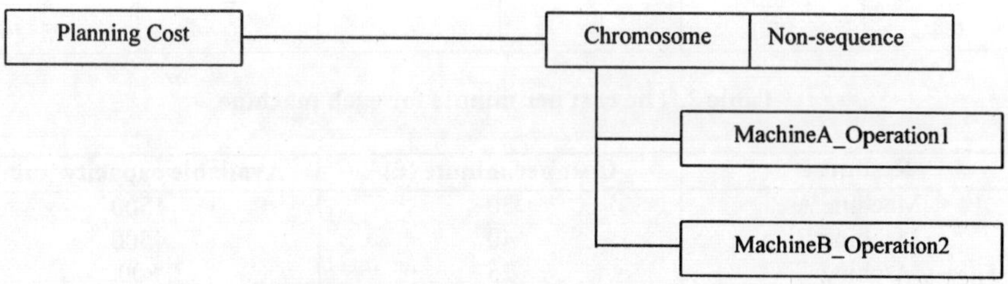

Figure 2. A non-sequence chromosome

3.2.2 Mutation

In non-sequence chromosomes, mutation is a random change in the value of a gene within the chromosome. In sequence chromosomes, mutation involves swapping the positions of two adjacent genes randomly selected.

3.2.3 Adaptation

Adaptation represents a local 'hill climbing' search in the vicinity of a solution. Like mutation, adaptation is a random change to the value or order of genes within the chromosome. However, unlike mutation, the changes are only retained if they improve the cost (fitness) of the chromosome. As such, Adaptation can be seen as a 'wise' mutation which is a deviation from the evolutionary selection process.

With incremental adaptation, each adaptation cycle works on the result of the previous cycle, regardless of whether it represented an improvement in cost. This will often produce better results, when the genes have a high degree of interaction in the way that they impact on the cost.

4. SIMULATION STUDY

Consider the typical example of machine planning and scheduling in manufacturing. A factory has four machines A, B, C, D, these machines are of different ages and capabilities. Their respective maximum available capacity in a week is 1500, 4500, 800, and 1700 minutes. The products manufactured involve four operations O1, O2, O3 and O4. Some products require O1, O2, O3, O4, and others O2, O3 and O4 and so on. The requirements for a week have been summed up as orders of 1400 units, 1200 units, 900 units and 900 units for operations O1, O2, O3 and O4 respectively. The times in minutes per operation and the cost per minute for each machine are shown in the Table 1 and Table 2 (8).

Table 1. The times per operation processed with different machines

Operation	Predicted weekly demand	Unit production time (minutes)			
		Machine A	Machine B	Machine C	Machine D
O1	1400	2	3	1	1
O2	1200	3	4	2	1
O3	900	1	7	4	3
O4	900	1	6	2	5

Table 2. The cost per minute for each machine

Resource	Cost per minute (£)	Available capacity (min)
Machine A	30	1500
Machine B	40	4500
Machine C	25	800
Machine D	20	1700

The problem is how many units of each operation should be manufactured in each machine in order to fulfill the requirements at a minimum cost, without exceeding the capacity of the four machines, using a simple job scheduler, which works on a first come first scheduled basis.

Simulation runs were carried out with *XperRule KBS Package on a P166 32M computer. Which take few minutes to converge to the optimal cost. The simulation results are shown in Figure 3, minimum and mean cost for optimised scheduler are £275000 and £288350.8, and minimum and mean cost for optimised machine output are £319465 and £531120.8. The advanced parameters are illustrated in Table 3.

Table 3. The advanced parameters for optimal solution

Population size	50
Maximum number of generations	50
Crossover rate	0.6
First mutation rate	0.050
Second mutation rate	0.500
Adaptation rate	0.50

*XpertRule KBS is marketed by Attar Software

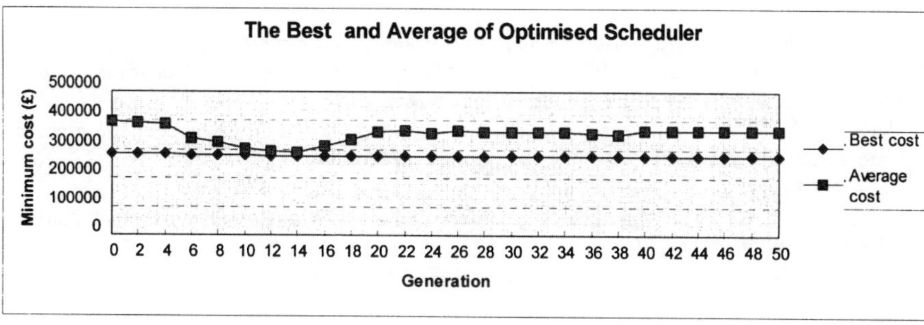

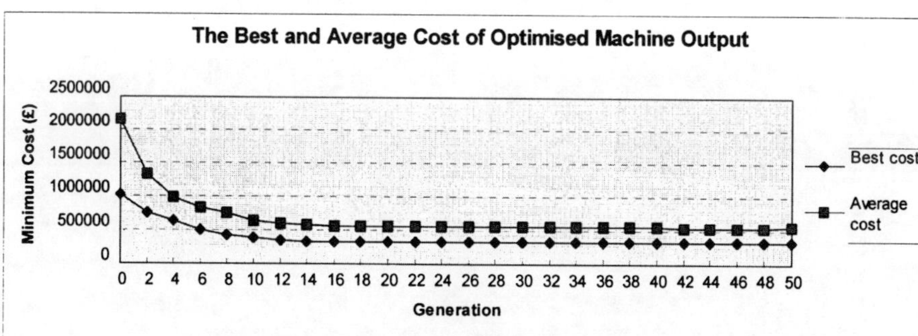

Figure 3. The results of the proposed genetic algorithm

5. CONCLUSION REMARKS

The simulation results indicate that this new planning and scheduling scheme based on a genetic algorithm is an effective and efficient technique to tackle the problem of cost. The algorithm also shows a performance robustness to machine breakdown, problem expansion and inability of machines to process all operations situations.

Further work will concentrate on the development of algorithm to be applied in a more complex manufacturing environment.

6. REFERENCES

1. Proth, J. M., "A comprehensive approach for planning and scheduling on-line systems", *IEEE International Conference on Systems, Man and Cybernetics*, San Antonio, Texas, pp. 1340-1344, Vol. 2, 1994.
2. Morad, N. and Zalzala, A., "Integrated production planning and scheduling in cellular manufacturing using genetic algorithms", *Second International Conference on Genetic Algorithms in Engineering Systems: Innovations and Applications*, IEE Press, Sept 1997.
3. Pinedo, M., "Scheduling: Theory, Algorithms, and Systems", Englewood Cliffs, NJ: Prentice-Hall, 1995.
4. Panwalker, S. S. and Iskander, W., "A survey of scheduling rules", *Operations Research*, **25**, 45-61, 1977.
5. Al-Attar, A., "A Hybrid GA-heuristic search strategy", *AI Expert*, Sept 1994.
6. Adeli, H. and Hung, Shih-Lin., "Machine Learning-Neural Networks, Genetic Algorithms, and Fuzzy Logic", John Wiley & Sons, 1995.
7. Lee, C. Y., Piramuthu, S. and Tsai, Y. K., "Job shop scheduling with genetic algorithm and machine learning", *International Journal of Production Research*, Vol. 35, No. 4, 1171-1191, 1997.
8. Anderson, E. J., "The Management of Manufacturing: Models and Analysis", Addison-Wesley Publishers Ltd, 1994.

Adaptive fuzzy logic scheduling for just-in-time flexible manufacturing systems

J D TEDFORD and **C LOWE**
Department of Mechanical Engineering, The University of Auckland, New Zealand

ABSTRACT

Production scheduling for flexible manufacturing needs to satisfy multiple and often conflicting criteria. The multi-criteria problem is generally difficult to solve using existing heuristics without a means of combining rules and adapting them to changing circumstances.

A scheduling system incorporating an adaptive fuzzy logic system is proposed which can be tailored to suit the characteristics and objectives of an organisation and maintain its performance by constantly adapting to changes in the production environment. The system developed, indicates that its application shows benefits in terms of the resulting schedule performance and with respect to the simplicity of the scheduling algorithms required. The computational overhead of the fuzzy system is also shown to be insignificant thus permitting real-time operation in a manufacturing situation.

1. INTRODUCTION

The advent of flexible manufacturing systems operating "just-in-time" has placed increasing demands on the production scheduling process. The importance of meeting traditional performance criteria such as makespan and flowtime, etc., need to be weighted to reflect other important operating requirements. Specifically, lateness is unacceptable in situations where an assembly line running under JIT relies on the timely production of components to supply it.

Much of the research undertaken for flexible manufacturing systems has focused on the adaptation of existing rules (1). These rules however, display significant drawbacks due to the assumption that the systems operate in steady state conditions and because the rules usually seek to optimise a single criterion. This generally results in schedules with inherently higher risks of failure.

Fuzzy logic has been shown to be ideally suited to solving the multi-criteria problem through its facility to combine and weight multiple inputs. This enables fuzzy systems to emulate much of the decision logic demonstrated by human schedulers (2).

A hybrid scheduling method is proposed here which incorporates adaptive fuzzy logic to generate intelligently reasoned schedules which perform better than traditional heuristics.

2. FUZZY SCHEDULING

Fuzzy systems, although mathematical techniques, define their behavioural framework through a compact linguistic rule-base. They are capable of clearly defining most manufacturing environments and are able to simultaneously consider multiple criteria. Furthermore, fuzzy systems do not generally present severe computational overheads (3).

Consequently, a fuzzy system is able to make reasoned decisions based on many inputs thus lowering the decision risk and increasing the confidence in the output. Fuzzy systems may infer a plausible result even in the absence of complete input data and the fuzzy membership functions and rules can be readily modified to allow adaptability.

Several fuzzy logic based scheduling systems have recently been developed (4,5,6,7), although direct comparisons between them is difficult due to their different implementations and objectives. An alternative approach to the scheduling problem is presented here, it replaces traditional scheduling rules and iterative simulations with an artificial intelligence based approach. Key features of the system are:

- *Schedule from the task not from the resource*
- *Generate a schedule through reasoned decision making*
- *Fully utilise available modern manufacturing information systems*
- *Build real-time capability into the system*
- *Build adaptability into the system*

A fuzzy system incorporating these features was developed. The attributes of each job in the system awaiting processing were assigned both a numerical *task index* and a *priority index* [0..100], based on the following factors:

Value	- the cost of the component
Run Length	- std. cycle time * run quantity
Slack	- lead time
Type	- whether the component was produced for inventory, consignment or as an expedited order.

In a similar manner, the performance and reliability of the available resources were assigned a *resource index* [0..100], based on:

Cost	- the cost of processing
Run Length	- actual cycle time * run quantity
MFT	- mean failure time
MTBF	- mean time between failures
Quality	- typical reject rate

The system then attempts to match a job's *task index* with a resource that has a complementary *resource index*. For example:

If the production run is *lengthy* and *urgent* and *difficult* and the task index evaluates to 90

Then select a resource with a resource index as close as possible to 90.

i.e. a resource (machine) with *good reliability* and *fast performance*.

Components are then backwards scheduled and queued for the selected resource as long as the lateness criterion can be met. Order within the queue for a resource is determined by the *priority index*. If there is insufficient capacity for the component at the preferred work centre, the part with the lowest priority index will be transferred from this work centre to an alternate resource with the next closest resource index. This promotes clustered production of related components and reduces the likelihood of a subsequent assembly line product missing one or more key components.

The fuzzy logic functions were implemented as run-time libraries which were able to be called by the scheduling software. Design of the fuzzy system was facilitated by the use of FuzzyTech® fuzzy logic development software and later compiled to C++ code. These resulting libraries perform only the fuzzy processing functions. A diagram outlining the structure of the scheduling system is shown in Figure 1.

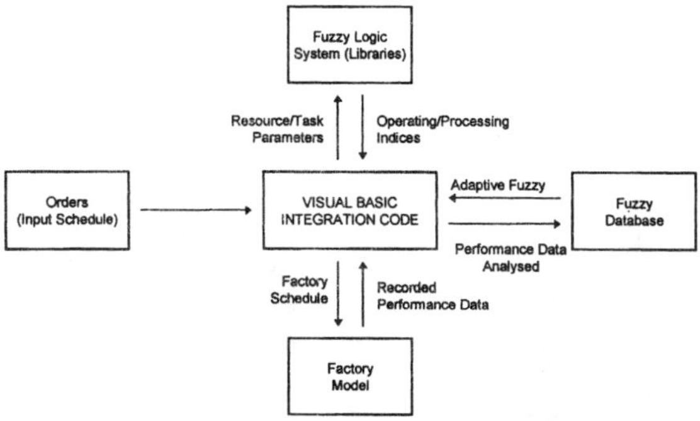

Figure 1 - Scheduling system information flow

The membership functions and rule base of the fuzzy system are stored externally in a relational database and dynamically read into the fuzzy system at run-time. This allows the definition of the fuzzy system to be readily accessed and changed in real time thus enabling adaptability. The interface between the schedule input/outputs, database and fuzzy system is handled by front-end software written using Microsoft Visual Basic®.

The computational overhead of the system is minimal as tested on a standard Pentium® class personal computer operating under Microsoft Windows NT®. The fuzzy system described, can be loaded and 100,000 fuzzy calculations performed on each of the operating and performance indexes in less than fifteen seconds.

3. SIMULATION MODEL AND RESULTS

An initial simulation model was constructed using Arena® simulation modelling software. The model for analysis was based on a subset of the injection moulding shop of a medium sized manufacturing organisation. The simulated model consisted of five injection moulders with twenty-five different components to mould. A weekly master schedule was created for the assembly line and individual component orders derived from it. Within the simulator a module varied set-up times, cycle times, failures, etc. using distributions derived from acquired factory data. Each simulated week processed 90-150 orders.

The simulator is also able to schedule production using the FIFO, SPT, and SLACK dispatching rules in addition to the Fuzzy method proposed here. Each production scenario was simulated using each of these rules in turn. The randomness of work centre failures etc. was repeatable ensuring as far as possible that a similar scenario was being modelled for comparable performance evaluation between the different scheduling methods.

Fifty replications were run equating to one year of production. The performance criteria included, the number of orders completed late, average lateness, makespan, and flowtime. Figures 2 to 5 present comparative histograms of the results, where the worst performing dispatching rule is given a normalised value of 100%. It can be seen that flowtime, using Fuzzy, was reduced by 28.8%, while makespan was reduced by an average of 11.5%. The average lateness and the number of orders completed late criteria benefited the most from the Fuzzy method. The number of orders completed late fell to 45.6% using Fuzzy, while the lateness of individual orders was, on average, 62.1% less when compared with the other rules - a significant reduction in average lateness.

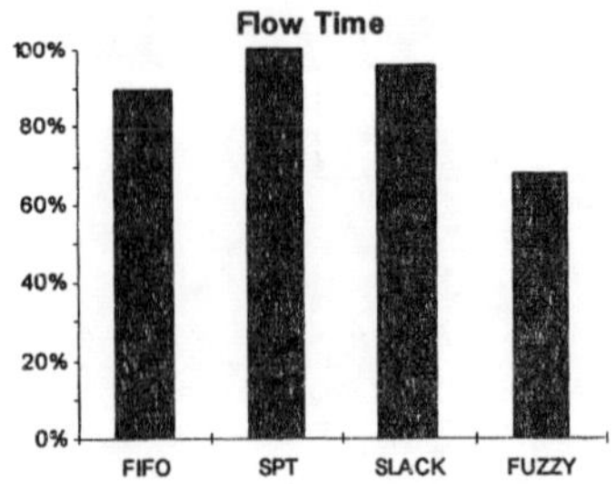

Figure 2 Comparison of Flow Time

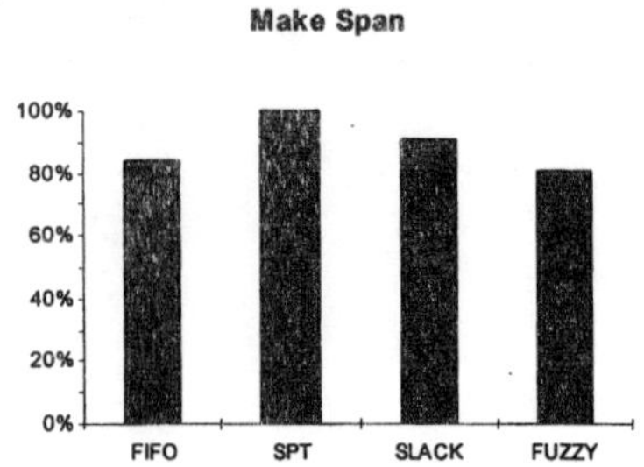

Figure 3 Comparison of Make Span

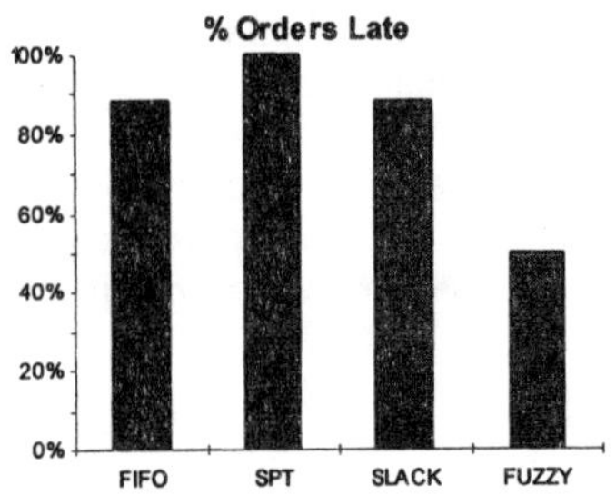

Figure 4 Comparison of Late Orders

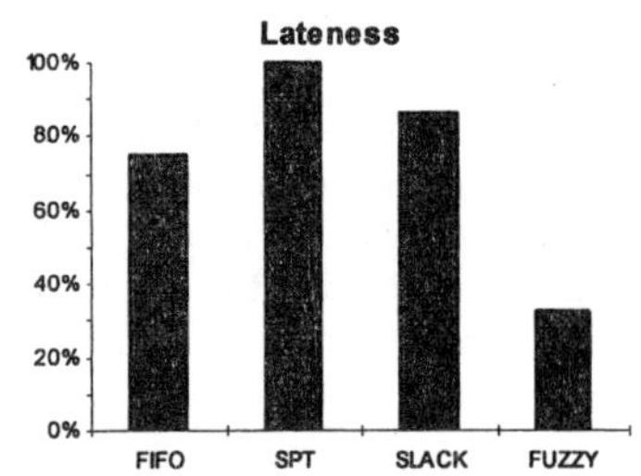

Figure 5 Comparison of Lateness

Following on from these results, a more complex simulation model was developed using Microsoft Visual Basic. Although lacking the graphical displays afforded by Arena, the development of a more flexible simulator was possible, having superior simulation performance and with virtually limitless model variations.

As configured for the subsequent analysis of the fuzzy scheduling technique, the simulator featured:

- 100 machines arranged in 20 groups of 5 machines
- 1000 tasks comprising a mixture of 200 components with process plans requiring up to 5 different machines per component
- Variable cycle and set-up times for each part and machine combination
- Variable cost of processing for each component at different machines
- Variable reject rate for each machine-tool combination
- Variable placement of machines to reflect a real factory layout
- Variable time between failure and failure time profiles, for each machine - able to be set for constant, random, or trending (gradual wearing out of tooling for example)
- Run-time was for one simulated month of production.

The fixed parameters were:

- Constant velocity for transfer between machines

- Invariant set-up costs for machines
- Fixed batch sizes i.e. components were only moved to the next stage of processing after all components were completed.

This model featured real-time adaptive fuzzy scheduling. Each task when moving from one process stage to another was re-evaluated with respect to its task index in real-time and the resource index of each machine was re-evaluated at the end of each week to account for trends in machine performance.

The results from this model are shown in Figures 6 to 9. Overall the results reflect a similar pattern to the earlier model with a clear improvement through the use of adaptive fuzzy scheduling. However, the increase in task variety and process plans requiring up to 5 machines had a slightly adverse effect on the degree of superiority of the fuzzy scheduling technique. This was due to some tasks being moved to less than optimal machines for processing in order to maintain reasonably balanced machine utilisation at the first stage of component production. This was compensated for however, by the better resource-task matching able to be achieved at each subsequent step in the production of a component.

The percentage of orders late under adaptive fuzzy scheduling, rose from 50% to 65% relative to the worst performing rule and average lateness of orders similarly increased from 33% to 47%. This was largely due to the flow-on effects of the lateness incurred in the early stages of processing for individual components. Further research will determine if a change in scheduling strategy to curtail or abandon production of some components may be beneficial.

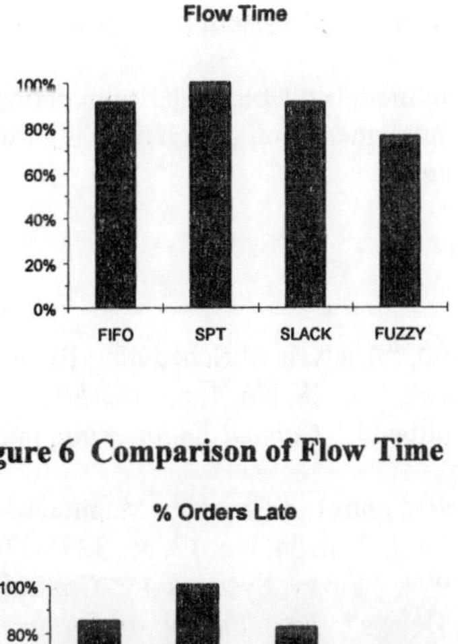

Figure 6 Comparison of Flow Time

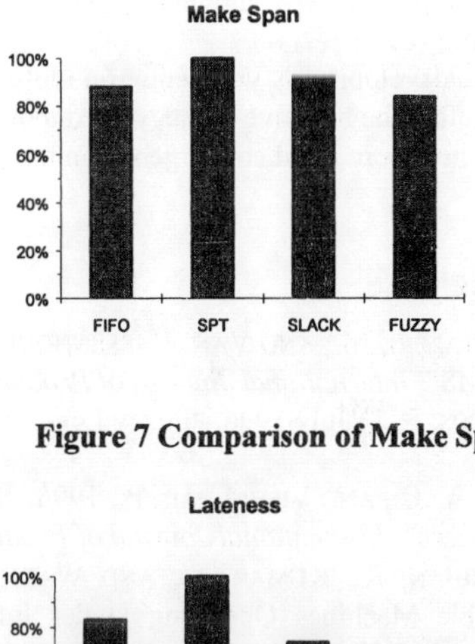

Figure 7 Comparison of Make Span

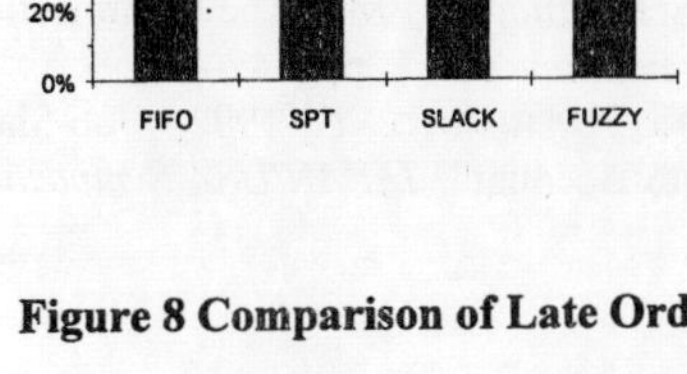

Figure 8 Comparison of Late Orders

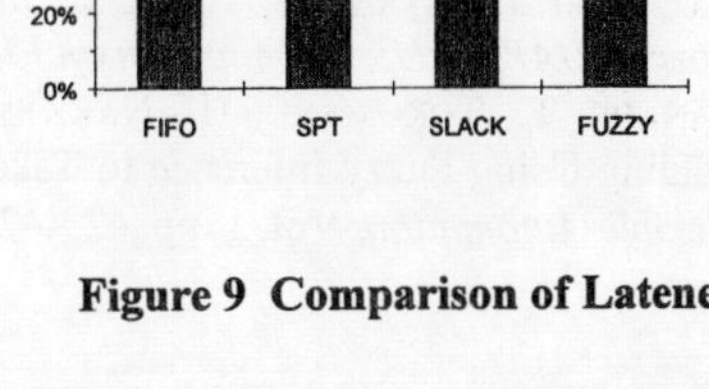

Figure 9 Comparison of Lateness

Overall, the fuzzy scheduling technique proposed was able to make, on average, schedules with at least as good if not better performance than the other commonly used scheduling rules in terms of flowtime and makespan. The performance of the fuzzy scheduling technique however, was consistently superior with respect to lateness which is an important criterion for just-in-time production.

The overall computational overhead was more than acceptable, requiring 1-2 minutes to build and derive a series of initial schedules on a Pentium class computer.

4. CONCLUSIONS

Through the inherent tolerance in fuzzy systems, varying levels of detail can be incorporated as required by any particular manufacturing organisation. This aids in the practical implementation of reliable scheduling where there is missing or incomplete information available to the scheduler.

The results to date indicate that the fuzzy scheduling technique presented, is able to make schedules of improved performance over other commonly used scheduling rules. Its adaptivity provides the ability to modify its decision making in response to changes in the behaviour of the system in real time, thus maintaining its performance superiority over other commonly used heuristics.

Although adaptability was implemented, it is felt that under the simulation model tested here, the run length was not sufficiently long enough to realise its full benefits. Additional simulation runs spanning much longer time periods would likely demonstrate these benefits more fully since they would enable greater long term trends in the system performance to be observed.

Future developments will centre on more variations in models over a longer time period to further refine the adaptive strategy used along with the implementation of a genetic algorithm to assist in a more rapid convergence on the best schedule.

REFERENCES

1. MONTAZERI, M., AND VAN WASSENHOVE, L. N., 1990, "Analysis of Scheduling Rules for an FMS", *International Journal of Production Research*, Vol. 28, No. 4, pp. 785-802.
2. BARTOS, F. J., July 1996, "Fuzzy Logic Reaches Adulthood", *Control Engineering*, pp.50-56.
3. SHIN, Y. C., AND VISHNUPAD, P., 1996, "Neuro-Fuzzy Control of Complex Manufacturing Processes", *International Journal of Production Research*, Vol. 34, No. 12, pp. 3291-3309.
4. CAPRIHAN, R., KUMAR, S., AND WADHWA, S., 1997, "Fuzzy Systems for Control of Flexible Machines Operating Under Information Delays", *Int. Journal of Production Research*, Vol. 35, No. 5, pp. 1331-1348.
5. GRABOT, B., AND GENESTE, L., 1994, "Dispatching Rules in Scheduling: A Fuzzy Approach", *Int. Journal of Prod. Research*, Vol. 32 , No. 4, pp. 903-915.
6. SOMLO, J., AND WATANABE, T., 1992, "Intelligent Scheduling of FMS: The Optimisation Approach", *JAPAN/USA Symposium on Flexible Automation,* Vol. 1, pp. 447-455.
7. WATANABE, T., TOKUMARU, H., NAKAJIMA, Y., AND HASHIMOTO, Y., 1992, "Job-Shop Scheduling Using Fuzzy Inference to Take Profit into Account", *JAPAN/USA Symposium on Flexible Automation,* Vol. 1, pp. 423-427.

An approach to assembly line balancing problem using genetic algorithms

G XU, R PENSON, G LITTLEFAIR, and **R CALLAN**
Systems Engineering Faculty, Southampton Institute, UK

ABSTRACT

The Assembly Line Balancing (ALB) problem (Deterministic Model) is an important aspect when many flow lines of manufacturing systems are designed. To generate line-balanced and cost-effective assembly sequences; we proposed an algorithm using Genetic Algorithm (GA) in this paper. The sequence restriction of the ALB problem is described with the directed graphs and simple binary adjacency matrix and a feasible solution judge Theorem is offered. Then, an approach to the ALB problem using GA was put forward, which describes the solution to ALB problem in the form of the permutation matrix. It is easy to modify the algorithm to suit other form problems, such as flexible task-time, that depend on workstation and minimum the number of workstation.

NOTATION

A – The binary adjacency matrix, N by N matrix, which includes the information of tasks' sequence.

a_{ij} -- The element of the matrix A, $A = (a_{ij})$, a_{ij} denotes the element in row i and column j.

B -- The binary permutation matrix, N by M, which denotes the solution to the problem.

N – Maximum number of tasks

M – Maximum number of workstations

T_i -- The task i

W' – Task times vector (1 by N); it may be a matrix (M by N), if the task time value depend on workstations.

W_i' -- The ith task time: it is considered a constant or it can be considered a variable depend on workstation.

CT -- Cycle Time given by designer

i – Task number $(1 \le i \le N)$;
j – Workstation number $(1 \le j \le M)$;
Px – The rate of crossover
Pm – The rate of mutation

1. INTRODUCTION

1.1. ALB Problem and Genetic Algorithm

The Assembly Line Balancing (ALB) problem is known as one of difficult combinatorial optimisation problem since the search space explosive as task number increase. Various types of the constraint conditions and methods to generate assembly sequences have been reported [10]. ALB problem (Deterministic Model) is a important aspect when many flow lines of manufacturing systems are designed in order to increase productivity and reduce assembly cost. There are many separate and distinct processing and assembly operations to be performed on the product in flow line production. The assembly steps or the sequence of processing is restricted. These restrictions are called precedence constrains.

The ALB problem is to arrange the individual processing and assembly task at the workstations. The total time required at each workstation is approximately the same we call it predict line balancing when the task can be grouped so that all workstation times are exactly equal; but it is very difficult to achieved. In most practical situation, the workstation times are unequal; the slowest workstation determines the overall production rate of the assembly line.

A major feature of GA is the ability to take care of a variety of objective functions [4]. It has been used for seeking a set of diverse optimal solutions for different ALB problem [5]. An approach to ALB Problem using GA is determining the minimum cycle time, subject to precedence and given number of workstation [4]. GA can generates multiple solution to the ALB problem. Good results are achieved by combining the genetic approach with a simple local optimisation procedure. On other hand, in the random search, the average cycle value does not converge to the best cycle time even after 1000 generation's [3]. The most random search is effecienless. But the random search can help to escape the local minimum.

The different crossovers and mutations used in the ALB problem are tested by Rubinovitz and Levitin [3]. The GA required longer computational CPU time. It may be tolerable since the ALB problems are not demanding a real time solution [5]. Therefore we do not put main attention at the reducing computational CPU time in this paper.

1.2 The Framework of the Methodology

The overall framework of the algorithm in the paper is shown below.

1. Initial an arbitrary population P (gen = k), k=1, P is a constant and gen denotes the generation number, in which each chromosome stands for a feasible solution, using the feasible solution judgement theorem.
2. Select k parents from the population P (gen = k) after Fitness is evaluated for each individual of the population.
3. Apply genetic operators, crossover and mutation, over the parents' chromosomes chosen in last step and produce offspring.
4. Add offsprings to P (gen = k) and delete redundant chromosomes to generate new population P (gen = k+1) according to the fitness.
5. Repeat step 2 to step 4, until a termination condition is met. Usually the termination condition can be a number of the generations or a global objective value and so on.

This paper is divided into four major sectors as follows; after this sector the ALB problem formulation is described, which discussed feasible solution criterion, coding and crosssover. A 12-task-element example is taken as case study. Then the experimental results and its design using the proposed algorithm were given in the third sector. The conclusion were put forward in the forth sector.

2. PROBLEM FORMULATION

The ALB problem, as an example will be discussed in the most popular form, the objective is to minimise the cycle time CT, given the number of station.

2.1 Precedence Diagram

The example below is from Groove Fig 1. [8].

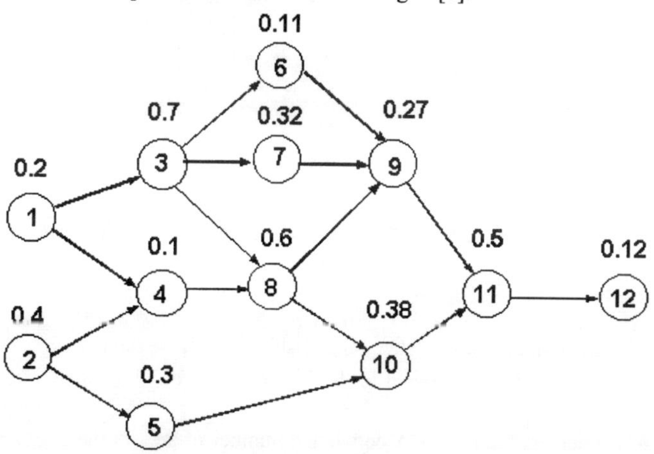

Fig.1. 12-element example

Using nodes to symbolise the tasks, above which task times are recorded, with arrows (edges) connecting the nodes to indicate the order in which the tasks must be performed. Using 0, 1 variable when task *j* has immediate predecessor task *i* then the element $a_{ij} = 1$, in the matrix A, Otherwise is 0. We will assume in the development of the procedure the nodes of the network are numbered such that if task *m* precedes task *n*, *m*<*n*. Therefore, the adjacency matrix can be upper triangle and sparse one. Its main diagonal element is zero. The diagonal elements are zeros since no task start from and terminate to itself.

Since the directed graph is been used to represent the relationship of the task sequence, we have the definition as follows.

I_i -- the indegree of node *i*, i.e. the number of immediate predecessors (tasks) which the task *i* has, e.g. $I_1 = 0$ for the task 1, $I_4 = 2$ for the task 4.

O_j -- the outdegree of node *j*, i.e. the number of immediate follow-up task the task *j* has. e.g. $O_1 = 2$ for the task 1, $O_4 = 1$ for the task 4.

2.2. Matrix Representation

The precedence relationships are indicated with a square matrix which shows the precedence relations among the tasks where cell a_{ij} above the main diagonal of a matrix A, where element a_{ij}=1 if task *i* precedes task *j*, and a_{ij}=0 otherwise.

The matrix for Fig.1 is as follows.

$$A = \begin{bmatrix} 0 & 0 & 1 & 1 & 0 & 0 & 0 & 0 & 0 & 0 & 0 & 0 \\ 0 & 0 & 0 & 1 & 1 & 0 & 0 & 0 & 0 & 0 & 0 & 0 \\ 0 & 0 & 0 & 0 & 0 & 1 & 1 & 1 & 0 & 0 & 0 & 0 \\ 0 & 0 & 0 & 0 & 0 & 0 & 0 & 1 & 0 & 0 & 0 & 0 \\ 0 & 0 & 0 & 0 & 0 & 0 & 0 & 0 & 0 & 1 & 0 & 0 \\ 0 & 0 & 0 & 0 & 0 & 0 & 0 & 0 & 1 & 0 & 0 & 0 \\ 0 & 0 & 0 & 0 & 0 & 0 & 0 & 0 & 1 & 0 & 0 & 0 \\ 0 & 0 & 0 & 0 & 0 & 0 & 0 & 0 & 1 & 1 & 0 & 0 \\ 0 & 0 & 0 & 0 & 0 & 0 & 0 & 0 & 0 & 0 & 1 & 0 \\ 0 & 0 & 0 & 0 & 0 & 0 & 0 & 0 & 0 & 0 & 1 & 0 \\ 0 & 0 & 0 & 0 & 0 & 0 & 0 & 0 & 0 & 0 & 0 & 1 \\ 0 & 0 & 0 & 0 & 0 & 0 & 0 & 0 & 0 & 0 & 0 & 0 \end{bmatrix}$$

Fig. 2 The adjacency matrix for 12-element problem

From the definition of the matrix A and the precedence relationships Fig. 1, we have:

1. The number of the arcs in the Fig. 1, *Ks*, is equal to the sum of indegree, *ΣIi*, as well as outdegree, *ΣOj*.
2. The number of non-zero element in the matrix A equals the number of arcs in the Fig. 1, *Ks*.
3. The sum of non-zero element of the column *i*, matrix A , is the indegree of the task *i*, *Ii*.

$$\sum_{j=1}^{N} a_{ij} = I_i$$

, The sum of non-zero element of the row *j* is the outdegree of the task *j*, *Oj*.

$$\sum_{i=1}^{N} a_{ij} = Oj$$

2.3 The Representation of the Solution

Suppose the solution is represented as the matrix Bs and unit matrix I_M;

$$I_M = \begin{bmatrix} E_1 \\ E_2 \\ \vdots \\ E_{M-1} \\ E_M \end{bmatrix} \quad B_S = \begin{bmatrix} E_{k1} \\ E_{k2} \\ \vdots \\ E_{N-1} \\ E_N \end{bmatrix}.$$

Bs is the linear combination of the E_i. i.e.

Where E_i is the unit vector of M elements, i.e., the row vector and the ith element is 1, the others are 0s. In this way, the solution shows that each task is allocated to one and only one workstation.

When it is shown as matrix, $B_S = [b_{ij}]$, $b_{ij} = 1$ means the ith task is assigned to the workstation *j*. Furthermore, chromosome CH is a string of 0s and 1s, i.e. it is a simple binary string, it looks like $CH = [E_{i1} \quad E_{i2} \quad \cdots \quad E_{iN-1} \quad E_{iN}]$, Where the *E* is the unit vector.

The solution to the ALB problem is described with permutation matrix B, which's element *bij=1* if task *i* is allocated to workstation *j*, and *bij=0* otherwise.

All the task time is a matrix $W' = [W'_{ij}]$, the element *Wij* is the time units of the *j*th task performed in the *ith* workstation.

2.4 Feasible Solution Judge Theorem

When the solution B is feasible, it satisfies the theorems below. The theorem guarantees it satisfies precedence constraints.

Feasible solution judge Theorem:

If the definition of the binary matrix $A = \lfloor a_{ij} \rfloor$ and $B = \lfloor b_{ij} \rfloor$ are described as above, they stand for the precedence constraints and the solution separately. When and only when the following condition is satisfied, the matrix B is the solution that satisfied the precedence constraints.

$$\sum_{i=1}^{N}\sum_{x=1}^{M-1} b_{ix}\left(\sum_{y=x+1}^{M}\left(\sum_{j=1}^{N} a_{ij}^{\tau} b_{jy}\right)\right) = 0 \qquad \text{Where } a_{ij}^{\tau} = a_{ji}$$

Proof: It can easily be derived from the definition of the matrix A and B.

2.5 Crossover Procedure

There are many crossover operators; they must be utilised according to the feature of the problem and the characteristics of algorithm. So some of the conventional can not used in the case of ALB problem. The crossover is based on the principle of best survive. They inherent the parents good merit. There are some argument on which is better in the chromosome coding using real value or 0~1 variable [9]. The real value chromosome coding is used in the papers [3,4,5]. We use 0~1 variable in this paper. Since the search is only in the feasible solution space, each chromosome is satisfied the precedence constrains. In other words, it satisfies the feasible judgement theorem. The crossover procedure is as follows.

1. Look for the minimum number of the workstation Wa in the left sector. For the left sector of parent 1, it is Wa=1, that means the 1st station has been allocated.
2. The allocation begin from the (Wa+1)th workstation; For the left sector of parent 1, it is (Wa+1)=2. Keep it was in the sector. Then $P_{1L} \oplus P_{2R}$ is a new chromosome (see Fig.4), but generally, it is not a feasible solution judged by theorem. Therefore the chromosome should be modified as following steps.
3. Manipulation on the matrix A in order to get the tasks that will be allocated. Set the row 1, 2, 3, 4, to zero, and the column 1, 2, 3, 4 to 1, since the allocated tasks are Number 1, 2, 3, 4. They are in parent 1 left sector, See Fig. 4.
4. Get the columns in which the sum of the elements is zero. The number of column corresponds the number of the tasks that will be allocated. In our case, they are the task 5, 6, 7, 8.

5. Consider the constraint of CT, i.e. $CT \geq Wt * B_i$, delete the tasks they can not satisfied the constraint. Since the row index in B denotes the index of the tasks. The B_i is the column i of the matrix B. the elements in the column are 1s corresponding the index of the allocated tasks and an allocating task. In our example, they are 2,4 and a k ($k \in [5,6,7,8]$). Other elements are 0s.
6. Allocate a task to the workstation Wa (here Wa=2), the task in the P_{2R} have priority. Otherwise they are randomly allocated according to uniform distribution. It is the task 6 in our example is allocating to the workstation 2.
7. Suppose the task T_i is allocated in this step. In order to add the tasks that will be allocated, Set the row i, to zero, and the column i to 1.
8. Go to the step 4 and repeat these step until all tasks are allocated that generate the first offspring.
9. Apply the above steps 1 ~ 8 to P_{2L} and P_{1R} to generate the second offsprings.

Let us take the 12-element problem [8] as example.

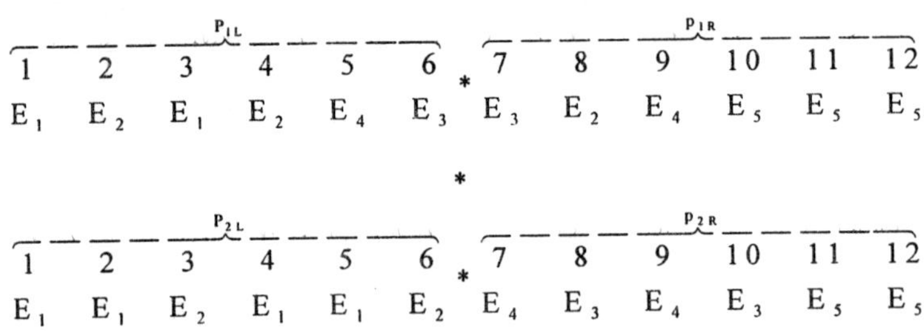

Fig.3. Crossover point and chromosome

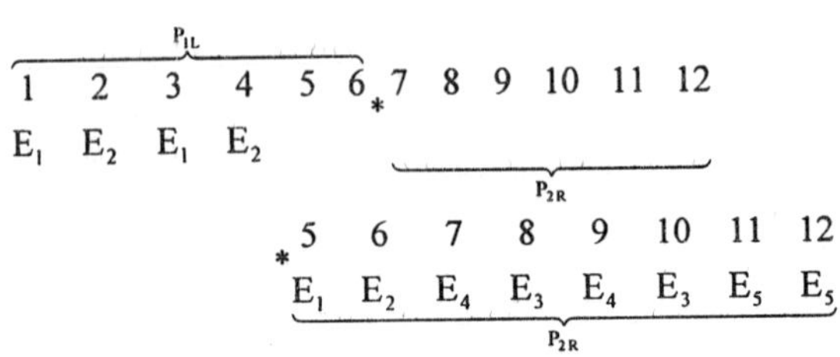

Fig.4 modified crossover point parent sectors

Note: * -- the crossover point.

P_{1L} -- The left sector of the crossover point in parent 1.

P_{1R} -- The right sector of the crossover point in parent 1.

P_{2L} -- The left sector of the crossover point in parent 2.

P_{2R} -- The right sector of the crossover point in parent 2.

The first row in the Fig is the number of the task.

2.6 Mutation Procedure

It is important to apply mutation operator even the crossover operator play a main role. It usually gets into the local minimum when the crossover operator is only used in the GA without mutation operator. In our formulation, chromosome (see Fig. 3) is represented with 0~1s code strings. The mutation operator is that,

1. Selecting randomly K points in the binary chromosome and simply change a 1 to 0 or vice versa. After that, the E_i in the chromosome may have more than one 1 or all 0s.
2. In the first case, suppose the first 1 is at the position ik and the last 1 is at the position iL, here $ik, iL \in [1, M]$, the E_h, which will replace E_i, is been chosen from $[E_{ik} \quad \cdots \quad E_{iL}]$ .in the second case, suppose the original 1 position is j. the E_h is chosen from $[E_1 \quad \cdots \quad E_M]$ except E_j.
3. Then go to crossover procedure step 4 to step 8.

3. EXPERIMENTAL DESIGN RESULTS

A feasible solution is one that satisfies the precedence constrains requirements and does not cause the sum of the time values at the station to exceed the cycle time CT. search is in the feasible solution space can speed convergence. However, A crossover alone, which leads to local minimum, is not sufficient in finding good solution, and mutations, to which the premature convergence rarely occurs, are needed to work with the crossover to obtain good result [5]. The mutation operator makes random changes to one or more element in a solution string and play only a secondly role, but without mutation, some potentially useful material may be lost. It can be seen form fig. 7, where *Px* is the rate of crossover; *Pm* is the rate of the mutation. It is for the 21-task problem, symbol '+' stands for the average of the first 10 best objective, and the 'o' for the average of all objective.

The objective is as follows

$$SI = \sqrt{\sum_{k=1}^{M} (S_{max} - S_k)^2} \qquad k = 1,2,\cdots,M$$

Where S_{max} is the maximum station time and S_k is the individual station time.

In order to handling irregular search spaces, it is necessary to allow infeasible solutions in the population [4]. In our formulation, the infeasible solutions come from the mutation. In particular, the different ratio of *Px* and *Pm* is tested in our experiments. The experiments show that the good match of the $Px \sim Pm$ is *Px* and *Pm* in the area of 0.5~0.8.

The experiments for the GA were carried out on the 21-task and 70-task problem from Tonge [7]. The population size in our experiment was 36 and 72 separately. Fig. 5 and 6 is typical trend of convergence.

4. CONCLUSION

In order to operation easily, the sequence restriction of the ALB problem is described with the directed graphs and simple binary adjacency matrix. The binary major advantage in the method is that it deals with the binary matrix in which only the simplest matrix operations are needed. A concise matrix expresses and manipulates the ALB problem in this paper. It can be using in the case where task times depend on the workstation. It can also offer a set of solution when the real line design needs to investigate alternative solutions, where preferences for work allocation to station constraints are considered.

From the results of the experiments, it is concluded that the proposed method can generate good result for assembly line balancing. It is necessary to combine the GA with the random search in order to speed convergence.

Make use of the precedence relationship and its matrix representation, the feasible solution judge theorem are applied, shrink the solution space searching time, these are valuable in generating good offspring to ensure both the speed of the algorithm and the quality of the solutions. The experiments show that the good match of the $Px \sim Pm$ is Px and Pm in the area of 0.5~0.8.

REFERENCES

[1] Baybars I. 1986, A Survey of Exact Algorithms for the Simple Assembly Line Balancing Problem. Management science vol.32, no. 8, 909-932

[2] Wai-Kai Chen, Graph Theory and its engineering applications. world scientific publishing Co. Pte. Ltd. ISBN 981-02-1859-1.

[3] Rubinovitz, J. and Levitin, G.. 1995. Genetic algorithm for assembly line balancing. International journal of production economics, 41, 343-354

[4] Suresh, G., Vinod, V. V. and Sahu, S.. 1996. A genetic algorithm for assembly line balancing. Production planning & control, Vol, 7, No, 1, 38-46

[5] Yeo, K. K., Yong, J. K. and Yeongho, K.. 1996, genetic algorithms for assembly line balancing with various objectives, computers ind. Engng., Vol. 30, pp. 397-409

[6] Groover, M. P. 1987, Automation, Production Systems, And Computer Integrated Manufacturing. Pretice-Hall, Englewood cliffs, New jersey. ISBN 0-13-054652-6.

[7] Tonge F. M. 1961, A Heuristic Program For Assembly Line Balancing, Prentice-Hall, Inc.

[8] Goldberg D. E. 1989, genetic algorithms insearch, optimization, and machine learning. Addison Wesley, reading, MA.

[9] Reeves C. R., 1995, Modern Heuristic Techniques For Combinatorial Problem. McGraw-Hill Book company Europe ISBN 0-07-709239-2.

[10] Soumen G. and Roger J. G., 1989, A Conprehensive Literature Review And Analysis Of The Design, Balancing And Scheduling Of Assembly Systems. INT. J. PROD. RES. Vol. 27, No. 4, 637-670.

[11] Moodie, C. L. and Ben-Tuvia S., 1966, A heuristic method of assembly line balancing for assumptions of constant or variable work element times. Journal of industrial engineering, Vol. 16. 23-29.

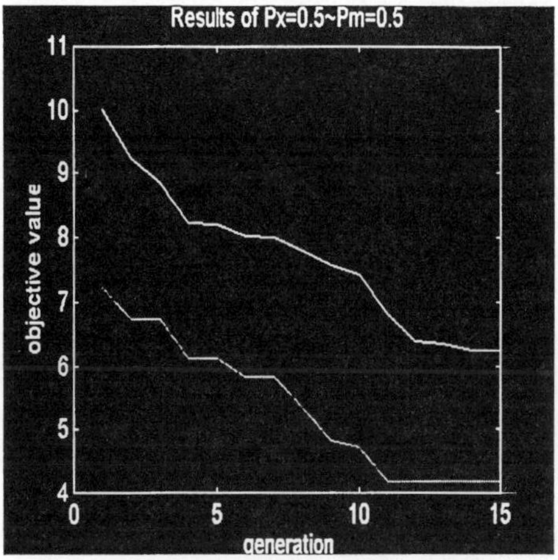

Fig. 5 The result of the 21-element problem

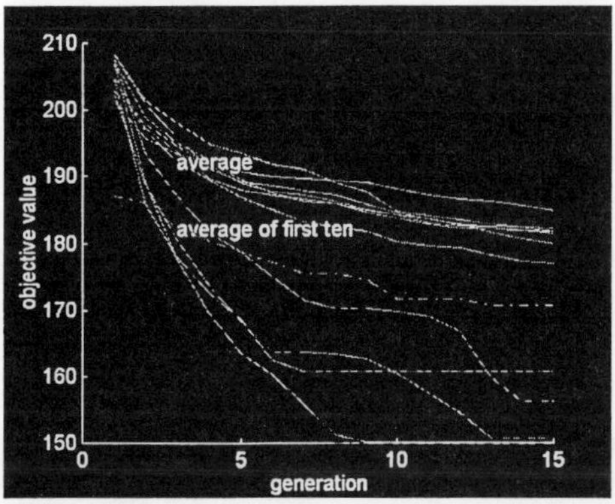

Fig. 6. The convergence trend in 70-element problem experiment

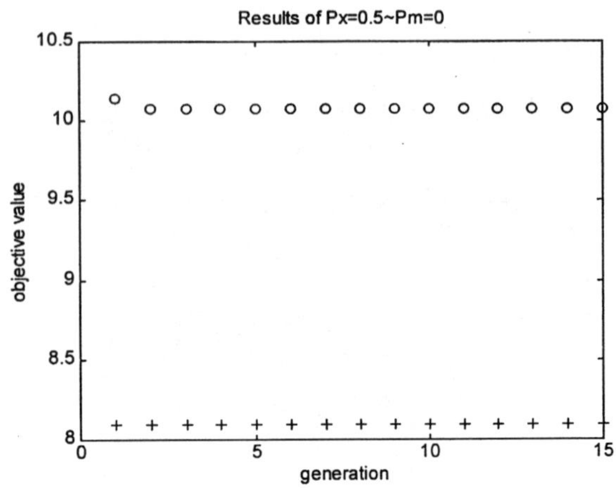

Fig. 7. The 21-element problem experiment without Mutation

Virtual Reality

Virtual vertical integration as a strategy for manufacture

L ARGUMENT and **C E R WAINWRIGHT**
The CIM Institute, Cranfield University, Bedford, UK

Abstract

Historically, businesses have changed their organisational structure in order to adapt to factors including the constantly changing environment in which they operate. For industrial organisations, change may include the move from integration (where design, manufacturing and other associated functions are performed internally within the business) to disintegration, involving the break up of organisations into smaller parts. Organisations who have pursued this type of strategy may contract out specific practices to outside parties as they no longer have the capability to perform certain processes. An alternative strategy to pursue is virtual vertical integration, combining the principles of the virtual corporation with those of vertical integration, in order to provide an agile environment.

This paper addresses the concept of virtual vertical integration by initially exploring the implications of an organisation considering virtual vertical integration. Subsequently, a conceptual framework for a tool, which may be used as an aid to decision making by an organisation wishing to pursue virtual vertical integration, is proposed.

1.0 INTRODUCTION

It is vital that organisations make changes to their organisational structure and hierarchy so that they may adapt to fluctuations in their operating environments in order to survive. Changes have included the move from being a fully integrated organisation (where all processes to produce and deliver a product are performed by one company) to disintegrated organisation (where one or more of the processes is outsourced for another company to perform).

This process has come almost full circle in some organisations that have undergone vast structural changes in recent years to cope with the demands of modern markets. This focus on structural change is also accompanied by changes to the business and manufacturing strategies pursued by some companies so that sustainable competitive advantage may be achieved. Strategies are constantly evolving and undergoing global promotion in order to cope with future generation manufacturing and business operations. The argument is that companies must be prepared to change or adapt any mindsets they may have in order to cope with the requirements of manufacturing in the 21st century. The pressures on the firm, and fluctuations in markets and product demand are emphasising a need for a reduction in time between an order being placed to receiving the payment for it. Taiichi Ohno (1) focused on the need for this time reduction in the Toyota Production System, (which eventually gave rise to lean production), and this is carried over into agile manufacturing (2), a concept which aims to build on what was achieved by lean production and to create a specific level of flexibility in a firm so that it may cope with the cornucopia of demands placed on it.

2.0 BUSINESS AND MANUFACTURING STRATEGIES

The difference between business and manufacturing strategy has been attributed to the different terminologies used by managers in the manufacturing function and managers at the corporate level. There is one other level of strategy between the upper corporate level and the bottom functional level - the business level, which resides between the corporate and functional levels, feeds into both strategies yet has it's own objectives and goals. All strategies should offer support to the ultimate objectives of the company

The need for strategies is ever increasing in order to guide companies towards achieving their short and long term goals and objectives. As companies have evolved so have the concepts that guide their business and manufacturing processes. As mass production gives way to mass customisation strategies will have to be adapted or developed in order to cope with the demands that customisation will place on companies. Mass customisation is assisted through the adoption of manufacturing concepts such as agile manufacturing but eventually this concept will age and give way to a new concept. It is proposed that virtual vertical integration, a strategy for manufacture combining principles of vertical integration and the virtual organisation, will be a suitable strategy for manufacturing in the next century.

The Product Model shown in Figure 1 shows a simple breakdown of the levels of activity in a company. At the very top level there is the Organisation which will have a corporate strategy to govern processes. The next level is the Product level which is governed by a business strategy allowing for marketing and logistical processes. The next levels, Assembly and Component, are governed by functional design and manufacturing strategies drawing on daily and monthly output levels to monitor progress against strategic goals.

Figure 1: The product model

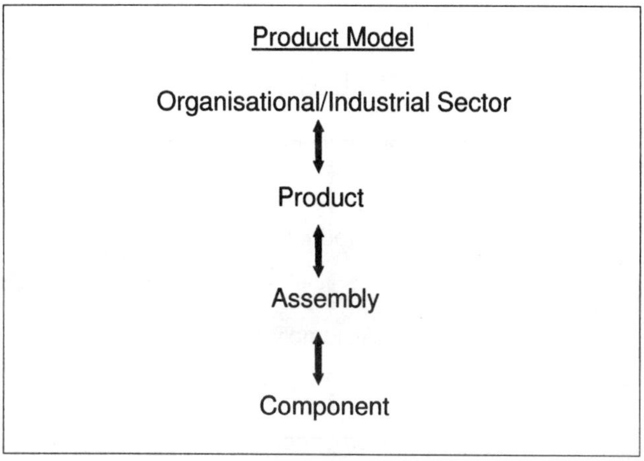

The manner by which the Product Model has been traditionally implemented in a company has been through the process of vertical integration. This approach has been the choice of companies wishing to retain control over their processes rather than outsource them. It may not be suitable for all companies but those using it may choose a degree of integration thus tailoring the approach to suit their business and manufacturing requirements.

Vertical integration has been described as the combination of technologically distinct production, distribution, selling and/or other economic processes within the confines of a single firm (3). It is an essential strategic move taken by a firm in order to guarantee sources of supply and secure bargaining leverage on vendors (4). When considering such a strategic option it is vital to consider the implications for the products offered by the company. There is little reason to pursue such a strategy if its effects will be detrimental to the success of the product and the company. It is often chosen as the means for executing a significant portion of business processes as opposed to contracting as firms believe it is cheaper, less risky or easier to co-ordinate (3).

Agility is the umbrella term used to denote the change in the way companies are approaching manufacturing in the next century. Changes in market characteristics are triggering structural changes within many companies so that they can best prepare themselves to cope with broader product ranges, fluctuating lot sizes, and shorter product lifetimes. After coming through a period of lean manufacturing, which some have argued has seen companies neglecting changing consumer tastes in favour of continuous improvement practices (5), a period of mass customisation is looming which may require further structural and strategic changes within manufacturing companies. It is no longer enough just to define customers and their corresponding market segments. It is necessary to understand the needs of the customers in terms of the products and services they require. The traditional mass production systems, where production processes and resources tend to define the product (6), will give way to systems geared to providing individualised products. Figure 2 shows how agile manufacturing can be mapped onto the product model.

Figure 2: Mapping agile manufacturing onto the product model

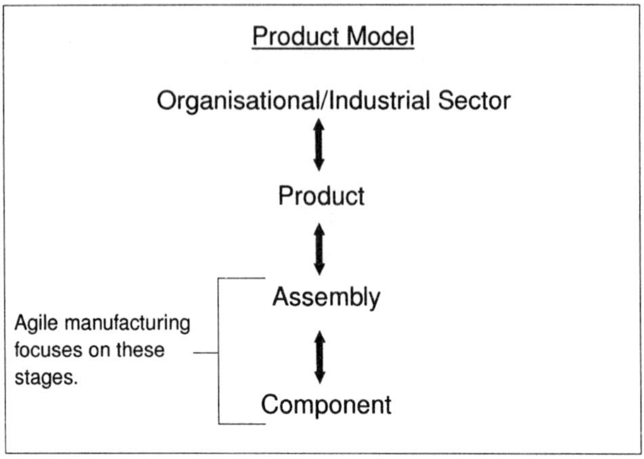

The virtual organisation has been identified as a feature of agility (7). For some it would seem that the virtual organisation defies consistent definition (8) but whatever its features or construct the virtual organisation brings alliances, joint ventures, extended information technology links, collaborations and other partnerships into the spotlight. As a feature of agility the virtual organisation emphasises internal and external co-operation, the aim being to bring agile products to market in the minimum time by leveraging resources through co-operation (7).

The virtual organisation has been described as a temporary network of independent companies - suppliers, customers, linked by information technology to share skills, costs, and access to one another's markets (9). In relation to the product model its effects are primarily top down and this is illustrated in Figure 3. It has also been said to be characterised by an opportunity pulled and opportunity defined integration of core competencies among a number of real organisations. This integration of the core processes is usually based on electronic means and a deep trust between the partners (10). The purpose of the virtual organisation is to band together to rapidly respond to arising market opportunities, to meet any opportunities by exploiting the relevant skills of the parties involved, and then to disband once the job has been done.

3.0 VIRTUAL VERTICAL INTEGRATION

As we move into the next century manufacturing concepts will evolve to account for the advances in technology and other changes taking place in the external environment. An emerging paradigm known as virtual vertical integration may be examined as a next generation strategy for manufacturing in the 21st century. Virtual vertical integration refers to the integration of companies involved in the production, distribution and selling of a product using electronic links as opposed to having full or part ownership. This combination of the principles of vertical integration and the virtual organisation requires that the partnering companies combine their resources and processes but does not require ownership of firms up or down the supply chain to achieve this. The companies involved in this relationship use information technology links to conduct their business with all the companies involved in the integration thus having access to one another's technologies, data, and processes.

Figure 3: Mapping the virtual organisation onto the product model

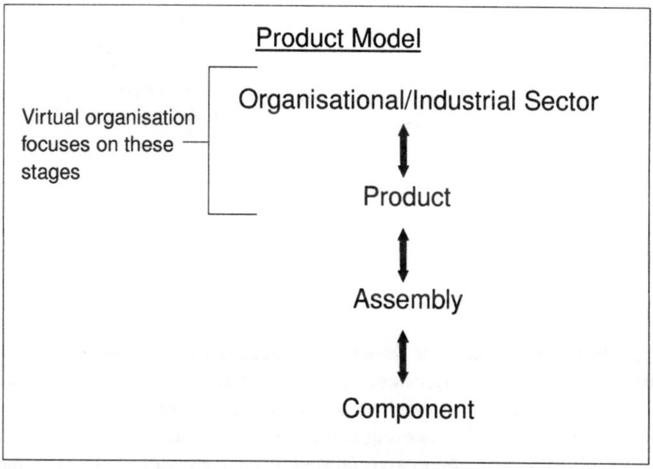

Virtual vertical integration requires a great deal of co-operation internally and externally between the companies involved if they are to successfully co-operate with each other on a more permanent basis. The difference between virtual vertical integration and the virtual organisation is that the relationships are extended over a wider field to incorporate not only the head organisations but also the smaller companies in the chain. Figure 4 shows how virtual vertical integration may be mapped onto the Product Model.

The argument supporting the need for research into this area is simple. The product model in Figure 1 shows the stages in the product development process and strategies that may be pursued by companies in order to achieve the greatest flexibility at various stages in the concept to grave cycle. The first section, named in the model as the Organisational/Industrial Sector, is the last stage in the development process and will be referred to as level 4. At this level in the product model it is possible for companies to form partnerships of some sort to improve their product development process and sustain their competitive advantage. The next level in the model, which shall be known as level 3, leads down from level 4. This level refers to the final assembly, or the final product requested by the customer. This level leads down into level 2 which is the assembly level. At this level in the model various components are assembled to make the sub-assemblies which are eventually combined to form the final, fully assembled product. Level 1, which is known as the Component level, feeds into level 2.

Component parts are produced at this level which are then used to form sub-assemblies at the next level.

Figure 4: Mapping virtual vertical integration onto the product model

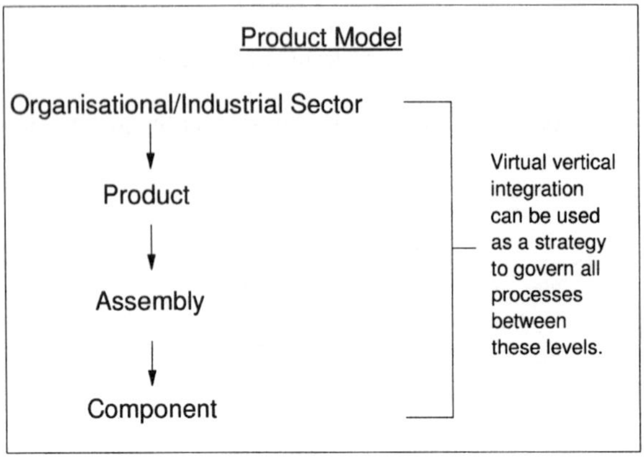

4.0 CONCLUSIONS

For many years, European companies have been searching for a concept to help them achieve the competitive success that the Japanese automotive industry has enjoyed. Such concepts have included the extended enterprise, lean production and concurrent engineering, amongst others. These concepts are still governing manufacturing today, in one guise or another, but eventually they will give way, or evolve, allowing newer concepts to take manufacturing through the next century. Agile manufacturing is a manufacturing concept, or strategy (as it has also been referred to), that is predicted to govern next generation manufacturing processes. The virtual organisation is also predicted to become a strategy for the future. This paper has examined the implications of these concepts for companies wishing to pursue them and proposes another concept, namely virtual vertical integration, which combines principles of vertical integration, the virtual organisation, and also agility in one strategic approach to next generation manufacturing.

It is proposed that there is insufficient research to support the creation of virtual vertical integration which involves all levels of the organisation, from the very top corporate level to the bottom functional level. Strategies can be formed and implemented at the corporate, business and functional levels but with increasing levels of technology being used to integrate companies across all levels it may not be possible to ensure that these strategies are aligned and supportive to the ultimate corporate strategy. As virtual vertical integration is a strategic approach to production then it is necessary to conduct research to operationalise the Product Model framework to assist companies to integrate effectively and align their strategies.

References

(1) Womak, J, Jones, D, Roos, D, (1990): The Machine that Changed The World
Rawson Associates, 1990

(2) Kidd, P, (1996): Agile Manufacturing - Forging New Frontiers
Addison Wesley, 1996

(3) Porter, M, (1980): Competitive Strategy: Techniques for Analysing Industries and Competitors
Macmillan Publishing, 1980

(4) Galbraith, Kazanjian, (1986): Strategy Implementation – Structure, Systems, and Process
2nd Edition, West Publishing Company

(5) Esmail, K, Saggu, J, (1996): A Changing Paradigm
Manufacturing Engineer, December 1996

(6) Goldman, S, Nagel, Preiss, K, (1995): Agile Competitors and Virtual Organisations
American Society of Mechanical Engineers, Van Nostrand, 1995

(7) Harrison, A, (1997): From Leaness to Agility
Manufacturing Engineer, December 1997

(8) Ward, J, (1997): The Lure of the Virtual Corporation is Flexibility
Photonics Spectra, July 1997

(9) Charles, F, (1983): The Virtual Corporation
Business Week, February 8, 1993

(10) Muller-Wunsch, M, (1995-1996): The Execu-desk Project: Managing the Virtual Corporation
Journal of Computer Information Systems, Winter 1995-1996

The application of immersive virtual reality for layout planning of manufacturing cells

B KORVES and **M LOFTUS**
School of Manufacturing and Mechanical Engineering, University of Birmingham, Edgbaston, UK

ABSTRACT

Immersive Virtual Reality (VR) is a new form of human computer interaction, which has the potential to support a range of engineering applications. A recent review of the methods of planning and implementing Manufacturing Cells indicates that the visual capability of VR could assist the planning process. Current planning methods are restricted by 2D and 3D modelling systems, which provide visual representation of the Cells but lack the full user interaction. This paper describes the framework of an Immersive VR approach to the planning and implementation of Manufacturing Cells in which the planner enters the modelled world of the application.

INTRODUCTION

Layout Planning is not a new problem in the manufacturing environment. For centuries, craftsmen such as blacksmiths have arranged their working area according to the process requirements and the movements of work. Project layout, functional layout and line layout schemes have emerged in response to the growing complexity of the operations and the physical size of the organisations.

Muther (1) was one of the first to apply a systematic methodology to the planning process and his Systematic Layout Planning (SLP) approach is still a valid procedure. Muther considers the planning process as a loop, which has to be executed twice; once for the general overall layout and a second time to detail the layout plans. Since Muther, a great number of computer tools have emerged to aid the planner through the planning process and evaluate layout alternatives to suggest an optimal solution. According to Sly (2) the decision between the alternative layout scenarios has to be based on the defined features, such as travel frequency, distance and the physical attributes of the parts, devices and operators. The values associated with the features can be analysed and comparisons between the different schemes evaluated. Solutions with minimal transportation are often chosen for the final layout. Although the procedure proves to be satisfactory on the first stage of planning the general layout, it is less successful when the detailed layout has to be determined. Muther attributes this difference to the decreasing influence of quantitative features and the increasing importance of qualitative features as the level of planning detail increases. For example, aspects such as safety and employee preference have to be considered in the final design.

IMPLEMENTATION

The introduction of Manufacturing Cells (MCs) in the 1980s highlighted the need to satisfy all the requirements on a layout plan. Burbridge (3) considered that the MCs combined the advantage of direct short material handling distances of the line layout approach with the greater flexibility of a functional layout system. Wemmerlov (4,5), who conducted two major studies on the implementation of MCs in 46 US plants, reports that the formation of MCs were not especially problematic, but the actual placement of the equipment within the cell had a high impact on the performance of the cell. Significantly, the latter aspect was often not given enough attention during the planning and implementation stages. These reports suggest a lack of operator participation in the early stages of planning countered by subsequent individual machine placement in a process of iterative overall improvement in the final layout.

Gallagher (6) considers the problem of shopfloor layout to be bounded by constraints, *"...the actual alteration of layout of the shop floor does not present difficulties, as there are usually few possible alternatives"*. Sekine (7), on the other hand, devotes an entire book to layout case studies of MCs without giving a clear methodology for the process. Proth (8) and Irani (9) also describe mathematical algorithms to minimise inter-cellular transport, but intra-cellular layout problems are rarely mentioned. This view is supported by Hilger (10) who states *"...there is no satisfactory procedure available for this last step"*. Finally Harhalakis (11), who applied an inter-cellular material flow algorithm to intra-cellular material flow noted: *"Although the intra-cell layout algorithm may serve as the basis for the solution of this problem, major enhancements are necessary. In this case actual machine and cell shapes and dimensions must be considered. Furthermore, walls and other physical objects on the shop floor, which impose traffic obstacles, must be accounted for in both the location of cells and the evaluation of traffic distances"*.

The authors believe that these differences could be attributable to the lack of adequate tools to support intra-cellular layout planning. At present, whilst the inter-cellular layout planning is done using classical methods (pen and paper, or computer aided systems), the intra-cellular layout is mostly done through the intuition of the planner. Current CAD based tools are not well supported at shopfloor level and may form a barrier between the planner and the machine operators. An interactive visual tool for analysing layout scenarios for MCs, which can enhance the communication between the planner and the operator, would be desirable. Wemmerlov (5) supports this view and argues that the difficulties with planning a layout are associated with the qualitative criteria. Until now, there is no way to build a quick and inexpensive mock-up of a cell. Some companies describe the use of scale model blocks, which are placed, on a model floor plan, to analyse a proposed cellular layout. This paper reports the steps taken to provide a visual planning tool based on Virtual Reality (VR) technology. The tool, which would address the last step in cell layout planning, has the capability to model alternative layouts and analyse their effectiveness. There would be no need for physical scale models and disruption on the shop floor would be avoided.

VIRTUAL REALITY TECHNOLOGY

Loeffler and Anderson (12) have formed a working definition for VR: "*...VR is a three-dimensional, computer generated, simulated environment that is rendered in real time according to the behaviour of the user*". The aim of VR is to achieve a maximum sense of reality for the user within a computer generated environment. In a unique way, a sense of presence has to be established between the user and the computer. A visual update rate of 15 frames per second (fps) or higher and a system lag time of < 250msec are generally regarded as pre-requisite for a VR application. At present VR systems can be categorised into four different setups:

(1) Desktop System
(2) Wide Screen Projection System
(3) Immersive CAVE System
(4) Immersive VR System using Head Mounted Displays (HMD).

Desktop VR resembles the typical Human Computer Interaction (HCI) most closely. It uses the monitor as the viewing device, and a mouse, keyboard or glove for interaction. Together with a computer, which is capable of displaying the virtual model in real time and a suitable VR software package, this arrangement often forms the entry system to VR applications.

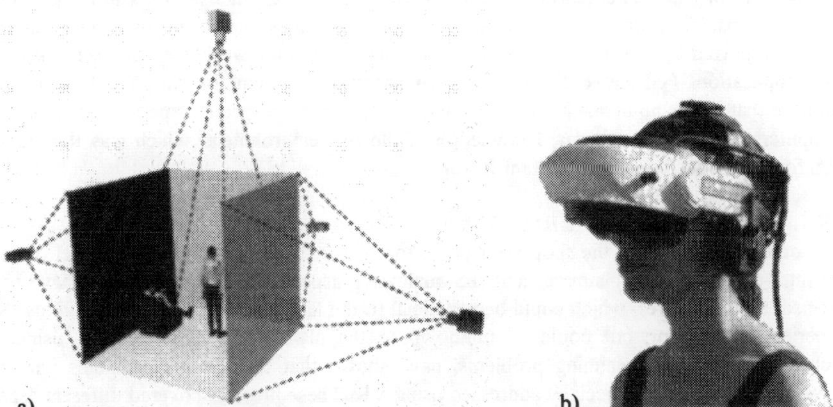

a) b)

Figure 1: Immersive VR using a CAVE System and an HMD

Wide screen projection is used to increase the Field Of View (FOV) from 20°-30° with a monitor, to over 100°, or if multiple projection screens are used, the projection can fill the entire viewing area. A setup with multiple projection, usually three walls and the floor, is called a CAVE, see Figure 1a, and is by far the most expensive VR platform. Shutter glasses, which provide stereo vision, complete this very effective immersive VR system. VR, using an HMD offers a much more affordable immersive system. Two displays, see Figure 1b, are mounted into a helmet, which provides the user with a stereoscopic image of the model. The HMD is position tracked in real time and the images are updated according to the users position and orientation. A keyboard and mouse can not be used because they are invisible to the user but a six degree of freedom (6DOF) interactive device is used instead.

Immersive VR, with a position tracked HMD, has the following advantages over desktop or single projection systems:

(1) Provides a sense of scale, from motion parallax and the stereo view.

(2) Ease of interaction with complex components using 3D input devices.
(3) Skill acquisition is improved.
(4) Unique way to study complex interactions.
(5) Improved understanding through direct interaction.

The visual advantages of motion parallax and stereo viewing and the improved understanding through direct interaction make it the preferred choice for this research. An Indigo2 Maximum Impact workstation, a VR4 HMD and a DIVISION 6DOF mouse support the VR activities in the School of Manufacturing and Mechanical Engineering at the University of Birmingham.

PREVIOUS RESEARCH

Previous research by Banerjee (13) clearly focused on the assistance to the layout planner. He used a CAVE setup to provide a viewing platform for the proposed layout. The interaction with the model was not of particular interest. The project has shown that the CAVE approach can be an extension to an existing layout planning tool, based on quantitative design criteria. Kreitler (14), on the other hand, focused on the interaction with the model. His model was based on a desktop system allowing the user to manoeuvre large pieces of equipment through a crowded model shop floor. He clearly stated the need for an immersive system as a future requirement for easier interaction. The input with a standard 2D mouse was said to be difficult and time consuming for the 3D task and required specially designed menus. The Fraunhofer project, reported by Peil (15), aimed at the normal shop floor worker as the prime user for their application. Peil stated the need for an immersive system for natural interaction and reported that it was no obstacle that the standard users had no special experience in the use of computers, or VR in particular. Knowledge of the real environment, which was the building plan for the model, was the important factor.

VR LAYOUT PLANNING CRITERIA

A good understanding of the shop floor operation is essential to satisfy the high level of detail for intra-cellular layout planning, and the qualitative aspects affecting the layout have to be defined. Some features, which could be beneficial to the layout of a cell, may be obvious to an experienced operator, but could be unknown to the planner. Existing projects, using VR technology for layout planning problems, have shown that the specific problems with MC layout planning can be especially addressed using VR. These projects covered different aspects of support for cell layout planning and provide the following guidance on the development of a VR strategy:

(1) Make use of the <u>superior 3D viewing</u> capabilities of VR to enable the planner to quickly assess a layout alternative visually.
(2) <u>Reduce the required skill level</u> for the planning activity, by including <u>natural interaction</u> with the model. For example an <u>expert system to assist</u> the layout planning process, and <u>user feedback on potential layout constraints</u>.

Certain general layout rules can be implemented in the model framework, which enables the user to conduct a visual analysis. Burbridge listed the following points that ought to be considered when planning the final cell layout:

(1) Allocate space for the temporary storage of materials issued to the cell, and for finished components.
(2) Arrange machines for first operation near the materials storage area.
(3) Arrange machines for last operations near the finished component storage area.
(4) Place machines, generally used in the same sequence, in line (I or U).

(5) Place machines with automatic cycles and short material load/unload times, near to each other, so that one man can load/unload more than one machine.
(6) Place facilities for manual operation, such as inspection and deburring, next to machines with long automatic cycles, so that the machine loader can do them in the intervals between loading work.
(7) Plan convenient positions for group services, such as swarf removal, tool storage, and pre-setting.

All these points should be included in a generic framework of the system, where the particular size and shape of the objects, equipment, parts, and materials can be easily adapted for each planning problem.

THE VR LAYOUT PLANNING SYSTEM

At the present stage the model includes standard shop floor equipment, which can be selected from an immersive menu and placed on a model shop floor. The functionality of the machines include the axes travel and loading / unloading procedures. Equipment can be moved on the shop floor according to realistic behaviour and be arranged to the users preference. Feedback is given, when predefined constraints are violated, for example a machine is placed too close to a wall or another piece of equipment. Specific floor properties can be simulated. Some machines, due to weight and vibration, can only be placed on certain specifically supported floor areas, whilst others can be placed more freely. Constraints relating to under floor access points, power supply, etc. have to be observed and the user is automatically alerted when one of these restrictions has been violated, see Figure 2. At the end of a user session, a final two-dimensional floor plan can be saved into any standard CAD file format.

Figure 2: Visual Representation of a Lathe and Displayed User Feedback

Future work will optimise the immersive menus and continue building a library for sample objects and equipment. Simulated machining operations are being prepared which will enable the user of the system to get an indication of the pace within the cell. This will allow the seven principles set out by Burbridge to be considered, e.g. long cycle machines next to manual operations.

CONCLUSION

People, who have had some experience with VR, often express a positive view about the potential of this interesting technology, but they are not so certain about specific areas of application. Due to the existing technical limitations of VR (haptic feedback is poor or non-existent, resolution and speed is limited, tracking systems have limited range and accuracy, etc.) it has also been called 'a solution in search of a problem'. Successful applications with VR will only be possible, where the unique advantages of VR are central to the problem, and the current limitations of VR are of minor importance. The layout planning of MCs offers such an opportunity because the key problem areas of the current planning tools coincide with the particular advantages of VR technology.

Initial investigations have shown that the functionality of the VR software can be used to trigger user feedback. The collision detection capabilities of VR can be used to test spatial constraints and the model can create a natural feel to it by including machine travel and machining operations. Performance related issues such as size and distance estimation or task performance times in relation to graphical update rates have still to be investigated but the primary findings are encouraging that a VR layout planning tool is feasible.

REFERENCES

(1) Muther, R, (1973), Systematic Layout Planning (2d ed), Cahners Books, Boston, ISBN: 72091983;

(2) Sly, DP, (1996), "*Systematic approach to factory layout and design with FactoryPLAN, FactOPT, and FactoryFLOW*", Winter simulation conference proceedings, pp.584-587, IEEE, Piscataway, NJ, USA, ISBN: 02750708;

(3) Burbridge, J L, (1989), *Production Flow Analysis for Planning Group Technology*, Oxford Science Publications, ISBN 0198591837;

(4) Wemmerlöv, U, and Johnson, D J, (Jan 1997), "*Cellular manufacturing at 46 user plants: implementation experiences and performance improvements*", International journal of Production Research, Vol: 35, pp.29-49, ISBN: 0020-7543;

(5) Wemmerlöv, U, and Hyer, N L., (1989), "*Cellular manufacturing in the U.S. industry: a survey of users*", International Journal of Production Research, Vol: 27, pp.1511-1530, ISBN: 0020-7543;

(6) Gallagher, C C, and Knight, W A, (1986), *Group Technology Production Methods in Manufacture*, Ellis Horwood Limited, Chichester, ISBN: 0853126097;

(7) Sekine, K, (1992), *One-Piece Flow: cell design for transfoming the production process*, Productivity Press, Cambridge, Mass, ISBN: 091529933x;

(8) Proth, J M, (1991), "*COALA: A New Manufacturing Layout Approach*", Design, Analysis, and Control of Manufacturing Cells, pp.15-29, PED, ASME, New York, ISBN: 0791808564;

(9) Irani, S L, (1991), "*Design of Cellular Manufacturing Systems*", Design, Analysis, and Control of Manufacturing Cells, pp.47-70, PED, ASME, New York, ISBN: 0791808564;

(10) Hilger, J, (1991), "*Design of Manufacturing Cells: The Inter-Cell Traffic*", Design, Analysis, and Control of Manufacturing Cells, pp.197-204, PED, ASME, New York, ISBN: 0791808564;

(11) Harahalakis, G, and Minis, I, (1991), "*Cell Formation for Minimal Traffic under Capacity Constraints*", Design, Analysis, and Control of Manufacturing Cells, pp.83-98, PED, ASME, New York, ISBN: 0791808564;

(12) Loeffler, and Anderson, (1994), *The VR Casebook,* Van Nostrand Reinhold, New York, ISBN: 0442017766;

(13) Banerjee, A, Banerjee, P, and Mehrota, S, (1996), "*An Enabling Environment For Inputting Qualitative Information In Manufacturing Systems Layout Design*", Proc. of VR in Manufacturing Research and Education Symposium, UIC, The University of Illinois at Chicago, Oct. 7-8, 1996,

(14) Kreitler, M., Heim, J., and Smith, R, (1995), "*Virtual Environments for Design and Analysis of Production Facilities*", Proc. of Working Conference on Managing Concurrent Manufacturing to Improve Industrial Performance, University of Washington, Sep. 11-15, 1995, Published on the Internet on "http://weber.u.sashington.edu/~jheim/VirtualManufacturing/vrPaperIFIPS.html"

(15) Pleil, T, (1997), "*Virtuelle Realität zeigt den optimalen Arbeitsplatz*", VDI nachrichten, 29-08-1997,

A fast three-dimensional modelling approach for virtual reality environments

Q J PENG and **M LOFTUS**
School of Manufacturing and Mechanical Engineering, University of Birmingham, Edgbaston, UK

Abstract

This paper focuses on a quick way to extract geometry information from an object and transfer the information into the VR system. A 3D surface reconstruction system, based on image analysis, forms a new approach to data acquisition and promises to be a competitive technique for VR applications. This system integrates the binocular stereo principle and the shape from shading technique. Experimental results are encouraging and the performance of the new system has been evaluated by simulation procedures.

1. INTRODUCTION

The exciting new technology of Virtual Reality (VR) is being increasingly employed to improve and reduce the costs in design and manufacturing. The great potential of VR is that it can provide a virtual environment in which a user can experiment with different design and manufacturing techniques. Unlike most simulation systems which replay a pre-determined sequence of the design or manufacturing process, the VR system allows the user to follow their instinctive idea to complete a design or manufacturing process, and allows the operations to be performed on a workpiece in real time. Therefore, rather than being just a simulation of the process, VR can provide much more support for the design and manufacturing process. However, this places a heavy demand on the geometric modeling task in the virtual design and manufacturing system, and building an environment around a particular system continues to prove one of the most difficult development problems. For instance, a replica of a factory floor, its machine tools and supporting equipment could be modeled on a CAD package. Each item would have to be individually created as a 3D model and merged together to form the whole manufacturing environment. This accounts for an enormous period of the overall process, especially, when CAD databases of object models are not available.

Most of the current Virtual CAD/CAM systems use a 'synthetic' virtual world as their virtual environment, which is based on advanced computer graphics and real-time simulation technology [1, 2]. The first step to set up a virtual 'synthetic' environment is to make a copy of the real world. Current commercial VR software provides powerful VR functionality but may fall short of creating the new VR environment. The shape information of a new object has to be built by other CAD software, such as AutoCAD, in a standard CAD format and then translated into a virtual object in a VR system. The most general demand of a VR system is to make the user feel that they are a part of the environment and can easily interact with it, i.e., provide an immersive capability.

There are several methods being considered to provide a quick means to construct VR environments. The feature extraction and recognition method analyses the geometry and topology of the existing CAD model and attempts to classify geometric features of the model as different manufacturing entities [3]. This method is limited to the existing CAD model, and restricts the user using new features to define the model. An alternative way is to use image-based approaches. 3D shape reconstruction from vision information has received considerable attention and various methods have been used to capture the 3D depth information of objects from their 2D images [4]. These methods can be divided into two groups: photometric stereo-based and binocular stereo-based. The photometric stereo-based techniques use images of the same scene with different directions of the light sources. If the surface reflectance properties of the object are known, the local surface orientation and the height of illuminated points can be computed by the techniques of shape from shading. One of the important advantages of this method is that it does not suffer from the correspondence problem, which will be referred to later in this paper. It has been used effectively by the authors in reverse engineering applications. Unfortunately, as the images are captured in different time, this approach can not meet the need of real time in a VR system. The binocular stereo-based techniques extract the depth information from intensity images acquired by two cameras in the same time, the two cameras being displaced from each other by a known distance. The key to this method is to determine the point in one image which corresponds to a given point in the other image, namely the correspondence problem. A significant amount of work has been carried out to solve this challenging problem, but the results have not been satisfactory [5], because the time delays hinder real time applications. The technique is also limited by the sparse (feature points) depth information, which it can provide.

The authors have overcome these problems by integrating the information from the shape from shading process with the binocular stereo-based technique. The 3D shape information of an object is recovered from the key feature points extracted by the binocular stereo-based method and the shape from shading method reconstructs the surface of the object.

2. THE SHAPE FROM SHADING METHOD

Given a Lambertian image, with the constant diffuse reflectivity k_d and the incident intensity of the point light source I_n, the image irradiance equation relates the intensity of an image point (x,y) with the surface normal **N,** as follows [4]:

$$I(x,y) = k_d I_n \mathbf{L} \bullet \mathbf{N} = k_d I_n \frac{1 + pp_s + qq_s}{\sqrt{1 + P^2 + q^2}\sqrt{1 + p_s^2 + q_s^2}} \qquad (1)$$

Where p and q are the partial derivatives of height Z with respect to image coordinates; p_s and q_s are the partial derivatives of the light direction.

The shape from shading algorithm by a single image can not produce accurate results. The surface normal at each point is represented by two variables, p and q, which can not be computed by using one equation. Boundary conditions and local constraints based on reasonable assumptions support the most common reported solutions. A direct formulation of the shape from shading problem, with a smoothness condition on the surface, is stated in the following minimization problem:

$$\iint_{\Omega}((I(x,y)-k_d I_n \frac{1+pp_s+qq_s}{\sqrt{1+P^2+q^2}\sqrt{1+p_s^2+q_s^{\ 2}}})^2+\lambda\left(p_y-q_x\right)^2)dxdy \rightarrow \min. \quad (2)$$

Where λ is a Lagrangian multiplier used to enforce the constraint of surface smoothness, $p_y = q_x$. Therefore, p and q can be computed by the iterative scheme from equation (2).

Unfortunately, even for ideal Lambertian diffuse reflectance, this model has practical difficulties in achieving an accurate solution for 3D shape recovery. However, the shape from shading process can convergence to a unique solution when the boundary information is available at the occluding contours [6]. The binocular stereo-based techniques can be used to find the boundary information of the object.

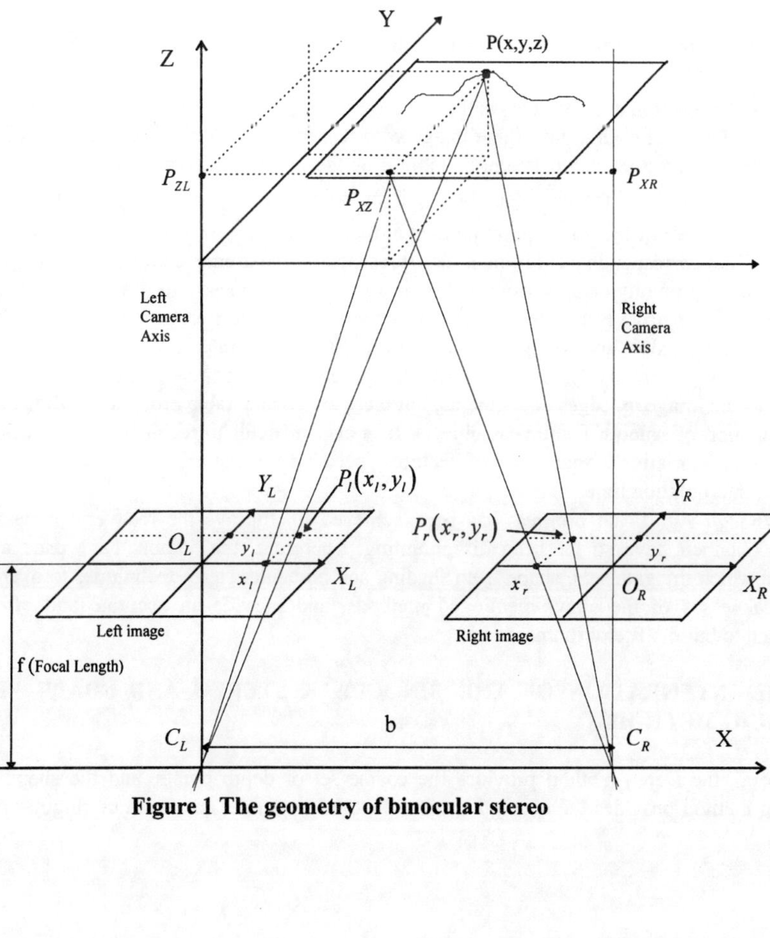

Figure 1 The geometry of binocular stereo

3. THE SHAPE FROM BINOCULAR STEREO METHOD

The binocular stereo method is an effective means of extracting a 3D data set from image information. This technique simulates the human visual system which uses the difference between the images in the left and right eyes to judge depth. The brain processes the two separate 2D images and forms a 3D shape. Similarly, when a 3D object is photographed by a pair of cameras, with the same focal length f, the depth information can be surmised (Figure 1):

$$\frac{X}{Z}=\frac{x_l}{f} \quad (3) \qquad \text{and} \qquad \frac{X-b}{Z}=\frac{x_r}{f} \quad (4)$$

Combining the equations (3) and (4):
$$Z=\frac{bf}{(x_l-x_r)} \quad (5)$$

Substituting (5) back into (3):
$$X=\frac{x_l b}{(x_l-x_r)} \quad (6)$$

Similarly :
$$Y=\frac{y_l b}{(x_l-x_r)} \quad (7)$$

Where: b is the displacement between the two cameras used for obtaining the image pair, and (x_l-x_r) is a shift in pixels at a point of an object between the two corresponding images.

It is obvious from equations (5), (6), and (7) that, if the quantity (x_l-x_r) can be known , the values of X, Y, and Z in point P can be obtained.

Before the position and depth at each point on the images can be determined, it is necessary to know the shift of each point in one image with respect to the other. Thus the images must be matched (known as stereo image matching, or the correspondence problem) in order to obtain the position of the corresponding pair of points (x_l, y_l) and (x_r, y_r). In a discrete space, the number of the image point pairs is limited to the number of pixels in the two given images. The correspondence problem has presented an extreme challenge to the image processing community, and various methods have been proposed to find a set of stereo matches to reconstruct a surface from the matched information. For example, the area-based method, which is slow and costly, matches the images with the intensities of each pixel or the local window. Alternatively, the feature-based approaches are popular,which find the match feature of the image by edges or segments, but there are considerable problems with the shape determination of smooth featureless objects. It is also difficult to reconstruct the surface of object with a relatively sparse set of features, because the feature-based algorithms need massive interpolation time.

Although significant progress has been achieved to improve the recovery process, the results obtained have so far been disappointing. Therefore, the authors have used a new method which integrates the shape from shading and binocular stereo techniques to overcome the weaknesses of the above-mentioned methods, and provide an accurate and effective approach to build VR environments.

4. THE INTEGRATION OF THE BINOCULAR STEREO AND SHAPE FROM SHADING METHODS

In general, the stereo method provides the coarse set of depth points, and the shape from shading method provides the detailed local surface information. The depth of singular points and feature points, extracted by the binocular stereo-based method, will attempt to force the resultant reconstruction towards the true depth value by the shape from shading method. Where the singular points are the brightest points in the image, they are the local minima, local maxima and saddle points of the surface. The object can be divided into the piecewise smooth surfaces according to the singular points and feature points to satisfy $p_y = q_x$ in equation (2). The integral method can be used to recover the whole surface.

In Figure 2, the integrated module has been used to produce a VR model of a gear train. A pair of digital cameras capture two images of the gear train, which are detected for edges, interesting points and singular points. The correspondence process is performed at the interesting points and singular points, and the height at these points are calculated by the binocular stereo algorithm. This information is fed into the process of shape from shading, where the complete set of 3D surface points can be obtained. Finally, the required VR environment is realised from the geometry modeling process.

5. DEVELOPMENT OF THE VR ENVIRONMENT AND RESULTS DISCUSSION

Experiments and simulations have been employed to evaluate this approach. The two images of a model car, which taken by the binocular stereo method, are shown in Figures 3 (a) and (b). Figures 3 (c) and (d) show their edge detection results, with the recovered shape of the car shown in Figure 3 (e). The geometry model for the VR application is shown in Figure 3 (f).

This approach is relatively fast, and it is particularly useful for recovering the 3D depths of smooth surfaces, which have previously been difficult to perform by the binocular stereo-based technique.

6. SUMMARY AND CONCLUSIONS

Many of the ideas and technologies of VR are still being researched, but progress suggests that VR technologies will play a key role in future design and manufacturing applications. The approach described in this paper is still under development, and more experimentation will be done to evaluate its efficiency and accuracy. The initial results are encouraging and support VR environments to be constructed with little expense and effort.

REFERENCES

[1] S. Kim, R. Sonthi, and R. Grdh, Virtual Engineering Environment Applications: An Overview, De-Vol. 67, *Design for Manufacturability*, ASME, pp. 1-11, 1994.

[2] C. P. Chu, T. H. Dani, and R. Gadh, Multi-sensory user interface for a virtual-reality-based computer-aided design system, *Computer-aided Design*, Vol. 29, No. 8, pp. 709-725, 1997.

[3] J. Kraftcheck, T. Dani, R. Gadh, State of the Art in Virtual Design and Manufacturing, *VR News*, Vol. 6, Issue 4, pp. 16-22, 1997.

[4] B. K. P., Horn, and M. J., Brooks, *Shape from shading*, London: MIT Press, 1989.

[5] Z. Zhang, R. Deriche, and O. Faugeras, A Robust Technique for Matching Two Uncalibrated Images Through the recovery of the unknown epipolar geometry, http://www.inria.fr/rapports/sophia/RR-2273.html.

[6] A. Blake, A Zisserman and G. Knowles, Surface descriptions from stereo and shading, *Image Vision Comput.* Vol. 3, pp. 183-191, 1985.

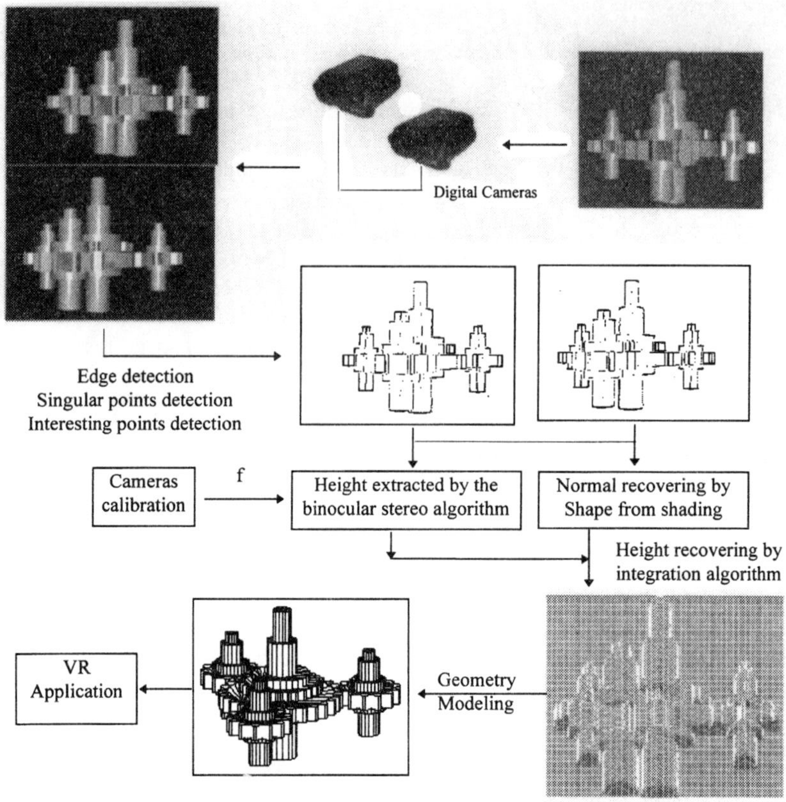

Figure 2 Integrated binocular stereo and shape from shading methods

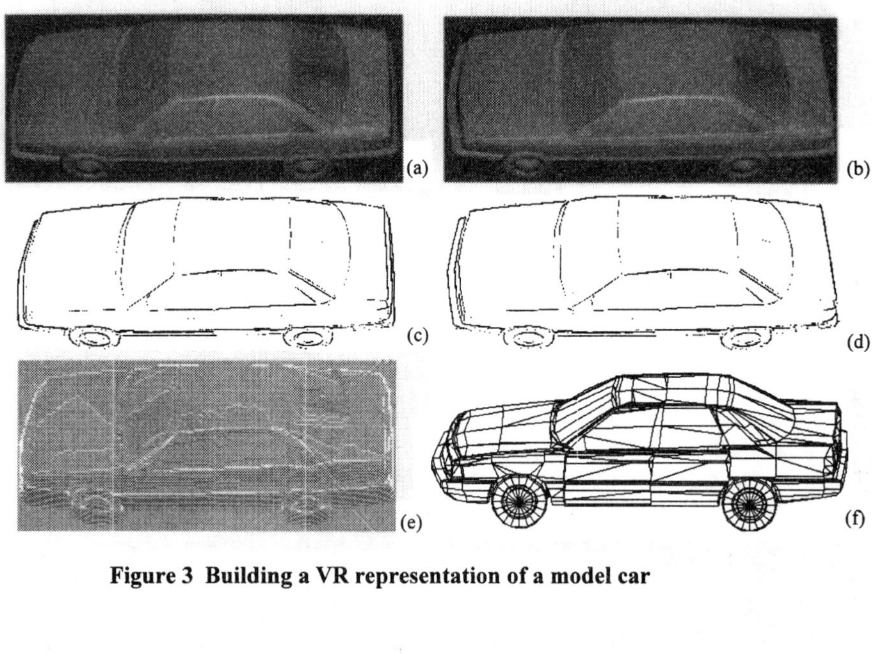

Figure 3 Building a VR representation of a model car

Management of complex virtual manufacturing environments for manufacturing applications

Z XU, Z ZHAO, and **R W BAINES**
School of Engineering, University of Derby, UK

ABSTRACT

The construction and management of complex virtual reality (VR) environments is a bottleneck for applying virtual reality technology to manufacturing problems. Current bottom-up construction approaches are tedious and inflexible. This paper presents a variant approach to managing and constructing virtual manufacturing environments. The approach is established on the management of systematically stored virtual manufacturing environments. Each environment is associated with manufacturing tasks defined in the data base. Based on the approach, a virtual manufacturing environment management system is created. This system is currently developed as a framework for manufacturing knowledge acquisition. In the framework, the manufacturing knowledge stored in the knowledge base is represented and recognised within virtual worlds. A knowledge interpretation scheme is defined to determine the real world manufacturing facilities and processes data in terms of virtual world data format. The data can then be visualised into virtual manufacturing environments and used for manufacturing planning and control.

1 INTRODUCTION

Virtual Reality (VR) is an emerging human-computer interface (HCI) technology that aims to provide users a computer generated 3D digital environment where users can "walk-in" and realistically interact with its inside objects (1). VR has already demonstrated the application potential in manufacturing in terms of product design, process planning and resource and facility management (2)(3). VR technology is more informative and realistic than conventional simulation for acquiring cognitive knowledge, enhancing scenario knowledge representation, understanding sophisticated problems and eliminating risks in complex processes.

The applications of VR for manufacturing processes can be justified as follows:

(i) Manufacturing knowledge is in general difficult to represent and formalise in computer terms. A large proportion of manufacturing data and information is based on experience, which is empirical and non-generic. Traditional or even conventional computerised manufacturing systems and the techniques behind those systems such as modeling, planning, simulation and AI (artificial intelligence) expert systems have employed rather formal and simplified models to represent such knowledge in forms of specified formula, text, logic and rules. In most cases, they have failed to solve the difficult problems when empirical knowledge is involved and have to rely on human intervention.

(ii) The theme of VR is not only to bring automatic computerised system functions to engineering applications, but also to incorporate human abilities into those computerised systems in a way that human cognisance and senses to the real world can be fully utilised in a virtual environment.

(iii) Most manufacturing systems or programs only work well if based on well structured databases that are populated with realistic data and information. This data and information has to be captured and represented from various knowledge sources. Manufacturing knowledge acquisition is an old but still challenging problem, VR provides more effective techniques for collecting, refining, storing and utilising the non-generic and empirical information.

2 CONVENTIONAL METHODS OF CONSTRUCTING A VIRTUAL ENVIRONMENT

Different types of VR systems with various developing tools and environment exploring tools are available on the market. According to their VR world creation methods, those systems can be put into three categories: (i) those that use complete virtual world modeling toolkits to allow users to construct their virtual worlds and then to navigate the worlds from within the package or to export the file to other formats. Typical examples of this type are Superscape VRT and Division dVS (4), (ii) those that use independent description languages, such as Virtual Reality Modeling Language (VRML) which has an ISO standard file format that describes 3D virtual objects and virtual worlds on the Internet, (iii) those that use the VR API (Application Programmer Interface) which typically provides a large number of programming functions for users to take advantages of writing 3D graphics applications.

However, the construction of a large-scale, complex VR environment is still a pain-taking process for three reasons (i) the diversity of VR systems demands extensive learning of individual toolkits and programming skills before a virtual world can be created (5), (ii) most VR software builds the synthetic environment from a tedious process similar with the CAD modeling process, and (iii) the created virtual environments are generally insufficient in geometric accuracy (6).

3 VARIANT APPROACH TO CONSTRUCTING VIRTUAL WORLD

In contrast to the conventional methods of VR world creation, this paper provides a variant method to construct a virtual environment by modifying the ready-made master environments rather than building one from scratch. Through a virtual manufacturing environment management system, the built environment is used to acquire and represent manufacturing knowledge that can be applied to manufacturing planning and control.

3.1 The concept of a variant approach

The variant concept of constructing virtual manufacturing environments allows for a set of similar manufacturing tasks (for example, machining a part of certain geometric features) to be represented by one typical or representative task. If a specific virtual environment (for example containing a lathe machine and a robot) exists to accomplish this representative task, then it should be able to accomplish marginally all similar tasks. Therefore, a VR world can be created based on a representative environment called a 'master environment' and the representative task is called a 'master task'. The master task and the master environment are related and stored together in a certain format in the data base.

Currently, a number of master environments and master tasks that have been built in the data base, each representing a group of similar VR simulation worlds for a family of similar simulation tasks (which are represented by a master task). If a specific simulation task is required, then the task will be described, by a coding programme, in the same format as that of the master task in the data base. To construct the required VR simulation environment, the index of the formatted task is used to search the data base by performing following logic;

(i) If a master task exists in the data base that is exactly the same as the given task, then the master environments attached to that master task will be retrieved as the required simulation environments.

(ii) If, in the data base, a master task that is the most similar to the given task, then the master environments attached to that master task will be retrieved and modified into the required simulation environments. The amount of modification depends on the similarity between the two tasks. The more similar there is, the less modification is needed, thus avoiding creating it completely from scratch.

(iii) If there is no master task in the data base that is similar to the given task, then a new VR simulation environment should be created for the task. The given task and the newly created VR environment can be stored in the data base as a new master task and master environment.

3.2 System Structure

Figure 1 shows the structure of the system developed for constructing VR environments in three parts - the virtual world editor, the system manager, and the VE data base. The top layer of the system is a virtual environment editing tool used for creating and viewing master virtual environments.

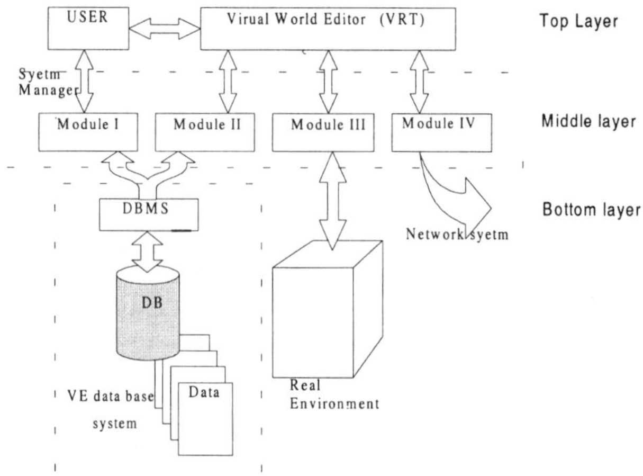

Figure 1 Variant approach of constructing a virtual environment

The middle layer is the System Manager designed for data exchanging between the top-layer virtual environment authoriser and the bottom-layer VE data base, and the physical manufacturing equipment. The System Manager has four modules, Module-I is a programming utility for building data base applications with real-time connections to a knowledge base. It manages the interface of different data base files and provides standard data base accessing tools. Module-II supports the real-time data exchanging between virtual world objects and the data base files through virtual world displaying vectors. A data converter - for translating the self-defined VME schema format into standard VRML has been developed and included. Module-III is a program for processing signal received from real manufacturing equipment - in the test applications, a PUMA 560, Lansing PWIOII industrial robots and CNC milling machines. Module IV is a tool to link the virtual worlds with networking system to support the sharing of virtual environment information through the Internet and has been developed using Superscape SCL, HTML and JavaScript.

The bottom-layer is a dedicated manufacturing data base, which includes a data base management system and domain-specific data files. The data base management system is for managing and querying data and managing the master virtual manufacturing environments and related manufacturing tasks. Unless the data base stores a large number of task families and each family has only one task, the retrieved master environment normally needs to be modified to form the required environment. The fewer the task families and the more tasks each family has, the more modification will be needed. However, after the system has been used for a period of time, new master tasks and master environments will be added to the data base, the amount of modification will reduce significantly. A commercial data base system can be directly integrated (in a plug-in style) with the system. Paradox data base was used to test this function however other stand-alone data base systems or server based data base systems such as Oracle, Informix can be supported.

4 KNOWLEDGE ACQUISITION AND REPRESENTATION FOR MANUFACTURING APPLICATIONS

The acquisition of manufacturing knowledge is to capture and represent the information for accomplishing the manufacturing task within a virtual environment.

4.1 Manufacturing knowledge acquisition

The novelty of using VR for manufacturing knowledge acquisition is due to the real-time interaction that can be achieved by developing a virtual manufacturing environment management program. The functions of this program include: (i) a user interface to randomly access and edit a schema file, (ii) ability to embed manufacturing knowledge into a schema file in a conventional IF - THEN format, (iii) communication between the schema and the virtual manufacturing environment data base, (iv) automatic translation of a schema file into standard VRML format, and vice verse.

To capture and represent manufacturing knowledge within a virtual manufacturing environment, the schema file has been developed to a so-called 'VME file format' to relate the manufacturing knowledge with the contents and contexts of the virtual environment and then to convert the knowledge into Superscape VRT format. Thereafter the knowledge can be visualised or displayed as stand-alone or Internet plug-in. Within the virtual environment, the knowledge acquisition process is simply the user's interaction with its contents and contexts and manipulating them to accomplish a task. The whole scenario of such an interaction and manipulation involves users high-level decisions, thinking and cognitive knowledge that are recorded by the data converter into the VME file. The knowledge in the schema definition can be fully visualised in VR systems or 3D VRML Web page.

4.2 Computing implementation

As for the variant system for constructing virtual manufacturing environments, the knowledge acquisition system that works in the constructed environments has also been implemented. The schema (VME file) was designed for two purposes. One for holding the acquired knowledge in formalised format - VME, the other to describe a master virtual manufacturing environment by illustrating (i) suitable manufacturing task(s), (ii) the list of master task(s) related manufacturing machines, tools and other devices, (iii) the relationship among those objects. In the research, the schema was written in ASCII format and nested in the Superscape VRT displaying vector T_STANDARD structure. The schema can be used directly in creating VR world objects and generating the virtual manufacturing environment.

The schema file contains different virtual object code sections, each section being formed by different fields that describe the static and dynamic characteristics of the virtual objects. The virtual manufacturing environment data base stores a master virtual environment by saving its schema. Each token in the schema is a data base field, and each of the object description forms a data base record. This arrangement facilitates the conversion of the VME file into Superscape VRT environment file format. Figure 2 shows the master virtual environment data base interface of the integrated virtual manufacturing environments construction and knowledge acquisition system.

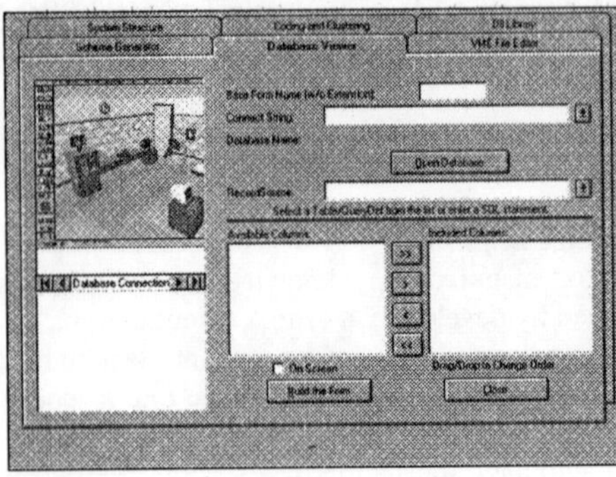

Figure 2 Integrated VE construction and knowledge acquisition system

5 CONCLUSION

A framework for manufacturing knowledge acquisition and the methodology of managing large-scale and complex virtual manufacturing environments has been developed. It has provided an effective approach towards solving the bottleneck problem of constructing complex virtual worlds. A schema for manufacturing knowledge representation has been designed in VR terms, its file format - *.VME - has been defined with in a data converter module. A virtual reality based knowledge base system has been implemented and integrated with a commercial VR system (Superscape VRT). At present, a workable VR based prototype system for manufacturing knowledge acquisition has been achieved and a generic methodology for managing virtual manufacturing environments established. Further research and development work will apply the developed software into a manufacturing planning system to obtain a flexible planning modules and optimise the information transfer for different manufacturing activities.

6 REFERENCE

(1) Rheingold, H., 1992, Virtual Reality, ISBN: 0749308893, MANDARIN, London.

(2) Hitchcock, M. F., Baker, A. D., and Brink, J. R., 1994, The Role of Hybrid Systems Theory in Virtual Manufacturing, IEEE, 0-7803-1800-5/94.

(3) Vance, J. M., 1996, Virtual Reality: What Potential Does it Hold for Engineering? International Conference on Mechanical Design and Production Proceedings. Volume (2).

(4) Peter, Y., 1996, Virtual reality systems, Survey of VR and VRML systems, University of Durham, February, 1996, Internet file. dcs.ed.ac.uk/~dcs3py/pages/work/documents

(5) Singh, G., Serra, L., Fairchild, K., and Poston, T., Visual creation of virtual design environments and virtual worlds research at ISS, Presence, Vol. 3, No. 1, Winter 1994.

(6) Mckenna, M., and Zeltzer, D., 1992, Three dimensional visual display systems for virtual environments, Presence, 1, 421-458

Rank-order-clustering (ROC) for construction of virtual manufacturing environments

Z ZHAO and **R W BAINES**
School of Engineering, University of Derby, UK

This paper presents a method of classifying virtual manufacturing environments, attempting to provide an effective data base by which complex virtual manufacturing environments can be built with less construction effort and shorter computing time. First the paper presents the concept behind the classification method. Then it describes the computing algorithm for implementing the method and proposes a system for managing virtual manufacturing environments. Finally the paper discusses the applications of the algorithm in constructing virtual manufacturing environments for manufacturing planning.

1. INTRODUCTION

Virtual Reality (VR) is very much a real-world grounded, business-oriented technology of particular interests to manufacturing (1),(2). The potential benefit of VR stems from the richness and flexibility of VR worlds, but the creation of those worlds is a challenging task. For manufacturing, a VR world is often a complex computerised environment made of a large number of different three-dimensional computerised models. It can be either created from scratch or derived from existing VR worlds (3),(4). While creating a completely new VR world is a tedious and pain-taking process, deriving one needs the support of an efficient storage and management of existing virtual objects or virtual environments.

1.1. Creation of virtual worlds

Most VR systems provide users with 3D modelling tools like VR world editors and shape creators. Some VR packages provide application development tools that enable users to create fundamental code to construct virtual environments and to navigate within them (5). Other VR systems support added functionality to build VR worlds on a set of 3D primitives provided. Despite these system utilities, building a virtual world is far from easy. Anyone who has spent time wrestling with the task of mastering conventional CAD software knows that even building a 3D graphic representation of a single object can only be made easy up to a point. To create a complex virtual manufacturing environment, even the design and planning process at the beginning still takes enormous thought and effort, not to mention the actual constructing process.

At present, researchers are exploring VR systems' realistic modelling and simulation which conventional technologies like CAD, CAM and manufacturing simulation cannot do. Construction and management of manufacturing environments seems to be treated as an application issue. Little work is reported in literature that specially addresses the automatic construction of large and complex virtual manufacturing environments. Currently one popular method of VR world creation is using Virtual Reality Modelling Language (VRML) (6) to down-load already existing 3D object models from other modelling systems or shape libraries. Another method is to design primitive shapes first within a CAD system in a standard data exchange format (for example, Autodesk's DXF) and later to translate the shapes into VR objects (7), (8). Although VRML has already become an industrial standard and using a CAD system to define the required 3D shapes can reduce substantial amount of shape editing work, neither methods alleviate the problem of editing and arranging the shapes in a real world's physical context.

1.2. Management of virtual worlds

Another problem that needs to be addressed is the management of already created VR worlds. This mainly concerns how to make the contents of an existing VR environment retainable and reusable to avoid building a new one completely from scratch. Storing individual large complex virtual manufacturing environments could consume a large amount of disk memory and may be difficult to reuse. Similarly storing constructive blocks or primitive shapes may loose valuable data about the physical relationships and simulation contexts that belong to individual VR worlds. It is therefore desirable that virtual environments can be both retained economically and managed effectively in order to recycle not only their physical contents but also their simulation contexts.

Relating to the above two problems, this paper reports on an effective methodology of constructing and managing virtual manufacturing environments. The emphasis is laid on the classification of VR worlds and their data base management. A method of classifying virtual manufacturing environments is presented that shows how VR worlds can be managed and retrieved to simplify the process of constructing complex VR manufacturing worlds.

2. RANK-ORDER-CLUSTERING OF VR MANUFACTURING WORLDS

Virtual manufacturing worlds can be classified by using a Rank-Order-Clustering (ROC) algorithm. The underlying concept of ROC comes from Production Flow Analysis (PFA) (9), (10). ROC algorithm classifies VR worlds according to (a) the virtual objects, i.e., the static contents such as machines, tools or other manufacturing equipment, (b) the virtual events, i.e., the dynamic animation context such as manufacturing operations, activities and processes. Accordingly two classification tools are derived. One is the world-content matrix which derived from PFA's part-machine matrix. The other is world-context matrix which is derived from PFA's part-operation matrix. An element in either of the matrices has a value of 1 if the corresponding content or context is owned by a virtual world, otherwise it has a value 0. Figure 1 (a) shows a world-content matrix where 5 virtual objects (machines, for example) are placed within 10 VR worlds. Figure 1 (b) illustrates a world-context matrix where 5 virtual events (manufacturing operations, for example) are to be animated in the same VR worlds.

Virtual Operations \ Tasks	1	2	3	4	5	6	7	8	9	10
1	1	0	0	1	1	1	0	1	1	0
2	0	0	1	1	1	0	1	1	0	1
3	1	1	0	0	1	0	1	0	1	1
4	0	1	1	0	1	1	0	1	0	0
5	0	1	0	1	0	1	1	0	0	1

(a) Task-operation matrix

Virtual Machines \ Tasks	1	2	3	4	5	6	7	8	9	10
1	1	1	0	1	1	1	0	1	0	1
2	1	0	0	1	0	1	1	1	0	1
3	1	0	1	0	1	1	1	0	1	1
4	0	1	1	0	0	1	0	1	0	0
5	1	1	0	0	1	1	0	0	1	0

(b) Task-machine matrix

Figure 1. Rank-order-clustering matrices with bit-wise code elements

With either of the two matrices, the 10 virtual worlds can be classified into families, each family having similar worlds in terms of their contents or contexts. Since the ROC algorithm is to deal with matrices, the resolving computation is often done using specific arrays or memory. Due to the unspecified size of the matrices, the length of the bit-wise ranking code can become considerably large for applications where a complex virtual manufacturing environment involves a large number of machines and operations. To overcome the problem, the ROC computation programme is developed based on data base files instead of buffer memory so that the length of the ranking code can be extended to as long as 2^{32} (number of data records that a data file can hold). Accordingly, the clustering is done by a file rotating mechanism that uses four data files ***f1***, ***f2***, ***f3*** and ***f4***.

File ***f1*** holds the original ranked codes, each code being a row of the ranking matrix and occupying one data record. The codes are automatically sorted in ascending order with the last data record holding the code of the greatest value. From ***f1***, new codes are automatically created, each being a column of the ranking matrix and occupying one data record in file ***f2*** which, in fact, is obtained by "rotating" file ***f1*** 90 degrees. Files ***f1***, ***f2*** together define the ranking matrix. Next rotating ***f2*** in the similar way to form file ***f3***, which, in turn, is rotated to form file ***f4***. Files ***f3*** and ***f4*** together define another ranking matrix which is better clustered than the previous one. Rotating ***f4*** to form ***f1*** and continuing the process until the matrix defined by files ***f1*** and ***f2*** and the matrix defined by files ***f3*** and ***f4*** are both ranked and are the same as each other, which means that all elements are clustered in families.

Figure 2 shows the result of the ROC clustering on the two matrices shown in Figure 1. The left table clusters the manufacturing tasks around operations and forms task families as {***task5***}, {***task3***}, {***task2***, ***task8***, ***task1***}, {***task4***, ***task9***}, {***task10***, ***task6***, ***task7***}. The right table clusters the task around machines and forms task families as {***task6***, ***task7***}, {***task9***, ***task1***}, {***task10***, ***task8***},{***task4***}, {***task5***}, {***task2***, ***task3***}.

ROC clustering can directly allocate manufacturing information such as machines, tools, operations, and processes into the clustered families, which is useful in designing virtual environments for an individual manufacturing task family. Its strength lies in rapid classification of existing VR worlds and thus avoiding the necessity of constructing new ones from scratch. It is particularly useful for virtual factory and workshop layout where there is a requirement for constructing different virtual manufacturing environments that have great diversity of manufacturing systems but each has specified machines and involves definitive processes.

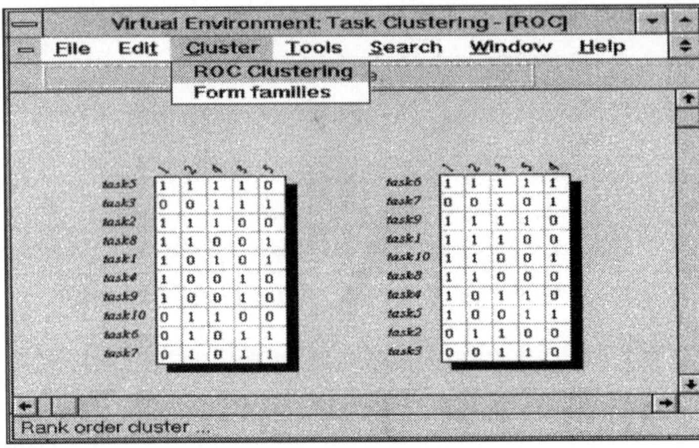

Figure 2. ROC clustering system

3. THE SYSTEM DESCRIPTION

The programme runs on IBM-compatible PC's under Windows NT or Windows95 and supported by VR systems such as Superscape VRT5.5 and 3D Website browsers that are compatible with VRML2.0 standards.

Figure 3 Snapshots of the detailed virtual environment

The programme takes existing virtual objects or virtual worlds as input from a data base or disk files and classifies them into clusters according to specific manufacturing tasks or processes. It then dispatches the clusters to a VR systems to construct the required virtual manufacturing environments. For example, if the manufacturing tasks are required for machining raw material to a finished component, the machines, tools, operations or processes used for each task have to be identified and assigned to the corresponding task to form a machine flow, tool flow, operation flow or process flow. Ideally, these flows need to be verified in a virtual manufacturing environment before they could be implemented in the real world. The clustering programme performs two tasks (i) generating task families, machine flows, tool flows, operation flows and process flows in real-time and (ii) providing required virtual machines, virtual tools and even virtual manufacturing systems for a VR system to assemble (or construct) rapidly the required virtual manufacturing environments.

Figure 3 shows a virtual machining environment constructed using the clustered virtual objects and their display in a VR system.

The virtual components shown in the left of Figure 3 are used as tasks and the snapshot in the right show a detailed representation of one of the lathes in the constructed environment for machining the components.

4. CONCLUSIONS

The ROC algorithm reported in this paper proved useful in reducing the effort required in constructing virtual manufacturing environments by avoiding the creation of similar environments from scratch. It provided a building tool for constructing large and complex virtual worlds that could be used to develop an object library for virtual manufacturing.

However, the clustering algorithm depends on human judgement to assign virtual objects to specific manufacturing tasks. The potential of the techniques described in this paper is yet to be fully assessed against a real industrial manufacturing background. Little literature reporting similar investigations is available. The development of virtual reality technology and its fast permeation into manufacturing applications will demand effective techniques for the fast and economic creation of large and complex virtual environments in the future.

REFERENCES

1 Blanchard, D. The reality of virtual manufacturing, *Intelligent Manufacturing, June* 1995 (http://lionhrtpub.com/IM6-Reality.html)

2 Sherard, D. C. and Narayanan, M. The aid of virtual reality in the industry, Conference Record on *Wescon/94, Idea/Microelectronics*, Anaheim CA, USA, September 1994, pp.374-377 (IEEE Press, USA)

3 Neugebauer, J. G. Virtual reality - more than just simulation, *Industrial Robot*, 1992, 19(3), 30-34

4 Kalawsky, R. S. *The Science of Virtual Reality*, 1993, (Addison-Wesley Publishing Co.).

5 Superscape UK. plc. *Superscape VRT for Windows: User Guide*, 1998, (Superscape Inc.)

6 The VRML Consortium Incorporated, *The Virtual Reality Modelling Language*, International Standard ISO/IEC 14772-1: 1997

7 Pierre, D. P. Applied virtual reality, *Virtual Reality in Engineering*, 1993, (IEE, London UK).

8 Carpenter, I. D., Dewar, R. G., Ritchie, J. M. and Simmons, J. E. L. Enhancing a virtual environment for manual assembly, Proceedings of *The 12th International Conference on CAD/CAM Robotics and Factories of The Future* (Eds. R. Gill and C. S. Syan), Middlesex University, UK, 1996, pp. 1067-1072 (Middlesex University Press)

9 Burbidge, J. L. *Production Planning*, 1971 (Heinemann, London)

10 Gallagher, C. C., and Knight, W. A. *Group Technology Production Methods in Manufacturing*, 1986 (Ellis Horwood, England)

Robots, Automation, and CAD

A method for dynamically producing CAD drawing files within a multimedia-based electronic catalogue

N AMIN and **D SU**
Department of Mechanical and Manufacturing Engineering, Nottingham Trent University, UK

Abstract

A novel approach has been developed for electronic cataloguing of mechanical products, using Visual Basic. Distinguished from the existing electronic catalogues, the new approach integrates hypermedia, CAD drawing and data processing techniques into a single environment, making this package more efficient, interactive and user friendly. A C++ program has also been written which creates a DXF file depending on the dimension that the user has chosen. The output DXF file can then be imported to CAD packages, such as AutoCAD and Pro-engineer, which provides a useful tool for the designers to produce CAD drawings.

1. INTRODUCTION

In this research, a prototype electronic catalogue has been produced for the 'Davall Gear Stock Company' using Microsoft Visual Basic version 5.0, C++ and other hypermedia techniques. With recent development of computing techniques, product catalogues in electronic forms, or so-called electronic catalogues have been increasingly emerging, for example (1, 2, 3). In comparison to the traditional product catalogues, i.e. the ones in the forms of hard copies, electronic catalogues have obvious advantages, such as compact size, large amount of information storage, data handling by the means of computers, etc.

Distinguished from the existing electronic catalogues, the new approach integrates hypermedia, CAD drawing and data processing techniques into a single environment, making this package more efficient, interactive and user friendly. In addition to providing all the information available in the traditional catalogue, the software is particularly beneficial for engineers working with CAD packages. This software would save hours of time creating the individual drawings. Using this software, a spur gear drawing with any dimension could be created as a DXF format.

The data is displayed in a scrollable tabular format. The user can select a spur gear with any number of teeth. The program then produces a DXF file of either a component type or assembly type, with appropriate parts being hatched. This file could then be copied to any directory, for example a floppy disk. The output DXF file can then be imported to any CAD package, such as AutoCAD or Pro-engineer. The software contains all the technical information about the company's product, spur gears ranging form module 0.5 to module 6.0. The related data includes gear type (A & B), number of teeth (from 12 to 127), geometrical dimensions and CAD drawings. The software incorporates some multimedia features such as images and sound, which improve the graphical user interface dramatically. The software is attractive and user friendly. It also provides a brief information about the company.

2. THE SOFTWARE DEVELOPMENT STRATEGY

Microsoft Visual Basic version 5.0 is an application development system, whose purpose is the creation of Microsoft windows programs (4). It is a powerful, flexible and inexpensive programming language that can be used to quickly build any application. The applications produced have the same features as other windows softwares. Visual Basic was utilised as a platform for the electronic catalogue.

Visual Basic is an object-oriented programming language that is used to build graphical user interfaces (GUIs). It has the capability of integrating multimedia features such as images, sound and animation. These features play an important role in the success of any software, making it more efficient and user friendly.

The data in the electronic catalogue is displayed in a spreadsheet format and the user is able to scroll through it and to select the desired record. The MSFlexGrid Control has been used to accomplish this. The MSFlexGrid Control is tied directly to a Data Control. This relation is shown below:

The first step is to create a file containing all the necessary data. Excel version 5.0 has been used for this purpose. Each table of data is inserted in one sheet of the Excel. The next step is to bound the Data Control with this file. The *connect* property of the Data Control is set to Excel 5.0. This tells the Data Control which type of database is being used. In the *DatabaseName* property the name of the database file is inserted. The visibility of the Data Control is set to False. This is due to the fact that it is not being used independently and its purpose is to provide data to MSFlexGrid. Therefore, there is no need for it to be visible to the user. The final step is to bound the data to the MSFlexGrid. To do this, the *DataSource* property of the MSFlexGrid is set to the Data Control. The information in the *Recordset* of the Data Control is now automatically loaded into the grid. It has to be mentioned that MSFlexGrid is read-only, i.e. the data could not be overwritten which is an excellent method to protect the data.

This feature makes the software very user friendly. The whole data is available to the user and the user can easily scroll through them. Figure 1 shows the layout of the screen displaying the table of data. The user is also able to change the value of the Module and Material, in

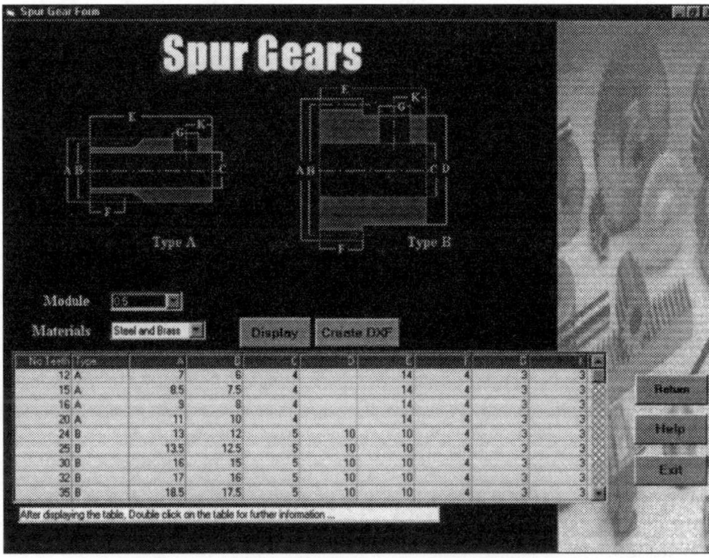

Figure 1 - Layout of the screen showing the table of data

order to display another set of data. When the user selects any record and clicks on the *Create DXF* button, the program then produces a DXF file of the selected dimension. The user can then import this file into any CAD packages that can read DXF format, such as AutoCAD or Pro-engineer.

3. METHOD FOR DYNAMIC PRODUCTION OF CAD FILES

The Drawing Exchange Format (DXF) was developed and is maintained by AutoDesk. It enables the interchange of drawings between AutoCAD and other programs. It was decided to write a program to create a DXF file for the Spur Gears. The reasons are listed below:

- The DXF file format is supported by many applications, and many CAD systems can import drawing files of this format.
- The DXF file could be in ASCII or binary format. The ASCII format could easily be viewed and analysed in DOS Editor.
- Only a single DXF file is needed for each type of gear. Every time the user selects a dimension, the data is overwritten over the original value. So, there's no need to keep all of the different drawings on file.

A DXF file consists of 5 sections (5): Header section, Tables section, Blocks section, Entities section, End of file. It is organised in these sections and not all of these sections need to be edited. Depending on the task at hand, some of these sections could be completely omitted. In relation to this, an investigation was carried through which it was found that the data for hatching was located in the *Blocks* section and the actual lines of the gear are situated in the *Entities* section. Therefore, only the data in the *Blocks* and *Entities* section needed to be changed. Each co-ordinate component of line has an identification tag. In order to create the drawings with new dimension, these co-ordinates needed to be overwritten. However, the data given in the table, could not simply be copied onto the DXF file. This is because the data specifies the length of the line and not the x and y co-ordinates. To overcome this problem, it was decided to define each point parametrically. In this way, the co-ordinates could be calculated from the data given in the table.

3.1 Calculating co-ordinates of the lines

To determine the parametric values of each point, it is first required to determine an origin on the diagram. The left co-ordinate of the centre line of the gear is assigned to be the origin. This is shown in figure 2, the letters correspond to the dimensions given in the table of data.

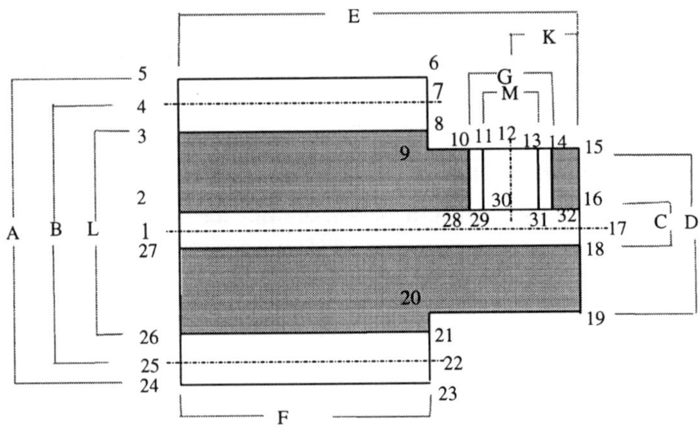

Figure 2 – Spur Gear with the co-ordinates being marked

The drawing is made up of 32 points. Each of these needs to be assigned an equation (a parametric value). Since point 1 is the origin, its value is (0, 0). Now, the value of point 2 needs to be calculated. Its x co-ordinate is the same as point 1, i.e. 0. Its y co-ordinate is +C/2. It is positive because it is located above the origin (point 1). So, its co-ordinate is (0,C/2). The co-ordinate of the point 27 would be negative, i.e. (0, -C/2). This is due to the fact that it is located below the origin. In this way, the values for all the co-ordinates are calculated.

3.2 Overwriting the original DXF file

The first step is to create a drawing in AutoCAD with one of the dimensions given in the catalogue. This drawing is then exported as DXF file. When the user clicks on any records on the table, the Visual Basic program saves the data on that row in a file. The Visual Basic then runs a C++ executable file. The C++ program reads the data which has been saved to the file. It then calculates the co-ordinates of each point using the appropriate equation and then overwrites the co-ordinate values of the lines on the DXF file that has been created. Only the *Blocks* and *Entities* sections of the DXF file needs to be edited.

4. HATCHING

The information about the hatching is located in the *Blocks* section of the DXF file. The lines that construct the hatching are defined exactly as the lines that construct the outline of the drawing. However, a lot of calculations and conditions are needed to be taken into account in order to determine the co-ordinate of the hatching lines. When the dimension of the gear changes, the co-ordinates of the lines that construct the hatches, should be changed relatively. The program holds the information on all the points that construct the gear. This information is used to calculate the co-ordinates of the lines that define the hatches.

For a better understanding, hatching the top right section of the gear is going to be explained here. This section is shown in figure 3. The marked points on the diagram are those that are needed in the calculation for hatching this section. The program starts hatching from the top left corner of this section, i.e. point 14. The lines are drawn from the top edge to the left edge. Initially, the co-ordinates of the two points of the lines are set to be the same as point 14. The spaces between the hatches are set to be 1.0.

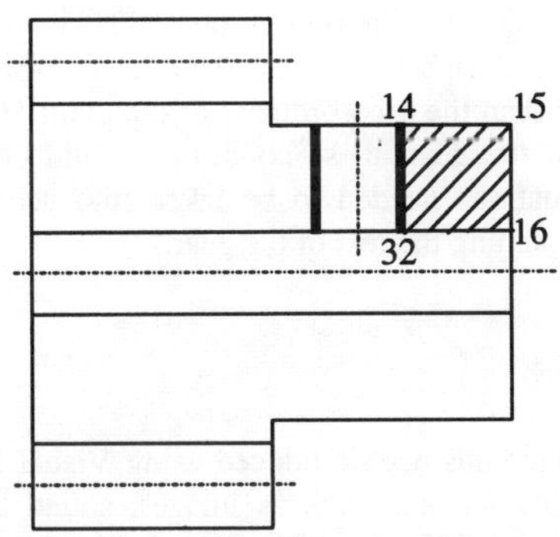

Figure 3 – The top right section that needs to be hatched

Let us assume that (*ax, ay*) is the co-ordinate of the first point of the line and (*bx, by*) is the co-ordinate of the second point of the line. The program first starts by incrementing *ax* and decrementing *by*. *ay* and *bx* stay unchanged. The program then checks if *ax* has passed the top right corner of the gear, i.e. point 15, and also if *by* has passed the point 32. For the first condition, the program then starts decrementing *ay* and sets the value of *ax* the same as the x co-ordinate of the point 15. In this way, the co-ordinate (*ax, ay*) will lie on the right edge. For the second condition, the program starts incrementing *bx* and sets the value of *by* the same as the y co-ordinate of the point 16. The co-ordinate (*bx, by*) will then lie on the bottom edge of this section.

A problem that arises here is that when *ax* reaches point 15, if *ay* is decremented by 1.0, the hatching will not be consistent and the space between this line and its previous one does not seem to be 1.0. The same problem applies when *by* passes point 32. Therefore the distance of *ay* to the bottom of the point 15 must be calculated. The equation that calculates this, is shown below:

$space = 1.0$ *(the distance between the hatches),* $\alpha = 45^\circ$

$$\tan 45^\circ = \frac{T}{S} \Rightarrow 1 = \frac{T}{S} \Rightarrow T = S \therefore T = ax - 15x$$

$$ay = 15y - T \Rightarrow ay = 15y - ax + 15x$$

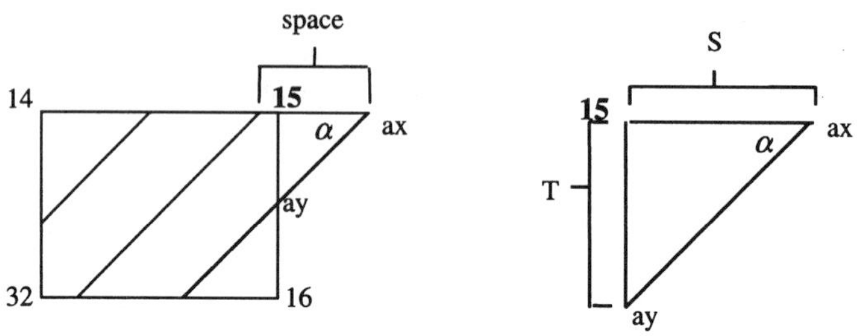

This value of *ay* is used only once. From the second time onwards, the value of *ay* is decremented by the value of space, i.e. 1.0. A similar calculation needs to be done for the value of *bx*, i.e. the distance to the right of point 32. The program then stops drawing the lines when either the value of *ay* becomes less than the y co-ordinate of the point 16, or the value of *bx* becomes more than the x co-ordinate of the point 16. It has to be noted that depending on the dimension of the gear, these conditions could occur simultaneously or at different times. All these situations needed to be taken into account in the C++ program. Similar method is used for hatching the rest of the gear.

5. CONCLUSIONS

An electronic catalogue has been produced using Visual Basic AND C++ which included a number of multimedia features such as images, sound and animation. Research has been carried out in the area of DXF files. The program is able to create a DXF file drawing for a spur gear with various dimensions. The DXF file can then be imported to CAD packages as an assembly or component drawing. The main feature of the electronic catalogue was the scrollable database. The data is displayed in a scrollable tabular format which is an outstanding feature compared to other electronic catalogues. The user can easily view all the data on the screen and select the appropriate dimension.

REFERENCE

(1) RS Catalogue on CD-Rom, RS Component Ltd, U.K, 1998
(2) Lee Spring Electronic Catalogue, Lee Spring Ltd, U.K, 1997
(3) SKF Electronic Handbook Version 2.0, SKF (U.K) Limited, 1996
(4) Harrington, J., Spenik, M., Brumbaugh, H., Diamond, C. , "Visual Basic 5 interactive course", The Waite Group, Inc., USA, 1997
(5) AutoCAD, "AutoCAD release 13, Customization Guide", Autodesk, Inc., USA, 1996

Surface visualization and assessment using geometric parameter curves

R J CRIPPS and **R E HOWE**
School of Manufacturing and Mechanical Engineering, University of Birmingham, Edgbaston, UK

Ensuring the quality of a CAD generated surface model is of paramount importance to the follow-on activities within CADCAM(1). Current quality assessment methods either rely on the mathematical representation or are difficult to interpret geometrically. We briefly review these problems then introduce geometric parameter (gp) curves(2) as a geometric method for visualising and assessing the quality of a surface. Surface gp curves based on arc length, tangent angle and normal angle are illustrated on a range of surfaces. It is expected that quality surfaces have smoothly varying gp curves. The method is applied to an industry generated surface.

1. INTRODUCTION

Before the introduction of CADCAM, the quality of surfaces was assessed by skilled craftsmen looking at large scale drawings of section curves and the physical models or patterns. With a CADCAM generated definition, it is now possible to consider its quality on a workstation screen. However, interpreting the results of some interrogation methods is difficult especially at a screen. Further, some methods relate to the parametrisation not the geometry. To understand why, we need to consider how surfaces are represented. Free form surfaces in CADCAM are typically defined by a vector valued parametric equation of the form:

$$S(u,v) = [\, x(u,v),\, y(u,v),\, z(u,v) \,] \quad 0 \le u\,,\, v \le 1,$$

The most popular representations are Bézier and B-spline forms(3). For example, the rational Bézier form can be represented as :

$$S(u,v) = \frac{s(u,v)}{w(u,v)} = \frac{\sum_{i=0}^{n}\sum_{j=0}^{m} \mathrm{b}_{n,i}(u)\mathrm{b}_{m,j}(v) w_{i,j}\, \boldsymbol{v}_{i,j}}{\sum_{i=0}^{n}\sum_{j=0}^{m} \mathrm{b}_{n,i}(u)\mathrm{b}_{m,j}(v) w_{i,j}} \qquad 0 \le u\,,\, v \le 1;\ w_{i,j} > 0,$$

where $\mathrm{b}_{n,i}(u)$ and $\mathrm{b}_{m,j}(v)$ are the Bézier basis functions, $w_{i,j}$ are scalar weights which are normally restricted to being positive, and $\boldsymbol{v}_{i,j}$ are the vertices of the characteristic polyhedron.. We note that when the weights are constant we have the non-rational Bézier form. Each surface is then a rectangular grid of (rational) polynomial patches of the form $S(u,v)$ and a typical definition is created as a collection of one or more surfaces. However, the surface representation should not affect the way in which we are able to visualise its form. For example, a shoe last or a car roof (half shown because of symmetry about the centre line along the bottom right-hand edge) (Figure 1) should have the same shape, to within some tolerance, whatever its representation. It is the geometry of the object that is important not its computer-based representation.

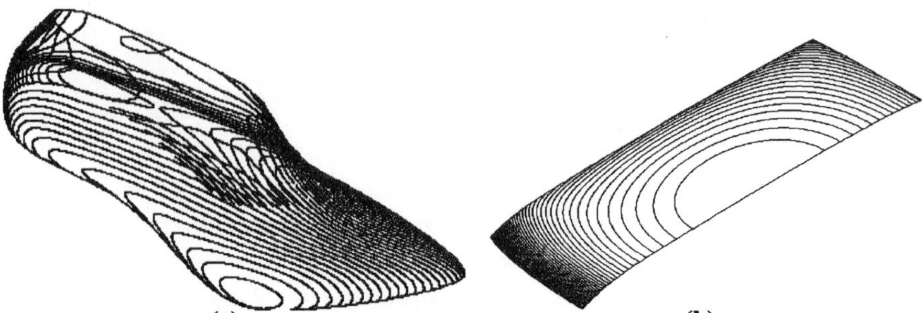

(a) (b)

Figure 1. Computer-based definitions - contours for shoe last(a) & car roof(b).

Before actual production of tools, dies and moulds, etc, interactive design methods allow the surfaces to be assessed for quality and manufacturability. Modification is usually carried out by direct manipulation of the $\boldsymbol{v}_{i,j}$. This can be time consuming since it is difficult to relate changes to the design handles, i.e. the polyhedra, back to the geometry. Once a quality definition has been generated, it can then be used to generate a sequence of points and normals. These are used to define numerically controlled (nc) tool cutter paths to drive an nc milling machine, producing the dies and moulds for the components(4). Thus with CADCAM we have the capability of matching the computer based definition and the production tool to machining tolerances, i.e. what we define is what we get. This has resulted in the need to create high quality surface definitions. This quality is not just for mathematical rigour or computational speed and robustness, nor just for the aesthetics imposed by the stylists or designers. Small surface imperfections can have large consequences in down-stream activities such as nc tool manufacturing and robotic assembly (1).

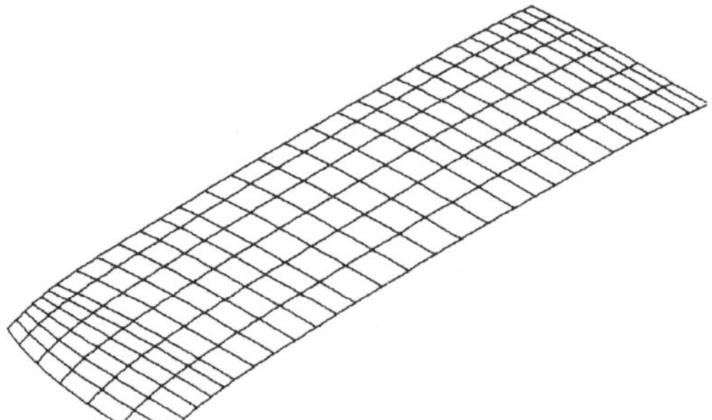

Figure 2. A regular polyhedron net indicates a quality surface.

CADCAM systems offer a range of interrogation tools to aid in the production of high quality of surface definitions. Some are computationally equivalent to the familiar manual techniques, for example isoparameter curves and planar intersects. Others are new techniques with no manual equivalents, for example curvature based analysis. These procedures have been well documented(5-8). With increasing order of complexity these are isoparameter curves, planar intersects and curvature analysis.

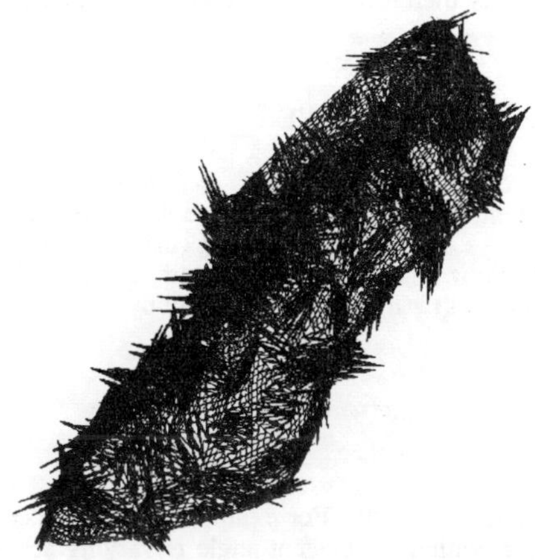

Figure 3. Irregular polyhedra does not necessarily imply a poor surface.

The problem is that the simple techniques, whilst easy to interpret, can give misleading results. A good net of isoparameter curves (equivalently the polyhedron net) virtually guarantees a quality surface. This is illustrated in the car roof where the polyhedron net is highly regular (Figure 2). However, an oscillating polyhedron net does not necessarily imply a poor surface. This phenomenon is well illustrated by the polyhedron net for the shoe definition (Figure 3) and its parametric grid (Figure 4).

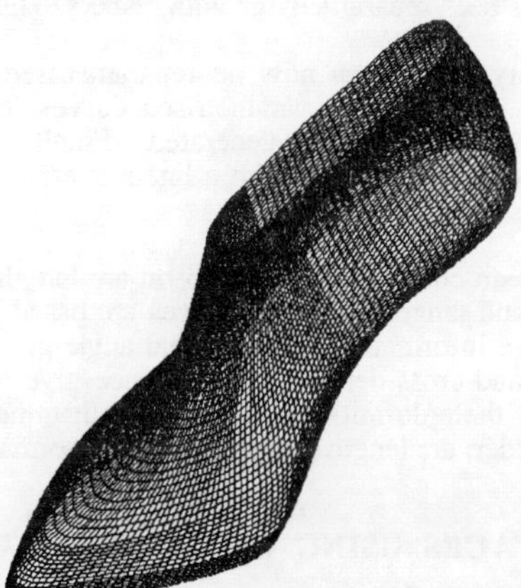

Figure 4. Isoparameter curves for the shoe last in figure 3.

This suggests that we should look at the isoparametric grid. However, the disadvantage of this approach is that the isoparameter curves are dependent on the representation and are therefore not geometric in nature. Thus if we were able to detect an irregularity in the isoparameter curves, relating it back to the geometry is by no means straightforward. An alternative approach is to obtain more information about the geometry of the definition. This leads to more complex methods of interrogation, for example curvature analysis techniques.

Unfortunately their added complexity makes them difficult to use and interpret(5-8). The aim is to develop a method of interrogating a surface which has the property that it is geometric and simple to interpret. This could then be used in addition to existing surface quality assessment methods. Our starting point is therefore isoparameter curves and our aim is to generate gp curves.

2. GEOMETRIC PARAMETER CURVES

Given a surface boundary curve of the form $\boldsymbol{r}(u) = \boldsymbol{S}(u,0)$, where $0 \le u \le 1$, with some underlying parametrisation, u, which does not necessarily relate to the geometry of the boundary curve or surface. We can reparametrise the curve such that our new parameter $g(u)$ say, relates to some geometric invariant, for example arc length. Thus our reparametrisation takes the form $\boldsymbol{r}(u) = \boldsymbol{r}(g(u))$ with $g(u)$ given generically by:

$$g(u) = \frac{\int_0^u G(\alpha)\mathrm{d}\alpha}{\int_0^1 G(\alpha)\mathrm{d}\alpha},$$

where $G(\alpha)$ is some geometric invariant. For example, to reparametrise the curve so that the parameter relates to the rate at which the tangent angle is varying across the total length of the curve, $G(\alpha)$ is given by:

$$G(\alpha) = \frac{\left|\frac{\mathrm{d}\boldsymbol{r}(\alpha)}{\mathrm{d}u} \times \frac{\mathrm{d}^2\boldsymbol{r}(\alpha)}{\mathrm{d}u}\right|}{\left|\frac{\mathrm{d}\boldsymbol{r}(\alpha)}{\mathrm{d}u}\right|^2} \left|\frac{\mathrm{d}\boldsymbol{r}(\alpha)}{\mathrm{d}u}\right| = \frac{|\boldsymbol{r}_u \times \boldsymbol{r}_{uu}|}{|\boldsymbol{r}_u|^2}\boldsymbol{r}_u = \tau(\alpha)\boldsymbol{r}_u,$$

and we say that our curve has been reparametrised with respect to tangent angle.

Each of the four boundary curves can now be reparametrised with respect to the same geometric invariant. Along each of the reparametrised curves, equally spaced points with respect to the geometric invariant can then be generated. Finally, we connect corresponding parameter values across opposite boundaries with a linear interpolant which approximates the true geometric surface curves.

The gp curves that have been considered are based on arc length, tangent angle and normal angle. Both the arc length and tangent angle gp curves are based purely on boundary and at most first geometric derivative information. The normal angle gp curves are based on second order geometric derivatives and cross-derivatives and hence give more information regarding the surface interior. Given their definitions, it is likely that their sensitivity to geometric irregularities will be in the order: arc length < tangent angle < normal angle (3).

3. VISUALISING SURFACES USING GEOMETRIC PARAMETER CURVES

Since the gp curves are a function of the underlying geometry, we would expect that standard geometric features, for example an inflection, will give characteristic patterns in their gp curves. To illustrate we consider two geometric features, namely a mound and a surface containing an inflection.

The resulting gp curves for the mound surface (Figure 5) show that the isoparameter curves are very regular and give no indication of the geometric nature of the surface. The gp curves however show distinctive patterns. The arc curves become closer as we approach a corner, whilst the two angle based curves become closer towards the centre, reflecting the symmetry of the surface. Both patterns reflect the changing rate of the geometric measure as we traverse the boundary curves. However, neither pattern indicates whether the mound is a hump or a hollow.

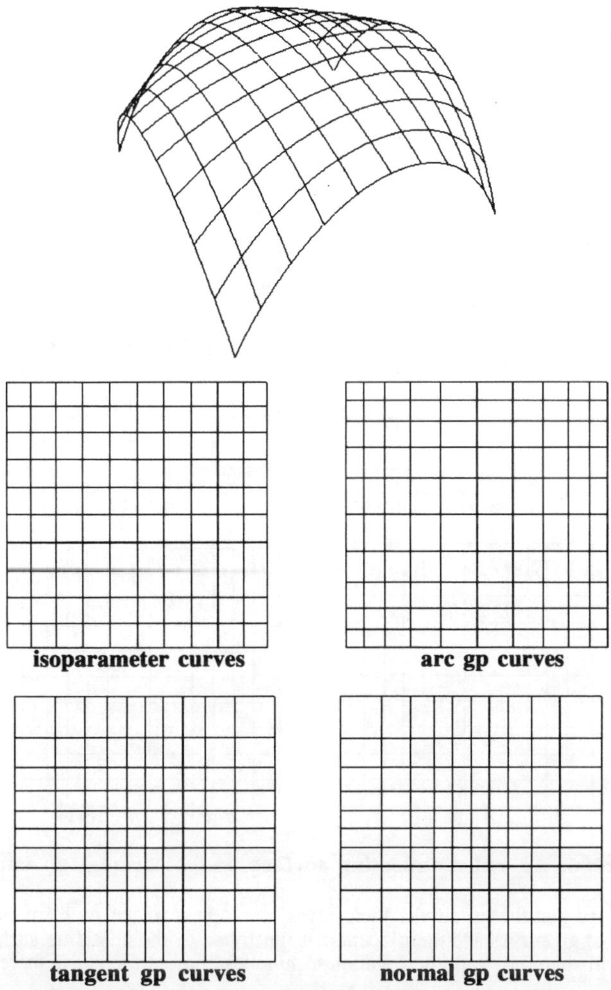

Figure 5. A 'mound' surface and associated gp curves.

The resulting gp curves for the inflecting surface (Figure 6) show that the isoparameter and arc gp curves are similar in nature and give little information regarding the geometry of the surface. The two angle based gp curves characteristically diverge away from the inflection point, the amount being related to the degree of the inflection. This pattern reflects the changing rate of the geometric measure as we traverse the boundary curves with the gap indicating the location of the inflection - opposite boundaries have geometric features which are in sympathy with each other. The location could be sensed better by increasing the density of the gp curves.

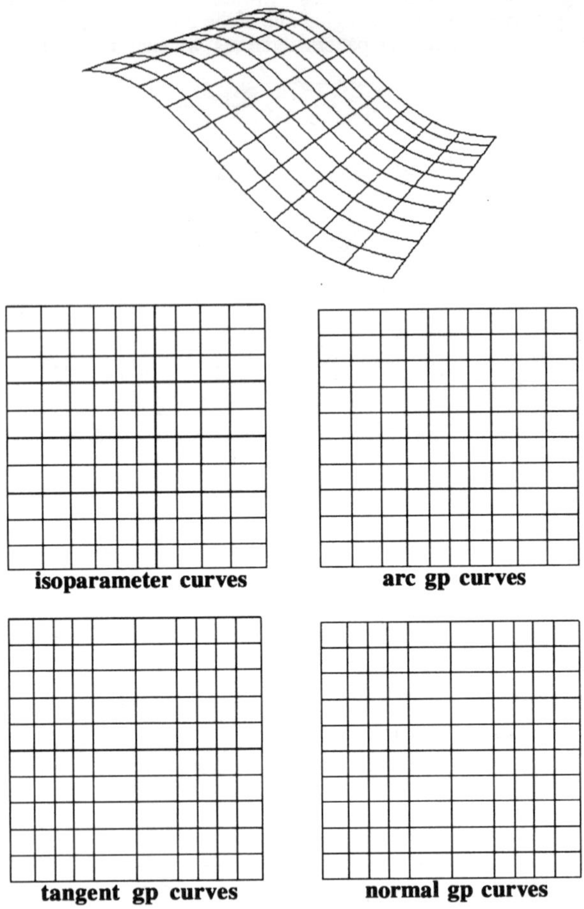

Figure 6. An ruled inflecting surface and associated gp curves.

It is recognised that the gp curves themselves are subject to the problem of choosing the correct number of gp curves and their location as illustrated by the inflecting surface gp curves. This is a problem shared by all interrogation techniques and we recognise there are no simple solutions.

4. QUALITY IMPROVEMENT BASED ON GEOMETRIC PARAMETER CURVES

It is also expected that the regularity of gp curves will indicate quality surfaces just as in their non-geometric equivalent isoparameter curves. Again we illustrate by considering two surface patches taken from the shoe last definition in the heel region. The isoparameter curves are shown in Figure 7 and the gp curves for the original patches are given in Figure 8. Here the arc length gp curves look fairly regular, however the tangent and normal angle gp curves have irregular spacings on opposite boundaries. This suggests the underlying geometry of the boundaries could be made more consistent.

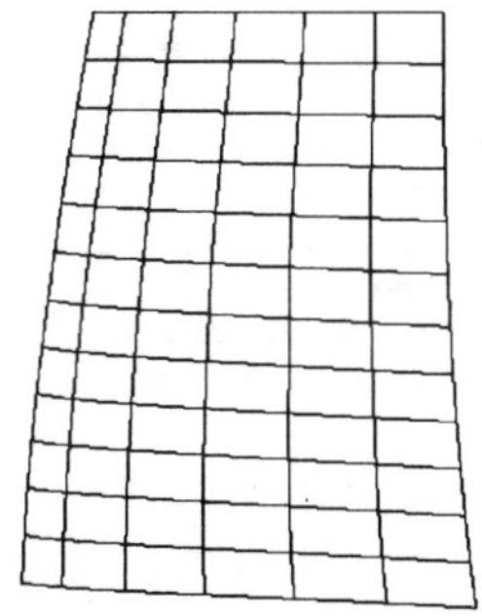

Figure 7. Isoparameter curves for shoe last patches.

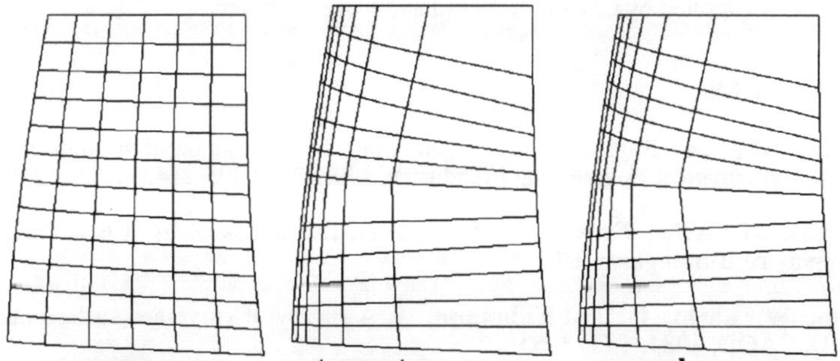

arc gp curves tangent gp curves normal gp curves

Figure 8. Original gp curves for the shoe last patches.

Minor modifications to the shoe patches were carried out manually to tidy the tangent and normal gp curves. These modifications were achieved with movements to the original vertices within a tolerance of ±0.1 mm with a corresponding surface movement within ±0.037 mm. The resulting gp curves are given in Figure 9 and show an improvement in terms of their distribution.

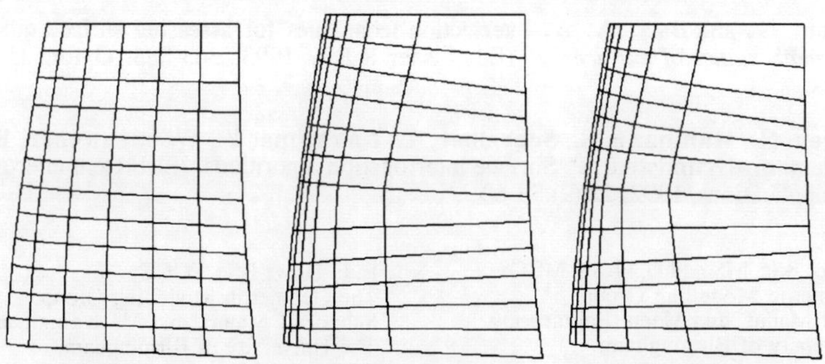

arc gp curves tangent gp curves normal gp curves

Figure 9. Modified gp curves for the shoe last patches.

5. CONCLUSIONS

Geometric parameter curves have been introduced as a means of visualising and assessing the quality of free-form surfaces in CAD. Work is continuing on their characterisation and in developing automatic procedures for improving the regularity of gp curves as a means of improving the overall quality of a surface definition. The ability to predict problems in down-stream activities in the CADCAM process by assessing the quality of the underlying surface geometry will have a positive input in the development of a fully integrated CADCAM system. Relating quality improvements directly to the design handles, $\boldsymbol{v}_{i,j}$, will help in both the integration of CADCAM and also improve the interactiveness of the computer-based total design process

6. ACKNOWLEDGEMENTS

The authors are pleased to acknowledge the support of the CDP Directorate of EPSRC (Grant No GR/K 31268) and Delcam International plc. The definitions of the shoe last and the car roof were kindly supplied by Clarks Shoes Ltd and Rover respectively.

7. REFERENCES

(1) **Cripps, R.J.** and **Barley, S.A.** A geometric characterisation of springback in drawn panels. Mathematical Engineering in Industry, 1992, **3** (3) 205-214.

(2) **Czerkawski, A.M.** *Fitting procedures for curves and surfaces*. 1996, Ph.D. thesis, University of Birmingham, UK.

(3) **Böhm, W. Farin, G.** and **Kahmann, J.** A survey of curve and surface methods in CAGD, CAGD, 1984, **1** (1) 1-60.

(4) **Marciniak, M.** *Geometric modelling for numerically controlled machining*, 1991, Oxford Science.

(5) **Cripps, R.J.** and **Ball, A.A**. Visual assessment of free-form surfaces in CADCAM. *Proc. Inst. Mech. Engrs. Part B*,1998, 212, 207-214.

(6) **Farouki, R.T.** Graphical methods for surface differential geometry. In: *The mathematics of surfaces II*, (Ed. Martin R.R.), 1987, 363-385, Oxford University Press.

(7) **Liden, G.** and **Ball, A. A.** Intersection techniques for assessing surface quality. In: *The mathematics of surfaces V*, (Ed. Fisher R.B.), 1993, 363-385, Oxford University Press.

(8) **Hagen, H.**, **Hahmann, S.**, **Schreiber, T.**, **Nakajima, Y.**, **Wördenweber, B**. and **Hollemann-Grundstedt, P**. Surface interrogation algorithms. IEEE Computer graphics and applications, 1992, **12** (9) 53-60

R. J. Cripps BSc, MSc, PhD, FRSS, MBCS,
The Geometric Modelling Group,
School of Manuf. and Mech. Engineering,
The University of Birmingham,
Edgbaston,
Birmingham, B15 2TT, U.K.

R. E. Howe BSc, PGCE
The Geometric Modelling Group,
School of Manuf. and Mech Engineering,
The University of Birmingham,
Edgbaston,
Birmingham, B15 2TT, U.K.

A comparison of techniques for modelling robot dynamics

J U DOLINSKY, G COLQUHOUN, and **I D JENKINSON**
School of Engineering, Liverpool John Moores University, UK

An industrial robot can be programmed in two ways either, by writing the program online using teach programming or by generating the program off-line with the aid of a CAD based Off- Line Programming system (OLP). An OLP uses simulation to imitate the robot activities. This simulation is based on a model (for computing, a mathematical model) which sufficiently describes the robot kinematics, dynamics and the controller strategy. In general, analytical models are used. However, some investigators have proposed models based on neural networks.

In this paper the advantages and disadvantages of neural network based models are explained and compared with general analytical modelling methods. Furthermore possibilities for the refinement and adaptation of the analytical model of the respective robot are discussed. This includes investigations regarding the applicability of automatic mechanisms for parameter identification etc. in order to automate model generation.

1. INTRODUCTION

Offline Programming Systems allow the development and test of robot programs without using the physical robot. With a CAD system a sequence of tasks is designed (3D data) and thereafter applied to a simulator, which is based on kinematic, dynamic and controller model of the respective robot.

In general the problem an industrial robot controller has to solve, may be simplified and expressed by the following formal notation:

$$T = f(X_{Start}, X_{Target})$$

Between two given points X_{Start} and X_{Target}, which are vectors in the 3 dimensional cartesian space, a trajectory T (a series of configurations → end effector positions or joint angle vectors) has to be found, which models the special kinematic and dynamic properties of a robot. Therefore the controller f has to realise a mapping between the 3D data of the end effector position and the joint angle configurations of the robot. Such a mapping is inaccurate due to the manufacturing tolerances of robot components and wear.

It is essential to identify the individual physical robot parameters and the controller strategy used for determining trajectories in order to build accurate controller models. It is obvious that both procedures are learning processes, which have the potential to be implemented e.g. using Artificial Neural Networks (ANN).

2. ARTIFICIAL NEURAL NETWORKS

In recent years various types of artificial neural networks have been developed, studied and thereafter applied to practical and theoretical research fields related to the computational learning of natural systems like pattern and speech recognition, machine learning etc. The success of ANNs may be explained by the fact, that they can be simply implemented either in software or hardware.

A neural network provides a method of mapping between high dimensional input/output spaces. The information held in the ANN (In this context knowledge about mapping joint angle- and Cartesian space) is stored in the form of weighted connections between artificial neurons, which are generally organised in layers [4]. Adjusting the values of the weights during what is known as the training process configures the ANN.

During the training process the network is fed with example pairs of joint angles and respective Cartesian end effector vectors. In order to provide a proper mapping, the weights of the net have to be adjusted using an appropriate learning rule. Learning rules vary according to the type of network and learning strategy, examples include:

i. The *Hebbian rule* and the *Delta rule* (a gradient descent method) are applied to unsupervised learning in for example single layer perceptrons;
ii. The *Backpropagation* rule used in *supervised learning* of multi-layer perceptrons.

After finishing the training process the ANN is be able to classify incoming patterns (joint angles vectors) and to provide the proper mapping. The ability of ANN to give reasonable output for inputs not contained in the training set is known as *generalisation*. This property makes an ANN in general a useful approximation tool [12] e.g. for learning the behaviour and modelling of highly non-linear mechanisms [5][8] such as robot arms [13].

Neural models are pure numerical computational models. To convert the information contained in the weight matrices of the net into a readable symbolic form to extract information about the learning state of the network, is a difficult task [11].

However, the key problem a designer of an ANN has, is in selecting an appropriate net structure for the mechanism to be learned. This means the right choice of the network type (either a *feed forward* or a *recurrent network*), the number of the neurons, layers, suitable activation functions e.g. *hyperbolic tangent* or *sigmoidal* within the neurons, all have to be determined empirically [15]. Another problem is the initialisation of the weight matrices of the ANN. Tests have shown that non-optimal weight configurations lead to a considerable longer training time especially when applying backpropagation [10].
Furthermore during the training of the net with pairs of example patterns, which is an iterative optimisation process, the algorithm may run into a local minimum [9]. This means, that the weight values of the ANN have not been optimally adjusted to the new example. The example has then not been correctly learnt by the ANN. In this context a point of concern is also the capacity of the ANN, which has to be analysed in order to avoid over-training errors [1]. It is also acknowledged that ANNs provide an undetermined mapping beyond the range they have been taught (locality). For an application it means, that a possible critical value that exceeds the admissible range of mapping has to be considered and the ANN has to be taught with respective examples.

1. ANALYTICAL MODELS

The more traditional engineering approach is to build analytical models of the robot using system equations. The pre-condition therefore is knowledge of the respective kinematic, dynamic and controller parameters. Whereas the kinematic (length and twist of links etc.) and dynamic (mass of links etc.) parameters are mostly supplied by the manufacturer of the robot, the controller algorithms are in general not documented. This makes it generally difficult to find a closed analytical description for the entire mechanism.

However, using the kinematic parameters (pref. in *Denavit- Hartenberg (DH)* notation) a direct forward kinematic model (coarse geometric model) can normally be generated. To work towards more accuracy the manufacturing tolerances of the physical parameters of the individual robot have to be integrated into the model as parameters that have to be identified experimentally. Therefore new models or refinements of the DH model have been proposed [14][16]. To cover non-linearity such as joint boundaries etc. respective constraints have also to be added to the model. The growing number of parameters raises model complexity and may impair the solvability of the inverse transformations[1] of the model to compute end effector positions into the joint angle vector space.

However, a complete or partial analytical model (provided it exists) offers many advantages. The generation of the nominal kinematic models can be conveniently performed with the aid of computer algebra systems like *Maple* or *Mathematica* [17]. Using such tools the equations can be manipulated exclusively symbolically, which allows flexible modelling[2] and guarantees maximal stability and minimises the overhead of a subsequent numerical computation.

Several software toolboxes like *robotica* [7] provide predefined methods for calculating

[1] mostly iterative solution of non-linear equations

[2] system equations can be dynamically expanded with parameter terms

dynamics e.g. *Euler- Lagrange* equations of motion of the model. However, in general these packages are straight-forward implementations (based on matrix algebra) of the standard robot modelling concepts without really taking advantage of the possibilities of dynamic structuring and programming combined with symbolic representation and computation provided by the computer algebra system.

2. CONCLUSION

In this paper the advantages and disadvantages of analytical and artificial neural robot model have been discussed and the essential features of both approaches may be summarised as in table 1:

	Analytical Model	Neural Model
Limitations of modelling methodology	- many non-linearities - high complexity in general	- training examples must be provided - extrapolation
Effort involved in Model generation	- initially relative simple: generation of a nominal kinematic model with DH-parameters	- a suitable architecture (numbers of neurons and layers) has to be chosen empirically - high training effort
Task variability	- covers all modelled (described with the equations) situations	- models only trained tasks
Complexity	- mostly high: many kinds of model configurations like singularities have to be considered - complexity rises with complexity of the robot (non-linearities, Degrees of Freedom) and the number of parameters to be identified	- only connectionist complexity (between neurons)
computer internal representation	- symbolic descriptive	- numerical
accuracy	- depends on accuracy of the model	- depends on number of training examples
learning capacity	- dynamic growing	- in general restricted, except by referring an architecture such as in [2]

Table 1 : A summary of analytical and neural model characteristics

In general analytical modelling of robot mechanisms is more difficult because of the need for a mathematical description of complex non-linearities. The quality of a neural model depends on the choice of the network architecture and the number and quality of trained examples.

Although artificial neural nets have been widely studied and successfully applied to numerous specialised research projects, until now they have been in more empirical disciplines. With better computing facilities and the aid of dynamic programming and computer algebra techniques it has become worthwhile to investigate in general analytical models rather then separated solutions of subtasks modelled by artificial neural nets.

1. REFERENCES

[1] S. Amari, N. Murata, K.-R. Müller, M. Finke, H. Yang
Asymptotic Statistical Theory of Overtraining and Cross-Validation
METR 95-06 August 1995

[2] Fritzke, Bernd, *Growing Cell Structures - A Self-organizing Network for Unsupervised and Supervised Learning*. TR-93-026, May 1993, International Computer Science Institute Berkeley, California

[3] Hsinchun Chen. *Machine Learning for Information Retrieval: Neural Networks, Symbolic Learning, and Genetic Algorithms*
Journal of the American Society for Information Science, 1994, in press.

[4] Kröse, B.J.A. and Smagt, P.P. van der. *An Introduction to Neural Networks*.
University of Amsterdam, Amsterdam, The Netherlands, 1994.

[5] F. Lange and G. Hirzinger, *Learning to Improve the Path Accuracy of Position Controlled Robots*, IEEE/RSJ/GI Int. Conference on Intelligent Robots and Systems IROS'94, München, Germany, Sept. 1994

[6] Lippmann, R.P., *An introduction to computing with neural networks*. IEEE Acoustics Speech and Signal Processing Magazine, 4(2):4-22, April 1987.

[7] J. F. Nethery, *Robotica: A structured environment for computer aided design and analysis of robots*, Master's thesis, University of Illinois, Urbana, IL, 1993. Department of Electrical and Computer Engineering.

[8] Poggio, T. & Girosi, F. (1989), *A theory of networks for approximation and learning*, Technical Report AIM-1140, Artificial Intelligence Laboratory and Center for Biological Information Processing, Whitaker College, Massachusetts Institute of Technology.

[9] T. Poston, C. Lee, Y. Choie, and Y. Kwon. *Local minima and backpropagation*, in International Joint Conference on Neural Networks, vol.2, (Seattle, (WA)), pp. 173-176, IEEE Press, July 1991

[10] Shavlik Jude W., Raymond J. Mooney, Geoffrey G. Towell.
Symbolic and Neural Learning Algorithms: An Experimental Comparison. TR 857,

Computer Sciences Department, University of Wisconsin-Madison, June 1989.

[11] Shavlik Jude W. *A Framework for Combining Symbolic and Neural Learning*. TR 1123, Computer Sciences Department, University of Wisconsin-Madison, WI, November 1992.

[12] Smagt P. van der, and F. Groen. *Approximation with neural networks: Between local and global approximation*. In Proceedings of the 1995 International Conference on Neural Networks, pages II:1060-II:1064, 1995. (Invited paper).

[13] Smagt, P. van der and Schulten, K. *Control of pneumatic robot arm dynamics by a neural network*. In World congress on neural networks pg III/180--183. Lawrence Erlbaum Associates, Inc., Hillsdale, NJ, Portland, OR, USA, 1993.

[14] Stone, Henry W. *Kinematic modelling, identification, and control of robotic manipulators. Dissertation.* Carnegie Mellon University, 1986

[15] Vysniauskas, V. and Groen, F.C.A. and Kröse, B.J.A. *The optimal number of learning samples and hidden units in function approximation with a feedforward network.* CS-93-15, 1993.

[16] Whitney, D. E. and Lozinski, C.A. ,*Industrial Robot Calibration Methods and Results*. In the Proceedings of the international Computers in Engineering Conference, pages 92-100. ASME, New York, August, 1984. Las Vegas, NV.

[17] S. Wolfram, *Mathematica: A System for Doing Mathematics by Computer*.
New York, NY: Addison-Wesley Publishing Company, Inc, 1991.

Skill acquisition for robotic assembly via integration of vision and force sensing

C D'SOUZA, I LOPEZ-JUAREZ, K SIVAYOGANATHAN, M HOWARTH, and **V BALENDRAN**
Department Mechanical and Manufacturing Engineering, University of Nottingham, UK

This paper describes a scheme to integrate vision and force sensing with a PUMA robot to acquire manipulative skills for autonomous mechanical assembly. Recognition and location estimation of the assembly components can be done by using two cameras in a fixed stereoscopic set-up or by a camera mounted on the robot arm. In the latter case (eye-in-hand option), two stereo views of the object can be obtained by moving the robot arm. Using continuous path modification, the arm grasps the part and moves it to the insertion point. At this stage the process is guided only by force feedback via a Neural Network Controller (NNC), which recognises and classifies the different force contact states during the operation. These clustered states are then used to complete the mapping to incremental arm motions and guide the component towards the end-condition. The testbed hardware design is described and some initial results obtained for both, vision and force sensing are presented.

1 INTRODUCTION

The success of a robotic assembly technique depends on the accuracy of the robot's sensing systems. Vision and force sensing systems are helpful for grasping the correct component and performing operations such as peg-in-hole insertions. At the start of the operation, visual information is more important since the right component must be recognised and located. As the process progresses, the importance of force feedback increases. This is the case when the assembly component is already grasped and placed in readiness for assembly. At this time the robot has only to "sense" the forces acting on the component being inserted and to decide on the next direction of arm motion that favours the insertion.

This paper describes the research progress on the integration of force and vision with an industrial robot. The aim is to perform assembly operations autonomously. In simple words, the robot has to be able to see, pick and assemble components intelligently, with a minimum of instructions. The robotic system forms part of an intelligent cell with different sensory capabilities currently being investigated within the Manufacturing Automation Research Group. These areas include language understanding, object recognition [1] and force control [2].

A connectionist approach based on the Adaptive Resonance Theory (ART) [3] has been used for both object recognition and the implementation of the Neural Network Controller (NNC). Initial results have demonstrated the ability of the algorithm to recognise simple objects [1]. In this paper, the scheme to integrate the vision system to effectively recognise 3D objects is described and results of the preliminary study are presented. The testbed system and the NNC are described. The results show the incremental learning capability of the NNC to recognise force contact states on-line. Finally, some conclusions and plans for future work are given.

2 SYSTEM DESCRIPTION

The testbed basically consists of a 6 DOF PUMA 761 robot, a robot controller and a supervisory host computer as illustrated in Figure 1.

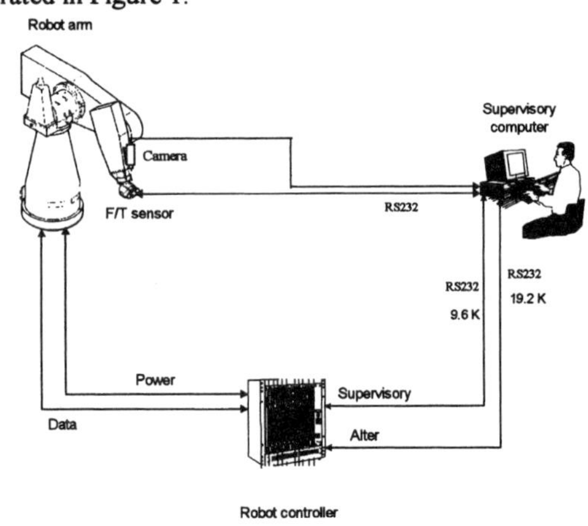

Figure 1 The overall system

The two main units of the PUMA robot are the controller and the robot arm. Power and data are transmitted between the two units through two interconnect cables. The controller houses the components that control and power the robot arm. The VAL II operating system is used for control of the robot. The robot controller is only used as a slave to move the arm, since the VAL language neither allows for development of complex programs nor to implement the force and vision interface boards. Instead, a supervisory computer is being used. It communicates with the controller in two ways: via the "supervisory" mode, in which commands are sent to the controller, and the ALTER mode, for incremental motion of the arm.

Two signals close the control loop to the supervisory computer, the video signal and the Force/Torque (F/T) serial data. The vision system is to be mounted just above the wrist of the robot arm. It consists of a PULNIX TM-460 CCD camera connected to the PC through a MATROX Meteor frame grabber which uses MIL library functions. Images are digitized by this board and subsequent image processing is done for object recognition. The force sensing system consists of a 6-axis JR3 wrist F/T sensor and a DSP based board implemented within the supervisory computer.

3 OBJECT RECOGNITION AND LOCATION ESTIMATION

The vision system is used for two purposes: object recognition and location estimation. Object recognition is done by using Artificial Neural Networks (ANNs). Preliminary work on this in classifying and recognising objects has been completed[1]. Efforts are underway to develop a recognition system that is invariant to rotation by using a neural network called Gaussian ARTMAP[4]. It is able to perform incremental learning of vectors (patterns) and associate them with another vector. The connectivity of surfaces and edges of the object is used for this.

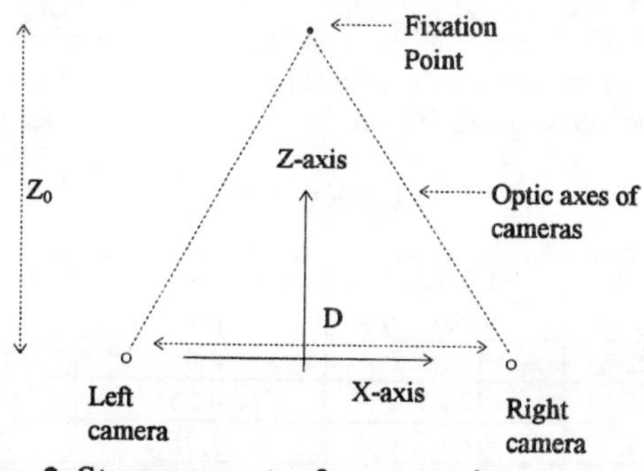

Figure 2 Stereo geometry for converging cameras

Stereo vision is used for location estimation. In stereo vision, two slightly shifted views of the same object are taken (Figure 2). The disparity between the two views is proportional to the depth of the object from the camera. The two cameras normally converge on the object of interest and the angle by which they converge is called the *vergence* angle. The two angles should be equal and the fixation point (the point at which the optical axes intersect) should be orthogonal to the baseline. This allows us to make use of the stereo equations [5]. Figure 2 shows the top view of the cameras. The distance between the cameras D is called the baseline.

There are two options available for using stereo vision with the robot:

(1) Fixed camera configuration.
(2) Eye-in-hand configuration.

In the first case the stereo setup can be placed away from the robot just above the table where assembly is to be performed. Thus two cameras can be placed in a fixed calibrated setup where the fixation point and vergence angles are known. In this case, the camera coordinates need to be transformed into the tool coordinate system of the robot.

The second configuration has been chosen for implementation. There will be a single camera mounted just above the flange near the end-effector. The robot arm will be programmed so that a precise movement of a distance D, the baseline, will be made at a specified angle. Thus, a single camera can be used to obtain both the stereoscopic views. This will allow us to determine the location of the object. The recognition can be done with input from a single view. Thus once the object has been recognised and its co-ordinates known, the robot can move to the desired location. This approach compares favourably with research done recently at Purdue University [6] for the vision-based bin-picking problem. Continuous visual feedback is not used but the image is additionally processed twice when the robot is on its way to pick the object.

3.1 Experiments and Results

Experiments with the fixed camera configuration have already been made where the two cameras were mounted on a tripod stand. Results obtained in estimating a depth of a point with a fixation distance Z_0 of 73 cm and a baseline of 19.5 cm have been tabulated below.(Table 1)

Table 1

Calculated depth (cm)	Measured depth (cm)	Absolute Error found	Percentage error
114.22	114.3	0.08	0.07%
139.45	139.7	0.25	0.18%
163.32	165.1	1.78	1.08%
186.94	190.5	3.56	1.87%
212.09	215.9	3.81	1.76%

4 ASSEMBLY

Several peg-in-hole assembly operations were carried out to determine the requirements of the force control system and the NNC. These experiments have been presented earlier [2]. Based on this work, it was found useful to decompose temporal force signal patterns into spatial patterns as shown in Figure 3a. This was achieved by taking the force readings after each incremental motion, which eliminated any transient response in the sensor due to the inertia of the arm. These type of patterns are encoded and then fed to the NNC. The structure of the NNC is illustrated in figure 3b.

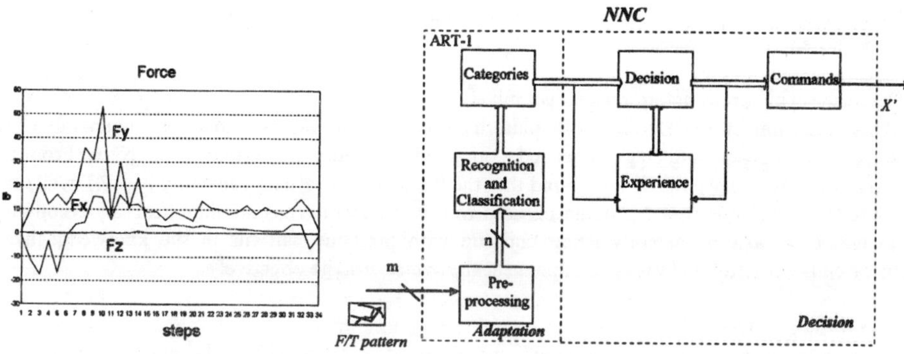

(a) (b)

Figure 3 Structure of the NNC

The first stage, (*adaptation*) has an initial section where the m-dimensional F/T value vector is pre-processed to provide an adequate representation of the contact forces. This pre-processing section may include some discriminant function in order to eliminate redundant contact forces, hence producing an n-dimensional input vector. The *recognition/classification* and the *category* sections are essentially the counterparts of F1 and F2 layers in ART 1 [3] that are in charge of recognising contact states and classifying them. If new force patterns arrive, the adaptation stage will allocate space for these new patterns into the current categories. When a category already exists it will become active (or in resonance according to ART theory) , so the system still preserves the previous categories, hence providing the required flexibility for the knowledge base.

During the *decision* stage, the corresponding category contains a group of contact states with similar pattern features. However, due to the geometry of the mating pairs there might not be a unique motion direction, hence this information has to be associated through the experience section (past events and geometry) in order to make a decision about the optimum motion. Finally, the output *X'* is a motion relative to the previous position of the arm, that produces a reduction in contact forces as well as a motion towards the end condition.

Results on the implementation of the NNC have shown that the algorithm is able to classify all contact force patterns during insertions in real-time. The number of patterns is only restricted to the computer's memory. Patterns such as the one shown in Figure 3a were learnt by the NNC after they were presented twice. The ability of the network to generalise these force patterns can be varied according to the network's vigilance parameter.

5 CONCLUSIONS AND FUTURE WORK

This paper describes the current state of research on the integration of vision and force sensing systems with a robot arm. For the vision system, the fixed camera stereo configuration was implemented to locate the object and the results obtained were presented. The mounting of the camera on the robot arm as described above is envisaged for future work. Preliminary work on object recognition has been done using ANNs and research is progressing to further enhance the vision system.

The analysis of forces acting during assemblies was important to specify the requirements of the NNC. The number of contact force patterns occurring during assemblies is innumerable and therefore a preprocessing stage is needed as pointed out in the description of the NNC. From the study of temporal patterns, it was found that the incremental learning capability of ART while not "forgetting" previous patterns could be used to recover from gross misalignments by comparing the temporal pattern currently being built up with previous patterns in the knowledge base. Research is currently underway to implement the *decision* stage effectively.

Acknowledgements

The authors wish to thank Rolls Royce and Associates for providing the manipulator for this project. The support of The Nottingham Trent University and The Consejo Nacional de Ciencia y Tecnologia (CONACyT) to Mr Lopez-Juarez in his research work is also acknowledged.

REFERENCES

(1) C D'Souza, K. Sivayoganathan et al; Machine Vision for Robotic Assembly: Issues and Experiments, *Proceedings of the Thirteenth National Conference on Manufacturing Research,* Glasgow, pp 114-118, 1997.

(2) I. Lopez-Juarez, M Howarth, K. Sivayoganathan. Using Force Data to Control Robotic Assembly., *Proc. of the 13th National Conference on Manufacturing Research,* Glasgow, pp 300-304. September, 1997.

(3) Gail A. Carpenter, Stephen Grossberg; A Massively Parallel Architecture for a Self-Organizing Neural Pattern Recognition, *Computer Vision, Graphics, and Image Processing,* pages 54--115. Academic Press, Inc., 1987.

(4) Williamson, J. R.; Gaussian ARTMAP: A Neural Network for Fast Incremental Learning of Noisy Multidimensional Maps. *Technical Report CAS/CNS-95-003,* Boston University, Center of Adaptive Systems and Department of Cognitive and Neural Systems, 1995

(5) R.D.Boyle, R. C. Thomas; *Computer Vision -A first course,* Blackwell Scientific Publications, 1988.

(6) Krisnawan Rahardja, Akio Kosaka; Vision-Based Bin-Picking: Recognition and Localization of Multiple Complex Objects Using Simple Visual Cues, *Proc. of the IEEE/RSJ International Conference on Intelligent Robotics and Systems,* Osaka, Japan, 1996.

CAD-based recognition – a neural network approach

P MENON, V RAJEEV, K NATARAJAN, and **P RADHAKRISHNAN**
PSG College of Technology, Coimbatore, Tamil Nadu, India
S LOKE and **S ARUNACHALAM**
Department of Manufacturing Systems Engineering, University of Hertfordshire, Hatfield, UK

ABSTRACT

A neural network approach to CAD based object recognition for components is presented in this paper. Regular and irregular shape holes are used as entities of the component in this investigation. Images of 2D projection of component are analysed. This paper focuses on the identification of the component with reference to the template pattern of the component.

1.0 INTRODUCTION

The vision inspection system that is widely used in the industry today mainly processes and correlates data collected by the computer vision system, with the actual data obtained from the measurement of the component. Based on this correlation, the component is accepted or rejected. In some systems the data from the CAD model of the component is transferred to the vision system and the measurement is carried out based on the strategy specific to the CAD database. All these have a disadvantage that for different components the strategy developed is different. Thus if several components are passed through the same assembly line it utilises considerable computational resources. There is also a possibility of speed being compromised. Another important problem is in the complexity of the identified part. That is if the part image obtained is not complete as per CAD model programmed, the system may not identify the component at all. Charles and Thomas[1] proposed integration of CAD model with vision system to generate a strategy tree for automatic identification of the component. In another method[2] a CAD model of the object to be recognised is first made and features and their attributes are extracted. Following this, cluttered image of objects is captured. Based on this a search is carried out on the cluttered image for object identification and orientation.

Recently, neural networks are increasingly being made use of in classification and recognition of components. Neural networks are used in recognition due to their adaptive nature and robustness to noise and deformation of images. This ability is radically different from the standard software techniques because it does not depend on the programmer's prior knowledge of rules. They can reduce development time and can solve problems that have inputs, which are not well defined. Several neural network models have been proposed for recognition like the third order neural network based on invariant recognition system [3] and Higher-Order Neural Net (HONN) with invariance to translation, rotation, scaling [4].

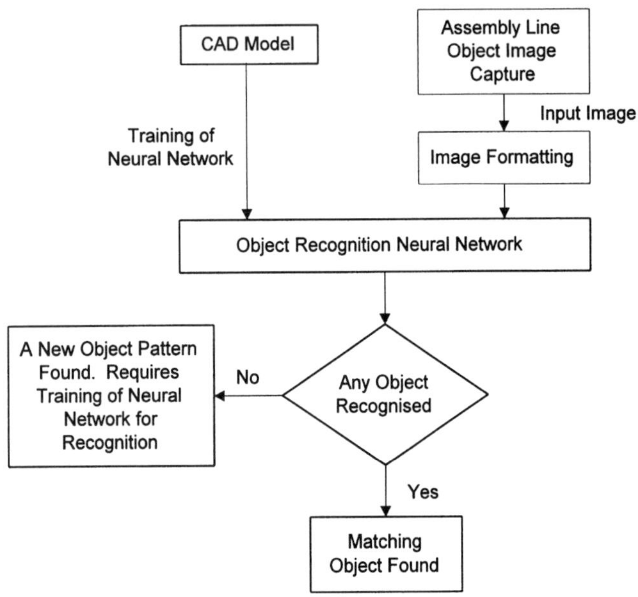

Fig. 1 Object Recognition Module

2.0 PROBLEM DEFINITION

This paper deals with relating CAD database and vision inspection for the ultimate aim of integration of these in a flexible manufacturing system. The focus of this work is in the recognition of the component using adaptive resonance theory-2 (ART 2). Since the components are differentiated based on their hole features an assumption is made that the component contains multiple holes. The major objectives in this CAD based approach are:

1) To recognise a component with the help of a template CAD model and a Neural Network without use of master component.

2) To incorporate real time calibration of the image in the software using a model component.
3) To achieve scaling, translation and rotation invariance for recognition of the component using some selected features.

The problem of correct identification is significant due to the fact that, several components may be passing through the same inspection station. Hence recognition becomes an important requirement. For this a Neural Net model would be appropriate. A template CAD model can be used to generate the entity database for all components. The neural net would then be trained with a CAD model of the proposed components, as the template. The image of the component would then be captured and extracted features formatted in a manner that the net can understand. The recognition strategy involves measurement of position of holes and other features with respect to one another, which could be unique for a particular component. The recognition scheme proposed in this work is shown in Fig 1.

3.0 EXTRACTION OF CAD MODEL DATABASE

AutoCAD based template model has been proposed for creating object database. It gives details about the entities or features as a set of indexed data with the index representing the type of data in the set. Raw data that is obtained from the system is of not much use since some of the attributes associated with entities are solely used for internal representation in AutoCAD and has no relevance to the physical model. AutoLISP is used to process the raw data into a meaningful one and provides an ideal method for automated database creation. The parameters extracted are area, perimeter, moment of inertia, compactness, and distance between holes. In this instance, compactness is defined as the ratio of perimeter / area .

4.0 EXTRACTION OF VISION DATABASE

The image processing process is divided into three stages. The first two are pre-processing of the grabbed image and the calibration of the component. The third stage deals with selection of features of a component and the extraction of these features.

4.1 Pre-processing

Pre-processing of the image is initialled with a *smoothing* operation followed by *thresholding / binarising* of the gray level image to a binary image. The approximate threshold value is obtained from the histogram of the gray level image. With the aid of morphological operations, the unwanted minute holes on the foreground (component) image are obliterated. Back lighting is used in the experiment resulting in black binary model and component images.

4.2 Real time calibration

In the Real Time Calibration method developed, analysis of component starts with the calibration of the camera. For this both model and the component are captured in the same image space. The model chosen is a circular object of known dimension (the diameter. The calibration (scale) factor is computed from the ratio of the physical dimension to image dimension of the model. The image is then suitably scaled (either up or down). The result of this will be, image co-ordinates getting transformed into real world co-ordinates and they will have a one-to-one mapping.

4.3 Feature selection and extraction

The derived features are area, perimeter, centroid, central moment of inertia, compactness, distances between features, etc. Selection of these features for a component is done keeping in view that it is translation and rotation invariant. The extracted parameters e.g. area, perimeter, etc. of the features are calculated and stored as groups of data.

5.0 OBJECT RECOGNITION USING ADAPTIVE RESONANCE THEORY (ART) 2

The main problem of any artificial Neural Network is the stability of the network in classifying the input patterns. Neural Network essentially classifies patterns according to the weights associated with its connections. But often when new patterns are presented to the network it affects the previous training. In the real world the Net is subjected to constantly changing patterns, which has to be classified appropriately. There might be cases in which the Net would not come across the training vector twice, and the Net would be asked to classify the input patterns correctly. But if all the weights are modified continuously the Net may never reach convergence. In this context the Network like back-propagation etc. fail miserably. ART-2 network and algorithm maintains the plasticity to learn new patterns, while preventing the modification of the stored weights which has been learned previously. The mathematics behind ART-2 is complicated but the fundamental implementation and idea are not.

Neural net interface that is developed takes inputs from two sources. The first one is from the CAD module whose database is used to 'train' the network. The next is from the vision system that grabs image of physical model to be recognised. The six extracted parameters are taken into consideration for the Net to ensure that the object is the right one. Each set of parameter form a layer resulting in a six layer Neural network. For recognition of each of the parameters only the weights of the connections are changed. Since the weight changes, the Net gets configured for the corresponding type of input vector. Both top-down and bottom-up weights for each parameter are stored in separate files and they are loaded when the corresponding parameters are invoked for recognition.

5.1 Formatting and Normalisation

Since the user can build the training and vision database in any order, before the parameter values are passed to the neural network they are normalised and formatted. This is done in order to standardise the procedure for inputting,. Formatting of values involves normalising the values and arranging these in an ascending order. Normalising is performed with the expression:

$$\text{Norm values} = \frac{x_i}{\sqrt{x_1^2 + x_2^2 + x_3^2 + \ldots}}$$

where x_i is the ith component parameter.

5.2 Training of the Neural Network

The training of the network is done using the formatted CAD data values. Each element of a respective set becomes input for a particular node. The output will be in the form of a neuron registering 1. While in training, for each given input vector, the same neuron is made to fire so that the sum total of all the excitation will be six.

5.3 Object Recognition

Recognition of the object is said to be perfect if the same neuron fires for a particular component. Thus the cumulative index of the firing neuron must be six i.e. same neuron must win for each set of input vector group given in the order as explained above. If the components taken for recognition have the same value for one parameter or if the distortion in the component is above a certain limit then cumulative index may not be six. Therefore, it is better to accept a match with the highest cumulative index and indicate to the user the level of match. Thus the reason for choosing a six-layer network for recognition so that a slight mismatch in parameters will not affect the recognition of the component. The formatted vision data values are used for recognition.

6.0 RESULTS AND DISCUSSIONS

Training is carried out with data derived from a CAD model built using AutoCAD R13. The following parameters are derived from the object image holes: area, perimeter, compactness, central Moment of Inertia (X & Y), and centroidal distances between holes. The values are obtained from AutoCAD, formatted and normalised before being used for training the neural net. The CAD model used is shown in Fig 2.

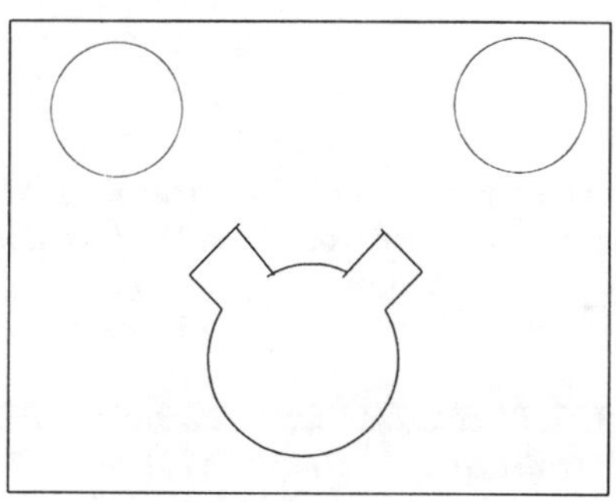

Fig. 2 Template of CAD Model Image

The gray scale image is processed using the MATROX imaging series and the image-processing module is implemented in Borland C++. The image is then ready to be used for extracting the parameters of the component. All analysis are carried out on the hole features.

The network is trained using the CAD database and three other components CAD databases. The vision database generated is used for recognition. The network not only recognised the component but also gave the cumulative index as six. The network is tested on four different components. Another five components having the same features are also tested. The ART 2 network identified each component according to the hole features.

The procedure for pattern recognition that was discussed earlier is implemented. Results of one such implementation are shown below:

Feature	Area		Perimeter		Compactness		Central M. Inertia X		Central M. Inertia Y	
	CAD	Net	CAD	Net	CAD	Net	CAD	Net	CAD	Net
Hole 1	113.10	130.00	37.7	45.21	1	1.25	1017.8	1354.7	1017.8	1342.9
Hole 2	113.10	131.00	37.72	44.14	1	1.18	1017.8	1443.6	1017.8	1292.4
Hole 3	344.06	403.00	73.84	86.77	1.26	1.49	9587.6	13998	9747.4	13142

Feature	Centroidal distance between holes	
	CAD	Net
Hole 1 - Hole 2	49.5	50.34
Hole 1 - Hole 3	34.5	34.58
Hole 2 - Hole 3	34.5	33.50

7.0 CONCLUSION

From experiments it has been found that the system response is very stable to scale, translation, and rotation of the image. As calibration is real time, height of the camera from the object is not critical. However, the camera must be placed perpendicular to the face that has to be sensed. This gives the image profile of holes in its true dimensions and can be easily compared with CAD database using neural net. It has been verified that the neural network designed is capable of identifying the component even though small amount of distortion is present in the component image. One of the limitations is the inability of the net to classify the component if it contains a single hole what ever may be the profile.

REFERENCES:

1. C. Hensen and T. C.Henderson, "*CAD-Based Computer Vision*", IEEE Trans. on Pattern Recognition and Machine Intelligence, vol. 11, no 11, November 1989,
2. F. Arman and J.K.Aggarwal,"*CAD-Based Vision: Object Recognition in Cluttered Range Image Using Recognition Strategies*",CVGIP Image Understanding , Vol. 58 No. 1, pp33-48, July1993.
3. S. J.Perantonis and P. J. G. Lisboa, *Translation, Rotation and Scale invariant Pattern Recognition By Higher-Order Neural Network And Moment Classifiers*, IEEE Transaction on Neural Network, vol. 3, no 2, March 1992, pp 241-251.
4. L. Spirkovska and M. B. Reid, "*Coarse-Coded Higher-Order Neural Network for PSRI Object Recognition*", IEEE Transaction on Neural Network, vol. 4, no 2, March 1993, pp276-283.

Development of a methodology to support a robotic assembly cell

S B MOHAMED
Sultan Zainal Abidin Religious College, Malaysia
D J PETTY
TIME Research Institute, University of Salford, UK
D K HARRISON
Department of Engineering, Glasgow Caledonian University, UK

There has been considerable success in automating machining processes. Flexible automation of the assembly process however has proved to be more challenging. Notably, flexible/robotic assembly cells have only been applied infrequently. One of the key problems is that setting up flexible assembly cells is difficult. Moreover, changing between different products, even when these are relatively similar, is time consuming. This means that product can only be manufactured in relatively large batch sizes.

This paper discusses the on-going development of a methodology to support a flexible assembly cell. The objective is to reduce the batch size necessary for flexible assembly by changeover time reduction. This is achieved by a combination of computer based tools with integrated fixturing allowing off-line set-up during cell operation. A pilot cell, based on a Sysmo TS1300 Gantry robot with a Robox 186PM controller, will be used to test the approach. Also, initial research has suggested that there are large numbers of transactions needed to link the assembly process with the organisation as a whole. Thus, the software design also accommodates interfaces to link the cell to other business systems. This increases the economic attractiveness of flexible assembly, further reducing the feasible batch size.

1. Introduction

1.1 Background

After World War II, with the requirements of high production, better quality and low cost, dedicated automatic lines replaced manual assembly in many industries. Because of the nature of the equipment, the automation of assembly at that time was limited to large production runs. Today, market needs mean that many versions of the same product are required. This has resulted in a need for assembly systems capable of dealing with mixed production in low quantities [1]. Moreover, as product lives have become shorter with frequent changes in design, more flexibility is required from assembly systems. Generally however, as more automation is applied, lower flexibility is the outcome. Thus, producing small volumes with large variety with automated cells is a major challenge. To accomplish this, it is likely that the use of industrial robots in assembly environments will become common [2, 3].

This paper describes on-going research at Salford University on the development of tools to support a flexible, robotic assembly cell. At present, a conceptual design for these tools have been created with the creation of a pilot robotic assembly cell currently underway.

1.2 The Assembly Task

Assembly typically represents 30% - 40% of total product cost [4] and involves picking, orientating, placing and securing of a part. There are three main assembly methods:-

Special Purpose Assembly. Special purpose assembly machines are those that have been built to assemble a specific product. They consist of transfer devices with single-purpose workheads and parts feeders at the various workstations. These special-purpose machines are costly, inflexible and require considerable engineering development before they can be commissioned.

Robotic Assembly. Because of the need to cope with large product varieties, there is a need for assembly systems capable of dealing with mixed production in low quantities. Robotic assembly systems offer tremendous promise for the flexible assembly automation.

Manual Assembly The most common method of assembly is mechanically-assisted manual assembly. Manual processes have considerable flexibility and adaptability. In some instances, it is economical to provide the assembler with mechanical assistance in order to reduce assembly time.

There are several factors that determine which is the most appropriate method for a given product [5]. In general terms however, the key issues are variety and complexity. This is illustrated in Figure 1:-

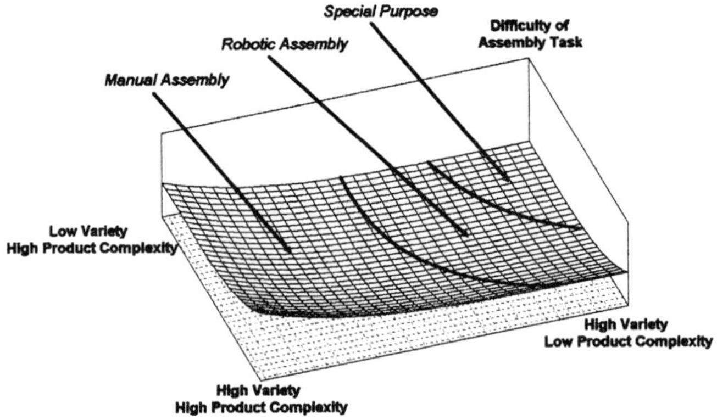

Figure 1. Difficulty of the Assembly Task Vs Variety and Product Complexity

Thus special purpose assembly machines are only applicable to low complexity tasks for a restricted variety. Manual assembly can cope with high product complexity for high variety. Between these extremes, robotic assembly is applicable. At present however, the region over which this technology can be used is relatively small.

2. Improvement of Assembly Flexibility

2.1 Economics of Robotic Assembly

The principal restriction for the limited use of robotic assembly is changeover time between products. This has two aspects: first, the time to introduce new products. Second, the time required to change between existing products. In both of these cases, the changeover time results in the need to manufacture in lots in order to amortise set-up costs over a large number of products [6]. To increase the use of robotic assembly, the key task is to reduce changeover times.

2.2 Research in Robotic Assembly

Because of the perceived potential of robotic assembly, there has been considerable research in this area. This research can be categorised into three basic areas:-

Workholding and Grasping. Several authors [7, 8] are examining ways in which assembly can be increased by the application of sophisticated grasping and workholding devices. Changeover times are reduced in this case by eliminating elements of a set-up. In many respects, this stream of research can be characterised as an attempt to emulate the flexibility of the human hand.

Sensors. The CMS will have the provision to be linked directly into a number of sensors. In order to avoid the complexity associated with the intelligent assembly systems research described earlier, the sensors will not be integrated into the robot controller, but rather directly into the CMS PC. Standard interface cards will be used to reduce cost and flexibility of application. This approach is more limited however than the previous work on intelligent assembly systems. This is because the information captured from the sensors cannot be used to modify the assembly process. Because the sensors are not linked into the robot controller, it is only possible to use this data within the CMS. For example, the CMS system will be able to provide warnings that components parts in feeders are close to being exhausted. The virtue of this approach is simplicity and because it exists quite separately from the cell controller, a generic system can be created.

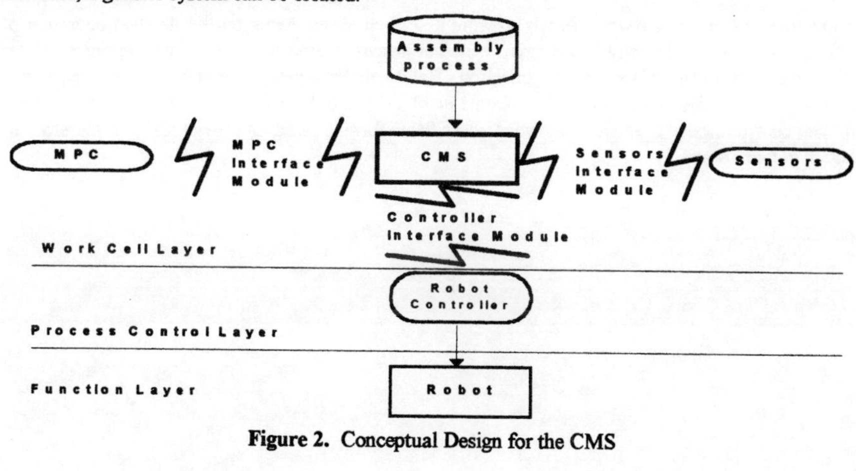

Figure 2. Conceptual Design for the CMS

Manufacturing Planning and Control (MPC) Systems. Considerable effort can be expended in making the transactions necessary to support MPC systems [11]. Within an assembly process, it is necessary to update business systems with the number of finished products completed and the number of components consumed. The operator responsible for the cell must also interpret instructions from MPC systems on the appropriate priorities for orders being manufactured by the cell. To minimise this administrative effort, the CMS will be linked via a standard interface to an MPC system.

The interfaces can be considered to be analogous to a compiler for a high level programming language or a post processor for a Computer Numerical Controlled (CNC) machine tool. The interfaces allow a standard, generic system to be used in conjunction with different cell hardware.

3.3 CMS Operator Support

It is intended that the CMS will support the personnel responsible for a robotic assembly cell. The CMS design incorporates a database that will be synchronised with the external environment with the interfaces described above. The database will enable the following functionality to be provided:-

Priority Status. Cell operators will be provided with a display showing orders to be assembled. This display will be updated via the link with the MPC system. By selecting particular orders, the cell operator will be able to access all of the relevant information associated with that job.

Tooling and Component Checking. As previously discussed, to minimise changeover time, it is essential to maximise the amount of work that can be completed while plant is still running. The CMS will allow tooling and component availability to be checked prior to the start of the actual changeover.

Intelligent Assembly Systems. Another restriction of robotic assembly is the lack of intelligence in the process. For simple robotic assembly, misalignment or incorrect orientation of components will invariably cause an uncoverable problem for the system. Several authors [9, 10] have investigated error detection and recovery methods in robotic assembly.

Systems Integration. The two research streams described above focus on the detailed operation of specific elements of the assembly process. Another research area has been the integration of the assembly cell into the manufacturing organisation. Relatively little research has been conducted in this area, though Erevelless and Aithal [11], Onari *et al* [12] and Najjari and Steiner [13] have made significant contributions. Systems integration is the focus for this research project and will therefore be discussed in more detail.

3. The Research Project

3.1 Basic Philosophy

The research project has the objective of reducing the costs associated with robotic assembly. This has two main elements which will be integrated into a coherent methodology:-

Reduction of Changeover Time. As discussed previously, one of the key restrictions in the application of robotic assembly is the time required to change between different products. The best known methodology for reduction of changeover time was developed by Shingo [13]. The basic premise of this approach is to differentiate between internal and external elements of a changeover. Internal elements are those parts of a changeover that can only be completed while the plant in not in operation. External elements on the other hand, can take place in parallel with production. Shingo argues that if some of the internal elements can be converted to external, significant savings in set-up time can be made. Within this project, this will achieved in two ways. First, by the use of standard assembly pallets that can be configured while the robot is still in operation. Second, the development of a Cell Management System (CMS). The CMS is a Personal Computer (PC) based system developed to support the robotic assembly process.

Integration into the Manufacturing Organisation. Previous research [11, 12, 13] has demonstrated the benefits of integrating a robot cell into the broader manufacturing organisation. The CMS will provide standardised tools for linking into other business systems.

By reducing these costs, the range of products that can be manufactured economically using robotic assembly techniques can be increased.

3.2 Conceptual Design of the CMS

To date, the conceptual design of the CMS has been formulated (see Figure 2). The CMS is an add-on to a robotic assembly cell. It does not attempt to replace the cell controller, but rather is intended to provide overall support for the process. Fundamental to the design is the use of standard interfaces. The CMS design is generic, but to operate in different environments, the interfaces need to be tailored to the particular situation. The CMS has interfaces to the following elements:-

Robot Controller. The CMS will link directly to the robot controller. In this case, the link is in the form of a standard RS232 serial interface. The CMS does not replace the cell controller, but provides top level support for controlling which part is to be manufactured. Thus the interface allows the CMS to start and stop the cell, download a assembly program or receive error messages.

Changeover Instructions. To minimise changeover time, it is also essential that the operator is provided with clear instructions on set-up procedure. The CMS will provide this information on-line. It is hoped that this facility combined with the availability of standard assembly pallets will provide significant reduction in changeover time.

4. Conclusions

This paper describes an on-going piece of research at the University of Salford. To date, the conceptual design of the CMS has been completed. A pilot cell using a Sysmo TS1300 Gantry robot with a Robox 186PM controller is currently being created. Some of the basic interfacing work has already been completed, in particular the links between the controller and the CMS. The next stage is to complete the CMS software and demonstrate its use within the pilot cell.

5. Acknowledgements

Mr. Mohammed's research at the University of Salford is funded by Sultan Zainal Abidin Religious College, Malaysia. Dr Petty is funded by Federal Mogul.

6. References

1. Shukri M.I., *Computer aided analysis and planning of flexible assembly system*, PhD Thesis, University of Salford, 1991.
2. Redford A.H. and Lo E., *Robots in assembly*, Open University Press, 1986.
3. Redford A.H., *Is there hope for robots in assembly?,* Assembly Automation, The Int. Journal of Assembly Technology and Management, Vol. 11, No. 1, pp. 3, 1991.
4. Schepacz C., Brand A., and Lancour M., *Design for automation: A Case Study*, Proc. of 3rd Int. Conf. of Product Design for Manufacturing & Assembly, Newport, Rhode Island, 1988.
5. Boothroyd G. and Dewhurst P., *Design for assembly: Selecting the right method*, Machine Design, November, 1983.
6. Lotter B., *Planning and implementation of flexible assembly cells,* Proc. of the 7th Int. Conference on Assembly Automation (ICAA), Zurich, Switzerland, IFS Publication Ltd., Springer-Verlag, UK, 04-06 February, 1986.
7. Taylor P. M., *Presentation and gripping of flexible materials, assembly automation*, The Int. Journal of Assembly Technology and Management, Vol. 15, No. 3, pp. 33-35, 1995.
8. Gorce p. and Fontaine J. G., *Design methodology approach for flexible grippers*, Journal of Intelligent and Robotic Systems: Theory and Applications, Vol. 15, No. 3, pp. 307-328, 1996.
9. Borchelt R.D. and Alptekin S., *Error Recovery in Intelligent Robotic Workcells*, Int. Journal of Production Research, Vol. 32, No. 1, pp. 65-73, 1994.
10. Gujar R.K. and Sanders J.L., *Analytical models to study the impact of error detection and recovery on the performance of a robotic assembly cell.* Int. Journal of Prod. Research, Vol. 32, No. 4, pp. 769-785, 1994.
11. Ereveless W.F. and Aithal S.K., *Development of an interface between a supervisory shop controller and MRP system in a CIM environment,* Journal of Material Processing Technology 61, pp. 120-123, 1996.
12. Onari M.A., Langbeck B. and Grondahl P., *The MARK III flexible assembly cell,* Robotics and Computer-Integrated Manufacturing, Vol. 13, No. 3, pp. 193-202, 1997.
13 Najjari H. and Steiner S. J., *Integrated sensor-based control system for a flexible assembly cell*, Mechatronics, Vol. 7, No. 3, pp. 231-262, 1997.
14. Shingo S., *A Revolution in manufacturing: The SMED system.* New York: The Productivity Press, 1985.

Modelling and Simulation

Computer-aided modelling of the human spine

K CASE and **D C XIAO**
Department of Manufacturing Engineering, Loughborough University, UK
B S ACAR
Department of Computer Studies, Loughborough University, UK
J M PORTER
Department of Design and Technology, Loughborough University, UK

ABSTRACT

The human spine is the main structure to support human body weight and external loads, to allow the torso to reach to a variety of positions and to protect the spinal nervous system. Lumbar back pain and disorders may be related to spinal curvature and disc pressure, and it is an ultimate objective of the work reported here to include consideration of these issues in computer aided ergonomics design systems for evaluating a wide range of situations including manual handling and car seat design.

Several methods from structural analysis have previously been used to model the human spine, principally lever and beam structures, but these have frequently shown discrepancies when compared with experimental data. As an alternative, an arch representation for the spine is considered here and allows the establishment of a criterion for the failure of the spine that may be useful in determining absolute maximum loading conditions. However, the main interest is in sub-maximal loading conditions where damage or discomfort are the concerns rather than fracture. It is proposed that the location of the thrust line in relation to the centre line of the spine is a useful predictor, and optimisation techniques have been developed to find the 'best-fitting' thrust line for the statically indeterminate structure.

Further work is concerned with adding muscle and ligament forces to the loading system of the model, extension of the 2D model into 3D, validation against experimental data and integration with the SAMMIE computer aided ergonomics design system.

1. INTRODUCTION

The spine is a three dimensional mechanical structure that transmits, allows for physiological motion and protects the spinal cord from damage. The spinal column consists of cervical vertebra, C1 to C7, thoracic vertebra, T1 to T12, lumbar vertebra, L1 to L5, the sacrum and intervertebral discs. The lumbar spine is designed to bear great loads and provide truncal mobility while the discs resist compression and determine mobility. The spinal column system also includes ligaments, attached muscles and the rib cage. The ligaments take tensile loads and the muscles supply forces to maintain spine stability.

An understanding of the spine is important in the study of its normal function conditions such as osteoporosis, scoliosis and kyphosis, but current understanding is limited by the structural complexity of the spine and the difficulty of in vivo experiments (i.e. with human subjects). Hence many mathematical models of the human spine have been developed, and these are principally classified as *levers*, *simple beams*, *cantilever beams* and *arches*.

Lever models typically describe the whole spine as a rigid lever without consideration of curvature, and loading of the spine is balanced by reaction forces at the sacrum. Simple beam models describe the whole spine as a straight slender elastic beam, and Euler's theory of elastic buckling is used for analysis. Cantilever models describe the whole spine as a cantilever beam rigidly built into the pelvis with applied forces and moments. Arch models describe either a part or all of the spine as an arch. Stability under a variety of loading conditions can then be determined by compliance with the criterion that the thrust line should be completely located within the "core" of the arch spine.

Intra-vertebral disc pressure increases with intra-abdominal pressure (1), but lever models predict the reverse effect. and can produce unrealistically large reaction forces at the sacrum (2). For example, with these models a 77 kg man holding a 90.7 kg weight in a stooped posture leads to a predicted reaction force at the sacrum of 6.6 kN (2) which is comparable with the fracture strength of the end-plates of the vertebral body (whose bearing strength is about 5-8 kN). Clearly this situation, although well outside normal working practices, is sustainable without damage in the extreme by weightlifters. This discrepancy is considered to arise from the way in which lever models ignore internal forces and spinal curvature. In contrast, modelling the spine as an arch could be a more realistic representation which considers both spinal curvature and internal forces. The increase in disc pressure with increasing intra-abdominal pressure in standing and upright postures (2) can be explained by the introduction of lumbar lordosis and intra-abdominal pressure which acts on the convex surface of the arch lumbar spine. The fact that the disc pressure reduces with increasing intra-abdominal pressure in stooping and leaning (1) can be explained by the intra-abdominal pressure acting on the concave surface of the arch spine. These factors can be included in an arch model and result in a reaction force at the sacrum of 1.3-1.5 kN for the loading situation described above.

2. STATICALLY INDETERMINATE STRUCTURE OF THE SPINE

The loading system of the human spine can be simplified as body weight, external loads, muscle forces, ligament forces and abdominal pressure as shown in figure 1. The spine structure may be approximated as a tower consisting of independent block layers. Each layer is fixed by a number of strings to the base. This structure is statically indeterminate as the number of unknowns is greater than the number of equations available for their solution. The tension in the strings, the frictional forces between the layers and the connection conditions between the strings and the base all are unknown.

A loaded arch (figure 2(a)), is statically indeterminate, as the reaction forces H1, H2, R1 and R2 cannot be found using equilibrium equations such $\Sigma f_x = 0$, $\Sigma f_y = 0$ and $\Sigma M_A = 0$.

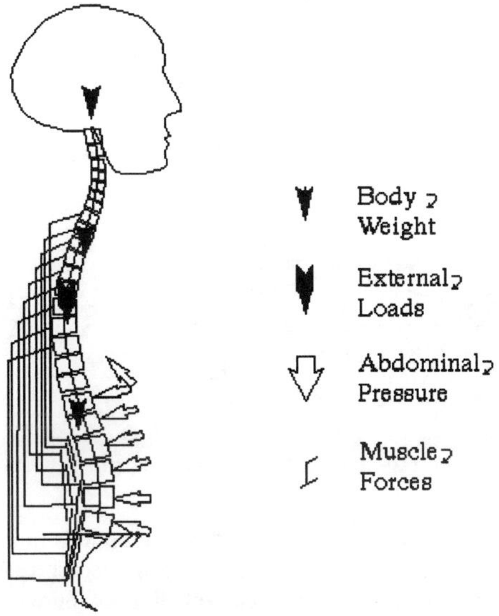

Figure 1. Loading system of the spine (not all muscle forces shown)

The funicular polygon is an analysis tool (4) that generates alternative solutions (figure 2(b)). The problems of the hanging string in figure 2(b) and the arch which is modelled as hinged rigid rods in figure 2(c), are the same in statics. The rods work in compression while the strings are in tension. Compressive force, thrust, is transmitted along a line called a thrust line. The thickness of the arch voussoirs which contain a thrust line, and the funicular polygon determines the stability of the arch.

If the horizontal component H (H1 = H2 in this case) is known, then the funicular polygon in figure 2(c) can be drawn from the force polygon. The pole O could be placed anywhere, so an infinite number of thrust lines corresponding to different pole positions could be found.

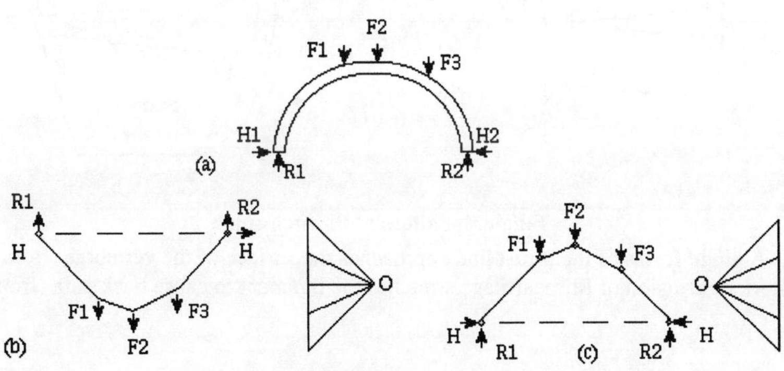

Figure 2. String and arch in equilibrium (from (4)) (a) Masonry Arch, (b) String in Equilibrium, (c) Arch in Equilibrium

Loads applied in general directions are treated in a similar way to loads applied in the vertical direction (figure 3). The position and shape of the funicular polygon again changes with the pole O of the force polygon. According to the "Middle-Third Rule", if the applied load (thrust line) stays within a “core" of the section, stresses across the whole section will be compressive and the arch is safe. For a rectangular section the core has a depth of one-third of the total depth. The existence of one satisfactory thrust line ensures that the arch cannot collapse, whereas thrust lines outside the arch cause collapse due to the formation of hinges.

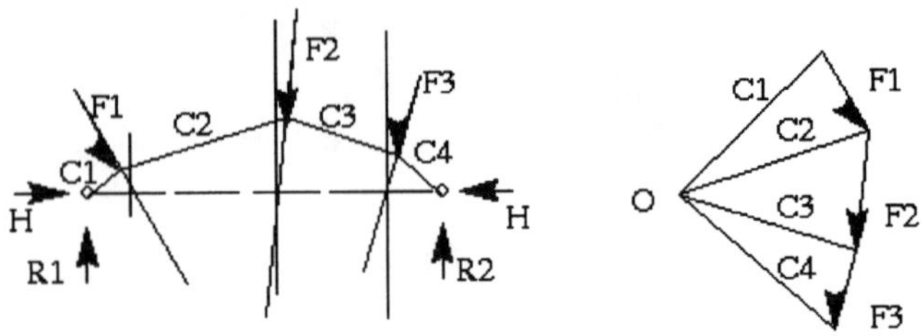

Figure 3.Arch in equilibrium with loads in general directions

The spine can be modelled as an arch with the assumptions that (a) loads are transmitted by compressive forces along the spine length, (b) normal compressive forces are lower than the crushing strength of vertebrae and discs and (c) sliding failure cannot occur.

The spine is similar to the masonry arch but the factor of safety may be different. Also the centre line of the “core” may not be located at the central line of the vertebrae of the spine, and the diameter of the “core" may increase due to the actions of ligaments and muscles. A hinge forms as a thrust line approaches the surface of the vertebrae as shown in figure 4(a). Posterior extrusion of fibrocartilage from the disc, caused by hydraulic wedging pressure and the stretching of ligaments may be a cause of back pain as shown in figure 4(b). See (5).

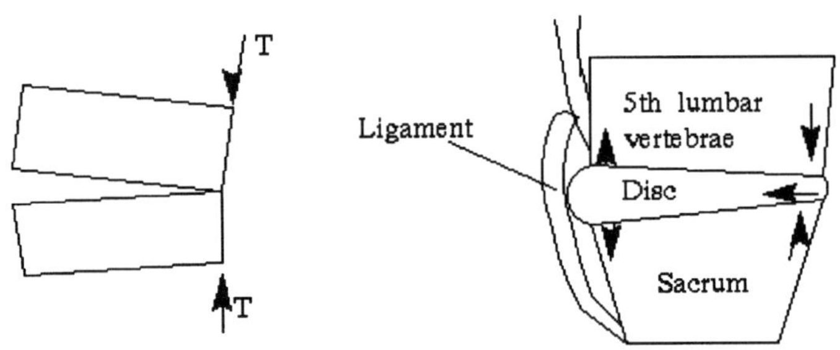

(a) (b)

Figure 4. Failure of the arch spine

(a) A hinge forms as the thrust line approaches the surface of the vertebrae (from (4))

(b) Posterior extrusion of fibrocartilage stretches the ligament to cause back pain (from (5)).

3. FAILURE CRITERION AND THRUST LINE OPTIMISATION

An infinite number of thrust lines for an arch exist as it is a statically indeterminate structure. However, there is one thrust line which is the best, and finding it is an optimisation problem. A definition of "the best-fitting" is that the resultant thrust line is as close as possible to the central line of the arch. A criterion of failure for the spine is then: **If**: *the best-fitting thrust line amongst all thrust lines in a spine can be found but it is not located within the "core" of the complete arch spine*; **Then:** *the spine fails or disorders occur.*

To find the best fitting thrust line it is necessary to find the best position of the pole O of the force polygon. This is a 2-D non-linear optimisation problem, and four objective functions to achieve this are presented as follows:

1) $f_1 = \min [\max |d_i|]$; i=1, n; 2) $f_2 = \min[\sum d_i^2]$; i=1, n;

3) $f_3 = \min [\sum (w_i * |d_i|)]$; i=1,n; 4) $f_4 = \min[w_1 * |d_n| + w_2 * \max |d_i|]$; i=1,n-1

where |d| is the distance between the thrust line and the spine central line, w is a weighting factor which forces the thrust line to be as close as possible to the central line and n is the number of forces. The best result of several locally optimised positions of the pole O in force polygon from the above four objective functions is chosen as the final optimisation result.

4. CASE STUDY

The forward bending, heavy load problem in reference (6) was taken as a calculation example. A person in a stooped position carrying a 900 N load in both hands and which acts on T6, a head and shoulder weight of 130 N acting on T2, a trunk weight of 230 N acting on T12, intra-abdominal pressure of 70 N acting perpendicularly to each of the vertebrae L1 to L5 were modelled as shown in figure 5. The body weight, intra-abdominal pressure and additional load carried are treated as the external loads, and the ligament and muscle forces are treated as internal forces. The optimisation calculation results are shown in figure 5. A comparison between a optimisation result and the result from (1) is shown in figure 6. It should be noted that the optimised thrust line is closer to the spine central line.

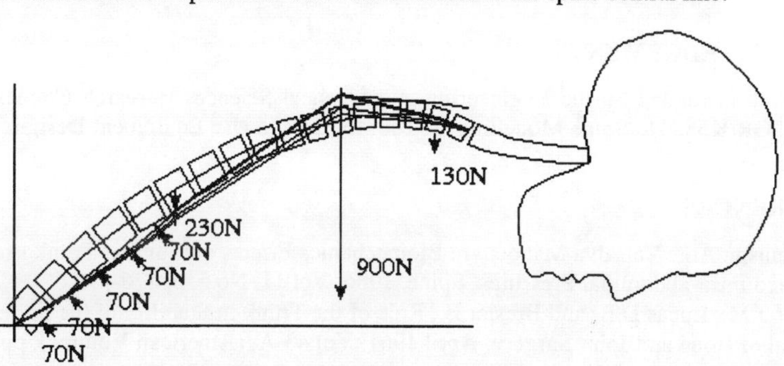

Figure 5. Optimisation of the thrust line.

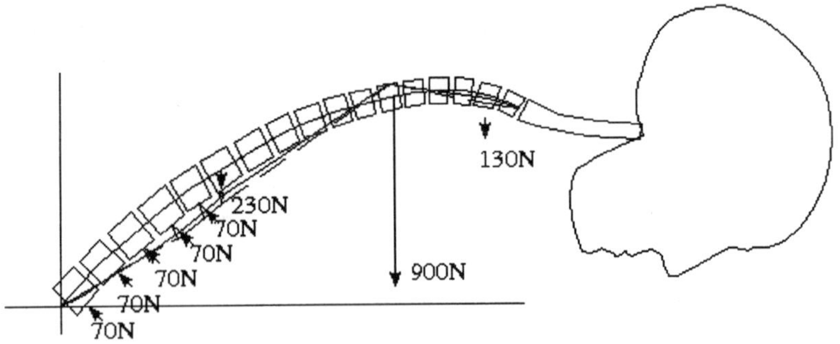

Figure 6. Comparison of thrust line for arch model in (6) and the optimised arch model

5. CONCLUSIONS

The reaction force at the sacrum calculated by the elastic lever model was about 6.6 kN which is comparable with that capable of rupturing the end-plates and causing crushing of the vertebral bone (5-8 kN). Compared with the lever model, the predicted reaction force calculated by the arch model with optimisation is approximately 1.3-1.6 kN which is a more realistic evaluation of the situation.

A criterion of failure of the spine is generated that if the best-fitting thrust line among of all thrust lines in a spine can be found but it is not located within the "core" of the complete spine, then the spine fails or disorders occur. A better, or even the best-fitting, thrust line which is as close to the reference line of the spine as possible can be obtained using optimisation techniques and is a better predictor compared with those reported in the literature.

6. ACKNOWLEDGEMENTS

The research is funded by the Engineering and Physical Sciences Research Council under Grant No. GR/K58241, 'Spine Modelling: A Tool for Safe to Use Equipment Design'.

7. REFERENCES

1. Nachemson A.L. Valsalva Manoeuvre Biomechanics Effects on Lumbar Trunk Loads of Elevated Intra-abdominal Pressures, Spine, 1986, Vol 11, No.5, pp476-479.
2. Morris J.M., Lucas D.B. and Bresler B., Role of the Trunk in Stability of the Spine, The Journal of Bone and Joint Surgery, April 1961, Vol 43-A, (American Volume), pp327-351.
3. Macintosh J.E. and Bogduk N., The Attachments of the Lumbar Erector Spinae, Spine, Vol 16, No. 7, 1991, pp783-792.
4. Heyman, J., The Masonry Arch, Ellis Horwood Ltd, New York, 1982.
5. Keegan J.J. Alterations of the Lumbar Curve Related to Posture and Seating, The Journal of Bone and Joint Surgery, 1953, Vol 35-A, No.3, July, pp589-603.
6. Aspden R.M., The Spine as an Arch: A New Mathematical Model, Spine, 1989, Vol 14, pp266-274.

Rule-based Petri-net modelling and scheduling of manufacturing systems

N KONSTAS, S LLOYD, and **C R CHATWIN**
School of Engineering, University of Sussex, Brighton, UK
H YU
School of Computing and Mathematical Sciences, Liverpool John Moores University, UK

Abstract--- A Rule Based Petri net (RBPN) modelling and scheduling scheme is proposed in this paper. The proposed method merges the simple If-Then rules (rule base) which describes the higher level decision into the PN modelling. The rule base represents the static aspect of the system, while the PN represents the dynamic aspect. To model a more realistic manufacturing system and to solve a general scheduling problem (e.g., zero-waiting [5]), a modified timed PN definition is introduced. PN modelling of manufacturing systems and the development of the RBPN software tools are presented.

1. INTRODUCTION

In a flexible manufacturing system which belongs to the family of the discrete event dynamic systems, modelling and scheduling are very important and complex issues. There is not a unique tool, like the differential equations in modelling a continuos system, to model those discrete even dynamic systems. PNs [1] presents a possible tool to model such systems and to carry out the performance analysis. However, the conventional PN has many limitations which constraints its modelling power. To overcome these limitations, a new concept must be introduced. The modular modelling methods recently proposed in [6,7] provide a promising method to handle a big manufacturing systems modelling using PNs. The effort of combining the AI base techniques with the PN already begun [2,4,8]. The research reported in [2-5] studied the scheduling problem when the number of input products equalled the number of output products. However, in practice manufacturing systems are far more complex. Apart from having a specific behaviour, manufacturing systems also have a set of rules that are commonly used to control the overall operation of the system and cannot be modelled using PN models. Usually these rules are associated with a number of external parameters although this isn't necessarily true for every system. Recent advances in artificial intelligence (AI) have created significant new opportunities in manufacturing scheduling. From the AI view, scheduling could be considered as a search with constraints through a problem space which allows to generate automatically a schedule with required characteristics. The PN provides the natural constraints of the manufacturing systems.

In this paper, the concept of the RBPN model is introduced and a general case is considered. A new timed PN definition which can be used to model more realistic manufacturing systems is discussed in Section 2. The development of the software tools required for producing Rule Based PNs is described in Section 3. An example are used to illustrate the modelling techniques. The scheduling algorithm for Rule Base PNs is described in Section 4 and conclusions and some future research lines are presented in Section 5.

2. PN MODELLING OF MANUFACTURING SYSTEMS

2.1 PN Definition

The PN can be used to analyse the qualitative *performance*, such as *boundedness*, *liveness*, and *reversibility* of the discrete event dynamic system. However, it is difficult to perform quantitative analysis. To perform quantitative analysis and increase the PN modelling power, time must be introduced. The two main ways of adding time to a PN are the use of either timed places or timed transitions [1]. In this paper, the timed place model is used to preserve the convention of transitions as instantaneous events and avoid the self- loops. To increase the modelling power, a modified timed PN is proposed as follows.

A timed PN PN=(N, h), where

N is a normal PN as defined in the usual way [1,2];

h: P→ (C, t_{min}, t_{max}), C∈PClass, $t_{min} \in R^+$, $t_{max} \in R^+$ is the *time delay vector* (or *life-time*) associated with the relevant state. *PClass* is the set of the different place classes:

PClass = { HasMinTime, HasMaxTime, HasMinAndMaxTime, NoTime}

The class of a place determines how the minimum and maximum time attributes will be interpreted. A token in a timed place PN may have two possible states: available or unavailable. In any time interval h, the present marking $M=M^a+M^u$, such that M^a is made of available tokens and M^u is made of unavailable tokens. A transition is enabled if and only if $m^a(p_i) \geq I(p_i)$, where $m^a(p_i)$ is the available tokens in the places. When a token arrives in a timed place p_i, depending on the class of the place the following rules apply:

- *C = HasMinTime:* t_{min}=c and $t_{max}=\infty$, i.e., the token must remain in the place for a time interval c to become available and can stay in the place for the infinite time.
- *C = HasMaxTime:* t_{min}=0 and t_{max}=c, i.e., the token is immediately available and it cannot remain in the place for more than c time unites. If the token remains for more than c time unites then a deadlock occurs.
- *C = HasMinAndMaxTime:* $t_{min}=c_1$ and $t_{max}=c_2$, i.e., the token must remain for a time interval c_1 to become available but it cannot remain in the place for more than c_2. The precondition for this class of places is that $c_1 \leq c_2$. If $c_1=c_2=c$ then the token must be moved out of the place exactly after time c (*zero waiting place*)[5]. If the token remains for more than c_2 then a deadlock occurs.
- *C = HasNoTime:* t_{min}=0 and $t_{max}=\infty$, i.e., the token is always available and can stay in the place forever.

The timed PN can be employed in a quantitative analysis, such as *scheduling* to either determine the minimum cycle time, or to locate *bottleneck machines* [1].

2.2 Rule Based PN Modelling of Manufacturing Systems

It is often the case that the processing of a product does not depend upon the availability of a specific resource but rather on a higher level decision that has to do with the progress of the overall production. In this case the PN model is very difficult - and in some cases impossible - to construct. This can be illustrated using the following simple example.

Assume that the production of two products A and B takes place in a single processing machine M and that the condition for machining either product is that the current cost of the product is lower than the current cost of the other. The current costs for products A and B are defined as:

costA=(numberOfPartsA/hour)*UNitCostA

costB=(numberOfPartsB/hour)*UNitCostB

A PN model for this is impossible to build especially when the unit cost for each of the products changes dynamically. It would be much more efficient to be able to "merge" logical expressions and PN structures to form hybrid models, allowing the above system to be modelled as shown in Fig. 1.

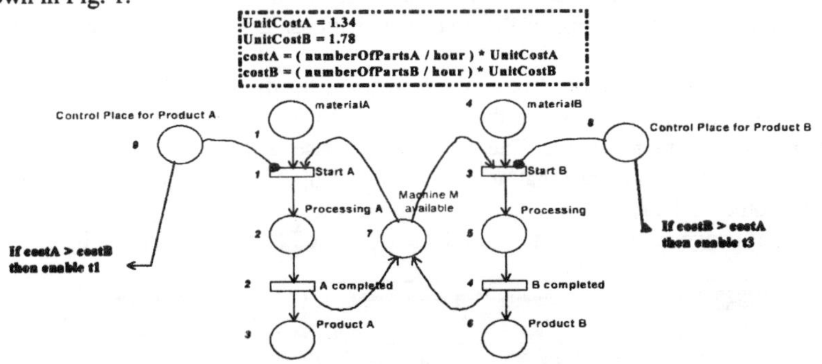

Figure 1 Hybrid Model

PNs of this type are referred as Rule Based PNs (RBPN) since the execution of the models depends upon the evaluation of a set of rules. The set of rules for a PN is also referred as its rule base. The PN model can trigger specific events in the rule base and the rule base can directly access and modify the PN model. There is a two-way communication between the PN model and the rule base.

2.3 Execution Model

In a similar way to the traditional PN model, the RBPN has two distinct aspects:

- a *static* aspect used during the development of the model, where the important factor is the flexibility of the modelling tool and
- a *dynamic* aspect used during the execution of the PN model, where the important factor is the performance of the PN model.

The two aspects define two different problems that require two different solutions. The static model has to be designed so that it is easy to build a consistent user interface while the dynamic model is designed to optimise the static PN model to be executed.

The conversion of the static model to a dynamic model is a complex procedure for the rule based PN. The stages for converting a static rule base into a dynamic piece of virtual code are:

- *Lexical Analysis:* During this stage the input rule base in text form is converted into a chain of *tokens*. Comments are ignored and an early symbol table is generated based on the identifiers found in the input text. An error at this stage will end the conversion of the rule based PN.
- *Parsing:* The chain of tokens alongside the symbol table are used as the inputs to this stage. The output is the complete symbol table and a block of intermediate code. During the parsing the syntax of the rule base is examined according to the Backus-Naur form (BNF) specification. An error at this stage will end the conversion of the rule base.
- *Code generation:* The code generator accepts as inputs the final symbol table and the intermediate code block and generates the final virtual code that it is designed to "run" on a predefined virtual processor. An error at this stage could only indicate that an extension library is missing or that an external function cannot be found. Again an error will end the conversion.

The above three stages are also referred as compiling the rule base. Before the rule base is compiled, the dynamic PN "sub-system" is generated. The generation involves the following stages:

- *PN validation:* The PN structure is first validated. If any errors are found the conversion is stopped. The degree of consistency required can be specified by the user.

- *Accelerated Model Construction:* As mentioned above the static PN model has to be optimised for dynamic execution. This involves the re-mapping of the places and transitions so that it takes less time to obtain the enable transitions. Errors cannot occur during this phase.
- *Schedule Database Initialisation:* One of the features of the dynamic PN is that it has built-in support for schedule retrieval. Once the dynamic PN has been constructed any previously stored schedules can be recovered. The schedule database is designed as an integral part of the PN model. This also means that any structural changes of the PN model will force all existing schedules to be removed.

The process of conversion from a static model to dynamic PN is illustrated in Fig. 2.

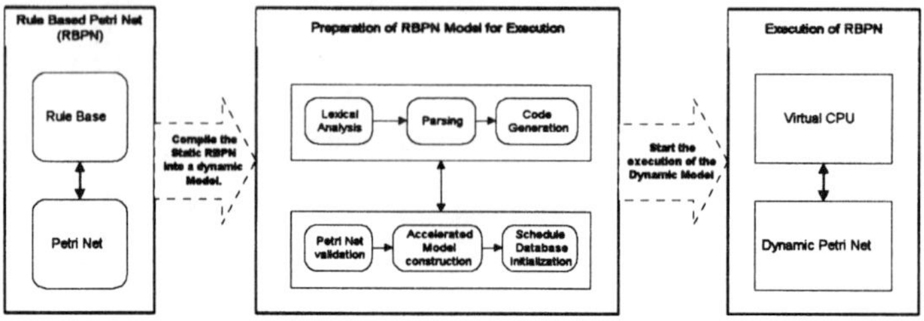

Figure 2 Conversion stages for Static to Dynamic PN

Once the dynamic PN model has been constructed the execution procedure is initialised. This involves the initialisation of the virtual CPU that will automatically load the generated virtual code and the required extension libraries. If any of the extension libraries cannot be loaded the execution terminates.

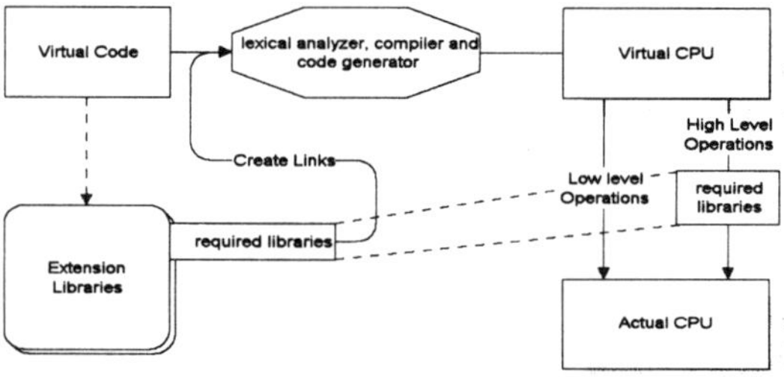

Figure 3 Execution procedure for the dynamic PN

Once the libraries are loaded they are "mapped" into the memory space of the virtual CPU by creating reference links from the virtual code to them. Note that the correct execution of the rule base also depends upon the correctness of the extension modules. Any errors in the extension modules might lead to abnormalities in behaviour of the model.

Finally, after the extension libraries mapping is complete, the virtual CPU links itself to the PN sub-system. This enables the virtual code to access PN primitives such as place, transition and arc information. If there are references to undefined (or non-existent) places, transitions or arcs the execution is aborted. The execution procedure is shown in block diagram form in Fig. 3.

4. SCHEDULING OF RBPN

All searching algorithms involve the basic nation of 'backtracking', i.e., the ability to abandon the current solution and continue the search by following a different path. For backtracking to work, the rule base and the PN have to 'save' and 'restore' their states. A PN model can easily implement backtracking by manipulating its current marking. For the rule base, two basic elements need to implement backtracking: the VP and the extension libraries. Saving and restoring the state of the VP is trivial and most of the extension libraries can be designed so that backtracking can be easily implemented. Thus RBPN can be scheduled like conventional PNs[7]. Additionally it is possible to add functionality to the rule base to improve the performance of the scheduling algorithm. E.g., the rule base can contain parameters for certain heuristic algorithms.

4.1 An Example: Adaptive Routing

One of the applications of rule base PNs is in the area of adaptive routing. Adaptive routing is important, because a high level of on-line control can be applied to the system. In the following example: material m can travel down two different paths. The flow is controlled using a set of *event handlers* and *functions*. The design of the rule based system is presented in Fig. 4.

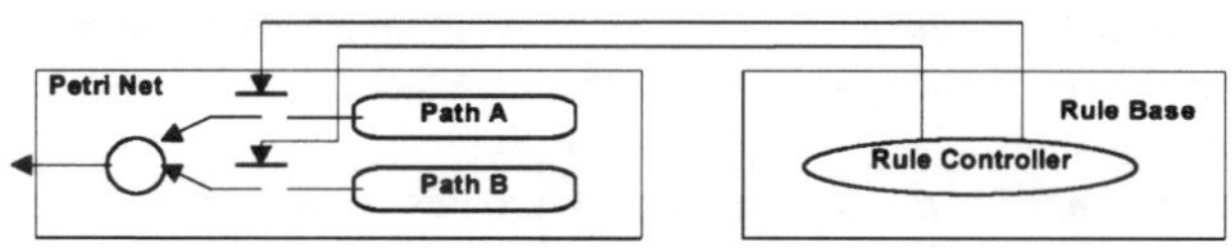

Figure 4 RBPN structure for simple adaptive routing

It can be seen from this that the routing functionality is implemented in the rule base and the actual system is modelled as a PN structure shown in Fig. 5.

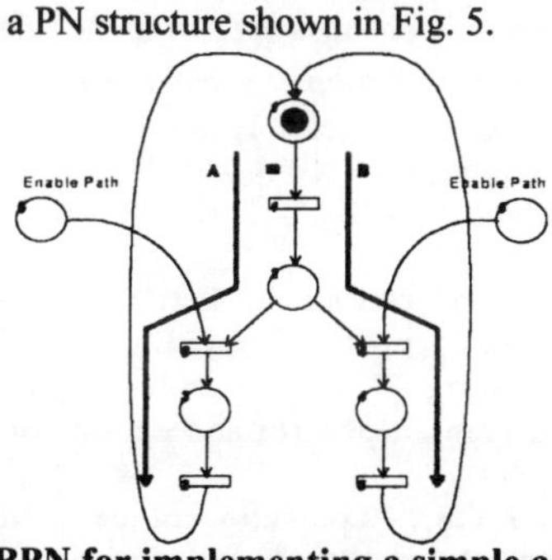

Figure 5 RBPN for implementing a simple adaptive routing

It is clear that the token in place 1 can follow two different paths depending on the availability of tokens in places 5 and 6 (controlPathA and controlPathB). It is clear that path A can be selected by placing a token in place 5 and at the same time making sure that place 6 has no tokens. Assuming that the function *tokens(n)* returns the number of tokens in place n then:

path A is selected if: *tokens(5) == 1* and *tokens (6) == 0*

path B is selected if: *tokens(5) == 0* and *tokens (6) == 1*

So by controlling the number of tokens in places 5 and 6 the desired route can be selected. The number of tokens in a place can be set by using the rule base function: *AddDynamicTokens(...)* . The function takes two arguments: the number (or *handle*) of the place and the number of tokens to insert. The complete definition can be stated as:

FUNCTION AddDynamicTokens(HANDLE place; INTEGER tokens)

The tokens in the control places must be set every time that material is about to be routed. In the PN of Fig. 5 transition 1 will fire when material is about to be routed. By using an even handler ('determinePath') the PN can trigger the required operation in the rule base so that the tokens in the control places are set accordingly. In this example the event handler will force the evaluation of the necessary rules (by calling 'GetPath') to determine which path to enable and it will set the tokens accordingly. The actual rules to be evaluated are insignificant but their complexity can vary from simple algebraic calculations to database queries.

5. CONCLUSIONS

The main goals of RBPNs are:

- *To increase the modelling power of PNs:* There are discrete event systems that cannot be modelled using *PN*s. This could be either because of the nature of the systems or because of some complexity that makes the modelling impossible. Any discrete event system that has some kind of high level adaptive routing is almost impossible to be modelled using *PN*s. The addition of rules, functions, event and state change handlers gives the *PN* model a procedural aspect that makes the models more realistic.
- *To simplify and enhance the modelling process:* Models can be constructed more easily because of the power of the modelling paradigm leading to a reduced modelling time.
- *To improve the scheduling performance:* Rule-base expressions are more compact and take less time to be evaluated than the corresponding *PN* structure. Also the search space is reduced since the number of places, transition and tokens is reduced. The overall effect is a shorter scheduling time and possibly a better solution to the problem.
- *To add high level control to the PN structure:* In production systems and specially in an FMS a high level of control is vital for the correct operation of the production system. The ability to model high level control is missing from traditional *PN* modelling. RBPNs can be used to add high level control and experiment with different control schemes. This can be done with basically the same *PN* structure with only changes applied to the rule base "sub-system".
- *To add interactive capabilities to the PN model:* Finally, a *PN* model is a static entity that is used only as an input to various analysis blocks. However the rule base extension gives the ability to create *PN*s that can be interactive.

REFERENCES

[1] Al-Jaar and A. Desrochers, Applications of Petri Nets in Manufacturing Systems: Modelling, Control and Performance Analysis, IEEE Press, New York, 1995.

[2] Lee and F. Dicesare, "Scheduling Flexible Manufacturing System Using Petri Nets and Heuristic Search", IEEE Trans. Robotics and Automation, vol. 10, no. 2, pp. 123--132, 1994.

[3] Shen L., Chen Q., and Luh J.Y.S., "Truncation of Petri Net Models for Simplifying Computation of Optimum Scheduling Problems", Computers in Industry, vol. 20, no. 1, pp. 25--43, 1992.

[4] Lloyd, S., H. Yu and N. Konstas, "Flexible Manufacturing Systems Scheduling Using Petri Net Modelling and a Branch & Bound Search," Proc. IEEE International Symposium on Assembly and Task Planning, Pittsburgh, USA, pp. 141--146, Aug. 1995.

[5] Lloyd, S., H. Yu, and C.V. Balakrishnan, " Petri Net Modelling and Scheduling of Flow Shop Systems with Different Storage Policies", IFAC Symposium on 'Information Problems in Manufacturing', October 4th-6th 1995, Beijing, China.

[6] Yu, H., S. Lloyd, A. Yusuf, and C.V. Balakrishnan, "Synthesising and Analysis of a Batch Processing System Using Petri Nets" Proc. of the Twelfth International Conference on Systems Engineering, Coventry University, UK, 9-11 September 1997.

[7] Proth, J.M., L. Wang and X. Xie, "A Class of Petri Nets for Manufacturing System Integration," IEEE Trans. on Robotics and Automation, vol. 13, no. 3, pp. 317--326, 1997.

[8] Gusikhin, O. and Kulinitch A. S., "Animated AI-Based Simulation in Production Scheduling," Elsevier Science Publishers B.V. (North-Holland), IFIP, 1993.

Rapid data collection in the simulation of manufacturing systems

K LIYANAGE and **T PERERA**
School of Engineering, Sheffield Hallam University, UK

ABSTRACT

This paper summaries the research work undertaken on input data modelling for the simulation of batch manufacturing system. It describes the development of a data reference model using IDEF1X data modelling language and a step-by-step approach through the simulation life cycle to identify and collect required data for a model more quickly. The purpose of this reference model is to describe the interaction between various data groups relating to items such as parts, resources and some aspects of system logic. This model has been developed and refined using data collected from real industrial situations. The models shows the major entities (data groups) with their attributes and relationships. This model was then translated into a normalized relational database. Having developed the data model, the appropriate sequence of collecting data was identified. A step-by-step approach has also been devised to support the simulation practitioners. The reference model was extended to link it into real data sources so that the practitioners could identify the most suitable source of data for their requirements.

1. INTRODUCTION

This paper focuses on how to carry out data modelling for the simulation of batch manufacturing systems. First part of the paper describes briefly the literature survey and a questionnaire survey conducted at the 1997 Winter Simulation Conference to explain data collection problems which can lead to longer data collection time. Second part of the paper explains the our research approach to identify and collect appropriate data for the batch manufacturing simulation projects quickly. The focus of this approach is that the input data of the simulation are gathered and analysed & manipulated outside the simulation environment and again export required data to the simulation software.

2. FACTORS INFLUENCING INPUT DATA COLLECTION

Most simulation practitioners argue that collection and analysis of input data takes a considerably long time, typically, more than one third of the project time. The analysis of literature and a questionnaire survey conducted at the 1997 Winter Simulation conference identified major seven reasons which can lead to longer data collection time (5).

2.1 Wrong or poor problem definitions

The first stage of any simulation project is the definition or formulation of a problem. Some situations in manufacturing systems simulation projects may be a collection of poor or wrong definitions problems. "Many millions of dollars are spent developing exotic solution to the wrong problem" (11). Due to this situation, information used to define the situation may be inaccurate (12). Therefore, the modeller tends to identify and collect invalid data and also a significant amount of the project time is spent on collecting this data.

2.2 Lack of clear objectives

One of the most important, but often neglected, parts of a simulation study is the clear definition of project objectives [(4) & (10)]. All projects needs objectives. But many project start without clear objectives. This reason will have an enormous impact on the scope of the model and the level of details to be incorporated and hence, the data that needs to be collected (7). It is impossible to decide an appropriate level of model details and required data for the system without clear objectives. This may cause the modeller to identify wrong and unnecessary model data which are not required for the model. Also a significant amount of project time is spent on collect these data.

2.3 Complexity of the system under investigation

When the system is too complex, the simulation modeller tends to identify and collect data in an ad hoc manner. It has been noticed that even in very similar projects, different practitioners adopt different approaches to gather and analyse data.

2.4 High level of model details

The level of model detail has clear implications on data collection. The level of detail depends on project objectives, data availability, creditability concerns, computer constraints and the opinion of system "experts" (4). The higher the level of model detail does not necessarily lead to a higher accuracy and understandibility but a can lead to longer data collection time.
A typical entity (data group) may have many attributes which naturally include some core attributes which are regarded as essential data elements for operation of the system. For example, process time is a core attribute. Collection of data relating to core attributes would be less time consuming and would not require much effort. As the model detail increases, more and more non-core attributes may be introduced. Collection of data for these non-core attributes may take a longer time. As an example, data items relating to a workstation such as breakdown data and allocation of labour require more effort and time.

2.5 Poor data availability

Two factors have been identified for poor data availability during the literature survey. When model details are increased, it may be difficult to find quality data for new attributes [(10) & (2)]. Simply the required data may not be available.

Simulation models are also built for the systems which do not currently exist. In such case, the modeller may not be able to collect the required data, due to the unavailability of past operation data and there is no chance to collect them [(8), (3) & (9)].

2.6 Difficulty in identifying available data sources

Sufficient amount of the accurate input data must be available in a required quantity and form within the company data sources. Simulation model needs many data from numerous data sources in the manufacturing company as shown in figure 3. These data sources can vary from simple manual systems to sophisticated computer based systems. These systems often can provide the required data. Model builders however find it difficult to identify reliable data sources due to;

- Existence of multiple data sources for the same data type. As an example, processing times of parts may be found in both MRP II and process planning systems. Due to lack of integration, these two sources may provides different values for same types of data. This uncertainty in data may force the model builder to seek the third party opinion to identify more accurate source of data.
- Indirect existence data. For example, the data required to model machine breakdowns (mean time between failures and mean time for repairs) may not be directly available. The maintenance department or maintenance contractors may have required data in very crude form. Hence a considerable time need to be collect and analyse this data.

2.7 Limited facilities in simulation software to organise and manipulate input data

Many simulation software packages have been developed for modelling of manufacturing systems but they have limited facilities to organise and manipulate input data. They do not have proper database management systems facilities to collecting and documenting data. Often the collated data is stored in propriety formats. When the core data is altered, it may be necessary to regenerate the data for simulation models. In large scale projects, this can be very time consuming. At the moment, the quality of input data handling of the simulation software is not very high. However, it can be seen that few simulators enable integration with spreadsheet packages and modern databases.
The participants of 1997 Winter Simulation Conference were asked to rank the impact of these factors on data collection (See Table 1).

Major reasons	Rank
Poor data availability	1
High level model details	2
Difficult in Identifying available data sources.	3
Complexity of the system under investigation.	4
Lack of clear objectives	5
Limited facilities in simulation software to organise and manipulate input data	6
Wrong problem definitions	7

Table 1. Impact of seven pitfalls (1 - most influential, 7 - least influential)

Poor data availability was considered as the major reason for long data collection time.

It appears that most of research undertaken to date has been focused upon statistical data analysis techniques and no or little has been carried out to develop methods to accelerate the collection of input data need in simulation models. It is obvious that a more methodical approach is required to identify and collect data. Perhaps, the best way is to develop a guided system which could take model builders through a series of tasks to ensure complete and efficient data collection. The development of a such guided system relies on a complete and accurate information model of the system. Once the core data types are identified using the information model, they can be linked to potential data sources. Depending upon the availability of data sources the model builder can be directed to collate the necessary data quickly.

3. A SUITE OF METHODS FOR IDENTIFICATION AND COLLECTION OF DATA

3.1 Overview of methods

This method consists of four integrated stages which identify and collect data requirements more effectively. It consists of development of functional modelling diagrams, reference data model, mapping tables between the functional modelling diagrams & reference data model and database (see figure 1).

3.2 IDEF0 functional modelling

The IDEF0 Functional modelling methods is design to model the decisions, actions and activities of the system (6). IDEF0 is designed to allow the user to "tell the story" of what happening in the system. The methodology permits a system to be described in as complete a level of details as desired. We have developed series of standard IDEF0 Functional Modelling diagrams for the system elements in generic batch manufacturing system which are often used in simulation. For an example, Figure 4 shows one of our standard IDEF0 functional modelling diagram which describes the required machine activities concerning the part processing.

Figure 1. An integrated approach

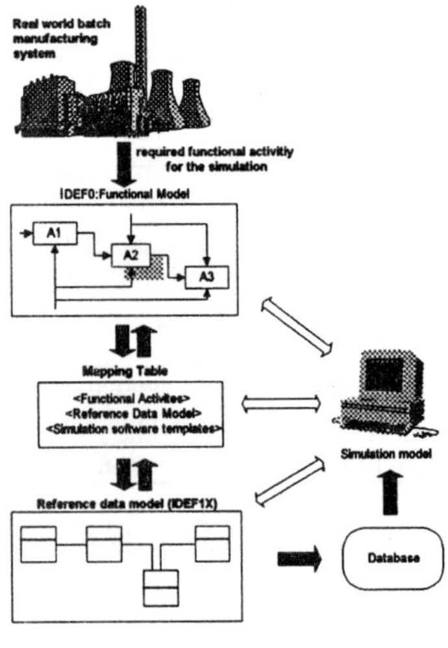

3.3 Reference data model

The reference data model has been developed using IDEF1X [(1) & (6)] data modelling language. The purpose of this reference data model is to describe a integration between a various component such as Parts, Resources and a Logic of a manufacturing system into a single cohesive system to describe a conceptual database implementation. The reference model currently contains 24 entities and over 75 attributes. For a relational DBMS, each entity in the reference model becomes a relational table, and each attribute become a column. Primary and foreign keys are declared for each table, and referential integrity constraints are declared for each relationship.

As an example, entity relationships between part, machine and operations are described on the reference model as shown in figure 2. i.e. . machine can perform many operations and an operations can be performed on many machines. An operation relates to many products and product can undergo many operations. In the Reference model, these many-to-many relationships can be resolved in three one-to-many relationships with associative entity as the standard pattern of IDEF1X Language. The figure shows that a MACHINE_ OPERATION represents a three-way association among PART (Identified by Part-ID), MACHINE (Identified by Machine-ID), and OPERATION (Identified by Operation-No). Due to paper length reason, every entity-relationships on reference model can not be described.

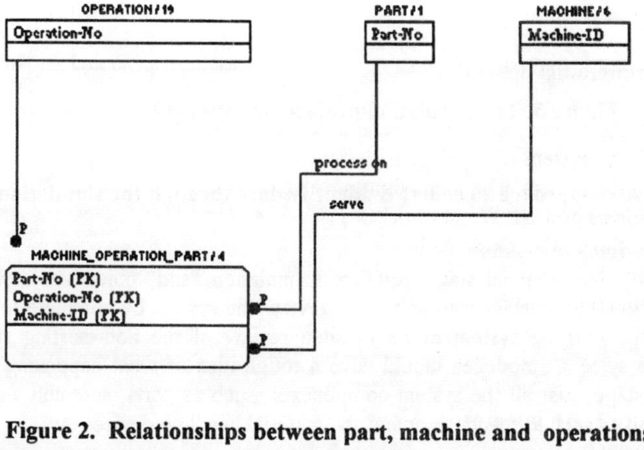

Figure 2. Relationships between part, machine and operations

3.4 Mapping table

The purpose of mapping table is to integrate both functional modelling diagrams and reference data model to identify modeller to system activities and corresponding information and data quickly. The sample of mapping table shown in Table 2 maps the manufacturing activity functions for the part processing against the entity and attributes represent in the reference data model with corresponding Arena simulation software package modules. This mapping table can be extended to represent the modules of other simulation packages as well. The development of this mapping table to Arena and other simulation packages will assist the novice modellers.

4. LIFE CYCLE OF A SIMULATION STUDY

The life cycle of a simulation project can often be presented in the series of steps. For the ideal simulation model building exercise, each of the step in the life cycle is completed before the next is started. However, Latest literature indicates that this not the actual situation. However, It can be seen that many activities take place simultaneously and influence each other as described in figure 3.

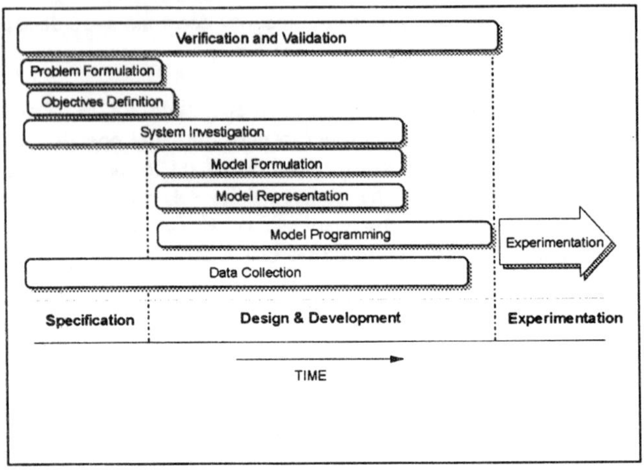

Figure 3. The simulation project activities (12)

4.1 Step by step approach to collect & identify data through the simulation life cycle.

Step 1: ***Investigate the system***
It can be seen that at initial stage, problem formulation and objectives definition become mixed together. At this stage, modeller can review the system by complete tour through the facility for the existing system or a detailed review of the non-existing system. Having reviewed the system, modeller should have a rough idea of what happening in the system. During this stage, List all the system components such as parts, assembly components and system resources such as machine, operators, material handling devices etc.

Step 2: ***Select appropriate functional modelling diagrams from the diagram database.***
Having listed all system components, retrieve relevant functional modelling diagrams from the our diagrams database. This functional modelling diagrams summaries operational activities of system elements (such as part, resources , material handling devices, etc.) of manufacturing systems in hierarchical manner and the interconnections between these functions and the data needed to support the functions.

Step 3: Once system components and functional modelling diagrams are studied, he or she will be able to prepare important questions to meet the project objectives. At this stage, modeller can decide how much level of model details will include for the model to achieve project objective. The mapping table is used to identify the data required to simulation model according to the level of model details.

Step 4: ***Collect data.***
At this stage, start the collect data using developed reference model. The reference data model or the database will guide the modeller through the right steps to collect required data. The collated data is stored in the database and it is possible to link data directly into simulation model at later stages.

CONCLUSION

This integrated approach has the following benefits to a simulation data collection exercise.

- Give a guideline to collect data at initial stage of a simulation study
- Minimise the chance of missing data items.
- Functional modelling diagrams database can provide significant contribution to model representation stage or totally eliminate that stage.
- Guideline to collect data in appropriate sequence (Hierarchical manner).
- A database management system can provide a valuable contribution to the data collection procedure. It can be used to accurately maintain all the collected data and also manipulate the data using SQL according to the user requirements.
- Within the database system, actions such as boundary definitions, input mask definitions for the data items can be defined as data validation & verification activities.

REFERENCES

1. Bruce A T, (1992), "Designing Quality Databases with IDEF1X Information Models", Dorset House Publishing, New York.
2. Hatami S., 1990, "Data requirements for analysis of manufacturing systems using computer simulation", Proceeding of the 1990 Winter Simulation Conference, 632-635.
3. Law A M, McComas M G and Vincent S G (1994), The crucial role of input modelling in successful simulation studies, Industrial Engineering, July, 55-59.
4. Law A. M 1990, "How to build valid credible and appropriately detailed simulation models", April, 16-17.
5. Liyanage K. and Perera T., 1998, Design and Development of a Rapid Data collection Methodology, International Conference of Simulation, University of York (Paper submitted).
6. Mayer R J, 1992, IDEF1X Data Modelling/IDEF0 Functional Modelling, Knowledge Base Systems, Inc. Texas.
7. Novel M D 1992, "Modelling the shape of the future", Professional Engineering, June.
8. Pegden C. D., Shannon R. E., and Sadowski R. P., 1990, Introduction to Simulation Using SIMAN, System Modelling Corporation.
9. Robinson S. and Bhatia V.,1995," Secrets of Successful Simulation Projects", Proceeding of the 1995 Winter Simulation Conference, 61-67.
10. Robinson S., 1994, Successful Simulation, McGraw-Hill International (UK) Limited.
11. Shannon R (1986), Presentation at General Electric company Simulation Symposium.

12. Tye B. and Perera T., 1997, Simulation Model Design Handbook (unpublished), Sheffield Hallam University

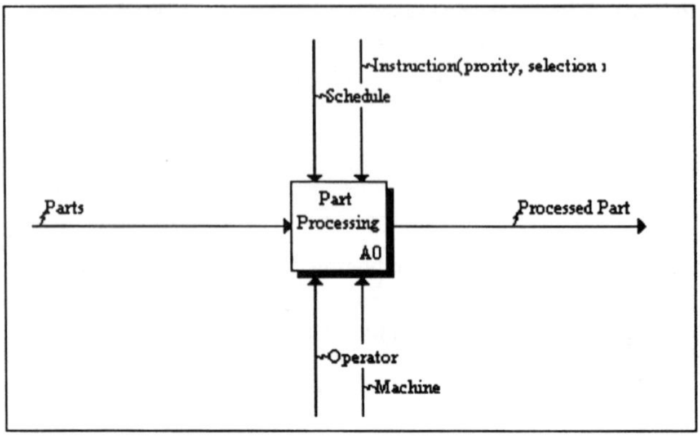

Level 1

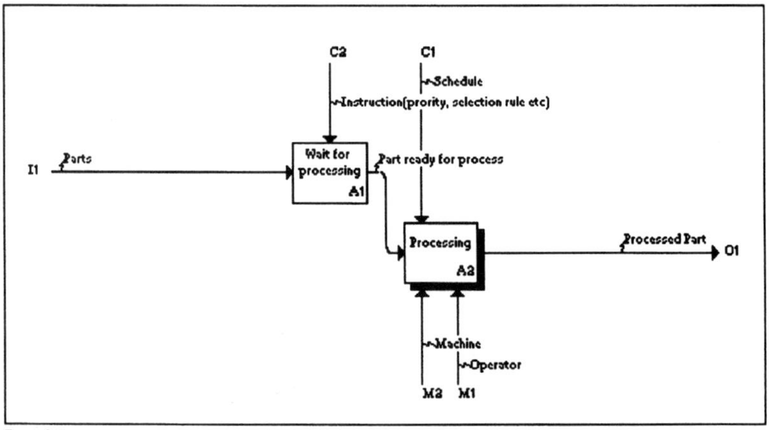

Level 2

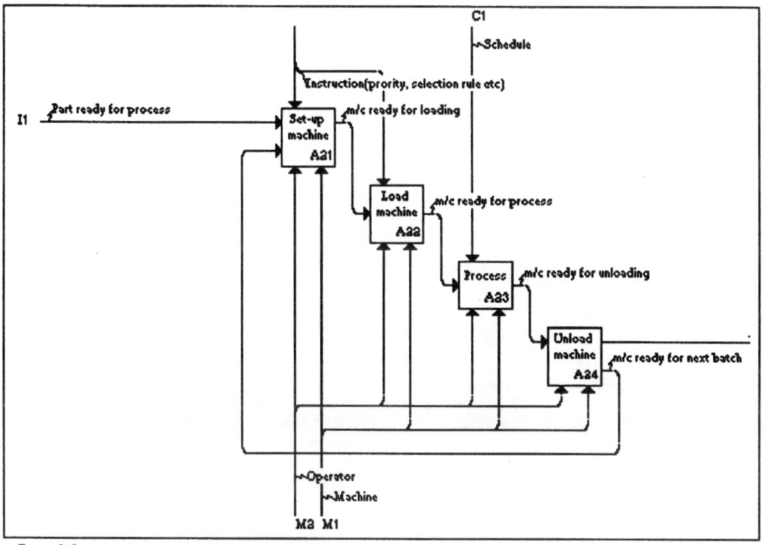

Level 3

Figure 4. IDEF0 functional mode diagrams for machine activities concerning part processing

Level of decomposition	FunctionalModelling Element	Required Data	Reference Model Path (Corresponding RM Entity)	Corresponding ARENA Template
1	Part	Part ID	PART	CREATE
		Part Description	PART	CREATE
		Batch Size	MACHINE_OPERATION	CREATE
		Max Batches	MACHINE_OPERATION	CREATE
		Inter Arrival Time	MACHINE_OPERATION	CREATE
	Machine	Machine ID	MACHINE	RESOURCE/SEIZE
		Machine Description	MACHINE	RESOURCE/SEIZE
		MTBF	MACHINE_GROUP	RESOURCE
		MTTR	MACHINE_GROUP	RESOURCE
		Input Buffer Capacity	MACHINE_GROUP	RESOURCE OR STORAGE
		Output Buffer Capacity	MACHINE_GROUP	RESOURCE OR STORAGE
	Operator	Operator ID	OPERATOR	RESOURCE/SEIZE
		Operator Description	OPERATOR	RESOURCE/SEIZE
		Efficiency	OPERATOR	
		Skills	OPERATOR	
		Learning Curve Effect	OPERATOR	
	Part	Machine Operation Desc.	MACHINE_OPERATION	
	Processing	Machine Time	MACHINE_OPERATION	DELAY
		Operator Responsibilities	MACHINE_OPERATION	
	Schedule	Schedule Name	SCHEDULE	RESOURCE-->Schedule block
		Schedule Description	SCHEDULE	RESOURCE-->Schedule block
		Schedule Time	SCHEDULE	RESOURCE-->Schedule block
		Duration	SCHEDULE	RESOURCE-->Schedule block
		Machine available time	MACHINE_SHEDULE	RESOURCE-->Schedule block
		Machine unavailable time	MACHINE_SHEDULE	RESOURCE-->Schedule block
		Operator available time	OPERATOR_SHEDULE	RESOURCE-->Schedule block
		Operator unavailable time	OPERATOR_SHEDULE	RESOURCE-->Schedule block
	Instruction	Priority Rule	MACHINE_OPERATION	SEIZE

Table 2. Sample Mapping Table

Integrated performance measurement – a reference model and an audit method

S LUTZ
Institute for Production Systems, University of Hanover, Germany
U S BITITCI, A S CARRIE, and **T TURNER**
Centre for Strategic Manufacturing, University of Strathclyde, Glasgow, UK

ABSTRACT

The article describes a Reference model developed for the analysis of performance measurement systems. This Reference Model provides a framework for the evaluation and design of integrated performance measurement systems. Furthermore, an approach for the analysis of current performance measurement systems is given testing the integrity and deployment of a performance measurement system against the Reference Model. The audit method provides guidelines to identify gaps in performance measurement systems and to rank the gaps in terms of importance for the company's success. The Audit outlines those areas of the business with a lack of suitable performance measures and focuses on the improvement of the performance measurement system.

1. INTRODUCTION

Today's industries are confronted with a changing economic environment determined by demanding customers, global competition, fast pace of change and unpredictable surroundings, social needs, and environmental constraints. To remain successful it is not longer sufficient for an enterprise to pursue excellence in productivity or efficiency only. A company has to be able to react on changes in its environment and to anticipate them. This requires that the company employs a correct set of objectives reflecting actual demands on the business from its stakeholders. The development of appropriate objectives and the control of the objectives' achievement, can be accomplished by a performance measurement system crossing boundaries and stimulating the integration of functions throughout the entire business.

2. REFERENCE MODEL FOR INTEGRATED PERFORMANCE MEASUREMENT

As part of an EPSRC Research project on performance measurement, a Reference Model for Integrated Performance Measurement has been developed by the Centre for Strategic Manufacturing at the University of Strathclyde in Glasgow. The aim of the Reference Model which is based on past and present academic work as well as industry's experiences is to describe an integrated, effective, and efficient performance measurement system which enables an organisation to manage its performance in line with corporate and functional strategies and objectives [1]. The Reference Model integrates various approaches of performance measurement design [2, 3], systems thinking [4], and business strategy models such as Business Classification [5] and order winning and order qualifying criteria [6].

The Reference Model considers an organisation to consist of four different organisational levels. These are the Business Level, the Business Unit Level, the Business Processes and the Activities.

The Business Level represents the entire Business and defines the scope of the Audit against the Reference Model. Each Business is expected to consist of a number of Business Units. A Business Unit is defined as the logical portion of the organisation which serves a particular market segment with particular competitive requirements. Each Business Unit incorporates several core and support processes which represent the operations of the Business Unit. A Business Process consists of different Activities. These layers of an organisation may exist physically or logically. "Logically" means that the organisation is not organised in this four layers but they exist already and need only to be recognised.

At each of its four levels the Reference Model breaks down into four elements. These are the stakeholders, an External Monitor, the objectives, and the performance measures.

Each level is required by the Reference Model to recognise its stakeholders and to understand their requirements. Possible stakeholders are the owner, employees, the society, or the environment. Customers are considered as stakeholders at the Business Unit level and at the levels beneath only. By the External Monitor the organisation is required to monitor its

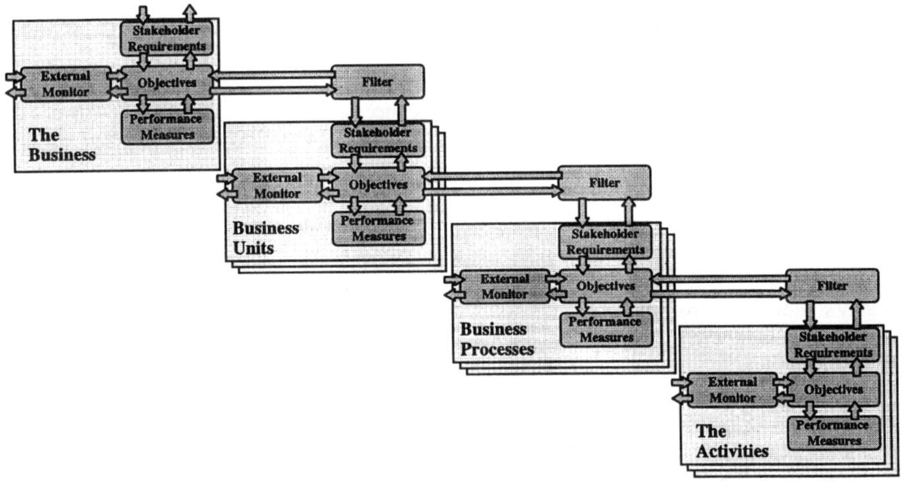

Figure 1: The Reference Model for Integrated Performance Measurement Systems

external position with respect to its stakeholder requirements against competitors as well as against Best Practice and World Class performance in order to identify improvement needs. The Reference Model requires the organisation at each level to employ a set of objectives based on implications and the criticality of development needs together with appropriate targets and timescales. By setting objectives due consideration should be given to stakeholder requirements and competitive gaps. Against each objective should exist at least one performance measure which is capable to report, monitor and review the achievement of the objective. In this context performance measures are used for both, the deployment of objectives throughout the entire organisation and for the control of the achievement of objectives. The combination of improvement and control measures creates a closed loop system for performance measurement.

The four hierarchical levels of an organisation and the four elements at each level form the complete Reference Model as depicted in Figure 1.

Resulting from the principles of deployment and management control, the higher level of the Reference Model becomes a stakeholder of the level beneath. Hence, the objectives of the higher level need to be considered as stakeholder requirements at the lower level.

Another key feature of the Reference Model is „transduction“ which means that the objectives of the higher level should be expressed in a locally meaningful way at the lower levels. The term „resource bargaining“ refers to a process of negotiation between the levels to ensure that set objectives and requirements are realistic and achievable.

Since only those objectives critical to the success of the business should be deployed downwards, two different levels of the Reference Model are linked to each other through the Filter. This element allocates the superior level's objectives to the different objects of the lower level with respect to their criticality and importance for particular parts of the lower level.

3. THE AUDIT

The following paragraphs describe an Audit Method for the investigation of an existing performance measurement system against the Reference Model which also allows the assessment of its integrity. In general, the Audit consists of two main phases: Data collection and data analysis. Figure 2 depicts the steps of the Audit.

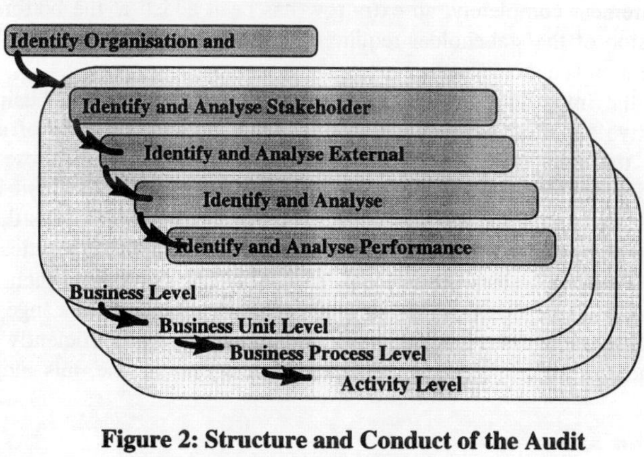

Figure 2: Structure and Conduct of the Audit

3.1. Data Collection

For the collection of relevant data, a questionnaire and a workbook have been developed. During a first phase of the Audit, general data about the company, its scope and business, the products and processes is captured. Based on this information, the organisational structure of the enterprise can be identified.

The questionnaire is used to collect data about stakeholders and their requirements, both categorised in terms of importance.

The examination of the External Monitor is split into two parts: The external position of the business for each stakeholder requirement with respect to competitors and with respect to Best Practice companies. The position of the company is scored.

For obtaining information about objectives, they are listed and targets and timescales are specified. The importance of the objectives for the business is also evaluated.

Finally, the performance measures employed and the frequency of measurement are listed.

3.2. Data Analysis

The main aim of the data analysis is to reveal inconsistencies and gaps in the current performance measurement system. To achieve this, the Audit investigates whether a set of objectives reflecting the stakeholder requirements exists and whether suitable performance measures for the appropriate monitoring of objectives are employed.

Starting with an analysis of the business' organisation and structure, the completeness of stakeholders and their requirements has to be investigated. The company should recognise its stakeholders at each level of the Reference Model. In particular, attention has to be paid to the complete deployment of objectives as stakeholder requirements to the lower levels.

An external monitor should exist at each level of the organisation. The absence of an external monitor can constrain a business' ability to anticipate changes and developments in its environment. The external monitor identifies areas in which improvement is needed. Especially in unstable environments, the monitoring has to take place regularly in short intervals.

Following, the Audit examines the existence of appropriate objectives which adequately address the stakeholder requirements. For the analysis of the relationships between objectives and stakeholder requirements a Quality Function Deployment oriented approach has been used. Objectives and stakeholder requirements are listed in a matrix; symbols („++“,“+“) indicate the degree of relationship. Since in many cases existing objectives do not address a stakeholder requirement completely, an extra row has been added at the bottom in which a sufficient expression of the stakeholder requirement by objectives is denoted „Yes“ and an insufficient expression is denoted with „No“.

According to the importance of the stakeholder requirement for the business and the external competitive and Best Practice position, a value for the criticality of a stakeholder requirement for the company can be calculated. Most critical are those stakeholder requirements which are very important for a very important stakeholder and in which the company additionally has a poor external position. This categorisation means that a business should pay special attention to the accomplishment and monitoring of very critical stakeholder requirements. Based on this classification gaps can be ranked in terms of their criticality by listing the criticality in the matrix. This simplifies the identification of dangerous gaps, as shown in Figure 3. For example: If a stakeholder requirement is not sufficiently addressed by the objectives and this stakeholder requirement is a very critical one, this indicates a very critical gap.

It may occur that an objective is only indirectly related to a stakeholder requirement or expresses the requirement not sufficiently. It also occurs that objectives are not related to any stakeholder requirement at all. This should be marked in the right column. If an objective is not meaningful or explicit enough, it should be optimised by the organisation.

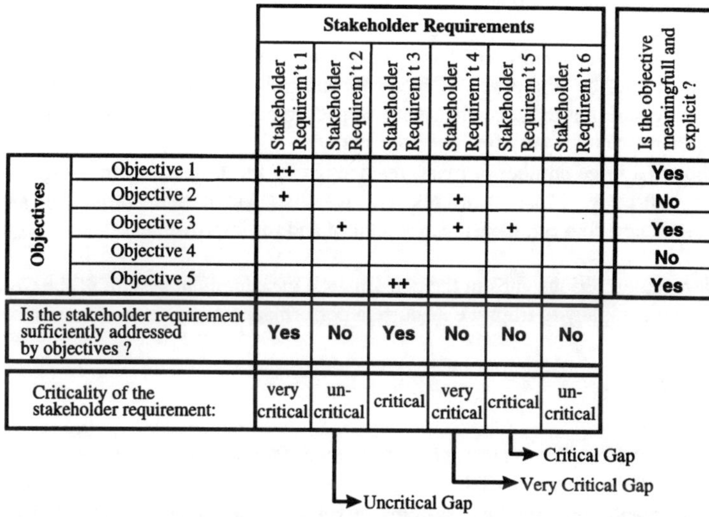

Figure 3: Stakeholder Requirements and Objectives Analysis Matrix

Another criterion for the integrity of the objectives is their consistency with the requirements from the classification models and the theory of order winning and order qualifying criteria. Furthermore the priority of objectives has to be in line with the criticality of the stakeholder requirement related to the objective.

For the analysis of the performance measures, provided data has to investigated whether a set of data is a performance measure or only descriptive data without further information about the quantification of the efficiency and effectiveness of an action. Similar to the analysis of objectives, matrices in which the relationships between objectives and performance measures are used for the analysis. By the comparison of the complete addressing of the objective by measures and the criticality of the objective, critical gaps can be identified. The criticality of an objective derives from the highest criticality of an stakeholder requirement addressed by the objective.

4. CASE STUDY

Several Audits against the Reference Model have been conducted in different industrial organisations. In this paragraph, the results of an Audit carried out at Company „D“, an major textiles manufacturer in the UK, should be presented. The main operation of „D“ is the conversion of cloth to gents and ladies garments. „D“ consists of two different Business Units with three core processes. Relevant data was collected by interviews with managerial staff and by questionnaires.

The results of the Audit can be summarised as follows:

- The company is aware of its stakeholders and their requirements.
- An external monitor is absent in large areas of the enterprise. Information is informally acquired based on contacts of managerial staff. A regular review of the competitive position does not take place.
- In many areas of the organisation, the objectives do not reflect the stakeholder requirements sufficiently. Objectives are not explicitly enough related to requirements and need to be sharpened. Objectives are missing against intangible issues. The objectives are not distinguished with respect to capability and potentiality. Targets and timescales exist for very few objectives only.
- Although a large number of measures does exist, most measures measure the objectives only indirectly. Most measures are reactive rather than proactive and are not differentiated between measures for control and for improvement.

The Audit revealed serious gaps in the performance measurement system of the Company „D" and helped the company to improve its current performance measurement system.

5. CONCLUSION

The Reference Model for Integrated Performance Measurement System integrates various academic theories and concepts. It provides an effective and simple tool for the analysis of current performance measurement systems in various types of organisations. The Reference Model helps to direct improvements and to maximise benefits from improvement efforts by rationalising and simplifying existing performance measurement systems.

The Audit Method described, facilitates the conduct of an investigation of industrial performance measurement systems against the Reference Model. The Audit identifies lacks in the deployment of stakeholder requirements and objectives. Furthermore, the Audit investigates the appropriateness of targets and objectives and reveals the absence of measures, surplus or meaningless measures, and wrong emphasis on measures. A number of Audits has been conducted successfully in several industrial organisations proofing the capability of the Reference Model and the Audit Method.

6. REFERENCES

[1] Bitici U. S., Carrie A. S., McDevitt L., Turner T.: „Integrated Performance Measurement Systems - A Reference Model", Proceedings of IFIP WG 5.7 Working Conference, 15th - 18th Sept. 1997, Ascona, pp. 35 - 44.

[2] Neely A., Richards H., Mills J., Platts K. Bourne M.: „Designing performance measures: a structured approach", IJOPM, Vol. 17 No. 11 (1997), pp. 1131 - 1152.

[3] Kaplan R., Norton D.: „The balanced scorecard: translating strategy into action", Harvard Business School Press, Boston / MA, 1996.

[4] Beer S.: „Diagnosing the system for organizations", John Wiley & Sons, Chichester, 1985.

[5] UK Department of Trade and Industry: „Factory for the future - final report", DTI, London, 1994.

[6] Hill T.: „Manufacturing Strategy", Macmillan Press, London, 1993.

Modelling manufacturing systems using GRAI and decision science techniques

M MENICOU and **I McCARTHY**
Department of Mechanical Engineering, University of Sheffield, UK

ABSTRACT:

The future organisation must be organised for the systematic abandonment of whatever is established, customary, familiar, and comfortable. In short, it must be organised for constant change. An organisation receptive to constant change must be structured in such a way so that signs of environmental and internal alterations reach promptly its decisional infrastructure and decisions are taken quickly with maximum insight gained.

The aim of this integration is to provide a common means of understanding the decisional network of manufacturing organisations and using well established Decision Analysis techniques to provide guidance to individual decision makers in order to enhance the quality of their decisions.

1. INTRODUCTION.

Years ago when change was slow, markets were concentrated in a handful of countries, and stability was the rule rather than the exception, an organisational approach that emphasised top - down hierarchy, rules and regulations, and authority in the hands of executives dominated (1, p.9). This kind of organisation, shaped since the outbreak of the industrial revolution, lasted up to the Second World War. During that period, competition was unknown to organisations and practically they could sell whatever they could produce.

After the end of the Second World War production overtook demand, and therefore competition among organisations became gradually more severe. Competitive advantage was sought through means such as cost, quality, service, customised design, flexibility and overall business agility. But the market pushes for new changes at an increasing rate and companies found themselves trapped in a never-ending race.

Drucker (2) claimed that the future organisation "must be organised for the systematic abandonment of whatever is established, customary, familiar, and comfortable. This might be a product, service, or process; a set of skills; human and social relationships; or the organisation itself. In short, it must be organised for constant change. The organisation's function is to put knowledge to work - on tools, products, and processes; on the design of work; on knowledge itself. It is the nature of knowledge that it changes fast and today's certainties always become tomorrow's absurdities" (p.96).

Organisational decisions should be thought of those mechanisms, which enable an organisation to respond effectively to arising problems. A problem could be defined as a deviation between what should be happening and what actually is happening, that is important enough to make someone think the deviation should be corrected (3).

Every decision is the outcome of a dynamic process that is influenced by multiple forces. Although there have been many attempts to model this process, this has proved to be a difficult task due to the fact that decision making is not a fixed procedure. However, an organisation receptive to constant change must be structured in such a way so that signs of environmental and internal alterations reach promptly its decisional infrastructure and decisions are taken quickly with maximum insight gained. This requires an organisation with high degree of flexibility, decentralisation, and empowerment.

As a result, the aim of this work is: *"to provide a common means of understanding the decisional network of manufacturing organisations and using well established Decision Analysis techniques to provide guidance to individual decision makers in order to enhance the quality of their decisions"*.

2. THE GRAI ENTERPRISE MODELLING TECHNIQUE.

Manufacturing enterprise modelling and analysis could be defined as the study of a manufacturing company to determine its activities or functions and to investigate how these are carried out (4). Enterprise models show the interrelationships of the people, objects, and information involved in the various activities.

There is no single ideal modelling technique capable of satisfying all requirements (functional, informational, decisional, financial, etc.). In general, the selection and use of a method is subject to the purpose of the modelling exercise, the nature of the system being investigated and the experience of the analyst (5). The GRAI modelling technique, developed during the late 70's by the University of Bordeaux I in France, aims to meet the needs of analysis and design of Production Management Systems (PMS) and particularly the decisional network (6).

2.1 The GRAI conceptual background.

The GRAI method consists of two tools: the GRAI grid (macrostructure) and the GRAI net (microstructure). By clarifying the nature of activities; time aspects of decision - making and the functional nature of the decisional activities, the macrostructure aims to achieve an overview of the PMS. Its activities are taken to be of the following three types: physical, decisional and informational. Physical activities are responsible for converting raw materials into finished products, carried out by men or machines. Decisional activities aim to control the physical resource subsystem in order to reach the economic targets set by the business and its environment while constraints are taken into account. Informational activities aim to help the decisional activities by both providing internal and external information. The concepts of Horizons and Periods are introduced as the time interval through which decisions are valid and the time interval at the end of which decisions are revised respectively. The functional nature of decisional activities within a PMS is identified in order to be associated with Periods and Horizons.

The microstructure model aims to depict the operation of individual decision centres both in routine and contingency situation. The parameters to be modelled are identified as the various activities of a decision centre; the decision frame (variables and decision limits); decisions made by the decision centre and information used by a decision centre (7).

2.2 The graphical formalisms used to build the conceptual models.

The GRAI Grid (Fig. 1) gives a hierarchical representation of the whole structure of decision centres within a production management system (7). In detail the various parts of the GRAI grid represent the various functions of the PMS (Columns); the different levels of PMS, which normally are strategic, tactical and operational (Rows); Decision centres (individual square) and Horizons and Periods. Double arrows show the decisional transmission or links between decision centres at different levels (always flow from upper level to lower level or same level, never from lower level to upper level). Single arrows show informational links between decision centres (flow in any direction).

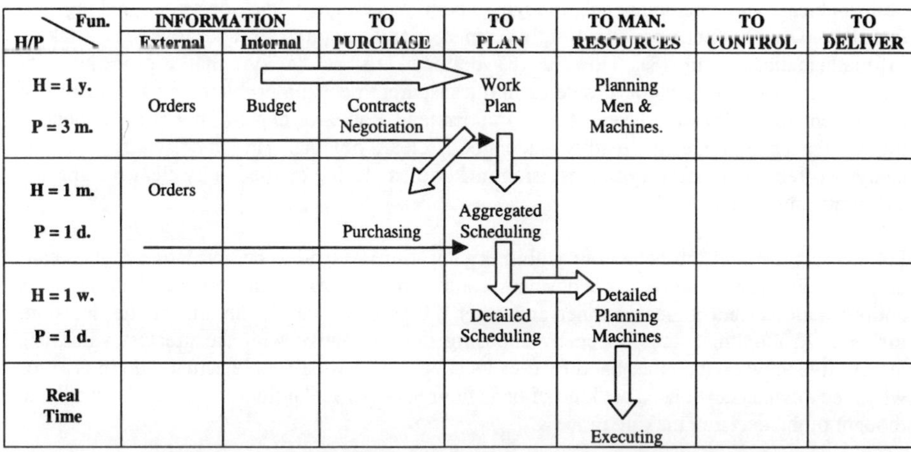

Figure 1: A sample GRAI Grid (Source: Doumeingths, 1987).

GRAI nets are used to describe the various activities that exist within the PMS decision centres. An activity, flowing horizontally, appears as a verb in the infinitive and it is written within a special frame as shown in Fig. 2. Decisional activities are drawn within the same frame but flowing vertically instead of horizontally. Information and means to perform the activity, as both may be the result of preceding activities, are written in frames linked to the activity with an arrow. The results (one or several) of the activity are written in rectangle(s) close to the end of the activity.

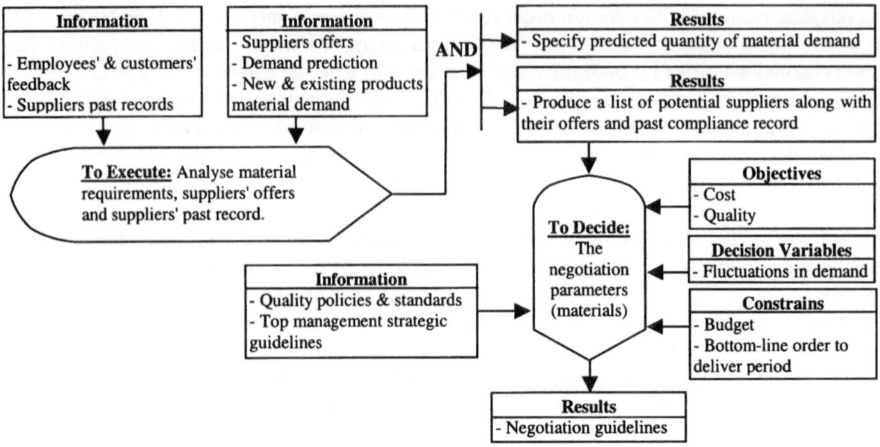

Figure 2: A generic material purchasing decision centre for an SME (H = 1y, P = 3m).

3. DECISION ANALYSIS SCIENCE.

Decision Analysis (DA) has experienced an evolution of hundreds of years based on philosophical and mathematical thought about uncertainty and decision making. Its foundations can be found early in 1700's when Bernoulli captured attitudes toward risk-taking in mathematical form (8). Howard (8) defined DA as "a systematic procedure for transforming opaque decision problems into transparent decision problems by a sequence of transparent steps. Opaque means 'hard to understand, solve, or explain; not simple, clear or lucid'. Transparent means 'readily understood, clear, obvious'. In other words, decision analysis offers the possibility to a decision maker of replacing confusion by clear insight into a desired course of action" (p.680).

DA expects the decision behaviour within organisations to tend to be normative. This means that in a decision situation, in a business context or at work, human beings behave in an entirely rational and logical manner and other things being equal, the aim of the decision-maker is to maximise satisfaction by choosing the alternative with the greatest value (9). Descriptive view on the other hand focuses its attention on what people actually do in practice when they make decisions. This kind of behaviour is of particular interest of the philosophical thought of the decision making process.

3.1 The decision environment.

Multiple objectives, Risk and Uncertainty, Decision maker's attitude towards Risk and Value Trade-offs have been the foundations, many of the DA theorems have been based on. These are presented below as follows:

It is highly unlikely to encounter a decision, which does not aim to satisfy a number of objectives. The objectives to be satisfied are contradictory to the vast majority of the cases. As a single alternative can not satisfy all objectives, it is important to evaluate the degree to which each objective is achieved by each competing alternative.

Decision making under certainty occurs if the outcomes of all choices are precisely known in advance. Predicting the consequences of each alternative course of action could prove a non-achievable task due to inherent risk and uncertainties. Decision-making under risk occurs if the probability of each possible outcome of a choice is known whereas decision-making under uncertainty occurs where these probabilities are completely unknown (10).

Important decisions involve critical value trade-offs to indicate the relative desirability among multiple influencing parameters (11). These value trade-offs have been the subject of many theorems including Utility Theory, Multi-Criteria Decision-Making (MCDM), Simple Multi-Attribute Rating Technique (SMART) and Analytic Hierarchy Process (AHP).

3.2 The process of Decision Analysis.

Decision analysis, by definition, constitutes a sequence of steps aiming to simplify a complex decision problem. Raiffa (12), Raiffa and Keeney (13), Keeney (11) and Howard (8) give evidence of these steps. Keeny (13) decompose the decision analysis process into four core steps as shown in figure 3.

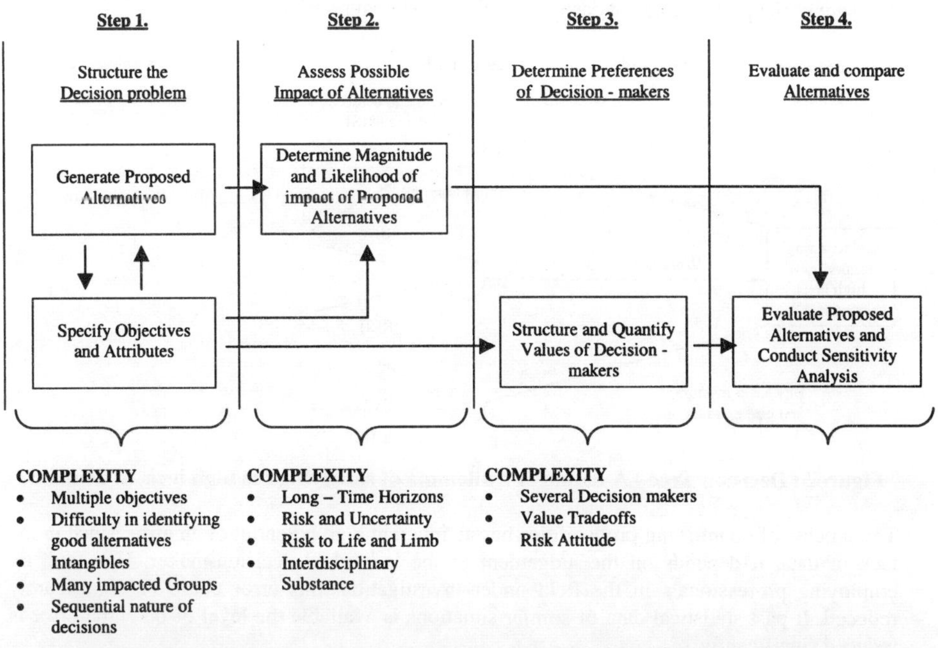

Figure 3: Schematic representation of the steps of DA (Source: Keeny, 1982).

The first step of a decision analysis is the specification of the objectives to be satisfied and the realisation of the alternative courses of action. Objective specification comes in a hierarchical form. The general objectives are specified first and then decomposed to lower level objectives in order to be measured more easily. Value Trees (Fig. 4) are used to represent this decomposition of objectives. The means used to identify the level of fulfilment of each lower level objective are called attributes (e.g. money, number of orders, etc). The Decision Tree technique is a very powerful way of describing the existing alternatives, as shown in Fig. 5.

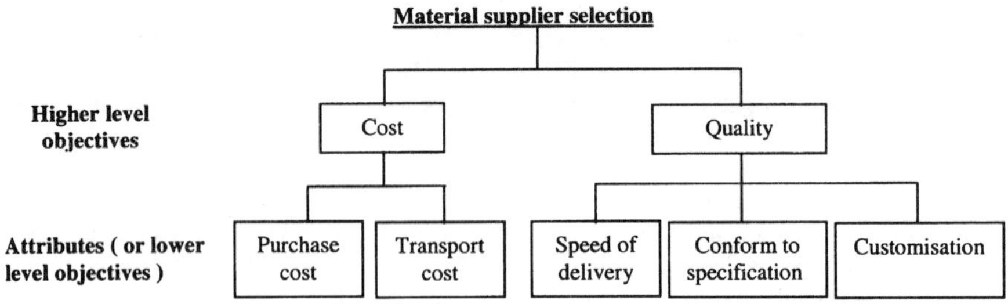

Figure 4: A typical value tree for a material supplier selection.

As the various alternatives have been defined, the magnitude of their consequences is determined. The problem is not so simple as each alternative might have a number of potential consequences as a result of inherent Risk and Uncertainty. For that reason, for each alternative, the various consequences are determined as well as the probability of occurrence. The sum of the probabilities of each alternative must sum up to one.

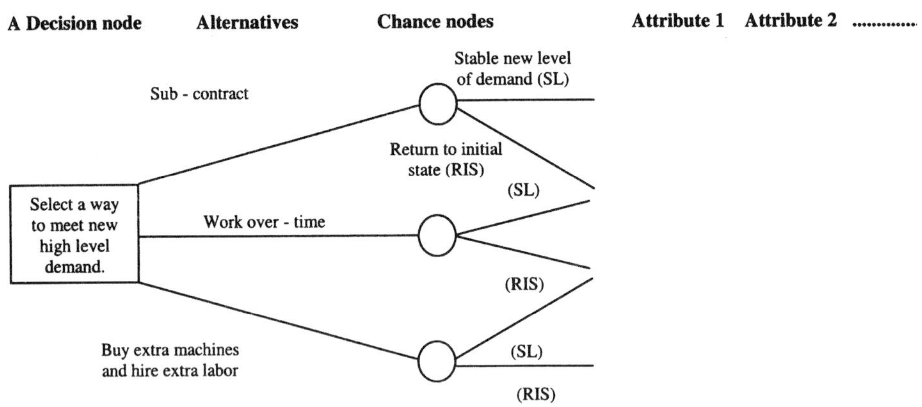

Figure 5: Decision Tree - A manager's dilemma of satisfying the high level of demand.

The process of quantifying probabilities has an inherent high potential of an error, since in the lack of data, it depends on the judgement of the individual decision-maker. However, by employing professionals in the field under investigation, this error could be significantly reduced. If past statistical data of similar situations is available the level of potential error is reduced significantly.

The preferences of Decision-makers are determined by creating a model depicting the values and risk attitude of individual decision-maker(s). Utility theory (utility functions) is used to determine these preferences. This is used to evaluate the various decision alternatives. A different approach constitutes the Expected Monetary Value (EMV) criterion, which simply states that the alternative to be chosen is the one with the highest EMV.

Each alternative identified is evaluated and compared with the other alternatives. Sensitivity analysis is used to examine how robust the choice of an alternative is to changes in the figures used in the analysis. This might come as a variation of the probability figures or the utility figures.

3.3 Simplified DA techniques, where the factor of Risk and Uncertainty is overlooked.

Simple Multi - Attribute Rating Technique (SMART) is a simple method for multiple attribute decision making, introduced by Edwards (14). The central idea of this technique is to split the problem into logical sequential steps so as to ease the decision-maker processing the available data. In this way the decision-maker will be forced to come across and evaluate various aspects of the problem and at the end to acquire a better understanding of the problem. Goodwin and Wright (15) give evidence of the main steps of this methodology.

Analytic Hierarchy Process (AHP) is another method acting within this area of DA techniques. Thomas Saaty developed AHP when he was acting as an adviser to the US government. From its axioms to its procedures, the AHP has turned out to be historically and theoretically a different and independent theory of decision making from utility theory. In less than a decade since its introduction, the AHP has found its way into various decision areas. These include areas like: economics and planning, energy policy, material handling and purchasing, flexible manufacturing systems, project selection, microcomputer selection, budget allocations and forecasting and many others (16).

It is widely evident that AHP has a large number of implementations in many diversified areas. Since its introduction, it has created a big controversy over its relative strengths and weaknesses. However, as AHP is a decision analysis technique its aim is not to give a correct answer to a decision problem but to provide insight and understanding into it. As a result one could argue that techniques like SMART and AHP, designed to deal with decisions with multiple objectives and limited uncertainty, provide a better insight into the problem as they closely guide the decision -maker deep into the insights of the decision problem.

4. COMMENTS.

GRAI method is been well established with over 50 documented industrial applications. GRAI users include several large engineering firms, as well as academics. Computer support software has been made available lately to assist its implementation. In spite of the fact that this modelling technique has been developed to facilitate the process of organisational change, emphasising the description of the decisional network, it does not provide any guidance to the individual decision – maker in structuring the decision problems or improving the quality of the decision making process.

These limitations could be a result of the following:

- The application of the method is mainly based on interviews, but these are rarely structured. The interview process is formulated according to the experience and knowledge of the analyst leading to inconsistency of results, depending on who is carrying out the GRAI exercise (This could especially be witnessed during the construction of the GRAI nets).
- Strategic, tactical and operational decisions follow a different decision making process. As their importance rises, the decision making process becomes more elaborate and the factor of risk and uncertainty becomes a major parameter in the final decision outcome. GRAI nets are relatively good when describing trivial or minor decisions, but when it comes to

strategic decisions they prove to be insufficient. This might be due to the fact that GRAI was designed to describe the PMS of a company and therefore strategic decisions are excluded from it, being the responsibility of top management.

Within this respect, Decision Analysis techniques constitute a natural continuation of the GRAI method. As soon as the decisional network of the organisation is realised (GRAI Grid) and the main function of each Decision centre is defined, DA techniques offer a means of:

- A coherent way of structuring the decision problems, along with appropriate notation;
- A formal way of evaluating the various alternatives by taking into consideration factors like multiple objectives, means of quantifying their importance, value trade-offs between them, and the attitude to risk of individual decision – makers;
- A sensitivity analysis, examining the robustness of the selected course of action to changes of the contributing parameters.

The application of various DA techniques with GRAI can contribute to a formalisation of an integrated methodology. Its aim would be to provide a common means of realising the decisional network of manufacturing organisations and a coherent, detailed guidance to individual decision-makers in order to enhance the quality of their decisions.

REFERENCES

1. Tompkins, J. (1995). *The Genesis Enterprise*. New York: McGraw - Hill, p.9.
2. Drucker, P. F. (1992). The New Society of Organisations, *Harvard Business Review*, September - October 1992, 95 - 104.
3. Kepner, C. H., Tregoe, B. B. (1965). *The Rational Manager.* New York: McGraw - Hill.
4. Colquhoun, G. J., Baines, R. W. & Crossley, R. (1996). A composite behavioural modelling approach for manufacturing enterprises, *International Journal of Computer Integrated Manufacturing*, 9(6), 463 – 475.
5. Colquhoun, G. J., Baines, R. W. & Crossley, R. (1993). A state of the art review of IDEFo, *International Journal of Computer Integrated Manufacturing*, 6(4), 252 – 264.
6. Doumeingts, G., Chen, D, & Marcotte, F. (1992). Concepts, models and methods for the design of production management systems. *Computers in Industry*, 19, p.p. 89 - 111.
7. Doumeingts, G. & Dumora, E. (1987). Use of GRAI method for the design of an advanced manufacturing system, *International Conference on Flexible Manufacturing Systems*, IFS (Conferences) Ltd and authors, ISBN 0-948507-60-8.
8. Howard, R. A. (1988). Decision Analysis: Practice and Promise, *Management science*, 34(6), 679 – 695.
9. Cooke, S. & Slack, N. (1991). *Making management decisions.* London: Prentice Hall.
10. De Neufville, R. & Stafford, J. H. (1971). *Systems analysis for engineers and managers.* London: McGraw-Hill.
11. Keeney, R. L. (1982). Decision Analysis: An Overview, *Operations Research*, 30(5), 803 – 838.
12. Raiffa, H. (1968). *Decision Analysis: Introductory Lectures on Choices under Uncertainty*. London: Addison – Wesley.
13. Keeney, R. L. & Raiffa, H. (1976). *Decisions with Multible Objectives: Preferences and Value Tradeoffs.* London: Wiley.
14. Edwards, W. (1971). Social Utilities, *Engineering Economist*, Summer Symposium Series, 6.
15. Goodwin, P. & Wright, G. (1998). *Decision analysis for management judgement.* Chichester: John Wiley & sons.
16. Zahedi, F. (1986). The Analytic Hierarchy Process - A Survey of the Method and its Applications, *Interfaces*, 16(4), 96-108.

Using simulation as a tool in implementing cellular manufacturing in a project-based organization

M M MUSBAH and **T THARMASEELAN**
School of Engineering, Coventry University, UK

ABSTRACT

This paper documents a case study that investigates challenges concerned with cellular manufacturing in a project-based organisation and uses data already gathered at the site of an industrial partner who is part of the rail industry manufacturing network. Problems arise in the form of large fluctuations in both total workload and mix of component types and possible solutions lie in careful selection of the scope of flexibility of each cell. The paper highlights on the benefits of using a simulation model for the cells and demonstrates the effects on the system during the fluctuation of projects, cells capacity, batch size and client's schedule change. The aim was to develop a model that simulates a manufacturing system with cellular layout to help to evaluate the systems behaviour for given sets of conditions.

1. INTRODUCTION

Over the last thirty years Western industry has expended considerable effort to improve flexibility and responsiveness to meet customer needs. Cellular manufacture has emerged as one of the major techniques being used in the drive for improved manufacturing competitiveness (1).

Different definitions are used to describe cellular manufacture (2)(3)(4), however all broadly describe a customer-focused operation containing all the resources required to complete the product. Methods of cell design confined to optimising material flow and space are still utilised by organisations despite the limitations recognised (5)(6)(7). Even where the higher level systems engineering based processes have been used (8); limitations on the development of fundamental engineering characteristics have been expressed (9). These methods also only consider 'static' aspects whereas in reality for some industries the production system is highly influenced by 'external' and 'dynamic' effects.

One such industry is that of rail transportation in which the original equipment manufacturer (OEM) can be either the main project contractor for a client or one of a number of first tier MTO suppliers to a contractor. Traditionally the 'dynamic' aspects have been absorbed through the use of a functional process approach to manufacturing, encompassing a wide variety of in-house manufacturing capabilities.

2. BACKGROUND

Coventry University has been working with an organisation which has been associated with the railway industry and is a long established renowned world wide supplier of Braking Systems and equipment, it can be described as a project driven specific make-to-order business of which engineering design responsibility is that of the original equipment manufacturers (OEM's), with high variety of products with low to medium volume and a high level of standardization.

The core business at the organisation is the design and manufacture of products as diverse as Train Braking Systems, Door Systems and Sub-contract Machinery, together with the provision of spares. To be competitive within the global market place, expectations have been raised whereby customer's nowadays demand shorter lead-times, high delivery reliability with keen prices and high quality.

During 1995 and 1996 the company recognised the growth signs and the traditional approach to manufacturing could no longer satisfy modern requirements, hence, they adopted a new business strategy with an improved manufacturing system and introduced cellular manufacturing.

3. THE CHALLENGES IN PROJECT-BASED ORGANISATION

Make to order companies are in the business of supplying products in response to customer orders in competition with other companies, on the basis of price, short lead times and reliability in meeting delivery dates (10).

Therefore, there is a lot of influence on the manufacturing process to meet the client's delivery date. At this organisation, the machine shop is the centre of all manufacturing activities for the four major core of business, which are characterised in table 1.

Table 1. The characteristics of the main business

	Core of Business	Main Characteristics of the Business	
1	Braking Systems	• *Mainly Projects* • *Large Product Range* • *Design and supply*	• *Short lead time* • *Variable Schedule* • *Variable Quantities*
2	Door Systems	• *Install at the Site* • *One off Delivery* • *Tight Lead Time*	• *Design/ Supply Installation.* • *Product Process Group*
3	Sub-contract Machines	• *Manufacturing Only* • *Mainly no design Required*	• *Good Forward Schedule* • *Fixed Quantities*
4	After Sales	• *Large Product Range & Varieties* • *Small Quantities*	• *Short Lead Time* • *One Off*

The traditional manufacturing process in the machine shop could be described, for example, as a sink into which the four main core business are placed, at the bottom of the sink is a tap which represents production control, once the tap is opened, a mixture of different products are produced, these products have different lead times and can take up to 12 weeks, operation methods, manufacturing and assembly routes can also be difficult to control within this type of environment. (As figure 1. illustrates).

A more flexible manufacturing system is required due to the fact, that in this type of environment, clients' demands and schedules change from week to week.

The present manufacture method in the machine shop is process orientated, e.g., separate sections for milling, turning, drilling, etc. Some 55 items for the Brake Actuators are manufactured within the machine shop on a variety of machines, both CNC and conventional, components need to be transported around the machine shop to various sections of machining, this involves a considerable amount of non value added activities, i.e. transport, storage and inspection.

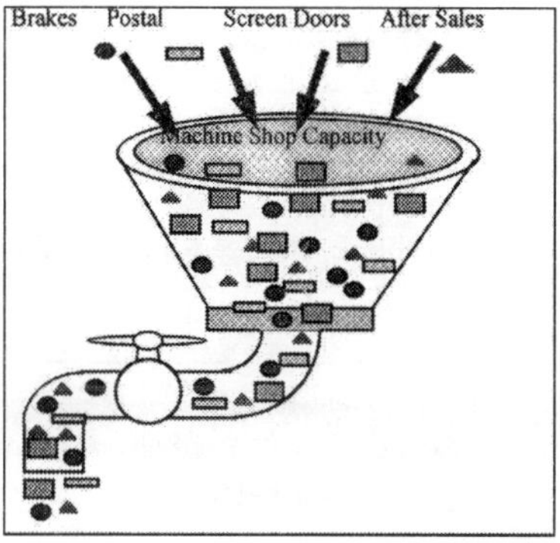

Figure 1. The bucket sink analogy

The lack of focus on these components can often lead to a bottleneck situation within the machine shop causing delays in manufacturing lead-time, longer throughput time and possible shortages within the assembly area. The most common way of dealing with this problem in the past has been to work excessive over time or to sub-contract the excess hours.

The introduction of cellular manufacturing will simplify work movement and responsibility for completion of jobs by placing all the resources required completing the job together in a cell. It will also simplify management and the production control by dividing the business into small self autonomous units that can be more easily controlled, cellular manufacturing will work by filtering the production control which will help lead to de-centralise the production control responsibility, work performance, output and scheduling, to those individual cells which can also be controlled more effectively (10).

4. USING THE SIMULATION

The definition of the simulation is simplified as a broad collection of methods and application to mimic the behaviour of a real system (11), simulation allows for the understanding and solutions of problems to be achieved by a trial and error approach (12), Simulation is the art of experimentation using models. It provides intelligent and cost effective platform to perform interesting and valuable experimentations to help making important decisions. But, it should be noted that the simulation must not be taken as a crystal ball. Usage of simulation is excellent in aiding the decision making process, but it cannot make the decisions. Simulation will help to envisage potential problems in a system and can help to solve the problems but it cannot automatically solve the problems by itself.

The simulation will also provide a number of results for long-term implementation of cellular manufacturing in a project-based company and the results will be shown in the form of project lead-time, cells utilisation etc. In such environments, introducing an optimum cell operation (i.e. batch size, number of employees, cells, reducing work in progress, priority rules), and synchronising the activities between the cells so the product will manufacture at the right time, the right quantities in the right place.

4.1 Model construction

The Discrete Event Simulation software ARENA was used to build the simulation model. By using appropriate features of the ARENA software the features of the manufacturing system concerned were modelled. A number of cells with a varied number of machines processing various projects (jobs) with different batch sizes and different cell visits were taken into account. Buffers for jobs awaiting processes were accommodated accordingly. Jobs process times at different cells were assigned appropriately. Figure 2 shows the model animation, and some of the results are shown in Table 2, the lead-times; Table 3, the cells utilisation and Table 4, the queuing time.

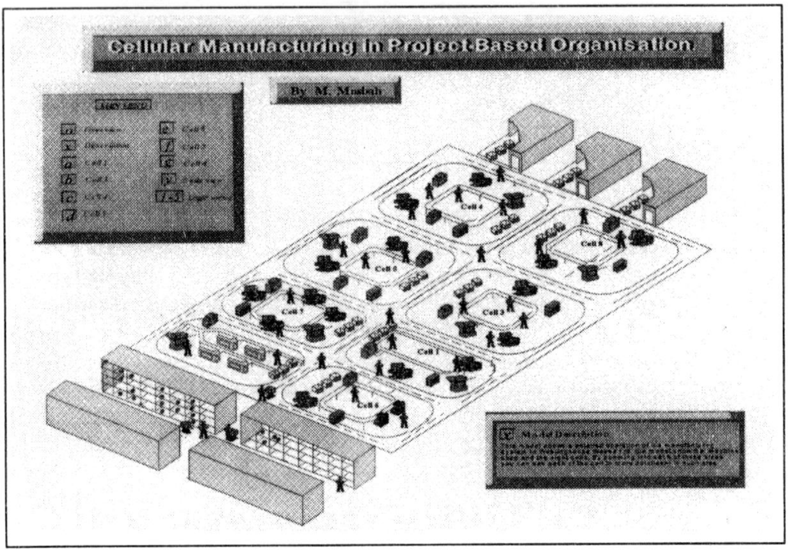

Figure 2 The Model animation area

Table 2 Lead time result

	Project Name	Lead Time/Hrs
1	Jubilee Brakes Lead Time	108
2	Northern Brakes Lead Time	160
3	Jubilee PED Lead Time	79
4	Jubilee DET Lead Time	17
5	Euro Shuttle Lead Time	85
6	Heathrow Lead Time	85
7	Class 92 Lead Time	85
8	Jubilee Contract Lead Time	85
9	Sundries Lead Time	100
10	Spares Lead Time	157

Table 3 Cells utilisation result

	Cells Utilisation	%
1	CELL 1 Utilisation	41%
2	CELL 2 Utilisation	57%
3	CELL 3 Utilisation	100%
4	CELL 4 Utilisation	63%
5	CELL 5 Utilisation	81%
6	CELL 6 Utilisation	26%
7	CELL 7 Utilisation	97%
8	CELL 8 Utilisation	13%

Table 4 Queuing time result

	Cells Queuing Time	Hrs
1	CELL 1 Queuing time	3.1
2	CELL 2 Queuing time	0.8
3	CELL 3 Queuing time	1.7
4	CELL 4 Queuing time	1.5
5	CELL 5 Queuing time	0.8
6	CELL 6 Queuing time	0.5
7	CELL 7 Queuing time	0.6
8	CELL 8 Queuing time	0.2

4.2 Collection of statistics

Relevant features were included in the model in order to collect statistics during and at the end of simulating the model. Lead times of the projects and the utilisation percentages of cells were decided to be two sets of important statistics to be produced from the simulations.

4.3 Model verification

In order to ensure that the model performs as expected and intended the model was tested by using facilities provided in the software, such as: the interactive debugger, tracing the textual descriptions of events and visualisation of events through animating the simulation.

4.4 Validation of the model

In discussing with the experts who are closer to the real system it was decided that the model adequately represents the real world system which was modelled and the model-generated behavioural data characterises the real world system's behavioural data.

4.5 The experiments

Several experiments were performed on the model by simulating the model for various conditions. A selection was made among these experiments to produce the tables of some results. The conditions of the experiments are described below:

EXP 1: This experiment shows the effect on the system by removing or stopping one of the projects as (Northern Brakes project).

EXP 2: This experiment shows the effect on the system by deleting one of the cells (i.e. breakdown or no longer needed) such as cell7.

EXP 3: This experiment shows the effect on the system by adding a new project.

EXP 4: In this experiment we tested the system by removing 2 cells (cell2, cell7).

EXP 5: This experiment shows the effect on the system by eliminating cell1.

EXP 6: This experiment tests the system by decreasing the order of the spares by 50%.

EXP 7: This experiment tests the system by increasing the order of the spares by 50%.

EXP 8: This experiment combines the cells into 3 major cells (cell1 is product-focused cell, second cell cubic cell and the third cell is rotational cell). Also one of the projects is taken out

EXP 9: This experiment combines the cells into 3 major cells (product cell, cubic cell and rotational cell). But in this case with an extra project.

EXP 10: This experiment combines the cells into two major types of cells (rotational cell and cubic cells) without one of the projects such as Northern Brakes.

EXP 11: This experiment combines the cells into the major two types of cells (rotational cell and cubic cell), and adds an extra project such as Northern Brakes.

5 RESULTS DISCUSSION

The experiment result showed that, the lead-time is inversely proportional to the number of cells, as figure 3 shows.

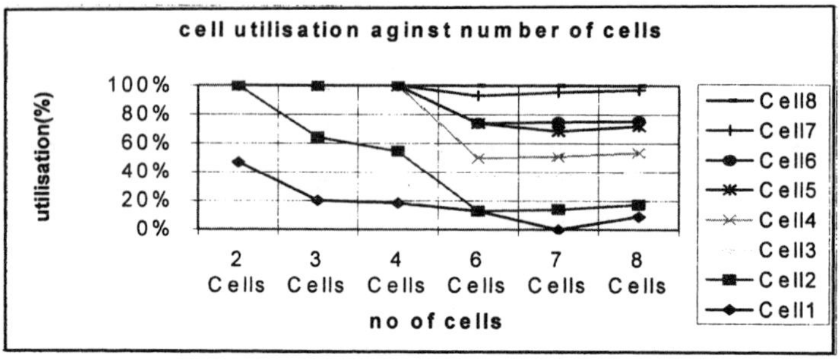

Figure 3 Cell Utilization

By bypassing cells 2 and 7, the two peeks in lead-time for the spares were generated. This is directly related to the specialism of these two cells to produce spares.
Lead times being measured in weeks was reduced down to being measured in hours.
Halting the production of Northern Brakes had the most significant affect on reducing the lead-time across the other products, especially the Jubilee Brakes, and to a lesser extent the other projects. The lead-time for the Spares remained relatively stable despite the change.
The data from this type of simulation forms the input to strategic decisions concerning the prioritisation of the cell loading.
In a project-based environment, as examined in this study, reference has to be made to customer requirements and the constraints of the workshop capacity. Critical path analysis combined with line-of-balance analysis, provides additional information to these decisions. The assumptions on which the decisions are made are the subject of further research.

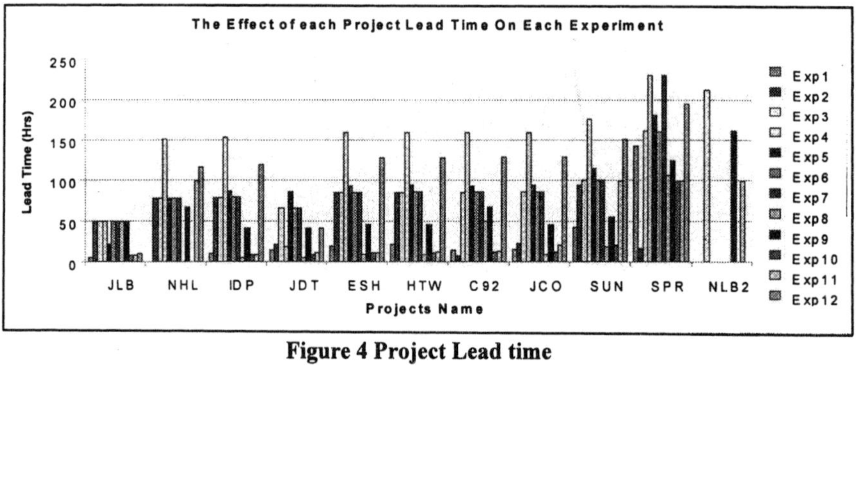

Figure 4 Project Lead time

6 CONCLUSION

From this study, the usefulness and appropriateness of Discrete Event Simulation can be seen. Cellular manufacturing has potential benefits for organizations where wide fluctuations in product workload and mix are normal. However, further investigation is required to obtain the much-needed benefits of quick service and flexibility, the design criteria for the cells and their operational rules. The developed simulation model gives successful results to the cellular manufacturing system concerned in this study. It must be seen that the data generated by the simulation forms a valuable contribution to the information required for successful project management in this environment.

Simulation is increasing in popularity for this type of environment. It is used to depict and analyse possible scenarios in manufacturing systems. Simulation packages are especially useful in experimenting on effect on various conditions (what if ? conditions) and control of the systems. By further experimenting with, understanding and analysing this simulation study, it will help to make management decisions on cells design and implementation. Project based manufacturing companies considering, enquiring into or having already in place a cellular layout will undoubtedly find simulation modelling extremely valuable in assessing possibilities; thus enabling experiments outside the costly real systems; helping to visualise various scenarios and forming the basis in decision support systems.

REFERENCES

1. Kirton. J and Brooks. E, (1994), Cell in industry "*Managing Team and Profit*"(McGraw-Hill, England)
2. Jones. D, (1993), Cell Manufacturing "The Quiet *Revolutio*n" (Ingersoll Engineers Rugby).
3. Johnson. P, (1992), Unlocking the Rewards of Cellular Manufacturing "*Machinery and Production Engineering*" 21 Feb. 1992. PP 32-35.
4. Schaffer et al, (1992), The Consideration of human Factors in the Design of Group Technology Manufacturing System, "*International conference on Flexible Automation and Information Management*
5. Burbidge. J, 1971, *Group Technology in the Engineering Industry*, MEP Ltd. London.
6. King. J, (1980), Machine Component Grouping in Production Flow Analysis "*An Approach Using Rank Order Clustering Algorithm*", (International Journal of Production Research), 18, No 2, PP 213-232.
7. Goodyer. J, et al, (1994), Job Design "*Work Force Empowerment in Cell Manufacturing to Achieve World Class Quality*", Proceeding of MANSA Conference, Cape Town South Africa.
8. Parnaby. J, (1988), System Approach to the Implementation of JIT Methodologies in Lucas Industries, "*International, Journal of Production and Research*", VOL. 26, No. 3. PP 483-492.
9. Higgins. J and Spraggett. S, (1995), Optimising Cell Performance Through Simultaneous Process Design, Pro. of APICS World Symposium of Integrated Resource Management, (APICS Australia). PP 63-67.
10. Musbah. M, and Martin. S, (1997) "Cellular Manufacturing In Project-Based Organisation" The 14th International Conference on Production Research ,Osaka, Japan
11. Kelton et al, (1998). System Modelling "*Simulation with Arena*" WCB/McGraw-Hill.
12. Szgmankiewicz. J, et al43. (1977). "Solving Business Problems by Simulation" McGraw-Hill

Automatic model generation of batch manufacturing plants using WITNESS

J ROGERS and **K D BARBER**
Total Technology Department, UMIST, Manchester, UK

ABSTRACT

Discrete Event Simulation packages, such as Lanner Group's Witness, have been successfully used by manufacturing specialists and researchers to model manufacturing systems. However, constructing the model is time consuming and tedious and requires specialist knowledge. Small to medium sized engineering companies frequently have a requirement to model their plant but they do not have the specialist skills within their organisation. This paper outlines a method for creating models of batch manufacturing plants using the work centre file, the part master file, the part routing file and the shop schedule from the companies manufacturing database (MDB). The companies MDB is imported to a Microsoft ACCESS database which is used to run WITNESS via an OLE link. The ACCESS database constructs the witness model using the work centre file. Part routing is achieved by using WITNESS parts with attributes specifying the process route. The model is designed so that it can be re-configured to operate as either a traditional process layout or as cellular manufacturer. The user creates the model via an ACCESS form and as such needs minimal computer skills.

1. INTRODUCTION

Over the past few years, the UK manufacturing industry has faced a constant battle for survival and growth in the face of fierce global competition. Both domestic and global customers are demanding and getting high specification and quality products at keen prices. This is compounded by shorter product life cycles where competitors are replacing their product ranges with increasing frequency.

The effect of this change on small and medium sized batch manufacturing companies (SME's) in the UK, is the requirement to introduce new products and their attendant processes and manufacturing reorganisation quickly and in a rational controlled manner. However, SME's, in general, do not have the skilled personnel or the time available to optimise their response to change and plan the change in the detail that is required.

Consequently, there is a requirement to provide tools which will aid SME's to rapidly evaluate the effect of changes, such as new product introduction, increased product variety, shortened lead-times and etc., on their plant. This paper outlines a means of using manufacturing simulation, with models constructed using data from the company's manufacturing database, to rapidly evaluate the effect of change on a manufacturing organisaion.

2. SIMULATION

Science and engineering are littered with examples of the use of experiments on models, which simulate the action of real world phenomena and systems. Actual manufacturing plants and systems are difficult to experiment with. A production manager would be very reluctant to allow any experimentation with his plant. Also experiments could take a long period of time and be subject to uncontrollable random variations (for example absenteeism, machine breakdowns, product volume-mix changes, etc.). Consequently, computer simulation is an attractive option to both researchers and manufacturers. Some of the earliest computer simulation studies were concerned with the operation of Flexible Manufacturing Systems (FMS) (Carrie, 1988)(1). Hoover and Perry (1989)(2) define computer simulation as:

The process of designing a mathematical or logical model of a real system and conducting computer-based experiments with the model to describe the behaviour of the real system."

Simulation holds many attractions for academic researchers. A model can be created of any plant. All factors can be controlled and experimental noise eliminated. Differing operational strategies can be applied to a single model of a plant. The literature is scattered with many examples of simulation based research. Just in Time is well represented (Yavuz and Satir, 1995)(3) (Lovell, 1992)(4). Comparison between operational policies is also popular (Cook, 1994)(5). Cellular manufacture is also well represented (Morris and Tersine, 1994)(6) (Shafer and Meredith, 1992)(7). Wu et al. (1994)(7) have tested the applicability of Optimised Production Technology using simulation.

Simulation is attractive to industrial users for similar reasons to those that attract academic researchers. In industry, the primary use of simulation is to design manufacturing plants and justify their financial viability. Farhadi (1994)(8) outlines where a manufacturer of automated warehouse systems used simulation to convince a client of the viability that their system met the client's requirements. Stapleton and Levary (1994)(9) extend simulation to the actual preparation of bids for contracts. Crowley et al. (1995)(10) describe the use of simulation to design a medium volume manufacturing facility. Chan and Smith (1993)(11) use simulation to convert an assembly line to JIT. Hwoke and Nelson (1993)(12), Ardhaldjian and Fahner (1994)(13), Koelsch (1992)(14) and Mabrouk (1996)(15) provide further examples of the general acceptance of simulation as a powerful tool for industry.

Both academic and industrial users of simulation have demonstrated the applicability and usefulness of simulation to solving problems. However, model building is a time consuming process and necessitates skills in computer programming and an in-depth knowledge of a simulation package. These skills and time are not normally available to a manager in a SME with an urgent need to adapt his manufacturing facility to exploit new market opportunities, operating polices, and etc. However, providing the manager with the means to rapidly model a plant and alter the model should enable simulation to be used as a management aid.

3. GENERIC MODEL OF A BATCH MANUFACTURER

In batch manufacture, the plant processes parts in batches under the authorisation of works orders and process routes. Consequently, any model only needs to process an individual batch and not its constituent parts. This is significant in terms of the computational efficiency of a model. If a batch had 100 parts, then a work centre would exhibit 100 events and would have to allocate memory for all 100 parts. However, by representing this as a batch, only one event and memory location is required. In this way, a complete model of a batch manufacturing plant is created and can run with an adequate speed.

Figure 1 outlines the basic components of the generic model; the model input section takes in a shop schedule (a part and associated attributes) and distributes it to the work centres; each work centre processes a batch and pushes it to the transport module; the transport module pushes the batch to the next work centre or the shipping module if the batch is finished; the shipping module ships the part and updates the stores file.

Figure 1 Components of Generic Model of a Batch Manufacturer

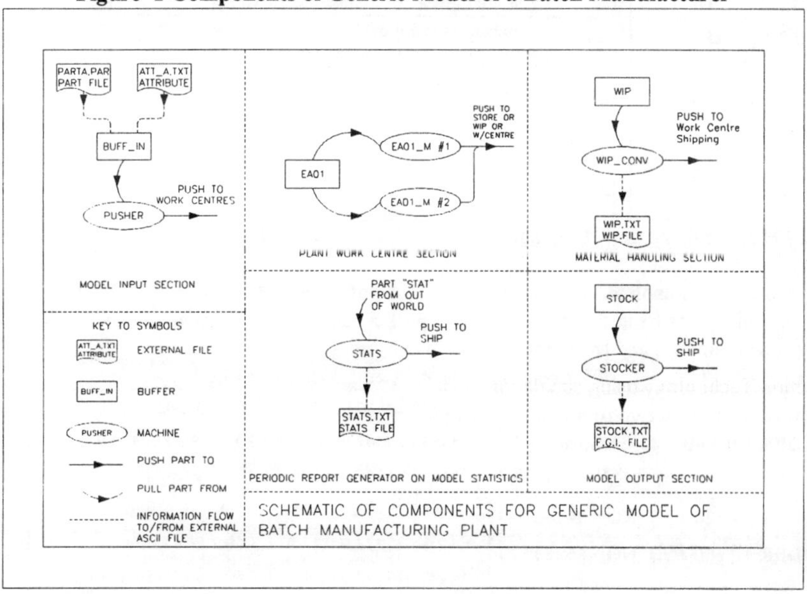

3.1 Part Description

In the generic framework, the Witness entity, part, is used to represent a batch of components. The works order information, such as batch size, process route and etc., are represented by part attributes as illustrated in Table 1.

3.2 The Control Modules

Figure 3. (b) illustrates the witness elements of the shop scheduling module. The shop schedule is contained in a text file. The parts are loaded into the buffer, input_b, by the witness element part file. The machine, SCH, pulls the part from the buffer and reads in an text file with the batches attributes and assigns them to the part. The part is then pushed to the buffer of the first work centre. Figure 2 illustrates a sample part and attribute file.

Once a work centre has processed a batch it pushes the batch to the material handling module. Figure 3. (a) illustrates the Witness elements. The material handling time can be applied as either a buffer with a delay or as a process time for the machine, forklift. The machine forklift pushes the part to the next work centre or to the shipping module if the batch is finished. The forklift module can also write a text file recording the material movement of parts.

Table 1 Description of Part Attributes

Attribute	Type*	Array	Description
JOB_NO	I	1	Job Number / Works Order Number
STK_CD	I	1	Part Number
BATCH	R	1	Batch Size
PTIME(1)	R	3	Start Time of Batch
PTIME(2)	R	3	Finish Time of Batch
CURROP	I	1	Current Operation Number
N_OPS	I	1	Number of Operations
BATCH_NO	I	1	Batch No used for transfer batches
SU	R	10	Set Up Time
PT	R	10	Process Time
DEST	N	10	Work Centre Name
WCN	I	10	Work Centre Number
M_COST	R	1	Material Cost
A_VALUE	R	1	Value added to Part.

* R,I,S,N are real, integer, string and name variables respectively.

If a batch is complete, it is pushed to the shipping module (Figure 3. (c)). The machine FGI_PP pulls parts from the buffer FGI and pushes the parts to shipping and writes a text file with the job number, and the model entry and exit times.

Figure 2 Sample Part and Attribute File

```
                                      45338,12510,0,0,300,1,1,60,4.2,PC08,0,0,0,0
ORDER 1 0 JOB_NO= 45338               44373,24747,0,0,300,1,1,30,3.2,PC07,0,0,0,0
ORDER 1 4320 JOB_NO= 44373            44392,24947,0,0,300,1,1,30,3.2,PC07,0,0,0,0
ORDER 1 4320 JOB_NO= 44392            44986,25972,0,0,1250,1,3,0,0.5,SH01,0,0,0,0
ORDER 1 5760 JOB_NO= 44986            44986,25972,0,0,1250,1,3,60,2,PC15,0,0,0,0
                                      44986,25972,0,0,1250,1,3,60,1.8,PC15,0,0,0,0

        Sample Part File                        Sample Attribute File
```

3.3 The Work Centre Module

The work centres are modelled as illustrated in Figure 3. (d). The first four digits refer to the work centre code of the work centre file. In this case, PC08 is the Mori Seiki 2 axis chucking lathes. The FORKLIFT pushes a part (the model equivalent of a batch of components and its associated documentation) to the buffer PC08 which adds the total batch processing time to the variable PC08_QT and the current cost of the batch to the variable PC08_QV. The variables PC08_QT and PC08_QV store the current work centre queue length and working capital tied up in the waiting batches respectively. The machine PC08 pulls the part from PC08 and pushes the part to either a designated machine buffer PC08_QM(#N) or to the buffer PC08_REP. The decision is based on the attribute WCN which denotes a specific numbered work centre e.g. PC08#1 or where WCN is zero the part is pushed to PC08_REP. This logic enables the model to emulate a process based or cellular based batch manufacturing where the cells are defined in the MDB routing file and not in the model structure. The machine PC08_PP(#N) pulls a part from PC08_QM(#N) or PC08_REP and sets-up and processes the batch according to the part attributes su and pt. Once PC08_PP has finished it adds its added WIP value to the WIP value of the batch and pushes the part to the buffer tran_buf.

Figure 3 Illustration of Witness Elements of Generic Modules

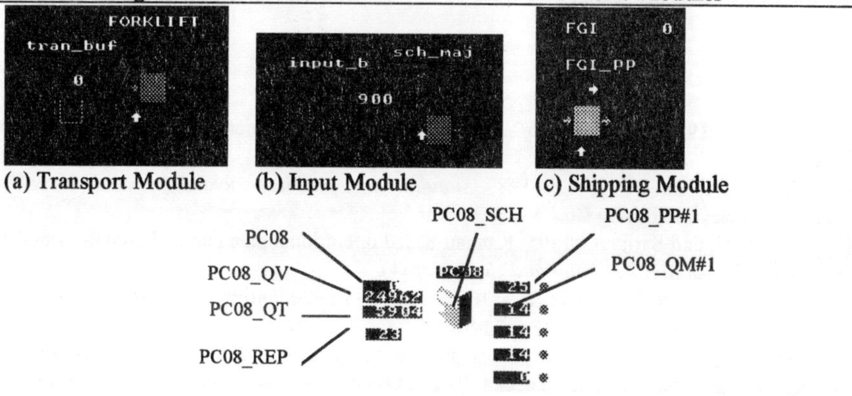

(d) Work Centre Modules

4. AUTOMATIC MODEL GENERATION

The Witness package provides a method for using Witness as an OLE server. Microsoft ACCESS, though a relational database package, has the ability to control OLE servers. Consequently, the manufacturing database of a SME can be exported to ACCESS. The key tables are the work centres (work centre names, quantities, shift patterns, etc.), the process routings (part name, set-up, process times, work centres), the part master (part number, description, batch size, etc.) and the works order table (works order number, part number, due date, batch size, etc.).

A model template is created in Witness which consists of the input, transport, shipping modules and shift patterns. A model is created of a sample work centre module and saved as a text file. Witness defines the work centre with three sections; the DEFINE section which defines the names of the witness elements; the DISPLAY section which lays out how the elements are displayed; the DETAIL section describes the logic of the model elements. In the programme, the listing of the work centres is converted to a parametrized string. For example the work centre's buffer is defined with *"wcl_stat = wcl_stat & "BUFFER: " & wk_name & ",1,10000;" & Chr$(13) + Chr$(10)"* where the variable wk_name is from the work centre file. Consequently, as the work centre file is looped through, the work centre is specified by the string wcl_stat and is input to Witness by the OLE WIT command.

Once the basic programme is written, probably by a simulation expert, then the user only needs to manipulate the work centre file, the routings file and part master file in order to experiment with the model. For example, cellular manufacture could be tested by assigning work centre numbers in the routings file and testing their effect.

5. Discussion

This paper outlines a method for automatically creating a model of a batch manufacturing facility. It should be possible for researchers to further refine the framework to represent differing manufacturing facilities. However, the key concept is to allow production managers to experiment with a model of their manufacturing plant by interacting with a database rather than actually constructing a model from scratch

6. References

(1) Carrie, A., 1988. *Simulation of Manufacturing Systems.* Chichester, UK: John Wiley and Sons.

(2) Hoover, S.V. and Perry, R.F., 1989. *Simulation: A problem-solving approach.* New York: Addison-Wesley Publishing Co.

(3) Yavuz, I.H. and Satir, A., 1995. Kanban-based operational planning and control: simulation modelling. *Production Planning and Control.* 6(4).

(4) Lovell, M.C., 1992. Simulating a 100% Just-In-Time economy. *International Journal of Production Economics.* 26 (1992) pp71-78.

(5) Cook, D.P., 1994. A simulation comparison of traditional, JIT, and TOC manufacturing systems in a flow shop with bottlenecks. *Production and Inventory Management Journal.* First Quarter, 1994.

(6) Morris, J.S. and Tersine, R.J., 1994. A simulation comparison of process and cellular layouts in a dual constrained environment. *Computers and Industrial Engineering.* 26(4).

(7) Schafer, S.M. and Meredith, J.R., 1993. An empirically-based simulation study of functional versus cellular layouts with operations overlapping. *International Journal of Operations and Production Management.* 13(2).

(8) Wu, S.Y., Morris, J.S. and Gordon, T.M., 1994. A simulation analysis of the effectiveness of drum-buffer-rope scheduling in furniture manufacturing. *Computers Industrial Engineering.* 26(4).

(9) Farhadi, J., 1994. Simulation. *Industrial Engineering.* July, 1994, pp 14-15.

(10) Stapleton, L.E. and Levary, R.R., 1994. Using XCELL+ in the preparation of a contracting bid. *Industrial Engineering.* April, 1994, pp 44 - 49.

(11) Crowley, D.J., Bard, J.F. and Jensen, P.A., 1994. Using flow ratio analysis and discrete event simulation to design a medium volume production facility. *Computers and Industrial Engineering.* 28(2).

(12) Hwoke, B.U. and Nelson, D.R., 1993. An overview of computer simulation in manufacturing. *Industrial Engineering.* July, 1993, pp 43-45.

(13) Chan, F.T.S. and Smith, A.M., 1993. Simulation aids JIT assembly line manufacture: a case study. *International Journal of Operations and Production Management.* 13(4).

(14)Ardhaldjian, R. And Fahner, M., 1994. Using simulation in the Business Process Reengineering Effort. *Industrial Engineering.* July, 1994, pp 60 - 61.

(15) Mabroak, K.M., 1996. Create your own low cost manufacturing environment. *IEE Solutions.* January, 1996, pp19-25.

(16) Koelsch, J.R., 1992. Prove it with simulation. *Manufacturing Engineering.* November, 1992, pp51-55.

Discrete simulation and heuristic search for decision support in manufacturing

M SCHULZ
Institut fur Produktionstechnic, Technische Universität Dresden, Germany
A BOYLE and **A KALDOS**
School of Engineering, Liverpool John Moores University, UK

Abstract

The paper demonstrates the potential benefit of using a heuristic search strategy combined with discrete simulation for operations planning and job scheduling in manufacture . For this purpose two notoriously difficult job shop scheduling problems are used in order to show the behaviour and performance of the applied search strategy. As a conclusion, the paper draws attention to the benefits of optimisation in both shop floor control and short term planning, where most applications in industrial practice still rely on well-known loading or priority rules.

1 INTRODUCTION

Discrete event simulation has been used as a system design and manufacturing planning tool for many years. Traditionally, simulation mainly plays the role of a decision support tool for long term planning tasks. In these cases the simulation analyst is seeking to find answers to "what if" scenarios, particularly when having to choose from a range of different possible configurations for a group of machine tools, a production line or a manufacturing system. The more general question however is how the selected operational and performance criteria can be achieved. In most cases the simulation analyst will rely on systematically varying the system parameters in the hope of finding the best parameter set or system configuration. Compared with these traditional applications for long term planning tasks, such as the support of plant layout design, simulation is being used more and more to facilitate decisions within a very short time horizon which take effect immediately on the shop floor. In such cases it is impossible to make a large number of simulation experiments in support of finding an acceptable or an optimum solution.

2 THE SCHEDULING PROBLEM

In general the task of job scheduling is to find a special but specific sequence for producing a set of jobs that satisfies a defined objective function. In many cases the objective is to minimise the makespan, which is the overall flow time for producing the jobs from the start of the first operation of the job processed first to the end of the last operation of the job processed last.

A scheduling problem usually involves several assumptions, namely:-

- each machine can only process one job at a time,
- once an operation is started, no change of order is allowed,
- no workstation breakdowns or other randomly occurring events are considered.

An indication of the complicated nature of scheduling problems is given by the fact that the famous 10×10 problem, representing 10 jobs in 10 machines, initially formulated by Fisher and Thompson in 1963, was exactly solved only in 1989 [1]. For this reason both this problem and the complementary 20×5 problem, representing 20 jobs in 5 machines, of the same origin [2] are used for the experiments.

3 AWARENESS AND HANDLING OF SCHEDULING

All manufacturing processes need a minimum level of scheduling and control in order to co-ordinate the flow of parts on the shop floor. If this condition is fulfilled and the system is running without any major disturbances or significant bottlenecks, the situation is often regarded as being satisfactory. In order to reach the increased productivity and output now being demanded of manufacturing systems, one tends to invest initially in new machinery and equipment instead of thinking about the underlying logic of the processes and the way the parts are scheduled through the processes.

State-of-the-art tools for scheduling and shop floor control mainly rely on well-known heuristic rules in order to find a schedule for a specified set of jobs. Typically, these tools provide both manual and automatic scheduling functions but generally only two characteristics of a job, namely due date and job priority, are considered as operational criteria for the automatic scheduling function. Based on these two attributes combinations of forward, backward and bi-directional scheduling are offered in many different tools. In some cases scheduling systems are integrated with simulation tools. This has the added advantage of allowing comparisons between alternative schedules, but it still does not answer the question of how best to generate the schedule.

4 MODERN HEURISTIC METHODS

Many practical problems encountered in areas such as production, transportation or telecommunications have prompted researchers to adapt ideas from other scientific fields in order to find new suitable approaches for these problems. Simulated annealing, for example, has its origin in thermodynamics, whereas Tabu Search represents a technique based on concepts from artificial intelligence.

Tabu search is a memory based search strategy [3]. At the starting point a possible solution is needed, in our case a sequence of n jobs as the representation of a possible schedule. Then a number of neighbouring solutions s' are generated. The set of all new neighbouring solutions represents the neighbourhood $N(s)$. Neighbouring solutions can be obtained by either an exchange of positions in the job sequence or by insertion. The new candidate solutions are analysed in a simulation run. The job sequence or priority allocation belonging to the best schedule is used as the starting sequence for the next iteration. The special operation which delivered the best solution is excluded during a number of subsequent iterations. The forbidden operation is then "tabu" for the specified tabu time and is written in the so called tabu list *T*.

Unlike the tabu search memory based approach, the acceptance of new candidate solutions can be controlled by means of an acceptance function. An acceptance criterion is found in the heuristic strategy of Simulated Annealing [4]. This strategy has its origin in a procedure to simulate the annealing of solids and can be applied for neighbourhood search. The simulated annealing strategy attempts to avoid being trapped in a local optimum by employing random selection and acceptance strategies. Based on certain probabilities, the random acceptance strategy allows the strategy to move to a solution worse than the previous one. These probabilities are calculated according to a parameter called "temperature". Starting with an initial temperature, after each iteration the new value is determined according to a function that defines the decrease of temperature. The temperature value is then used to calculate the probability *P* for the acceptance of the worse solution. This is done by means of the following rules, where $\Delta\ TF$ is the difference between the current and the previous value of the target function:

$$P\,\{\text{acceptance of } \Delta\, TF\} = \begin{cases} \exp\,(-\Delta TF\,/\,temp) & ; \quad \Delta TF > 0 \\ 1 & ; \quad \Delta TF \leq 0 \end{cases}$$

5 BEHAVIOUR OF TABU SEARCH AND SIMULATED ANNEALING

In order to get a picture of the performance of both heuristic strategies, the influence of the important parameters are analysed. For tabu search this is the tabu time or duration. When applying tabu search the initial solution has a certain influence on the outcome of the optimisation. For simulated annealing, the initial temperature has a major influence on the quality of the solutions. For each parameter value, 50 runs with 50 different initial solutions are carried out. In each case the run length is limited to 100 iterations. The search strategies and the job shop models are modelled in the simulation language SIMAN [4].

Figure 1 shows the results of the experiments with the 10×10 problem. Here the tabu search strategy proves to be better than the simulated annealing. With increased tabu duration the makespan results are getting better while the standard deviation is getting smaller, which reflects a higher reliability of the search in this parameter range.

The corresponding experiments have been carried out using the even more complex 20×5 problem consisting of 20 jobs being scheduled through a job shop consisting of 5 machines. Because of the higher number of jobs involved a neighbourhood $O(n)$ is used, instead of $O(n_2)$ which is used for the 10×10 problem. This is done in order to reduce the computation time. Figure 2 shows the results of the experiments with the 20×5 problem. Once again the tabu search shows better results than the simulated annealing. On the other hand both search strategies produce results that are more widely spread than for the 10×10 problem. This is due to the chosen neighbourhood.

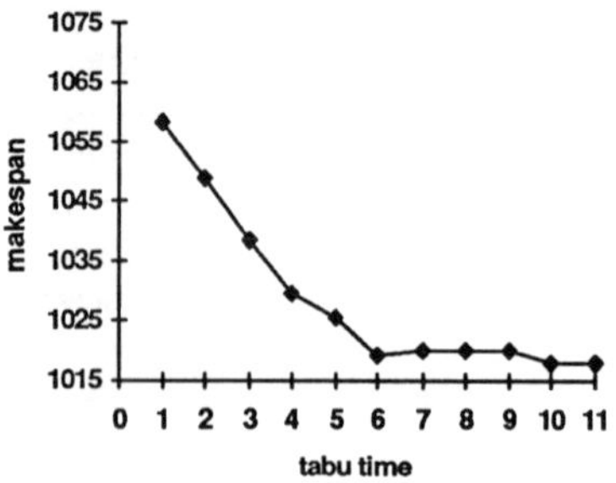

Figure 1a: Tabu Search applied to 10×10 problem

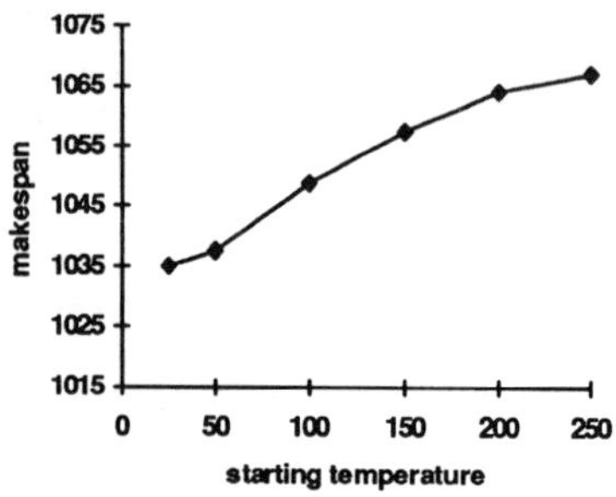

Figure 1b: Simulated Annealing applied to 10×10 problem

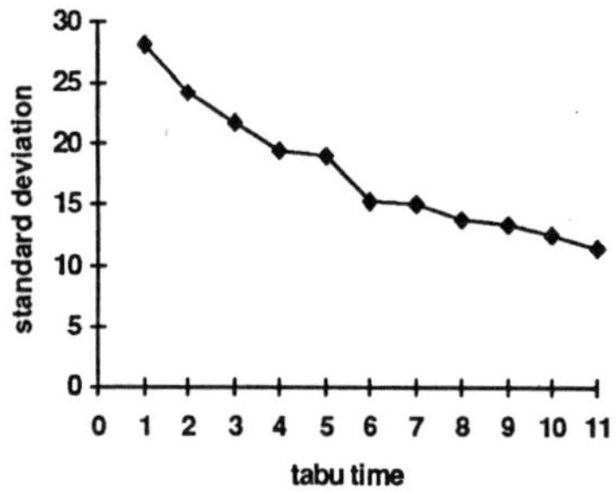

Figure 1c: Standard deviation of results from Fig. 1a

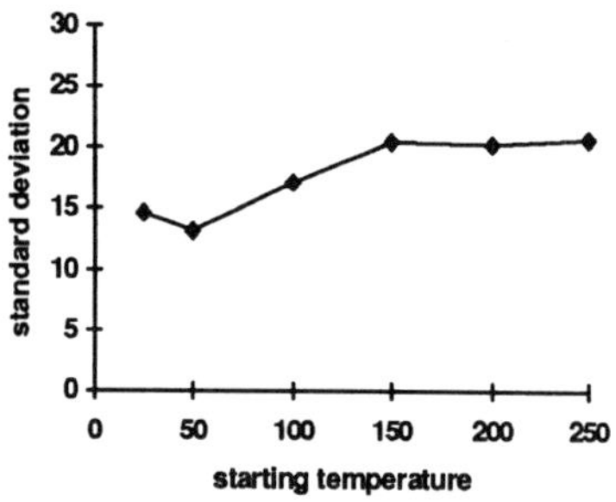

Figure 1d: Standard deviation of results from Fig. 1b

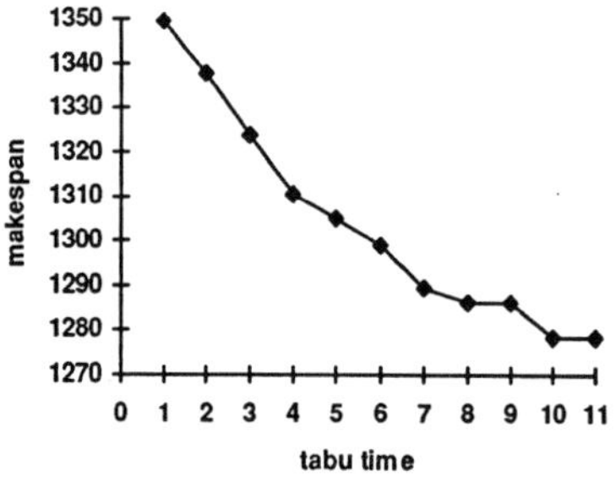

Figure 2a: Tabu Search applied to to 20×5 problem

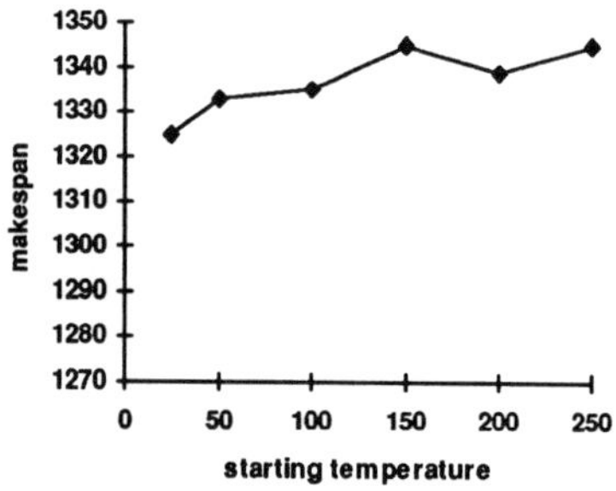

Figure 2b: Simulated Annealing applied to 20×5 problem

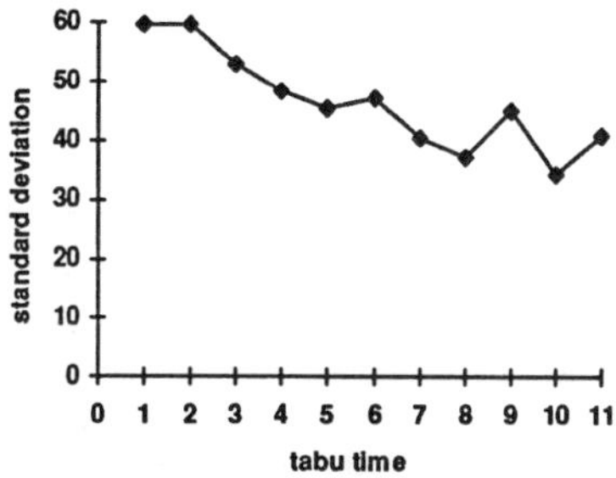

Figure 2c: Standard deviation of results from Fig. 2a

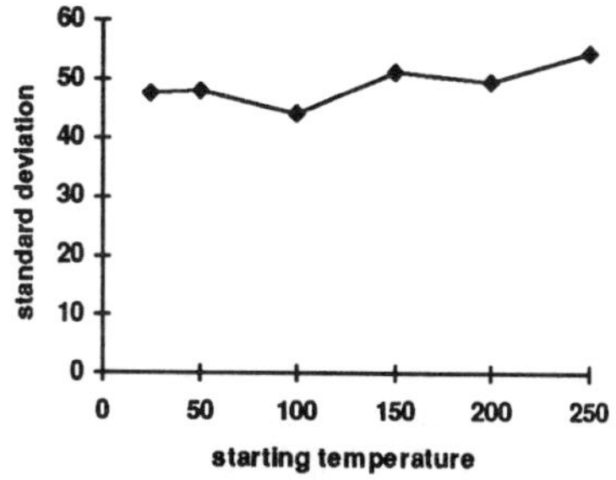

Figure 2d: Standard deviation of results from Fig. 2b

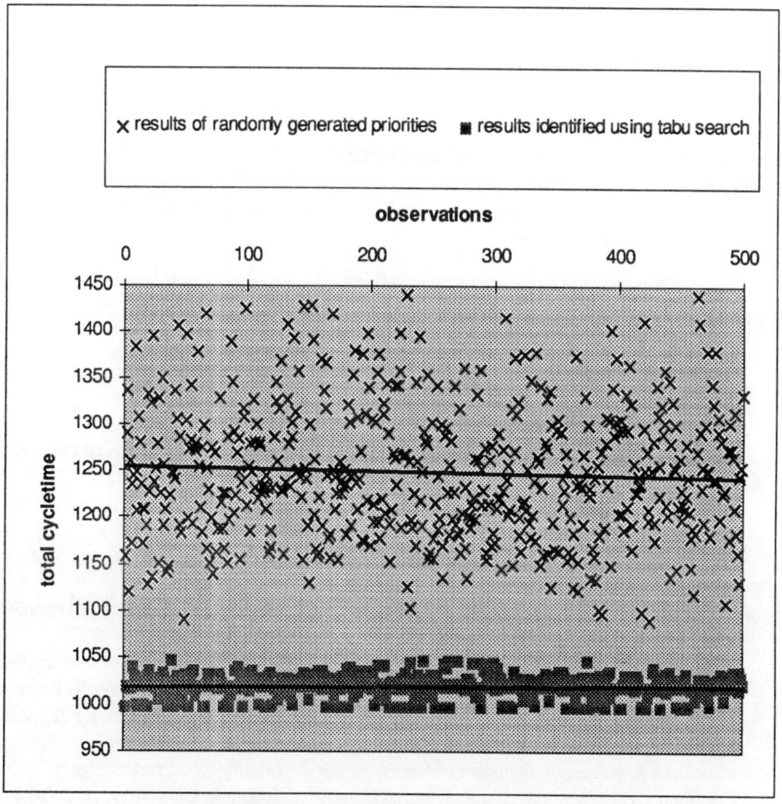

Figure 3: Optimised priorities by means of tabu search compared with the range of solutions covered by randomly chosen priorities

6 PERFORMANCE OF THE APPROACH

The approach used to deal with the job shop problem is to find priorities best suited to the jobs flowing through the system. This allocation of priorities remains unchanged during the whole process. Compared to the results that are achieved if the job priorities are allocated by random selection, which means the combinatorial problem is ignored, there is a potential for considerable improvement. Figure 3 shows the areas covered by the optimised solutions produced by means of tabu search when compared to randomly generated job priorities.The mean value of the total flow time for the 10×10 problem calculated from 10 000 observation is approximately 1250 time units. The effect of the well known SPT-rule is even worse according to Dell'Amico [1] with a result of 1399 time units. The tabu search approach using steepest descent and an $O(n2)$ - neighbourhood reaches an average result in the region of 1020 time units.

7 CONCLUSIONS

Both the tabu search and the simulated annealing search strategies allow very good solutions to be found within a relatively short run length [6,7]. Compared to the results attained by using priority rules [1], there is considerable potential to significantly improve the results [8]. In the experiments the results relating to the overall flowtime or makespan are improved between 10% in the worst case and approximately 25% in the best case. This improvement is achieved without using sophisticated tactical improvements of the search strategies. It is concluded that search strategies like those analysed here offer a worthwhile opportunity to support the experimentation phase in discrete simulation, which typically occupies a considerable amount of time, but is often neglected.

REFERENCES

1 Dell'Amico, M. Trubian, M. *Applying tabu search to the job-shop scheduling problem.* Annals of Operations Research, 41, pp. 231-252, 1993.

2 Muth, J.F. *Industrial scheduling.* Englewood Cliffs, NJ: Prentice-Hall 1963.

3 Glover, F. Taillard, E. de Werra, D. *A user's guide to tabu search.* Annals of Operations Research, 41, pp 3-28, 1993.

4 Fleischer, M. *Simulated annealing: Past, present, and future.* Proceedings of the 1995 Winter Simulation Conference, pp. 155-161, New York, IEEE, 1995

5 Pegden, C.D. Shannon, R.E. Sadowski, R.D. *Introduction to Simulation using SIMAN.* Mc Graw-Hill, New York, 1995.

6 Anderson, E.J. *The Management of Manufacturing - Models and Analysis.* Addison Wesley, Wokingham, Reading, p. 249, 1994.

7 Schulz, M. Boyle, A. Kaldos, A. *Optimisation of parts scheduling in flexible manufacturing systems using tabu search method.* Advances in Manufacturing Technology X. Proceedings of 12th National Conference on Manufacturing Research, University of Bath, pp 221-225, 1996.

8 Schulz, M. Boyle, A. Kaldos, A. *Reduction of experimentation in discrete simulation for resource allocation problems.* Advances in Manufacturing Technology XI. Proceedings of 13th National Conference on Manufacturing Research, Glasgow Caledonian University, pp 253-257, 1997.

The use of graphical modelling methods in the UK manufacturing industry

C A SMALL, R W BAINES, and **G J COLQUHOUN**
School of Engineering, University of Derby, UK

Synopsis
This paper reports the findings from a national survey into the use and status of static graphical modelling methods used in the UK manufacturing industry. The models considered are structured graphical models used to describe functions, activities, information, objects and behaviour in a manufacturing context for a wide variety of applications. Additionally, the paper reports on the criteria set by users to select a particular modelling method and identifies the most recognised and frequently used methods for a wide variety of applications. The paper concludes by discussing the need for a methodology to guide the effective use of modelling during change projects.

1 INTRODUCTION

A survey of $IDEF_0$ practitioners undertaken by Baines and Colquhoun in 1994 (1) revealed only 9 examples of manufacturing applications for the 23 returns (20% of the 125 surveyed) and a wide variation in the application for which modelling was used. In addition, limited literature publications were found to detail the use of modelling methods in the UK manufacturing industry with little research into their specific use as part of change projects. To explore this limitation further, a survey of UK manufacturing companies was conducted to examine the current state and level of implementation of modelling methods through a snap-shot study of their effectiveness, benefits, limitations, resource requirements, barriers to use, types of methods used, measures of success for application, application time scales, type of systems modelled, characteristics of industrial users and training needs.

2 QUESTIONNAIRE DESIGN AND IMPLEMENTATION

A target population comprising manufacturing companies from the (i) Times top 1,000 UK companies, (ii) companies known to the researchers, and (iii) Derbyshire databases was used under the précis that the most prominent/successful companies would have the potential to provide the greatest opportunity to study best practice (2) (3). A total of 1400 manufacturing companies were selected to provide an overall snap-shot of the use and knowledge of modelling methods.

The survey was conducted as a postal questionnaire, based on techniques used to gather wide scale information (4). Postage paid reply envelopes and 'tick' boxes were utilised to minimise completion time, optimise replies and facilitate analysis. The questionnaire comprised two sections, section A contained 14 questions addressing company structure and knowledge of graphical modelling methods, while section B contained a further 12 questions concerning practical modelling use. The modelling methods specifically addressed consist: GRAI, CORE, SADT, IDEF$_{(0,1x,2,3)}$, SSADM, DFD, Checkland Soft Systems, Petri Nets, Simulation Tools, and Object Oriented methods.

3 ANALYSIS OF RESULTS

The questionnaire had a response rate of 6.4% (89 respondents) and indicated that 67% of respondents were not formally using modelling methods for reasons ranging from associated cost and time to a limited awareness of modelling methods and their practical uses. Responses from 22 modelling companies indicated the most used methods were DFD, SSADM, SADT, IDEF$_0$, and Object Oriented methods for various applications including analysis of existing systems, design of new system, re-engineering projects, manufacturing systems, and business systems as described in a previous paper (1). The results of the survey are summarised below:

The principal activity of the organisation provided by 69 companies (table 1) showed a spread of responses throughout different manufacturing sectors with over 50% of the responses from metal goods manufacture, engineering, food and drink, textiles/clothing and footwear, plastics, and stationery and office equipment. Relating these responses to those from 22 modelling companies (displayed in the right hand column of the table) showed modelling was not limited to particular manufacturing sectors.

Types of changes carried out at 65 companies (figure 1) indicated that incremental and step changes have both been primarily conducted at a company wide level with slightly more incremental changes undertaken compared with step changes.

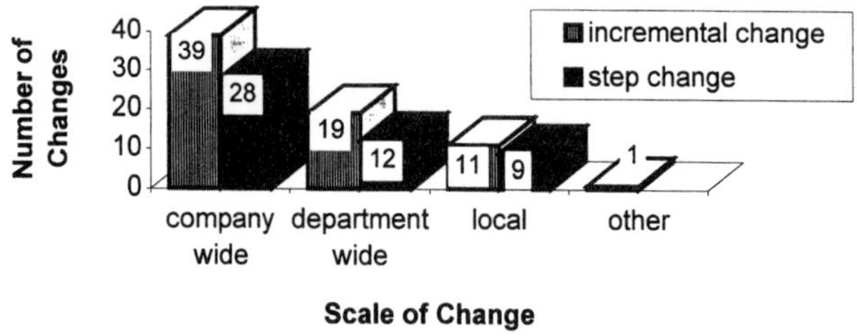

Figure 1: Nature and scale of changes carried out within companies (65 responses)

Table 1: Principal activity of manufacturing organisations (69 responses)

Principal Organisation Activity	No. of Responses	Percentage of Responses	No. of Modelling Method Users
Metal Goods Manufacture, Engineering	14	20.3%	5
Food & Drink Manufacture	8	11.6%	2
Textiles, Clothing & Footwear	6	8.7%	1
Plastics Manufacture	6	8.7%	-
Stationery & Office Equipment	5	7.2%	1
Automotive Manufacture	4	5.8%	3
Construction	3	4.3%	-
Healthcare Products	3	4.3%	1
Defence	3	4.3%	3
Energy & Water Supply	3	4.3%	-
Electronics Manufacture	2	2.9%	1
Mining/Chemicals/Metals/Minerals	1	1.5%	1
Computer Manufacture	1	1.5%	1
Government Department	1	1.5%	-
Other Manufacture	9	13.0%	3
Total	**69**	**100%**	**22**

Only, 26 out of 68 companies measured the effects of change primarily against manufacturing targets including productivity, stock reduction, and product quality and against business performance, improved customer service and financial parameters. No measures were made for the effectiveness of change processes or analysis and design methods used to progress projects. Forty companies (60% of 67 responses) indicated that operational management were primarily responsible for planning changes implying a need for effective communication between project teams and those staff with overall responsibility and approval of projects, which was additionally true for 21 model user companies (shown in the right hand column of table 2). Indeed, 4 companies additionally implied variable levels of influence on project teams from affected departments and individuals depending on the nature and scale of projects undertaken. This communication requirement was also stressed as an important factor in management and company participation during the change process as found in figure 2. Amongst these factors, new training posed the greatest concern during the change process to ensure effective work of new systems once implemented. Some significance was placed on standardisation of the changeover method in a company wide capacity suggesting a demand for formal standard project procedures within companies.

Table 2: Responsibility for planning changes (67 responses)

Planning Change Responsibility	No. of Respondents	Percentage of Respondents	No. of modelling method users
Primarily Project Team & Operational Management	9	13.3%	6
Primarily Operational Management	40	60.0%	10
Primarily Project Team	16	23.8%	5
Primarily CEO/Company Director	2	2.9%	-
Total	**67**	**100%**	**21**

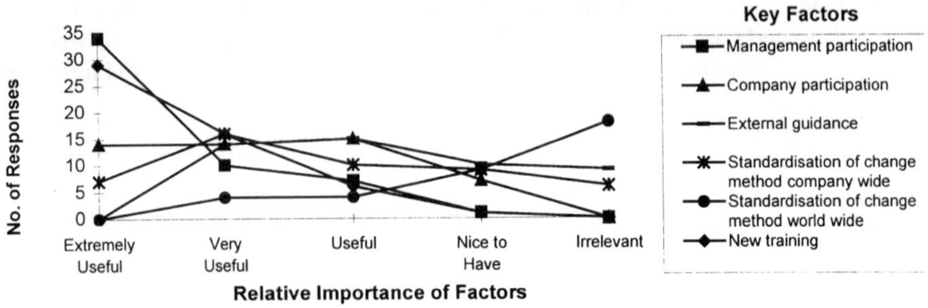

Figure 2: Factors considered beneficial to the change process (68 responses)

High percentages of the 89 companies had encountered Data Flow Diagrams (64%) and to a lesser extent SSADM (49%), Object Oriented methods (20%), and $IDEF_0$ (17%) (table 3). These findings are analogous with 22 modelling companies who indicated their use for various applications as shown in the right hand column of the table. Primary factors deterring the use of modelling by 50 respondents were in time taken to gain modelling skills, unawareness of methods available, and costs associated with modelling use (figure 3). Over 30% of the 50 responses cited more than one factor as a deterrent to modelling use with further factors including lack of senior management support and company culture.

Table 3: Modelling methods encountered (89 responses)

Modelling Method	No. of Encountering Companies	Encountering Percentage for 89 Companies	No. of Model Using Companies
GRAI	3	3.4%	1
CORE	1	1.1%	1
$IDEF_0$/SADT	15	16.9%	4
$IDEF_{1X}$	3	3.4%	0
$IDEF_2$	5	5.6%	1
$IDEF_3$	4	4.5%	2
SSADM	44	49.4%	10
DFD	57	64.0%	16
Checkland Soft Systems	7	7.9%	2
Petri Nets	8	9.0%	1
Simulation Tools	21	23.6%	9
Object Oriented Methods	18	20.2%	4

The triggers for modelling use cited by 19 companies have chiefly been through individual staff initiatives or parent company standards with some lesser triggers from academic/University recommendation (figure 4). The time taken to learn and develop necessary modelling skills and to create system analysis and design models varied depending on the modelling method used. A newly introduced modelling method meant that the process of creating models was more time consuming than in previously experienced cases because of the time spent in creating accurate models. All respondents agreed that previously known and used modelling methods allowed a quick model creation phase with the greater time spent in explanation to none modellers for modelling acceptance and to reduce misinterpretation.

Responses from 21 companies implied that ease of learning, user friendliness, and associated costs with modelling are the prime factors effecting the selection of a modelling method to use (figure 5). Eleven out of 20 (55%) companies cited lack of modelling expertise as a main barrier to modelling use, with a noticeable percentage (30%, 30%, 25%) citing company culture, management support, and staff resistance respectively. This agrees well with findings that relatively few staff are aware of modelling use and its benefits as an analysis/design and communication tool.

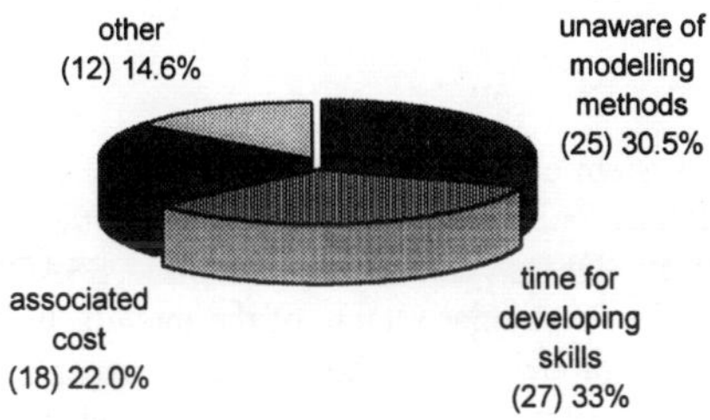

Figure 3: Primary factors that deter the use of modelling methods (50 responses)

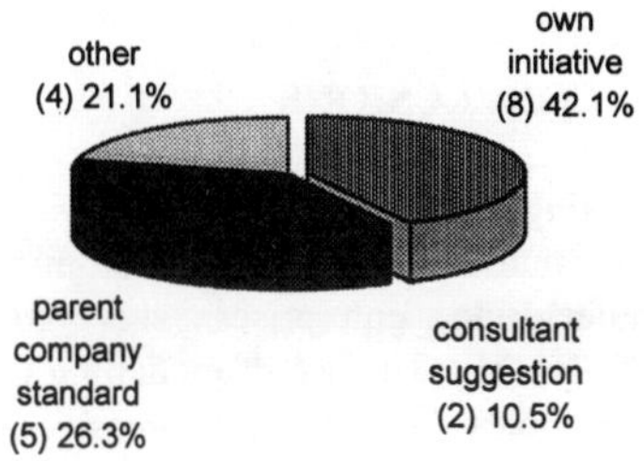

Figure 4: Initiation for using modelling methods (19 responses)

Other barriers specified included establishing modelling experts at an affordable cost and difficulties in interpreting created models by none modelling staff. Additionally, limitations experienced in using modelling during change projects included communicating ability of completed models to none modelling staff which also featured highly as a factor in the selection and use of a modelling method. Several companies cited difficulties in finding software to create models, project timing constraints on model creation times, and some management attitudes to associated costs in timing to learn and develop modelling skills required.

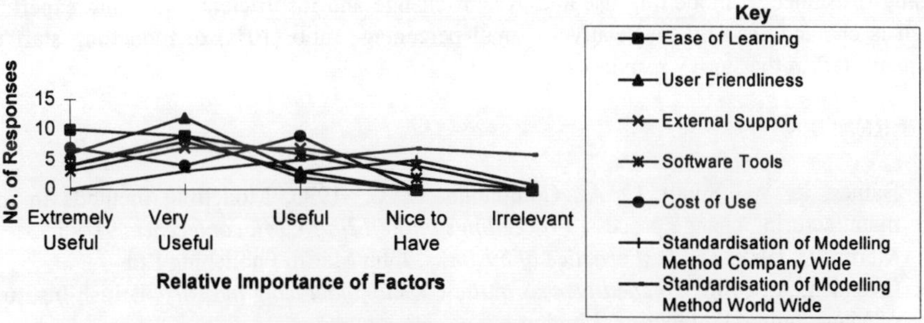

Figure 5: Factors influencing the use of modelling methods (21 responses)

Advise given by modelling respondents could be divided into the following points:

- understand the reasons for creating models and identify subjects and viewpoints
- know and understand the current system under review
- gain advice and support from other modellers and learn from their mistakes
- investigate modelling methods available and select a simplistic method to produce models that are easy to build and modify with an ability to get system users involved in the building process
- maintain completed models in line with current systems
- gain management support for the project and the use of modelling
- back up training for modelling with practical experience as quickly as possible

3.4 CONCLUSIONS

Modelling methods are used for a wide range of system change applications for analysis, and design purposes. The most prominent modelling methods recognised within the UK manufacturing enterprises surveyed comprise: DFD, Simulation Tools, SSADM, $IDEF_0$/SADT. The use of modelling is not limited to particular sectors of the manufacturing industry or to specific job areas. Since no one modelling method appears capable of effectively describing all manufacturing systems, the criteria for selection of a method should be based on user requirements. No formal measurements have been made by respondents to determine effects of graphical modelling methods used to complete system analysis and design.

Planning of projects and system change is the responsibility of both the project team and operational management suggesting a need for strong communication to effectively progress projects, where modelling may be implemented as a communication tool to describe systems efficiently between project team members, management and system experts and end users with little modelling experience.

Efficient use of modelling provides a clear, common graphical description of systems to enable interchange of ideas during both step and continuous system change projects. Model users demand simple methods requiring short skill development times, at a minimal cost to the company and which can produce simplistic easily amended models. Barriers experienced using modelling methods have included lack of management support, company culture causing resistance to modelling use and system change and insufficient modelling expertise which is characterised by the relatively small percentage ratio (10%) of modelling staff to company staff in the survey responses.

REFERENCES

(1) Baines, R. W., Small, C. A., Colquhoun, G. C., 1996, Modelling methods in the manufacturing change process, *Proceedings of the 12th NCMR conference*, 91-95

(2) Martin, J., 1978, *The best practice of business*, John Martin Publishing Ltd

(3) Ovenden, A., 1986, *Competitiveness in the UK manufacturing industry*, British Institute of Management Foundation, London

(4) Denzin, N., and Lincoln, Y., 1994, *Handbook of qualitative research*, Sage Publications, London

Modelling engine performance in the automotive industry

M H WU and **Z B LIAO**
University of Derby, UK
S Y DUAN
Rover Group, Birmingham, UK

ABSTRACT

It is becoming increasingly important for the Engine Control Unit (ECU) to control the multiple input variables such as ignition timing, Exhaust and Inlet valve timing, in order to achieve efficient combustion, lowest Break Specific Fuel Consumption (BSFC) and highest output toque, etc. at a variety of engine operating conditions.
It is important to set the optimum input variables using the ECU to meet the following requirement:

--Maximum toque output, minimum fuel consuming,
--Low temperature, Low pollution,
--ease of starting for a cold or hot engine,

To achieve the above requirements, experiment data that are from an Engine Test Bed under different constrains and conditions have been analysed. A multi-variable high power function has been created to describe the relationship and features between the variables and the experimental data.
The paper, first gives the background of the research project and introduces existing Engine Test Bed (ETB) procedure. The paper then discusses the modelling of experiment data. Weaknesses of existing optimisation models are discussed and the features of a new model are presented. Finally, the results are shown that the new model provides a better fit to existing test data.

1. INTRODUCTION

In view of Euro III and IV legislation, and the increasing customer demands for higher performance and lower cost engines, many new internal car engines currently under development are adopting a number of innovative technologies including Variable Valve Timing (VVT), Exhaust Gas Re-circulation (EGR) and Gasoline Direct Injection (GDI).

The development of such new generations of internal car engines is marked by increasingly stringent requirements including higher performance, fuel consumption, lower exhaust emission regulations, avoiding misfire and guaranteed stability. In order to meet these requirements engines must provide a stable combustion process at any operating conditions, thus requiring engines to be equipped with more and more controlling elements requiring a higher level of sophistication of electronic Engine Management System (EMS), and complexity of engine control.

In order to provide the optimum control parameters to EMS. The research works on the new engine test and modelling is necessary. This paper will discusses both engine test and modelling.

2. ENGINE TEST ON THE ETB's

2.1 Structure of ETB's

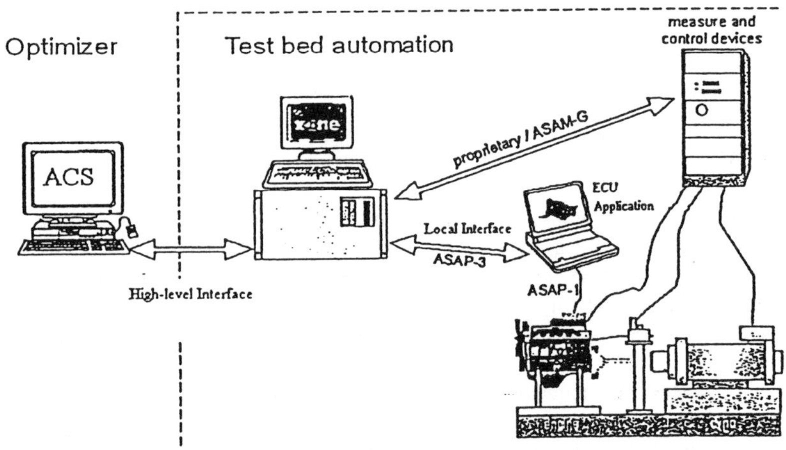

Fig.1 The structure of the Engine Test Bed

Fig.1 the structure of an Engine Test Bed which is used to test a newly designed injection engine.

The Test bed automation (TBA) is a system controller. Its functions are :

- control the whole system,
- send the adjustment parameters to measure and control devices for engine test,

- send operation parameters to desk top,
- monitor the ETB performance,
- save the test data into individual datafiles,
- communicate with distance PC computer for data exchange,
- etc.

The desk top is used to replace the Engine Control Unit (ECU) for engine operation. The function of it is the same as the ECU in any electrical inject car. For each test, the desk top will receive a set of data from the test bed, e.g. ignition timing length, inlet/exhaust valves opening degree, fuel inject time and length, etc. All of the parameters are employed to operate the normal engine on an ETB. The desk top still obtains the normal Engine operation information through the transducer fixed on the engineering, e.g. temperature, speed, etc. and sends this information back to the TBA.

The measure and control devices (MCD) are firstly used to physically adjust the engine parameters, e.g. valve's opening angles, engine loads, fuel injection amount, etc. according to the command and information from TBA. Adjustment status are measured and feedback to MCD to form the close loop element adjustment control. Secondly, the MCD still measures the engine operation status and then sends the information to the TBA for example Break Specific Fuel Consumption (BSFC) and output Toque information, etc.

2.2 Operation principle

In section 2.1, the system was introduced and each element's function in the system is explained. The operation principles are as follow:

The present engine test is a four-variable test system such that . It means that four input variables and two output variables are tested for engine operation. The input variables include ignition time, inlet valve opening degree, exhaust valve opening degree and engine speed.

Two output variables are Break Specific Fuel Consumption (BSFC) and output Toque. The constraints for the test is exhaust temperature and emission. No matter how the variables are changed, it has to be met that the readings of the exhaust temperature and the emission value have to be less than the Maximum limit values.

The remote computer will send a group of initial set up data to TBA. The data will be analysed and then send to desk top and MCD as they required. MCD will adjust the engine operation parameter physically to set up the machine. When the above actions are completed, engine will be started and waiting for about 5 minters, then the engine performance data will be measured by desk top and MCD.

The measured data is finally passed to a remote computer ACS for modelling. This model will be used to generate a few sets of optimum engine operation parameters. Those chosen optimum operation parameters are sent back to the ETB for further testing. After repeat test cycles, the optimum operation parameters will be chosen and finally install into the ECU for the appropriate type of engine.

3. ENGINEERING PERFORMANCE MODELLING

3.1 Current model used for engine parameter optimisation

The current mathematical model for engine optimisation are adopted by quadratical model denoted by [1] [2] [3],

$$y = x^T Q x + a^T + b \qquad \text{Eq.1}$$

Where:

y is an objective vector,
x is an input variable vector,
Q is co-efficient matrix,
a & b are constant vectors,

The coefficient of matrix Q and the vectors a and b are calculated by least-squares-method to minimise the deviation of measured values from the model values.

This model has the advantage of simplicity but is imperfect for the real world. Figure 2 shows examples which indicate the model's imperfect nature.

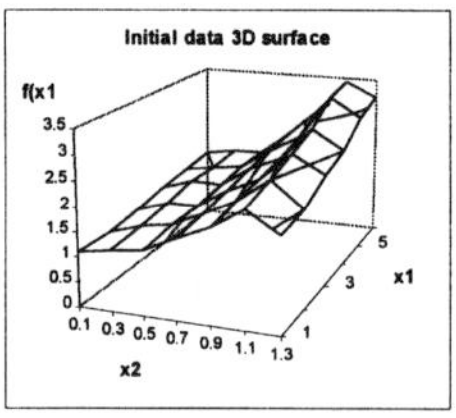

(a)

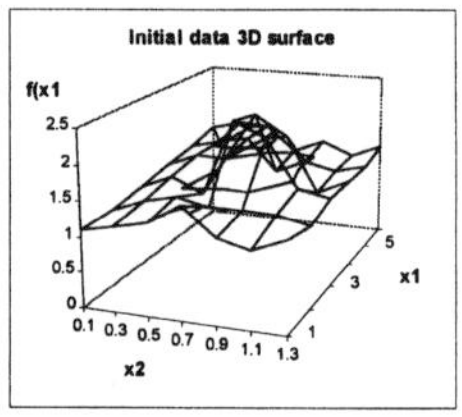

(b)

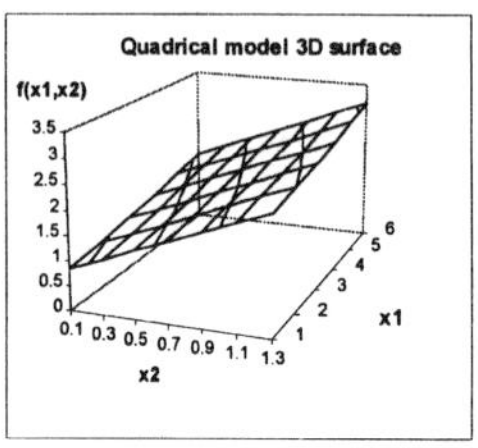

(c)

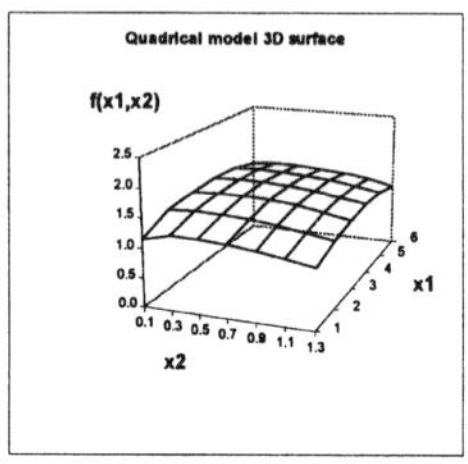

(d)

Fig. 2 3D views from a engine test simulation compared with calculated data by Eq.1

Fig.2 has been drawn according to the Engine Test Simulation Data. Figures 2 (a) and (b) are drawn by the simulation data. Figure 2 (c) and (d) are drawn by the data calculated by Eq.1.

It is obvious that the fit between two pair drawings (a)/(c) and (b)/(d) is very poor.

The relative errors are about ±30%. In (a) the maximum point is at (6, 1.1) but in (c) the maximum point is at (6,1.3). This is because the long convex peak has been lost in the model. In (b) there is a suddenly raised convex peak at (3, 0.7) but at (d) it become a smooth raised convex area. The signification feature has been lost.

The current model Eq.1 is not a suitable model for the engine performance parameters model, it gives the large errors and loss of data features. Another disadvantage of using quadratical equations is that they only can express one convex (or concave) data set.

3.2 A New model for engine parameter optimisation

After carefully analysing the engine test data, the new engine performance model has been created with three features: Constant, Liner characteristics and convex or concave peaks. The new maths model is:

$$f(x_1, x_2, x_3) = a + \sum_{i=1}^{3} b_i x_i + \sum_{j=1}^{p} h_j E(x_1)E(x_2)E(x_3)\,; \qquad \text{Eq.2}$$

where .

$$E(x_i) = e^{-8\left(x_i - m_{i,j}\right)^2 / w_{i,j}^2}$$

$f(x_1, x_2, x_3)$: break specific fuel consumption;

x_1 : ignition time variable;

x_2 : inlet valve opening degree variable;

x_3 : exhaust valve opening degree variable;

p : the number of peaks in 4D space; it assumes $p = 10$;

a : a constant;

b_i : a slope coefficient respect to input x_i ;

h_j : the peak height respect to jth peak in the concerned dimension;

$m_{i,j}$: the peak position respect to jth peak in the concerned dimension;

$w_{i,j}$: the peak width respect to jth peak in the concerned dimensional space but its value may go over its input measured range.

Fig. 3 shows the application of new maths model. In figure, the drawings (a) and (b) have the same meaning as it in Fig.2. The drawings (c) and (d) are constructed according to the data calculated by new maths model Eq. 2.

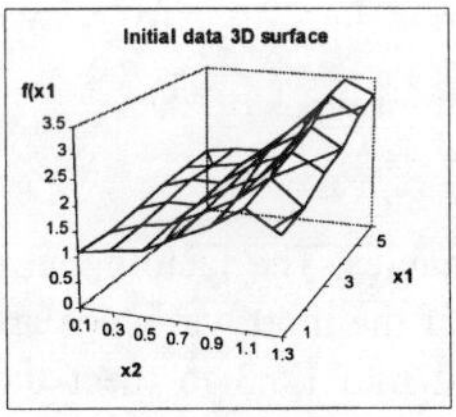

(a)

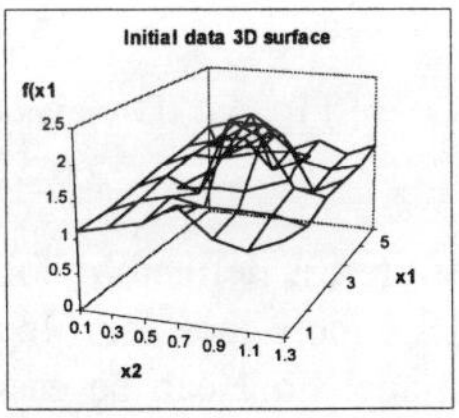

(b)

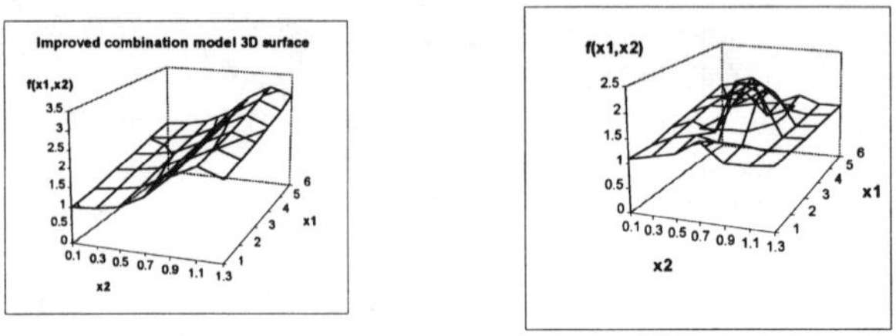

(c) (d)

Fig. 3 3D views of engine test simulation data and calculated data by Eq.2

The Figure 3 is clearly shown that the new maths model can represent the tested data's features with a relatively small error. It is a significant improvement on the current quadratic model.

4. CONCLUSION

The new maths model was successfully applied for real engine test data analysis and modelling. Fig.4 shows the results. The new maths model provides a close fit to the ordinary tested data.

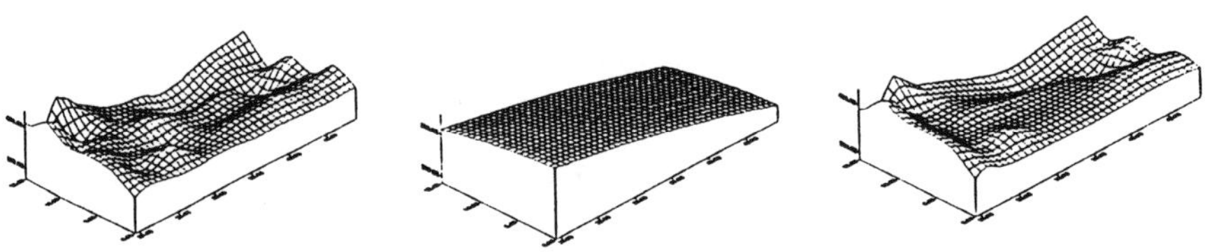

(a) Original data (b) Quadrical model (c) New maths model

Fig. 4 3D views of the different types of data

The current test is defined by four input variables. The factors that influences the car engine operation are more then four. In the future, if the input variables are more then four, the new maths model Eq.2 can be easily modified into Eq.3 to meet the requirement. In theory, there is no number limitation of n of the input variables.

$$f(x_1, x_2, \ldots x_n) = a + \sum_{i=1}^{n} b_i x_i + \sum_{j=1}^{p} h_j \prod_{i=1}^{n} e^{-8\left(x_i - m_{i,j}\right)^2 / w_{i,j}^2} \qquad \text{Eq.3}$$

REFERENCE

1. K.U.Voigt, (1993) "Model based on line optimisation for modern engine management system" *SAE Brazil.*

2. H.Lechner, (1996) "Automated system for optimised calibration of engine management systems", *Rover Group internal publication circa.*

3. H.Hochschwarzer, (1992) "Fully automatic optimisation of engine calibration" *IMechE proceedings pp 89-100.*

4. W.Mendenhall, (1995) "Statistics for Engineering and the Sciences", Prentice, Inc.,

5. K.Arczewski & J.Pietrucha (1993) "Mathematical Modelling of mechanical Complex Systems"; *Ellis Horwood*

6. P.E.Gill, (1991) "Numerical Linear Algebra and Optimisation", *Addison Wesley*

Manufacturing Strategy/Management

Organization culture for world class manufacturing

S P BARRY and **A W DUFFILL**
School of Manufacturing and Mechanical Engineering, University of Birmingham, Edgbaston, UK

SYNOPSIS: In this paper the authors outline the objectives of research and consultancy into world class manufacturing (WCM). They introduce a basic model for an organisation wishing to become world class; a model which was later enhanced and developed. In working with manufacturing organisations it soon became apparent that three issues had to be tackled very early on. The first was the issue of changing the organisational structure and developing a more open culture. The second was a top-to-bottom revision of employee's value and reward. And the third issue was the associated re-thinking of information flows and information understanding. The paper concludes by showing the need to satisfy the 3 issues and over-viewing the rewards the researched companies had achieved.

1 INTRODUCTION

This paper is based on some preliminary research and consultancy with manufacturing organisations. The work commenced as companies requested assistance/consultancy to implement a world class manufacturing (WCM) methodology. It soon became apparent that an exact definition of what WCM means was going to be difficult to find. World class often meant different issues to different organisations. In general 'becoming world class' is an initiative embarked upon by organisations that find they need to improve.

To a growing number of organisations world class means being the 'best' in the 'field' (i.e. market sector) and in the 'world' (1). In this context 'best' can be regarded in terms many issues (e.g. product design and performance, innovative designs introduced more quickly than competitors, shorter lead times and reliable deliveries, low manufacturing costs, excellent customer service, quality and reliability and so on). The list is not intended to be exhaustive neither is it suggested that all companies would need to investigate and improve in all of these issues. But, as improvements take shape, they have to beware of complacency. The pace of global change has become so great that, without constant efforts to keep on top, an organisation can find itself losing out to more dynamic competitors. With these issues in mind the authors searched for a comprehensive model which would assist organisations who wished to become world class. The model had to be capable of driving the organisation through the various world class methods and tools and, at the same time, be reasonably un-restrictive.

2 AIMS OF THE RESEARCH

The main aim of this research was to develop a model for a manufacturing company intent on becoming world class. The secondary aims, the main topics of this paper, were to overview the areas in which a company needed to develop its organisation, people resources and information. Finally the research needed to provide data on benchmarks and show how benchmarks had to improve if the company was to stay world class.

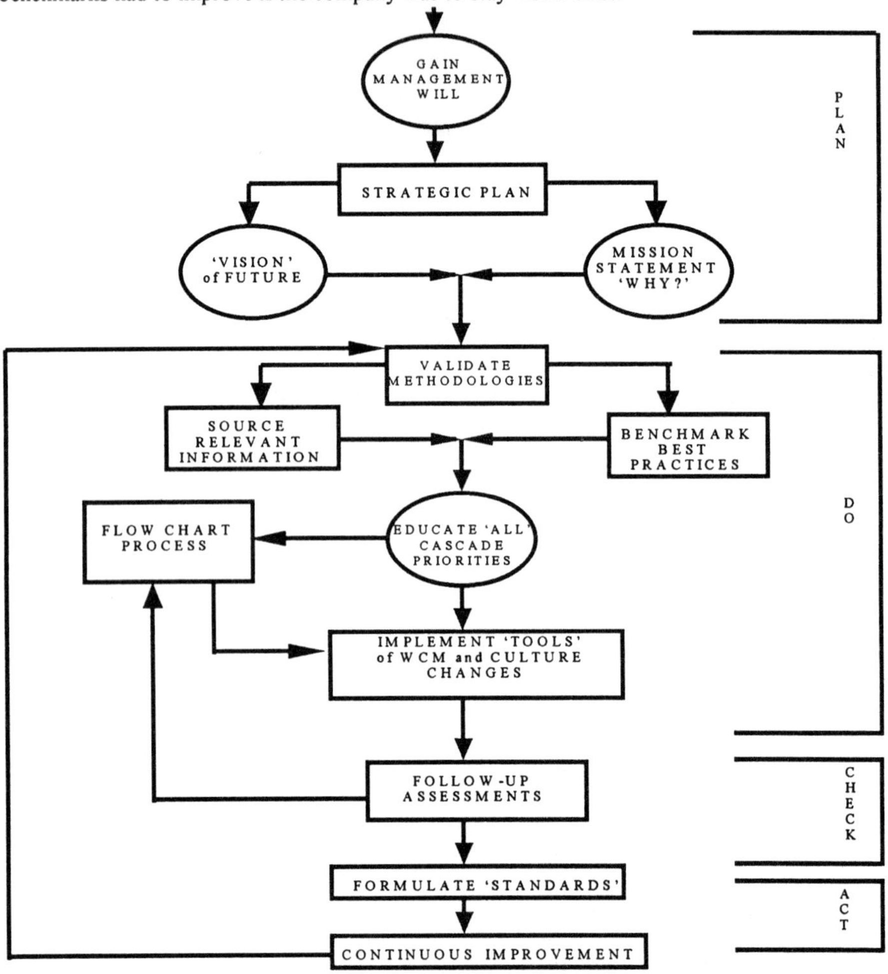

Fig.1 A Basic World Class Model

Fig.1 shows a basic model used to embrace the research aims for a WCM initiative. It indicates the key components of a WCM campaign which are broken down into:

i. Gain management will for without this no WCM principles will be possible. It will probably mean changes in organisational structure.
ii. Establish a strategic plan to provides purpose and direction. Management define what their vision of the future is and communicate it to all employees.
iii. Management and employee teams define the organisation's mission; management then need to reinforce the mission at all times.
iv. WCM methodologies need to be researched and validated. The teams source and benchmark successful practices and identify the relevant programmes.
v. All employee teams should be educated in the methodology, tools and language of WCM. The education process should be cascaded throughout the organisation.
vi. Relevant WCM enabling tools should be implemented along with the necessary resources and training.
vii. Follow up and assess progress, it is important not to let the momentum fade.
viii. Formulate and document standards, activities and responsibilities.
ix. Continuous improvement which means returning to start the cycle again. There is always room for improvement as it is impossible to achieve perfection.

As this basic model developed it soon became obvious that many manufacturing organisations were not structured for WCM. The 'plan' stage (Fig.1) often produced 3 key issues.

3 STRUCTURE AND CULTURE

Traditionally the role of organisational management is on clearly delineating departmental responsibilities and narrowly defined employee job categories. The organisation is divided into semi-autonomous departments linked together in a hierarchy composed of multi-levels of descending authority. The goal is to control and enforce behaviour and performance to fit departmental and individual pre-defined standards. According to Schonenberger (2) each department is treated by its manager as their domain whose responsibilities end at the departmental boundary. The resulting culture tends to promote parochialism, distrust, secrecy, and finger pointing.

In order to successfully implement WCM, companies must be prepared to change significantly their organisational structures and operating cultures. Manufacturers must be prepared to effect 4 broad organisational changes (3):

i. There will be a requirement for new organisation designs, new channels of information transfer and different distributions of authority. This usually means a flattening of the traditional hierarchy, simplifying the organisation, establishing integrated functions and participating as a team and supporting the strategic vision of the company.
ii. Employee roles and knowledge not found in the old organisation will be required. This usually means broader and more flexible job categories and on-going educational opportunities aimed at re-skilling employees at all levels.

iii. It requires a new operating culture concerned with teamwork, the elimination of waste, value added manufacturing, commitment to total quality and an endless search for an environment characterised by open communication and information sharing.
iv. Finally, it is essential that there is a radical change in the measurement of performance. There must be a management style where total productivity and total cost form the standards of success.

In the ensuing case studies the organisations which, broadly, adopted these type of changes produced the most effective world class type improvements.

4 PEOPLE RESOURCES

Until recently, manufacturing management did not see the human organisation as a source of competitive advantage. The culture of human resources in such a rigid hierarchical environment was oriented around gaining employee control and compliance. The purpose of management was to monitor and make routine behaviour, departmentalise and inhibit communications between functions, restrict the knowledge of individuals by narrow job training, and decrease the organisation's dependence on the individual by engineering away skills.

For example the application of WCM technology, without a radical change to the traditional model of human resources, has often proved to be disappointing. There are several comparative studies, in automotive and shipping industries, that show that the successful implementation of advanced technologies requires significant changes to the human organisation. Jaikumar (4), found American FMS systems to be characterised by 'an astonishing lack of flexibility compared with similar systems in Japan'. He attributed the difference to the fact that 75% more Japanese than American workers received technical training and that Japanese companies invested two-thirds more effort in upgrading worker skills.

The consultancy work, at a number of companies, showed that if WCM is to be successfully implemented, management needed to enhance and tap into the skills and experience of the whole work force. Simply to fit the old structure and people management methods into the new clothes of WCM will result in failure. What will be required is a top-to-bottom revision of employee value and reward systems. Manufacturers seeking to implement WCM needed to address the following points:

i. Implementation of participative management methods that promotes employee self-learning and commitment.
ii. Strong commitment to continuous training and education aimed at raising skill levels.
iii. Development of initiatives aimed at worker centred quality control.
iv. Continuous removal of organisational barriers that inhibit effective communication, mutual trust and information sharing.
v. Training workers to perform multi-functional tasks.

This will provide for flexibility in human resources and will position the employee to be able to search for continuous improvement across the organisation, rather than in a narrow job category.

5 INFORMATION

Closely associated with structures, cultures and people resources is data/information. Organisations implement new, often computational, systems to collect, assimilate, and transmit information. However it is accurate and timely information that permits purposeful planning, control and decision making. A manufacturing company's accessibility to information is critical to world-class performance.

Despite its importance, the role of information has been poorly understood by most manufacturing companies. Historically, management sought to gain meaning from the flood of homogeneous data by dividing the enterprise into departments from which the data could be extracted and summarised. In such an organisation each department defined the data and how it was arranged into information. Trouble arose when this data was passed from one department to the next. Often data meaningful to one department had to be selected, extracted and translated into the vocabulary of another.

The implementation of WCM techniques requires a re-thinking not only of traditional information flows but how information is understood. Linking the shop floor and office systems require manufacturers to formulate new approaches to information control and co-ordination. The new role of information will enable the creation of networked functions which are referencing and co-ordinated by a common system. In addition only the relevant information would be used by the teams and individuals. The foundations of this integrated information environment are here today with electronic communications. Organisations searching for world-class performance must be prepared not only to define information but to provide for a company wide open dialogue. The understanding and utilisation of information is a fundamental resource in WCM strategies.

6 CONCLUSIONS

The paper introduces a basic model which represents the main aim of the research. This model was to be developed in future work (5)[1]. The paper then gave details of the secondary aims (i.e. organisation, people resources and information) which were high-lighted as the research progressed and showed management had to make some early decisions (Fig.2).

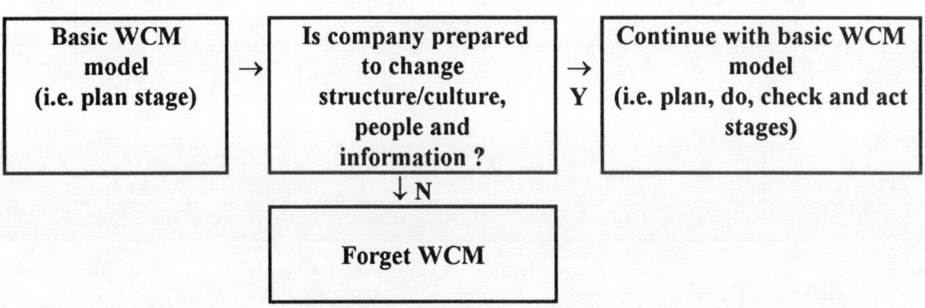

Fig.2 Simple Decision Flow Chart

[1] The enhanced model, using the software package Flow Charting PDQ, will be published later.

What was shown from this early work was that to achieve a world class organisation, it is not simply a matter of changing a few things (i.e. introducing enabling tools such as MRP II, total productive maintenance, CAD/CAM, statistical process control and so on); almost everything must be changed and changed dramatically. Ross (3) quoting Taylor stressed that the successful adoption of WCM required a complete mental revolution.

At any stage whilst working towards world class status any company may well be worried about the magnitude of the task facing it. Provided that it doesn't get carried away and try to do too many things at once. Provided it starts with the most appropriate improvement tasks, and using the right enabling tools, then it should soon get very encouraging results. The main thing to bear in mind is that small steps should be taken since these are less painful, easier to control and cost less.

In many companies, on the world class path, the rewards achieved within the first 6 months exceeded expectations. It provided a degree of improvement, in competitive edge, that would have taken several years to achieve in the days before they embarked on their world-class projects (1). These conclusions were supported by the results of the author's consultancy and research (5).

REFERENCES

1. Todd J, World-Class Manufacturing, McGraw Hill Book Company, 1994.
2. Schonenberger R J, World-Class Manufacturing, Free Press, NY, 1986.
3. Ross D F, Aligning the Organisation for World Class Manufacturing, Production and Inventory Management, pp 22-26, 1991.
4. Jackumar R, Post-Industrial Manufacturing, Harvard Business Review, pp 69-76, November-December 1986.
5. Barry S P and Duffill A W, A Model for a World Class Manufacturing Organisation, to be published.

Resource-based and network theories – a comparison

F E COLL and **U S BITITCI**
Department of Design, Manufacturing, and Engineering Management, University of Strathclyde, Glasgow, UK

SYNOPSIS

This Paper reviews the current literature that addresses the impact of interrelations between a company and the business network within which it is embedded. A conceptual model is then developed that focuses on the strategic development of horizontal and vertical linkages, the purpose of which is to provide a comprehensive approach to network strategy formulation. A case study in the textile industry is then outlined.

1 INTRODUCTION

Much has been published on the theory of industrial networks. Until the late seventies, the conduct of inter-firm business was regarded as a purely economic function in a market environment; the long-term 'relationship' effect was not formally recognised. Håkansson (1) provided a model of the 'interaction approach' to buying and selling between two firms. The tenets of the interaction model incorporated aspects from the theories of both the 'new institutionalists' (exemplarised by Williamson's transaction cost economics (2)) and inter-organisational theory. Currently, a variety of network concepts are receiving much research interest.

2 CLASSIFICATION OF INTERACTION THEORIES

Additional models have since appeared, dealing with similar interfirm issues. Harland (3) classified these hierarchically as different system levels:

Level 1: Relationships between internal processes in a firm- the 'internal chain'.
Level 2: Linkages between two parties- 'dyadic relationships'.
Level 3: The interconnections along an entire supply chain.
Level 4: An analysis of the relationships in an entire network.

The following table uses this method to outline some notable contributions to research in this area:

Table 1: Contributors to industrial interactions

Source	Concept	Level
Håkansson	The interaction model	2
Williamson	Transaction cost economics	2
Penrose	Resource based theory	2
Porter	Value chain & system	1, 3
Houlihan	Supply chain	1,3
Rijs	Networks	4
Burt	Structural hole theory	4

Although each level is a subsystem of the one above, the concepts themselves do not necessarily exhibit such recurring properties; e.g. Porter's model (4) is founded on traditional industrial organisation economics- 'structuralism', whereas that of Holmlund and Törnroos (5) is based on the tenets of network theory. Many of the above can be classified further, as New (6) did with a taxonomy of supply chains. The major differences are typified by contrasting the assumptions of the structuralist and network paradigms:

2.1 Network paradigm

1. The level of analysis resides at that of the actor.
2. The environment consists of a network of other organisations.
3. The actors' behaviour is determined by the relationships they have with other actors.
4. The properties of the network are dependent upon all the actors while constraining each individual.

2.2 Structuralist paradigm

1. The company unit is the system in focus.
2. The environment of the firm is undifferentiated.
3. Firms place themselves in their environment according to strategic considerations
4. Decisional behaviour is influenced by bounded rationality.

It is necessary to acknowledge the influence of the immediate customer and supplier as recognised in Porter's work, and also the existence of the internal value chain. When discussing the strategic ambit of a company, the linkages involved are nonetheless at the company level, and can lead to competitive advantage through "optimisation and co-ordination"- again referring to positioning rather than directly influencing strategy. The current impetus in research into business networks is directed towards enriching existing theory rather than refuting that of industrial economics, so there exists little empirical evidence to discriminate against either. The following two theories, resource based theory and industrial network theory, provide insights into the differing approaches.

2.3 Resource-based theory

Resource based theory (RBT) is founded in traditional industrial economics, and was devised to explain why similar firms in similar conditions performed differently. The theory views the firm as a bundle of resources that interact together to determine situational performance and help determine strategy. The resources may be intangible, and no distinction is made between those that have originated internally or those that have been acquired. In a review of major work in this area, Olavarrieta and Ellinger (7) categorise them as follows:

- Input factors are generic resources that can be acquired on the market- they include standard new equipment and raw skills. Once subsumed within the company, they become part of the firm's assets or capabilities.
- Assets are those visible resources owned or influenced by the firm which have been 'accumulated' over time, during which they have absorbed investment. Examples include dedicated capital equipment, patents, joint ownership of a proprietary logistics network, etc.
- Capabilities represent the potential combinations of skills, assets and knowledge that are indigenous to the company. They reside in and are inextricably linked to organisational processes. Finally, they are knowledge based in that they are dependent upon the interaction of personnel within the firm.

The interaction between these resources is emergent, and result in competitive advantage, which the firm uses to seek 'rent'- the return for the operation of the organisation. Monopolic rent derives from the firm restricting supply to maintain demand. Economic rent is the surplus return gained over the opportunity cost of the resources. Both forms of rent can be earned in varying combinations. A further characteristic of the resource based theory is that firms view rent as more important than profit, entailing that they seek to exploit emerging opportunities rather than solely relying on established ones. This elicits the notion of 'sustainable competitive advantage'.

2.4 Network theory:

Unlike the behaviourist assumptions of social network theory, there is no definitive description of an industrial network topology. The following three characteristics are, however, common to all:

- Activities represent the means to achieving the overall objective of the network and are delimited as such. They exhibit interdependence, and are functionally reliant upon actors. The most apparent example of a network of activities would be the transformation of raw materials into goods and services for end customers.
- Actors are responsible for controlling or performing activities and managing resources. Depending on the perspective from which the network is viewed, they may be individuals, groups or organisations.

- Resources are required to facilitate activities by actors. Examples include financial capital, knowledge bases and technology.

It is the cumulative interactions between these three network object types that both provide the network with its *raison d'être* and influence its characteristics. The network is in dynamic equilibrium; although stable it undergoes constant change. This occurs not only to the above characteristics, but to the interrelations between them. As the network evolves, it grows competent at achieving certain goals. This competence is generally resident in the combination of interrelations between the activities, actors and resources and in the properties of the characteristics themselves; therefore the possibility of transferral or replacement is reduced and this acts as a constraint. It is worth noting that implicit in the assumption that a network is an intermediate form of organisation between an open market and a vertically integrated industry is the equilibrium, again dynamic, between competition and collaboration. Much of the research into network relations focuses primarily on the latter.

3 SYNERGY?

If the two models are compared at the firm level some common elements are discovered, despite the conceptual differences outlined above:

3.1 Similarities

- Resources (network theory) and assets (RBT) are analogous, in that are both subject to the influence of the firm, and barriers exist to their removal. These are past-investment and the dependence of competencies in RBT, and the effects of interrelations with activities and actors in the other.
- Capabilities and actors / activities. Although capabilities are latent rather than actual, they are in many ways similar to the combination of actors and activities. Actors are responsible for controlling and enabling activities; and these developing interrelations enable the network to function efficiently. The concepts of tacit knowledge, social complexity, and causal ambiguity (the difficulty of analysing the links between resources) result in 'imperfect imitability' as described by Dierickx and Cool (8)- the barrier to imitation caused by the emergent properties of combined resources. This is similar to the explanation of interrelations as defined in a network.
- The dynamic nature of both models is emphasised as being central.

3.2 Differences

- There is no direct correlation in network theory to input factors. An equivalent situation would be the state of resources prior to being incorporated into a network.
- The concepts of rent are not evident in the network theory- the justification for a network is that it provides a more effective means of creating wealth for the constituent actors.

3.3 Vertical and Horizontal Linkages within the network

The superposition of the vertical and horizontal frame of classifying inter-firm relations can be useful in the network context. Horizontal linkages are typified by relationships with organisations at the same level in the industry, and are generally used to exchange various skills and key resources (e.g. franchising entry into a distribution network, or joint R&D) with a view towards increasing expertise and reducing investment costs. Vertical linkages follow the transformation of raw materials, and are concerned with improving logistics integration and effectiveness, collaboration on R&D of sub-assemblies or components.

The two are not independent, however, as the number and nature of the relationships constrain the choices of the firm when it comes to implementing strategy. Network theory conjectures that an individual linkage would best be described not by the dyadic relationship, but by focusing on all existing linkages. Elg and Johansson (9) created a two-by-two grid that highlights the most important relational issues for new and existing networks in the horizontal and vertical direction. The grid illustrates how pursuing a strategy which focuses on the core capability of the focus company can affect its existing inter-relations.

The economic and monopolic rent of RBT can be explained by networks that exist and are dynamically stable, and those which are in the process of being formed. The formers are competitive, and wealth is distributed evenly among members; the latter are led by some pioneering firms, with supernormal profits earned due to the lack of competition.

4 CASE STUDY

The combined approach of resource-based and network theory is currently being employed to create a model of a textile network, identify its core capabilities and develop some strategic scenarios which makes use of these strengths.

4.1 Background

The indigenous textile industry, in particular the part catering for the fashion market, typically consists of complex industrial networks. The products range from styles which change relatively little from fashion season to season, to those which are new, highly fashion orientated, and with limited shelf lives (typically six months). The average product lead-time from conception to final customer is approximately thirteen months.

The complexity of the network can be attributed to the traditional nature of the industry, and the high level of social interaction 'behind the scenes'. Vertically linked firms in the network include spinners, dyers, finishers, weavers, garment manufacturers and retailers. Horizontal linkages exist in the form of sub-contracted manufacture, franchising and licensing. The design of clothes is a unique, individual process. The properties of the various materials are subject to change, given a new fabric. The manufacturing firm (typically the largest body in the value chain) is organised along flow line principles, with utilisation rates and efficiency regarded as key performance measures by company staff.

The case study will identify the key characteristics of both approaches as outlined above, and compare them to discover how compatible they are in practise. In particular, the properties of the network interactions together with the actors will be compared to the capabilities as described previously.

5 CONCLUSIONS

The two approaches of network theory and resource based theory both stem from earlier attempts to provide a model of industrial purchasing and marketing that provides a strategic insight into the current situation. Reviewing the models reveal many characteristics capable of being tranduced from one to the other, despite different underlying assumptions. It is certainly true that the lack of stringent definitions for many of the characteristics (such as capability and interrelation) aid the comparative process; it is also true that little empirical proof exists that either is effective in improving the competitive advantage of a company. Nonetheless, the parallels in the models suggest an overlap that may help to strengthen them. It is also true that prescriptive theories (such as Porter's value chain) are regarded as sound from a strategic perspective.

The textile industry has been characterised by a complex business network, with a long supply chain, substantial lead times, and close personal ties. Changes in general consumer expectations, competition from abroad, and the imminent lifting of trade restrictions mean that the textile networks need to adapt, if they are to survive. The potential for market expansion is also there for firms successful in pursuing competitive advantage. A case study has been proposed which will examine the industry using the theories outlined above. The praxis of this study, further to such a comparison, will result in a model of an industrial network, and point towards some of the strategic considerations that should be entertained by the firms in question.

6 REFERENCES

(1) Håkansson, H. (1990) International Marketing and Purchasing of Industrial goods: An Interaction approach. International Marketing Strategy, 3rd edition, Pergamon Press.

(2) Williamson, O.E. (1979) Markets and Hierarchies, The Free Press.

(3) Harland, M. (1997) Supply chain operational performance roles. Integrated Manufacturing Systems Vol. 8 No.2, pp 70-78.

(4) Porter, M. (1995) Competitive Advantage: Creating and Sustaining Superior Performance. Free Press, New York

(5) Holmlund, M. Törnroos, J. (1997) What are Relationships in business networks? Management Decision Vol. 35 No. 4, pp. 304-309.

(6) New, S. J. (1996) A framework for analysing supply chain improvement. International Journal of Operations and Production Management Vol. 16 No. 4, pp. 19-34.

(7) Olavarrieta, S. Ellinger, A. (1997) Resource-based theory and strategic logistics research. International Journal of Physical Distribution and Logistics Management, Vol. 27 No. 9/10, pp559-587.

(8) Dierickx, I. Cool, K. (1989) Asset Stock Accumulation and sustainability of competitive advantage. Management Science, Vol. 5 No. 6 pp. 1504-1511.

(9) Elg, U. and Johansson, U. (1996) Networking when national boundaries dissolve. European Journal of Marketing Vol. 30 No. 2, pp. 61-74.

6.1 Recommended Bibliography

Burt, R. (1992) Structural Holes, Harvard University Press, Cambridge MA.

Croom, S. Batchelor, J. (1997) The development of strategic capabilities- an interaction view. Integrated Manufacturing Systems Vol. 8 No. 5, pp. 299-312.

Håkansson, H. Johanson, J. (1993) The network as a governance structure: interfirm co-operation beyond markets and hierarchies. The embedded firm: On the socioeconomics of industrial networks, Routledge, London and New York.

Stock, N. S. Greis, N. P. Kasarda, J.D. (1998) Logistics, strategy and structure; A conceptual framework. International Journal of Operations and Production Management Vol. 18 No. 1, pp. 37-52.

Case studies – learning in operations

V GILGEOUS
Department of Manufacturing Engineering, University of Nottingham, UK

ABSTRACT

An introduction to learning is provided and the characteristics of a learning organisation are explored. Using Garvin's five-indicator learning model, three international companies and one city council are studied to determine their extent of learning and the manner in which they pursue it. The models usefulness and faults as a means of contrasting learning in the four companies is examined. From this experience the author presents an action plan for learning. This considers competence acquisition, development programmes, schemes to facilitate learning, rewards and incentives to learn, a supporting organisation structure and processes for information/knowledge acquisition and transfer.

1 INTRODUCTION

Many management gurus have identified learning as the essential characteristic an organisation needs in order to progress effectively. This is due to the need of companies to create and manage the organisational cultures that foster commitment, responsibility and the development of individual potential for the corporate good. In other words, a learning organisation is needed to respond flexibly to the current business environment.

There are many definitions of the learning organisation and some disagreement between people. Straightforwardly Pedler, et al (1) describe a learning organisation as;

> *'an organisation that facilitates the learning of all its members and continuously transforms itself.'*

Peter Senge (2), who popularised learning organisations in his book *The Fifth Discipline*, described learning organisations as:

> *'places where people continually expand their capacity to create the results they truly desire, where new and expansive patterns of thinking are nurtured, where collective aspiration is set free, and where people are continually learning how to learn together.'*

Obviously, the learning organisation places a great emphasis on the development of the individual and the organisation together. It is important to develop the people in the organisation so that they want to, and have the ability to, learn, change and improve. In turn this will enable the organisation itself to learn, change and improve.

The importance of learning indicates the need for a much closer look at the human side of the organisation. In the process of changing an organisation, especially a manufacturing organisation, there is a tendency to let the hardware, the system, and the procedures dominate one's thinking and analysis. But when it comes to achieving improvement in performance, the people in the organisation and the way they link with these technical inanimate factors that will determine the success of the change. That is why emphasis needs to be placed on the learning element of people and the organisation itself.

2 CHARACTERISTICS OF LEARNING ORGANISATIONS

To understand better the forms of organisational learning, several interpretations can be given. Senge (2), recognises three characteristics of a learning organisation: it expands its capacity to create results, there is a collective aspiration, and people learn to learn.

Jashapara (3) recognises learning on several levels in the organisation:

- Organisational learning
- Team learning
- Individual learning.

Some of the above characteristics are quite self explanatory, but the term learning still remains vague. Rheem (4) describes several types of learning, which give more meaning:

- Competence acquisition;
- Experimentation;
- Continuous improvement;
- Boundary spanning (benchmarking).

Mills and Friesen (5) offer a similar variation on the above in the shape of three characteristics of an organisation which equip it for learning.

First it must make a *commitment* to knowledge. A very important aspect of this commitment involves the selection of people. Another aspect of commitment to learning involves the development of learning internally through a variety of means.

Second, a learning organisation must have a *mechanism for renewal* within itself. Departments and other units of firms continually fall into bureaucratic rigidity. They cease to adapt, to learn. They become impediments to the success of the firm. A learning organisation must be able to intervene in such situations to renew and revitalise them.

Third, a learning organisation possesses *openness* to the outside world so that it may be responsive to what is occurring there. For the firm's staff as a whole, openness to the changing needs of customers and suppliers is required.

2.1 Policies and Programmes for Learning

It is necessary for the management to define the policies and programmes which will enable the organisation to progress to becoming a learning organisation, and to specify the type of changes in behaviour are required. To assist this process, Garvin (6) proposes the following five building blocks of learning organisations:

- Learning from systematic problem solving
- Learning from experimentation
- Learning from past experience
- Learning from other's experience
- Transferring knowledge through the organisation.

Essentially, the organisation that can learn is the one will have high probability to success. Peters and Waterman (7) suggest that the control provided by strong learning culture is seen as the critical enabler of change.

For all of the above to occur a distinctive mind-set and pattern of behaviour is needed. Systems and processes that can support these activities and integrate them into the fabric of daily operations need to be in place. In this way companies can manage their learning more effectively.

3 FOUR CASES OF LEARNING

Three international companies and one city council were studied to determine their extent of learning and the manner in which they pursue it. Garvin's learning model was used as a basis for comparison across the companies. As mentioned the model has five learning characteristics and reference to it in the text is made in bold italics

3.1 The Pet food Company

The Petfood Company is a leading manufacturer in the UK of prepared Petfoods for cats, dogs, birds and fish. The management at the Company promotes a learning culture. Employees are kept well informed on the organisation plans, objectives and the role that they play in these plans. Overall the employees react well to new initiatives and this is helped by the financial incentives offered to them for performance improvements. Emphasis is put on employee training. Employees, called associates, are trained to National Vocational Qualification (NVQ) level 2 and all machine operators are trained to dismantle machines. Each department has their own training programme and a performance review which are documented and analysed by IIP assessors. However more emphasis needs to be placed upon ***systematic problem solving.*** To ***transfer knowledge*** regular meetings are held amongst the employees and shift team meetings are held at each change over. Regarding ***experimentation*** there is ongoing product development to ensure that the needs of the customer are met. To ***learn from others*** benchmarking is performed regularly and in extensive detail. There is close working with other companies to achieve mutual benefits. With regards to ***learning from past experience*** the Petfood Company are constantly looking at the successes and failures within their product range and redefining the range accordingly. Currently the company is working with Hero Drinks, Kodak Films and KP Foods. Overall the level of learning is very high at the Petfood Company and this is reinforced by the learning culture created by the management.

3.2 The Fruit Machine Company

The Fruit Machine Company, the largest gaming company in the Midlands, has the capacity to produce between 100 and 900 fruit gambling machines each week. The management style is very traditional. Management lead in all decisions, dictate what is to be done and how it will be done. There is little, if any involvement with the other employees. Regarding ***problem solving*** it is very rare that a shop floor decision is acted upon. Any team meetings are held in isolation to other departments and departmental barriers prevent ***transfer of knowledge*** between the departments. There are no training programmes ongoing in the company except when a new machine is purchased. Recently cellular manufacturing is being tried as a means of improving production efficiency. However, little ***experimentation*** is carried out in optimising the manufacturing processes since this is seen as time wasting if productivity is lost. A lot of emphasis is placed upon ***past experience*** within the company. They have found however, that this knowledge can be as much of a hindrance as a help since they have flooded the market with machines that are too similar and as a result they get played less. To ***learn from others*** market surveys are being carried out to find out what sort of game the customer wants to play. Benchmarking is an unpopular activity since they find it difficult to implement what they see in other firms, without going through the learning experiences for themselves. There is no ***systematic problem solving***. The employees are not very receptive to new ideas and half of them are on short-term contracts. In conclusion the environment created in the company is not suited for a learning culture. A major drawback is that the company structure is too traditional and they find it difficult to break away from the traditional processes and culture which hinders progress. Risks are unheard of and there is little employee motivation. If there are any changes to be made within the organisation the employee receptivity is very poor.

3.3 The Tractor Company

The Tractor Company manufactures a large product range of tractors in America, Germany, Mexico and Spain. They are sold throughout the UK. The management and team working inspire learning and ***problem solving*** is actively encouraged. The company ***transfers knowledge*** quickly and efficiently through written, oral and visual reports. Regular meetings are held to ensure that the employees are kept informed of the business objectives. Extensive staff training in the areas of customer care and the use of the computer system is provided. Regarding learning from ***past experience*** information is extensively recorded. When problems are identified they are recorded together with solutions that have worked, and those which have failed. Through customer records the company looks to identify influences that affected the customer in purchasing their tractor. To ***learn from others*** emphasis is placed on benchmarking and the company's policy is to visit at least eight companies a year. They are particularly interested in information on the salaries, benefits and packages other employers offer. Regarding ***experimentation*** ongoing research is conducted on product design and development of the manufacturing processes. To ***problem solve*** the company insists on collecting complete and accurate data but they do minimal statistical analysis. Employee receptivity to the management's decisions is high. Management rein down authority and instructions to the employees and they are encouraged to take an active part in the operation of the organisation. The management have created a learning environment where the employees are given free time each week to hold team meetings, discuss problems and seek ways of overcoming them. As a result there has been many changes on customer processing and on site product testing brought about by the ideas of the employees. In conclusion the degree of learning in the Tractor Company is high.

3.4 The City Council

The role of the City Council breaks down into two departments. The service and the leisure department's, both statutory departments are required by law. The responsibilities for services include Education, Training and Equality, Administration, Social Services, Electoral Registration and Technical Services. The responsibilities for leisure include supporting the service department through Finance, Law and Personnel. The Councils main aim is to increase customer satisfaction and fulfil the wants of the people who live within their boundary. They are also looking to reduce the queues in their service departments as well as dealing with problems more efficiently. The Council would like to increase the morale and performance of its employees. Currently there is low staff morale and performance and high absenteeism. To ***learn from past experience*** all training programmes and customer service developments are recorded and evaluated on a regular basis. Employees are encouraged to provide feedback on the programmes they attend. Customer suggestions and comments are also collected through suggestion boxes. ***Experimentation*** on form rationalisation and design and types of staff training is conducted to meet the Councils objectives and increasing customer service. Regarding ***learning from others*** work is done with other Councils to see if there are any practices which can be adopted, especially with respect to the use of computer software and the management of computer systems. ***Systematic problem solving*** is not carried out at all. However information that is collected has to be accurate and assumptions eliminated. To ***transfer knowledge*** courses are run all year round to develop the employees who work towards obtaining N.V.Q.'s. Most of the employees are receptive to training and an opportunity to talk directly with the line managers. In conclusion the City Council are keen to improve but there is a feeling that there is no great rush. Since there are no competitors, they are the only service providers.

4 CONCLUSIONS

A learning organisation can be described as one that facilitates the learning of everyone within it and is skilled at creating, acquiring and transferring knowledge. The literature shows that organisations, in order to progress, must learn and in so doing become aware of the need for change and how to manage it effectively.

Garvin's (1993) five-indicator model proved useful in ascertaining the degree of learning and the way that learning methods are being practised in the four companies chosen for study. The approach showed that the degree of learning within the companies varied greatly. At the Petfood Company there was a high degree of learning. The company strove for greater flexibility through employee training and good information flows; this was reinforced by the management who created an environment where learning could take place. There was a great deal of investment in their employees. In contrast the Fruit Machine Company had a traditional management style that inhibited learning. There was little experimentation, departmental barriers were present as was poor information flow; overall the degree of learning was low. Both the Tractor Company and the City Council exhibited a medium level of learning, but indicated a keenness to improve on their current situation.

Overall contrast of performance and use of Garvin's Model

Garvin's first indicator of ***systematic problem solving*** was not performed in any great detail in the companies studied. Further research would be useful to examine why this was the case, and establish if there are companies that actively perform the indicator. Garvin's second indicator, ***experimentation,*** posed several problems and could be more profligate in all the companies studied. The workforce in the Fruit Machine Company preferred not to experiment and take risks for fear for their job security. Organisations need to encourage experimentation and not punish employees when problems arise and management need to make allowances for lost productivity, trial production runs and wastage. Garvin's third indicator, ***learning from***

past experience, requires care in its application. As the Fruit Machine Company case showed it is easy to get trapped in producing old products and performing old processes that were once successful. The fourth indicator, ***learning from others experience***, has the benefits of gaining a new perspective and a source of new ideas. However just copying processes and ideas say through benchmarking, has its limitations since it prohibits companies experiencing learning for themselves. The fifth indicator, ***transfer of knowledge,*** showed that the information transferred must be relevant. This is not stressed in Garvin's model. The information must be spread quickly and to the relevant departments.

The literature and use of Garvin's model showed that systematic problem solving and experimentation should be encouraged. People need to be sufficiently trained and imparted with new knowledge, new ideas. An atmosphere where people learn from their past experiences and from each other should prevail. People should feel some kind of ownership concerning the outcomes of actions taken. Mechanisms for transferring knowledge and measuring learning should be in place. An action plan for learning should include competence acquisition, development programmes to facilitate learning, rewards and incentives to learn, a supporting organisation structure and processes for information/knowledge acquisition and transfer. It should also ask: to what extent are you a learning organisation? What are your learning objectives and policies? What framework or structured approaches you intend to follow? What are the main actions that must be taken? Who would be responsible? When will each action take place? How would you go about measuring learning?

5 REFERENCES

1. Pedler, M. Burgoyne, J. and Boydell, T. (1991) *The Learning Company: A strategy for sustainable development*, McGraw-Hill, Maidenhead, UK.
2. Senge, P. (1990) *The Fifth Discipline, The art & practice of the learning organisation,* Doubleday Dell Publishing Group Inc, New York, USA.
3. Jashapara, A. (1993) The competitive learning organisation: a quest for the holy grail, *Management Decision*, Vol.31 No.8, p.52-62.
4. Rheem, H. (1995) The learning organization, *Harvard Business Review*, March-April, p.10.
5. Mills, D. Q. and Friesen, B. (1992) The Learning Organisation, *European Management Journal*, Vol 10, No. 2, June.
6. Garvin, D.A. (1993) Building a Learning Organisation, *Harvard Business Review*, July/August, pp78-91.
7. Peters, T., and Waterman, R.H. (1992) *In Search of Excellence,* Harper & and Row, New York.

Modelling integrated process planning and scheduling for agile manufacturing

D M HALL and **W ZHAO**
School of Engineering, University of Huddersfield, UK

ABSTRACT

A programme of research is described, which aims to provide tools to simplify the profit maximisation task in the integrated process planning and scheduling activities. Process planning and scheduling operations in a number of companies have been modelled using formal business process modelling techniques. This has resulted in the formulation of a generalised model, which serves as a "benchmark" for case study models. Based on real data the results of work on the generalised model indicate the nature and primary specifications of a CAPP package and provide a solution capable of implementation in larger engineer-to-order small or medium sized enterprises.

1. INTRODUCTION

The current trend towards increased agility in manufacturing is causing increasing interest in the concept of integrated process planning and scheduling. This is not a new concept but as yet there is little evidence of good practice in industry, despite the substantial amounts of research that have been carried out. The benefits of this approach have been recognised for some time[1] and one of the authors has also proposed a novel approach to scheduling based on flexible, reactive process plans[2]. However, the bulk of research work in process planning has concentrated on generative process planning systems[3], architectures for integrated process planning and scheduling[3,4] and process plan optimisation[3,5]. Seen from the view point of industry much of this work does not seem sufficiently focused on immediate day to day needs or on implementation in mature manufacturing organisations. This is even more so in the case of engineer to order (ETO) companies to whom an integrated approach would be most beneficial. Most proposed solutions seem to be aimed towards a (somewhat distant) green field implementation.

This programme of research is aimed primarily at the derivation of a detailed specification of the necessary elements of an integrated process planning and scheduling method to facilitate an increase in agility. Furthermore, it is intended that the resulting system can be incorporated into existing operations with minimal disruption to other functions. The target market is the larger ETO companies within the small to medium sized enterprise (SMEs) sector where planning and scheduling is complex but where resources preclude an expensive off the shelf solution.

The approach is to model process planning and scheduling in a number of case study companies and to compare the models with a theoretical model. This will enable the creation of a reference model from which a detailed specification for improvements can be derived, particularly with regard to computer aided process planning. More importantly, by incorporating into the reference model the concept of reactive process planning, indicated for improvements in agility should result.

2. THEORETICAL MODEL

2.1 General Model

There is currently a range of good publications relating to methods and techniques in process planning[6,7,8] and it is sensible to draw upon this in the creation of a theoretical model. With the addition of local and in-house expertise, it was possible to derive a theoretical model of an efficient process planning and scheduling process. It is clear that there could be many variations to such a model and that much of the model could be considered to be a matter of opinion. However, this is acceptable in early stages as the purpose of such a model is to provide a 'benchmark' against which to compare the case study models. It was intended that the theoretical model should be as detailed as possible but should also evolve and be developed as and when necessary.

A number of popular modelling tools and methods were considered but it was clear that the most powerful tool currently available is ARIS (Architecture of Integrated Information Systems)[9]. In addition, ARIS is currently being used for other related projects in the Manufacturing Systems Research Group (MSRG). ARIS is an integrated environment for modelling, analysing and evaluating company models. It consists of a modelling part, an analysis part and a navigation part, which are integrated to enable user interface, functionality, development database and process control. ARIS allows for modelling in very great detail and it should be appreciated that to reproduce the entire model in paper form is not practicable.

The theoretical model was based on the initial structure shown in Fig.1, the majority of which is a fairly standard description. The main difference is the addition of Activity X, which in simple terms is the novel part of the model which requires to be defined and specified. Activity X is the activity which enables increased agility or 'reactiveness', although during the early stages of model building was not defined in any more detail than that. It was felt that initially the other areas of the model should be defined in more detail and that the priorities should be the process planning activity and the scheduling activity. It was clear that the model would undergo continual refinement during these initial phases.

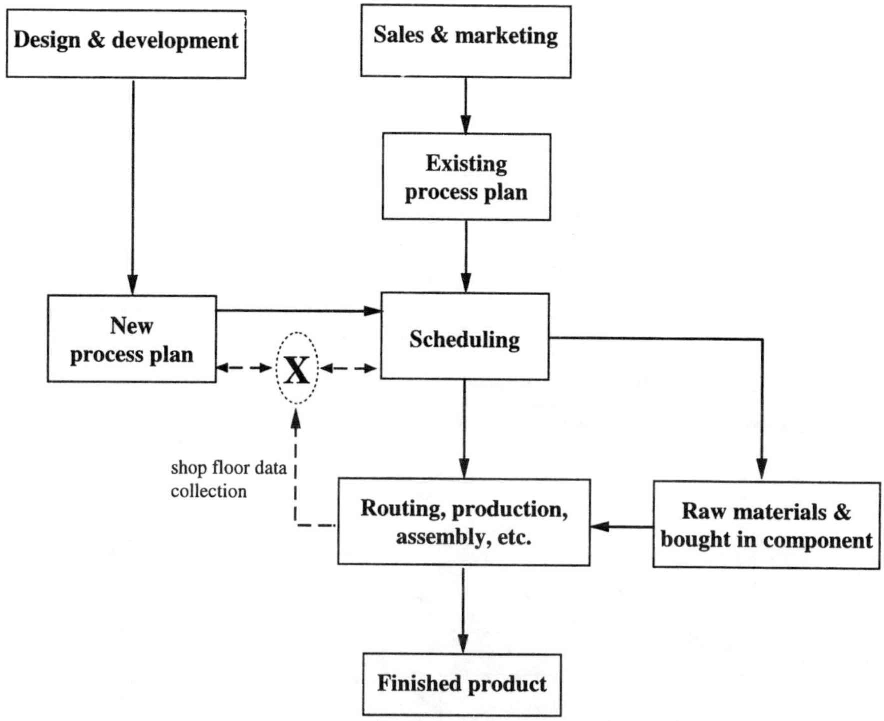

Fig.1 Theoretical Model of Process Planning and Scheduling

2.2 Modelling the Key Activities

As previously stated, it was clear that further decomposition was required for both the process planning and for the scheduling activities. A further development of the process planning activity is shown in Fig. 2 and is primarily a derivation from other models[8]. A number of derivations for the scheduling activity exist but it will be appreciated that there are more possibilities in this area and no particular approach has been identified. It is envisaged that a range of "plug in" modules would be a useful approach.

The notable aspect of the process planning activity model is the high degree of 'loop-back' flexibility required and the reliance on access to a large amount of existing data relating to the manufacturing system and its component parts.

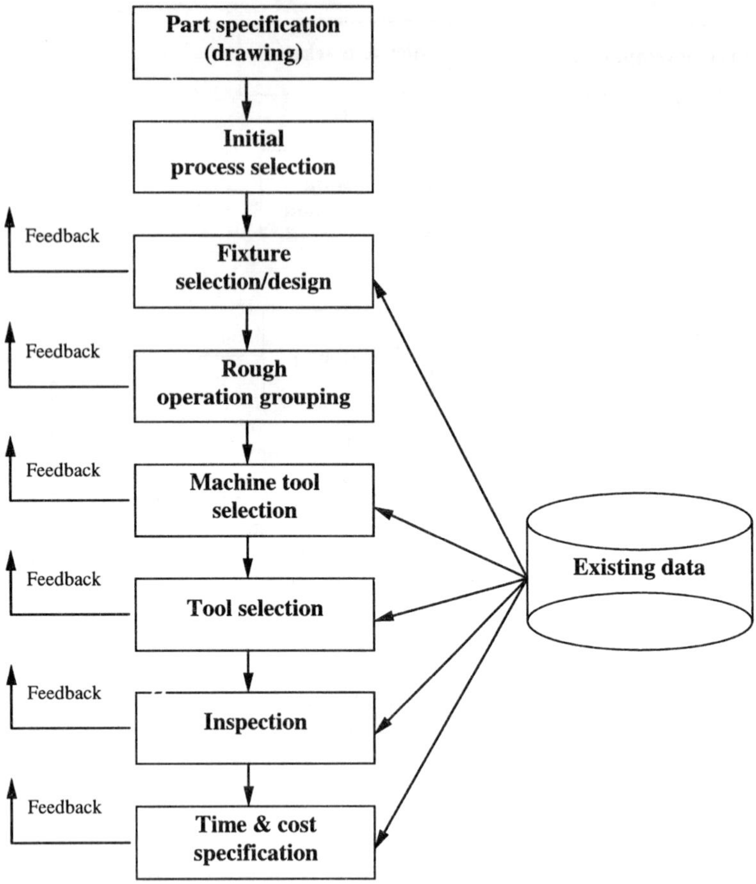

Fig. 2 Expanded Process Planning Activity

3. CASE STUDY MODELS

3.1 Case Study Companies

Three case study companies were initially identified. These were all companies in the larger SME ETO sector but all with different products exhibiting different production and planning characteristics.

Company A manufactures a small range of CNC machine tools, each tailored to customer requirements. This can be summarised as low product quantity, high component quantity and high component variety.

Company B manufactures a range of turbochargers, with short lead times and fast product turnover. This can be summarised as high product quantity, low component quantity and low component variety.

Company C manufactures a range of high quality gauges. This can be summarised as high product quantity, low component quantity and high component variety.

None of the companies had any form of integration between process planning and scheduling but all thought it to be highly desirable.

3.2 Case Study Method

The case studies followed a procedure established by other projects in the MSRG. This briefly comprises discussion and opinion gathering at different levels in the organisation, followed by detailed tracking and mapping of information and of component flow with a final discussion to clarify any problems. If necessary further mapping work would then be carried out.

ARIS models have been prepared for companies A and B whilst the model for C is not yet complete. Models A and B would benefit from the addition of some detail and a further refinement. As previously stated it is not practicable to include these models in paper form. However the extent of modelling is already sufficient to be able to draw some preliminary conclusions.

3.3 Model Comparison

A major strength of ARIS is its ability to compare models and identify similarities and differences. In general, the similarity of these case study models is that little awareness has been paid to Activity X. Also actions have not been fully taken to maximise net profit, i.e. bottleneck or low efficient work flow are retained and non-optimum process plans/schedules have been used. The difference between these case study models lies in the variety of product and quantity of component. The company with the lower variety of product and lower quantity of component normally performs better. This indicates that companies with high uncertainty and high variety of product will benefit more from the integration of process planning and scheduling.

3.4 Results

There are obviously major differences between the theoretical model and the case study models although some of these differences may not actually exist and may be a function of inexperience in modelling. Refinement of the models and growing experience should resolve these issues. Activity X is clearly absent from the case study models, but by placing it into these models it has been possible to develop a more detailed description of this activity. In brief terms this description is:

Activity X

- receives relevant shop floor data (machine status; component/batch location; etc.)
- reworks process plan on selected components based on a range of factors (as yet, specifies only loosely)
- reworks schedule based on a range of factors (again, only loosely specified)
- assesses consequences of new schedule
- implements new schedule, if appropriate

An important aspect currently under investigation concerns the determination of the frequency of replanning and rescheduling. This depends upon a number of factors including typical operation time, typical lead time, typical batch size, algorithm processing time and uncertainty of manufacturing (machine breakdown, bottleneck, change of customer demands, etc.)

Inclusion of Activity X in the reference model also enables the formulation of a specification for an improved process planning activity, the major improvement being that it will have the addition of computer aid. It is only when this specification is known fully that the next phase of the research programme will commence, either creating the package specified or matching the specification with one of the many existing research programmes elsewhere.

4. CONCLUSIONS

A representation of *real* process planning and scheduling operations is being developed in the form of a reference model. This has enabled the initial definition of features necessary to enable increased agility in the manufacturing system (Activity X). Definition in this way is more useful than the popular approach of defining and creating the ideal, but then having no implementation route. In addition, the means have been created to derive a specification for a computer-aided process planning system, which will fit into an existing manufacturing system. This is a major strength of the research in that the specification of the computer aided process planning activity and the objectives in producing it have definite consequences in existing process planning and scheduling operations currently in use in industry. The likelihood of implementation of such a system is thus greatly enhanced.

The development of the model still requires a substantial amount of work but the benefits of driving original research from a factual, realistic base are already clear.

REFERENCES

(1) Halevi G., The Role of Computers in Manufacturing Processes, John Wiley and Sons, 449 ff., 1980

(2) Hall D.M., Reactive Process Planning, 6th Int. Conf. Irish Manufacturer's Committee, Sept. 1989, Dublin City University, Eire

(3) Kiritsis D., Review of Knowledge Based Expert Systems for Process Planning, Int. Journal of Advanced Manufacturing Technology, 10, 240-262, 1995

(4) Gu P. and Norrie D., Intelligent Manufacturing Planning, Chapman and Hall, 1995

(5) Palmer G.J., An Integrated Approach to Manufacturing Planning Optimisation in Process Planning and Job Shop Scheduling. Ph.D. Thesis, University of Huddersfield, 1994

(6) Chang T.-C. and Wysk R.A., "An Introduction to Automated Process Planning Systems", Prentice-Hall, Inc., 1985

(7) Chang T.-C., "Expert Process Planning for Manufacturing", Addison-wesley publishing Company, 1990

(8) Halevi G. and Weill R.D., "Principles of Process Planning A logical Approach", Chapman & Hall, 1995

(9) Scheer A.-W., Business Process Engineering, 2nd Edn., Springer Verlag, 1994

Investigating the adoption of lean manufacturing at supplier companies

C R HERRETT, N S McBAIN, and **K WALLIS**
CAMM, University of Sunderland, UK

The paper investigates the adoption of Lean production among supplier companies in the North East of England. Using Lamming's Four phase model of customer-supplier relations and Porter's Five forces theory as the basis for a survey, information was collected that indicates conformance to accepted models and analyses industry circumstances that begin to explain resistance to the Lean production paradigm.

1. INTRODUCTION

It is suggested that Lean production is a concept mostly applied by large customer companies and ignored by the majority of supplier companies in the North East of England. This contradicts the conventional wisdom, which recommends Lean production as 'best practice', regardless of industry or company type(1). We believe that there are reasons, rooted in technology and the market, which enable supplier companies in the North East of England to resist 'best practice'.

The paper associates Mass and Lean production with particular types of customer-supplier relations contained within Lamming's Four phase model of customer-supplier relations (2). This model was used to survey supplier companies from the North East of England to investigate customer-supplier relations and to give an indication of the production paradigm employed. The investigation was extended using the Five forces model(3) to gain insight into the technological and market conditions which enable the supplier company to resist a change to Lean production.

The survey methodology contained in the paper has, in our opinion, great potential for producing insights into actual practice among supplier companies. However, it is still very much at an experimental stage.

2. MANUFACTURING BEST PRACTICE

Womack et al.(1) are frequently cited in the academic literature and alongside Monden(4) they have become arguably the most authoritative work on Lean production and the root of current 'best practice'. Theoretical studies related to the evolution of new manufacturing paradigms preceding (1), e.g. Doll and Vonderembse (5), have by default largely facilitated its success, because of the devotion of previous studies to abstract theory and their lack of empirical evidence. Probably the greatest strength of (1) and the source of its authority is the depth of empirical knowledge it demonstrates of both the Mass production paradigm and the emerging practice of the Lean production paradigm. However, the depth and focus upon the industry and country origin of the Lean production paradigm has also proven its weak point (see below), despite the transfer of the concept to other continents and industrial sectors.

An increasing number of academic papers critically evaluate the original success of Lean production and indicate problems associated with its implementation. O'Shea(6) contends that Lean production has had to undergo a period of conversion to compensate for large differences in social systems between Japan and the West. Katayama and Bennett(7) imply that Lean production has enjoyed success because of its use within the hitherto prosperous Japanese economy and that economies in decline would find true Lean production difficult to maintain. Duguay et al.(8) discuss the benefits of both Mass and Lean production, indicating that both manufacturing paradigms can be relevant depending upon the pace of the market context.

We believe that Lean production cannot be generalised to every manufacturing company in the West to conform with the original ideal. However, Lean production principles should not be ruled out. It may be that particular manufacturing companies are best served by drawing upon elements of different paradigms, which correspond with their specific business environment.

3. CUSTOMER-SUPPLIER RELATIONS

When customer-supplier relations are discussed in the supply chain management literature, reference is often made to two types of customer-supplier relationship: Traditional and Partnership(9, 10 and 11). Sako(12) departs from this convention by locating her definitions of customer-supplier relations at opposite ends of a continuum of trading patterns. Unfortunately, possible customer-supplier types that could occur between the poles are not defined. A more detailed understanding of possible types of customer-supplier relationship occuring along a continuum is found in (13).

The Four phase model of customer-supplier relations developed by Lamming(2) describes periods when particular types of customer-supplier relations were influential in the automotive industry. The characteristics of customer-supplier relations related to each period are contained within the Traditional model (before 1975), the Stress model (1972-85), the Resolved model (1982 onwards) (Mass production) and the Partnership/Japanese model (1990 onwards) (Lean production). The final phases of the Four phase model of customer-supplier relations, the Resolved model and the Partnership/Japanese model, depict the current state of customer-supplier relations respectively associated with Mass production and Lean production.

In his Four phase model of customer-supplier relations Lamming indicates the possibility that customer-supplier relations associated with Mass production are still influential, alongside customer-supplier relations associated with Lean production. This leads us to believe that although Lamming supports Lean production, he doubts that it will be broadly adopted in the automotive industry outside Japan.

4. METHODOLOGY

4.1 Origin of data

To facilitate access to company data the authors have been able to gain the co-operation of the Supplier Development Team (SDT) project at the University of Sunderland. The SDT project was developed to provide services to supplier companies in the North East of England. Supplier companies were not randomly chosen from these supply chains, but were selected for development by customer companies. The criteria for selection was that customer companies designated for development activities should be key to the customer companies' operations.

The supplier development team has employed a three-stage approach to improving supplier companies. These three stages are: 1) a survey of the supplier company; 2) the production of a development plan, which is agreed to by all parties, the supplier company, the customer company and the SDT; 3) a pilot project to 'kick-start' the improvement process.

This paper has used quantitative and qualitative information collected during the first stage of the improvement process, 1) a survey of the supplier company.

4.2 The Survey tool

The SDT consultants were surveyed for supplier company information using a tool that consisted of two parts.

1) Four phase model of customer-supplier relations; Lamming's (2) model indicates what he would describe as 'best practice' as far as customer-supplier relations were concerned and also provides an indication of the manufacturing paradigm of the company which is tested. It must be noted that this is primarily a model of the automotive industry. It is held that the automotive industry has evolved more quickly than many other industries. Hence this model will provide evidence of differences between industries and for those industries that are not as far down the evolutionary course as automotive companies it provides a target for the development of customer-supplier relations.

2) Supplier power within the customer-supplier relationship; Porter's Five forces model(3) was used as a framework for this section of the survey tool. Questions were developed for this part of the survey tool, which would provide information about the instrumental forces in the business environment and the techno-economic forces that would enable supplier companies to resist Lean production or force them into adopting its principles.

4.3 Description of the sample

Information was collected from ten supplier companies regarding the relationship with their main customer company. The supplier companies were involved in the following supply markets: Clothing/fashion, Plastic components, Engineering components, Injection mould tooling and Pressed metal components.

5. RESULTS

5.1 Survey tool: 1) Four phase model of customer-supplier relations

In the first part of the survey tool Lamming's original table was converted into a series of questions arranged by what we describe as the determinants of customer-supplier relations (see Table (1)). The consultant was expected to define the relationship between a supplier and its main customer by placing a tick in one of four boxes, which related to the four models that Lamming originally organised along a horizontal axis. The results were presented in table format using a label system.

Each of the models that Lamming presented were numbered as follows: Traditional model = 1, the Stress model = 2, the Resolved model = 3 and the Partnership/Japanese model = 4. Table (1) presents the results of the first part of the survey.

Table (1) The Four phase model of customer-supplier relations

Determinants of the customer-supplier relationship	Supplier company									
	A	B	C	D	E	F	G	H	I	J
Nature of competition	1	1	3	1	1	3	1	3	3	1
Basis of sourcing decisions	3	4	3	3	4	4	3	4	3	1
Role of data/information exchange	1	1	3	1	3	4	3	4	4	1
Management of capacity	2	2	2	4	3	4	3	3	3	2
Delivery practice	4	3	3	4	3	4	3	2	2	1
Dealing with price changes	1	4	2	4	4	4	3	4	3	1
Attitude to quality	3	1	2	2	4	4	1	3	3	1
Role of R&D	1	1	1	1	1	3	1	3	2	1
Level of pressure	3	1	3	4	1	3	3	2	2	2

If supplier companies adopted what the conventional wisdom considers 'best practice' it could be assumed that their customer-supplier relations would be dominated by the Partnership/Japanese model of Lamming's Four phase model of customer-supplier relations. If supplier companies were not practising 'best practise' it would be expected that they were at the stage of the Resolved model working towards 'best practice'. This would mean that supplier companies were moving towards 'best practice' production principles. Bearing this in mind we interpret the results shown in Table (1).

The characteristics of customer-supplier relations shown in Table (1) did not conform with the underlying implication made by Lamming that the relationship between the customer and supplier can be defined within the confines of one model. As can be seen from Table (1) the characteristics of the relationship between the suppliers and their main customer draw from three of the models in some cases, and from all of them in other cases. The closest that any of the supplier companies surveyed come to being confined within one model is demonstrated by supplier companies F, G and J. Based on the information contained in Table (1), we suggest that the only supplier close to demonstrating the principles of Lean production is F. The other suppliers remain divided between the three models that describe different types of customer-supplier relationship in the Mass production paradigm.

5.3 Survey tool: 2) Supplier power within the customer-supplier relationship

The second part of the survey tool, inspired by Porter's five forces model, consisted of seven sections of questions. Every question required a response of yes or no from the consultant. These questions were answered with reference to the same customer-supplier relationship described in the first section of the survey tool.

The sections of this survey tool contained an unequal number of questions. Therefore, to avoid an imbalance, an equal weighting was given to each section of questions. Following this system Table (2) was produced.

Table (2) Supplier power within the customer-supplier relationship

Determinants of power within the customer-supplier relationship	Supplier company									
	A	B	C	D	E	F	G	H	I	J
Supplier's products	6.8	6.8	6.8	2.3	4.5	6.8	4.5	4.5	6.8	0
Characteristics of supplier's market	2.6	5.1	1.3	3.8	1.3	0	3.8	1.3	0	2.6
Customer characteristics	3	3	0	4.5	3	4.5	7.5	4.5	1.5	4.5
Techno-economic structure of the supplier industry	5	1	4	5	2	2	5	4	2	4
Supplier's characteristics	6	5	1	6	4	6	6	8	4	7
Competitors	6	3	6	3	9	3	9	6	0	0
Substitute products	9	9	9	9	9	9	9	9	9	9
Total score	38.4	32.9	28.1	33.6	32.8	31.3	44.8	37.3	23.3	27.1
Rank	2	5	8	4	6	7	1	3	10	9

The purpose of this part of the survey tool is to provide the Four phase model of customer-supplier relations with more depth through a description of technological and market forces, which potentially enable a supplier company to avoid the transition to Lean production.

Using the rank provided in Table (2) it is possible to arrange the supplier companies into three groups according to the level of power that they control in the customer-supplier relationship: high = G, A and H, medium = D, B, E and F and low = C, J and I. We suggest that the higher the level of power that a supplier company controls, the more capable it is of avoiding the demands of its customer company to make the transition to Lean production. It is assumed that customer companies are Lean manufacturers and would want to influence their supplier companies to become lean.

Referring back to Table (1) it is possible to see how the power of the supplier company has influenced a transition towards Lean production. As far as the supplier companies G, A and H are concerned it could be argued that their strength in the customer-supplier relationship has enabled them to avoid the transition towards Lean production, although A and H do practice some of its principles. The supplier companies with a medium level of power within the customer-supplier relationship tend to demonstrate Lean principles more than the stronger supplier companies. The supplier companies that have a low level of power within the customer-supplier relationship do not support the idea that weakness would force them towards the Lean production paradigm. In fact the weak companies are further away from adopting Lean production than the companies that have a high or medium level of power within the customer-supplier relationship.

It is noticeable that the supplier companies with a low level of power within the customer-supplier relationship scored zero points, whereas only F of the other supplier companies did so. The zero score is an indication of weakness. Or is it that the particular sections of the survey that scored zero were not important to the supplier company and provided no power within the customer-supplier relationship?

6. DISCUSSION

As mentioned in the Introduction the Survey methodology is still at an experimental stage. We acknowledge that the number of companies used in the sample is not statistically significant, but a major objective of this experiment was to see how useful the survey tool might prove in practice.

Part one of the survey tool, Lamming's Four phase model of customer-supplier relations, performed better than part two of the survey tool, the Porter inspired questionnaire upon supplier power within the customer-supplier relationship.

Our claim that Lean production was being ignored by supplier companies in the North East of England was largely upheld by the Four phase model of customer-supplier relations. It was hoped that the second part of the survey tool would provide evidence of the distribution of power within the customer-supplier relationship and indicate the possibility of a change by the supplier company towards Lean production.

On reflection we feel that our assumption that customer companies practiced Lean production principles probably helped to diminish the richness of the picture that could have been derived from the second part of the survey tool. Given our suggestion that supplier companies with a high level of power within the customer-supplier relationship are able to resist Lean production and the evidence that the weaker suppliers have remained in the Mass production paradigm, it can be assumed that a driver towards Lean production does not exist. This may be an outcome of the customer company manufacturing paradigm, i.e. Mass manufacturers will not push for Lean production.

It is also accepted that more work needs to be done on the individual questions contained in the second part of the survey. There is a high possibility that some questions need to be given more weighting than others. This would obviously change the percieved strength of the supplier within the customer-supplier relationship and may lend more accuracy to results produced by the second part of the survey tool.

7. CONCLUSION

The main hypothesis of this paper, that Lean production is ignored by suppliers in the North East of England, has been supported through the first part of the survey tool. However, the second part of the survey tool has not provided enough evidence for us to confidently explain the reasons for supplier companies remaining with Mass production principles.

The evidence provided suggests that the survey methodology has potential, which at the present time remains untapped. Areas for improvement definitely lie in the second part of the survey tool, where more information needs to be collected about the customer company and the importance of each question needs to be examined in more depth. A larger sample size would also provide a more reliable indication of the usefulness of the survey tool.

8. REFERENCES

(1) Womack, James, Daniel T. Jones and Daniel Roos (1990), The Machine that Changed the World, Rawson Associates, Macmillan Publishing Company, New York.

(2) Lamming, Richard (1993), Beyond partnership - strategies for innovation and lean supply, Prentice Hall, Hemel Hempstead, Great Britain.

(3) Porter, Michael (1985), Competitive Advantage, The Free Press, New York, United States of America.

(4) Monden, Y. (1983), Toyota Production System – Practical approach to Production Management, Industrial Engineering and Management Press, Atlanta, GA.

(5) Doll, William J. and Mark A. Vonderembse (1990), "The Evolution of Manufacturing Systems: Towards the Post-Industrial Enterprise", Business Research Center, College of Business Administration, The University of Toledo, Toledo, Ohio.

(6) O'Shea, T., N. S. McBain and A. C. Morris (1998), "New Product Development Process-The Mediating Effect of the Behaviour and Belief", Accepted for the Fifteenth Irish Manufacturing Conference: University of Ulster at Jordonston. 2 - 4 September.

(7) Katayama Hiroshi and David Bennett (1996), "Lean production in a changing competitive world: a Japanese perspective", International Journal of Operations and Production Management, Vol. 16 No. 2, pp. 8-23.

(8) Duguay, Claude R., Sylvain Landry and Federico Pasin (1997), "From mass production to flexible/agile production", International Journal of Operations and Production Management, Vol. 17 No.12, pp. 1183-1195.

(9) Dyer, Jeffrey H. and William G. Ouchi (1993), "Japanese-style partnerships: Giving companies a competitive edge", Sloan Management Review (Fall).

(10) Stuart, F. Ian (1993), "Supplier Partnerships: Influencing Factors and Strategic Benefits", International Journal of Purchasing and Materials Management, Vol. 29, Issue 4, (Fall).

(11) Watts, C.A., K.Y. Kim and C.K. Hahn (1995), "Linking Purchasing to Corporate Competitive Strategy", International Journal of Purchasing and Materials Management, (April), 3-8.

(12) Sako, M. (1992), Prices , Quality and Trust – Inter-firm Relations in Britain and Japan, Cambridge Studies in Management, Cambridge University Press, Great Britain.

(13) Spiers, John (1997), The Theory and Practice of Buyer-Supplier Relationships: A Sectoral and Unipart Perspective. In: Cox, Andrew and Hines, Peter (Eds.) Advanced Supply Chain Management, Earlsgate Press, Great Britain, pp. 107-137.

A methodology for examining organizational values

M J HOOPER and **C N WINTERS**
Manufacturing and Supply Chain Management Centre, Coventry University, UK

ABSTRACT

Although consistent and ongoing change has been identified as a requirement for world-class organisation a significant number of companies have been identified within the UK, compared to competitor nations, as seriously under performing. Cultural and human factors are identified as the major barriers to the instigation of change, focus being placed on the values held within companies. The relationship between corporate values and mission statements are considered, allowing the identification of a framework to map values through an organisation.

1.0 INTRODUCTION

Fundamental to the continued existence of all manufacturing organisations is the ability to instigate and sustain continued innovation and change in terms of products, business processes and organisational structures. This is underlined by Cordon and Vollmann (1) who report that 90% of European executives interviewed expect to restructure their business units continuously over the next ten years. Agile manufacturing and the wider concept of World Class manufacturing all call for sustained organisational development focused on customer requirements and expectations. The real impact of change failure can be seen within smaller organisations. As size decreases the importance of the success of any implemented change increase, such that within SME's a lack of success can have a significant impact on organisational viability.

2.0 THE UK UNDER PERFORMANCE - MANIFESTATIONS AND CAUSES

Economically the UK has the lowest Gross Domestic Product (GDP) per head within the G7 group of countries (2). Although the phenomenon of deindustrialisation is common to advanced manufacturing nations (exemplified by Britain becoming in 1983, for the first time in 150 years, a net importer of manufactured goods), the decline of UK manufacturing has been identified as a relative weakness of the UK economy (3). Taking a manufacturing management perspective, the UK has a comparable numbers of World Class companies but has a larger number of under performers when compared to other countries (4,5). The concern relating to the competitiveness of UK manufacturing in relation to other economies is not a new phenomenon, indeed it was identified as a problem as long ago as 1852 (6).

This under performance has been attributed to the reluctance of companies to make investment in manufacturing technology and a short term based financial system (7). Comparisons of manufacturing technology with advanced manufacturing countries should, therefore, reveal a measurable gap. When comparisons are made, as shown in Table 1, no significant gap is revealed. Since investment in manufacturing technology relies both on the willingness of companies to make the investment and financial resources being made available for the technology a less simplistic answer is required.

Table 1 -: The UK's Investment Performance

Industry	Countries Compared with UK	Relative Age of UK Machinery	Relative Technical Sophistication
Machine Tools (8)	US	Slightly Younger	Similar
Metal working (9)	Germany	Similar	Less NCMs Employed in the UK
Metal working (10)	Germany France US and Japan	Slightly younger than France, Germany and US. Slightly older than Japan	Similar
Fitted Kitchens (11)	Germany	Similar	UK not as advanced
Clothing (10)	Germany	Older	Similar
Food Processing(12)	Germany Netherlands France	Similar	Similar
Engineering (13)	Germany		UK not as Advanced
Chemicals (12)	Germany		UK at Least Equal
Precision Engineering (14)	US Netherlands	Similar to US but Older than Netherlands	Less NCMs employed in UK

Note:- NCM refers to Numerically Controlled Machines

The earlier work of Hitchens and Birnie (15) support this view contesting that any performance gap is a result of the human inputs of labour and management rather than the quality of capital, the level of variability of demand, plant or firm size, industrial structure and the degree of foreign ownership. This performance gap is reflected in the poor success of organisations in implementing lean and world class practices (16).

3.0 ORGANISATIONAL INCLUSIVENESS

Oliver (17) in his study of Japanese manufacturers within the UK, found the most common barrier to organisational change was the recruitment of labour with the necessary skills in terms on behaviour within the organisation. The next most commonly reported problem referred to 'different ways of thinking', manifesting in the difficulties in relations between local managers and their Japanese counterparts. These results are echoed in the survey of manufacturing companies carried out for the Coventry and Warwickshire World Class Partnership (18) which found that 47% of companies attributed the main barriers of change within there organisation to be the human input elements of manufacturing.

These results support the view that aspects of organisation can influence the performance of activities within companies. The culture within organisations can be defined in narrow terms as shared values within an organisation (19) or can take on the broad context of the shared philosophies, ideologies, values, beliefs, expectations, and norms (20) within groups.

The linkage with these fundamental foundations and the higher level structures within companies gain greater importance in organisations seeking to attain agility. Agile structures can be viewed as internal and extended dynamic networks of individuals and companies brought together to implement customer solutions. Such structures are essentially fragmented, leaving little that is consistent or constant within its boundaries.

The study by Kotter and Heskett (21) of the performance of US firms between 1977 and 1988 found that the 12 outstanding performers, having an average profit growth of 756% over 11 years, were characterised by the strong emphasis they placed on the relationship with their customers, employees and shareholders. This conclusion is further supported by Stanford Universities "Built to Last" study (22) and the Gallup organisations Q12 studies (23).

Using a benchmarking tool developed from the European Quality model (EQM) (24), assessment took place of 5 manufacturing companies engaged in the production of automotive rubber components. The results for both leadership and business policy and strategy are shown in Figure 1 and Figure 2 respectively.

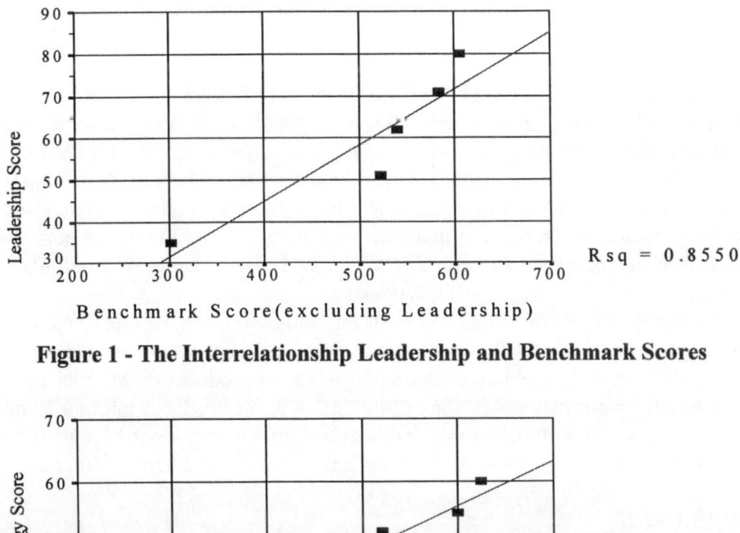

Figure 1 - The Interrelationship Leadership and Benchmark Scores

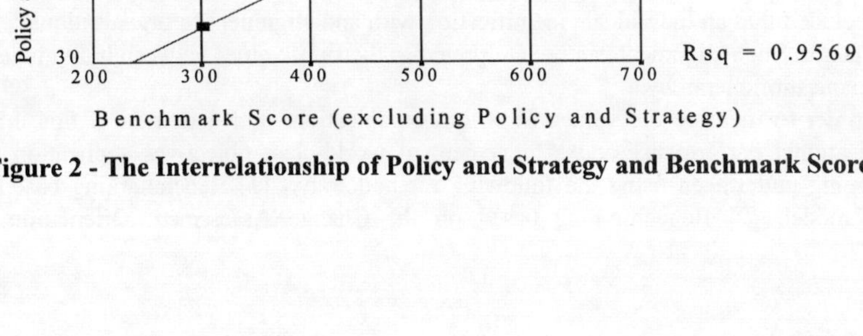

Figure 2 - The Interrelationship of Policy and Strategy and Benchmark Scores

The correlation shown between the leadership and policy and strategy employed within the organisation clearly demonstrate the relationship between inclusiveness and the level of World-class performance demonstrated by companies.

These studies taken together indicate that the aspect of culture involving values is particularly significant.

3.1 MISSION STATEMENTS AND VALUES

Pearce (25) identifies five primary components of company mission statements, [1]Product or service, market, and technology, [2] Company goals: survival, growth, and profitability, [3] Company philosophy, [4] Company self-concept and [5] Public image. David (26) broadens these components to include customers, location and concern for employees. Klemm et al (27) suggest that mission statements are used first and foremost in an internal role facilitating the communication of managerial beliefs and to indicate standards of expected behaviour, with lesser use externally. The Ashridge model (28) synthesises these views into four points of a network, consisting of purpose, values, behaviour standards, and strategy.

A consistent theme within the literature is the inclusion of values, and aspects of culture within the structure of mission statements. This suggests that within the structure of statements lie the elements that define the values that exist or those which management would like to prevail within the organisation. The development of a tool to examine values within an organisation can therefore use mission statements as its basis, relying on no underlying precepts.

3.2 IDENTIFYING VALUES WITH MISSION STATEMENTS

Organisations have wide and diverse forms depending on the markets they serve, manufacturing technology employed and business processes followed. This variety will be reflected in the value systems contained within each company. This results in a large number permutations and combinations of value types. In developing a generic framework to map organisational values the problem of including this variety within the model must be solved. One solution to this problem is to examine a wide and varied corpus of mission statements, thereby including a wide verity of value systems.

Abrahams (29) analysis of a large corpus of mission statements has identified, on the basis of occurrences, a series of key words. Each word can be associated with a particular organisational element. Examining the resulting word list produces meaningful groupings or value sets based on six major functions (Table 2). The six headings selected form a broad umbrella of organisational function and are therefore sufficient to describe any operation that an organisation must perform in order to satisfy either internal or external demands.

4.0 CONCLUSION

The power of values within organisations can only be developed into operational practice and behaviour by the clear communication of the prevailing or desired values(30). It can therefore be concluded that an individuals identification with and alignment to organisational values is a key factor. Such alignment can be interpreted as the fit of values between individuals within the organisation hierarchy.

In order to measure the values fit within organisations and determine if this influences organisational performance and achievement of world-class objectives application work is now being undertaken using the following methodology:- [1] Benchmarking based on the EQM model, [2] Benchmarking based on the Change Assessment Orientation Survey

(Ch.A.O.S.) (31), [3] Interviews with a cross section of employees using the value sets identified here.

During each interview, individuals are presented with the value word sets, organised as two identical columns. They are instructed to select the thirty words on the list that they most associate with the company in a positive way.

Results have now been obtained from seven manufacturing and service related companies within the UK and the results will be published in further articles.

Table 2- Mission Statement Derived Value Sets

Direction	Control	Market Orientation	Integrity	Involvement	Development
Security	Harmony	Unique	Conscience	Joy	Tomorrow
Accountability	Communicate	Accomplished	Reliable	Exciting	Solution
Strategy	Dedication	Environment	Ethics	Passion	Initiative
Mission	Return on Equity	Customers	Dignity	Enthusiasm	Ability
Direct	Productivity	Service	Principles	Pride	Diversity
Risk	Profit	Success	Trust	Employees	Potential
Responsibility	Shareholders	Long term	Fair	Empower	Future
Strengths	Growth	Serve	Respect	Communities	Change
Leadership	Value	Best	Goodwill	Mutual	Innovation
Values	Quality	Excellence		Dedicated	
Support	Asset			Corporate citizen	
Goal	Performance			Fun	
Commitment				Relationship	
Vision				Life	
Leader				Teamwork	
				Individual	
				Team	

REFERENCES

(1) Carlos Cordon and Thomas E. Vollmann, Perspectives for Managers Report, Number 8, July 1997, IMD Lausanne

(2) Pocket Britain in Figures, The Economist, 1997 Edition

(3) Tony Charles, Rosemary Charles and Alan Roulstone, New Technology and Social Organisation in British Manufacturing, EC/FAST: European Assessment Project Final Report, CEE-FAST Programme, Bruxelles, December 1990.

(4) Made in Europe - A Four Nation Best Practice Study, IBM Consulting Group and London Business School, 1994.

(5) Robert Collins, Carlós Cordon and Denyse Julien, Lessons from the 'Made in Switzerland' Study: What Makes a World-class Manufacturer, European Management Journal, 1996, Volume 14, Number 6, Page 576 to 589.

(6) Nichols, T., The British Worker Question, Routledge, London, 1986

(7) Will Hutton, The State We're In, Vintage, 1996

(8) Beacon, R.W. and Eltis, W.A. (1974), The Age of US and UK Machinery, NEDO

(9) Daly, A., Hitchens, D. and Wagner, K., Productivity, Machinery and Skills in a Sample of British and German Manufacturing Plants: Results of a Pilot Survey, National Institute Economic Review, February, 1985.

(10) Prais, S., Some International Comparisons of the Age of the Machine Stock, Journal of Industrial Economics, 1986, Volume 34, Number 3, Page 261 to 277.
(11) Steedman, H., and Wagner, K., A Second Look at Productivity, Machinery and Skills in Britain and Germany, National Institute Economic Review, November, 1987
(12) Mason, G., van Ark, B. and Wagner, K., Productivity, Product Quality and Workforce Skills: Food Processing in Four European Countries, National Institute Economic Review, February, 1994.
(13) Mason, G. and Wagner, K., Innovation and the Skills Mix: Chemical and Engineering in Britain and Germany, National Institute Economic Review, May, 1994
(14) Mason, G. and Finegold, D., Productivity, Machinery and Skills in the United States and Western Europe: Precision Engineering, NIESR Discussion Paper, Number 89, 1995
(15) D.M.W.N. Hitchens and J.E. Birnie, The United Kingdom's Productivity Gap: Its Size and Causes, OMEGA International Journal of Management Science, 1989,Volume 17, Number 3, Page 209 to 221.
(16) P.E. Waterson et al, The Use and Effectiveness of Modern Manufacturing Practices in the United Kingdom, ESRC Centre for Organisation and Innovation, Institute of Work Psychology, University of Sheffield, UK, 1997.
(17) Nick Oliver, Making it in Britain? Japanese Manufacturers in the UK in the 1990s, Judge Institute of Management Studies and the Engineering Department Cambridge University, June 1993.
(18) Survey of Manufacturing Companies Project Report, Coventry and Warwickshire World Class Partnership, The Management Consulting Group, July 1997
(19) Kevin H. Pratt and Brian H. Kleiner, Towards Managing By a Richer Set of Organisational Values, Leadership & Organisational Development Journal, Volume 10, Number 6, 1989, Page 10 to 16
(20) Rohit Deshpandè and A. Parasuraman, Linking Corporate Culture to Strategic Planning, Business Horizons, Volume 29, Number 3, May-June 1986, Page 28 to 37.
(21) Kotter, J., Heskett, J., Corporate Culture and Performance, The Free Press 1992
(22) Collins, J., Porras, J., Built to Last, Century, Random House, 1995
(23) Curt Coffman and Jim Harter, A Hard Look at Soft Numbers, The Gallup Organisation, 1997
(24) M.J. Hooper and S. Martin, Benchmarking for Best Practice, Coventry University, 1996
(25) John A. Pearce II, The Company Mission as a Strategic Tool, Sloan Management review, Volume 23, Spring 1982, Page 15 to 24
(26) Fred R. David, How Companies Define Their Mission, Long Range Planning, Volume 22, number 1, Page 90 to 97, 1989
(27) Mary Klemm, Stuart Sanderson and George Luffman, Mission Statements: Selling Corporate Values to Employees, Long Range Planning, Volume 24, Number 3, Page 73 to 78, 1991.
(28) Andrew Campbell and Sally Yeung, Creating a Sense of Mission, Long Range Planning, Volume 24, Number 4, Page 10 to 20, 1991.
(29) Jeffrey Abrahams, The Mission Statement Book, Ten Speed Press, 1995
(30) John Humble, David Jackson and Alan Thomson, The Strategic Power of Corporate Values, Long Range Planning, Volume 27, Number 6, Page 28 to 42, 1994.
(31) Richard Hale and Peter Whitlam, Towards the Virtual Organization, McGraw-Hill Book Company, 1997.

The development of a user-oriented information systems development strategy

M INCE and **J KAY**
School of Industrial and Manufacturing Science, Cranfield University, UK

The research described here examines the input of potential users in the process of developing information systems. The investigation has shown that most companies leave the process of capturing user requirements in the hands of the developer, which is highly dependent on past experience. This could mean that a less experienced developer may obtain lower quality information. Further investigation has led to the conclusion that the involvement of the user plays a large role in the final success of the system. The steps necessary to develop and test an improved method of including user involvement are described.

1. INTRODUCTION

There has been a great deal of research that assesses the potential relationships between the systems development team and the end users of the systems. However, there is not as much detail directed towards methods of actually extracting the information from the end user. The lack of expertise in this area means that there is a small gap in the existing collection of knowledge. However, there are methods being developed that acknowledge the need for user involvement and show how the user could play a larger role in the process of systems development. These methods are identified in this paper together with their good points and weaknesses in relation to their method of capturing user requirements. The methods present some useful aspects that can relate to user involvement. These will be investigated and the most favourable areas will be used to make the new method benefit potential users.

This research aims to identify organisational attitudes towards user involvement in relation to the development of Information Systems. It will utilise previous literature within this area and make intelligent assumptions about methods that have been used in the past. The key objective of the research is to identify a method for achieving successful user involvement. This proposed method will aim to achieve an increase in the quality of information gained from the user, leading to higher levels of satisfaction and more efficient usage of the information systems.

1.1 The need for the new method

Many academics have been attempting to prove that system development methods are not as successful as they may seem. Since the 1980s, there has been a recognised need for improving the way that systems developers interact with the end users of the system. Cheney and Dickson (1) state that according to their research findings the end user has become crucial to the success or failure of a system. This type of evidence leads to various assumptions. The first is that the end user is a very important factor. Therefore methods that are developed to make the job of interacting with the user easier is vital to the success of a system. The second being that once the end user is considered the company can work towards improving various elements that effect how productive a worker can be. For example, Yaverbaum (2) states that the computer affects worker satisfaction, worker motivation and job content. These will eventually have an effect on the use of information and the resources of the company, which in turn leads to the level of success gained by the organisation. Yaverbaum (2) also states that if the organisation hopes to realise the optimum productivity promised by high technology, there is an urgent need to focus attention to the user. Howard and Smith (3) have created a list of goals, which would lead to an increase in involvement for the end users, which should eventually increase worker satisfaction, these include: (1) Increasing user participation in the design of a system, (2) Understanding user characteristics that foster system acceptance and (3) Understanding organisational and task factors which effect user needs.

Folkes et al (4) state that the traditional methods did not satisfy the business needs of a company and that it took too long and had too many compromises to make it acceptable. Therefore, this study aims to develop a method of involving the user, which ensures that the requirements of the business are placed in the system.

1.2 The methodology adopted to develop the new method

Figure 1 demonstrates the route that will be taken towards developing the new method.

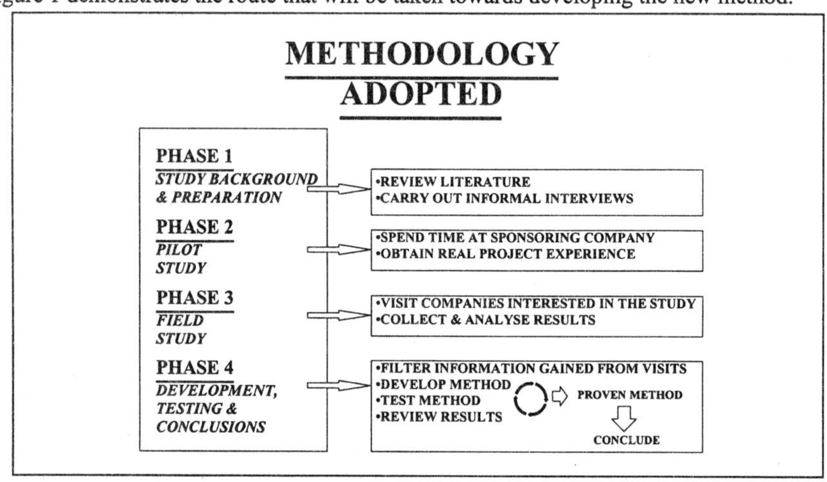

Figure 1 - Methodology adopted for the research

1.3 Where the new method fits into the Systems Development Life Cycle (SDLC)

The SDLC aims to formalise the process of systems development leading to beneficial implementations within a company. The five main stages are Requirements analysis -

identification of the needs of the system, Specification - shows the technical modelling for designers, Design - programme development, Implementation - inputting the system into the organisation and Maintenance - Upgrading and changing requirements where necessary, as shown in figure 2. The proposed method will fit within the region of the first three stages.

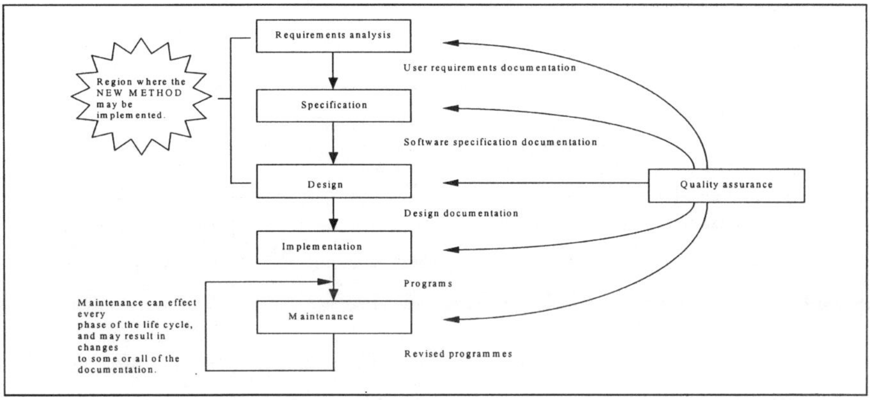

Figure 2 - New method positioned within the SDLC (Open University)(5)

2. METHODOLOGY INVESTIGATION

The current structured methods were studied in detail to discover how they can benefit companies and if they satisfy the requirements of this area of research. From table 1 it is clear that they do have many worthwhile attributes. It is also possible to see that the methods do not fully adopt a strategy to show the analyst / developer specific details of how to capture user requirements. However, it would be unfair to say that the methods do not offer factors that would help in the production of the new method. Therefore, the new method will adopt various good points which will help to satisfy the need that has been identified by the study.

2.1 Field study

In order to complete phase three of the research which is shown in figure 1 visits were made to different companies. The companies were from the manufacturing, services and Information Technology sectors (company names have been omitted for confidentiality). The results obtained so far have been analysed in order to create a general view from the companies regarding the area of capturing user requirements. A summary of the general problems are identified in figure 3.

2.2 Discussion of summarised results

The methods that were used in the companies visited included brainstorming sessions, Flow diagrams, group discussions and one to one meetings. These methods, once used in the right way may have the potential to produce better quality requirements. For example, if group sessions are used to produce the list of members to go into a brainstorming session this may be a good basis to identify the users needed for one to one sessions.

Table 1
Table identifying the various structured methods investigated, emphasising their relation to capturing user requirements

METHOD	DESCRIPTION	GOOD POINTS	USER INPUT WEAKNESSES
SSADM Structured Systems Analysis & Design Methodology	Uses various diagrammatic techniques, to show the processes that are relevant to the required system.	Uses detailed rules and standards to help structure the development process.	The technique can be too rigid and needs tailoring for different types of systems.
DSDM Dynamic Systems Development Methodology	Uses categories of well-defined prototypes for use within the development life cycle.	Well structured method that encourages user participation and prototyping for better results.	No exact structure for capturing user requirements. However, guidelines are used for involving the user.
RADD Rapid Application Data Development	Uses group sessions to extract information from users.	Brings users together. Interactive process, which uses software to represent results immediately.	Hard for participants to understand. Need for extra training which strains resources such as time and money.
RAD Rapid Application Development	Group sessions are used. Encourages speed in the analysis process.	Very fast method of development.	Harder to get very specific information. Better used as a brainstorming session.
JAD Joint Application Development	Groups session and CASE tools are used to represent the processes.	Good approach to getting the user involved. Brings the user and the developer together	Problems could come about due to technology overpowering the whole process.
GERAM Generalised Enterprise Reference Architecture and Modelling	Identifies sets of components necessary for integration of enterprise architectures.	Life cycle approach shows a specific beginning and end to the definition process.	No real user consideration. Also lacks a prototyping stage, which is necessary for testing purposes.
CIMOSA Computer Integrated Manufacturing Open Systems Architecture	Architecture used to streamline systems to the needs of the enterprise.	Separates the system into various sections. This helps to produce building blocks which can be worked on individually.	No user participation identified. It is also very complicated and requires a great deal of expertise in systems development.
GRAI Graphical Results and Activities Interrelated method	Tool used for the analysis and design of planning and control systems	Uses a structured grid to identify the importance of information. Good for analysis purposes.	Users are not seen as a major factor. The processes are shown as very broad areas. Very complicated method.
IDEF$_0$ ICAM Definition Method	Uses boxes and arrows to represent activities, inputs, outputs and controls.	Very good for modelling processes. Has set boundaries on diagram sizes.	Lack of specific user input. However, is very useful for representing basic process information.
YSM Yourdon Systems Method	Used to break down a system into easily definable components.	Very good for defining the present system and is not complicated. Not very rigid.	No specified method for involving the user. Very brief structure. Needs more controls.
OOA Object Oriented Analysis	Defines areas of systems in a way that information is related to a specified area.	Good for identifying where information originates from.	No major user strategy. It aims to give developers a method for use within their own techniques.

Figure 3 shows the type of complaints that users and developers have with regards to the system and its development. These problems could be dealt with if there was a formalised method, which documents each step to enable the end result to be produced. This will enable the systems analyst to break down the system and identify where specific problems actually come from. For example, if the users find that they cannot find the functionality, this may be a result of not enough time given for the requirements capturing process. This could also mean that there was a lack of understanding due to developers thinking that they know what users want and it is a waste of time asking potential users.

Major Problems Identified	
USERS	DEVELOPERS
• System is Slow and Old • Cannot find the functionality • Getting results from the system slows down the workflows • Lack of understanding • Errors are created by the system • Hard to get the information that is required • Lack of end user support • Users have not had enough input	• Time Factors • Missing requirements • Users not knowing what they want

Figure 3 - Brief summary of the major problems identified from the visits

2.3 How the method will be developed

Huang (6) has proposed a general methodology for the design of a development framework. This method will be used as a guide for this particular research. Figure 4 shows how this will be achieved.

2.4 How the method will be tested

The proposed method will be tested within a particular manufacturing company. This process will be used to generate results from actually implementing the method in a project. Several parameters are proposed for the evaluation of the results obtained to see which is the best for the proposed method. These include:

- Time - which identifies results obtained from using different percentages of the whole process. e.g. 5%, 10%, 20%
- Size - which looks at the number of end users using the system. e.g. 50%, 70%, 100% of all potential users.
- End user experience - This tests the method with different levels of end user experience.

The results obtained from the proposed approach will be compared to those obtained by methods used within the company. The method will be used to develop a prototype where test data may be implemented. Various points will be used to determine the success of the method. These include overall company recognition, user satisfaction, developer satisfaction and correct results created from the prototype. These factors can be obtained by using a questionnaire asking developers, users and managers questions comparing methods used in the past and the new method.

Stages	Considerations in relation to the current study
1 Requirements analysis	• Gather facts about the present situation • Identify strengths / weaknesses
2 Modelling for decision making activities	• What is needed to make a good method • What components relate to each other • What is important about their relationships
3 Modelling for process analysis	• What are the basic characteristics necessary • How should the process be represented
4 Selecting performance measures	• How should the method effect current performance
5 Compiling a manual	• Factors to determine the development of a user guide
6 Compiling workbooks	• Paying attention to easy of use • Use of a systematic procedure • Ability to get the required results
7 Verification	• What should be tested • How will the verification process take place • How can the method be improved

Figure 4 - Development process, adopted from Huang (6)

2.5 Initial Findings

There are a few important observations that have been produced by the study so far. These are as follows:

- Systems are failing to satisfy the needs of the users.
- Companies are spending excess time and money on information systems.
- Systems produced in the past are not very easy to change.
- There is no real structure used for capturing user requirements. This means that there is no basis for repeating the process even if things went well.
- Developers are quite clear about how they feel that the current methods should be improved. For example, developers have said that "There is a need for more user participation", "There are not many standards or guidelines to say what should go into the requirements", and "There is a need for a stricter documentation procedure". Further investigation is needed to determine why things have not changed. However, companies are recognising the need for structure and it is likely that things will change in the future.
- It is also clear from the study that user involvement does improve the willingness of users to use the system. This is shown from the answers to the question, "Would the fact that you were involved increase your willingness to use the system?" Every member who was involved said YES.

The results obtained here show that users are getting involved. However, if we look at questions relating to functionality there are still problems especially with users not being able to find what they require and also that there is an overall lack of confidence in the actual use of the systems.

3. CONCLUSION

The study has produced some results that will be used to define a formal structure for use within companies who are involved in developing information systems. It is clear that there is a need for something that would assist the process and could benefit the production of more efficient and maintainable systems.

ACKNOWLEDGEMENTS

I would like to thank BICC Cables and the companies who have been involved in the study for their technical support within this research.

REFERENCES

1. Cheney, P and Dickson, G. Organisational Characteristics and IS: An Exploratory Investigation. *Academy of Management Journal*, March 1982.
2. Yaverbaum, G. The technical revolution: Organisations must take positive actions to reduce stress caused by computers. *Journal of Systems Management*. December 1982.
3. Howard, J and Smith, R. Computer Anxiety in Management: Myth or Reality? *Communications of the ACM*. July 1986.
4. Folkes, S. 1992 *Accelerated systems development*. Prentice Hall 2nd
5. Open University. 1994. *Software Engineering*. Open University
6. Huang, G 1996. *Design for X. Concurrent Engineering Imperatives*. Chapman and Hall.

The nature of performance measurement systems and measures in UK industry

D MEDORI
Manufacturing and Supply Chain Management Centre, Coventry University, UK

This paper details results from empirical research carried out between 1995 and 1997. This research was based around developing and testing an integrated performance measurement framework and assessing the nature of performance measurement systems and measures.

The research led onto an investigation into 12 organisations' performance measurement systems paying particular attention to non-financial measures.

This paper will outline and discuss the breakdown of the performance measurement systems, highlighting certain measures which were termed as being common, and so illustrating the nature of measurement systems and measures, in use and in industry today.

1 INTRODUCTION

Performance measurement which is the process of quantifying the efficiency and effectiveness of actions has become a topic of considerable interest through almost all of the 1990s. Performance measures are individual metrics which are contained within a performance measurement system. A performance measurement system can be defined as a set of metrics, while an integrated performance measurement system is a set of metrics that are derived from a variety of competitive priorities.

Eccles (1) had identified in the '90s that a performance revolution had been launched. Kaplan and Norton (2) stated that a performance measurement system is an integral part of the management process; Vitale and Mavrinac (3) claimed that a measurement system is a strategic investment in itself, and Drucker (4) stated that establishing performance measurements is one of the five basic principles of management.

1.1 The purpose of performance measurement systems

Performance measurement systems have a number of purposes:

- Monitoring of performance so identifying areas in need of attention.
- Influencing behaviour so affecting the implementation of strategy (4). A measurement system promotes self diagnosis, with measures indicating whether strategies are being achieved.
- Tracking the implementation of change and acting as a catalyst for change. Once a company recognises that change is needed, it first has to identify what needs to be changed and what is the target. Performance measures can then be used effectively as a means of setting goals and quantifying the progress.
- Fostering teamwork by linking business functions together and focusing on results of the entire organisation.
- Identifying and adding focus to early measures that are correlated to desired end results.
- Improving communications whilst stimulating learning and enhancing motivation.

2 HISTORY OF RESEARCH PROGRAMME AND PROBLEM

The research programme was formed through both academic and industrial interest. Performance measurement incorporating both financial and non-financial measures has gained much interest and has been advocated by many academics, consultants and professionals, for example (5,6,7,8).

While the importance of performance measurement and measurement systems incorporating non-financial measures had been well documented, there were problems associated with establishing measurement systems. A major problem in establishing performance measurement systems especially incorporating non-financial measures is the selection of measures from the sheer plethora of measures that can be used. The solution to the problem resulted in frameworks to guide companies to set up performance measurement systems.

Davis and O'Donnel (9) had noted that many companies have problems trying to establish effective performance measurement frameworks. Dumond (10) observed that while there is much written, there is little consensus regarding definitions, methods of measurement, or even what should be measured, and White (11) stated that there appears to be little, if any agreement on precisely which measures to use.

A major literature review identified that there are only a few intellectual frameworks for tackling performance measurement, for example (5,7,12). Each framework or system has its advantages and disadvantages with there being no clear framework, consensus or industry standard relating to the content or use of non-financial measures. Neely et al (13) identified that no guides of how to improve or upgrade measurement systems exist, and a major report and survey undertook by CIMA (14) found that there is no optimal set of measures in practice or theory to monitor the performance of manufacturing companies. What was identified by CIMA is that a measurement system approach to establish a range of measures is needed.

The research programme which evolved from the literature review and associated problems, was split into two fundamental areas:

- Development of an integrated performance measurement framework for medium to large-sized manufacturing organisations to use in order to audit and improve current performance measurement systems in line with world class manufacturing; and

- To assess the nature of current non-financial performance measures and measurement systems in medium to large-sized organisations.

3 RESEARCH METHODOLOGY ADOPTED

The first area which consisted of the development of an integrated measurement framework, was developed in a deductive action research mode. The framework was applied and tested with great success in 5 medium to large-sized manufacturing organisations, where 4 of the company measurement systems were improved as a result of using the framework.

The remaining part of the paper will concentrate on the second area of assessing the nature of measures and measurement systems in medium to large-sized manufacturing organisations. This area of research was based around an inductive ethnography approach in which 12 measurement systems from 12 manufacturing companies in the UK were observed.

The 12 companies that participated with the research (shown below) were all manufacturing based, but from different manufacturing sectors in order for a cross section of measurement systems to be assessed. Indeed, over half of the companies were perceived as being World Class Manufacturing (WCM) companies.

The common denominator with all the companies were that they were following WCM policies. These policies are in fact ingredients that WCM companies possess. These policies being: Total Quality Management, Just-In-Time, Employee Involvement and Teamworking, Flexibility, Close Supplier Links, Total Productive Maintenance, Continuous Improvement and Integrated Performance Measurement.

Sector and size of collaborating companies

Company	Manufacturing Sector	Employees	Turnover
Company A	Body in White Assemblies	800	£120 Million
Company B	Electronic Automation	250	£20 Million
Company C	Diesel Engines	2,000	£600 Million
Company D	Automotive Engines & Gearboxes	4,500	£1,500 Million
Company E	Automotive Polymer Components	557	£70 Million
Company F	Food Process Equipment	900	£80 Million
Company G	Agricultural Equipment	1,500	£200 Million
Company H	Automotive Mechanical Components	450	£80 Million
Company I	Electronic Automation	200	£15 Million
Company J	Pneumatic Control Equipment	400	£30 Million
Company K	Fibreglass	500	£45 Million
Company L	Tyres	900	£170 Million

4 RESULTS

The company performance measurement systems were broken down into six competitive priorities: quality, cost, flexibility, time, delivery and future growth. These priorities were widely accepted by the companies that collaborated with the research; as competing effectively in these priorities were seen as critical in achieving world class manufacturing status. The cost competitive priority although financial based, involved only manufacturing or product cost measures, and did not contain higher level financial accounting measures which are standard to all companies. The rest of the measures in the other competitive priorities were all non-financial based.

The research work carried out in the companies was in the segregation of individual measures (if not already done so by the company) by competitive priority. The results shown in the table below identify that all the companies have a balanced set of financial and non-financial performance measures.

Break-down of company performance measurement systems by competitive priority

Number of Measures According to Competitive Priority							
Company	**Quality**	**Cost**	**Flexibility**	**Time**	**Delivery**	**Future**	**Total**
A	4	7	2	2	2	1	**18**
B	9	8	2	1	3	2	**25**
C	6	9	4	3	4	6	**32**
D	4	6	2	4	6	5	**27**
E	8	4	2	1	1	3	**19**
F	8	13	2	3	7	4	**37**
G	9	9	2	4	4	8	**36**
H	13	12	3	5	3	4	**40**
I	5	10	6	7	5	5	**38**
J	8	7	3	1	3	2	**24**
K	9	12	4	6	3	2	**36**
L	6	9	4	2	3	4	**28**
TOTAL	89	106	36	39	44	46	**360**
MEAN	**7**	**9**	**3**	**3**	**4**	**4**	**30**
% MEAN	**23%**	**31%**	**10%**	**10%**	**13%**	**13%**	**100%**

From the above results it is clearly illustrated that there is a variety in the number of measures employed by each company and in each competitive priority. This result was not surprising as most of the companies were engaged in different manufacturing sectors having different manufacturing strategies, hence having different performance measurement systems.

What became apparent from the results is the spread of measures from each competitive priority: on average most measures fell under the cost competitive priority, closely followed by quality measures, with there then being a fairly large gap in relation to the other four competitive priorities.

In total from the 12 companies, 102 different types of measures were employed. Although a wide variety in terms of type and number of measures were used between the companies it was found that 27 of these measures were similar or common to a number of companies. The table overpage illustrates 2 measures from each competitive priority which were determined common; how they were determined common was from the number of companies which utilised the same measure, which is also illustrated in the table.

Common measures in 12 companies with WCM policies

Ranking	Competitive Priority	Type of Measure Utilised	Companies Utilising Measures	Total
	Quality			
1		Supplier Quality	A,B,C,D,E,F,G,H,I	9
2		Warranty Claims	B,C,D,G,H,I,J,K,L	9
	Cost			
1		Absenteeism	A,B,C,D,E,F,G,H,I,J,K,L	12
2		Scrap	A,B,C,D,E,F,G,H,K,L	10
	Flexibility			
1		Cross-Training	A,D,E,F,G,H,I,J,K,L	10
2		Engineering Changes	B,C,F,G,H,I,K,L	8
	Time			
1		Lead-Time	A,B,C,D,E,F,G,H,I,K,L	11
3		Material Availability	C,D,G,H,I,K	6
	Delivery			
1		Schedule Adherence	A,B,C,D,E,F,G,H,I,K	10
2		Customer Late Delivery	C,D,F,G,I,J,K,L	8
	Future Growth			
1		Number of New Product Introductions	C,E,G,H,I,L	6
2		Market Share	G,I,J,K,L	5

5 DISCUSSION

Initially from the results it can be observed that all the companies had a balanced set of performance measures. This was as expected, as most of the companies were perceived to be world class manufacturers, and a policy of a WCM company is the adoption of an integrated measurement system - incorporating both financial and non-financial measures. This equates to the system being balanced.

The results also illustrate that medium to large-sized manufacturing organisations have on average 30 measures with most measures being cost (31%) and quality (23%) based. The remaining measures then fall fairly evenly under the following priorities: flexibility (10%), time (10%), delivery (13%) and future growth (13%).

It was discussed that these measurement systems had evolved over time and that the companies ideally look to reducing the number of measures within the system. Quality and especially cost based measures tend to be universally emphasised well beyond their significance; it is measures based under these priorities which are likely to be reduced in the future. It was noted that as non-financial measures gain in popularity, then cost based financial measures will decrease in popularity, hence leaving measurement systems to be more equally weighted in terms of number of measures in each priority.

Finally from the results, it was shown that three quarters of the observed company measurement systems' employed what was termed as common measures: absenteeism, lead-time, cross-training, scrap, schedule adherence, supplier quality and warranty claims.

6 CONCLUSION

From an in depth analysis of the 12 companies' measurement systems it was shown that the systems were balanced, albeit somewhat unevenly. This illustrates that companies wishing to proceed along the path of WCM will need to adopt such systems. From the research it was found that there were no worst or best set of measures, rather companies have developed their own judicious permutation of measures applicable to their requirements.

A number of measures were identified as common in the research and these measures could be seen as an ideal base for companies setting up or modifying a measurement system, as the measures are structured around six essential competitive priorities.

6.1 Acknowledgements

The author is grateful to the manufacturing organisations that collaborated with the research, and the Economic and Social Research Council (ESRC), for the award of a research grant.

7 REFERENCES

1. Eccles, R.G. (1991), "The performance measurement manifesto", *Harvard Business Review*, January-February, pp. 131-137.
2. Kaplan, R.S. and Norton, D.P. (1993), "Putting the balanced scorecard to work", *Harvard Business Review,* Vol. 71 No. 5, September-October, pp. 134-147.
3. Vitale, M.R. and Mavrinac, S.C. (1995), "How effective is your performance measurement system?" *Management Accounting (USA),* August, pp. 43-47.
4. Drucker, P. (1994), *Management: tasks, responsibilities, practices,* B'worth-Heinemann.
5. Maskell, B.H. (1991), *Performance Measurement for World Class Manufacturing,* Productivity Press, Cambridge Massachusetts.
6. Hall, R.W., Johnson, H.T. and Turney, P.B.B. (1991), *Measuring Up - Charting Pathways to Excellence,* Irwin, Homewood, IL.
7. Kaplan, R.S. and Norton, D.P. (1990), "The balanced scorecard - measures that drive performance", *Harvard Business Review,* Vol. 70 No. 1, January-February, pp. 71-79.
8. Drucker, P.E. (1990), "The emerging theory of manufacturing", *Harvard Business Review,* May-June, pp. 94-102.
9. Davis, A. and O'Donnel, J. (1997), "Modelling complex problems: systems dynamics and performance measurement", *Management Accounting (UK),* Vol. 75 No. 5, pp. 18-20.
10. Dumond, E.J. (1994), "Making best use of performance measures and information", *International Journal of Operations and Production Management,* Vol. 14 No. 9, 16-31.
11. White, G.P. (1996), "A survey a taxonomy of strategy-related performance measures for manufacturing", *International Journal of Operations and Production Management,* Vol. 16 No. 3, pp. 42-61.
12. Cross, K.F. and Lynch, R.L. (1989), "The 'SMART' way to define and sustain success", *National Productivity Review,* Vol. 8 No. 1, pp. 23-33.
13. Neely, A., Mills, J., Platts, K., Gregory, M. and Richards, H. (1994), "Mapping measures and activities: a practical tool for assessing measurement systems", *Proceedings of the First International Conference of the European Operations Management Association,* University of Cambridge, Manufacturing Engineering Group, UK, pp. 313-318.
14. CIMA. (1995), *Performance measurement in the manufacturing sector.*

Development and implementation of manufacturing strategy for small-medium enterprise (SME)

L MITCHELL
University of Hertfordshire, Hatfield, UK

ABSTRACT

The development and implementation of manufacturing strategy for small-medium enterprise (SME) has been aimed at developing a proper conceptual base by an action research methodology. This has led to a low-cost competitive advantage for manufacturing operations. The manufacturing strategy includes a unified modelling framework and utilises a 10-module manufacturing plan. The author using an engineering orientated definition and a detailed top-down approach has developed the framework. The development of the manufacturing strategy has led further to the design of a modular integrated systems strategy (MISS) programme, and its implementation at several SME cases with proven efficiency gains. The development process of the manufacturing strategy has been based on a body of applied knowledge gained from academic research, strategic models, engineering experience and management practice. The manufacturing strategy has also been planned and designed with the focus on control of manufacturing processes, maintenance criticality of plant and machines, and low-inventory manufacturing and material-flow movement as employed in many SMEs.

1. INTRODUCTION

MISS is a manufacturing framework with far-reaching implications. The basic concept is relatively simple: identify and re-design the major/minor sub-systems to take advantage of the total manufacturing system process. The various modules of MISS are considered with the resultant effect on company organisation with emphasis placed on chosen Production and Operations (POM) concepts and techniques which facilitate the rationalisation process. It is a technological improvement and it is from this class of change that SMEs gain the largest manufacturing productivity increases; as referenced in Skinner (1), Hill (2), Parnaby (3), Mitchell (4)(5). The initiative in the eventual design of the modular integrated systems strategy involved selected production and operations (POM) performance measurement studies at several SME cases of: plant and machine maintenance criticality, material-flow movement, how factories were loaded, how orders were placed, and how work was scheduled through facilities i.e. low-inventory manufacturing and material-flow movement as researched by Mitchell (5)(6)(7).

Conceptually the exercise was deceptively simple i.e. analyse the current and expected future situation for competitive advantage, determine the direction of the SME cases and design the integrated systems strategy for achieving the implemented manufacturing strategy. In reality this was to be an extremely complex process, which demanded a systematic approach for identifying and analysing factors external to the SME cases and matching them with their own capabilities.

2. THE MODULAR INTEGRATED SYSTEMS STRATEGY (MISS)

The flow-process of the modular integrated systems strategy process is shown in FIGURE 1 and systematically addresses market requirements, the performance measurements of existing manufacturing systems and current manufacturing practice, with a view to identifying key areas for improvement. Particular attention is given to the practical aspects of carrying out the process towards its implementation over a time-phase, and the action research covers the theory underlining the approaches. The 10-module manufacturing action plan is shown in TABLE 1, and includes the MISS framework which builds upon selected production and operations management performance measures and sets of results e.g. SWOT analysis and systems approach checklists, a sub-systems (activities) re-design and re-build process, and a time-phase analysis at the beginning of the manufacturing system design process. Only after feasible alternatives have been identified can the integrated systems strategy be designed necessary for full implementation.

At all of the SME cases real operational data-input and observations offered the most direct opportunities for full management decision analysis and support, as the detailed MISS framework was critical to the consistency of the overall manufacturing strategy.

3. THE DEVELOPMENT PROCESS AND THE 10-MODULE FRAMEWORK

Models that are sufficiently sophisticated to reflect accurately the real world are often large and complex. As a result models tend to be difficult to understand and to explain to others. Furthermore, inconsistencies, omissions and many other errors are difficult to detect, as discussed in the book by Dunn and Ramsing (8). One way to minimise these problems, was to use a detailed top-down approach and to ensure that where possible the MISS framework was built up of discrete modules, any of which may be amended or replaced without confusing the framework structure. Thus, the manufacturing strategic process was broken down into manageable parts, by dividing the MISS framework into a 10-module manufacturing action plan as shown in TABLE 1. The decomposition of the framework process is not only in agreement with engineering and management practice, but is closely related to the composition and design of any integrated systems strategy as identified in Law (9).

The MISS framework can be tailored and refined to suit the needs of SME manufacturing enterprises, as a systematic step-wise (incremental) approach to the analysis is used. This promotes the idea of solving problems through increased understanding, awareness and questioning. This leads to the implementation of the manufacturing strategy and the management of change on time, at least-cost, and with minimum disruption to the people involved and the work they perform.

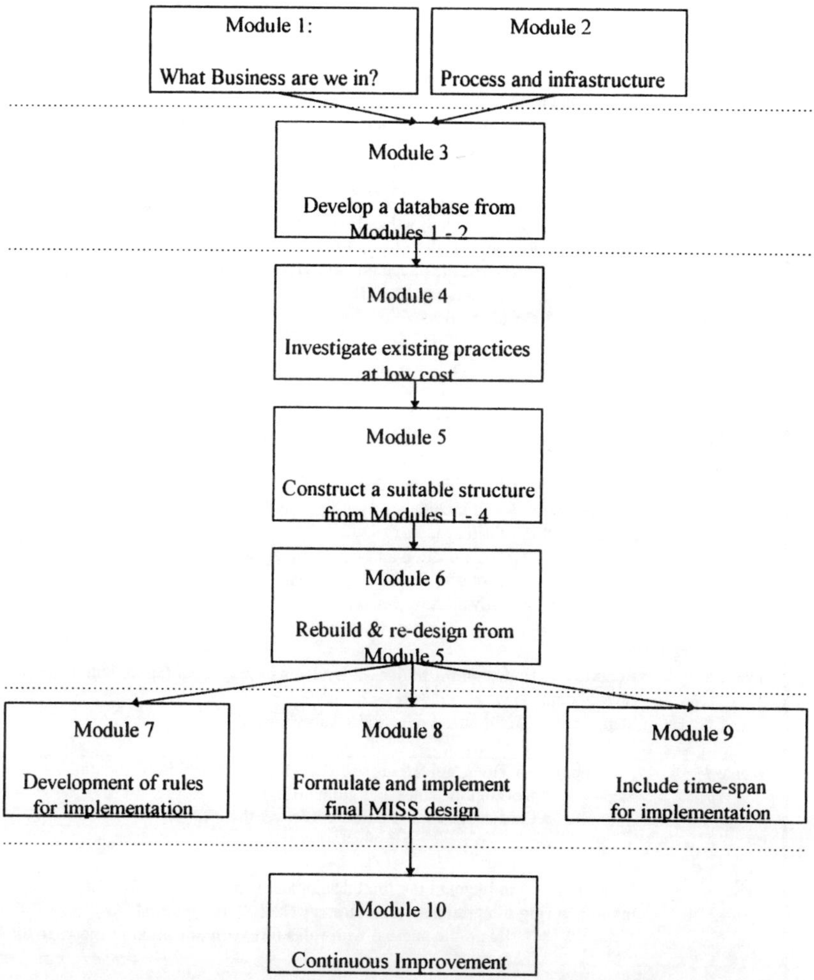

Figure 1 - The MISS Flow Process

Module 1	What business are we in? Steps * Use SWOT analysis to identify the major strategic problems. * Use the major/minor systems approach checklists to identify the manufacturing system problems and weaknesses.
Module 2	Process and infrastructure! Steps * Investigate the initial state of the company and measure all major sub-systems.
Module 3	Develop a database of analysed and refined information from ***modules 1-2***. Steps * Rationalise all existing data. * Introduce the development of knowledge systems for computer integration e.g. MRP II. * Develop rules.
Module 4	Investigate existing practice at low cost! Steps * Use selected production and operations management POM concepts and techniques to measure the identified weaknesses.
Module 5	Construct a suitable structure (rebuild and re-design) of integrated manufacture from ***modules 1-4*** to include any or all of the following; Steps * 1 - Factory layout 2 - Organisation and other management systems 3 - Production planning and control systems 4 - Advanced technology
Module 6	Rebuild and re-design the major sub-systems based on ***module 5***, with the new revised data. Step * Simulate the new major sub-systems.
Module 7	Development of rules, guidelines and checklists for MISS implementation! Steps * Develop the rules and guidelines. * Create a rules and guidelines checklist.
Module 8	Formulate and implement the final design and recommendations of the manufacturing integrated systems strategy (MISS) as a manual. Steps * Write up the manual with rules for communication purposes to all personnel.
Module 9	Include a time-span for implementation. Steps * Use Gannt chart, CPA or PERT techniques.
Module 10	Introduce a monitoring and feedback mechanism (CI) after implementation for the MISS framework. Step * Continuous Improvement for ***modules 1-9***.

4. TESTING AND APPLICATION OF THE (MISS) DESIGN PROCESS

The 10-module manufacturing action plan shown in TABLE 1, was developed and designed for all SME cases. The action plan included the process of integrated systems strategy formulation; the process was implemented and tested at several SME cases with proven and recorded gains as shown in Mitchell (5). The principal advantages of having a unified framework based on the selected use of production and operations management performance measures for the design of integrated systems strategy are summarised as follows:

1. It can support any discrete batch manufacturing SME or SBU.
2. The behaviour of the sub-systems (activities) represented, as data is simple to update.
3. Selective areas from the modules can be investigated even further.
4. The user can specify the individual module and the detail of performance measurements and results.
5. Integration is achieved from the modular re-design and re-build of the sub-systems (activities).
6. The ability to integrate the total manufacturing system with all sub-systems (activities) is possible.

5. RESULTS AND GAINS FROM THE (MISS) APPROACH TO FACTORY INTEGRATION

Each step in the MISS framework and the development of the manufacturing strategy for SMEs resulted in noticeable productivity gains, without prohibitive costs of larger-scale projects as proven and evidenced by Mitchell (5). The modular approach requires a hierarchical organisation of factory equipment, use of selected production and operations concepts and techniques and a management of change approach. Modules have been arranged within a hierarchy according to complexity. Sub-systems (activities) are re-designed and re-built from the chosen modules and the factory encompasses the integrated operation of all-existing components and sub-components. The modular approach involves total integration of the elements of the sub-system (activities), followed by integration of specific manufacturing cells chosen through the group technology (GT) concept. However, it differs from other approaches to factory system integration, in that it incorporates a greater degree of local control, database storage and manufacturing control into each individual module. This allows tremendous flexibility in the final manufacturing strategy. Thus, any SME can apply any degree of integration desired within its own manufacturing strategy. It can decide to integrate only one critical machine, or it may decide to integrate all critical plant and machines and processes affecting one type of product. Because, an entire factory can not change at once, the MISS framework suggests rationalisation of a vertical slice of the manufacturing operations i.e. choosing one group of products and integrating every aspect including the management decision making process e.g. people, production scheduling and control, and all manufacturing operations. Such an integrative approach is consistent with the modular approach presented.

6. CONCLUSIONS AND RECOMMENDATIONS

The 10-module manufacturing action plan is used to investigate changes in manufacturing systems and recommend a suitable manufacturing strategy for SME competitive advantage. The performance measurements within the modules can be monitored, divergences from the projected performance analysed and the modules thereby refined and tailored for applying to other European SMEs. The modelling triad concept from the book of Dunn and Ramsing (8), was chosen as the approach to a better understanding of the methods of problem solution, and the resulting potential for the decisions from the results. The MISS framework shown in TABLE 1, follows a sequence of modelling, modifying and applying for each sub-system (activity) within the manufacturing system. Where the use and application of the results and gains give a better understanding of the relationship of variables and problems that become more evident with practice and implementation. The MISS framework is based on strict time-scales and the concept and its related factors have shown to be of great significance in evolving total manufacturing strategy and control to achieve high business profitability, responsiveness and growth for SMEs looking towards manufacturing excellence and survival for the millennium and onwards.

7. REFERENCES

1. Skinner, W. (1978). "Manufacturing in the corporate strategy". Wiley, New York.
2. Hill, T.T. (1985). "Manufacturing strategy-the strategic management of the manufacturing function". Macmillan, London.
3. Parnaby, J. (1979). "Concept of a manufacturing system". Int. Jour. Prod. Res. Vol.17, No.2, 123-135.
4. Mitchell, L. (1985). "Advanced manufacturing technology at an electronics company". MBA (Ind. Mgt.) Thesis, University of Newcastle-upon-Tyne.
5. Mitchell, L. (1996). "Development of manufacturing strategy for the small-medium enterprise (SME): theory and practice". PhD Thesis, University of Hertfordshire.
6. Mitchell, L. and Spurgeon, D. (1990). "Manufacturing strategy within a sheet-metal manufacturing facility". OMA (UK) Conference and Proceedings.
7. Mitchell, L. and Spurgeon, D. (1991). "Work scheduling through manufacturing cells". Int. Jour. of Integrated Manufacturing Systems. Vol.2, No.1, 30-32.
8. Dunn, R.A. and Ramsing, K.D. (1981). "Management science-a practical approach to decision making". Chapter 2; (The Model Triad), Macmillan Publishing Co.
9. Law, A.M. and McComas, M.G. (1989). "Pitfalls to avoid in the simulation of manufacturing systems". Industrial Engineer, 28-31.

Benchmarking small and medium-sized manufacturing enterprises

G P NELDER and **A SKANDALAKIS**
The CIM Institute, Cranfield University, UK

Abstract: Benchmarking can be described as an alliance between partners to share information on practices, processes and measures to stimulate innovation improvement in corporate performance. Small and Medium-sized Manufacturing Enterprises are reluctant to participate in benchmarking studies due to lack of time, financial and personnel resources, and difficulty in selecting partners. A Diagnostic Benchmarking process has been developed to meet the requirements of smaller enterprises. The process involves a business Needs Analysis that provides participating companies with a fresh perspective on their strengths and weaknesses. The analysis connects to business performance, and is supplemented by a benchmarking study to help the company progress toward best practice excellence.

1 INTRODUCTION

In today's challenging business environment, Small and Medium-sized Manufacturing Enterprises (SMMEs) experience a plethora of problems affecting their growth, profitability, success and survival. Whilst some may command a niche market, many others are minor players at the end of the supply chain, and have to conform to the demands of Original Equipment Manufacturers (OEMs). Faced with increased competition at the local, national and transnational level and emerging pressures from their supply chain, most SMMEs remain uncertain about their longer-term future.

To combat these emerging pressures, SMMEs must continuously reduce their cycle time, introduce cheaper products more quickly, with higher quality and that better satisfy customer and market needs [1]. Realising and maintaining coherence between continuous improvement and day-to-day management is of increased importance for success. SMMEs that fail to embrace continuous improvement and development initiatives to enhance their performance and competitiveness imperil their sustainability.

The authors describe the Diagnostic Benchmarking processes developed by the ADAPT-CORE project to assist with the business development of Small and Medium-sized Manufacturing Enterprises. Important issues are the accessibility of benchmarking to SMMEs and the identification of better practices.

2 BENCHMARKING AND SMALL TO MEDIUM-SIZED ENTERPRISES

The importance of Small and Medium-sized Enterprises is reflected by the European Commission, which concluded that they constitute 99.8% of all companies, provide 66% of total employment and attain 65% of business turnover in the European Union [2]. Enterprises with fewer than 100 employees have been responsible for almost all of recent job creation in the European Union. However, 50% of small to medium sized companies cease operation within five years of their creation, which raises concerns that SMEs constantly under-perform [2,3].

Benchmarking provides recognition that profitability and growth come from a clear understanding of how the business is doing, not just against its own performance in the previous year, but against the best they can measure [4]. The gaps in performance between the best companies and the "also-rans" can be vast: manufacturing industry leaders typically generate products up to two and a half times faster than the industry average and at half cost [5]. Potentially SMEs have the most to gain from both internal and external benchmarking since analysis of the performance gap helps to keep a company competitive [6,7]

However SMEs do not seem to engage in benchmarking activities. 'Too expensive' and 'not enough time' are two of the most common reasons why small companies think they cannot afford benchmarking [8]. Small and Medium-sized Enterprises can be left behind, operating in the same way and designing similar products as they always have. SMEs are typically put off benchmarking by what they perceive to be the considerable effort involved [9].

3 BENCHMARKING AS A DIAGNOSTIC TOOL

Benchmarking is an instrument for improvement by providing a reference point [10]. Benchmarking can be defined as a formalised quality process that is used by business organisations to measure their products, processes, services and practices against competitors or best practice companies [11]. The American Productivity and Quality Centre [12], together with the International Benchmarking Clearinghouse [13], has defined benchmarking as: "a process of continuously measuring and comparing an organisation's business processes against business process leaders anywhere in the world to gain information which will help the organisation take action to improve its performance".

Benchmarking as a diagnostic tool, involves a process of continuously measuring and comparing an organisation's performance against superior performance anywhere in the world, in any industrial sector, to gain information and knowledge on *how* premium performance can be attained. Continuous improvement is a step process and diagnostic benchmarking provides the opportunity to achieve step improvements. The emphasis is not placed on where a company is positioned on the performance chart, but on how it can improve its performance.

Small and Medium-sized Manufacturing Enterprises require a positive and proactive method of improving their business activities. The CIM Institute has developed an Information Technology-based diagnostic benchmarking tool to assist small companies toward business improvement. The tool is based on the widely recognised European Foundation for Quality Management Business Excellence model.

The Business Excellence model has an important role to play in enhancing the competitive position of European companies in the world market by reinforcing the importance of quality in achieving competitive advantage and in stimulating and assisting the development of quality improvement activities [14]. It is based on the premise that customer satisfaction, people satisfaction and impact on society are achieved through leadership driving policy and strategy, people \(employee) empowerment, resources and processes, and that this leads ultimately to business results [15].

4 THE DIAGNOSTIC BENCHMARKING SCHEME PROCESS

The scheme involves a business needs analysis study that can give participating companies a fresh perspective on their strengths and weaknesses and how these affect business performance. The study is supplemented by a benchmarking study intended to help recalibrate the company towards excellence through expertise dissemination from other small companies who have demonstrated best practice in their functions or operations. The main strengths of the scheme are that it provides the management team with an immediate 'snap-shot' of the enterprise performance and allows an incremental approach to benchmarking with realisable benefits at each stage. By using the diagnostic benchmarking tool, a company can identify and prioritise business processes or functions that need enhancement, and seek available expertise in the field.

4.1 Needs Analysis

The companies that participate in the benchmarking process are targeted under the umbrella of ADAPT-CORE, a European Social Fund programme aiming to improve the efficiency and effectiveness of European Small and Medium-sized Enterprises.

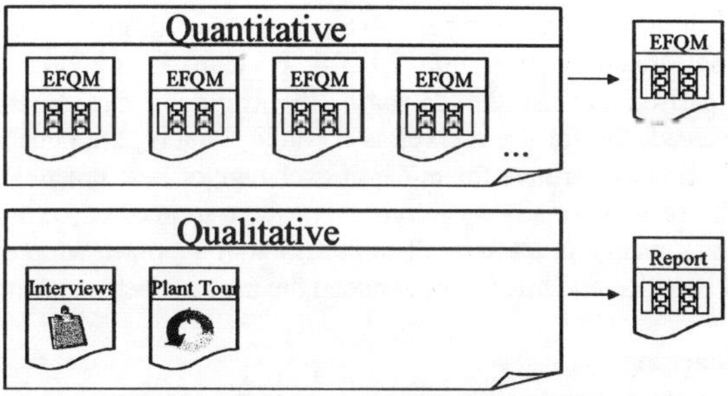

Figure 2: The Needs Analysis Process

The needs analysis is performed through the examination of qualitative and quantitative data obtained from the SMME. The quantitative data is collected using a questionnaire based on the European Foundation for Quality Management Business Excellence Model (EFQM model). Qualitative data is collected through a series of personnel interviews and a plant tour. The quantitative data questionnaire was tested for content and readability in a pilot study with a group of two hundred and forty Small and Medium-sized Manufacturing Enterprises in Spain.

To overcome the problem of objectivity, the EFQM questionnaire is distributed both horizontally and vertically within the organisation. Top and middle management datasets are essential for the analysis process, as they emphasise and actualise the basis for decision making through crucial internal information. The process also recognises that shopfloor workers are of high importance as they provide the foundations for the business operation. Ideally, returns from shopfloor workers will account for as much as forty percent of the received datasets from any given company. The completed questionnaires are subsequently analysed to provide a consensus set of data, which is employed to provide information about the company profile and is used in the diagnostic benchmarking process.

To provide a thorough review of the company, qualitative data is obtained through a series of interviews with key personnel within the company. The interviewees are chosen on the basis that their position provides or requires access to crucial internal business or production information. This process is employed to identify potential communication problems between the employees, and provide a viewpoint of the information flow within the company. A report is handed out to the company to provide recommendations and conclusions concerning their business processes.

The Needs Analysis process is carried out by experienced consultants in the fields of small enterprise management, and usually requires between two and five working days, depending on the company.

4.2 Step-by-step Benefits

The data collection and needs analysis process is integrated with benchmarking practice so that the participant company can realise benefits at every stage. This is an important reassurance for SMMEs that may not be confident of their ability to complete the whole benchmarking exercise. The horizontal and vertical distribution of questionnaires within the company allows analysis to pick up internal issues, which may not otherwise be recognised and which can often be addressed quickly and effectively. The second stage allows comparison of the company's results with both an objective standard, the EFQM, and a database of other companies, which permits a subjective comparison against market competitiveness. This may be sufficient to identify areas for improvement and motivate the company to initiate change. Repeating the analysis process after a period of time allows the company to track progress against actions taken and to measure improvement objectively against the EFQM benchmark. Importantly, by repeating the subjective comparison against the database, the company can assess whether its rate of improvement exceeds that of the market as a whole. Finally, the benchmarking process itself leads to the selection of partners for in-depth exchange of best practice knowledge. For many SMMEs, which are traditionally secretive about their sources of competitive advantage, this requires a massive change in attitudes. The incremental approach adopted allows companies to gain substantial benefits, and build confidence in the process, before taking the final step.

4.3 Benchmarking

The quantitative data is subsequently used to engage the company in cross sector benchmarking activity, aiming to disseminate best practice and share expertise from the field, and to give the company visibility of exemplars beyond their own immediate environs. Recommendations for

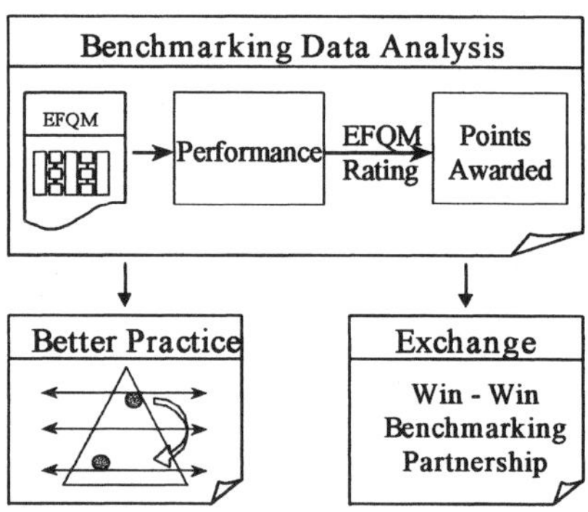

Figure 3: Diagnostic Benchmarking Process

improvement are given to the companies based on proven best practice or superior performance obtained from other companies in the database, regardless of business or industrial classification. The diagnostic benchmarking process is demonstrated in Figure 3.

5 DIAGNOSTIC BENCHMARKING IN PRACTICE

The data presented in Table 1 is based on eight participant companies from the United Kingdom, obtained since the launch of the benchmarking scheme in January 1998. The data shows that the highest priorities for companies were customer satisfaction and processes, which attained an average percentage score of 63% and 61% respectively. The lowest score was for people management with 37%. The results support the perception that SMEs focus on the short term needs of their customers and day-to-day performance, while neglecting the longer term gains associated with training and development. The majority of the participant companies had similar enablers and results scores, as predicted by the EFQM model.

Table 1 : Participant Company Data

EFQM Categories	Companies							
	1	2	3	4	5	6	7	8
Leadership	49	37	44	55	60	45	18	64
Policy and Strategy	38	19	30	57	51	42	34	71
People Management	32	23	34	48	52	36	10	59
Resources	63	38	41	58	64	51	38	61
Processes	59	64	68	66	71	54	36	69
Customer Satisfaction	71	61	64	71	69	55	45	68
People Satisfaction	56	13	44	48	56	50	37	69
Impact on Society	37	10	32	43	58	51	37	60
Business Results	79	16	43	55	37	55	43	61
Enablers	48	36	43	57	59	46	27	65
Results	63	27	48	56	55	53	41	65

6 CONCLUSION

The use of benchmarking as a diagnostic tool provides a realistic approach to assessing and improving the performance of SMEs. A benchmarking process for Small and Medium-sized Manufacturing Enterprises has been established and proven. Further work is required in the area of benchmarking, focusing on Small and Medium-sized Enterprises across Europe, which will include:

- Research into the data collection process. Quality data is essential to provide meaningful results. There is a need to assess different collection methodologies, the trade-off between quantity and quality, and the minimum data requirements for useful benchmarking.
- Research into the relationship between normative data on the company (sector, size, markets), the Benchmarking analysis, and best practice recommendations. An area of interest is to what extent can recommendations be formulaic or automated. (For example, by using Artificial Intelligence and a database of best practices)
- Longitudinal research should be carried out on the cause - effect analysis and the consistency of the recommendations. (For example, do recommended actions lead to the expected improvements in EFQM scores and can this be separated from other factors?)
- Longitudinal research into regular use of Diagnostic Benchmarking as a management information tool and its impact on business performance.

7 REFERENCES

(1) Fallah, M.H.; 'Process Audit: From Compliance to Self-Management', International Journal of Technology Management, Vol. 14, Numbers 2/3/4, pp. 277-286.

(2) The European Foundation of Quality Management (EFQM); Avenue des Pleades 19, B-1200, Brussels, Belgium. Internet Address: http://www.efqm.org

(3) Nelder G. P., Skandalakis A. and Martin V.; Integrated Business Support for SMEs, First International Stimulating Manufacturing Excellence in Small and Medium Enterprises (SMESME) Conference, Sheffield, UK, April '98.

(4) Dti; Best Practice Benchmarking, Managing in the 90's Series, Department of Trade and Industry, Amended Reprint, May 1995.

(5) The Economist; Find Your Bench, May 11th, 1991.

(6) Balm J.; Benchmarking and Gap Analysis: What is the Next Milestone, Benchmarking for Quality Management and Technology, Vol. 3, Nov 4, 1996.

(7) Monkhouse E.; The Role of Competitive Benchmarking in Small to Medium Sized Manufacturing Enterprises, Benchmarking for Quality Management and Technology, Vol. 2, Nov 4, 1995.

(8) Micklewright M. J.; Competitive Benchmarking: Large Gains for Small Companies, Quality Progress, Vol. 26, No. 6, June 1993.

(9) Dale B.; Benchmarking on Total Quality management Adoption: A Positioning Model, Benchmarking for Quality Management and Technology, Vol. 3, Issue 1, 1996.

(10) Riis O. J.; How to Implement Benchmarking, Report on Group Work, Benchmarking – Theory and Practice, Edited by Asbjorn Rolstadas, Chapman and Hall, 1995.

(11) Cabello C.; Benchmarking of Best Practices: A Perspective for Policy, The IPTS Report, No. 9, November 1996.

(12) APQC; Planning Organising and Managing Benchmarking: A Users Guide, Houston, American Productivity and Quality Control, 1992.

(13) International Benchmarking Clearinghouse (IBC); American Productivity and Quality Control, Internet address: http://www.ibc.apqc.org

(14) Hakes C.; The Corporate Self-Assessment Handbook for Measuring Business Excellence, Third Edition, Chapman & Hall, 1996.

(15) Davis J.; Khodabocus F., Obray C.; Self-Assessment: A Path to Business Excellence, Quality World Technical Supplement, March 1996.

A review of environmentally conscious manufacturing (ECM) strategy

K F S NGAN and **R W BAINES**
School of Engineering, University of Derby, UK
K CHAPPELL
School of Environmental and Applied Science, University of Derby, UK

Abstract

Increasing commercial pressures and environmental laws and regulations are having impacts across industry. In response to these pressures, the study of manufacturing strategy has generated considerable interest in Environmentally Conscious Manufacturing (ECM) strategies. This paper comprises a review of relevant published literature on the subject. The main categories of publications have been identified: life-cycle approach, Eco-design; and environmental management. The paper concludes with a discussion of issues raised as a result of the review, identifying general agreement that there is a need for a methodology that will help manufacturing enterprises to develop their ECM strategy that provides a basis for manufacturers to improve their environmental performance.

1 INTRODUCTION

The present and future adverse effects of manufacturing activities, their damage to our environment and to the earth's ecosystem - which cannot be expressed in money terms, and ultimately their effect on the quality of human life are by now well recognised by the public as well as governments (1). It challenges manufacturers to address environmental concerns at the level of their core competence (2),(3).

Environmental regulations (4),(6) are becoming more stringent, and their implementation has a major impact on the economic operation and financial health of industrial organisations (5),(6), (7). Regardless, and not withstanding arguments by some that environmental damage from these activities is overestimated, manufacturing engineers and management of companies have the responsibility for planning and implementing environmentally safe manufacturing operations to compete in the global markets (8).

This review is a result of a survey of published work carried out by the authors relative to Environmentally Conscious Manufacturing (ECM).

2 ENVIRONMENTALLY CONSCIOUS MANUFACTURING (ECM)

Manufacturing enterprises can improve their environmental performance by careful analysis of products, design, materials used, and the manufacturing processes and practices employed in making these products.

Espousing the environmentalist approach in manufacturing operations means becoming involved in a historic project to save and regenerate the natural environment. In 1995, Kalpakjian (6) described it as an attitude to work, which goes beyond the mere profit motive and can awaken powerful latent creative powers in people. In this respect, it is at the same time a sound basis for successful business management (3), (5), (6).

ECM goes beyond simple compliance with environmental regulations to drive a philosophy of pollution prevention. Manufacturing operations that involve paradigm shift from end-of-pipe control to meet environmental regulations, to avoiding pollution in the first place should be adopted by all manufacturing enterprises. This type of manufacturing operations are called Environmentally Conscious Manufacturing (ECM) (5).

Weissman *et al.* (5) and Sarkis (9) summarised that Environmentally Conscious Manufacturing involves planning, developing and implementing manufacturing processes that minimise or eliminate hazardous waste, reduce scrap, or are operationally safer. The benefits include safer and cleaner factories, reduced environmental and health risks, improved product quality at lower cost, and higher productivity (9).

3 ECM STRATEGY

Minimising waste can reduce waste-related costs. Optimising processes and procedures to more efficiently use material will reduce production and disposal costs, because one frequent component of waste is material that is not used in the final products (5).

Many approaches can be taken by manufacturing enterprises. The following is a partial list of general ECM strategies (6), (9), (10), (11).

- prevent/reduce waste of materials/pollution at the source;
- conduct research and development in environmentally safe products and manufacturing technologies;
- reduce the use of hazardous materials in products and processes and ensure proper handling and disposal of them;
- take preventive and correction environmental actions;
- measure environmental impact;
- reduce consumer scepticism by forming alliances with stakeholder.

Many individual companies have developed their own ECM strategy initiatives. Hewlett Packard, IBM, 3M, Volvo and Xerox are examples. The reviewed literature would indicate a consensus of the mail elements to be included in an ECM strategy, these are life-cycle approaches, waste minimisation strategies, remanufacturing, recycling and reuse, Eco-design, environmental management systems and green supply management.

3.1 Life-cycle Approach

Life-cycle approach involves viewing the whole life of a product. Life-cycle assessment (LCA) includes considering the flow of materials and their linkages and relationships with other concerned activities and processes. This is useful in examining the environmental impacts of a product. The life-cycle assessment approach assists in the selection of alternative product characteristics, manufacturing processes and technology (12). One example of a commonly used life-cycle analysis model is Porter's value chain (13).

Frosh and Gallopoulos (14) explained the life-cycle analysis assessment methodologies from a manufacturing point of view as follow: *'Like their biological counterparts, individual manufacturing processes in an effective industrial ecosystem contribute to the optimal function of the entire system. Processes are required that minimise the generation of unrecyclable wastes as well as minimise the permanent consumption of scare material and energy resources. Individual manufacturing processes cannot be considered in isolation. A process that produces relatively large quantities of waste that can be used in another process may be preferable to one that produces small amounts of waste for which there is no use.'*(14)

Within the life cycle of a product, reduction of waste generation is one of the major goals in ECM strategies (9).

3.2 Waste-minimisation

All pollution is waste whether solid, liquid or gas. Minimising waste will improve operating cost and energy efficiency. This includes *packaging material.* Emphasis of waste minimisation is placed on source reduction (15) or pollution prevention. It is a strategy that closely links with Just-in-time (JIT) and Total quality management (TQM) of manufacturing related processes and philosophies (16), (17). The relationship of quality at source and defect reduction has direct implications on waste reduction. Waste reduction can often result in significant cost reduction (5), (18). Companies that have achieved environmental excellence using continual improvement objective and do not deviate from their goals are recorded by Wheeler (1992) (19).

3.3 Remanufacturing

Remanufacturing is a strategy that is closely related to waste-minimisation. Is refers to repair, rework, or refurbishment of components and equipment for external sale or internal use. In a remanufacturing process, the worn-out parts and equipment are normally grouped into production batches, completely disassembled and cleaned. These parts are then replaced to the level of new product with the incorporation of any newer improvements. The product is assembled, tested, packaged, and distributed in the same manner as new products (20), (21).

3.4 Recycling and/or reuse

Recycling (3) and reuse are closely related to remanufacturing and the waste minimisation strategy. More than 70 percent of solid waste produced every year could be recycled (22). Technology has made the opportunities available to take discarded materials back and recycle into usable materials, therefore, recycling waste that leads to its minimisation.

Reuse refers to a strategy similar with recycling however minimal treatment of a material is normally described as reuse of a product. For example, it could mean others consider waste generated from a company to be of significant commercial value. This specific example has also been defined as "waste exchange" by Vadja (23).

3.5 Eco-design (or Design for Environment DFE)

Eco-design (24), also known as Design for Environment, (DFE) especially in the U.S. - evolved from the design for X principles (25), (26). Environmental concerns are addressed in the product and process design phase. Decision made during the design phase have a profound impact on the entire life cycle of the product, from design, through manufacturing distribution, installation, servicing and disposal.

Gatenby *et al.* (27) estimated that 80 to 90 percent of the total life cycle costs of a product are determined in the design phase (27), (28). Thus integrating environmental consideration in the design process will substantially reduce environmental problems.

Eco-design helps process designers to analyse the environmental ramifications of a new process before development and installation. By carrying out a material flow balance or waste audit on a process *before* it is implemented, designers can prevent expensive, intrusive fixes later (5). Eco-design incorporates design for reuse, recycle, remanufacturing, disassembly and disposal.

Weissman *et al.* (5) described DFE as a means to an end bigger than reducing factor waste streams. During the design stage, environmental concerns associated with the front end (of the supply chain) and the after-sale phase of the product life cycle can also be addressed. It is this idea of influencing the entire product life cycle that make DFE a potent concept and more than just a waste minimisation approach.

3.6 Environmental Management and ISO14001,

Today's competitive business environment requires that environmental considerations be integrated into a corporate culture and business strategy. ISO 14001 provides a basic system for environmental management. The ISO14001 standard, environmental management systems is part of the overall management system for developing, implementing, achieving and maintaining the environmental policy. The standard is based on the concept of ISO 9000 series and has evolved from of British Standard 7750 (29), (30).

3.7 Green Supply Management

It must be acknowledged by manufacturing companies that many, if not most, of their impacts occur outside their factory gates, from an environmental life cycle and quality perspective it is important to consider the supply chain of the business. It maybe possible to influence supplier's environmental programme and ability to meet environmental specifications (3), (31), (32).

4 CONCLUSION

A number of references to ECM strategies have been reviewed in this paper. The magnitude of environmental issues requires an approach that has a clear definition of goals to be achieved through Sustainable Development. If priorities are wrong and the techniques used to address waste minimisation ineffective then there is a danger than ECM could add to waste and not its minimisation.

The review of ECM strategies shows that the technology developed for ECM can have ancillary benefits for manufacturers (5), (33). There are many approaches a company can take, however, a combination of ECM approaches is necessary for short- and long-term improvement in environmental performance. Ansoff (34) and Sarkis (9) proposed the need for a systematic strategic approach to relate to its environment to assure continued success (34). From the study of the literature on ECM strategies, it is clear that many large organisations have developed their own ECM initiatives but there is a significant research opportunity for a 'best practice' methodology for manufacturing enterprises to develop their own ECM initiatives that allow them to compete in the global market and at the same time minimising waste. Further research by the authors is being carried to develop a ECM strategies that will stand along side companies' corporate, business and manufacturing strategies to meet its sustainable development in the future.

The environment movement will only be successful by communicating the positive aspects of an environmentally oriented lifestyle by companies.

5 REFERENCES

1 Henriques, I.. and Sadorsky, P. (1996). The determinants of an environmentally responsive firm: an empirical approach, *Journal of Environmental Economics and Management,* vol. 30, 381-395

2 Winter, G. (1988). *Business and the Environment: A handbook of industrial ecology with 22 checklists for practical use*, McGraw-Hill, Hamburg.

3 Richards, D.J. (1994). Environmentally Conscious Manufacturing, *World Class Design to Manufacture*, 1 (3) 15-22.

4 Bergland, R.L. (1991). *Quality Costs and Pollution Prevention*, ASQC Quality Congress Transactions, Milwaukee, WI, 602-609.

5 Weissman, S.H. and Janine C.S. (1991). Environmentally Conscious Manufacturing: A Technology for the nineties', *AT&T Technical Journal*, November/December 23-30.

6 Kalpakjian, S. (1995). *Manufacturing Engineering and Technology*, third edition, Addison-Wesley.

7 The Engineering Council. (1994). *Guidelines on Environmental Issues*, LR Printing Services Limited, UK.

8 Ghobadian, A., Viney, H. and Liu, J. (1995). The impact of environmentally conscious corporate strategies upon manufacturing management, *Advances in Manufacturing Technology IX - Proceedings of the 11th National Conference on Manufacturing*, September, Leicester, UK, 757-761.

9 Sarkis, J. (1995). Manufacturing strategy and environmental consciousness, *Technovation*, 15(2) 79-97

10 Ottman, J.A. (1992). Green consumerism: the trend is your friend, *Directors & Boards*, 16 (4) 47-50.

11 Pinchot, G. (1992). The ecology of success, *Executive Excellence*, 9 (10) 7-8.

12 Fouhy, K. (1993). Life cycle analysis sets new priorities, *Chemical Engineering*, 100 (7) 30-34.

13 Porter, M.E. (1980). *Competitive Strategy*, Free Press, New York.

14 Frosch, R.A. and Gallopoulos, N.E. (1989). Strategies for manufacturing, *Scientific American*, Sept., 144-152.

15 EPA. (1992). *Facility pollution prevention guide*, USEPA, Office of Research and Development, EPA/600/R-92/088, Washington, DC.

16 Isbell, T.S. (1991). The backside of TQM: How waste indicates haste, *Industrial Engineering*, Nov., 42-45.

17 Willig, J. (1994). *Environmental TQM*, McGraw-Hill, New York.
18 Tillman, J.W. (1991). *Achievements in source reduction and recycling for ten industries in the Unites States*, EPA/600/2-91/051, Cincinnati, OH.
19 Wheeler, W.A. (1992). The revival in reverse manufacturing, *Journal of Business Strategy*, 13 (4) 8-13
20 Lund, R.T. (1983). *Guidelines for an original equipment manufacturer starting a remanufacturing operation*, Technical Paper, Massachusetts Institute of Technology, Cambridge, MA.
21 Garvin, D.A. (1992). *Operations Strategy: Text and Cases,* Prentice Hall, Eaglewood Cliffs, NJ.
22 Rennie, C. and MacLean, A. (1989). *Salvaging the Future: Waste Based Production*, Institute for Local Self-reliance, Washington, DC.
23 Vajda, G. (1992). Integrated waste minimisation, *Manufacturing Systems*, 10 (1) 36-40.
24 Duckworth, A.P., Lewis, P.A. and Baines, R.W. (1997). A Framework for integrating Ecodesign into the Small and Medium Sized Enterprise Design process, *Advances in Manufacturing Technology XI - Proceedings of the 13th National Conference on Manufacturing*, September, Glasgow, UK, 119-123.
25 Sarkis, J. (1998). Evaluating Environmentally Conscious Business Practices, *European Journal of Operational Research*, forthcoming.
26 Fiksel, J. (1996). *Design for Environment: Creating Eco-Efficient Products and Processes*, McGraw-Hill.
27 Gatenby, D.A. and Foo, G. (1990). Design for X: Key to Competitive Profitable Markets', *AT&T Technical Journal*, 69 (3) May/June, 2-13.
28 Oakley, B.T. (1993). Total Quality Product Design - how to integrated environmental into the Product Realisation Process, *Total Quality Environmental Management*, 2 (3), Spring, 310
29 Lamprecht, J.L. (1997). *ISO 14000: issues &implementation guidelines for responsible environmental management*, AMACOM, American Management Association, New York.
30 Willits, S.D. and Giuntini, R. (1994). Helping your company go green, *Management Accounting*, 75 (2) 43-47.
31 Murphy, P.R. Poist, R.F. and Braunschwieg, C.D. (1994). Management of Environmental issues in Logistics: current status and future potential, *Transportation Journal*, 25 (2) 5-19
32 Sarkis, J. (1995). Supply chain management and environmentally conscious design and manufacturing, *International Journal of Environmentally Conscious Design and Manufacturing*, 4 (2) 43-52
33 A survey conducted by the Gallup organisation indicated that about 90 % of those surveyed would be willing to pay a premium for green products or packaging (Hayes, D. (1991). Harnessing Market Forces to Protect the Earth, *Issues in Science and Technology*, winter 1990-91, 46).
34 Ansoff, H. (1984). *Implanting strategic management*, Prentice Hall International, London.

Towards customer-focussed manufacture for SMES

P-L NOLAN and **M H JONES**
Department of Manufacturing Systems Engineering, University of Hertfordshire, Hatfield, UK

ABSTRACT

This paper is based on an industrial case study, carried out as part of a Teaching Company programme, developing JIT techniques within a flextime environment. Principal issues facing this company were the need to reduce lead-times to create competitive advantage in UK and World markets, while at the same time reducing stocks and WIP. The typical answer to this problem is to move from batch to flow techniques through the implementation of JIT and/or flexible manufacturing. For many small companies there is a large gap between the rhetoric of such approaches and the everyday realities of life.

This paper discusses how a TQM framework in place was utilised to identify key issues for the company and how they were solved. Such issues included creating human centred systems, stimulating multi-skilling and enabling JIT and flextime systems to operate together. This was carried out in an environment with a complex product mix, a variable demand rate within this mix and a typical "end of month" approach to production planning.

1. BACKGROUND TO THE CASE STUDY

This industrial case study is based on an SME located in Bury St. Edmunds, Suffolk. Its business is the design, development and manufacture of electro-mechanical and electronic equipment.

Within the last five years, changes in the SME's management style created the need to develop a people-centred company. The SME decided to embark upon the route towards TQM and used the vehicle of the DTI's Teaching Company Scheme to facilitate this. The TCS programme lasted 2 years and once finished, the SME decided to progress further in their quest for World Class Manufacturing status.

The company identified its major problem as being manufacturing procedures that have not developed fast enough and do not reflect best practice in modern markets. The effect of this shortfall incurs costs that are too high, delivery times that are too long and often missed, and stock levels that are too high. It therefore decided to set up a TCS Programme with the following objective:

"Contribute to the creation of a World Class Manufacturing company, geared to customer needs, by implementing most appropriate and effective manufacturing systems to suit a variety of product ranges and differing market requirements."(1)

The shortcomings of the Company's inefficient manufacturing techniques also led to apparent difficulties in the market. The Company's Sales department was convinced that these "problems" (long & missed lead times) meant that they could not fully exploit the market potential for the Company's products.

2. FLEXTIME, PATTERNS OF WORKING & DEMAND PATTERNS

The SME is situated in an area of the UK that has low unemployment, less than 2.5%(2). This factor, linked with the high percentage of female employees at the company, makes the application of flextime very desirable. As Ralston(3) has shown:

"flextime means that they [employers] can increase the size... of their labor pool by being able to include individuals who might otherwise be forced out of the full-time labor market".

The way in which the company applies flextime is based on a 37½ hour week. The factory is open for 10 hours a day, 5 days a week and employees can complete their hours largely as they wish. This system has resulted in two work patterns - the *predictable* and the *unpredictable*. Many employees complete their hours in the first 4 - 4½ days within a predictable pattern of work; the unpredictable pattern developed from those who apply the "freedom" element of flextime to its fullest extent.

Demand rates for Company products have historically been extremely variable with no discernible demand patterns. As shown in the polar diagram (Figure 1), the Company's perceived competitive advantage resides principally in the "product" category – they produce many product design "specials" attracting customers by offering products exacting to their requirements. Of course, the product category alone does not win them additional business, but a combination as shown in the polar diagram does.

3. EXISTING MANUFACTURING TECHNIQUES

To understand the setting of the case study, this section will give an insight into the manufacturing techniques used at the Company prior to the project.

All customer orders are processed through an MRP system that generates Works Orders and Work-To lists. These are administered through the Production Control department, which distributes them to the relevant departments.

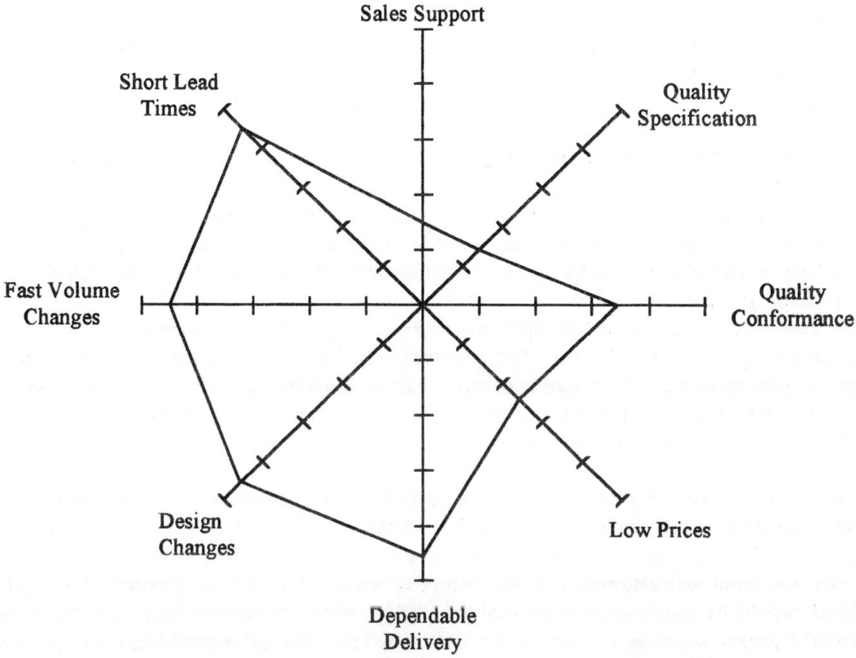

Figure 1 Polar representation of the Company [after Slack(4)]

Due to factors such as lack of training, lack of MRP housekeeping and minimal control of data input, supplementary activities and occasional by-passing of the system have been necessary to ensure the throughput of finished products. This has led to a highly disorganised system where "Peter robs from Paul" on kitted schedules with shortages, lead times are highly speculative as no capacity planning exists and there is "no time for process improvement as getting the product out of the door is priority!". Strangely enough it is people's commitment and care towards the Company and its Customers that created this situation, which from a WCM point of view is a positive point. The next challenge would be to channel this care and effort into the correct type of operations.

4. GETTING STARTED

Getting Started is all about being honest about where you are and what needs improving. The Company had experienced several years of a TQM-focused approach and it was understood that the next phase would be to use this TQM foundation to build up the Company's understanding of World Class Manufacturing. Via the route of Continuous Improvement, the Company would undergo several years of guided process improvements implemented by those who worked in each area.

We describe this approach as **"Here's a great cardigan, would you like to learn to make one?"** as the ideal operational system is described and the Company team is taught how to achieve it. Hence, the project would have to:

1. Bring the teams together and develop their ideas through the teaching of modern manufacturing techniques.
2. Direct the team and help implement their ideas through the application of project management.

In this case, the reality was somewhat different due to incorrect assumptions at the outset. These were:

1. The existing employee knowledge of modern manufacturing methods would be adequate.
2. A clear Company direction had been given to employees by the management team.
3. Employees had a Continuous Improvement attitude which had been achieved through the TQM programme.
4. There would be an internal champion to spearhead the project implementation.

The underlying reality was that the TQM programme had not infiltrated the employees' cynical, sceptical outlook that had developed over many years of aborted manufacturing projects. Despite this, the TQM programme was certainly not a failure as new initiatives had been introduced and a better "atmosphere" had been achieved.

However, the initial approach to WCM needed to be adjusted to create (or to start creating) an open, willing and challenging environment within which some of the cynicism and scepticism could be dispelled. To start this ball rolling, an initial success had to happen. The importance of this step (which was learnt with retrospect) should have been reinforced when time after time, the project slipped back by trying to transfer ownership too quickly. *They just were not ready.* The lack of an internal champion made the situation worse, and created the need for external help. This position requires the traditional perspectives of impartiality and action-orientation and the need to become an ally to the manufacturing teams. It is also advantageous if the position at the Company does not hold any hierarchical "power" to ensure that all steps of the project are accomplished via *teaching by doing* and not *doing by telling*.

The Getting Started stage of this case study showed that an initial success would be paramount to setting the correct environment for the implementation of the project and that an external champion was required to do this while "pulling" the team along. This initial success stage we call **"Look you can make a cardigan"**.

5. CREATING THE HUMAN CENTRED APPROACH

We believe that productivity improvements can be achieved through increased employee responsibility, multi-skilling and improved quality of working life. Creating a human centred approach to manufacturing must be a primary element in ensuring the success of World Class Manufacturing. It is why at this stage of the project, the facilitator has to ensure that everyone is involved and aware, even if the project becomes a single-handed implementation. At this stage of an SME manufacturing project it is very easy for the champion to stride ahead without the team, but with sometimes disastrous consequences.

This particular case study made use of Monday morning WCM training sessions that took 10 employees at a time to explain the Company's perspective on WCM and where it felt its goals lay. At the same time, a Manufacturing Game was held once a month to help disseminate Modern Manufacturing Techniques and to explain some of the jargon. At the same time the core group representing the first area to be affected by the new manufacturing methods was constantly updated as to what was happening. At this stage the employees remained very sceptical and little help was offered from any discipline.

6. OPERATIONS MANAGEMENT AND COMMON SENSE

The analysis of the production area singled out for improvements was undertaken using methods that have been written about in many Operations Management publications, e.g., (4) or (5). Analysis of all products and variants, sales history over the last four years, possible manufacturing combinations of sub-assemblies and line balancing were among the techniques applied.

The implementation stage involved many aspects of human centred design ensuring that the workstations were ergonomically designed and situated, that the correct tools would be available and every stage was implemented with the operators' help and guidance. One of the most poignant methods was allowing the implementation to happen at the employee's paces. For example, a 2-bin system was being implemented as a predecessor to Kanban to try to vary the way in which components are issued to line (previously issued in very large kits). Enabling the stores supervisor time to understand the system and implement the cards at her own pace allowed that crucial time for self development. Had the implementation been hurried and forced into place, the ownership of the new system would never have been transferred.

7. OUTCOMES

Despite knowing that trust is earned, which was shown by the way that the operators started to help with the physical implementation, it was quite a shock to experience a step change in the overall attitude that occurred after the initial implementation.

The first line to come under the World Class Manufacturing project reduced all back orders in two weeks and reduced lead times from 5 weeks to 3 days within its first three weeks. However, the change in attitude seemed to occur about 4 weeks after the line had been running. Our observations have been that:

1. The team leader had taken full responsibility for the operations of the line and was openly praising the project in terms of organisation of his work, ease of assembly and lack of "problems".
2. The Quality Manager had openly expressed his support for the project.
3. The success of the line in terms of lead times was being publicised through Team Briefing and the Managing Director.

Is it the "parents" that must express their agreement with the project before the "family" agrees to follow the idea? So much has been written about step change, particularly in larger organisations where Senior Management decides the main implementation stages. Could it be that in SMEs, an incremental change is essential to sustaining the trust that the "family" requires? Our experience shows this to be the case.

Another factor that plays an important part in the trust generated by this SME is flextime. How can the Company be sure that flextime will not pose a threat to their competitiveness? When the new line was implemented, a customer-focused approach was used to ensure that the line was only making what it needed to. One of the reservations by the management team (and ourselves!) was that if the line was more or less idle for four days and then a customer order was received on a Thursday night, would the operators be in work on the Friday, and when would the customer receive the goods? Over the last few months, we have seen marked changes in the patterns of flextime worked by operators (we are currently collecting data to quantify these changes). In some cases, when there are few or no customer orders (due to the unknown demand rates as discussed at the beginning of this paper) the operators take "flexi" days or half-days to "fit in" with demand. The SME's employees have reciprocated the trust that the Company showed by implementing flextime.

This approach to varying demand patterns is in contrast to some contemporary manufacturing ideologies. One encounters set parameters for implementing certain methodologies, such as:

> "some degree of demand stability is a prerequisite for effective JIT"(5)

With this fact in mind, it may seem strange that modern manufacturing techniques were considered as viable options for this project. As more data is collected and the story unfolds, the importance of setting parameters will show how modern philosophies are *just* as applicable in a flextime environment as in a traditional one.

8. THE FUTURE

During these first stages of this World Class Manufacturing project we have identified the importance of creating the correct milieu for change. In fact, creating the appropriate environment is as important as providing the enabling tools.

Due to the response that this first stage received, we are now progressing through the stage of the project called, **"Would you like to learn to knit?"**. At this stage the data analysis techniques and implementation stages are being taught to the teams, who are much more receptive and eager than 6 months ago. Upon completion of this phase we hope to hear the resounding cry of,

"WOW, WE'VE MADE A JUMPER!"

REFERENCES

1. TCS project proposal document.
2. Data from Table S30, Unemployment by Travel-to-work Areas, June 1997. *Labour Market Trends*, August 1997, DfEE publication.
3. Ralston, D. A. (1989) 'The benefits of Flextime: Real or imagined?'. *Journal of Organizational Behaviour*, Vol. 10, 369 – 373.
4. Slack, N., Chambers, S., Harland, C., Harrison, A. & Johnston, R. 1995. *Operations Management*. Pitman Publishing.
5. Wild, R. (1991). *Production and Operations Management*, Fourth edition, Cassell,London.

Teaching strategic management through simulation

B PARKINSON, R SENIOR and **P HUDSON**
Manufacturing Systems Centre, University of Hertfordshire, Hatfield, UK

ABSTRACT

This paper describes the development of BOSS (Business Organisation Strategy Simulation); a computer aided learning system, produced by the iDer (Innovative Design Engineering Research) Group based at the University of Hertfordshire. Undergraduate students of engineering and management related subjects may be required to understand the implications of operational and strategic decision making processes. Paper based case studies are often used to practice these decision making skills, but suffer the disadvantage that students cannot directly see the outcome of their work. By acting as an interactive case study providing continuous formative feedback, BOSS overcomes these limitations, allowing students learning to be reinforced.

1. INTRODUCTION

1.1 Paper based approach to decision making skills

Paper based Case Studies are often used by educationalists as a means of giving students opportunities to practice operational and strategic decision making skills. Typically this process involves students gleaning relevant information from within a Case Study, analysing it's significance and proposing operational and strategic actions designed to bring the portrayed situation into line with some pre-determined goals. Being remote from the working environment, the case study approach suffers the disadvantage that students cannot directly see the outcome of their recommendations. Furthermore, they are unable to try out different recommendations and do not have the opportunity to introduce corrective action if their decisions fail to have the desired effects. For these reasons opportunities to reinforce learning are lost.

1.2 iDer research

The iDer (Innovative Design Engineering Research) Group, a consortium based at the Manufacturing Systems Centre, University of Hertfordshire, was set up under the TLTP (Teaching and Learning Technology Program) to develop intelligent computer aided learning systems teaching concurrent approaches to engineering design. It became clear that the knowledge and programming skills acquired during the development of the group's 'ICE'(1) and 'Design Builder'(2) software (which introduce and allow students to experience concurrent engineering), may be used to facilitate an interactive alternative to paper based Case Studies.

1.3 Benefits of an interactive approach

Botto asserts that students interaction with multimedia is a key aspect of it's role in education, leading to as much as 70 percent recall(3). In the case of strategic decision making, this interaction makes it possible not only to discover what situation the organisation in question is facing, but more importantly, to actively alter the scenario by actually implementing operational and strategic decisions and experience the combined effects of these decisions through a simulated outcome.

2. BOSS - 'BUSINESS ORGANISATION STRATEGY SIMULATION'

BOSS is a CAL (Computer Aided Learning) software programme using multimedia techniques to simulate a manufacturing company. Taking the form of a Business Game, BOSS acts as an interactive case study providing students with continuous formative feedback.

2.1 Educational Objectives

The Computer Aided Learning system aims to :

- Introduce students to the nature of operational and strategic decision making processes.
- Allow students to experience these processes by interactively implementing operational and strategic decisions.
- Introduce students to a process type industry with realistic commercial data.

2.2 Nature of the programme

Whilst it would demonstrate the effects of example strategies and so illustrate the decision making process, a linear approach to programming was deliberately rejected, on the grounds that it would not satisfactorily represent the dynamic interactions occurring within and acting upon actual organisations. Rather, an attempt was made to simulate the workings of a manufacturing company in terms of it's organisational attributes, thereby bringing students decision making processes into the virtual working environment.

2.3 Nature of manufacturing organisations

iDer sees manufacturing organisations as 'socio-technical systems wherein individuals co-operate together to produce outputs from inputs.' Such systems exhibit static hierarchical structures which define and regulate dynamic processes(1). In order to create a computer model that adequately represents a manufacturing company, it follows that both structural and process attributes need to be included. Britton and Doake make the point that the complexity of many real world situations results in such models using techniques of decomposition, breaking the problem into manageable units; and abstraction, concentrating on important elements in order to bring about simplified understandable synthesis(4).

2.4 Structure model

In BOSS, the organisation is decomposed into functional departments which serve to define the organisational structure. Departments define and contain the sub-processes that make up the overall activities of the organisation. Each department has operational goals that it seeks to attain.

2.5 Process model

The system sees the entire organisation in terms of processes, including, but not confined to the manufacturing process itself. BOSS uses a simplified process model based on principles of 'throughput' and 'constraints' taken from OPT (Optimised Production Technology)(5), a philosophy primarily concerned with production, but in this case also applied to other operational areas such as sales and finance. Accordingly, processes are seen in terms of throughput, the rate of flow of inputs through the process which converts them into outputs. Throughput is limited by the slowest part of the process, known as a constraint, which can vary in nature according to circumstances. Throughput is also demand driven, being matched so far as possible to the rate of sales. Examples of inputs might be market demand or raw materials, outputs finished products or fulfilled orders and constraints production capacity or staff levels.

Strategies are pre-assigned certain consequences in certain areas, which may constitute any of the inputs outputs or constraints. In this way, strategies determine if departments meet their goals. An important feature of BOSS is that these consequential changes, and the corresponding formative feedback, are determined by the software in real time, depending upon the particular permutations of selected strategies and the order in which they were chosen. This approximates to the situation in real life, in that there are no 'pre-set' answers.

2.6 Use of BOSS

BOSS takes a fictitious confectionery company, 'A. C. Bow & Co.', as an example of a process industry.

By visiting the virtual company the programme allows students to discover the prevailing situation within each of the six operational departments, where the departmental managers provide information and ask questions. Further information is available by clicking on certain objects shown within the departmental views. iDer is fortunate in securing the help of a multi-national confectionery company willing to supply genuine manufacturing and marketing data, for inclusion within the programme. Having analysed the situation, students may consult Bow's managers who will suggest a number of different strategies to remedy problems and improve operational performance. After choosing a strategy, students immediately see the effects of it's implementation on the scenarios presented within each department.

Some strategies are more effective than others. Often, a strategy will improve one aspect of the company's operations at the expense of another. A thoughtful analysis by students is needed in order to reach their objectives (set by Bow's Managing Director) which are to deal with various short term problems, then in the long term, to increase A. C. Bow's market share.

3. FORMATIVE FEEDBACK

Formative feedback makes the system data concerning the current state of affairs available to students in a meaningful format. By objectively presenting the results of strategy implementation, this feedback allows students to see the effectiveness or otherwise of a chosen strategy and to decide if it was the most appropriate available.

3.1 Textural feedback

Text based feedback is the main way in which students see the workings of the virtual company. This is presented as if the managers were speaking to the students about the situation in their own department. Managers see circumstances from their own viewpoint, so will emphasise certain aspects at the expense of others. As BOSS represents a fictitious but typical manufacturing organisation, students need to enquire within different departments in order to obtain a balanced view of the situation within the company; in a similar way to if they were actually in a real world organisation.

3.2 Graphical feedback

In addition, BOSS generates a bar graph showing two superimposed sets of key data, the values of which are calculated in real time according to the company's current situation. This graph allows the results of strategies to be easily compared with the situation as it was immediately prior to their implementation.

3.3 Hypertext

In order to maximise it's potential use in different educational contexts, BOSS is 'self contained' and able to be used without reference to other information sources. For this reason, additional information pertaining to the company, it's markets and the nature of operational and strategic decisions is included in the form of textural links to objects found within the various departments.

4. GRAPHICAL INTERFACE

Multimedia techniques are used to depict A. C. Bow's company by a combination of visual and textural representations. These representations form an interface between the student and the mathematical model of the organisation contained within the computer programme. Since this interface essentially represents aspects of the company's organisational attributes, it is itself dynamic and is able to be controlled both by the computer model and the student allowing non-linear interaction to take place.

The graphical interface developed for BOSS uses a number of different screens to allow access to various parts of the programme. These screens are accessed by means of control buttons. Controls may function differently depending on the screen currently displayed. 'Department' screens represent functional departments within A. C. Bow's company. There is a short sequence of texts within each department, generally beginning with an introduction to the department's role within the organisation, explaining the current situation and asking the student a question. A 'Site Plan' screen shows a diagrammatic view of A. C. Bow's factory. It acts as a menu allowing access to the department screens. 'Managers' and 'Strategy' screens allow the student to make operational and strategic decisions, whilst

'Library' 'Hypertext' and 'Graph' screens all provide access to information. Figures 1 and 2 show example BOSS screens.

Fig. 1. An example 'Department' Screen

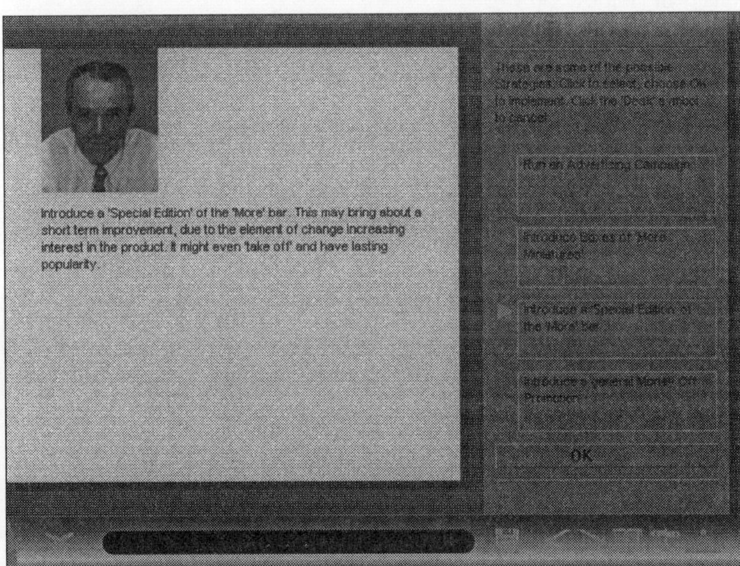

Fig. 2. An example 'Strategy' Screen

5. FUTURE WORK

At present the system is in a complete but concise form with the ability for expansion already built in. There is scope for expanding both the breadth of content and the detail in the organisational model, leading to a more realistic representation of the virtual company.

6. ACKNOWLEDGEMENTS

The iDer team wishes to thank and acknowledge the funding and support provided by HEFCE under the TLTP Program; and to thank Nestlé UK Ltd. for their valuable assistance.

Further details of the project may be found on the iDer web pages at :
http://www.mansys.herts.ac.uk/ider/

REFERENCES

1. Parkinson, B., R. Senior *et al.* "Virtual Manufacturing Organisation Modelled Using a Lexivisual Approach" in *Proceedings of The Second Working Conference on Engineering Education,* (Sheffield UK 24th - 27th March 1997). Sheffield Hallam University Press Sheffield UK pp 337-341.

2. Parkinson, B., P. Hudson *et al.* "Teaching Concurrent Engineering Design Using a Virtual Environment" *Proceedings of The Eighteenth Annual Design Conference Computer Aided Design Education,* (Bristol UK 26th - 27th June 1996). SEED University of Hull UK. pp 48-55.

3. Botto, F. 1993. *Multimedia, CD-ROM and Compact Disk - a Guide for Users and Developers.* Sigma Press, Wilmslow UK.

4. Britton, C. and J. Doake 1993. *Software System Development A Gentle Introduction.* McGraw - Hill, London.

5. Goldrat, E. M. and R. E. Fox 1986. *The Race.* North River Press, New York.

Data management – an overview of information requirements for small and medium sized enterprises (SMEs)

J G THOBURN and **C N WINTERS**
School of Engineering, Coventry University, UK

1. ABSTRACT

This paper describes research work undertaken as part of a 3-year programme, having the objective of defining implementation methodologies for organisations installing Product Data Management (PDM) systems. Although there is evidence of their utilisation and application in large enterprises, there is little evidence of this in Small- and Medium-Sized Enterprises. This paper explores the relationship between the information that exists in Small-and-Medium-Sized Manufacturing Enterprises and details its interaction with Management Information Systems.

2. INTRODUCTION

Today, every company must be able to process significant amounts of information. Without information, no business can properly perform any of the functions that are required of it. Every business must collect and blend a wide variety of information, distribute and use it throughout its operations, and provide accurate and timely outputs. In the widest sense, Information describes the trading environment of the company. More narrowly it is the basis of its management and operational systems.

There are substantial differences in Information Systems between companies. These differences are most marked between large and small companies. Highly formalised systems are evident in the majority of large companies, while, especially in most small companies, structured information system may not exist at all. Often, only isolated parts of an organisation have established systems. The most common example is the Financial Accounting System, whose formality may be imposed by such outside organisations as Tax Authorities and Lenders. There is usually little integration between different functional systems of the organisation, resulting in duplication of information and effort, errors and poor communication

The term "Information Systems" means many things to many people. Some companies view them with importance, others as purely business sub-systems. In smaller organisations, it is often a nebulous and undefined series of operations - simply a means to an end. To many, Information Technology and Information Systems are synonymous terms, seen as computer hardware and software.

Computers are now affordable for all but the smallest organisation. PC based systems such as financial accounting and relatively simple integrated applications that include word-processing, databases, spreadsheets and presentation software are now common. Yet, this has not solved all of the information needs of companies. Indeed, the opposite is the case. Information may be entered, processed and circulated by uncoordinated users with great ease. There may be considerable duplication of effort, with each user trying to focus information by function. This frequently results in multiple, often inaccurate, versions of information being used within an organisation. The same information may be labelled or described differently within different sub-systems making cross checking difficult or even impossible.

At the heart of any manufacturing company is its Operations Management (OM) system. This system may range from the highly formal to the extremely informal. Some techniques such as Manufacturing Resource Planning (MRPII) are widespread, and draw together much of the data flow of manufacturing companies, with its origins in the requirement to integrate the Accounting with Manufacturing. Other systems have integrated different functional areas within companies. More recent developments include Computer Integrated Manufacturing (CIM), Concurrent Engineering and Product Data Management (PDM), that seek to integrate design and manufacture. They may provide advances for some companies, but smaller companies have largely failed to adopt them. None provides a truly integrated Information System, which draws together all of the information available to a company.

3. "INFORMATION SYSTEM" DEFINED

Authors have defined Information Systems in different ways, though looking at both manufacturing and service companies.

> "An Information System may be defined as a system that serves to provide information within an organisation when and where it is needed at any managerial level. Such a system must take the information received and store, retrieve, transform, process and communicate it using a computer system or other means." (1)

> "One of the components (or sub-systems) of an organisation is the 'information systems'. The components of this system are people, hardware, software, data and procedures. The organisational information system thus collects, transmits, processes, and stores data, and retrieves and distributes information to various users in an organisation." (2)

> "An information system exists only to serve the business system of which it is a component. It keeps records and maintains the various facts and figures needed to run the business" (3)

"A system for the most efficient and effective means of identifying the 'real' needs of users, and developing information processing systems for satisfying these needs; ensuring that the resulting information processing systems continue to satisfy changing user needs by the most efficient means of acquiring, storing, processing, disseminating and presenting information; by providing facilities and a learning environment for users and information systems specialists to improve the effectiveness of their decision models; and by supporting operational, control and strategic organisational objectives." (4)

"An information system can be defined technically as a set of interrelated components that collect (or retrieve), process, store, and distribute information to support decision making and control in an organization. In addition to supporting decision making, coordination, and control, information systems may also help managers and workers analyze problems, visualize complex subjects, and create new products." (5)

There is a growing view, that Information is a strategic resource. Leading companies are today using information and information systems as tools for staying ahead of the competition. They have achieved this by developing Strategic Information Systems. These are defined as:

"Systems which change the goals, operations, products, services or environmental relationships of organisations to help them gain advantage over competitors." (5)

Many definitions assume some degree of computerisation, yet vital information is never held exclusively in computers.

4. INFORMATION FLOW IN AND AROUND A MANUFACTURING COMPANY

To be effective, managers must concern themselves not only with information flow within the company, but also with its external information environment. This presents the company with incoming information or make demands upon it, that require outputs. While the internal operating environments involve common or regular events, it is often the more irregular or widely spaced external occurrences that have profound and disruptive effects on companies.

No Company can afford to trade for long periods while ignoring its external environment. A Company that operates while ignoring, for example, provisions of the Health and Safety at Work Act could find itself at risk of closure on the visit of a Health and Safety Executive inspector. In common with many legislative and legal provisions, mere ignorance would be no defence. Similarly, ignoring external influences, such as relative strengths of currency, changes in demand, technological developments, or growing skill shortages can have profound effects. On the other hand, failure to communicate accurately and in time with, for example, Suppliers, Customers, Lenders or Tax Authorities can be costly and damaging.

While the final purpose of any company is clearly to make a profit, the *raison d'être* of a manufacturing company is to produce material goods which satisfy customer needs. Thus, the Operations Management System (OMS) is at the heart of such an organisation. Its other systems are arguably there to support the OMS in the optimum way.

5. INFORMATION CATEGORIES DEFINED

Based upon the work of Porter (6) and Taylor & Sparks (7), (Figure 1) the authors have defined eleven broad Information Categories that each company, regardless of its size, must take into account. Examples are given under each category, though the list is not intended to be exhaustive. They are:

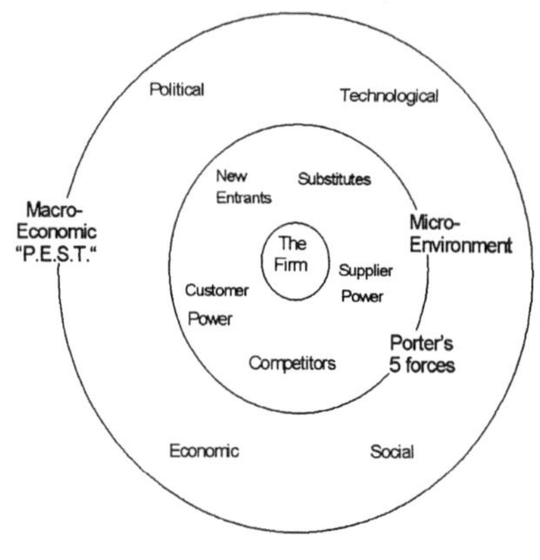

Figure 1 - Company Environment

1. **Legislative & Legal**: Tax Accounting, Statutory Provisions and Common Law
2. **Stakeholder Information for:** Banks & Lenders, Employees, Shareholder and Investors
3. **Business Environment**: Marketing & Sales Information, Competition, General Trading Conditions, Supplier Information and Employment Information
4. **Design Information:** Founded on Existing Products, Acquired and Developed
5. **Corporate Knowledge Base**: Founding knowledge, Experience, Acquired Knowledge and Research & Development
6. **Administrative Information**: Financial and General Accounting Information, Human Resources, Customer and Supplier Information, Business Resource Information, General Information.
7. **Technical Requirements**: British & ISO Standards (etc.), Machining & Process Requirements, Material & Physical Constraints, Universal Quality Requirements
8. **Specific Customer Requirements**: Price, Delivery and lead-times, Quality, Specification. Performance Requirements, Contractual Relationships and Project data
9. **Internal Operational Information Requirements**: Work instructions, Purchasing & Inventory, Manufacturing & Process Information, Quality Systems Information, Billing Information, Maintenance & Buildings Management, Capabilities, Restraints and Employee competencies
10. **Management Systems**: Company Culture, Structure, Communication, Attitudes, Financial Requirements, Strategic Factors, Internal Management Requirements
11. **Information Technology**: Hardware, Software, Personnel, Systems Integration

6. RELATIONSHIPS BETWEEN INFORMATION FLOWS

There are two broad flows of information within a company. The first goes through the organisation linking the support services. The second is through the OMS after intervention from support services. With each flow there are associated feedback systems.

The model (Figure2), shows these relationships within the Operating Environment. External Information is represented by both incoming and outgoing flows of data. Porter's (6) work was

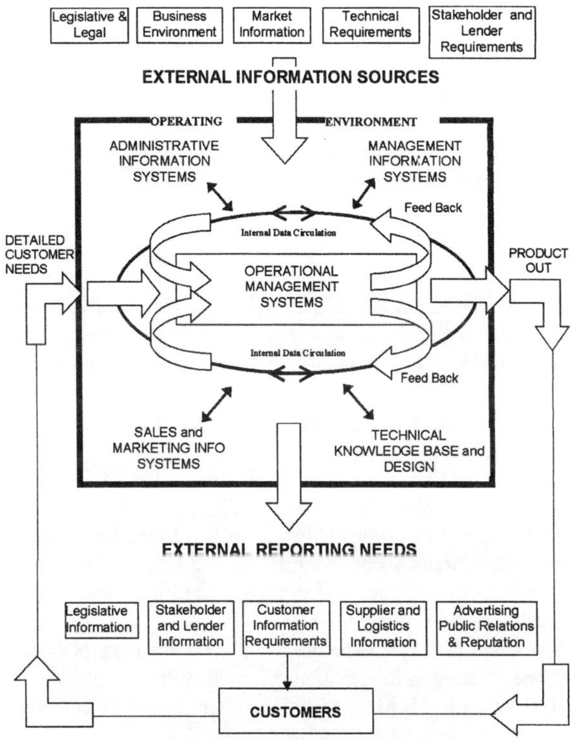

Figure 2 - Data Relationships

used to define Macro-trading and Industry Environments. This provided the basis for the identification of a number of information categories. Many authors (8,9,10,11,12,13) have described the importance of the Customer, and Customer Focus within the Company. The separately defined and continuous link between the Customer and the Organisation reflects this special relationship.

7. CONCLUSIONS

This paper has identified different categories of information that exist in manufacturing companies. Given the complexity of information it is necessary to provide a framework that will permit its optimum use within the company. In many SMEs, information systems are often evolutionary rather than planned. As a result, they are fragmented and inefficient, and do not meet the strategic or operational needs of the company in the best way possible.

Tools and techniques for achieving this have been developed for large companies. These have centred on the integration of the manufacturing operations with the company's data infrastructure. Many involve sophisticated, and expensive, software solutions, and a high level of investment, not least in skilled and specialised staff. Most of these solutions seem to be highly inappropriate, and commercially risky, for most SMEs. Further research is being undertaken on this subject, with the aim of devising methodologies for data and information management in manufacturing SMEs

8. ACKNOWLEDGEMENTS

The Economic and Social Research Council in collaboration with Omega Electrical Systems Ltd (Nuneaton) fund this 3-year programme.

9. REFERENCES

(1). Aktas, A.Z (1987) Structured Analysis and Design of Information Systems Prentice-Hall New Jersey USA

(2). Ahituv, N and Neumann,S (1990) Principles of Information Systems for Management, Wm C Brown Publishers, USA

(3). Kendall, P (1992) Introduction to Systems Analysis & Design – A Structured Approach, W.C.Brown Publishers, USA

(4). Jayaratna, N,(1994) Understanding and Evaluating Methodologies – Nimsad – A systematic Framework, .McGraw-Hill International(UK), Maidenhead

(5). Laudon,K.C. and Laudon, J.P, (1996) Management Information Systems – Organization and Technology (4th Ed) Prentice-Hall, New Jersey, USA.

(6). Porter, M.E.(1985) Competitive Advantage- Creating and Sustaining Superior Performance, Free Press, New York, USA

(7) Taylor, B. and Sparkes, J.(1977) Corporate Strategy & Planning, Heinemann, London

(8) Levitt, T (1983) The Marketing Imagination, Macmillan Press, London

(9) Schonberger R.J. & Knod Jnr, E.M., (1991) Operations Management: Improving Customer Service, Irwin

(10) Christopher M (1992) Logistics and Supply Chain Management, Pitman Publishing

(11) Chisnall, P.M.(1989) Strategic Industrial Marketing 2nd Edition, Prentice-Hall New Jersey USA

(12) Hill, T. (1991) Production/Operations Management: Text & Cases 2nd Edition Prentice-Hall New Jersey USA

(13) Sterling, J.U. & Lambert, D.M. (1989) Customer Service Research: Past Present and Future, International Journal of Physical Distribution & Material Management Vol 19 No.2

The gap between theory and practice in manufacturing management

B TROUGHT
Management Research International, Grantham, UK

ABSTRACT

This paper, based on recorded observational data of the work activities of a sample of manufacturing managers in the UK, reveals that in a substantial segment of manufacturing industry commonly taught theories of management do not seem to address problems with which managers are confronted. This gap between academics and practitioners can be widely seen in the use, or not, of computer planning and control systems and the subsequent decision-making methods employed by managers, but also more recently in the emerging theory of complexity. This paper sheds some light on this discrepancy and the reasons for it.

1. INTRODUCTION

Manufacturing management in its present form can be viewed as a chaotic practical activity, yet it is hemmed in by a body of logical theory which usually bears little resemblance to what goes on in a manufacturing system. It is not as if this discrepancy has not been pointed out over many years (by for example Rosemary Stewart in her many writings), it has, and certainly even more so from the early 1970's until the present day. The cry to researchers has often been for more applicability but sad to relate these cries have generally gone unheeded. Why this is so is an interesting conjecture but it could be an inevitable result of the nature of organisations, which, because of their diversity do not lend themselves to analysis through generalised theories. As late as September 1997 we can still read (1) that:

> 'Management research cannot subscribe to any ivory tower ideal of a disinterest in and a distaste for practice. Indeed the ability to develop ideas and put them into practice should be the distinguishing competence of the skilled management researcher.',

and this quotation highlights the point that management is a practice first and foremost whatever else it also is. Another real problem is that not only is supposed management theory not usefully prescriptive but that it is not normally sufficiently descriptive either. Close consideration of many text-books and of management course material will vouch for this problem. The study of (a) the theory of management and (b) the practice of management, still appears to be proceeding along two separate, perhaps parallel, tracks with little if any interconnection.

2. THE RESEARCH

Although the author has been collecting and recording observational data on the total work activities of a sample of manufacturing managers in the UK since 1975, the present paper mainly represents collection of data, interviews and statistical analysis over the past two years. About thirty managers from five companies were involved and the work is continuing.

3. RATIONALITY AND NON-RATIONALITY

3.1. Decision-making

The study of manufacturing systems has always left me with more unanswered questions than conclusions. Is this due to attempting to apply a rational mind-set to a non-rational scenario? Or perhaps more importantly as (2) notes:

> 'Good management research raises questions for practitioners as well as answering them.' (pg.7),

but this is only if practitioners actually have contact with research ideas and the difficulties of this are also well expressed in (2). The language used in research papers has often little meaning for practitioners. Researchers only seem to be talking to other researchers when they are discussing management theory. For me the idea which differentiates the gap between theory and practice in manufacturing management most markedly is the dual concept of rationality and non-rationality. The stated theories of management, whether they be concerned for example with decision-making, computer control systems or simulated plant layout are all sustained on a rational basis - the text books say so, the courses teach it. Computerised systems such as MRP2 have been advocated over many years in academia, extensively implemented in companies, but found to be difficult if not impossible to operate accurately in practise. Many of us have taught the step-by-step, systems orientated method of arriving at decisions, which, as the evidence shows (3) is rarely if ever used in managerial decision-making - the process is just too long for most manufacturing decisions. Previously it had been shown using statistical significance testing (4) that there was no difference in the way routine and non-routine manufacturing decisions were actually made. The way they were both made was non-rational. It is the way that managers choose to make decisions, by intuition based on shared company values.

3.2. Myths

Corporate myths often tend to drive the managerial processes in manufacturing companies or are used as an incorrect descriptions of these processes. Some, still prevalent myths which are apparent in the present sample and have been so for the last year or two, are:
(a) that manufacturing managers are in complete control of product-flow,
(b) that managerial work is proactive,
(c) that managerial decision-making is a logical process,
(d) that human conflict is the norm in manufacturing systems,
(e) that a competent manager can manage successfully in any organisation, in other words that managing is a general skill, and,
(f) that managers can be trained to manage.

In each case the converse is usually the reality. Managers only have partial control of product-flow and in some cases control is so diversified that it is difficult to see where control actually lies. Whilst managers may like to visualise themselves as proactive it is usually the case that their time is spent reacting to circumstances. The illogocality of decision-making processes has already been mentioned. Close observation of managerial activities will show that cooperative activities predominate. Real conflict, even if we had an agreed definition of the feature, is difficult to find contrary it many opinions. In respect of (f) it is maintained that managers cannot be trained in the generally accepted sense but that potential managers can learn to manage.

4. COMPLEXITY AND UNCERTAINTY

4.1 Complexity

For some people complexity theory has become the new 'holy grail' of organisational analysis. The theory has a mathematical base and is linked to manufacturing systems by an assumption that organisations can be considered as a sub-set of complex adaptive systems and so will adopt the characteristics of the wider set of complex adaptive systems. This is becoming an academic view (5) and (6). However the concepts are often difficult to understand, as reported in (7), even though in their case study the company was the Ford Motor Company and the people involved were all graduates. Furthermore in (8) detailed attention was drawn to what are apparent defects in the complexity argument for manufacturing systems. An important finding suggests that it is outside observers (eg. researchers) who see manufacturing systems as complex not internal managers. The view of a system from the outside is invariably different to the view from the inside. If this statement appears obvious it is perhaps worth noting that it is not a difference which is normally taken account of in academic organisational analysis.

From the academic complexity literature it appears that some authors see complexity and uncertainty as similar factors - two names meaning the same concept, for example (5) and (9). However, I am arguing that this is not so. For me complexity in practical situations suggests unknowability which cannot necessarily be managed to desired outcomes, whereas uncertainty does not suggest such intractability. Complexity appears to be the result of a myriad of interlocking relationships, large numbers of these relationships seem to be a condition for the label, complexity, to be attached. The state of complexity in the eyes of an observer also varies depending on which person is viewing the system or sub-system. This is because the degree of perceived complexity is a function of a person's depth of knowledge of the system. This raises the question of whether complexity is a view of the real world wholly residing, but differently, in the minds of observers be they internal or external to an organisation or is an unvarying feature of a manufacturing system or model of that system. With uncertainty in a manufacturing context a range of possible options seem to allow partial knowability within prescribed limits - a statistical view. In fact one could also argue that managers do tend to manage probabalistically rather than deterministically.

4.2. Uncertainty

My data suggests that managers of manufacturing systems see themselves surrounded by uncertainty rather than complexity. Large numbers of interlocking relationships do not appear necessary for uncertainty to be apparent. Simple, limited systems or small sub-systems can develop uncertain features such as those caused by theft or stock recording errors (10). Sometimes these uncertainties may be of a transient nature such as indeterminate delivery dates. Whether this adds to system complexity is an open question. In manufacturing systems uncertainty can therefore frequently be seen as an external impact on the system caused for example by the already mentioned indeterminate delivery of material. If complexity is noted at all it is more likely to be seen as a factor inherent in the system.

5. CONCLUSIONS

As in the case of rationality and non-rationality, the differences between complexity and uncertainty also seem to mark out more clearly the gap between academic theory and managerial practice in manufacturing management. Similarly, managerial strategies for coping with uncertainties in production also serve to underline the gap between what is frequently taught as management theory and what can be observed as management practise. Typical management work strategies for coping with uncertainty in a manufacturing environment are, intuitive decision-making rather than a blind following of computer simulations, overlapping work roles between managers and non-managers rather than the use of clearly defined job descriptions. This overlap is still widespread in manufacturing systems and can be observed occurring in about 60% of discrete managerial activities, extensive use of lateral managerial relationships to further the aims of manufacturing production rather than the use of the system hierarchy and an almost obsessive embracing of cooperative activities rather than activities which are embroiled in conflict. These are all work strategies which receive little attention in mainstream manufacturing, managerial literature and course material. One often wonders how curriculum builders arrive at their interpretation of managerial course material.

6. REFERENCES

(1) Tranfield. D. and Starkey. K. 1997. Abstract from a discussion paper in The Newsletter of the British Academy of Management. September, 1997, No.9.

(2) Watson. T.J. 1997. Theorizing Managerial Work: a Pragmatic, Pluralistic Approach to Interdisciplinary Research. British Journal of Management. Vol.8. pp.3-8.

(3) Trought. B. 1997. Comparing Decision Theory with Practise in Manufacturing Companies. 13th. NCMR, Glasgow Caledonian University, Glasgow, UK, September.

(4) Trought. B. 1990. Manufacturing Decision-Making: irrationality rules OK! 6th. NCPR, University of Strathclyde, Glasgow, UK, September.

(5) Dodd. M, Efstathiou. J, Schirn. J, Calinesci. A. and Berniejo. J. 1997. Measuring Complexity to Identify Problems: Case Study in the Metals Industry. Ed. Harrison. D.K. Proceedings of the 13th NCMR, Glasgow Caledonian University, Glasgow, UK, September.

(6) Stacey. R. 1997. The Applicability of the Science of Complexity to Organizations. Paper presented at a manufacturing complexity seminar, University of Sheffield, Sheffield, UK. Organised by McCarthy. I., Department of Mechanical Engineering. December.

(7) Kay. J. and Campomenes. I. 1997. Manufacturing at the Ford Motor Company. Paper presented at a Manufacturing Complexity seminar, University of Sheffield, Sheffield, UK. Organised by McCarthy. I., Department of Mechanical Engineering. December.

(8) Trought. B. 1998. Manufacturing Complexity: a dissenting view. Paper presented at the workshop on Complexity and Organization, European Institute for Advanced Studies in Management, Brussels, Belgium. June.

(9) Fan. I-S. and Gillies. J.M. 1997. An Initial Mapping of Complexity Theory on to a Model of Extended Enterprise Relationships. Ed. Harrison. D.K. Proceedings of the 13th NCMR, Glasgow Caledonian University, Glasgow, UK. September.

(10) Trought. B. 1986. Why Your CIM is Unlikely to Work. 2nd NCPR, Napier College, Edinburgh, UK. September.

A review of customer service as a strategy for competitive advantage

K VEITCH and **C N WINTERS**
School of Engineering, Coventry University, UK

ABSTRACT

This paper reviews the role of customer service as a strategy for attaining competitive advantage. The paper examines both the importance of competition and compares and contrasts existing competitive strategies. It then goes on to examine the development of customer service as a competitive strategy. The paper concludes by giving a brief outline of an industry based research case study.

INTRODUCTION

The objective of this paper is to examine the evolving role of customer service as a strategy for attaining competitive advantage. It will consider the developing nature of customer service both within the wider context of competition and existing competitive strategies.

The last quarter of this century has seen increasingly changing market places and rapid advances in technology. These factors have given rise to intense global competition making it difficult for organisations to sustain long term profitability (1). The need to address this issue has resulted in a growing recognition of the importance of competitive advantage. Schonberger & Knod (2) observed that, 'Getting to know the competition is vital'. This is confirmed by Christopher (3) who concluded that, 'seeking a sustainable and defensible competitive advantage has become the concern of every manager'. Addressing the need of achieving a competitive edge has seen the implementation of a range of business strategies. More recently customer service has developed from a marketing gimmick to a source of competitive differentiation. Vandermerwe (4) stated that, 'customers.... now have to become the focal point for corporate activities'. This is confirmed by Ohmae (5) who considered that if an organisation, 'consistently succeeds in serving customers more effectively than its competitors, profit will follow'.

The structure of the paper is as follows:

1. An evaluation of the importance of competition and its role in an organisation achieving commercial success.
2. An examination of existing competitive strategies, comparing and contrasting their characteristics, strengths and weaknesses.
3. An assessment of the evolution of customer service and its validity as a competitive strategy.
4. An outline of Covpak (Coventry, UK) a Small - Medium Sized Sheet Plant

1. AN EVALUATION OF THE IMPORTANCE OF COMPETITION

Although the importance of competition as a key factor in an organisation achieving commercial success has been a long held principle, the understanding of its nature and role has undergone a developing process with time. This development has been brought about by both the pressures of the market place and by businesses' responses to those demands. The changing nature of competition has seen the nature of competitive advantage transformed from a product development tool to strategic driving force, and has given rise to a range of competitive strategies. These will be considered in more detail in the next section.

Originally competition was defined in terms of direct competitors. The emphasis of companies being to maximise market share, reduce costs and increase profitability. Often product development was in direct response to the desire to offer a greater number of technical features or perceived benefits above those of the competition. Ansoff (6) observed that, 'product differentiation became an important competitive strategy'. This was confirmed by Drucker (7) who identified the need for a product to have market leadership qualities in order to gain an edge over its competitors.

More recently there has seen a change in emphasis from market dominance to new market creation as, 'in more and more markets the power of the brand has declined and customers are willing to accept substitutes', Christopher (3). Companies have sought to maximise their products' potential and increase long term profitability.

In spite of this changing market focus organisations have experienced more wide reaching strategic forms of competition (8) (9) and intense company rivalry, (10). This has been brought about by a, 'confluence of a variety of social, technological and economic forces' (4), which has resulted in organisations finding it increasingly difficult to sustain competitive differentiation, (11). This in turn has threatened companies' commercial success. Ohmae (5) concluded that, 'deterioration of a company's position relative to that of its competitors may endanger the very existence of the enterprise'.

The change in competitive environment to one of a strategic nature has influenced the development of a variety of competitive strategies, as companies, 'continue to search for new ways to improve the distinctiveness or uniqueness of their products / services', (12) and seek to spend, 'more time creating fundamentally new competitive space', (13). Although different in nature these competitive strategies are each shaped by the aim of attaining competitive advantage and thus providing superior performance for the organisation, (14). A comparison of these competitive strategies will allow an insight in to the nature and importance of attaining competitive advantage in the future.

2. A COMPARISON OF EXISTING COMPETITIVE STRATEGIES

Existing competitive strategies appear to fall into three broad areas :

- Manufacturing Strategies - those concerned predominantly with the structure or working practices of the manufacturing or production processes of an organisation.
- Organisation Management Strategies - those concerned predominantly with the structure and the working systems of the organisation as a whole.
- Marketing Strategies - those concerned predominantly with the marketing and sales activities of the organisation.

Due to the wide reaching strategic nature of competition, individual strategies may be found to overlap one or more of the above areas. In addition to this organisations may adopt a combination of different strategies in order to achieve their own competitive advantage. Each of the strategies can be seen to possess its own characteristics, strengths and weaknesses.

2.1 Manufacturing Strategies

This paper will consider three established Manufacturing Strategies: Lean Manufacturing, Agile Manufacturing, and World Class Manufacturing. Differing in nature and objectives, each of these strategies does encompass shared elements and manufacturing methodologies.

2.1.1 Lean Manufacturing

Established from the Toyota Production System, the main emphasis of Lean Manufacturing is the maximising of production efficiency. Efficiency gains are made by the reduction of waste from the manufacturing and support processes of the organisation, whilst still allowing full production demands to be met, (15).

Giving rise to both just-in-time (JIT) and Kanban production methods, Lean Manufacturing has received world wide acceptance and implementation. However more recently the limits of its success have been debated. Cusumano (16) concluded that, 'continually pushing for gains in manufacturing efficiency has resulted in a new set of problems'. Moreover Womack et al's research conclusions have been brought into question, (17), and these doubts may give rise to a disenchantment with Lean in the future.

2.1.2 Agile Manufacturing

Agile Manufacturing is a fusion of a range of accepted production and organisational methodologies and work systems, (18). Unlike Lean the main focus of Agile Manufacturing is not maximised production efficiency, but manufacturing flexibility. Agile Manufacturing seeks to provide a dynamic response to meeting customer requirements, and although a relatively recent development it has received wide acknowledgement.

2.1.3 World Class Manufacturing

World Class Manufacturing (WCM) creates a standard for excellence in manufacturing, and as such has received broad acceptance by industry. Like both Lean and Agile, WCM is a combination of several key elements, 'total quality control, just-in-time, total preventative maintenance and employee involvement', (19). With the aid of bench marking and best practice World Class Manufacturing aims to bring about continuous improvement both within individual organisations and the manufacturing industry as a whole.

2.2 Organisation Management Strategies

Within this broad area this paper will discuss two key competitive strategies: Total Quality Management (TQM) and Business Re-engineering. Fundamentally different in nature both TQM and Re-engineering are concerned with an organisation wide re-evaluation process and a complete commitment to the principles of the strategy.

2.2.1 Total Quality Management

Derived from the work of such theorists as Deming, Juran and Feigenbaum, quality has long been accepted as a critical element in the development of manufacturing methodologies. Concerned with the work practices and systems of the entire organisation, as defined by Feigenbaum (20) TQM has developed into a key competitive strategy. Juran (21) concluded that, 'market place pressures have their impact on manufacturers, who look for ways to improve their quality performance so as to make it easier to meet their competition'.

Total Quality Management is a long established management philosophy, which has received broad acceptance by industry. However it is for this very reason that the future of TQM as a competitive strategy is uncertain. Slater (1996) observed that, 'Quality may no longer be as strong a basis for competitive advantage as it was in the past. In many cases it has become necessary for competitive parity'.

2.2.2 Business Re-engineering

Derived from the ideas of Hammer & Champy, Re-engineering is concerned with the restructuring of an organisation in terms of critical business processes, (22). This restructuring process results in, 'quantum levels of transformation', (23) encompassing both the organisation structure and work practices. Re-engineering has found limited implementation due to its radical principles, however its critical successes to date may see its growing acceptance in the future.

2.3 Marketing Strategies

Based on the accepted principles of the Marketing Mix and concerned with all the interface activities required, 'to achieve the corporate purpose of attracting and holding customers' (24), Marketing Strategies can be subdivided into four areas : Product, Promotion, Distribution and Pricing Strategies.

Including such elements and activities as: market segmentation, portfolio development, branding, loyalty rewards and franchising, it is the combination of these strategies that establishes an overall competitive Marketing Mix. This mix is a direct response to the demands of the market place, and as such Marketing Strategies continue to receive wide acceptance. However in order to continue to gain competitive advantage, 'Marketers must constantly adapt their strategies to the rapidly changing competitive market' (25).

As can be seen from the above section a range of strategies exist for attaining competitive advantage. Each of these strategies possesses its own principles, strengths and weaknesses. In response to the need to create and sustain competitive advantage, brought about by an ever changing market place (26) the 1990's has seen the development of customer service as a competitive strategy.

3. AN ASSESSMENT OF CUSTOMER SERVICE AS A COMPETITIVE STRATEGY

Long Accepted as an important element of commercial success, in the last decade customer service has become a key factor for competitive differentiation (3). Evolving from a marketing concept, customer service has developed into the central driving focus for all competitive strategies. Levitt (27) concluded that, 'all energies should be directed towards satisfying the customer no matter what'. This is confirmed by Vandermerwe (4) who contested that service is, 'the most important competitive lever for the 1990's and beyond'.

Yet in spite of change in the nature of customer service to one of strategic importance, the adoption of customer service strategies has been limited to date, with most companies remaining unfocused on customer satisfaction, (1). However being the central emphasis of existing competitive strategies and growing in importance, the indication is that customer service will play a critical role as a strategy for attaining competitive advantage in the future.

4. COVPAK - A CASE STUDY

This paper is the outcome of an industry based research programme, funded by the Economic and Social Research Council and in conjunction with Covpak (Coventry, UK). Covpak is a Small - Medium Sized Sheet Plant, which is facing intense competition in a regional market place and is seeking new ways to create an edge over its competitors.

Having realised the penalties of poor service, loss of orders and long term profitability, Covpak is looking to customer service to create added value to a commodity product. It is envisaged that future research will incorporate an assessment of a re-engineering of Covpak's customer service operation and the development of a customer service strategy, including its generic applicability.

CONCLUSION

This paper has considered the role of customer service as a strategy for attaining competitive advantage. It has considered both the nature of competition and existing competitive strategies.

It has been seen that addressing of the issue of competition is critical to an organisation's commercial success. The need for creating and sustaining a competitive advantage has given rise to a variety of competitive strategies. Differing in nature each strategy has its own strengths and weaknesses.

Changing market place pressures continue to create the need for new competitive strategies and as a result the 1990's has seen the development of customer service as a method for creating competitive differentiation. Customer service continues to grow in importance and it is indicated that it will play a critical role as a competitive strategy in the future.

ACKNOWLEDGEMENTS

The authors would like to acknowledge the assistance of the Economic and Social Research Council in carrying out this research.

REFERENCES

1. Griffin A. et al, Best Practice For Customer Satisfaction In Manufacturing Firms, Sloan Management Review, Winter 1995, pp. 87-98.
2. Schonberger R.J. and Knod E.M., Operations Management 4th Edition, Irwin, 1991.
3. Christopher M., Logistics and Supply Chain Management, Pitman Publishing, 1992.
4. Vandermerwe S., From Tin Soldiers Russian Dolls, Butterworth-Heinemann, 1993.
5. Ohmae K., The Mind Of The Strategist, Penguin Books, 1982.
6. Ansoff H.I., Corporate Strategy, Penguin Books, 1965 & 1987
7. Drucker P.F., Managing For Results, Heinemann, 1964.
8. Porter M.E., Competitive Strategy, Free Press, 1980.
9. Porter M.E., Competitive Advantage, Free Press, 1985.
10. Hill T., Manufacturing Strategy 2nd Edition, MacMillan, 1995.
11. O'Shaughnessy J., Competitive Marketing 2nd Edition, Routledge, 1992.
12. Pitts R.A. and Lei D., Strategic Management, West Publishing, 1996.
13. Hamel G. and Prahalad C.K., Competing For The Future, Harvard Business School, 1994.
14. Slater S.F., The Challenge Of Sustaining Competitive Advantage, Industrial Marketing Management, 25, 1, Jan 1996, pp. 79-86.
15. Womack J.P. and Jones D.T., Lean Thinking, Simon & Schuster, 1996.
16. Cusumano M.A., The Limits Of Lean, Sloan Management Review, Summer 1994, pp. 27-32.
17. Williams K. et al, Against Lean Production, Economy & Society, 21, 3, 1992, pp. 321-354.
18. Gunneson A.O., Transitioning To Agility, Addison-Wesley Publishing, 1997.
19. Schonberger R.J., World Class Manufacturing, Free Press, 1986.
20. Feigenbaum A.V., Total Quality Control 3rd Edition, McGraw-Hill, 1991.
21. Juran J.M. and Gryna F.M., Quality Planning And Analysis 3rd Edition, McGraw-Hill, 1993.
22. Hammer M. and Champy J., Re-engineering The Corporation, Nicholas Brearly, 1994.
23. Hardaker M., Total Competitiveness, McGraw-Hill, 1995.
24. Levitt T., Marketing For Business Growth, McGraw-Hill, 1974.
25. Kotler P., Principles Of Marketing 3rd Edition, Prentice Hall, 1986.
26. Morehouse D.L., Essentials Of TQM, National Press Publications, 1996.
27. Levitt T., The Marketing Imagination, Macmillan, 1986.

Managing legacy systems in small manufacturing organizations

S C WILLOUGHBY
School of Computing and Information Systems, University of Sunderland, UK
N S McBAIN
Centre for Achievement in Manufacturing and Management, University of Sunderland, UK

ABSTRACT

This paper discusses an EPSRC funded project that aims to develop a generic model for assessing the risks facing manufacturing SME's that are considering changing their existing (legacy) IT systems and related incumbent processes. The paper focuses on the cases of three engineering firms that manufacture in small and medium sized batches. The project fieldwork used an inductive investigative tool called 'grounded theory', which enabled legacy systems in the firms to be evaluated in their organisational contexts rather than as stand alone systems. This found two types of IT implementation - integrated packages that supported all business processes, and piecemeal structures of stand-alone applications that just supported administrative rather than core manufacturing operations. Managers at the SMEs with the piecemeal structures favoured replacing them with an integrated package. The wisdom of this decision is questioned, particularly as the SMEs with integrated packages had found that these did not meet all their requirements. They had therefore either developed additional applications, or had replaced the integrated package with PC applications that supported particular processes. The paper suggests that a piecemeal approach might be more appropriate for SMEs to meet individual demand given the need for flexibility to adapt to changing external environments.

1. INTRODUCTION

The discussion of what to do with legacy IT systems and how to manage them almost invariably revolves around large, complex, software systems [1]. This perspective results in researchers concentrating on the development of technical solutions to technical problems.

However, there is, in our experience, a whole range of problems that are overlooked because they are situated in the context of small to medium sized enterprises (SMEs), defined as employing 250 staff or less. For these organisations, IT systems are used purely as a support mechanism rather than as an end in themselves.

This paper examines the issues associated with managing legacy systems used in SMEs, based on detailed case study fieldwork at three manufacturing SMEs. Since the IT systems are just used as a support mechanism, this issue is examined from a sociotechnical viewpoint, where the IT system is viewed in the wider context of the organisation. Therefore for the purposes of our work we define a legacy as *"Any existing state which impacts upon change"* and thus a legacy system as *""A group of interacting elements forming an entity where one or more elements impacts upon potential change"*. In this paper, we firstly outline the main legacy systems problems, including a detailed case study of the situation at one organisation. We then discuss issues for SME managers to consider when addressing the legacy systems issue.

The research was conducted using an inductive investigative tool called 'grounded theory' [2], that is oriented towards the generation of theories which are firmly grounded in empirical data and relies on the researcher systematically collecting, coding, categorising and analysing data to derive the features which belong to the theory (or model). In our research, use of this approach enabled the legacy systems to be evaluated in their organisational contexts rather than as stand alone systems. The work reported is part of an on-going project sponsored by the Engineering and Physical Sciences Research Council (EPSRC) through the Systems Engineering for Business Process Change (SEBPC) programme.

2. TYPICAL SITUATION WITHIN SMES

The three SMEs we examined were each considering the effect of a business process change exercise on their legacy system. The legacy systems problems that exist for SMEs arise as a result of their IT environment, and the manner in which they have acquired software systems and attempted to maintain them [3]. The strength of the legacy systems is that they encompass each SME's history and processes. They are often the basis of many of its operations, and support processes that have been shaped alongside the SME's culture. Our research therefore examined the development of the IT systems currently in use at the SMEs and what part of the business activities they were used to support.

One of the mechanical engineering SMEs we examined had introduced an integrated computer package that supported all their main operations. The only computing software at this company prior to this was an accounts package and some word processing packages. At the time of our fieldwork, the company was experiencing some problems with this integrated package. There were a number of technical problems, such as production of the material requirements report taking several hours rather than a few minutes, which meant that it was compiled on a weekly rather than a daily basis. The slow speed was because of a large amount of redundant data as computer records were not updated upon completion of jobs. Indeed, production was significantly faster after a recent exercise where this redundant data had been rationalised.

In addition, the integrated package was not suitable for all of the company's requirements. The accounts department still used a package dating from the early 1980s, though was considering changing to a more recent package at the time of our research, mainly because of concerns over compatibility with the year 2000. The purchasing department still ordered consumables using the database package from before the integrated system, and only used the integrated package to order production materials. This was partly because the integrated package required a code for each order, and the company had yet to develop codes for their large amount of consumables. This was annoying for staff, who had to swap between the two systems, and the purchase orders produced by the integrated package were not in a suitable format for the company.

This suggests that the accounts and purchasing departments' requirements were not considered when selecting the integrated package. Our research found that the decision to purchase it had been made on the basis of its ability to automate job card production, with an eventual aim of further development into a production control system. At the time of purchase, the company was unsure whether it could support certain activities, but were happy to proceed, even if the only outcome was automated job card production.

There were two views concerning the future of this legacy system - to either replace or to continue with it since it had only been installed for 2½ years. Those in favour of replacing it were from the accounts and purchasing departments that had experienced the problems outlined previously. Most other users thought the integrated system was satisfactory for activities that it currently supported; there was a belief that the perceptions concerning its inadequacy to support business requirements were because the company had not used it to its full potential. For example, employees used spreadsheets that they had developed to support their personal activities: they had no idea whether these facilities were supported by the integrated package. This factor, along with the continued use of separate software by the accounts and purchasing departments resulted in a piecemeal structure to the legacy system at this SME. There was some data transfer between these user applications and the integrated package which might have been unnecessary if the activity was also supported by the integrated package. Indeed, our fieldwork established that one of the main problems with the legacy system at this SME was that users had received inadequate training and technical support from the software manufacturer.

3. RESEARCH ISSUES

Subsequent analysis of the data collected at the other two mechanical engineering SMEs using grounded theory identified a number of similar findings to the organisation featured above. The emerging generic issues identified can be summarised under the headings of "IT acquisition and maintenance", "user skill" and "data issues".

3.1 IT acquisition and maintenance

The IT acquisition process has varied over time in each organisation, with the situation evaluated afresh each time: this has led to a lack of consistency which suggests an ad hoc approach to system development. Evidence of this is seen in the range of software applications: including off-the-shelf packages, in-house development of centralised bespoke systems, and end user application development. Although some of the software which supports the business has been developed by IT professionals (either internally or externally) a large amount of it has been developed by end users: typically within commercial spreadsheet or data base packages, to automate those tasks that are repetitive or mechanical. The hardware platforms, meanwhile, have varied, but what has been common to all is that although a central computer exists, the IT end users most frequently use a PC environment in which they are networked.

3.2 User skill

The second issue identified by our research is the varying IT skills of the users. It is apparent from the empirical evidence that very few SMEs have in place dedicated IT departments: the norm appears to be to assign an individual to the systems manager role and to dedicate few resources to IT support and maintenance. There was also a lack of IT training at the SMEs we examined, and most users were self-taught. This approach has led to applications of varying quality, since they are dependent on the ability and IT skills of the end user concerned, and was one reason for the failure of the integrated package mentioned earlier. The resultant nonconformity produces difficulties when staff are required to undertake tasks normally carried out by others.

3.3 Data issues

The third issue concerned the storage of data in the legacy system. In SMEs the value of the legacy system is often not in the functionality provided but in the data held. The stored data contains both current and historical data and, in effect, is the corporate memory of the organisation. Therefore it can provide the opportunity for managers to analyse previous performance, and plan for the future provided that the data can be relied on (and can be accessed effectively). Although discrete IT applications are often used within a small organisation the data used, produced, and stored is rarely unique to those applications. For example, the design and production departments will hold data on the same part of business activities. Some of the data they store may differ but large elements are duplicated. This issue is central where any concerted effort is made to rationalise the data held and eliminate data duplication.

Where data is transferred between the various components of the legacy system, the safest approach is to use electronic transfer. However, if the data is held in different formats this requires some intermediate IT solution to ensure a "fit" between applications is achieved. Misunderstanding of data can result in "acceptable" data being fed from one application to another (since the format fits) but with data values being incorrect. However, most problems occur when manual transfer is used (as at the SME described previously). For instance where the print out from one application is passed for subsequent input into another, perhaps situated in another department. Here human error in inputting data is a danger however scrupulous the individual is in checking the entries.

4. LEGACY SYSTEM PROBLEMS

We have observed in the SMEs examined that most of the IT has been developed by end-users to support their individual activities. An advantage of this is that individuals in the company know the applications and this can be adapted to meet changing business requirements. On the other hand, this has meant that software development is essentially functionally based, the precise functionality is unknown to anyone other than the originator and is undocumented, and there is no consideration for how software could be used to support business processes.

Hence it appears that the typical structure of the legacy systems at SMEs is fragmented, as shown schematically in Figure 1. Here a business process is shown crossing the departments which support it and areas in which there is IT support are denoted. In mapping such IT support the linkages between components and the nature of the linkages are identified.

There was the belief at each SME that such legacy systems were no longer adequate to support business requirements, as a result of considering the effects of various levels of business process change on the legacy system and vice versa. In addition to the previous example at the case study organisation, management at another SME thought that their current piecemeal structure would be best replaced by an integrated package. Meanwhile, the final SME was, at the time of our research, replacing the legacy system of a number of spreadsheets that were cross-referenced for reporting purposes with a database that would speed up the cross referencing between the various parts of the system. There was no desire to replace this piecemeal system with an integrated package, largely because of two previous unsuccessful experiences of implementing an integrated software package that supported all their main processes. Hence, there were two different perceptions amongst management at each of the SMEs over how best to resolve this situation - using a large-scale integrated package or redeveloping the piecemeal structure of stand-alone IT components.

5. RECOMMENDATIONS

The desire to replace the piecemeal structure with integrated systems to support all business processes appears to be based on two perceptions. Firstly, that it is seen as adopting "best practice" of large, successful, organisations, therefore transference into SMEs will be positive. Secondly, the managers can see problems with their existing operation of IT and such a solution is the "obvious" cure. However, the reality of the IT operation diverges from these two perceptions: although the true fit between the processes and systems is often ill understood. Therefore the strategy required to move from discrete systems to integrated systems is difficult for such organisations to formulate.

The use of grounded theory for this research revealed issues that would not otherwise have been identified. Seen from the perspective illustrated in Figure 1, the patchwork effect of different systems, which managers perceive as a weakness, may actually present a strength. Where a monolithic integrated system is in place, it may be very difficult to isolate the interaction of a particular process with it and to change it. The knock-on effects could be disastrous. The process is defined by the computer system and is shaped by it. However, in the case of the SMEs examined a process is served by a number of loosely linked, or completely disparate, computerised systems, therefore the risk in change is far less. If a computerised system has to be abandoned or radically changed due to process change, the ramifications are limited and the job of replacing it is comparatively self-contained. Where computerised systems represent only small portions of a task, it is more likely that it will be possible to retain them more or less intact. In the organisation examined in section 2, this flexibility could be achieved through allowing the accounts and purchasing departments to use separate packages that suited their requirements, and for the integrated package to focus on supporting activities in production and control where it was perceived that there is potential.

Therefore, the emphasis should be on facilitating communication between isolated programs rather than on integrating them. This could be achieved by developing software bridges between different programs to enable them to communicate, and ensuring that the same type of hardware and software platforms are used as widely as possible. This kind of integration allows better continuity of processing but doesn't set the current state of the business processes in stone, and is also beneficial given to need for SMEs to be flexible in adapting to changing external environments. In the environment of the three SMEs examined, people and processes are running the legacy systems rather than the legacy systems running the way that business is conducted: which in any business is a feature that should be aspired to (and retained where it exists).

6. CONCLUSION

This paper has identified a number of issues concerning legacy IT systems in SMEs through the use of grounded theory. This focused on one organisation where the legacy system did not support all their business requirements. The paper discussed how managers apparently favoured an integrated package to resolve the problem with legacy systems, particularly the interfaces between components. However, it is suggested from the evidence of this research that an IT system consisting of separate inter-related components might be more appropriate for the flexible environment within SMEs. Organisations where such a structure exists should consider its strengths when evaluating appropriate action for their legacy system. Each individual component can be changed when required, rather than replacing the entire system, though it is important there are appropriate software interfaces between the various components to ensure integrity of the data.

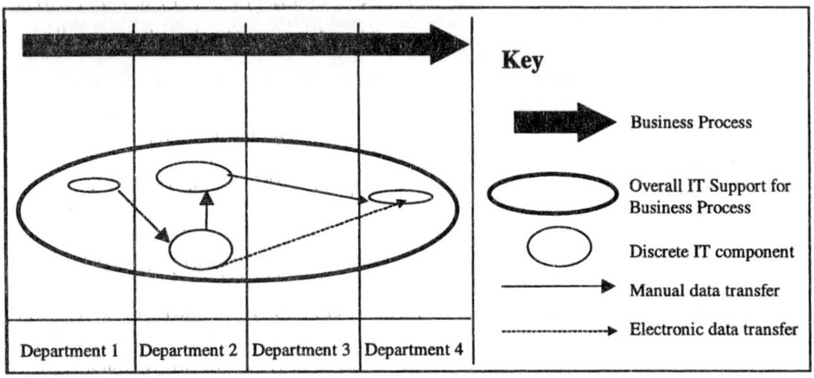

Figure 1 - Schematic Representation of IT Support for Business Processes

7. REFERENCES

[1] Bennett, K. *Legacy systems: coping with success*. IEEE software, Volume 12, Issue 1, pp. 19-23, 1995

[2] Strauss, A and Corbin, J., "Basics of Qualitative research: Grounded theory Procedures and Techniques". Sages Publications, 1990.

[3] Edwards, H.M., Harvey, C.F., McBain, N.S., *Alternative Futures - Constrained by the Past? The Impact of Legacy Systems on Business Process Change*, 7th Annual Bit Conference Proceedings on CD ROM, Manchester Metropolitan University, 1997.

Manufacturing management – interfirm relationships

C WYATT and **S EVANS**
The CIM Institute, Cranfield University, UK

SYNOPSIS

The fields of politics and economics have traditionally taken an approach to relationship management based on exchange theory, whereby companies engage in relationships only when perceived benefits outweigh perceived costs. Manufacturing supply chains consist of unequally balanced power-dominance relationships, where one partner (the customer) has the potential to achieve profit at the expense of the other (the supplier). This paper will expand existing work in exchange theory, by attempting to understand the mechanisms underlying the ability of organisations to manage their relationships in an environment of inequality. An example from the automotive industry will be given which focuses specifically on product development.

1.0 INTRODUCTION

Emerson (1962)[1] rejects the view that power is an attribute of an individual or an organisation, rather he describes it as a property of a relationship involving two or more parties. Simplistically, he describes power as a function of the dependency of one party on another. Dependence is proportional to how much one party desires the goals mediated by the other and the availability of these goals elsewhere. Power is how much one party can make the most of the assets that it holds to influence the other. Power dependence theory has often been used in the literature to describe interactions between customers and suppliers ([2], [3],) and traditionally the situation has been one of heavy dependence of suppliers on their customers. This has been particularly true of the automotive industry in the West where "adversarial" approaches to supplier management have led vehicle manufacturers to wield the favourable balance of power in order to force down component prices [4].

It is well documented that adversarial relations in the automotive industry are being replaced by moves towards "relationships" and so called supplier "partnerships" [5, 6]. This is happening for a number of reasons. The increasing pace of technological change, the impact of globalisation on the number and variety of end customers and the maturity and competition within the car making business itself have led to pressure for both cost and product development timescale reductions. European car companies are therefore focusing their resources on fewer core competencies, such as styling, assembly and some parts of engine design and increasing the proportion of component outsourcing [7]. Today, up to 80% of the costs of the end vehicle can be attributed to supplier products [8]. In addition as Lamming [9] states: "it is estimated that whereas the car market will grow at between 1% and 2% per annum until the end of the century, components should increase by 5-6% per annum. This is the result of two trends: the value of components is increasing with developments in technology, and the amount of value-adding work which the vehicle assembler outsources to the supplier is increasing in many cases." For a large number of first tier suppliers the value adding activity involves increasingly high levels of design responsibility [7, 9, 10]. Moves towards "partnerships" and "relationships" have therefore been brought about by the realisation of the increasing dependence of vehicle manufacturers on the capabilities of their key suppliers, particularly in the area of design and development.

This tilting power-dependence balance creates an environment in which the term "relationship" can now be usefully applied. In the social sciences and economics literatures, the most influential interpretation of the term relationship is based on the ability of the two parties involved to make decisions about the level of investment and return they are receiving from the situation [11-13]. From these assessments, either party is able to make decisions about their level of satisfaction with the relationship and therefore whether or not to stay with that particular partner. Within traditional "adversarial" contexts these decisions i.e. who to work with, when and how, were entirely the prerogative of the vehicle manufacturers and as a consequence the nature of the interaction was almost entirely one sided. This paper has argued that the situation is changing; in the context of power dependency theory [1] as vehicle manufacturers become more dependent on suppliers, suppliers gain more power. With more power comes the increased ability of suppliers to influence the decisions taken in their relationships with vehicle manufacturers. A key challenge facing the industry is therefore to develop new structures and processes that take into account fundamental shifts in relationships between vehicle manufacturers and their first tier suppliers

2.0 RELATIONSHIP THEORY

There are a number of key propositions highlighted in relationship theory as essential to the successful maintenance of relationships. The existence of these propositions has been proposed and empirically tested by social scientists and economists and synthesised into general theories of exchange. [12, 14], transaction cost economics [15] and equity theory [16, 17]. Application of these propositions to the business environment is dependent on the assumption that decisions made between organisations are likely to differ from those made at an interpersonal level in one key aspect. It is assumed that organisational decisions are based on conscious assessments of the situation and are irrespective of implicit thought processes. Although beyond the scope of the current work, valuable insights may be gained through the study of less easily accessible relationship based thought processes. The following is a summary of the key relationship propositions outlined in the social sciences literature:

Proposition 1: Both parties must be in "profit". Profit can be defined as the inputs into the relationship minus the outputs from the relationship and can include any type of benefit received from the relationship. For example a supplier may work with a vehicle manufacturer for comparatively low levels of monetary gain, but may perceive the opportunity to learn particular skills from that vehicle manufacturer as a more salient benefit.

Proposition 2: The level of <u>actual</u> profit received by both parties must be equal to or greater than <u>expected</u> profit levels. For example a supplier's component must match the quality and cost targets agreed at the outset by the customer.

Proposition 3: The level of actual profit received by both parties must be equal to or greater than that that could be achieved with an alternative partner. Either party will only be satisfied with the relationship if they know that they cannot receive greater benefits from an alternative partner.

Proposition 4: The levels of profit achieved by the other party in comparison to the organisation's own profit levels must be acceptable to both parties. For example it may appear to suppliers that customers are achieving substantially more benefits from the relationship than the suppliers themselves, however because of the intrinsic power position of the customer, this may be deemed as both acceptable and "normal".

If all propositions hold true the result will be satisfaction with the relationship, satisfaction is associated with commitment. However, where a discrepancy occurs either one or both parties will engage in equity restoring behaviours, for example reducing their own inputs or asking their partner to increase their inputs. Restoration of equity can be the result of both actual behaviours and the psychological alteration of perceptions. For example where actual benefits do not match up to expected benefits, the party concerned may convince himself that what he had originally been expecting was completely unreasonable and therefore the actual benefits received are quite acceptable. Where equity restoring behaviours fail to produce the required results, or even in extreme cases of perceived inequity the relationship may be dissolved [17].

3.0 A PROCESS FOR INTERFIRM RELATIONSHIP MANAGEMENT

The remainder of the paper will describe one approach taken by the European design centre of an international vehicle manufacturer to the development of a process for managing its relationships with suppliers specifically in the area of design and development. The process aims to align the new product development cycles of both parties through a series of regular joint gateway reviews prior to the kick off of a development programme. The success of the joint gateway reviews depends on the nature, the level and the quality of the information brought to the table by both parties. Fundamental to the design of the process is an assumption of the commitment of both parties to provide relevant, detailed and accurate data. Decisions are based on the degree of alignment between the two companies in terms of future development requirements, and the level of risk involved. A simple scoring system allows products to be assessed both individually and in relation to others. Progression decisions (for example target timing), and specific actions are recorded during the meeting and used as the basis for future meetings to assess product progress. The frequency of the joint reviews is determined jointly but on average is currently twice annually.

Each review is an opportunity for both parties to assess the level of investment required to successfully progress a new product idea. It is also an opportunity to identify the predicted benefits of the product development to their business. The term "relationship" can therefore be defined in this context as joint involvement in new product (component) development. The first criteria established for the process was that it should enable both parties to be explicitly aware of their relationship inputs and outputs in order to facilitate the conscious relationship decision making process. A standard operation for the review process was defined and documented. It currently includes a framework for the information to be presented by both parties at each review, a documented process for discussion of each product, a guide for assessment and simple scoring of products and documentation that captures the main outputs of the review, further actions and specific responsibilities of both parties. This document is signed by both the customer and the supplier as a firm indication of commitment to proceed. As far as is reasonably possible all aspects of the process are made transparent to both parties, providing the information on which adherence to the propositions outlined above can be assessed.

At the outset, both parties will be able to clearly identify their inputs into the process - at the very least for a minimum of six months. However the outputs and therefore the overall benefit levels of the relationship will be harder to define in the initial stages. A traditional product development funnel e.g. [18] was taken as a conceptual basis for the strategic review process, and modified to include the roles of two firms as opposed to one. During each review both technical and market information is exchanged by both parties and decisions are jointly taken on: a) whether or not the product is to be progressed - based on market need, development risks and the availability of the necessary resources from both companies and b) what actions are to be taken by whom in order to progress the product

A number of products can be in the funnel at any one time and all products compete against each other for resources and funding. Products may enter or exit the funnel at any point. Entrance will depend on whether the product is new, core or modified core technology and exit point will depend on how well it competes with other products. Outputs to be gained from particular projects will therefore depend on a number of factors and may not become explicit until the product has progressed through several gateway reviews. Assessment of profit levels from the relationship is therefore delayed at the outset, however experience with a number of products over a period of time will enable both parties to make predictions of expected outputs and therefore profit levels in future development projects.

It is only on successful progression of components into vehicle development programmes that assessments can be made regarding the match between expected profit levels and actual profit levels, and between actual profit levels and the level of profit that can be achieved in alternative relationships. The review process itself does not provide a formalised mechanism or context for the making of these comparisons, the rationale being that these are intra rather than inter company decisions. However, the transparency of the review process does enable both parties to make judgements about the relative profit achievable by their partner. Again the assessments of whether inequalities in profit levels are acceptable are likely to be made individually rather than jointly, however the underlying data will be common to both parties.

The continuous review process enables both parties to monitor any changes in the inputs or expected outputs from the development project, and any changes in the inputs and outputs of their partners. It therefore provides a basis for regular monitoring and review of the relationship. In addition, because of its formalised nature the strategic review process itself will serve as an explicit reminder to both parties to make these types of assessments on a regular basis. If the outcome of these assessments is less than optimal for either party, both parties have the opportunity to engage in equity restoring behaviours - i.e. reducing their own inputs, reducing their expectations or increasing the inputs of the other party. The relationship should therefore rarely be ignored, the reviews can be seen as opportunities for health checks and the taking of appropriate action to ensure the relationship is either nurtured or dissolved.

4.0 CONCLUSIONS

Arguments for the shifting balance of power and dependence between automotive manufacturers and suppliers suggest that the move towards closer more partnership type relationships by the manufacturers may be mediated by the increased decision making potential of suppliers. It is acknowledged that the position of the customer will always be stronger than that of suppliers, however the logical conclusion of current trends points to increasing supplier influence. Much of the current literature in the area emphasises the Japanese model of close, long term relationships however as the model of lean supply [19] suggests, suppliers may well become the technology leaders developing products independently of the requests of vehicle manufacturers, thereby increasing the range and variety of choices available to them.

This paper describes a process designed by one major vehicle manufacturer to increase their integration with suppliers in product design, thereby counterbalancing increased devolvement of design with increased investment in suppliers. The process has been described within the framework of equity and exchange theories to highlight how such processes can be developed as tools for the management of relationships, as opposed to traditional buyer supplier interactions. It has been suggested that the industry as a whole needs to consider how to cope with these new and unfamiliar situations, and to develop appropriate mechanisms and procedures. As has already been outlined, one major assumption of this paper has been that relationship decision making between organisations is a conscious process guided by a logical set of rules. However, organisations cannot have relationships, it is the people within the organisations that have relationships with people in other organisations. Decision making will therefore not necessarily be based on explicit information nor made in any logical and rational way. If suppliers do not want to take responsibility for design, if vehicle manufacturers are reluctant to relinquish their power, moves towards true “relationships” between two autonomous parties are unlikely to succeed.

REFERENCES

[1] R. Emerson, "Power-Dependence Relations," *American Sociological Review*, vol. 27, pp. 31-41, 1962.

[2] M. Bresnen, "An Organisational Perspective on Changing Buyer-Supplier Relations:A Critical Review of the Evidence," *Organization Articles*, vol. 3, pp. 121-145, 1996.

[3] J. J. Cronin, T. L. Baker, and J. M. Hawes, "An Assessment of Role Performance Measurement of Power-Dependency in Marketing Channels," *Journal of Business Research*, vol. 30, pp. 201-210, 1994.

[4] J. Morris and R. Imrie, *Transforming Buyer-Supplier Relations:Japanese Style Industrial Practices in a Western Context*. London: Macmillan Academic and Professional Ltd, 1992.

[5] S. R. Helper and S. M., "Supplier Relations in Japan and the United States: Are they converging?," *Sloan Management Review*, vol. Spring, pp. 77-84, 1995.

[6] J. Bevan, "What is Co-Makership," *International Journal of Quality and Reliability Management*, vol. 4, pp. 47-56, 1987.

[7] Boston Consulting Group, "The Evolving Competitive Challenge for the European Automotive Components Industry," Commission Des Communautes Europeennes July 1993 1993.

[8] K. B. Clark and T. Fujimoto, *Product Development Performance*. Boston: Harvard Business School Press, 1991.

[9] R. Lamming, "Strategic Options for Automotive Suppliers in the Global Market," *International Journal of Technology Management*, vol. 5, pp. 649-684, 1990.

[10] C. Leisk and J. Wormald, "Supplier Innovation:Building Tomorrow's Components Industry," The Economist Intelligence Unit, London June 1992.

[11] O. E. Williamson, "Transaction Cost Economics and Organization Theory," *Industrial and Corporate Change*, vol. 2, pp. 107-156, 1993.

[12] P. M. Blau, *Exchange and Power in Social Life*. New York: Wiley, 1964.

[13] G. Homans, *Social Behaviour:Its Elementary Forms*. New York: Harcourt Brace, 1961.

[14] M. Deutsch, "Equity, Equality and Need:What Determines which Value will be used as the basis of Distributive Justice," *Journal of Social Issues*, vol. 31, pp. 137-149, 1975.

[15] B. K. Pillig, L. A. Crosby, and D. W. Jackson, "Relational Bonds in Industrial Exchange:An Experimental Test of the Transaction Cost Economic Framework," *Journal of Business Research*, vol. 30, pp. 237-251, 1994.

[16] E. Walster, G. W. Walster, and E. Berscheid, *Equity: Theory and Research*. Boston: Allyn and Bacon, 1978.

[17] S. Sprecher and P. Schwartz, "Equity and Balance in the Exchange of Contributions in Close Relationships," in *Entitlement and the Affectional Bond:Justice in Close Relationships*, M. J. Lerner and G. Mikula, Eds. New York: Plenum Press, 1994.

[18] S. C. Wheelwright and K. B. Clark, *Revolutionizing Product Development*. New York: The Free Press, 1992.

[19] R. Lamming, *Beyond Partnership:Strategies for Innovation and Lean Supply*. London: Prentice Hall, 1993.

Business Process Re-engineering

Performance measurement identification – an application for maintenance

D J CARSON
Globe Research and Consulting, UK
R LEONARD
Total Technology Centre, UMIST, Manchester, UK

ABSTRACT

Identifying a comprehensive and manageable set of measures for an organisation is one of the fundamental challenges that face organisations. Being able to understand the interactions between measures, cross functionally within a corporation, is necessary to gauge strategic business objectives. This paper demonstrates the use of two techniques, subjective and objective, cross functionally within the maintenance division of a national utility. The techniques have been correlated, which have identified performance measures that were aligned and misaligned to the strategic intent of the organisation. It was demonstrated that the division showed strong degrees of alignment. The method is applicable to other sector scenarios.

1. INTRODUCTION

With increasing levels of service and customer satisfaction, organisations are confronted with the pressures for performing. This situation, together, with growing commercial pressures from emerging competitive markets, multi-national conglomerates and the need to enhance competitive advantage, means adopting rationalisation procedures. Measures must therefore be adopted to continuously improve performance, whilst maintaining the high standards upon which corporate identities are based.

There is a growing realisation that financial performance cannot be sustained unless the non-financial underpinnings of successful operation are measured as well. Financial measures may summarise past performance adequately well, but in isolation they predict future performance poorly. The problem is not the measurement as such, but knowing what to identify and measure for long-term growth. For an effective understanding of a firm, progress towards the attainment of strategic objectives is the principal aim.

In the quest for identifying appropriate performance measures for new industrial paradigms, the maintenance function of a national utility provides an example of this need. New paradigms within the service industries need to offer superior performance, not only in terms of tangible, quantitative financial measures, but also intangible, qualitative non-financial measures. Such techniques are capable of spearheading strategic initiatives into organisations that often find traditional measures inadequate within the contemporary climate.

2. PERFORMANCE MEASURES

Historic growth of industrial practices has been influenced by technological, socio-economic development and the need to meet competition in a globally market driven, capitalist economy. Meyer (1), cites "post war techniques involved only monitoring rejects and returns." Browne(2) develops this further, citing "that in the 1960's and 70's, the focus was on manufacturing and on the costs of production. In the 1980's, the emphasis shifted to quality measurement. Today the appropriate measures are less clear." The climate is therefore very confused, with a plethora of authors and techniques each demonstrating the ramifications of their own methods, which may or may not adopt to other situations.

Hanson(3) develops this view further, citing "every tool in our management kitbag goes through a familiar cycle. This starts with an initial phase of being conference hyped, and finishes with the concept beginning to be discredited by its successors initiatives. At neither end of the spectrum is the critique likely to be correct."

Indeed, trends between both public and private sectors are seemingly just as confused. Nutt & Backoff(4), state "that one of the significant factors of difference between the public, the third/non-profit and the private sectors is performance expectations. In the private sector, they are clear and fixed long-term, creating urgency; in the third sector, action taking has many interpretations until a consensus emerges; and in the public sector performance expectations are suffering from a vague and constant flux, changing with electoral appointments."

Osborne & Gaebler(5) have produced guidelines upon the art of performance measurement. Eilon(6) denotes a methodology for the analysis of the factors affecting performance, shown in **Figure 1.**

Which direction should we take? Meyers (1) perhaps holds one of the keys to the many doors that can be opened. His approach recommends the principles upon which organisations must develop in order to attain long-term sustainability. He cites "that finding an informative and manageable set of financial and non-financial measures for an organisation, is one of the principal challenges facing management, only when they have understood the differences between financial and non-financial measures can they be instigated." There are of course numerous non-financial measures, that at first glance have no meaningful relationships between one another, unlike financial measures, which are often portrayed as ratios and linked to one another through accounting conventions.

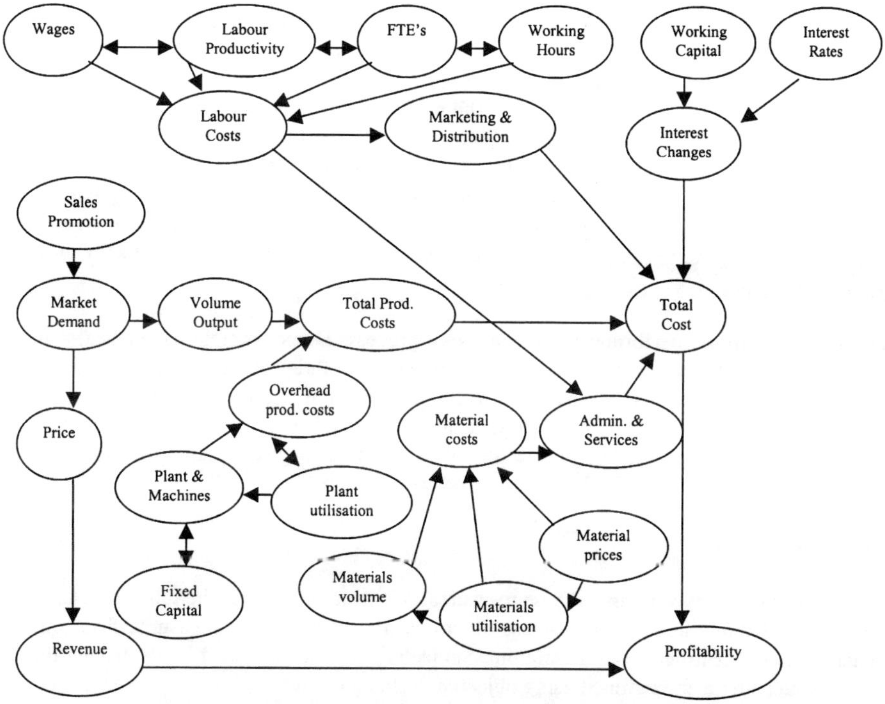

Figure 1 - Performance Factors, Eilon(6).

It is even more difficult to establish links between these financial and non-financial measures. Meyers(1) gauges the number and combination of measures, citing "that there should be more than one or two measures gauging progress towards strategic objectives, but fewer than five or six. There must be a balance between the financial and non-financial measures, of which the non-financial measures must meet the criteria that:

i) there must be room for variance, and
ii) they must be under the managers control, and
iii) there must be a clear path from non-financial to financial performance results."

2.1 STRATEGIC MEASUREMENT IDENTIFICATION

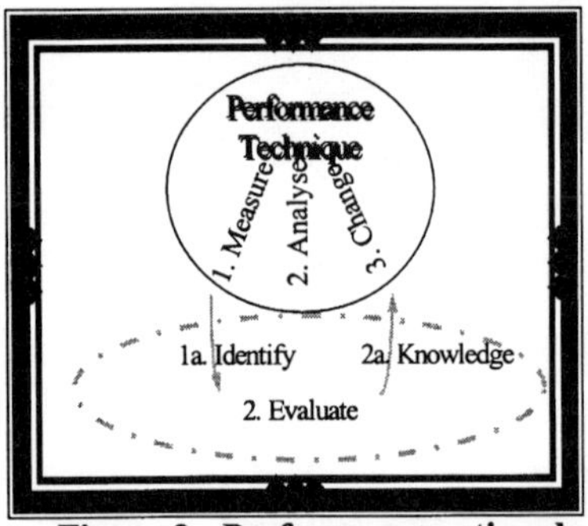

Figure 2 - Performance rationale.

It is a fact that no two industrial situations are completely alike, and yet the more measurements and approaches that are used to categorise them, the more they might seem to be the same. Whilst it is in everyone's own interests to perform well and then upgrade the performance, the techniques used to record performance need adaptability and flexibility. Not only must the techniques adopted become better suited to the real situation, but the measures identified for systematic improvement must be strongly related to the governing determinants of real success that are inherent in most organisations.

The raison d'être of performance is a six way process. Demonstrated schematically in **Figure 2**, most performance techniques are composed of three processes, measurement, analysis followed by change. What the techniques do not explore are the areas of identification, evaluation and the knowledge factors achievable before measuring, analysing and then changing. This area is key in selecting what measures the organisation monitors, and their alignment with the strategic objectives of the organisation.

3. METHODOLOGY

So exactly how do we determine these measures? We need to begin at the top and cascade down across all dimensions of a business situation, and compare the measures from this approach to those identified from the bottom. Subjective methods are used in identifying the measures, which are later examined using objective techniques.

3.1 IMPLICATION GRIDS

The subjective approach is derived from industrial psychology and the use of repertory analysis, Bannister(7). A facilitator is required to formulate the measures that are perceived as being instrumental to the organisation's strategic intent. These are augmented across four echelons, namely '*the doing work*', '*what work to do*', '*business*' and '*constraints*' levels. These measures can be derived through individuals or groups of key members of an organisation , which is the subject of the analysis.

From this stage, an Implications Grid (ImpGrid), Kelly(8), is constructed, **Figure 3**. Resembling a symmetrical matrix, the ImpGrid will have a leading diagonal, though its symmetry will be dependent on the implications perceived between the measures, when subject to resistance to change methods. From such a matrix, derivation of the super and sub-ordinate measures is possible, according to the ranking of the row implications.

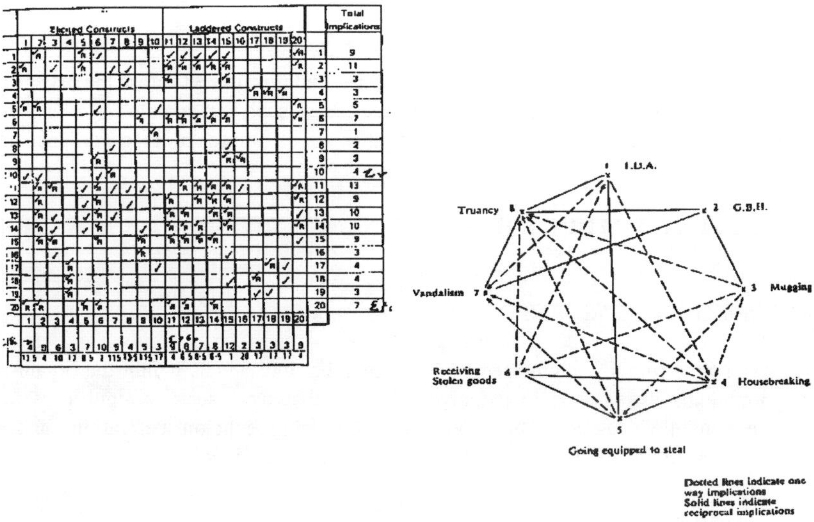

Figure 3 - Implications Grid, Kelly (8).

3.1.2 Polygon Representation

The ImpGrid can be represented schematically, using polygon notation, Bannister(7). This is achieved by segmenting the implication scores into the four echelons. These can then be displayed as polygon diagrams, with each node representing the appropriate measure. From such diagrams, the implications can be derived in uni-directional or reciprocal relationships, and ranked from the scoring profile set for each type of relationship depicted at each node. This analysis is now available in a software version, RAPID[3], which performs the calculations for the user.

3.2 MULTI-VARIATE ANALYSIS

Measures derived from the subjective top-down approach are compared to results taken from an objective bottom up perspective. The results from this analysis are assimilated across the organisation, and when compared to the RAPID© results indicate common qualitative and quantitative measures that are ultimately pertinent to the situation under investigation. The analysis uses survey results which are delivered across the organisation, and to those individuals that are most likely to have any impact upon the function being observed.

From the survey, statistical software packages such as SPSS© can be used to correlate relationships between the measures. This is achieved using factor analysis, which derives a set of factored equations relating the highest correlated set of measures. Once this is derived,

[3] RAPID - Carson-Mee, DJ (1997) - Repertory Analysis for Performance Identification of Determinants, software available for PC applications.

a method of clustering those measures examines the highest correlation's, which can be compared to the results of the RAPID© analysis.

Comparing the results reveal those that are aligned and misaligned across the organisation. The next stage is to set a range for each of the measures, such that monitoring over a period of time will enable the measurement performance to be improved in accordance to the organisational objectives.

4. CASE STUDY WITHIN THE MAINTENANCE DIVISION OF A NATIONAL UTILITY

4.1 TOP-DOWN MEASURES

Measures were initially augmented by 7 key member individuals within the organisation. These individuals were asked to perceive measures that were seen as significant for the maintenance of their assets. These were derived using echelon derivation, and were transferred to an Implications Grid. Thirty three measures were identified, as shown in **Figure 4**.

		assetinv	assetloc	assettyp	assetuti	assetvar	climate	comp	contract	culture	custsati	economic	emerg	empskill	eqptqty	flexibil	freq	fte	innovate	knowled	labcosts	mainmats	mainplan	mainpoli	matcosts	orders	politics	qa	quality	replace	social	sysacces	technol	terrain				
		1	2	3	4	5	6	7	8	9	10	11	12	13	14	15	16	17	18	19	20	21	22	23	24	25	26	27	28	29	30	31	32	33			row total	Ranking
assetinv	1			1		1	1						1		1		1					1	1	1	1	1				1					1	assetinv	12	24
assetloc	2				1		1	1	1						1						1		1	1	1							1	1	1	2	assetloc	12	24
assettyp	3		1		1	1	1	1	1						1		1					1	1	1	1					1		1		1	3	assettyp	15	14
assetuti	4		1	1		1					1				1		1				1	1	1		1					1		1		1	4	assetuti	13	20.5
assetvar	5		1				1	1	1						1									1	1			1	1	1		1	1	1	5	assetvar	13	20.5
climate	6		1																															1	6	climate	2	33
comp	7		1	1	1	1			1		1				1		1				1	1			1			1	1				1		7	comp	14	16.5
contract	8		1	1	1	1	1	1			1	1			1		1				1	1	1	1	1			1	1	1		1	1	1	8	contract	21	2
culture	9		1		1	1	1	1			1	1		1		1		1	1	1	1		1	1			1		1		1		1	1	9	culture	20	5
custsati	10		1	1	1			1	1					1		1		1	1	1	1	1	1	1	1			1	1	1		1	1	1	10	custsati	21	2
economic	11						1	1		1																	1						1		11	economic	5	29.5
emerg	12		1	1	1	1	1								1		1				1	1			1	1				1		1		1	12	emerg	14	16.5
empskill	13			1		1	1	1		1			1		1	1	1	1	1	1			1	1					1			1	1	1	13	empskill	18	9
eqptqty	14	1	1	1	1	1	1															1	1		1				1	1		1		1	14	eqptqty	13	20.5
flexibil	15		1	1	1	1	1	1		1			1	1	1		1	1		1		1	1							1		1		1	15	flexibil	18	9
freq	16		1	1	1	1	1				1		1		1						1		1	1	1			1	1	1		1	1	1	16	freq	18	9
fte	17	1	1	1	1	1	1						1	1	1	1	1		1	1	1		1	1					1			1	1	1	17	fte	20	5
innovate	18					1		1		1	1			1						1									1		1		1		18	innovate	9	26
knowled	19			1		1							1	1	1	1															1		1		19	knowled	8	27
labcosts	20		1	1		1	1	1	1			1		1	1	1	1	1														1		1	20	labcosts	14	16.5
mainmats	21	1	1	1	1	1	1	1	1		1		1		1		1	1					1	1	1	1				1		1	1	1	21	mainmats	21	2
mainplan	22		1	1	1	1	1	1				1			1		1	1			1			1	1			1				1	1	1	22	mainplan	17	12
mainpoli	23		1	1	1	1	1										1			1	1				1			1	1				1	1	23	mainpoli	13	20.5
matcosts	24		1	1	1	1	1	1			1	1	1		1		1					1	1	1					1	1		1	1	1	24	matcosts	19	7
orders	25	1		1		1	1								1		1					1	1	1	1				1	1			1	1	25	orders	14	16.5
politics	26							1	1	1		1																			1				26	politics	5	29.5
qa	27			1		1	1			1	1			1	1		1			1				1					1				1		27	qa	12	24
quality	28			1		1	1	1		1	1		1	1	1	1	1			1		1		1				1					1	1	28	quality	17	12
replace	29	1	1	1	1	1	1			1		1	1	1		1	1		1	1		1	1	1	1							1		1	29	replace	20	5
social	30					1	1		1	1	1									1									1						30	social	7	28
sysacces	31		1				1																											1	31	sysacces	3	32
technol	32		1	1	1	1	1	1			1	1		1		1				1				1	1		1	1	1					1	32	technol	17	12
terrain	33		1				1					1															1								33	terrain	4	31
		1	2	3	4	5	6	7	8	9	10	11	12	13	14	15	16	17	18	19	20	21	22	23	24	25	26	27	28	29	30	31	32	33			449	
		assetinv	assetloc	assettyp	assetuti	assetvar	climate	comp	contract	culture	custsati	economic	emerg	empskill	eqptqty	flexibil	freq	fte	innovate	knowled	labcosts	mainmats	mainplan	mainpoli	matcosts	orders	politics	qa	quality	replace	social	sysacces	technol	terrain				
col total		5	22	22	17	25	26	17	9	9	12	9	10	11	20	9	18	7	5	12	11	13	16	18	17	3	4	9	16	13	4	17	19	24		449		

Figure 4 - ImpGrid for Maintenance taken from National Utility.

The implication grid results were ranked and divided into the four echelon categories. From these categories, polygon structures similar to those shown in **Figure 3** were produced.

4.2 BOTTOM-UP MEASURES

A survey was issued across the division, of which, 64 respondees replied. Using the objective analysis, factoring these results, correlated the most significant measures to fifteen, as shown in ***Figure 5.*** These measures, when clustered, identified the measures most strongly related, as shown in ***Figure 6.***

Factor Matrix:

	Factor 1	Factor 2	Factor 3	Factor 4
ASSETINV	.56027	-.32269	.37339	-.24469
ASSETTYP	.46078	-.70675	.15943	.22791
CONTRACT	.61544	.26127	-.37033	.20246
CUSTSATI	.47596	.42142	.12854	-.32179
EMERG	.38765	.03504	.44712	.04275
EQPTQTY	.52590	.23245	.55014	-.06088
FREQ	.61224	.47608	.11393	.14682
LABCOSTS	.74755	.08119	-.15850	-.37008
MAINPLAN	.65476	-.31866	-.44430	.12660
MAINPOLI	.76512	.04665	-.29217	-.01935
MATCOSTS	.77523	-.14766	-.07358	-.25615
ORDERS	.60063	.32681	.00900	.04194
REPLACE	.51408	-.37619	.29761	.32125
SYSACCES	.35798	.32486	.04602	.70965
TECHNOLO	.71066	-.24583	-.18745	-.07184

Figure 5 - Factor equations.

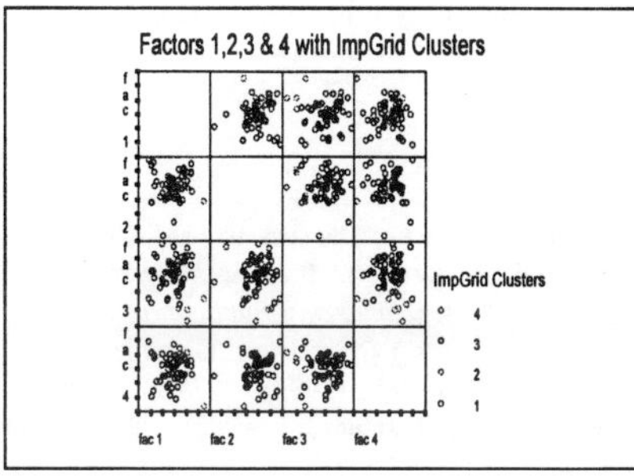

Figure 6 – Clustered measures

When both sets of results were compared, those which were found to align with one another were evaluated upon their merit and the determination of measures as indicated by Meyer. Measures which were found to be misaligned to one another were also evaluated and merited according to the differences found. Once these had been validated, then the measures and their gauges were determined and set out as shown in ***Figure 7.***

No.	Performance measures	Range
1	ASSETTTYP - the types of asset maintained	Variations between assets (keeping to minimum list parts). Keeping asset inventory at max/min levels. Establishing best practice per item.
2	EQPTQTY - the quantity of assets maintained.	Exact quantities and types of asset. Complete maintenance overview and breakdown into manageable systems.
3	LABCOSTS - the labour cost in maintaining assets	Taken from complete maintenance overview, for forecast basis and for direct and indirect cost allocations.
4	MAINPLAN - the maintenance plan	Directly proportional to nos.1-6.
5	MATCOSTS - the material costs in maintaining the assets	Quantity of assets, type of maintenance scheduled per item, implication on inventory and analysis of failures.
6	REPLACE - asset replacement	Ease of replacement, time, system outage and whether duty/standby is necessary.
7	TECHNOLO - the technological environment	Direct changes to items that have implications upon maintenance, their ROA
8	FREQ - frequency of asset maintenance	Times that assets are maintained, according to basic, intermediate and full overhaul periods. Resources necessary to conduct maintenance over various frequencies and optimisation.
9	MAINPOLI - maintenance policies	Establish optimum maintenance policy for asset base. Policy amendments monitored.

Figure 7 – Aligned measures

5. CONCLUSIONS

The method of determining performance measures presented in this paper has been demonstrated within the maintenance function of a national utility. With careful monitoring the measures have been shown to be manageable across all business divisions of the organisation. The measures have helped future long-term direction for the particular function since it is aligned towards organisational strategic objectives. The application demonstrated within this paper is considered to be indicative of such an organisation, and it is considered that the method would be applicable across most sectors and organisations that identify the need for strategic direction.

REFERENCES

(1) **Meyer, M (1996)** - "The secrets that lay behind corporate performance. *Financial Times, 9.2.1996.*

(2) **Browne, J. Sackett, P.J. & Wortmann,J.C. (1996)** - "The Extended Enterprise - a context for benchmarking". *Benchmarking Publication.*

(3) **Hanson, P. (1997)** - "Sharing achievement." *Manufacturing Engineer, April 1997, pp.62-63.*

(4) **Nutt,D & Backoff,R (1993)** – "Government measures" *Wiley, NY.*

(5) **Osborne, D. & Gaebler, .T (1992)** - "Re-inventing Government." *Williams Patrick Books.*

(6) **Eilon, S. (1993)** - "A methodology for analysis of corporate performance." *Omega International Journal of Management Science, Vol. 21, no 5, pp.551-560.*

(7) **Kelly, GA (1963)** – "A theory of personality" *Norton & Co. NY.*

(8) **Bannister, D (1977)** "A manual for repertory grid technique" *Academic Press, London.*

Managing change in the garment manufacturing industry – the devil in the detail

R GASSMANN
Daks-Simpson Limited, Larkhall, UK

Many garment manufacturers are moving their production functions abroad due to more favourable labour and union conditions. To compete with them on cost whilst fulfilling acceptable levels of product quality and continuing to manufacture in Britain is an immense challenge.

To meet this challenge, companies must simultaneously improve productivity as well as the quality and design of products. This is all very well and indeed most industries are facing the same or similar problems but specifically how do you fulfill these ambitions in the garment industry and what are the main obstacles?

This paper provides a detailed insight into these issues from the perspective of a British men's garment manufacturer focusing specifically on the product development process. The challenge faced by the various functions involved in this process is to accommodate the 'touchy-feely' processes of garment design in general and ambiguous retailer strategies in particular with the more quantifiable customer requirements and market drivers of cost, delivery performance and responsiveness.

The paper describes the 'good practice' methods used to establish formal systems throughout the product development process, the main obstacles encountered and the experienced benefits.
Keywords: garment manufacture, change management

1. INTRODUCTION

A characteristic of product development in the garment industry is the pervasive conflict between what the designers think the consumer wants, what the consumers believe they want, what the retailers are willing to sell and what the manufacturers are able to make.

Any of the former market forces can shift prior to and during a season for whatever unforeseen trivial or significant economic, political and/or social change. Typically in these cases, new pressures are placed on manufacturers and their suppliers to accommodate any shift in garment specification whilst maintaining original timescales and costs.

The least flexible of the participants in this picture are the manufacturers as a consequence of the relatively long styling design and component design-delivery lead-times. Any last minute changes to, for example, a pattern, the finish of a cloth, label details or button colour can put back development by anything up to six weeks. As a consequence, in striving to meet timescales there is a danger that shortcuts may be taken for commercial interests and garment integrity compromised.

DAKS-Simpson Ltd. has been a manufacturer of traditional men's and ladies tailored garments for over 100 years. The company employs 1600 people in its manufacturing site in Scotland and over 400 in its design, sales, marketing and retailing functions in London. The company has experiencing massive technological investment since Sankyo Seiko bought the company in 1992. Over the last six years DAKS has rapidly expanded its skills and knowledge base emerging from a de-skilling policy in the 1980s to embrace the latest information technology tools, manufacturing organisation methods and management techniques.

The company manufactures essentially for two customers. The first customer is the DAKS-Simpson Group itself who retails men's and ladies garments throughout Europe and the Far East. The second is Marks and Spencer (M&S) for whom DAKS manufactures men's jackets, trousers and suits for M&S stores throughout the world.

Process re-engineering work was carried out in DAKS-Simpson Ltd. over a period of twenty-six months focusing specifically on the M&S product development (PD) process. The overall objective of the 'project' was to eliminate the garments returned to DAKS by consumers from M&S stores (return to manufacturer garments - RTMs) that were directly the fault of deficiencies in the production and/or PD processes. This additional cost is absorbed completely by DAKS and has an obvious knock-on effect on the cost of garment manufacture in Scotland.

The paper firstly briefly describes the PD process itself and identifies the main sources of internal and external disruption. The experiences of the change process are then described rather than the work itself because the former has more universal relevance than the latter which is more specific to the industry and DAKS. Finally the practical lessons learnt as well as the key benefits of the work are summarised.

2. M&S PRODUCT DEVELOPMENT

At any one time, DAKS experiences four production seasons: the early development of ideas and experimentation for the season after next, the formal product development (PD) for the next season, the manufacture of the current season's garments in bulk and finishing off the bulk manufacture of last season's garments. The work described in this paper embraces the activities carried out during the second production season.

The PD process is unique in that no single person or department is wholly responsible for garment development. High task interdependency and the requirement for coordinated decision making has necessitated a strong concurrent product engineering environment. Groups of departments assume responsibility for completing each stage of the development process. As the various stages are completed for each style, responsibilities are gained and others transferred to other groups to fulfill the relevant tasks.

Typically about 28 garments are developed in parallel during a season. The PD process may be divided into 4 stages (greatly simplified):

1. Developments (0.5 months): DAKS' selected cloths and patterns are presented to the customer for selection.
2. Administration (1 to 2 months): Pre-production contracts are agreed and garment

specifications confirmed. Pre-production materials are ordered.

3. Incoming materials are tested against agreed criteria. Prototype batches are produced to trial materials and patterns (1 to 3 months).
4. From these batches, 'technically correct' garments are produced and submitted to the customer for inspection. If the garments are satisfactory production contracts are signed for bulk production. Production materials are ordered (0.5 to 1 month).

The control, visibility and measurement of this PD process has improved as a consequence of the change activities referred to in this paper. Despite this, the following 'internal' and 'external' obstacles to product development may still occur:

1. **Developments**
   a) Internal:
      Delays in developing a range to meet customer requirements
      Late changes to fabric make-up, shade and/or finish
   b) External:
      Delays in cloth buying by remaining non-committal on cloth selection
      Difficulty in developing the latest cloths on-time and/or to specification
2. **Administration**
   a) Internal:
      Delays in agreeing contracts with mills
      Delays in determining internal components for newer cloths
      Verbal specification changes
   b) External:
      Last minute changes to garment make-up
      Ambiguous and/or delayed final garment specification
      Verbal specification changes
3. **Prototype Batch**
   a) Internal:
      Incorrect production set-ups due to poor early communications between design, production and laboratory personnel
      Poor 'hanger appeal', i.e. a garment may look good on a person but not on a hanger
      Poor make-up and 'feel' of garment
   b) External:
      Delivered materials manufactured using a different method from which bulk material will be made, so when used is not representative of bulk manufacture
      Late material delivery
      Late requests for alterations to patterns
4. **Technical Garment Assessment**
   a) Internal:
      Manufacture of technically correct garments not to specification
      Damages to garments during transportation between Scotland and London
   b) External:
      The same garment may be required to be made by more than one supplier. It is difficult without many iterations to develop identical technically correct garments from different suppliers especially when in direct competition
      Inconsistent criteria used for assessing technically correct garments
      Customers inspecting garments may vary at each (re)assessment

Successfully fulfilling the project's objectives (refer to the previous section) required the implementation of a PD process that was robust and flexible enough to cope with the many sources of uncertainty and change internal and external to the company. Because these and many other sources are beyond DAKS' immediate control, it was important for the project to ultimately gain the 'high ground' with the customer, i.e. developed systems will seek to eliminate all internal and cope with all external obstacles by fulfilling the requirements of the customer 'to the letter'.

This approach ensures that DAKS' systems are beyond reproach in quality system terms and business relationships with the customer are strengthened.

3. PROCESS RE-ENGINEERING

3.1 Preparing for Change

The two year project received full support and was frequently reviewed by DAKS executives. The project team membership, made up from PD stakeholders changed, throughout the period depending on project need. The author, acting in the best traditions as a project 'champion' at an operations level, was an original member of the project team and remained with it throughout the two year period.

The importance of executive support for strategy development and resource management as well as the inclusion of all product development (PD) stakeholders in project shaping and implementation was not in doubt. This represented the bare minimum of what should be in place to provide credibility to the project activities and give the project a chance of success.

Whilst this is the case, evaluating the role and degree of participation of stakeholders (in this case heads of departments and key staff) to ensure success was a complex and delicate task involving changes in workload, skills used, dependencies between departments and team structures. To initially determine stakeholders' reactions, enthusiasm and uncertainties towards the change process as well as to 'not leave anyone out' everyone tended to be involved with everything all the time. At an operations level this approach, although found to be useful initially for developing confidence and trust in the process, proved to be too simplistic and was done purely as a goal in itself to meet perceived 'best practice' methods. This activity as well as the moves to 'involve nearly everyone in nearly everything nearly all the time' were inefficient and impractical.

A strategic 'back step' was taken by the project team at an operations level to identify lessons learnt and propose methods of optimising stakeholder organisation at different stages in the change process. Although this episode lasted for only three weeks, in that time a number of clear, identifiable lessons were learnt that influenced the planning and organisation of subsequent project activities:

1. Be wary of using 'text book' best practice unsuitable for the application context. Do not only do things because you read it in a book
2. Develop totally unambiguous roles for all stakeholders
3. Be fully prepared to rapidly change direction and tactics to get the job done. Avoid 'throwing good money after bad' and assess future costs and benefits.

Project timescales, milestones, performance measures, review procedures, dependent tasks, resources and costs were clearly defined, budgeted for and agreed prior to the start. During the project they were modified to accommodate unforeseen events and unplanned disturbance.

Formally establishing these project elements is a basic project requirement because they provide a route-map for stakeholders to follow, focus the efforts of people, guide people to move between tasks and most importantly contributes towards project control. Although the latter can at times be illusory, perceptions of control are often just as beneficial for project credibility than its existence.

Formal IDEF0 model building of the product development process helped clarify the limits of stakeholder responsibility and authority as well as defining task interdependencies. The process of developing an 'AS-IS' model was far more valuable than the model itself. The AS-IS modeling and accompanying training workshops provided a formal, overt method of involving stakeholders in the change process.

TO-BE model development from the resultant AS-IS model to form the basis of an improved PD process was found to be unwieldy because of its complexity and inability to describe what happens because of the time independent nature of the method. The modeling process and modeling workshops did help, however, to develop a familiarity with the method sufficient for developing a TO-BE model with minimal reference to the perceived constraints of the AS-IS model.

3.2 Managing Change

The management of change is a nebulous activity that cannot be easily described or planned. Often it is described in chronological terms or project phases. These involve a series of interconnected planned tasks such that the completion of one phase prepares the way for the remainder of the project. This view of a change process is often provided in management brochures and at consultancy seminars and badly trivialises and mis-represents its true complexity and disruptive potential.

The true nature of the change process at the operations level is less like controlled phased activity and more like a busy web of activity defined within project timescales and costs. The primary reason for this situation is that the stakeholders involved in product development focus their efforts, not surprisingly, on developing new garments for forthcoming seasons. This is their primary task and remains so despite continuous emphasis on the benefits of the change process and the importance to fulfill longer term company objectives. The daily melee and the unestablished newness of change often demotes project activities into second place.

In progressing the project to combat this situation, the following methods were learnt and applied:

1. Emphasise the presence of third party (i.e. M&S) involvement in the change process. Reminding people of its presence helps focus effort and gives the process legitimacy.
2. Arrange for the stakeholders to attend relevant structured training courses conducted by a third party on for example: project management, auditing, model building, performance measurement and computing. Ensure that the trainers clearly re-emphasise the project messages with case studies. A common message re-iterated by a third party adds credibility to the project activities as well as the individuals facilitating them.
   The case studies help to broaden the world-view of the stakeholders beyond that of their own company experiences by emphasising the ubiquitous and prevailing nature of change as well as its past and current fads and band-wagons.
3. Ensure the fundamental tenets of the project never change, e.g. its purpose, objectives and importance. Timescales and targets may change but the high-level justification and content of the project must not. To do so will put at risk the credibility of the project and project members.

Conversely, the following methods were learnt and avoided:

1. Avoid using three letter acronyms (TLAs) if they are associated with failed attempts in the

past. If not then they can be used with great effect to enhance project credibility. Definitions of the TLAs must be unambiguous and couched in language familiar to the particular industry. Definitions through case studies help broaden the world-view of the stakeholders (as with 2 above) through examples of good and bad applications in a range of industries.

2. Avoid starting a piece of work that has not been asked for by its stakeholders even if part of the overall project plan. If the work is essential for say auditing purposes and/or if it will provide improved systems to the benefit of the stakeholder then the stakeholder will freely request and participate in it. This involves the clear communication by the project 'champion' or other change agent of 'what is at stake' if the work is not carried out and a comparison of the current system with the benefits of the new. Without being asked, the work will get done but the stakeholders may experience minimal ownership and demonstrate little understanding as well as negative behaviour.
3. Avoid inflexibility. The nature of change in a manufacturing environment with its daily crises and changing priorities necessitates the requirement to progress many project elements at once in parallel. Do not waste time cajoling stakeholders to do work because the daily plan says so. Consequently, a project plan that is too detailed is not useful and does not appreciate or reflect the nature of such a project.

3.3 Benefits

The benefits of the work have been experienced on many levels: The individual stakeholders have primarily benefited through the improvement of their hard and soft skills from the daily use of the latest information technologies, exposure to the experiences of other industries, participation in 'new' management techniques and a greater understanding of the work that they do in relation to the whole PD process. Creating this situation in the company was highlighted early as a key pre-requisite for success.

The organisation and work methods adopted in the PD process fulfills customer requirements. The movement towards seeking the 'high ground' in quality system terms has improved DAKS' relations with M&S through improved system controls and work disciplines. Financial benefits have been estimated at annual savings of £150,000 from reduced RTMs, material waste, rework and administrative costs.

4. CONCLUSION

The experience of change is unique to every situation. The experience of change described here was shaped in general by the traditional nature of the industry, the behaviour and characteristics of the materials used in the industry and more specifically by the stakeholders' unfamiliarity with the latest technologies and the process of change itself.

Key to the success of the project was the rectification of the latter situation and the adoption of the following lessons:

1. Encourage active, public executive support to maintain project credibility and momentum
2. Be flexible in carrying out planned tasks. Work within the constraints of the environment
3. Focus on change to the system not the individual - keep it impersonal
4. Stick to what is required to 'gain the high ground' with the customer
5. Use methods to suit the politics, personalities and/or mood of the company
6. Continually review project activities and change tactics to suit future situations
7. Beware management 'best practice' and stick to 'good practice' adapted to context

The arena of core competence in manufacturing effectiveness

V GILGEOUS and **K PARVEEN**
Department of Manufacturing Engineering and Operations Management, University of Nottingham, UK

ABSTRACT

The concept of Core Competence are examined to ascertain what they are; what do they do; how they work and what makes them do what they do so well. Also the different models of Core Competence are reviewed e.g. The Core Competence Engine, The Model of Congruency and the Hierarchical Models of Competence. The results of a postal survey across different industry types shows to what extent companies Core Competency requirements are *inward/efficiency* culture driven as against *outward/strategy* culture driven. From the conclusions of the postal survey the authors have presented a model to support the core competence building and development

1 INTRODUCTION

The terms "Core Competence" struck the management scene by Hamel and Prahalad in the ground breaking paper published in the Harvard Business Review May-June 1990. Since that moment onwards, organisations, firms small and large nationally and internationally have been mesmerised to gain core competence e.g. some leaders in possessing core competence are Canon, Kodak, 3M, Sony, Honda, Sharp and NEC. Core competence as a conception has evolved through the Resource-Based View theory (RBV). RBV is an alternative way at looking at business in the 1990s. It is based on resources, capabilities and competencies of organisations. It helps to explain why some firms are consistently better than others in the same market with similar resources (1). Core competencies are built on individual intangible or groups of intangible assets which constitute and embody the organisations capabilities, skills and knowledge, experience, people, resources and intellectual property. Core competencies cannot be easily duplicated by competitors; however they also would be difficult to replace if a company found them destroyed or damaged. They are the source of the company's ability to deliver unique value to its customers; thus, they are not to be mistaken with "leading-edge technologies, "world-class processes", or other "production-driven" definitions of distinctiveness. They are flexible enough to straddle a variety of business functions, product families, and technologies, thus they are not tied to existing ways of doing business but are platforms for growth and stimuli for growth (2). They also define the unique opportunity set available to the firm, being those market openings or knowledge gaps that the company is uniquely qualified to fill; thus, they serve to narrow the focus of the firm's forward strategy (3). It is unlikely that a company will build world leadership in more than five or six core competencies which will take the best part of 15 years to gain. If a company has a list that stretches to 20-30 then that is more likely to reflect the capabilities of the company rather than the core competence, although it is good practice to generate such lists.

To achieve core competence it is fundamental that the organisation sees itself as a portfolio of competencies not just businesses and products. Corporate vision is required to build them and the administrative means to reassemble resources spread across multiple businesses. 3M, Black and Decker, Canon, Honda and NEC build global brand umbrellas by proliferating products out of their core competencies. This has allowed their individual businesses to build image customer loyalty and access to distribution channels. Conceiving of the corporation in terms of core competencies widens the domain for innovation.

2 WHAT IS CORE COMPETENCE ?

A Core competence constitutes of three fundamental properties. Firstly it allows the organisation to access a wide variety of markets such as, competence in display systems, for example enables a company to participate in such diverse businesses as calculators, miniature TV sets, monitors for laptops computers, and automotive dashboards. Second the core competence should make a significant contribution to the perceived customer benefits. Finally a core competence should be difficult for the competitors to imitate (4). Core competence is the reason behind, some companies can diversifying and competing into apparently disparate market domains.
There are three main definitions presented for Core competence and they are:

> *A Core Competence is the collective learning in the organisation, especially how to co-ordinate diverse production skills and integrate multiple streams of technologies, they also provide potential access to a wide variety of markets (4).*

> *A core competence is a unique combination of technologies, knowledge and skills that are possessed by one company in the market. A core competence is usually the basis for a whole variety of end products and services, both now and in the future. Core competence has various attributes such as Complexity, Invisibility, Inimitability, Durability, Appropiability and Non-Substitutability (1).*

3 EXAMPLES OF CORE COMPETENCE

Organisations which have embarked on the path of core competence have seen the fruits of the efforts. Some examples of the worlds leaders in developing core competence are; Canon's core competence in the fields of optics, imaging and microprocessor controls has allowed it to dominate markets in seemingly diverse as copiers, desktop laser printers, and image scanners; It took Philips 15 years to perfect its optical media and laser competence; Casio have specialised in LCD colour small screen television; 3M developed the Post it notes, pressure sensitive tapes, magnetic tapes, photography film and coated abrasives which are an extreme diversified portfolio of businesses that are based on a few core competence and Honda's core competence in power trains and engines gives it distinctive advantages in car, motorcycle, and generators businesses.

4 CORE COMPETENCE MODELS

There have been models and frameworks presented by authors and practitioners of the core competence. The concentration of these models and frameworks have contained a spectrum of issues for example: *Competencies: The roots of Competitiveness*, emphasises the location of core competence and represents the organisation as a large tree (4); *A Hierarchical Model of Business Competence:* Illustrates the Transparent linkage of congruence between different functions and emphasises the need for congruence at both higher levels and lower levels between the functions of the organisation (5); *The Core Competence Engine:* Illustrates how core competence can be evolved and grown through six meta-skills which are: skill identification, learning, knowledge embedding, rapid development, re-structuring and innovation. (1); *The Intangible Asset and Security Process framework:* emphasises on the protection and security of the core competence and presents a way forward through the aid of measuring intangible assets (2); and finally *The Three domains of Competence that provide the potential for internal corporate differentiation* concentrates on three domains: Epistemic, Idiomatic and Logistic domain which encompass and represent the capacity to think, act and operate differently respectively (3).

5 OUTLINE OF SURVEY.

The postal survey was designed to ascertain information in two significant sections. The first being designed to establish the depth of knowledge and understanding manufacturing managers had in the field of core competence. The second section was designed to ascertain whether manufacturing managers operated in an environment where core competence could be built and enhanced. Such as, were they operating in cultures that were inward efficiency driven against outward strategy driven.

5.1 Design of Survey. Section 1 contained nine expressions relating to core competence. The manufacturing managers had to indicate whether they had heard of these expressions and understood them. The expressions that were used in the survey were: core products, core competencies, core capabilities, core rigidities, strategic architecture and distinct capabilities. Section 2 composed of 8 significant statements relating to the types of working environments the manufacturing managers operated in. It was possible to establish the extent to which the manufacturing industry was inward efficiency driven (e.g. Sony and Honda) as against outward strategy driven (GE and Chrysler who due to costs sold parts of business to competitors). The latter culture is the prerequisite for building and enhancing the organisation's core competence. The 8 statements were used to address certain issues in the organisation, e.g. do the manufacturing managers view their company as leaders in their industry or laggards? How influential is the company in terms of their ability to set the rules and competitive standards: is the level from senior management the same for re-engineering strategies as it is for re-engineering core processes?, What drives the improvement of the organisation? is it competitors setting the standards for them to follow and adhere to and does the company preserve the past or does it create the future for its industry.

6 RESULTS

Of the 1300 surveys sent to a cross section of industries 200 replied giving a response rate of 15%.

Section 1: The nine terms identified associated with core competence and the results are shown below:

		(%) Heard of		(%) Understood	
		yes	no	yes	no
a	Core Products	92	8	90	10
b	Core Competencies	80	20	79	21
c	Core Capabilities	77	23	73	27
d	Core Rigidities	22	78	22	78
e	Strategic Intent	75	25	64	36
f	Organisational Learning	79	21	69	31
g	Organisational Citizenship	21	79	15	85
h	Strategic Architecture	31	69	24	76
i	Distinct Capabilities	43	57	36	64

As the results illustrate terms core products, core competencies and core capabilities are very well known, with a high percentage of manufacturing managers having knowledge and understanding of the terms., although of 20% of the participating managers in the survey were unaware of the term core competence and its meaning!. Terms strategic intent and organisational learning had high percentages of being heard of with a lower percentage of understanding them. The terms core rigidities, organisational citizenship, strategic architecture and distinct capabilities all fell below the 50% of knowledge, of that, only 40% understanding the terms. Having a strategic architecture is fundamental to the administration and development of achieving core competence, companies such as 3M, Vickers, Canon, NEC, and Honda all have a strategic architecture implemented, only 31 % of the respondents have heard of the term and only 24% of them understood what it was. Organisational citizenship was also a poor performer.

Section 2. The set of statements applied in this section are:

1 *Conventional & Reactive* **or** *Distinctive & far sighted*
2 *Re-engineering Core Processes* **or** *Regenerating Core Strategies*
3 *Mostly a Rule-taker* **or** *Mostly a Rule-maker*
4 *Operational Efficiency* **or** *New Business Development*
5 *Mostly Catching Up with Others* **or** *Mostly New to the Industry*
6 *Largely Driven by Competitors* **or** *Largely Driven by Our Vision*
7 *Mostly an Engineer* **or** *Mostly an Architect*
8 *Mostly Anxiety* **or** *Mostly Hope*

The results of this section are represented in Figure 1 below.

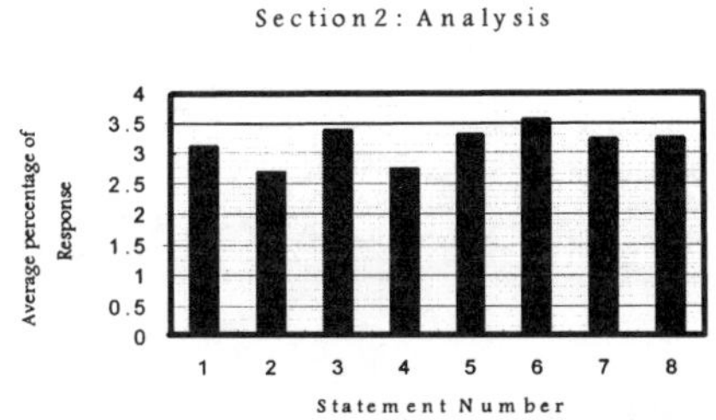

Figure 1 A Bar chart to illustrate knowledge of core competence related terms

The two extremely weak areas identified here were statements 2 and 4. Here manufacturing managers spend most of their time re-engineering existing core processes rather than concentrating on developing re-engineering core strategies. It was also evident that most manufacturing managers spend most of their time on operational efficiency rather than discovering new frontiers to their industries. Statements that performed just moderately were 1,5,7,8. Most of the managers view the company as conventional and reactive rather than having the ability see farther and distinctively than their competitors. It was also evident that most managers spent their time catching up to other competitors rather than creating new boundaries and frontiers for their industry. All this is further reinforced by the results that the majority spend too much time as laggards - accepting and adhering to competitors standards and rules rather than setting the standards for competitors to follow. This also reinforced the fact that they perceive themselves as mostly engineers maintaining today's present and ensuring everything runs smoothly instead of building and creating new businesses . Two statements 3 and 6 had the highest moderate average of mangers attempting to make significant changes in their organisations by building new future visions of their industry and to set the standard for competitors to follow instead of the reverse.

7 THE ENABLING CORE COMPETENCE LENS

The authors have presented a model (figure 2 below) to aid the manufacturing industry to build and enhance their Core Competence through looking at three fundamental systems: Enabling system, Disabling System and a Core Competence system. The enablers can either be soft or hard. Examples of the former are, Teams, Learning and examples of the latter are Information systems. The disabling system also constitutes of hard and soft issues, examples of the former communication and the latter are history, baggage, policies and regulations. These then work together to give the core competence of the organisation. Then these provide the nourishment for manufacturing effectiveness which ultimately results in three advantages. Firstly the organisation has a competitive advantage e.g. lower prices , secondly it gives the organisation a differentiating advantage e.g. difference in the product and finally it gives an organisation a market leadership and ultimately industry advantage which can lead to economies of scale for building and enhancing new core competencies.

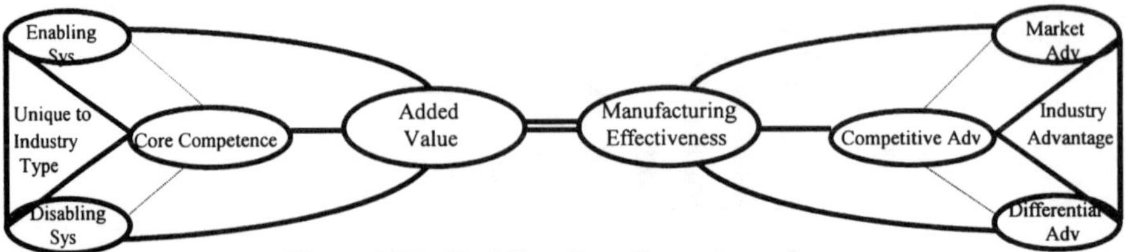

Figure 2 The Enabling Core Competence Lens

8 CONCLUSIONS

Core competencies are the sum of the intangible and tangible assets of the organisation. For a competency to qualify as a core competence it needs to fulfil three fundamental dimensions of firstly being inimitable, secondly transcending through different market boundaries or industries and thirdly to have significant customer perceived benefit. Together from the literature review and from the results of the survey it is also clear that organisations who are not core competence orientated need to oblige themselves to core competence however perilous the task is in order to qualify for the global competitive arena. From the results of the survey it is evident that manufacturing managers need to increase their knowledge and understanding of core competence and related issues in order to achieve the ability to enable the organisation to prosper. For example having a strategic architecture is fundamental to the administration and development of achieving core competence. Companies such as 3M, Canon and Honda all have a strategic architecture implemented. Also established in the survey is that manufacturing managers are spending too much time on preserving the past, concentrating on operational efficiency, accepting standards set by others, catching up with competitors vision rather than the opposite. Organisations need to see themselves as not only in terms of portfolios of businesses and services but also in terms of core competencies. Core competencies, unlike physical assets, should not diminish with use, they should be enhanced as they are shared and applied. The manufacturing industry needs to operate in an outward/strategy driven culture rather than inward/efficiency driven culture to succeed in achieving Core competence. The authors have suggested a model, *the enabling core competence lens* to aid building manufacturing effectiveness.

9 REFRENCES

(1) Nigel Petts (1997) Building growth on Core Competence's- A practical Approach *Long Range Planning* Vol. 30 No. 4, pp 551-561.

(2) Michael Harvey and Robert Lusch, (1997) Protecting the Core Competencies of a company: Intangible Asset Security, *European Management Journal* Vol. 15 No.4. August.

(3) Jules Goddard, (1997) The Architecture of Core Competence *Business Strategy Review,* Volume 8 Number 1 pp 43-52.

(4) G. Hamel and C.K.Prahalad, (1990) , The Core Competence of the Corporation, *Harvard Business Review* May-June, pp71-91.

(5) J.C. Baker, J. Mapes, C.C. New and M. Szwejzewski (1997), A Hierarchical Model of Business Competence *Integrated Manufacturing Systems* 8/5, pp 265-272.

Principles and methodology support system for re-engineering the product introduction process

R ROY, A P BEESLEY and **J E KEAST**
International Manufacturing Centre, University of Warwick, Coventry, UK

ABSTRACT

A number of reasons have been put forward for the failure of business process re-engineering (BPR) initiatives, and re-engineering the product introduction process (PIP) is a particularly complex task. A problem faced by an inexperienced team is the lack of easily available and comprehensive guidance. There are three types of literature, which provide assistance in different ways - the general methodology, the principles of re-engineering and process analysis tools. Little is directed at PIP, which has distinctive characteristics. In this paper we review the need for methodology, principles and tools, and describe work on the development of an advisory system that combines all three aspects to assist PIP re-engineering.

1. INTRODUCTION

BPR is concerned with the radical change of an organisation so as to achieve simultaneously substantial improvements in process cost, quality, lead time, flexibility, and other measures of success (1). Many of the ideas are common with manufacturing systems re-engineering and lean manufacturing (2), but applied to other functions and processes within the organisation. Although many companies have in recent years embarked upon radical change programmes, evidence suggests that 70% of corporate change programmes fail (3). Many reasons have been put forward for such failures. Lack of leadership or commitment from the top to radical changes in the organisation is often cited as a common cause, as is the general inertia to change that is usual, particularly in large organisations. Although many of the causes of failure cited are related to problems with implementation, many studies do not live up to expectations, take too long or fail even during the analysis phase. Even when the goals of the BPR initiative are clear, it is not easy to translate these into a change programme for improving performance of a complex process. In the case of manufacturing systems re-engineering, clear models of lean production systems have emerged and are documented, eg. (4); domain specific guidelines for change exist (e.g. cellular manufacturing, set-up time reduction, production and transfer batch size reduction, SPC, etc.), and a small number of modelling and analysis tools to assist the re-engineering team are well established (e.g. product flow analysis, string diagram, discrete event simulation, etc.). BPR, however, is concerned with change programmes in all spheres of business activities, from simple to complex processes, encompassing diverse functions. The BPR literature is full of material on generic methodology, eg. (5), tools and techniques (6) and, more recently, a limited number of publications attempting to define the principles of re-engineering, eg. (7). The BPR team may also need access to domain specific knowledge, e.g. on concurrent engineering in the product introduction process. For any complex process, it is not easy for an inexperienced team to assimilate information from so many disparate sources, choose appropriate tools and techniques, and formulate a change programme. The result can be long delays and, often, the whole exercise to be abandoned due to perceived lack of progress within a reasonable time. If the success rate of BPR studies is to be improved, an important requirement is to provide more direct and accessible guidance, combining knowledge of methodology, tools and techniques, principles and the domain.

The product introduction process in complex engineering industries is difficult to analyse since it encompasses many different, traditionally independent functions. Attempts at re-engineering it are still relatively immature, although in the last 10 years there have been considerable reductions in 'the time to market' within the automotive industry. In this paper, we discuss work currently in progress (in conjunction with two major industrial partners) on the development of an advisory system for the re-engineering of the product introduction process for such industries.

2. PRODUCT INTRODUCTION PROCESS

There are a number of reasons for the difficulty in re-engineering the product introduction process in technology driven industries with complex products. The process spans many years, consists of numerous activities, and knowledge of it is very dispersed in the organisation and not usually well documented or easily available. It encompasses many different functions - e.g. technology development, product definition, manufacturing engineering, supply chain definition, etc. The process depends on complex communications networks and reporting structures in geographically dispersed departments and organisations. The process is iterative in nature, the entity flowing through it is for the most part conceptual and ill defined, the activities and sub-processes are usually not documented and their durations have large variances. The necessary stage-gate management process requires explicit definition of purpose and strict control procedures to perform efficiently. There is a great deal of scope for concurrency of activities, but its extent and methods are not easy to define. The greater the degree of concurrency, the greater the need for strict change control procedures to maintain information integrity. The process is heavily dependent upon IT, but the development and implementation of such tools has often been driven by functional departments that have become bottlenecks to change.

3. BPR METHODOLOGY, TOOLS AND TECHNIQUES, AND PRINCIPLES

Assistance available to the BPR team from the literature takes three main forms:

- Methodology - Generic BPR methodologies found in the literature define, in detail, the stages in the study. However, they usually do not provide much guidance on the types of process analysis that may be required, and very little on the innovative thinking that is necessary to formulate a new, improved process.
- Tools and Techniques - Recently, a large number of process modelling tools and techniques have become available to capture and present process knowledge. A model, however, assumes a particular process structure, and does not suggest changes that may be necessary to achieve ambitious improvement targets. There is also a need for the classification of such tools, to provide guidance on the choice of tools and when they may be beneficially employed.
- Principles - Attempts to describe BPR steps as a set of formal principles are relatively new. They are valuable in helping with the necessary innovative thinking.

Case studies describing examples of BPR studies can also be found in the literature. Often these are used to illustrate particular aspects of a methodology, application of tools or principles. Some lengthier presentations of results of action research also exist, describing the approach taken, changes implemented and the benefits obtained (3).

All of these forms of assistance play an important role for the BPR team. The work described here, currently in its design stage, has set out to capture the essence of all of these methods found in the literature, and combine them into a single, easily accessible advisory system to support the team through the various stages of the project.

3.1. BPR Methodology

A BPR methodology describes a number of tasks that the team should carry out at various stages in the study. There are two distinct approaches to BPR that can be found in the literature. The methodology originally prescribed by Hammer and Champy (1) is a top-down approach, which suggests that the BPR team should focus on determining how the strategic objectives of the organisation can be met without letting its thinking be constrained by the existing process. The emphasis is on the 'to be' process, and is consistent with the step-change philosophy that the authors presented. The methodology is outlined in Fig.1A.

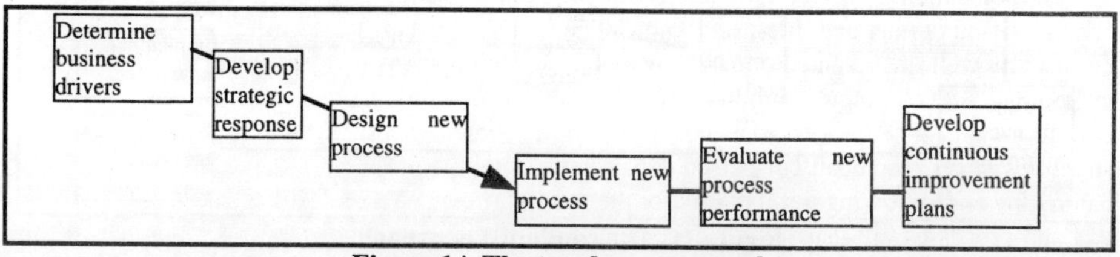

Figure 1A The top down approach

The more incremental change methodology outlined by, for example, Harrington (8) is a bottom-up approach which advocates modelling the existing process to gain understanding of it, and then streamlining it appropriately to meet the strategic objectives. The focus is on changing the 'as is' process by identifying opportunities for improving it. The methodology is outlined in Fig.1B.

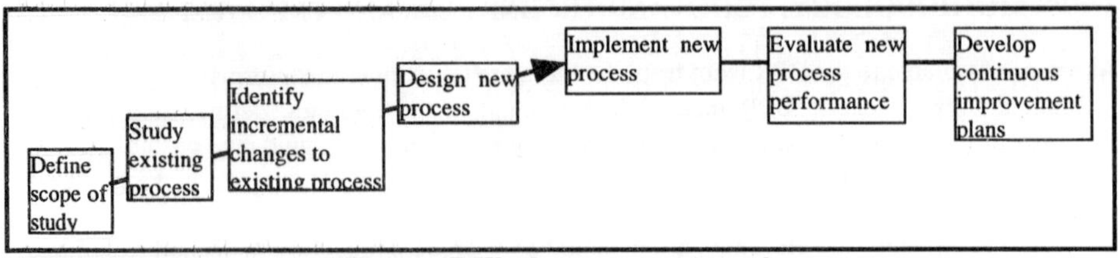

Figure 1B The bottom up approach

In practice, a BPR team will usually need to adopt a mixed approach (Fig.1C). If the top-down methodology is used as the basis, there is still a need to understand the current functionality, if not the detailed process, and to define carefully the transition path from the current to the preferred future process. With a bottom-up methodology, BPR teams can, and often do, spend too much time on detailing the current process, lose innovative thinking and, perhaps more commonly, run out of time to spend on exploring/defining all possible opportunities for major process improvements. A mixed approach would encourage the team to consider high level changes without being cluttered by the details of the current process, as well as detailed examination of the 'as is' (and/or the initial definition of the 'to be') process for incremental improvements and identify necessary transitional arrangements. It is important to recognise that an initial BPR study may lead to recommendations for a number of more detailed projects on improving sub-processes, which may only require relatively small changes (perhaps, to remove some bottlenecks). An advisory system should be flexible enough to cope with such different approaches and entry points. Domain specific knowledge should also be incorporated in the list of tasks outlined in the methodology, e.g. methods for maintaining data integrity must be defined for the 'to-be' process.

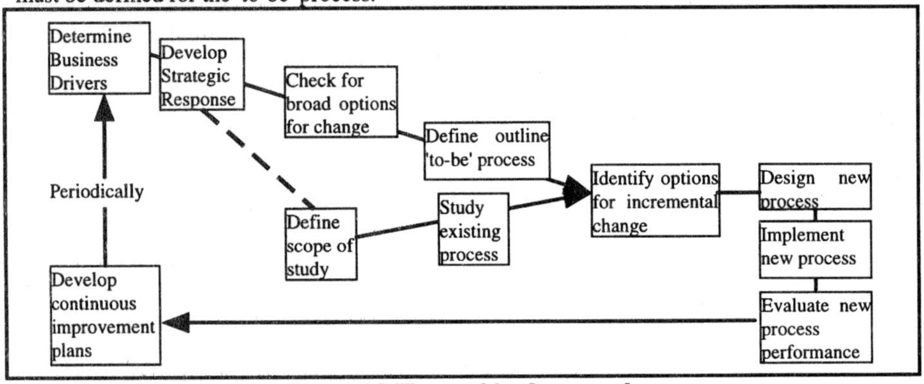

Figure 1C The combined approach

3.2. BPR Tools and Techniques

A large number of tools and techniques are referred to in the various BPR literature; some have been specifically developed, while others are existing tools and techniques considered useful in re-engineering studies. They can be broadly classified into the following categories:

- Soft systems methods - These are qualitative/brainstorming techniques, for formalising the thought process, and are often used in setting process/system goals (e.g. SWOT, QFD); problem analysis, for example, to identify causes of process failure (e.g. cause and effect diagram); risk analysis (e.g. FMEA), etc.

- Presentation tools - These are techniques for presentation of process views to understand and communicate the current process or the proposed future process. Examples are Role Activity Diagrams (9) to view the dependency of individuals/teams/departments within the process; process flow diagrams to view the activity dependencies; functional decomposition models, useful for viewing information dependency and Time Based Process Mapping (10) to present decomposition of process lead time into value added and non-value added components at various stages in the process.
- Analysis tools - These are tools which can be used to analyse process behaviour over time. Various tools exist with different levels of modelling capability, e.g. PERT/CPM, Petri Nets, Discrete Event Simulation.

An advisory system should point out the appropriate tools and techniques that may be used at the various stages of the BPR study, but should not also encourage unnecessary analysis. Experience suggests that BPR teams often go through a phase of 'paralysis by analysis'. Different views of the same process may all, in theory, help to understand it in some way, but usually do not enhance the necessary innovative thinking. Similarly an over-emphasis on modelling the 'as is' process can become a substitute for actual decisions. An advisory system should link the use of tools and techniques to the decisions that the team needs to make, and not simply some vague concept of gathering process knowledge.

3.3. BPR Principles

BPR teams also need principles to help with the innovative thinking necessary to change the process structure. The principles are mainly of value in what is usually the most difficult stage of BPR analysis, namely considering the options for changing or streamlining the process. The types of possible solution have been categorised recently into a number of change strategies. Carter and Handfield(7) described seven basic strategies. These are in sequence, simplification (which includes elimination of non-value added activities), standardisation, integration, parallelism, variance control, resource allocation and automation. The strategies are primarily directed at the objective of time compression, but also have relevance to cost reduction and quality improvement (11).

Carter and Handfield suggest that a BPR team should tackle steps (or strategies) 1 to 7 in a strict sequence. It would, for example, be wrong to attempt automating a process with an IT application without first considering its simplification; not only could simplification make automation redundant but the full benefits of automation may not be realised either. However, there is a danger in restricting the thinking process to a strict sequence. For example, integration of activities requiring different resources, into a single activity to be carried out by an individual, may sometimes become possible only with automation. An holistic view needs to be encouraged, while still conveying the fundamental message of Carter and Handfield through the advisory system.

The role of the support system is to explain, with examples, the principles, methods for applying them, and constraints that may need to be overcome (e.g. skill demarcation, lack of data integration). For example, a possible method for exploring the application of parallelism can take the form presented in Table 1. However, the basic principles could be usefully enhanced with the use of domain specific language, examples and composite strategies. Integration, for example, can take the form of combining two activities into a single task, which may require multi-skilling or team-working; the latter essentially combines features of integration with parallelism. Similarly parallelism could be achieved through simultaneous development of components of the product. This may require a design task to fix and release some critical information early enough for dependent tasks to start early (step 3 Table 1); or concurrent engineering, which would allow a dependent task to proceed in parallel without certain knowledge of the full information necessary for it to be completed. This requires strict procedures to maintain data integrity (steps 5-6 Table 1).

Table 1 Exploring opportunities for parallelism

1	*Map how the process operates, and identify activities on the critical path. Assess which activities are value adding.*
2	*Move non-value activities in parallel and therefore off the critical path, where possible.*
3	*Identify data and information flows between critical path activities. Determine whether enough information can be released by a partially complete activity to allow succeeding activities to start in parallel.*
4	*Determine any strategic or customer driven time constraints on the overall process. If the lead-time of the process meets the constraints then go to 7. Otherwise continue to 5*
5	*Assess the risk of starting each critical path activity before all the necessary information is available. Identify the overall process to try and meet the time constraints by increasing the concurrency of the process elements that present an acceptable level of risk to both the business enterprise and the customer.*
6	*Define method for managing risk to process and data integrity.*
7	*Assess benefits of new process structure and, if acceptable, define implementation plan.*
8	*Determine the reasons for the lack of parallelism in the existing process and the associated root causes to ensure that future process evolution is appropriately sequenced.*

The principles/strategies can be applied to either Hammer and Champy's top-down model (1) or to the bottom-up approach suggested by Harrington (8). The strict sequence advocated by Carter and Handfield would be more natural to Harrington's methodology, while Hammer and Champy's model would favour a more holistic approach. The mixed methodological approach advocated here would also favour a degree of free thinking, particularly when defining the initial outlines of the 'to be' process, but can build on the work of Carter and Handfield.

4. DEVELOPMENT OF A SUPPORT SYSTEM

Work on the development of a Principles and Methodology Support system (**PAMS**) is being carried out as part of a wider project to develop aids for a product introduction process re-engineering BPR team. PAMS is being designed to provide on-line advice on methodology, the tools and techniques that may be used and the circumstances in which they should be used, the information that should be collected to analyse the process and make decisions, and the principles that could be applied to improve process performance. Its domain of application is from the setting of goals for a BPR study through to the formulation of plans for process change. The key design requirements are that there must be sufficient depth of material to provide direct, practical help and pointers to more detailed public domain information; it must be rich in examples and case studies; have navigational flexibility; be easy to use; interactive in nature; provide context sensitive access to information/advice; be easy to deliver, maintain and customise within an organisation.

4.1 Architecture

PAMS will provide principles, tools and case studies at the appropriate stages of the BPR process. These key elements will be supported by a glossary of terms and references as well as links, which utilise any available information input from a previous stage (Fig.2).

To illustrate the nature of the online support, consider the integration, standardisation and parallelism strategies. A key domain specific composite strategy, combining all three, is the introduction of concurrent engineering. A context sensitive glossary will define the term, describe methods for achieving it and requirements and constraints; case studies will provide a more meaningful insight into the application of the concept. Another best practice composite strategy, the introduction of early supplier involvement within the product introduction process, is directly linked to the achievement of parallelism and also enables standardisation. As a subset of this, PAMS will highlight modularisation as a key consideration. If the user sees modularisation as appropriate, techniques will be listed to support consideration of the modularisation issues; for example, trade-off analysis will need to be deployed with consideration of certain attributes, some of which will be generic - design cost and time versus nature and cost of redundant capability, interface issues and distribution of risk between modules.

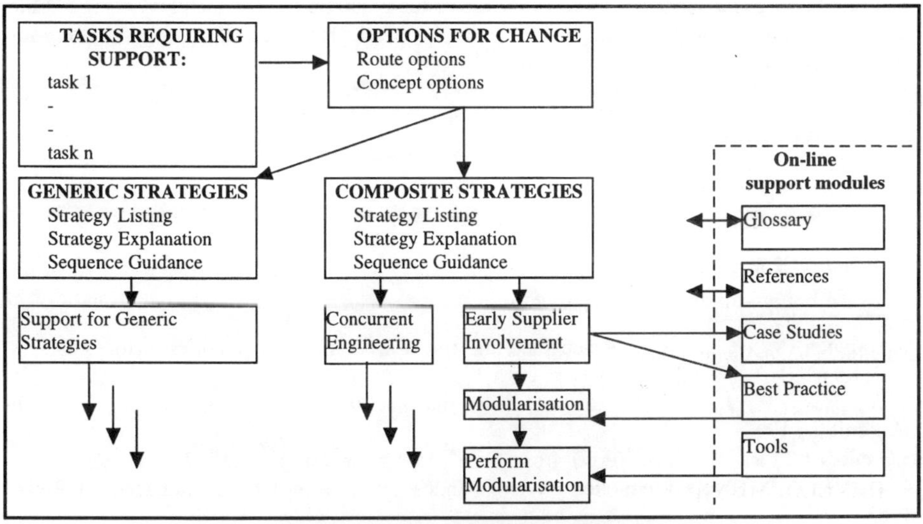

Figure 2 Outline of PAMS architecture

Such considerations and attributes will be very context specific. For example if we consider the product introduction process in the automotive versus the aerospace sector, the cost of redundant capability will be emphasised in the former whereas the distribution of risk of failure between modules will be paramount in the latter. The opportunity therefore exists to tune PAMS according to the domain of application.

4.2 Medium for Implementation

Possible IT technology that could satisfy some of the design requirements would include Windows help files and "Wizards", Multi-Media Authoring software or Internet based technologies. An approach based upon internet technology has been adopted for PAMS. This provides an open architecture within which information can be readily cross-linked and pages added, removed or edited. The browser provides a familiar context in which to operate the system. The rich set of programming languages, such as Java, Jscript and Vbscript along with the advent of DHTML allows pages to be highly interactive and context specific. Finally, internet technology provides a relatively cheap to implement client-server architecture which allows a single instance of the system to be created, maintained and used by any connected, browser enabled computer.

CONCLUSIONS

Re-engineering of the product introduction process for a complex product is a difficult task, and many initiatives fail to produce the expected results. Although much of this is related to problems with implementation, analysis of the product introduction process and formulation of an improvement programme is not an easy task for an inexperienced BPR team. The literature on the subject is somewhat fragmented, divided between emphasis on methodology, tools and techniques, and principles. There is a need to bring together all of these into a single, coherent on-line advisory system. Incorporation of domain knowledge will also be helpful in providing clear guidance. The paper has discussed some of the requirements for such a system, and the direction of work currently in progress aimed at the development of such a support system for the BPR team.

ACKNOWLEDGMENTS

The work described here is supported by an IMI/EPSRC grant

REFERENCES

1 Hammer M., Champy J."Re-engineering the Corporation: A manifesto for business revolution", Harper-Collins Publishers, USA 1993

2 Womack J.P., Jones D.T., Roos D.,"The machine that changed the world", Maxwell Maximillian 1990

3 Gregory I.C., Rawling S.B., "Profit from Time", Macmillan Press Ltd, 1997

4 Monden Y., "Toyota Production System", Industrial engineering and Management Press, 1983

5 V.Bach, L.Brecht, T.Hess, H. sterle "Enabling Systematic Business Change", Verlag Vieweg, 1996

6 J.Bal, S.N.Kota "An Appraisal of Process Analysis Tools for Process Improvement and Re-engineering", Proc. ISATA Conference, Dusseldorf, 2-5 June 1998

7 Carter M. & Handfield R., "Identifying sources of cycle time reduction", Re-engineering for Time based Competition, Quorum Books, 1994

8 Harrington H.J., "Business Process Improvement" McGraw-Hill, USA 1991

9 Murdoch J., McDermid J.A., "Modelling Engineering Design Processes with Role Activity Diagrams", Proc. Intergrated Design and Process Technology Conference, Berlin, 6-9 July 1998

10 Beesley A.P., "Time Compression, A new source of Competitiveness in the Supply Chain", Industrial Management & Data Systems Journal, Issue 1, MCB University Press, 1996

11 Stalk G. Hout T.M., "Competing against Time", Free Press, 1990

Design Management

Bridging the gap between product specification and manufactured output: a spreadsheet aid to parametric modelling

W CHAN and **B PARKINSON**
Manufacturing Systems Centre, University of Hertfordshire, Hatfield, UK

SYNOPSIS

This paper outlines research undertaken to develop a methodology utilising parametric modelling as an aid to achieving design requirements as the manufactured outcome. The project has resulted in development of software to link tabulated product specification criteria with geometric form through employment of a parametric modelling mechanism. Adoption of the system enables product form to be defined at a stage when a specification is being initially established through discussion with the customer. Production of outline drawing details at this early stage reduces the risk of error through misinterpretation of customer "needs" whilst enabling early agreement to be reached upon final manufactured outcome, thus supporting a "right-first-time" approach.

1 INTRODUCTION

The process of design plays a major role in the development of a product, it identifies the characteristics of the product and specifies the manufacturing procedures for its production. The degree of clarity of both requirements criteria and design detail is important since it has an influence on other aspects within the development process. The competitive market environment that exists at present demands efficiency during the design process and the capability of responding quickly to market changes in an effort to meet the increasing demands made by customers. The increasingly global market and resulting growth in competition by manufacturing concerns has emphasised the need for shorter design-to-manufacture lead-times, reduced costs and product "fit for purpose". Project managers are concerned at eliminating waste and significant savings have been made as a result.

The main objective of the research work discussed in this paper is to provide tools to assist efficiency in design. Development work has resulted in building a parametric modelling application, Requirement Options Modelling Parametrics (ROMP), as an aid to achieving a "right-first-time" to design for manufacture. ROMP enables changes in a products geometry to be visualised as specification criteria are being established thus encouraging rapid response to customer demands whilst minimising production costs.

2 PROBLEMS ASSOCIATED WITH A CONVENTIONAL APPROACH

Establishing product requirements is a crucial element in the design to manufacture process. Whilst it is important to define such requirements clearly, it is just as important that their interpretation is concluded in a fashion that leads to identified needs being satisfied. Conventionally, design requirements are expressed by the customer using a natural language, the fullness of which can lead to problems of interpretation and possible customer dissatisfied. In cases of customer requirements being initially misinterpreted or changes being requested during mid-process, modification to design decisions will need to be implemented, increasing lead time and associated costs in manufacture. The identification of component families can contribute towards minimising such efforts by enabling modification of established component criteria to provide the manufacturing specification for new products. Whilst such approaches may contribute towards improving manufacturing effectiveness, the relationship between needs and component geometry remains vague and hence advantages of such an approach are not fully realised. Since the cost of making essential design changes escalates steeply with advances in lead-time, the ability to make crucial changes during the design phase translates into significant savings over making changes during manufacturing process (1), (2).

3. PARAMETRIC MODELLING

Parametric modelling reflects the concept of design automation in which the conventional dimensional description are replaced by variables identified as constraints of the model. Applying different values to the geometry variables of the product model will result in the representation being changed accordingly. The advantage of this methodology is that a single representative model can rapidly be manipulated and an end result examined by the customer and associated manufacturing personnel. Being able to rapidly produce required manufacturing specification details has significant benefit in reducing lead-time of a product whose form has been agreed with the customer.

3.1 A parametric modelling aid

Application of parametric modelling software enables the designer to supply key component dimension values to a generic model to produce a unique set of associated drawings and related information. The construction of a representative geometric model is not a straightforward task and represents a high investment in time, even when using sophisticated solid modelling software systems. It is therefore of considerable advantage if a modelling environment supports the reuse of existing modelling information in the production of new design documentation. A considerable number of design projects are achieved by adapting an existing basic design to suit new requirements. Parametric modelling considerably aids variational design approaches whilst also assisting in conforming to a standardised approach of design documentation. As an aid, CAD modelling can be categories into approaches; variational, parametric and command recording (3). Whilst each approach has defined benefits, the most common is the combination of parametric with command recording (4) (5) as may be found in leading commercial products such as Euclid© and Solid Edge©

4. REQUIREMENT OPTIONS MODELLING PARAMETRICS (ROMP).

Requirement Options Modelling Parametrics (or ROMP) is a software application designed for reducing repeated effort towards the creation of drawing to represent a customer's specification. The application allows modification of these drawings with ease, thereby enables the content of the drawings to match closely the design requirement expressed. This aims to encourage a "right-first-time" approach and eliminate duplication of effort as a means of achieving a reduced design-to-manufacture lead-time.

4.1 The system architecture

The application development consists of three main component elements as illustrated in Figure 1 below.

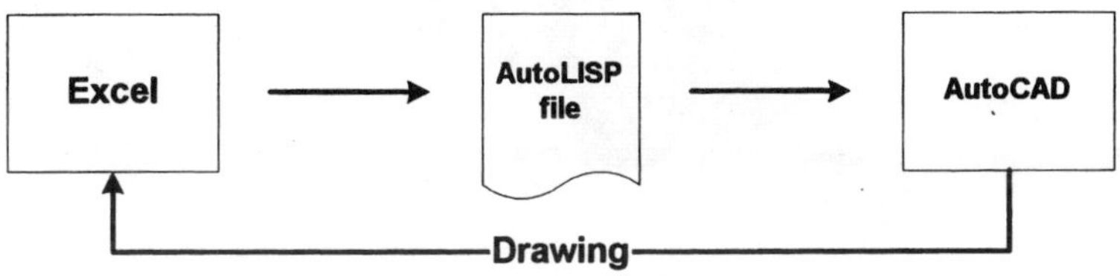

Figure 1. Architecture of the Parametric Modeler

ROMP employs AutoCAD for the creation of 3D drawings using base primitives and constructive solid geometry (CSG) approach. CSG is a solid modeling technique of combing primitives using Boolean operation as demonstrated by the example in Figure 2. The electric motor consists of twelve cylinders; eight for creating the motor profile, the remainder for defining the through holes. The main objective in developing the ROMP software was to reduce repartition in the design process by executing a process of directly converting customer specification details into a final geometric representation of the product. Using AutoCAD, automating the design process is achieved through the generation of an AutoLISP file. A typical geometry file consists of procedural drawing instructions for the creation or modification of entities (lines, circles, etc.) and numeric values which define the location and constraints of those entities. With ROMP, these entities are replaced with primitives and other related drawing instructions including the reinforcement of Boolean operations and the topological geometry of the primitives eg trimming commands etc.

Using the required AutoLISP geometry file a drawing may be produced, however, the content of that file cannot be changed at this stage. In order to enable variable geometry to be represented it is required to update the content of the AutoLISP geometry file with new values. In order to redraw an object with geometry entity changes, the numeric values specifying component dimensional properties of each entity needs updating. This may be achieved by changing the content of the AutoLISP geometry file, however, this is both time consuming and requires a working knowledge of the file structure. The approach adopted was to enable modification of the geometry file contents to reflect new requirements without need for user interaction. The solution adopted was to incorporate using a spreadsheet interface, in this case the spreadsheet adopted was MicroSoft's Excel. An AutoLISP file is text editable but a standard text editor does not have the capability to update content automatically, Excel, however, does not work as efficient as a text editor. Use of the "Visual Basic for Application"

(VBA) language, which is a feature of Excel, provides the capability to automatically modify the AutoLISP geometry file in an efficient manner.

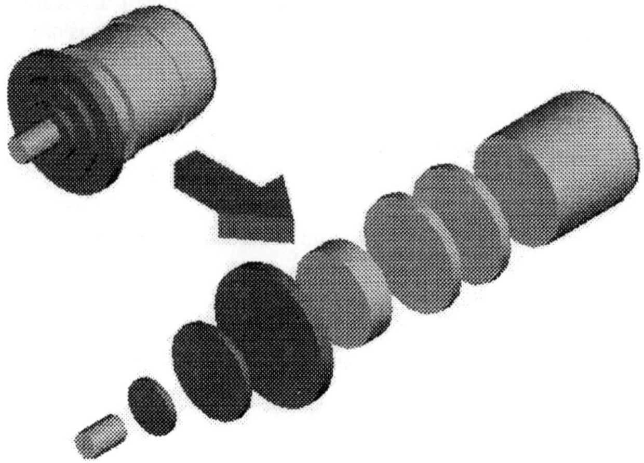

Figure 2. CSG model of product

The content of each individual instruction within an AutoLISP geometry file are arranged such that the numeric values are embedded within each instruction. The geometry of each primitive used to construct a final representation changes with respect to editing of its corresponding dimension values. Since a combination of these primitives is used to represent an object, the numeric values specifying position of each primitives needs updated, i.e. if the thickness of the mounting plate has increased, the main body has to be moved relatively for attachment. The Excel spreadsheet programme is used to break each individual instruction in the geometry file into part-instructions that occupy a single cell. A formula is written for this particular cell so that modification can perform automatically. For example, for the motor illustrated in Figure 2, the position of the shaft depends upon the height and the position of the mounting plate. In order to find out the new position of the shaft, a formula is applied for adding the original position of the shaft with its associated values with out necessitating the user's interaction. The ROMP user interface consists of a number of menus to enable information of the variable to be communicated to the system. The input values are then extracted and stored in the corresponding cells where the associated parametric form is written to rejoin the part-instructions together to form the final part geometry. The file defining the component that meets the criteria provided in the new specification is then saved. In the form of an AutoLISP file, the information provided may be used to produce the required detailed information to enable manufacture.

4.3 An example of a system application

The flow chart in Figure 3 outlines the method of operation adopted by ROMP. Initially the interface prompts the user to select the required product category for which the necessary parameter values are provided. The system then checks the validity of the values provided upon which, provided confirmation is agreed, a 3-D representation of the final design is displayed. Figure 4 illustrates an application of the system used to aid design of electric motors to parameter values specified by a customer. The motor is constructed using twelve geometric primitives with an average of time of two minutes being required to create each individual specification. The user is able to carry out further design options by means of specifying a new set of values.

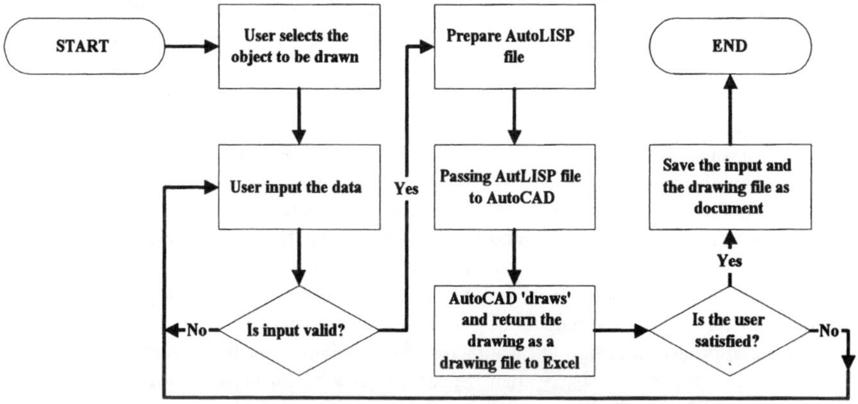

Figure 3. Flow diagram for the Parametric Modeler

5. THE ADVANTAGES OF USING PARAMETRIC MODELING

The use of a simple parametric modelling system as demonstrated by ROMP has indicated the improvements that may be gained during the design to manufacture process. The advantage of applying ROMP stems from the ability to rapidly convert customer requirements into design documentation. Further more, the drawings produced may be modified and redrawn within short periods of time as may result from further consultation with a customer . The application of ROMP has indicated that it is of considerable benefit in clarifying to a customer the product details being requested. The ability of being able to show a graphical image of a product that meets the customers needs confirms requirements and reduces the risk of misunderstanding between sales and the customer.

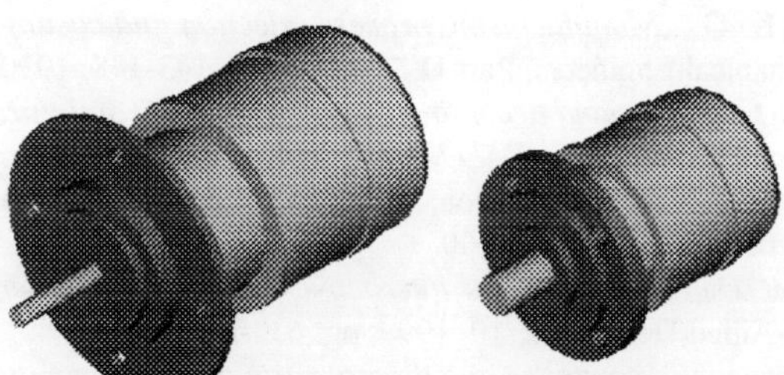

Figure 4. Example of parametric models

Benefits to be obtained by employing parametric systems such as ROMP both enhance customer relations and manufacturing efficiency. Employing existing information held withing design departments can be a great asset in assisting the reduction of lead-time when compared to conventional approaches, even if assisted by CAD systems. Valuable support is also provided to a user seeking identification of a solution to a customer request, especially for a company representative who is visiting the customers premises. With modification, ROMP could provide a quotation for the component being requested by the customer aiding the competitive nature of the business.

5.1 Feasible approach of adopting PM

Although criticism has been raised questioning the ability of parametric modelling to support innovative design, there are advantages to be gained.(6). In his paper, Oshuga (7) comments that the requirements for manufacture are not always fixed at the beginning of the process but are often changed during it. A source of needing modification is misunderstanding and misinterpretation of customer requirements, clarification of which ROMP can assist. The main benefits to be obtained from adopting such parametric systems are to be found where similarity in component geometry is evident. There are a significant number of products which fall into this category and employing a conventional approach of design to manufacture for supplying specific customer needs is both wasteful in both time and costs. System capabilities can be further enhanced by employing component libraries from which an extended range of product geometries may be obtained.

6. CONCLUSION

The iterative nature of the design process contains elements that restricted the smooth flow of the development process incurring unnecessary costs and extending time limits. A method has been introduced for bridging the gap between product specification and manufactured outcome by means of applying a software application incorporating the methodology of parametric modelling. This method supports creation of conceptual drawings at an early stage in the design to manufacture process which can be used to clarify needs with a customer and hence encourage a "right-first-time".

REFERENCES

1 Whitney D. E. : *Designing the design process*, Research in Engineering Design, 2, pg. 3-13, 1990.

2 Allen A. J., Swift K. G. : *Manufacturing process selection and costing*, Proceedings of Institution of Mechanical Engineers, Part D, 204, No D1, 143-148, 1990.

3 Girard P. and Pierra G. : *Command Recording Versus Parametric and Variational Systems, an Old/New Way of Parametrizing CAD Models by End Users*, Proceedings: Computers in Design, Manufacturing and Production, IEEE International Conference, 7th European Computer Conference, 1993, pg. 194-200.

4 Kondo K. : *PIGMOD: Parametric and interactive geometric modeller for mechanical design*, Computer-Aided Design, 22, 10, 1990, pg. 633-644.

5 Roller D. : *An approach to computer-aided parametric design*, Computer-Aided Design, 23, 5, 1991, pg. 385-391.

6 McMahon C. A., Lehane K., Sims Williams J. H. and Webber G. : *Observations on the application and development of parametric-programming techniques*, Computer-Aided Design, 24, 10, 1992, pg. 541-546.

7 Oshuga S. : *Towards intelligent CAD systems*, Computer-Aided Design, 21, 5, 1989, pg. 315-337.

Achieving synergy within the product introduction process of an extended enterprise

B T CLEGG, J T BOARDMAN, and **A J BOARDMAN**
De Montfort University, Leicester, UK

ABSTRACT
This paper is an account of the formative thinking on a current EPSRC project sponsored by companies from the UK aerospace industry. The project is concerned with the product introduction process (PIP) used to define a propulsion system; the realisation of this system requires a chain of suppliers; for example, 70% of the value of a Trent 800 engine lies outside of Rolls-Royce. The project seeks to establish an optimal process that will create the supply chain within a single 'extended enterprise' (EE). This paper provides visibility into the research methodology and some of the modelling, simulation and optimisation methods being used.

1. INTRODUCTION
Pressures in the market place continue to drive down the lead time for introducing new products. Manufacturers of high technology complex systems report various experiences in efficiency improvement initiatives geared towards product cycle-time reduction (PCTR). The reductions in many cases are impressive, but all agree that lead times are still far too long. A different approach is required to achieve radical improvement in PCTR. Rather than looking within an autonomous organisation, such as the prime contractor, and demanding more savings, companies such as Rolls-Royce recognise that it is not 'simply' its own design and manufacturing processes that introduce the product to market, it is the web of processes throughout the entire collection of their supplying companies: what is now being called the 'extended enterprise' (EE) [Browne, (1)], [Konsynski , (2)]. Naturally, most if not all of these suppliers have applied their own PCTR initiatives. However the opportunity for radical gains lies not within the parts of the supply chain, but in the interfaces between them. The challenge is to create a product introduction process (PIP) for the supply chain that not only combines, but integrates the individual PIPs, yielding an EE process that achieves strategic synergy between the contributors.

2. ACTION FOR IMPROVING THE EXTENDED ENTERPRISE
A recent study by the IMI (3) into PIPs of the aerospace industry found that research topics which ranked highly were: methods for simulating and modelling the PIP; process simulation for risk reduction and time compression; methods to improve the PIP within the supply chain; and, advanced digital environments to simulate and control all aspects of design, supply and

control throughout the EE. Another report by the OST (4) placed emphasis for success in UK manufacture on structures such as the EE (managing teams across a supply chain) and the virtual enterprise [Raulefs (5)] (core competencies across a network of companies). In doing so the report underlines the importance of team building, team working, and communication of business understanding. PIPSEE (Product Introduction Process Simulation of the Extended Enterprise) is a collaborative project that addresses these topics.

The project will propose an optimally efficient process and its supporting means, based on a scientific approach to data extraction, activity modelling, simulation of the EE, process definition, and hypothesis testing in the real world. The performance criteria to determine optimisation will be: minimum cost, minimum time and maximum customer satisfaction, all of which relate to the propulsion system that is created by an EE. The project objectives are:

- to develop a shared understanding of the EE as an operationally coherent whole.
- to understand the behaviours of constituent elements of the EE and how these behaviours depend on: heritage, capability, knowledge, working relationships, data & knowledge transfers, IT legacy systems, and business drivers.
- to determine an optimally efficient process and supporting means for the EE to work together as a coherent whole in order to produce new products.
- to provide a comprehensive set of deliverables to: facilitate the implementation of the optimally efficient process and its supporting means; and, maximise dissemination of process understanding amongst other EEs.

The critical success factors identified by the aerospace industry which this research programme addresses are: 'reduction of time to market', 'increase in responsiveness to customer needs', and 'increase in effectiveness in supply chain co-operation'. Time to market is influenced by co-ordination of effort and timeliness of knowledge possessed by those who provide the effort. By modelling the PIP in the EE, we will provide means for analysing the integration of roles & responsibilities. PIP simulation will aid the understanding of how to compose and schedule knowledge for process agents. Therefore, the EE will be able to respond to customer needs more readily and effectively by better understanding and owning *its own* PIP.

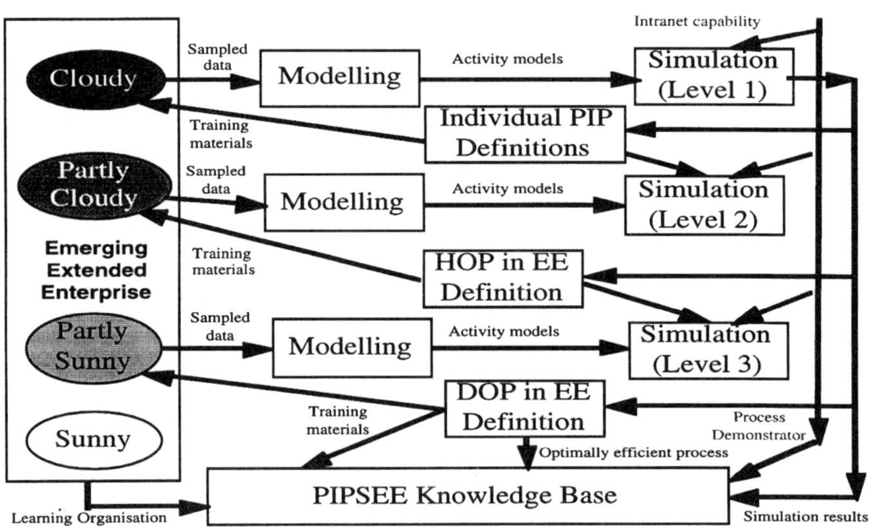

Figure 1: PIPSEE Project Process Logic

The project will: model and sample 'live' data from a pre-designated subset of individual PIP activities; develop a simulation capability to analyse process behaviour; propose a 'hypothetical optimal process' (HOP) for the EE; prepare materials to facilitate the adoption and testing of the HOP; sample similar 'live' data based on the HOP and then conduct further simulations in order to determine a 'defined optimal process' (DOP): this is depicted in Figure 1.

The value of simulation for process re-engineering is widely recognised [Evans (6)]. This project will apply simulation to the EE in a similar way to permit analysis of process and agent behaviour which will be 'observed', 'measured' and 'controlled'. The EE simulator will use a client-server architecture, a local area network and hypermedia technology. It will allow people from the industrial partners to undertake real data & knowledge transfers in accordance with improved definitions of the PIP. By sampling the EE PIP, lessons can be learnt to: improve process understanding; task ownership and commitment to radical improvement in the PIP; and, its interfaces with other processes. Applications of hypermedia technology for this purpose is discussed by Boardman (7), Foundyller (8), O'leary (9), and Taylor (10).

3. DESIGNING A PIP FOR THE EXTENDED ENTERPRISE

The challenge for the process analyst is to take a vantage point that permits a view of this heterogeneous grouping as a totality as well as from a variety of partial aspects or viewpoints. Various systems methodologies and tools will be deployed including the 'Boardman Soft Systems Methodology' (BSSM) described by Boardman (11); and, 'Business Systems Engineering' (BSE) tools developed by the Towill (12) in order to create a comprehensive view of the EE. Other forms of process capture and task analysis methodologies such as Yourdon (13), IDEF0 [US Air Force (14)] and Petri-nets [Murata , (15)] may also be deployed. The PIPSEE project team have initially modelled the PIPs for each of the collaborating companies; this effort has majored on defining the company interfaces to see how the EE is operating in its undesigned state (referred to as 'cloudy' in Figure 1). The need for information and process modelling as an aid to improving supply chains is well recognised and is discussed by Slats (16).

To perform data capture for the 'as-is' models the team has used a variety of data sources: company process / procedure manuals; interviews and workshops with company personnel; presentations from company personnel; and, by the researchers 'walking the process'. Some of the collaborating companies have comprehensive and widely propagated documentation about their PIP and so these manuals were used as a primary data source, backed up by interviews. For other companies, it was more appropriate to use interviews and workshops as a primary data source and then reconcile these findings with what was written in various procedures and guidebooks.Whilst capturing information about the process logic, it was also important to collect comments relating to why the process doesn't always perform as it should and opinion on how the process could be improved; particularly at the interfaces between the customer, the prime contractor and the supplier base. These comments will be used to contribute to the definition of the HOP and then the DOP, which should move the EE towards a 'sunnier' state.

From the process logic data, process models are being created for the individual company PIPs. The set of models needs to fulfil two requirements: firstly, to provide definition of the dataflows for the simulation; and secondly, they need to be used for promulgating and soliciting PIP knowledge to the collaborators. Interestingly, the different utilities place a different set of requirements on the model form. For example, the simulator requires that the models have complete logical integrity which means that they cannot have any 'dead ends' or 'floating artefacts' etc. However, the collaborators who wish to learn directly from the PIP models as a totality require the models to be readable and presentable at high *and* low level views. This requirement is important to stimulate proposals from the industrial collaborators for morphing the individual 'pipeline' PIPs into an amalgamated PIP for the Extended Enterprise. Initially this will be a 'hypothetical optimal process' but should eventually become the vision for a 'defined optimal process' for the PIP in the extended enterprise.

3.1 Exploring The 'Fuzzy Front End' Of The PIP

This section describes a case study which uses the BSSM to explore some of the process re-definition opportunities for the EE PIP at the early stages. The BSSM has been documented by Boardman (17) Bulbeck (18), Clegg (19), Ramsay (20), and Sherman (21). It has been derived from the Checkland (22) Soft System Methodology specifically to facilitate *process* improvement. It uses a unique diagram known as the *Systemigram* (**systemic** dia**gram**). This project will test and develop the BSSM with regard to the broader cultural influences of the EE which is characterised by being: a geographically dispersed, culturally diverse, and contractually bound amalgam of autonomous enterprises.

The emerging discoveries of the PIPSEE project indicate: a poor alignment of the individual PIPs; fuzzy interfaces; a slow start, and lack of design concurrency especially in the earliest stages. This confirms the previous findings of Burchill (23) when he states that the, '... front end of the product development process is not well understood ...' and Cooper (24) when he states that, '... the weakest rated activities [in terms of product development] were the "up front" or pre-development activities, namely initial screening, preliminary market assessment and detailed market study'. Due to similar factors there was a general feeling amongst the PIPSEE collaborators that there was most scope for lead-time compression within these stages. To explore these emerging issues primary data was obtained through: company presentations on the PIP; discussions with senior managers; and, from company procedures and documented improvement initiatives. Secondary data from published case studies [Backhouse (25)] were also referred to.

- The PIP 'phase under observation' was named the, 'Opportunity Evaluation and Preliminary Concept Definition'. The systemigram of this phase is shown in Figure 2. The phase modelling was 'picked-up' at the point where the *'business development / opportunity owners formulate a propulsion system product / problem statement'* and was 'dropped off' at the point where the *'preliminary propulsion system concept definition is referred to by the detailed concept designers'*. Between these two points in the process the systemigram describes the high level view of the activities that occur with the EE. This preliminary work raised a number of questions:
- How is a product opportunity initially identified?
- How defined is the Preliminary Concept Design Team?
- How is the interface between the customer, prime and sub-contractors handled?
- How can misdirected effort be reduced?
- What are the lead time pressures on each company in the opportunity scenario?
- How much research and development is done 'at risk'?
- How effective is the use of 'supplier databases' to develop 'partnering' and 'preferred supplier' relationships?
- Is the practice of competitive tendering truly helpful in developing partnerships?
- What impact would increased electronic commerce facilitation have on the PIP?
- What is the extent of knowledge exchange between the end-user and the subcontractor?
- How can design concepts be shared but intellectual property maintained?
- Is the balance right between technical and commercial issues; are they co-ordinated?

The project will investigate such issues within this phase of the PIP and other issues that arise from subsequent phases. In particular the project will extract issues from each of the individual companies that impact on the whole EE. This will not only involve modelling the PIP; but also identifying measures of performance and understanding the top level product architecture.

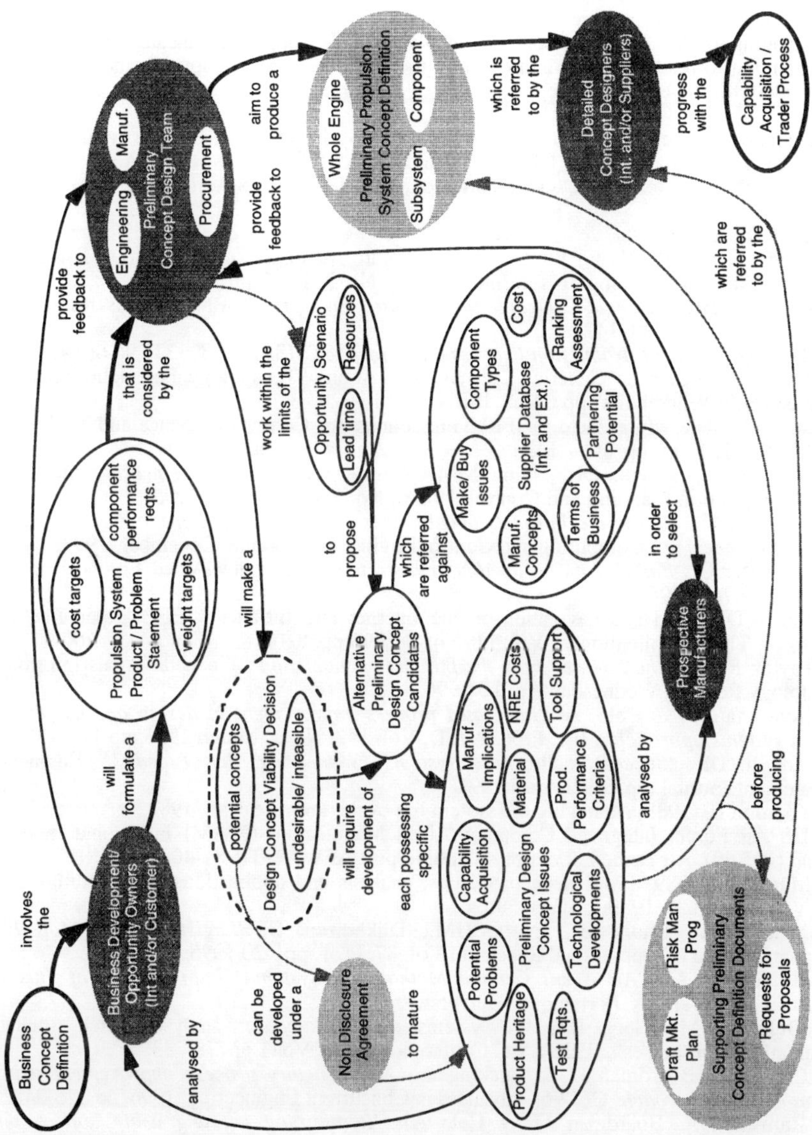

Figure 2: A Systemigram of the 'Opportunity Evaluation and Preliminary Concept Definition' Phase of the EE PIP

4. CONCLUSION

Creating a PIP for the definition of a propulsion system, one that transcends organisational boundaries, will challenge existing mind-sets. Existing company-specific paradigms need to be explored for their fragility to serve the whole, and a stimulus is required to generate innovative ideas about PCTR which need to be shared, explored and tested. Systemigrams and intranet-based simulation provide these opportunities. The approach to the simulation is original and seeks to: demonstrate the integrative nature of the EE as a dynamic entity; act as a platform for PIP optimisation; and, provide a metaphor for the EE. The project aims to develop a shared understanding of the EE as a coherent whole; provide supporting means for this shared understanding to be disseminated throughout the companies participating in the project; and, more widely for the benefit of the UK aerospace industry.

The authors are very grateful for the support given by De Montfort University, the LSDG at Cardiff University, the EPSRC (GR/L91030 and GR/L90118), Rolls-Royce Aerospace, Lucas Aerospace, Weston Aerospace, Kidde-Graviner and RHP Bearings Aerospace.

5. REFERENCES

(1) Browne J, Sackett PJ and Wortmann JC, "Future Manufacturing Systems - Towards the Extended Enterprise", Computers in Ind., Vol. 25, No. 3, 1995, pp 235-254.
(2) Konsynski, BR *'Strategic Control in the Extended Enterprise'*, IBM Systems Journal, Vol. 32, No. 1, 1993, pp 111-142.
(3) IMI, *'The Product Introduction Process for Aerospace'*, ISBN: 1 899371 96 6, 1997.
(4) OST. 1995. Scientific Generics Ltd. and WS Atkins "Technology Foresight Programme Sector Specific Workshops" pp 64-68. ISBN: 0 7115 0280 3.
(5) Raulefs P, 'The virtual factory' IFIP Transactions A - Computer Science and Technology, Vol. 32, No. 1, 1993, pp 111 - 142.
(6) Evans, GN., Towill, DR, Naim, MM, *'Business Process Reengineering the Supply Chain'*. Production Planning and Control. Vol. 6, No. 3, pp 227-237. 1995.
(7) Boardman JT, *'Corporate Intranets and Business Process Management: A Challenge for Systems Engineering'*. Computing and Control Engineering Journal. December, 1997.
(8) Foundyller C, *'Forming an Intranet/Internet strategy'*, Computer-Aided Engineering, Vol. 15, No. 6, p.76, 1996.
(9) O'leary DE, "AI and navigation on the Internet and Intranet", IEEE Expert-Intelligent Systems & Their Applications, , Vol. 11, No. 2, 1996 pp. 8-10.
(10) Taylor, P *'First the Internet: now the Intranet phenomenon'*, Financial Times Information Technology Review, Wednesday April 3 1996.
(11) Boardman, JT & Cole, AJ *'Integrated process improvement in design and manufacture using a systems approach'*. IEE Proc. Part D, Vol. 143, No. 2, March 1996, pp 171-185.
(12) Towill, DR, *'Successful Business Systems Engineering: Part 1 and 2'*. Engineering Management Journal. pp55-95. April 1997.
(13) Yourdon E, (1988) Managing the system life cycle, Prentice-Hall, 1988.
(14) US Air Force, Integrated Computer Aided Manufacture (ICAM) Functional modelling manual (IDEF0), Air Force Laboratory, Ohio, Report AFWAL-TR-81-4023, 1981.
(15) Murata T, (1989) 'Petri nets: properties, analysis and applications', Proceedings of the IEE, Vol. 77, No. 4, 1989, pp 541-580.
(16) Slats, P.A., Bhola, B., Evers, JJM., Dijkhuizen, G. *'Logistic Chain Modelling'*. European Journal of Operational Research. Vol. 87. Pt. 1. pp1-20 1995.
(17) Boardman, AJ, *'An Agent Centred Method for Systemic Improvement of Business Processes'*. PhD Thesis. Portsmouth University. 1997.
(18) Bulbeck JM, Clegg BT, "A systems methodology for analysing and simulating manufacturing processes", IEEE SMC Conference, 1996. Vol.1 pp778-783.
(19) Clegg BT, Boardman JT & Buckingham AD, *'Factory process improvement using a human-centred approach'* CE' 96: Advances in Concurrent Engineering, 1996, pp 326-342.
(20) Ramsay DA, Boardman JT & Cole AJ, *'Reinforcing learning using soft systemic frameworks'*, Int. J. of Project Management, 1996, Vol. 14, No. 1, pp 31-36.
(21) Sherman, D, Boardman, JT & Cole, AJ *'Assisting cultural reform in a project-based company using systemigrams'*. Int. J. of Proj. Mgt. Vol. 14, Number 1, Feb. '96, pp23-30.
(22) Checkland, P.B., 1990 'Systems Thinking, Systems Practice' Wiley. Chichester. UK
(23) Burchill, G., Fine, CH, *'Time Verses Market Orientation in Product Concept Development'*. Management Science. Vol. 43. No.4. pp465 - 478. 1997.
(24) Burchill, G. Fine, CH, *'Time Verses Market Orientation in Product Concept Development: Empirically-Based Theory Generation'*. Management Science. Vol.43. No.4. pp465-478. '97.
(25) Backhouse, C.J., Brookes, N.J. 'CE: What's Working Where'. Gower. 1996

The SMME design process

A P DUCKWORTH and **R W BAINES**
School of Engineering, University of Derby, UK

Abstract

The following paper describes the findings from a postal survey undertaken on the Small and Medium Sized Manufacturing Enterprises (SMME's) design process, as part of a larger research project on developing an eco-design methodology for SMME's. The postal questionnaire was sent out to 200 companies in the Derbyshire area. The questionnaire inquires about the nature and structure of the design process, the design team and the use of best practice design tools. The findings from the survey are discussed, compared to best practice and conclusions are reached on the process of design in SMME's

1. Introduction

A significant proportion of design research focuses on the larger company with best practice tools and methodologies being subsequently recommended, often with little thought of their application and suitability within the smaller company. Recent studies have shown that 94% of manufacturing companies in the UK employ less than 100 people (1). It is widely recognised that these companies are a key elements of the industrial base (2). SMME's therefore represent a significant research opportunity. Research undertaken by the authors into the development of an eco-design methodology for SMMEs has required a study into the method by which these companies design their products. This has been necessary to enable the development of a practical eco-design methodology and to determine how best to include environmental criteria into the SMME design process (3,4). It was hypothesised that the best practice design process as described in theory (5,6,7) would differ from what SMMEs perform in practice. A postal questionnaire sent to 200 companies in the Derbyshire area has been used to obtain the information.

2. Postal questionnaire

The questionnaire consisted of 29 questions used to obtain data relating to; the organisation, the structure of the design process, the design team and design characteristics and the use of design tools. The questionnaire was designed in a tick box format to reduce the time taken to complete the questionnaire and to aid the analysis. Some open ended questions were used to accommodate a wide range of answers, however the questionnaire was kept as concise as possible in an endeavour to increase the response rate.

2.1 Questionnaire findings:

2.1.1 The Organisation

The response rate for the questionnaire was approximately 20%, or 40 companies, with the largest proportion of responses coming from the manufacturing/engineering sector, who accounted for nearly half of all the companies surveyed, see figure 1.

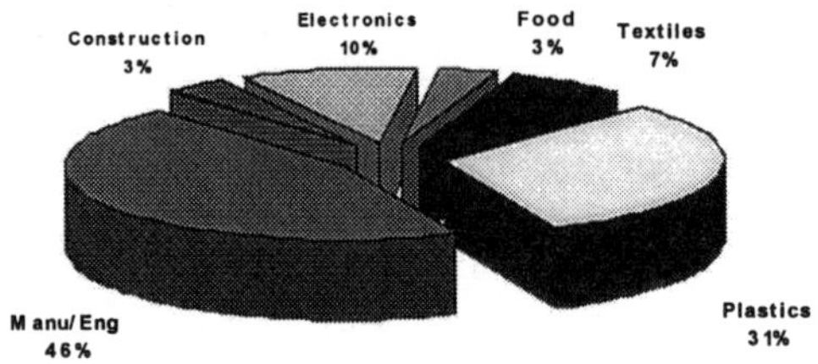

Figure 1. Response rate by sector

97% of the companies surveyed were found to perform both design and manufacture, see figure 2. 78% were found to employ between 10 and 99 employees and be categorised by Levy as small (2), 12% of responses came from medium sized companies employing between 100 and 250, and 10% of responses came from micro sized companies, employing less than 10 employees.

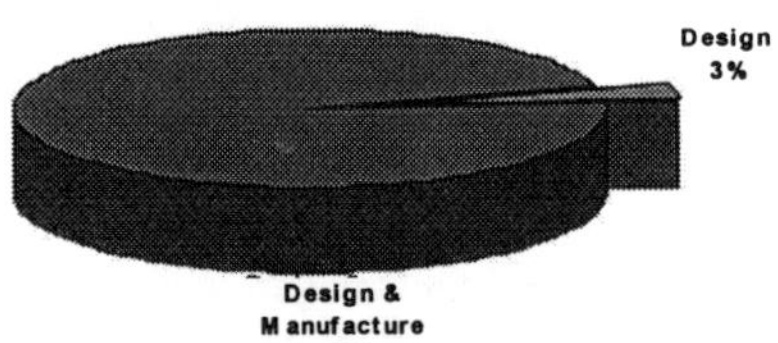

Figure 2. Percentage of companies performing design and manufacture

2.1.2 Design team

The survey revealed that the majority of SMME's had small design teams of less than four people. Companies also indicated that the majority of designers had responsibilities other than design, implying that the design role in SMME's is not as clearly defined when compared to a larger company.

2.1.2 The structure of the Design Process

This set of questions focused on the complexity of the design process, the generation of ideas and the involvement of customers, the amount of concepts produced and the success in getting the product right first time. To asses the complexity of the design process and to determine what steps SMME designers go through in the development of products, companies were given the choice of four models of the design process and asked to indicate which one most resembled their design process, companies were also free to add to or delete anything from the models. The models were presented in increasing complexity with model A being the simplest and model D being the most complex. The following diagrams detail the models A to D in increasing complexity.

A)

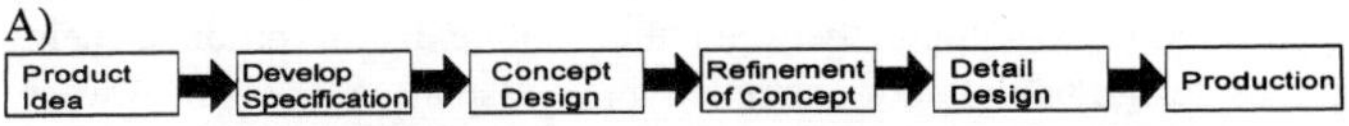

In this model a product idea is changed into a product specification, then developed into a working concept, refined and then bought up to a suitable level of detail for production.

B)

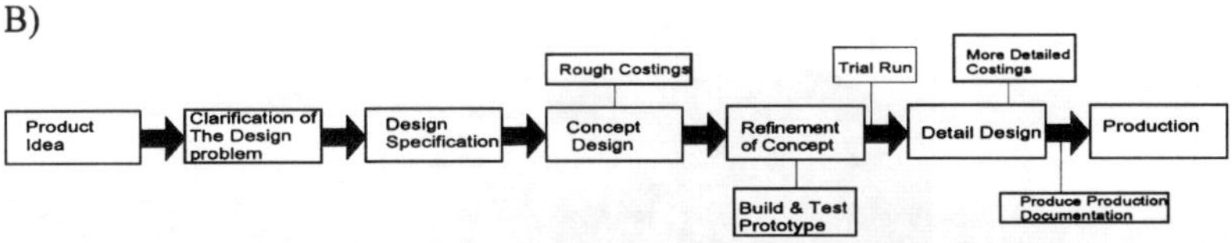

This is the same as A) apart from the product idea is further developed before the design specifications are produced. Rough costings take place around the initial design concept. To help in refining the concept a prototype is built which is then tested and undergoes a trial run to highlight problem areas in the design. The design is then brought up to a suitable level for production.

C)

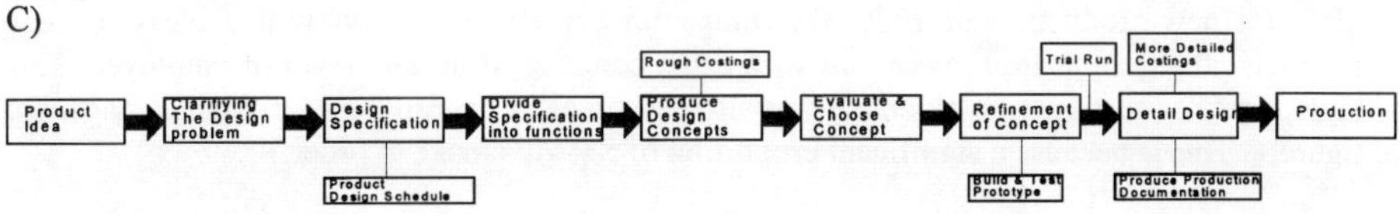

This is the same as B) except, due the complex nature of the design, the specification is divided into functions which are worked on separately and then brought together to form a product concept. More than one concept is produced, which is evaluated before the concept selection.

D)

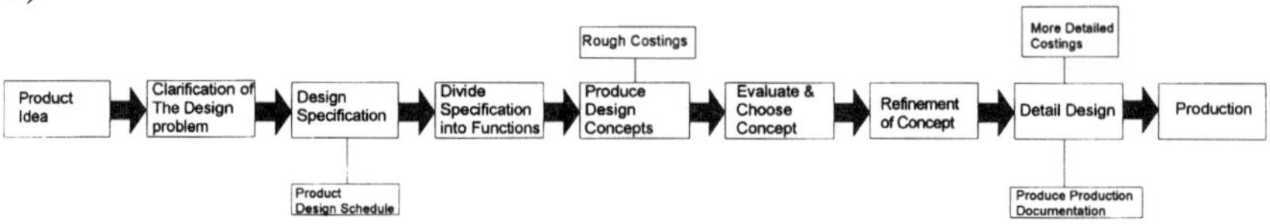

This is the same as c) apart from no prototypes are developed.

The findings show that 80% of companies expressed that their design process resembled the more simple models, figure 3. This indicates that the SMME design process is conceivably less detailed than best practice prescriptive models (6,7). This could be due to the simple nature of the products being designed and manufactured and the informal approach taken by the designers. A loose correlation was found between the complexity of products being designed and the selection of design models, for example typical responses to A and B where from companies performing simple plastic extrusions, sign manufacturers and the food industry, whilst responses to C and D came from companies designing vehicle transportation equipment and drilling machinery.

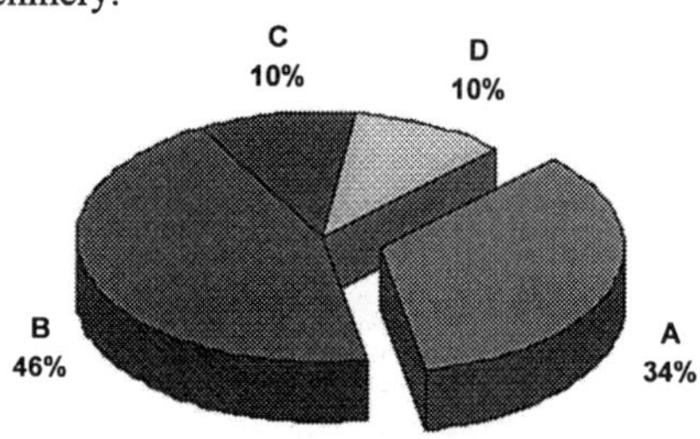

Figure 3. Models of the design process

It was found that customers and managers play a significant role in the development of ideas for new products, with only two companies expressing their view that ideas for new products could come from anywhere within the company from any level of employee. This high degree of customer involvement was also seen to carry over to the specification stage, see figure 4. This is because a significant proportion of SMMEs make to order.

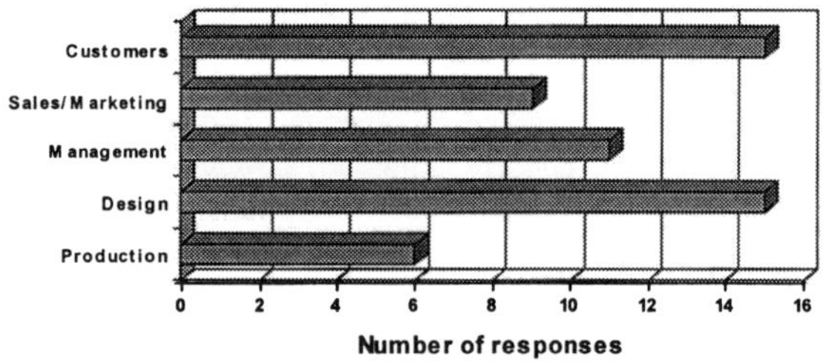

Figure 4. Responsibility for producing design specifications

It was found that 92% of companies produced less than four concepts for a product design, figure 5. Few concepts can be attributed to either the designer having a high level of certainty that the concept is right or the level of customer activity in setting concept designs reduces the need for a significant amount of concepts to be generated. The fact that SMME's typically suffer from limited capital (2), and therefore small design budgets, must also reduce the number of concepts being produced. The nature of the business should also be taken into account as it costs less for a textile company to produce a large number of visual concepts than it does for a manufacturer of drilling and mining equipment to produce a functional concept.

Figure 5. Concepts produced for a design

Further to this, when companies were asked to indicate if they often produced the concept right first time, 63% indicated that this was not the case, figure 6.

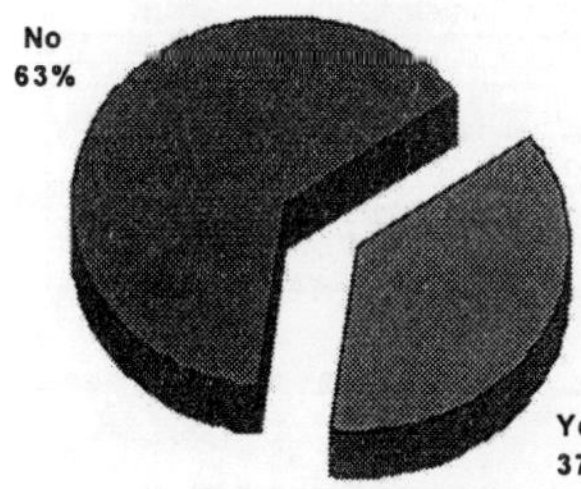

Figure 6. Companies producing the product concept right first time

This could be due to an inadequate exploration of concept ideas during the conceptual design stage resulting in the need for a significant number of subsequent design modifications. It was found that the majority of companies producing three or more concepts were producing right first time, however it was also noted that the majority of those companies who produced less than three concepts did not produce right first time. Further to this some companies added that they were under significant pressure from their customers to keep product development times low and design right first time, over 50% had product development times of less than 6 months, time, therefore, must also be a limiting factor when developing product concepts.

When companies were then asked to confirm when these further modifications were taking place it was found that most design modifications took place during the embodiment or prototype stage of the design process and in general in the later stages of the design process, figure 7. 18% of companies replied that most modifications were taking place after the product had been manufactured, indicating that design flaws were also being discovered by their customers. The results could indicate inadequate design work at the earlier stages of the design process or that some design problems may not be apparent until a physical prototype is produced.

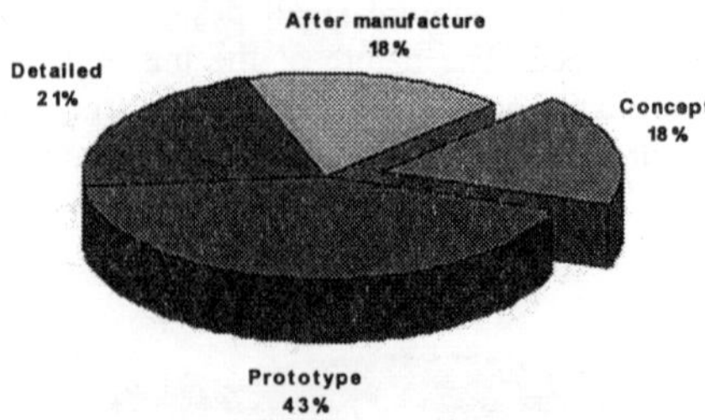

Figure 7. When do most of the design modifications take place?

2.1.3 Design characteristics and use of design tools

Companies were asked to give a score to which product characteristics are important, with 1 indicating high importance and 5 being unimportant, when designing a product. The lower the overall score the more important the characteristic.

Product Characteristics	Score
Cost	**65**
Reliability	**61**
Ease of Assembly	**107**
Ease of Manufacture	**101**
Appearance	**96**

Table 1. Product characteristic scores

As a whole it was found that most companies are concerned with producing a reliable product at the lowest cost and characteristics such as how the product looks and how easy it is to manufacture and assemble are less significant issues, table 1. However, it was found that some product characteristics were more important to certain industrial sectors, for example the appearance of the product was found to be very important to companies in the food and textiles sectors whilst the cost and reliability of the product was found to be more important to companies from the manufacturing/engineering sector.

When companies were asked to reveal what best practice design for 'X' (DFX) methods were being used to design those qualities into a product design, a low level of knowledge of best practice DFX methods was found. Figure 8 shows that Design For Manufacture (DFM) is the most widely used design tool with Taguchi being the least widely used. The survey also showed that most companies hadn't heard of DFX tools, with nearly 90% not having heard of quality function deployment (QFD).

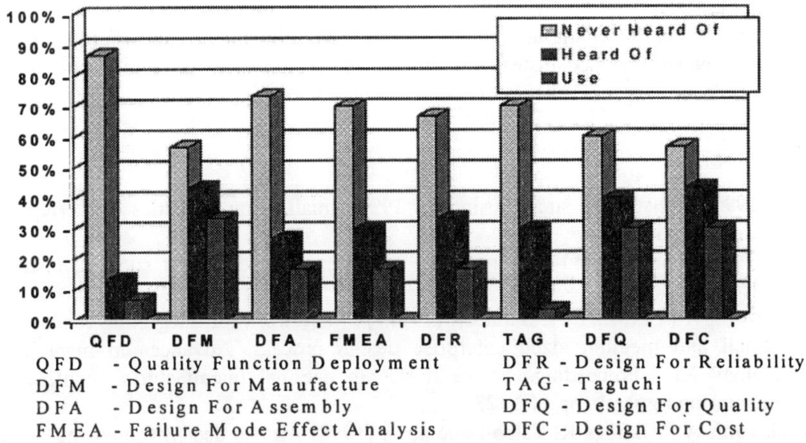

Figure 8. SMME knowledge of DFX

The nature of the business must influence the use of such tools, for example a designer in the textiles industry would have little practical use for some DFX tools. However, further studies of the SMME design process, that have been performed by the authors, in four case studies have revealed that, although DFX could have been used, designers either lacked the necessary knowledge to utilise DFX or the use of DFX was considered impractical to use because of short product development times. This infers that designers must be using other methods to design these characteristics into their products, and it was found from the case studies that designers in SMMEs rely heavily on their own experience, instead of using a design tool.

2.2 Survey conclusions

From the 40 returns of the questionnaire it can be summarised that:

- Most SMME's regard their design process as relatively simple.
- SMMEs were found to have design teams of less than four.
- Most SMMEs have short product development times.
- SMME develop few design concepts.
- If a company is making to order, the customer has a high degree of involvement in the design process.
- Most SMMEs were found not to design right first time.
- Most SMMEs are not familiar with DFX and do not use best practice DFX methods.

3. Conclusions

To increase the validity of the results from the questionnaire a more in-depth analysis of the SMME design process has taken place in four case studies, this has further highlighted the low level use of best practice design methods within SMME's. This suggests therefore that the suitability of current best practice design methods and tools for SMME's needs to be addressed. Research currently in progress on an eco-design methodology, by the authors, will take these factors into account and therefore use design guidelines which would help support the designers own experience, instead of complicated design tools, to integrate eco-design into the design process.

4. References

1. DTI (1996) Statistics on small businesses 1994, Small Firms Statistics Unit: Department of Trade and Industry
2. Levy J (1993) Small and medium sized enterprises: A recipe for success, Institute of Electrical Engineers
3. Duckworth A P, Lewis P A, Baines R (1997) A framework for integrating ecodesign into the small and medium sized enterprise design process, Advances in manufacturing technology XI - Proceedings of the thirteenth national manufacturing conference on manufacturing research, pp 119-123
4. Duckworth A P, Baines R, Taleb-Bendiab A (1998) An eco-design framework for small and medium sized manufacturing enterprises, 2nd international symposium on tools and methods for concurrent engineering - TMCE 98, pp132-141
5. French M J (1992) Form structure and mechanism, Macmillan, London
6. Pahl G, Beitz W (1995) Engineering Design - A Systematic Approach, Springer, London
7. Cross N (1989) Engineering Design Methods, John Wiley & Sons, Chichester

Design and manufacture – developing an integrated approach with undergraduate product design students

A J FELTON
School of Engineering and the Built Enviroment, University of Wolverhampton, Telford, UK
E BIRD
School of Art and Design, University of Wolverhampton, Telford, UK

ABSTRACT

This paper is based upon research and development undertaken over the last eight years for the BSc (Hons) in Computer Aided Product Design, at the University of Wolverhampton. Conceived in 1989 as a joint venture between the Engineering Division of the School of Engineering and the Built Environment and the Product Design Section of the School of Art and Design, it aims at developing an employable graduate possessing a greater awareness of the relationship between Design and Manufacture in the specialist field of Computer Aided Product Design.

This paper looks at how these and other elements have been integrated in the course and traces their development in undergraduates Design Projects. It analyses the development of the learning experience and will be illustrated on presentation with slides of project case studies.

(1.0) INTRODUCTION

The research focuses on the BSc Honours Course in Computer Aided Product Design (CAPD) within the School of Engineering at the University of Wolverhampton. The course may be completed in either 3 year Full Time or 4 year Sandwich mode and aims to produce a broad based educational programme of study in the technical, creative and commercial aspects of Product Design. The essential skills of a designer are developed through a balanced blend of creativity, analysis, synthesis and judgement with a strong emphasis placed upon manufacturing and modern high technology aids to design, particularly in the use of advanced computer graphics.

(2.0) AN INTEGRATED APPROACH TO CURRICULUM DEVELOPMENT

(2.1) Background to the course

During 1989 following research (1), (2) and feedback from Industry it was ascertained that a new type of designer was required who could combine the skills and theoretical knowledge of an engineer and the conceptual idea's of the industrial designer. The degree programme was based on this premise, the aim was to move away from what was loosely termed "Over the wall design" i.e. the designer who produced concept designs and passed them over the wall to the production / manufacturing engineers to sort out and produce. What was required was a designer who could work at both ends of the design spectrum, be able to produce design concepts but also had the technical knowledge to communicate and liaise effectively with manufacturing, materials and production engineers to achieve the product goal.

The new CAPD graduate would have a broad range of transferable skills to enter a new and fast changing job market that did not offer a traditional career for life, but required adaptability. We needed to offer a richer blend of skills between Industrial Design as then offered in the Three Dimensional Design Subject area of the School of Art and Design (SAD) at Wolverhampton and traditional Mechanical Engineering offered in the School of Engineering and the Built Environment (SEBE). This is one of the first close collaborative ventures between schools within the institution and aimed to bring together design, engineering, manufacturing and computer skills.

Integration is the key word both for our students and our programme which is drawn from expertise across the two schools and offers the broad range of inter-related skills outlined in Table 1. Four main skill areas are identified:-

i) Design Awareness and Visual Communication, supported by Product Design Staff from the SAD.

ii) Engineering Design, supported by Mechanical Engineering staff from the Engineering Division of the (SEBE).

iii) Manufacturing and Technology, supported by Manufacturing Staff from the (SEBE).

iv) Computing and Information Technology supported by Design engineering staff from (SEBE).

The other integrating vehicle for these skills is the practical design projects that occur in each level (year) of the course and have multi-staff supervision. These projects increase in complexity with each level (year) and ranged from being introductory and diagnostic at level 1 to a comprehensive test of all the skills outlined at level 3 where students undertake a self initiated project that occupied the final half of their completing year.

(2.2) Student profiles

Student profiles entering the 1st year of the course are of mixed educational backgrounds and qualifications these include 'A' Level students, students from Foundation courses in Art and Design, BTEC National, GNVQ and mature Access students. The prescribed programme of study (3) has therefore been developed on a progressive structure, establishing a foundation of study in Year 1 and then developing the subject disciplines through the subsequent semesters.

(3.0) PROJECT BASIS OF THE COURSE

(3.1) Level 1, Year 1 Project.

The first year project occurs in either the module Product Design Studies II or Product Design and Development depending on the option route the student has selected. The modules have an introductory design and make project which is diagnostic and supported with a series of lectures that introduce the main elements of practical design such as Design Management, Morphology of Design, Design Specification, Design Protection, Ergonomics and Anthropometrics. It is diagnostic because of the broad intake of students on the course. With a differing range of prior experiences it intends to identify individual strengths and weaknesses that can be picked up on in Level 2. It also tests the students knowledge accumulated from the other seven modules of study at Level 1. The project subject matter has varied over the years and has included:-

A mechanically powered toy.
A toy that flies.
Lifestyle projects for an ageing society.
Portable office for a designer on the move.
In flight airline meals tray.
A child's briefcase.

For many of the students it is their first attempt at a design brief and because of its diagnostic nature we are more interested in the process than the outcomes. The design brief is highly prescriptive and outlines in detail what is expected and the importance that we attach to the different elements through the marking scheme. In a report students are expected to outline and relate design and manufacturing issues and account for the management of their time.

(3.2) Level 2 Projects

Having had the opportunity to develop and rectify deficiencies at Level 1 through module feedback and Level 2 modules, the learning process is tested again in a Design and Make project which forms part of the Design Practice Module. This project is tackled by students working in pairs, working to their strengths as opposed to their weaknesses. Work load is shared for all aspects of the project. Each student takes responsibility for 50% of the design work., however, the manufacturing responsibility is switched. This is intended to allow students to experience the problems of communicating design details to others responsible for their successful manufacture. Projects have included:-

A **device for transporting a Windsurfer** and associated rig from the car, across a beach, to the waters edge.
A **cycle carrier** for carrying at least two cycles in a safe manner which can be bolted to the tow bar bracket of a car, taking into account that the towball maybe needed to be fitted at the same time.
A small portable or hand held **Wind Anemometer** for yachting and outdoor activities.
A weight lifting bench / Exercise bench.
A **Folding chair** for the Leisure market.

As well as the manufactured artifact a comprehensive project report is submitted by students to include research, concept designs, dimensioned CAD working drawings, 3D CAD modelling including some Finite Element analysis, costing details, packaging, intended usage, materials selection and details of manufacture.

(3.3) Level 3 Projects
Half of the final year, that is one semester, is devoted to project modules. These include Project Planning, Design Enterprise and a self initiated major Design and make project. Project planning allows the student to select and research their chosen topic, draw up a management plan for design and make in the Project modules.

The major project is expected to be developed to first working prototype stage which involves considerable time being spent in workshops. Project topics have been wide ranging and examples will be presented in slides.

(3.4) Design Competition
The module Design Competition has become an important part of the CAPD experience using an actual Design Competition as the background for this module, it is a test of the student's ability to work within the professional constraints and demands of outside generated criteria. On the other hand, the successes achieved in winning or being placed in both national and international events is a boost to both student and staff moral and confirms the professional integrity of what is being done. Since 1996 successes have included:-

- a placing in the finalists, **Plastics on the Road Competition.**
- winner of the **Phillips section of the National Lighting Awards**.
- second and third winning places in the **1997 BASF National Plastics Design Award.**
- a placing in the six finalists of the **1998 BASF Award** - To be finally judged in the summer of 1998.

(3.5) The Industrial Experience Year
This optional year which occurs between level 2 (Year 2) and Level 3 (Year 3) is very much dependent upon the students. It is based around their initial aims and ambitions and their personal tenacity to find a place. Obviously they are given considerable staff support in doing this and students who have gone out on placement have found the experience beneficial for a number of reasons.

- they learn the importance of working as part of a team.
- they develop a professional maturity from the experience.
- many gain a placement abroad and during the year become reasonably fluent in a European language.
- There is often the offer of a job when they finish their degree.

(4.0) CONCLUSION

The curriculum has been developed to educate undergraduates with broad based knowledge of design together with an understanding of manufacturing processes, materials, engineering science, and business. These disciplines are integrated through the use of modern aids in design computing and thus allow the students to develop and experience professional design practice within a high technology environment.

The field of product design is rapidly changing due to the increasing use of computer based technologies (4) Design for manufacture, design for assembly / disassembly and design for recyclability have resulted in the development of new approaches to design, the new approaches to include feature modelling, concurrent / simultaneous / parallel engineering concepts. The adoption of new methodologies to meet these new concepts indicate, industrial requirements for Designers coming from an engineering background to have broad perspectives and greater interactive viewpoints.

Skill levels and knowledge bases are changing therefore the Academic world needs to recognise this and develop new teaching / learning strategies (5) to promote and foster the culture of the integrative philosophy.

(4.1) First career destinations

The success of the CAPD course is demonstrated by the employability of its graduates. Apart from many of the industrial experience students being offered jobs when they leave the University the majority of the other graduates have gained employment in a wide variety of companies by the time they return to the graduation ceremony in September / October of the year of completion. Those completing in 1997 are a good example of those at the graduation ceremony, the profile of employment was:-

Company	**Position**
KPMG Health Systems	Designer
Polar Cup Ltd	Product Designer
Micro Focus	Quality Engineer
Quality Industries	Design Engineer
Thorn Security	Designer
Auto Design Ltd	Designer
Kylmar (KMC)	Designer
Vivid Image Development	Designer
Ishida Europe Manf Ltd	Design Engineer
Elliott Group Ltd	Designer, CAD
BAE Westlands	Design Engineer
Quest Business Technology	Designer
JDP CAD Bureau	Designer,CAD
Valor Heating	Designer
Omron	Designer
OYB London Bridge	Designer
Design Co	Design Engineer
Static Systems	Systems Engineer
Internet Design (Budapest)	Designer
ALPS Electric (Ireland)	Designer

REFERENCES

(1) *Training Designers For The 1990's*, Heads of Industrial Design Conference Report, 23rd Nov 1983, Royal Institution London: The Design Council.

(2) Ewing, Paul.D. Author, 1987. *Curriculum Development Report On Industrial Design Engineering* . London: The Design Council.

(3) *Award Handbook (1997-98)*, BSc (Hons) Computer Aided Product Design. University of Wolverhampton, School of Engineering and the Built Environment.

(4) Felton, A. J. *The Evaluation of Parametric CAD as a Teaching Aid in Product Design*, 2nd National Conference on Product Design Education, Coventry University 10th-11th July 95.

(5) Felton, A. J. *The Evaluation of Video Conferencing in a Distance Learning Environment for Product Design,* Thirteenth Conference of the Irish Manufacturing Committee, (Re-Engineering for World Class Manufacturing), University of Limerick 4th-6th September 1996.

Table 1. BSc (Hons) in Computer Aided Product Design (CAPD) Curriculum.

1. Design Awareness Visual Presentation Skills.	2. Engineering Design Skills	3. Manufacturing and Technology Skills	4. Computing and IT Skills	5. Design Projects. Testing of Areas 1,2,3,4
LEVEL 1 (First Year) **CM 1106 Product Design Studies 1** Design Awareness **CM1109 Graphics** Introduction to visual presentations skills both manually and via computer.	**CM1107 Design** Communication Studies. Introduction to numeracy and technical drawing skills in an engineering context.	**MA1103 Product Manufacturing Studies.** Introduction to manufacturing processes. **CM1104 & CM1105 Fundamentals of Technology 1& 2** Eng principles, Mat'ls, Electric's and Electronics.	**CM1108 Computer Aided Engineering.** Introduction it to information technology in an engineering context.	**CM1101 Product Design Studies 2.** Theory and practice of a design project. Students tackle their first design project alongside a series of lectures which take them through the design process.
LEVEL 2 (Second Year) **DS2229 Design Principles**. Development of visual presentation skills through assignments. **DS2228 3D Realisation.** Development of model making skills.	**ET2233 Engineering Design.** Technical aspects of designing Engineering Products. **CM2206 Design Enterprise Studies.** Design in a company structure.	**CM2205 Design Analysis.** Analytical methods for product component strength including Finite Element.	**CM2201 Principles of CAD.** Introduction to the principles of computer aided design 2D form. **CM2203 Graphical Modeling.** Extends CAD experience into 3D forms.	**CM2207 Design Practice.** Practical experience of the Design Process with a Design and Make Project.
LEVEL 3 (Third Year)	OPTIONAL	INDUSTRIAL	PLACEMENT	YEAR
LEVEL 3/4 (Final Year) **3rd Year direct course for students, or 4th Year students returning from Industrial Placements.**	**CM3302 Management Design.** Factors involved in managing design in an engineering context.	**CM3303 Design Assurance.** Quality Control in the Design and Manufacturing process.	**CM3304 Design Information Systems**. CAD Technology in the design context. **CM3306 Design Enterprise project**. Students work in teams and set up a company to manufacture a product.	**DS3310 Project Planning.** Planning process for the final Design Project modules. **DS3311 Design Project.** Double module, individual final project. **CM3305 Design Competition** The experience of entering an actual design competition.

Web-based design tools

G Q HUANG and **K L MAK**
Department of Industrial and Manufacturing Systems Engineering, University of Hong Kong, Hong Kong

ABSTRACT

Formal design methods and techniques such as Quality Function Deployment (QFD), Value Analysis (VA) and Failure Mode and Effect Analysis (FMEA) are, and should be, used as team tools. This paper proposes to exploit the Web technology to implement formal design tools on the Internet/Intranet. Team members can use standard browsers to access these services at different locations. In comparison with standalone computerised design tools, Web-based systems are effective in facilitating teamwork in product development.

1. INTRODUCTION

Significant progress has been achieved in developing and applying a variety of formal design techniques or methodologies. Examples include Quality Function Deployment (QFD), Value Analysis (VA), Failure Mode and Effect Analysis (FMEA), Design for Manufacture and Assembly (DFMA), and Morphological Charts (MC). They are usually packaged into pro-forma worksheets together with a systematic procedure of instructions, often assisted by computer software.

The importance and disappointments of using formal design tools become evident as a result of various industrial surveys (1-3). One way of overcoming the disappointments and maximising the potential benefits is to computerise these design tools. Reinders (4) discusses a number of computer based design tools or design assistants as he calls. However, computerised design tools do not completely remove these difficulties and in fact introduce some other overheads such as familiarity with operating a computer software package. While recent development in multimedia computing fortunately enables the ease of use, most computerised design tools are packaged as stand-alone systems. Stand-alone design tools support teamwork only in a limited way. Quite to the contradictory, they often hinder teamwork (5). This phenomenon can be described as isolated “Islands of Expertise”. It is not consistent with their tenet of facilitating communication, fostering teamwork, and enabling collaborative product development environment. This paper proposes to resolve the issue of “Islands of Expertise” by exploiting the rapidly developing Internet/Intranet technology.

2. WEB-BASED COLLABORATIVE PRODUCT DEVELOPMENT

A web-based virtual framework has been established for collaborative product development (6). There are five important constructs. First of all, the framework is aimed for multiple users involved in a product development project. They are individually or collectively supported by web-based design tools. The second construct of Virtual Consultants is introduced to refer to web-based design tools. They may maintain their own working databases as well as sharing common database on the Internet. They provide services on the Internet relevant to product development projects. Typical Virtual Consultants include web-based QFD, FMEA, DFMA, Morphological Charts, and Value Analysis. Third, Virtual Consultants are organised into Virtual Teams. A Virtual Team is basically a shared web page providing hyperlinks to various Virtual Consultants concerned. They are logically configured to solve a class of product development problems. The shared web page is specially designed to allow easy and structured access. Typical Virtual Teams include Engineering Change Management, Design for X, and Conceptual Design. Finally, both Virtual Consultants and Virtual Teams are registered with an Internet web site referred to as a Virtual (Design) Office. In addition to shared web pages of Virtual Teams, this web site contains a registration database for Virtual Consultants and Virtual Teams and associated facilities. Capabilities and HTTP addresses of Virtual Consultants and Virtual Teams are captured in the database. Such information enables the potential users to choose the most appropriate Virtual Consultants and Virtual Team to solve the problem at their hands.

Research and development have been conducted to experiment with the framework summarised above. One such experiment is concerned with web-based collaborative conceptual design. The importance and process of conceptual product design have been widely discussed and illustrated in a number of major textbooks on Engineering Design (7-9). Although there are minor variations among these conceptual design models, three major stages are in common. They are functional analysis, concept generation, and concept evaluation. The general framework proposed above has been instantiated for conceptual design. The resulting framework is shown in Figure 1. Its major components can be summarised as follows:

1. *Virtual Office*. During the experiment, one computer is set up as a Virtual Office. The corresponding home page is the startup page. A relational database is developed, containing data tables for storing information regarding Virtual Teams and Virtual Consultants, including their names, brief description of their capabilities, and their HTTP addresses, and their HTML file locations and names. Associated web pages are created for registering Virtual Teams and Virtual Consultants.
2. *Virtual Team*. A Virtual Team on conceptual design is attached to the Virtual Office startup homepage. It is basically a shared web page providing hyperlinks to the three Virtual Consultants involved. The user interface of this page is designed to allow easy and logical access to the three Virtual Consultants involved in conceptual design.
3. *Virtual Consultants*. Three Virtual Consultants are selected in the Virtual Team, corresponding to the three major stages of conceptual design. They are Functional Analyser, morphological Concept Generator, and morphological Concept Assessor. Functional Analyser is installed on the same computer as that of the Virtual Office. Concept Generator and Concept Assessor are installed on another computer. The experiment is conducted at a third computer. In the subsequent three sections, the design and implementation of the three Virtual Consultants are discussed.

3. WEB-BASED FUNCTIONAL ANALYSER

Functional Analysis is usually the starting point of developing new products or improving existing products. It is responsible for defining and clarifying the functional requirements or goals that the product under development aims to achieve. In fact, other design tools such as Value Analysis, FMEA, Morphological Charts, and QFD all include some elements of functional analysis. For this reason, it has been decided to develop a web-based Functional Analyser for common uses.

However, a problem has been encountered regarding which function analysis method should be adopted/adapted for the Functional Analyser. There have appeared a number of methods for functional analysis. Functional flow charts, functional block diagrams, and functional logic diagrams are only three examples. Each has its advantages and disadvantages and none is considered perfect (10). Lucas Design for Assembly proposes an entirely different approach to Functional Analysis (11).

After examining the functional analysis elements in various design tools, it has been found that the formal method used in Value Analysis is most generic and appropriate (12). That is, the Functional Analyser is now based on FAST (Functional Analysis Systematic Technique) diagrams. The FAST method and diagrams are familiar to trained engineers because they have been adopted in major textbooks and Value Analysis is practised in industries. In addition, the method is relatively simple because functions are linked together to form a hierarchical tree where only why/how relationships between them are represented. Other functional relationships and details, if necessary, may be incorporated into the analysis by exploiting other techniques. Although limited in a number of ways, the FAST method is considered sufficient for defining the functional goals.

Functional Analyser is the first Virtual Consultant included in the conceptual design Virtual Team. It is basically a web-based FAST diagram editor, providing a number of database operations. For example, the user can follow the procedure below to define a new functional goal:

1. Choose a functional goal already defined in the functional tree.
2. Create a sub-functional goal or a functional goal at the same level as desired.
3. Specify the general description of the newly created goal.
4. Add details of the newly created goal as necessary.
5. Repeat until all the functional goals are defined.

The above procedure follows a top-down approach, namely proceeding from the highest top level to the lowest bottom level. A bottom-up approach is also possible with this Functional Analyser. A flat list of functions is first defined. Its drag and drop facility is used to build the tree. Which approach, top-down or bottom-up, to use is beyond the scope of this paper. However, as a general guideline, the top-down approach is easier to start with and the bottom-up approach is then used to modify the tree structure.

4. WEB-BASED CONCEPT GENERATOR

Concept Generator is the second Virtual Consultant included in the conceptual design Virtual Team. It is responsible for concept generation - a divergent and creative stage of conceptual design. The purpose is to conceive as many solutions to sub-functions and to produce as many feasible solution alternatives to the overall functional goal as possible. The input to this stage is the hierarchical tree of functional requirements as established in the preceding functional analysis stage and the output from this stage is a set of solution alternatives to be forwarded to the next stage for in-depth evaluation. The conversion from the input to the output makes full use of the Concept Base which is the working database maintained by Concept Generator.

The basic mechanism used in the Concept Generator is a morphological generation chart. The vertical dimension of the matrix represents a list of basic functional goals, namely the functional requirements at the lowest level. This list is therefore flat, no longer hierarchical. The horizontal dimension of the matrix represents all the conceivable means that are relevant to achieve the corresponding basic functional goal. Different goals may have different numbers of conceivable means. Therefore, the numbers of cells of the rows in the chart may be different. Based on this morphological generation chart, the designer can make further analysis and choices regarding which means is more applicable than others for a given goal.

Concept Generator accomplishes three general steps. This first step is mainly for building up the morphological generation chart from the Concept Base as possible. The second step is to short-list conceivable means based on their feasibility. Those relevant but infeasible means are excluded at this step. The third step is to combine short-listed feasible means into complete alternative solutions for the overall functional goal. As a result, a fair number of alternative overall solutions or concepts are produced for further investigation.

Mathematically, a vast number of combinations is possible during this generation process. The user must apply his/her professional experiences and judgements to avoid such combinatorial explosion. Alternatively, sophisticated intelligent algorithms may be incorporated into this process for the purpose of avoiding the combinatorial explosion.

5. WEB-BASED CONCEPT ASSESSOR

Concept Assessor is the third Virtual Consultant included in the conceptual design Virtual Team. It is responsible for concept evaluation or assessment - a convergent and selective stage of conceptual design. The purpose is to narrow down the feasible alternative concepts from a fairly large number to a few more manageable number for further investigation. The input to this stage is the list of solution alternatives obtained from the Concept Generator and the output from this stage is a rank order of the inputted solution alternatives to be forwarded for further development. The conversion from the input to the output is based on the set of criteria against which alternative solutions are evaluated.

The basic mechanism used in the Concept Assessor is a morphological evaluation chart. The vertical dimension of the matrix represents a list of evaluation criteria. The horizontal dimension of the matrix represents all the solution concepts that are considered feasible to achieve the overall goal. This solution list can be directly transferred from the Concept Generator. The performance measurement of a solution concept against a criterion is entered into the corresponding cell in the matrix.

The Concept Assessor accomplishes four general steps. The first step is mainly for establishing the criteria against which alternative solution concepts are to be evaluated. The second step is to prepare a list of alternative solution concepts generated from the Concept Generator for evaluation. The third step is to work out specific measurements of each alternative concept against each criterion. The final step is to work out the overall measurements for all the alternative concepts and identify and shortlist a few most promising concepts for further investigation.

It is time-consuming and data intensive to assess alternative solution concepts against individual criteria one by one. The human user plays key roles in establishing performance criteria (indicators) and providing performance measurements. Compromise between fully quantitative and qualitative measurements is often necessary.

6. CONCLUDING DISCUSSIONS

This paper has discussed a framework within which web-based design tools can be developed and applied. A number of virtual constructs are proposed for establishing the framework. Web-based design tools are referred to as Virtual Consultants providing remote services from the Internet. They are registered with one or more Virtual (Design) Office and then configured into various Virtual Teams according to the nature of design problems. The Virtual Office provides a database for keeping information about individual Virtual Consultants and Virtual Teams so that they can be selected to solve the most appropriate design problems. Experiment is made to demonstrate the proposed framework with a simple example of conceptual product design.

The work reported here is based on a typical application scenario on conceptual design involving only three web-based design tools. In the future, more web-based design tools should be developed, including FMEA, VA, QFD, DFA, Engineering Change Request, Engineering Change Notice, Product, Process and Resource Data Management. Once these Virtual Consultants are developed, they would be configured to form other Virtual Teams for application scenarios such as Engineering Change Management and Design for X. The proposed framework would be further evaluated and improved after more insights are gained from future experiments.

REFERENCES

1. Andreasen, M.M. (1987) "Design strategy", *Proceedings of ICED 87*, Boston, MA, ASME, 171-178.
2. Norell, M. (1993) "The use of DFA, FMEA, and QFD as tools for concurrent engineering in product development processes", In: *Proceedings of ICED 93*, The Hague.
3. Aranjo, C.S., Benedetto, H., Campello, A.C., Segre, F.M., Wright, I.C. (1996) "The utilisation of product development methods: A survey of UK industry", *Journal of Engineering Design*, Vol 7, No 3, 265-277
4. Reinders, H. (1995) "Design information deployment in 'Design Assistants'", In: *Proceedings of ICED 95*, Praha, Czech Republic, 1339-1344.
5. Cutkosky, et al (1993) "PACT: An experiment in integrating concurrent engineering systems", *IEEE Computer*, January 1993, 28-37.
6. Huang, G.Q., Mak, K.L. (1998) "Web-Based Collaborative Product Development", to appear in *Collaborative Technologies in Product Design and Manufacturing*, Edited by Nanua Singh, to be published by Chapman and Hall.
7. Pahl, G., Beitz, W. (1984) *Engineering Design*, Design Council, London.
8. Cross, N. (1994) *Engineering Design Methods: Strategies for Product Design*, 2nd Edition, A Wiley Series in Product Development: Planning, Designing, Engineering, Chichester, UK.
9. Ulrich, K.T., Eppinger, S.D. (1995) *Product Design and Development*, McGraw-Hill, Inc., New York, USA
10. Sturges, R.H., O'Shaughnessy, K., Reed, R.G. (1993) "A systematic approach to conceptual design", *Concurrent Engineering: Research and Applications*, Vol. 1, No. 2, 93-105.
11. Lucas Design for Assembly Practitioners' Manual
12. Miles, L.D. (1965) *Techniques of Value Analysis*, 2nd Edition, McGraw-Hill, New York.

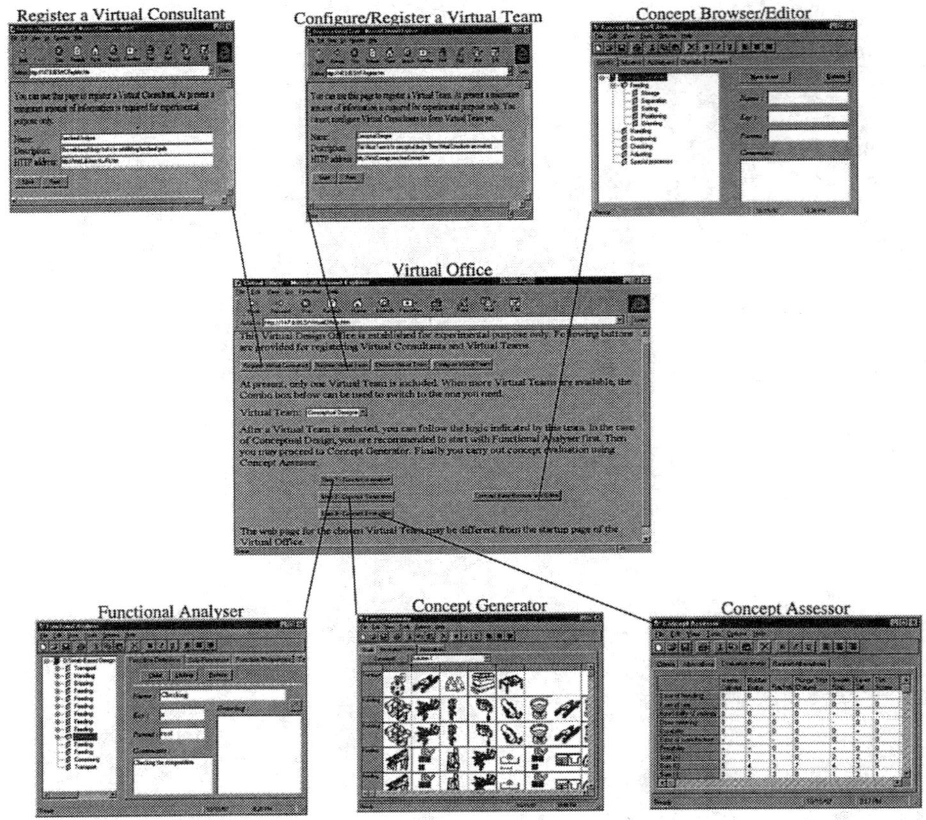

Figure 1 Web-based architecture for collaborative conceptual design

The implementation of shared workspace systems in design – common barriers and success factors

W J ION and **A I NEILSON**
Department of Design, Manufacture, and Engineering Management, University of Strathclyde, Glasgow, UK

ABSTRACT

This paper, initially discusses the application of shared workspace technology to the design process. Despite the potential offered by such systems, and the fact that they are commercially available and affordable, uptake within Small, and Medium sized Enterprises, (SME's) has been low. Furthermore, there is evidence that those companies who have adopted the technology are not fully exploiting its potential. A number of technological and organisational issues are responsible for restricting the adoption of this technology. This research identifies the barriers that are likely to face SME's during the introduction and use of shared workspaces and presents suggestions as to how they should be addressed.

SHARED WORKSPACES AND THEIR USE IN THE DESIGN ENVIRONMENT

The design process is characterised by the use of multi-disciplinary teams, integrating the knowledge and skills required to produce a successful product. In many cases the need for specialist skills has developed a trend towards a high level of outsourcing of design capabilities, either internal or external to an organisation. Such a scenario can lead to the creation of long supply chains and a wide geographic distribution of design specialists. As a consequence a high level of interest in the use of computer based collaborative working technologies which facilitate the creation of virtual design teams has been generated (1).

Many distributed design teams communicate electronically using groupware systems, such as Lotus Notes (2), that provide an efficient approach to asynchronous communication and the management of files and project data. Some sectors of industry are adopting a more advanced approach based upon the concept of shared workspaces. With shared workspaces parties that are separated geographically can share data in real time over computer networks through a common computer interface, the shared workspace. The shared workspace typically allows the following tools and facilities to be shared in real time:

- audio and video conferencing;
- data conferencing
    - shared whiteboards;
    - shared applications such as CAD.

Application of shared workspaces to design

The facilities offered by shared workspace technology allows designers who are separated geographically to share and discuss text, calculations and graphical information synchronously. For example, two designers located at different sites could share, in real-time, a CAD drawing and communicate using audio and video facilities to resolve a design problem. The network utilised could be Integrated Services Digital Network (ISDN), Local Area Network (LAN) or Internet based. Using conventional means of communicating such as fax, electronic-mail, phone or post, interaction between both parties would tend to be much slower and involve a lesser degree of interaction. Shared workspaces, offer the potential to significantly improve product quality and to reduce the cost and timescale of design projects involving parties with a widespread geographic area. Benefits typically achieved by companies adopting shared workspace technology within the design process include (3, 4):

- Improvements in the flow of work allow companies to move and react faster;
- Product development lead time, time to market and costs reduced while maintaining or improving quality;
- Quality failure costs reduced;
- Sharing of information and expertise between organisations/departments improved;
- Relationships between manufacturers and suppliers strengthened;
- Efficiency of day to day dealing with customers improved;

Shared workspace adoption and use

Despite the benefits and the fact the technology is commercially available and affordable the uptake of shared workspace systems, particularly within Small and Medium sized Enterprises (SME's), has been low. A number of technological and organisational issues have tended to restrict adoption to large multinational companies, such as Ford and Boeing (4), who have the resources and expertise to tackle and overcome them. Even within those companies that have introduced the technology for use within the design process, usage is often restricted to video conferencing (5). Data conferencing has not been widely adopted and as a consequence the full potential and functionality is not exploited. Research has shown (6), that designers, and others actively involved in the design and manufacture of new products could benefit significantly by utilising the full functionality available.

IDENTIFICATION AND CLASSIFICATION OF BARRIERS

The benefits that the adoption of shared workspace technology can bring to the design process have been discussed above and are now widely recognised. It is unfortunate that a higher level of utilisation of the technology is not in evidence, particularly, within the SME sector where the opportunity exists to shorten the communication gap between client and sub-contractor. Undeniably, technological and managerial barriers exist to the adoption of the technology and as a consequence uptake has been slow. There is a need, therefore, to identify and understand the barriers to the adoption of shared workspace technology so that appropriate implementation and management guidelines, maximising the potential offered by the technology, can be produced.

This paper henceforth presents a detailed account of the barriers encountered by SME's introducing shared workspace technology within a design environment and presents solutions for overcoming them. The barriers presented were identified during research into the implementation, management and use of shared workspace systems within a design context[1]. The research has involved monitoring the introduction and use of the technology on a series of live company based projects and interviews with actual and potential users and questionnaire surveys. Further details of the methodology adopted for this research are provided in (7). The barriers identified can be classified within five main groups. These groups and associated barriers are shown in Figure 1.

Management • Short sightedness - not looking beyond current project; • Lack of attention to innovation in management of projects; • Working to short deadlines 'constant firefighting'; • Lack of ownership, knowledge, attention to system location; • Lack of champion with clout; • Access/availability of remote party; • Poor self management in working synchronously or asynchronously; • Management commitment; • Location.	**Psychological** • Resistance to change e.g. way of working, new technology; • 'Big brother' syndrome. **Perceptions** • Perception to be expensive; • No recognition of benefits beyond reduced travel - project timescales; costs; product quality; • Lack of knowledge relating to capability; functionality; how best to use it; • Perception that it is difficult to use.
Technology Related • Technology glitches relating to network adopted, echo and audio; • Set-up problems; • Company infrastructure. • Integration of tools and the output - with management systems and existing computer tools;	**Training** Lack of: • training; • recognition of need for training; • guidelines and user information.

Figure 1 - Barriers to the adoption of shared workspaces

Each of the barriers shown in Figure 1 can be related to a stage in the process of introducing and using shared workspace technology and to staff at different levels of responsibility within the management hierarchy. Such an analysis helps in the understanding of the relationship between these three parameters (barriers, process stage and staff responsibility) and in the development of strategies to enhance the process of shared workspace introduction. Figure 2 shows in diagrammatic form the typical relationship between these three parameters.

The horizontal axis of Figure 2 represents the typical stages in the introduction of new technology into an organisation. These stages may be summarised as, firstly, a recognition of

[1] Integration of Design Specialists Through Shared Workspaces, a research programme being undertaken at the University of Strathclyde sponsored by The Design Council.

the need, followed by the preparation stage, involving specifying, procuring and planning for the introduction of the technology. This is then followed by the introduction and eventual wide spread adoption of the technology. The management hierarchy, spanning senior management, middle managers and designers/engineers, is represented on the vertical axis. Often, it is middle managers who initially recognise the need for the implementation of new technology. However, middle managers tend not to have the power to make decisions regarding its procurement, therefore, recommendations are made to senior management amongst whom the decisions are made. Designers and engineers will ultimately be the users of the technology that middle and senior managers procure. This process explains the general trend of Figure 2, where barriers in the recognition of need stage affect mainly senior management. Moving towards the preparation stage, barriers also start emerging within middle management. During the later introduction and adoption stages, designers and engineers encounter the majority of barriers when they become involved in the use of the technology.

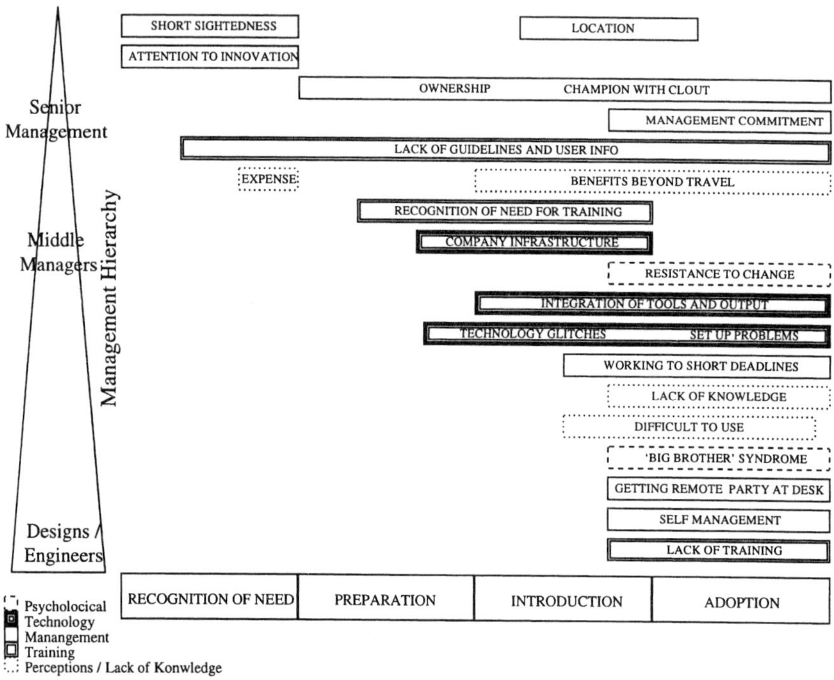

Figure 2 - Typical Barriers in the Shared Workspace Introduction Process

DISCUSSION

An analysis of the relationships shown in Figure 2 allows appropriate strategies to overcome the barriers to be developed. In the following section each of the main barrier groupings are briefly discussed in turn, together with indications as to how they may be addressed.

Psychological: These barriers are confined to users of the system, that is mainly designers and engineers. Resistance to change appears to be universal within organisations, with many individuals reluctant to adopt new technologies, particularly if it changes the way that they work. The general opinion being 'why should anything change when the current approach works'. 'Big brother syndrome' also affects system users and relates to a fear that management can employ shared workspace technology to 'check up' on them. To overcome these difficulties it is important that designers and engineers are aware of the system capabilities and of the benefits that the technology has to offer in comparison to the currently adopted methods.

Technology related: Technology related problems start to arise in the preparation stage and predominantly affect designers due to the fact that they are the main users. Technology problems are generally related to equipment being set-up incorrectly. Technical problems such as poor sound quality and slow screen refresh rate can be extremely off putting particularly to new users, and detrimental to the further use of the system. It is therefore crucial that equipment is set up correctly in the first instance and that appropriate IT support is provided.

Management: Management barriers occur in all areas of the management hierarchy and implementation stages. In the recognition of need stage senior management may not always recognise the benefits and may lose competitive initiative. A champion, with the appropriate authority, must direct and drive the project throughout the introduction process and beyond. This is particularly true of the latter stages of the introduction process where there is a tendency for management commitment to tail off following the commencement of system usage. Designers, particularly within SME's, tend to be continuously working to short deadlines, therefore, it is crucial that appropriate time is allocated to training to ensure that users feel confident with the capabilities of the system. Another difficulty is co-ordinating times where both parties are at their desks within their respective offices. Unless properly co-ordinated this can lead to designers reverting to conventionally employed asynchronous modes of communication such as e-mail, and fax.

Perceptions: Perceptions exist at all stages of implementation and throughout the management hierarchy and are generally the result of lack of knowledge caused by appropriate information and guidelines not being readily available. There is a need for clear, concise and readily available information and guidelines for computer based collaborative working in the design process.

Training: Designers and other system users generally suffer from a lack of confidence when proper training has not been administered. Correct training is a crucial success factor in the introduction and effective use of the system. Failure to administer appropriate training is often a direct result of management overlooking the need during the preparation stages. Both barriers are a result of the fact that guidelines and user information are sparse.

RECOMMENDATIONS - OVERCOMING THE BARRIERS

This paper provides an overview of the barriers that will generally be encountered during the introduction and use of shared workspaces within a design environment. It is based on findings from an ongoing research programme that aims to generate data that will enhance the management of the introduction and use of shared workspace systems within the product development process. In conclusion, the following key messages and recommendations are made.

Key messages and recommendations

1. Purchasing and installing a system will not alone ensure the benefits associated with shared workspace technology. A critical success factor is the management of the introduction and use of such systems. As an initial step companies should consider performing a self audit to determine their design information sharing requirements and readiness for such technology. Such an auditing tool is being developed as a part of the ongoing research described in this paper.
2. Companies should be aware of each of the barriers they are likely to encounter and how they can overcome them. An understanding of the factors discussed in this paper by middle and senior management would assist in the development of an appropriate implementation strategy.
3. Companies need to develop an implementation plan that addresses potential barriers and company requirements as identified through the audit process. A generic implementation framework is being developed as a part of the ongoing research, the basic stages are illustrated in Figure 3.

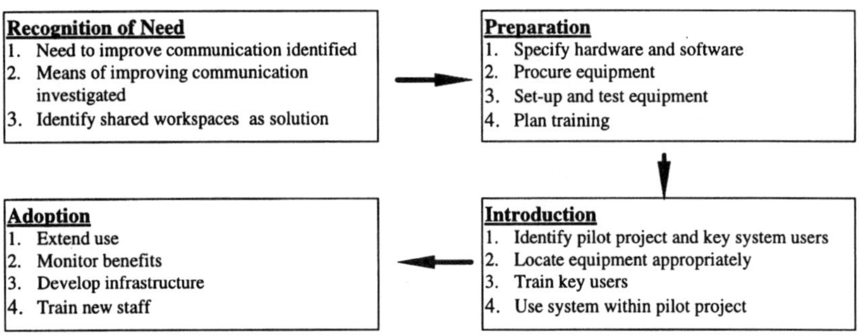

Figure 3 - Implementation framework

4. There is a strong requirement for more guidance to help companies achieve the maximum potential of shared workspace technology through its effective introduction and use. Implementation guidelines are being prepared as a part of the ongoing research, these will require extensive dissemination and assimilation amongst companies.

REFERENCES

(1) Orsak, G.C. & Etter,D.M. 1996. Connecting the Engineer to the 21st Century Through Virtual Teaming, IEEE Transactions on Education, Vol 39, No.2.

(2) Coleman, D & Khanna, R.(Eds). 1995. Groupware: Technology and Applications, Prentice Hall.

(3) Petrie, C. Madefast Home Page. http://www.madefast.org/

(4) Top Gear. 1996. Ford Motor Company Case Study, Computer Weekly, March 7th

(5) Collier, A. 1997. The End of Business Travel as we Know It, The Herald, February 27th

(6) Maher, M.L. 1997. A Model of Synchronous Collaborative Design Using CAD and Database Management, Research in Engineering Design , Springer-Verlag.

(7) Ion, W.J., MacCallum, K.J. & Neilson, A.I. 1997. The Use of Shared Workspaces to Support the Product Development Process. Proceedings of the International Conference of Concurrent Engineering, Detroit, Michigan, USA.

On representing industrial design brief requirements explicitly

K C LIANG, S A R SCRIVENER, and **L BALL**
Design Research Centre, University of Derby, UK

Abstract

Many authors have argued that establishing the design specification, a design requirement list, before beginning design leads to a better understanding of the solution possibilities and better design solutions. Typically, industrial design begins with a narrative form brief, in which design requirements are embedded implicitly. This paper describes a study designed to investigate how the explicit representation of design brief requirements influences designers' behaviour and the quality of the design solution. The results suggest that it does matter how design brief requirements are represented and presented to designers. The explicit representation of design brief requirements in the form of a checklist encouraged increased voicing of those requirements and their increased implicit consideration in solution and evaluation focused utterances. Furthermore, overall, the designers with access to the checklist judged their solutions to be a better match with the design brief requirements than those without the checklist.

Introduction

It has been argued that customer requirements can be serve as an important source for new product ideas (15), (28). While others have argued that the key factor which distinguishes new product success from failure is the consideration and understanding of customer needs (5), (9), (17), (18), (27).

Given their perceived importance, design researchers have investigated methods for capturing (4), (21), (6), (22), (24), (25) and integrating customer needs into operational definitions of design requirements during product development (1), (10), (11), (14), (21), (25), (26).

Typically, industrial design begins with a design brief, a narrative-form mission statement, in which design requirements are embedded implicitly. In the product and engineering design, the translation of brief information into a requirements list, or design specification, is recommended. Walsh et al. (27) state that the design specification provides the designer with a set of guidelines from which to create feasible design concepts and a set of criteria against which to evaluation alternative product solutions. Ulrich and Eppinger (25) also argue that the product development team should establish the target specification immediately after identifying the customer needs early in product development and then proceed to design and engineer the product to exactly meet this specification. Furthermore, various authors (20), (21), (25), (27) have suggested that setting down the requirement specification in the early design stage will help the designer to attain a better understanding of the solution possibilities and will positively affect the design solution.

However, the benefits of the design specification are largely anecdotal. Furthermore, although the value of design requirements is well-argued and the need for their clear representation in the material used to brief designers is recognised little attention has been given to how design requirements should be represented for design purposes and whether or how such representations benefits design.

In this paper, we describe and present the initial findings of a study designed to investigate whether or not it makes a difference if the design requirements in an industrial design brief are represented implicitly or explicitly, and if so what are the benefits or otherwise.

The Hypothesis

Design is regarded as coextensive with the domain of problem-solving (2). Typically, design development is viewed as an iterative process involving repeated activities of problem exploration, solution generation and selection which transforms the brief via many possible solutions to produce a particular design specified in sufficient detail to enable it to be manufactured (27). Cross (6) stresses that one of the most important design activities is the evaluation of design proposals prior to the acceptance of a final version. This evaluation occurs in an iterative process in which a design solution(s) is refined to meet certain criteria and requirements from clients, customers, industry standards, government regulations, *inter alia*, Figure 1.

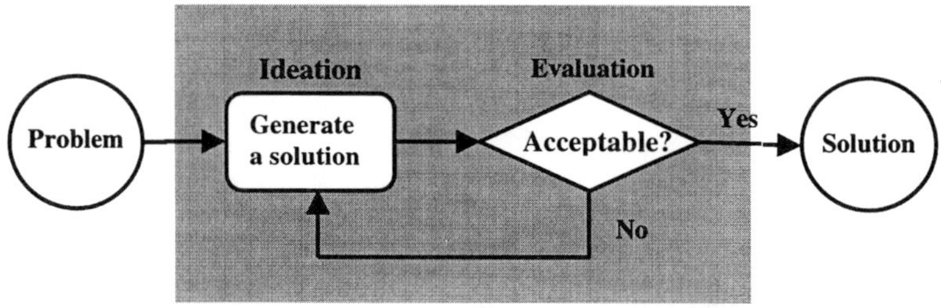

Figure 1. Design problem-solving process Source: Kim (13)

Overall, the aim is to develop a design concept which is 'optimum' in the sense that it can be measured against established criteria of success (16). However, this 'optimum' design solution cannot be identified at the outset of concept design and it is difficult to know what problems are relevant and what information will be useful until a solution is attempted. Consequently, the designer must be constantly aware of the design requirements to measure the success of the emerging design concept throughout the iterative problem-solving cycle.

We reasoned that the explicit representation of the design requirements contained in a brief would serve as an external memory aid heightening the designer's awareness of them during solution generation and evaluation thereby producing a more satisfactory design concept than would otherwise be the case.

The study

To test the hypothesis that the explicit representation of design requirements influences designers' behaviour, we conducted a study in which two groups of four design dyads worked on the same design task but under different conditions:

Condition 1 – The design dyads were given a brief in which a set of predetermined design requirements were implicitly represented in conventional narrative form.

Condition 2 – The design dyads were given a brief in which a set of predetermined design requirements were implicitly represented in a conventional narrative form, together with the same requirements listed in a checklist format.

1. The Domain

The domain of the study was industrial design. The Corfield Report (19) defines product design as including 'both engineering design and industrial design'. Industrial design is defined by the Industrial Design Society of American (IDSA) (12) as the 'professional service of creating and developing concepts and specifications that optimise the function, value and appearance of products and systems for the mutual benefit of both user and manufacturer.' This definition locates the focus of attention of industrial design upon the function, value, appearance and usability of products.

2. The Task

The task used in this study was the same as that developed by Cross, Christiaans and Dorst (7) for an international workshop on the use of protocol analysis in design research. The designers worked for two hours on the design of a fastening device which would allow a backpack to be carried on a mountain bike. We chose this task for a number of reasons including the availability of relevant information sources, the identification of requirements in the workshop papers and the possibility of comparing our results to those documented in Analysing Design Activity (8).

3. The Presentation of Design Requirements

The design requirements to be written into the brief for the study were derived by analysing (i) the original Delft brief, (ii) the papers reporting the study (8) and (iii) by using the IDEO Scenario Method (24), a visual confrontation technique. The brief for Condition 1 was then produced by editing the original Delft brief with the addition of requirements derived from (ii) and (iii).

While it is argued that writing a good specification is a matter of experience and depends on the type of product, it also recognised that one of the most useful aids to producing specifications is a check-list of points to be covered (27). We therefore modified Pugh's (21) product-design-based requirement checklist framework to emphasise industrial design issues (e.g. function, value, appearance and usability of products). Some authors (6), (27), (22) have argued that it may be necessary to distinguish between high-intensity requirements that are 'demands' and low-intensity requirements that are 'wishes'. However, to avoid biasing the designers towards particular requirements we did not prioritise the design requirements in this way.

The requirements determined by (i), (ii), and (iii) above where then organised in a checklist format to comprise what we call here the explicit representation of the design requirements of the task. Participants in Condition 2 received both the narrative brief provided to Condition 1 participants together with the checklist. The participants were all final year BSc Industrial Design students studying at Loughborough University.

4. Procedure and Apparatus

Each design dyad undertook the task in a controlled laboratory-like environment. In addition to the design brief and requirement checklists (for Condition 2) they were all provided with other information and resources such as a backpack, technical drawing of the backpack, mountain bike, mountain bike information, and technical drawings of the mountain bike, *inter alia.*

They were allowed to work on the task for two hours to develop a product concept throughout this period audio and video recordings were made for later analysis. At the end of each experimental session the design team was asked to self-evaluate the extent to which their design concept satisfied the design requirements using a rating scale.

5. Coding the Data

First the verbal discourse of each design dyad was transcribed for protocol analysis. The discourse was then coded in terms of discourse progression (23). As we were interested in the extent to which requirements were mentioned, interacted with the generation of solutions, and were involved in the evaluation of solutions we then focused on analysing and coding the discourse in terms of explicit requirement, solution and evaluation utterances, and in the case of the latter two those instances where specific requirements were implicit in the utterance.

An utterance was coded as a requirement if it matched a predetermined design requirement (embedded in the brief or checklist) or expressed a need or want generated by the designers themselves. A design proposal utterance was coded as a solution. An utterance was coded as an outcome if it expressed a problem arising from the appraisal of a solution against a requirement, for example, "perhaps it is too big to put inside"; or a constraint on what the designers may do arising from interaction between one requirement and another, for example, cost and material; or acknowledged that a solution met a given requirement(s), for example, "Using the frame of the backpack will make it easier to position the bag".

This coding scheme captured the flow and the evolution of the discourse over time. It contained information about what solutions were proposed, what requirements were considered, how the solutions interact with the requirements and how these solutions were positively and negatively evaluated against requirements. Thus, from the coding, we could identify explicit requirement utterances and those implicit in solution and outcome utterances.

6. The results

Table 1 presents the results of counting the occurrences of each category identified in the coding scheme for each design dyad, in each of the conditions. "R(Ex)" indicates explicit requirement statements. "Rxs" identifies those explicit requirement utterances where a proposed solution is implied, "V" and "X" signify whether the implied solution was considered negatively or positively with respect to the requirement. "S(Ex)" records explicit solution statements, "Sxr" identifies explicit solution statements where a requirement is implied, "V" and "X" signifying whether the solution was considered negatively or positively with respect to the requirement. O(Ex) identifies explicit outcomes utterances, and "Orxs" outcomes evaluating an implied solution against an implied requirement, while "V" and "X" identify whether the solution was considered negatively (presenting a problem) or positively (acceptable) with respect to the requirement.

Table 1 – Sum of the explicit requirement, solution, outcome and requirement related utterance

Condition 1

	R(Ex)	Rxs	V	X	S(Ex)	Sxr	V	X	O(Ex)	Orxs	V	X
G3	19	0	0	0	62	26	26	0	61	56	33	23
G7	8	3	1	2	78	40	40	0	44	37	19	18
G8	28	1	1	0	81	30	30	0	76	58	32	26
G10	9	4	3	1	51	18	18	0	51	46	19	27
SUM	64	8	5	3	272	114	114	0	232	197	103	94

Condition 2

	R(Ex)	Rxs	V	X	S(Ex)	Sxr	V	X	O(Ex)	Orxs	V	X
G2	39	5	3	2	90	41	41	0	70	55	12	43
G4	16	2	1	1	46	18	18	0	54	44	17	27
G5	19	0	0	0	51	32	32	0	61	58	32	26
G9	30	5	1	4	77	34	34	0	73	61	32	29
SUM	104	12	5	7	264	125	125	0	258	218	93	125

From inspection of Table 1, in total more requirements, solutions and outcome utterances are voiced in Condition 2 (i.e. the checklist condition) than in Condition 1. Figure 2, which

Figure 2. Proportion of explicit utterance

represents these categories as percentages of total explicit utterances, indicates a qualitative difference between Condition 2 and Condition 1 in that in Condition 2 there is relatively more voicing of design brief requirements and relatively less voicing of solutions than in Condition 1. When considering those utterances where requirements are considered either explicitly or implicitly (i.e. R(Ex), Sxr, and Orxs), again overall Condition 2 yields a higher volume than Condition 1, with the main qualitative difference being the greater relative proportion of explicitly voiced requirement utterances in Condition 2 compared to Condition 1. The main difference, then, between Condition 1 and Condition 2 is the higher number and proportion of explicit references to the design brief requirements in the latter.

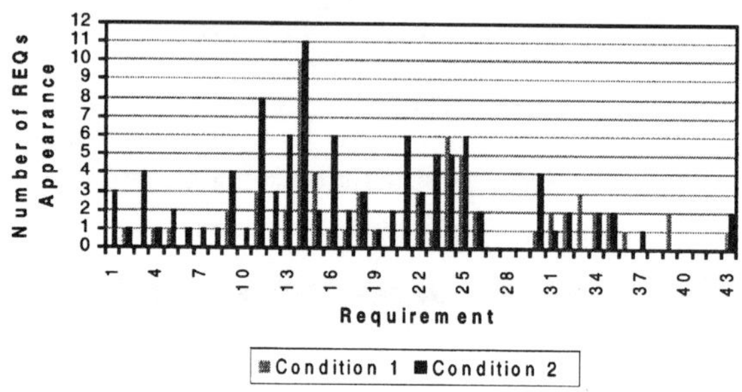

Figure 3. Frequency of explicit requirement

Figure 3, shows the frequency of explicit voicing of design brief requirements of which a greater number are voiced in Condition 2 (76%) as compared to Condition 1 (63%). Finally, Figure 4 shows the frequency of occurrence of rating scale values (i.e. 1 to 5) in Conditions 1 and 2. Overall, the solutions produced in Condition 2 yield higher rating values than those in Condition 2, indicating that designers in this condition felt their solutions matched better the design brief requirements.

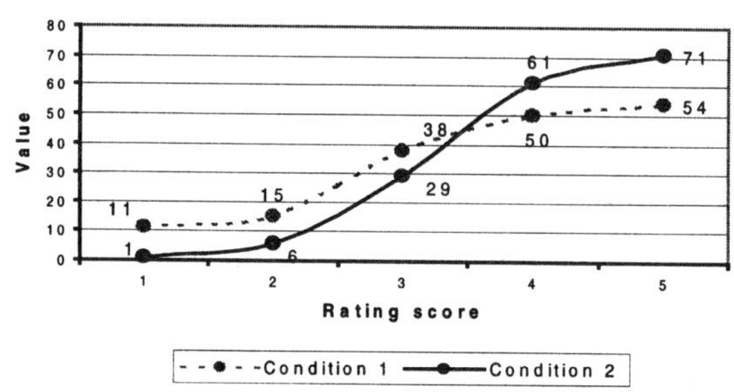

Figure 4. Sum of solution evaluation in each rating in both conditions

Conclusion

Overall, the evidence suggests that the explicit representation of design brief requirements in the form of a checklist encourages increased voicing of those requirements and their increased implicit consideration in solution and evaluation focused utterances. Furthermore, overall, the designers with access to the checklist judged their solutions to be a better match with the design brief requirements than those without the checklist. The evidence would suggest that it does matter how design brief requirements are represented and presented to designers.

However, although there are systematic differences between Condition 1 and Condition 2, the data set in each condition is relatively small and these differences could be the result of noise in the data or experimenter error. The data are insufficient to conclude with any certainty that there is or is not an effect. One way round this problem is to add to the data set by running more experimental sessions with new designer dyads in each condition.

While it would appear that those with the checklist rated their designs more highly than those without, this doesn't establish a causal link between the checklist condition and the quality of design. It could be that because the designers talked about the requirements more they somehow felt that their solution met the requirements. Alternatively, perhaps they found it easier to rate their solution against the requirements because they had seen them represented in a similar form in the checklist. We are currently considering how we might independently assess the design solutions against the design brief requirements.

Nevertheless, we conclude that the results are encouraging. Although, the designers in Condition 2 had assess to the checklist we did not interfere with their work. For example, they were not instructed specifically to read the checklist, nor were they reminded of it or encouraged to use it in any way during the design session. Whether or not they used the checklist was a decision left entirely to the designers. If systematic behavioural changes and positive benefits can be demonstrated with such little intervention, then we might expect these effects to be magnified through judicious instruction and intervention operationalised in educational and design methods.

In this respect much work remains to be done. For example, can benefits be achieved by prioritising some design brief requirements or sources of requirements (i.e. user or client) over other in the representation, or design specification? Does it matter when the requirements are presented and in what form they are represented? The goal of this work is to develop methods for representing industrial design brief requirements which can be employed effectively and efficiently by the designers in the early, conceptual stage of the design process.

Bibliography

(1) AKAO, Y.(1990) *Quality Function Deployment: Integrating customer requirements into product design*, Boston, MA, Productivity press.

(2) BAILETTI, A.J. and LITVA, P.F. (1995) Integrating customer requirements into product designs, *Journal of product innovation management,* vol.12 (pt.1), pp.3-15.

(3) BSI (1989) *Guide to Managing Product Design, BS7000*, London, British Standards Institution.

(4) BYRNE, J. G. and BARLOW, T. (1993) *Structure brainstorming: A method for collecting user requirements*, In proceedings of the human factors society 37th annual Meeting, Santa Monica, Human factor society, pp.427-430.

(5) COOPER, R.G. (1987) New products: what separates Winners from losers? *Journal of product innovation management*, vol.4, pp.169-184.

(6) CROSS, N. (1994) *Engineering design methods: Strategies for product design*, Second edition, Chichester, Wiley.

(7) CROSS, N., CHRISTIAANS, H. and DORST, K. (1996a). Introduction: The Delft Protocols Workshop, In N. Cross, H. Christiaans and K. Dorst, *Analysing Design Activity*, Chichester, Wiley, pp. 1-16.
(8) CROSS, N., CHRISTIAANS, H. and DORST, K. (1996b) *Analysing Design Activity*, Chichester, Wiley.
(9) DAVIDSON, J.H. (1976) Why most new consumer brands fail. *Harvard business review*, vol.54, pp.117-122.
(10) GRIFFEN, A. and HAUSER, J.R. (1991) *The voice of customer*, Working paper, MIT Sloan school of management, Boston, pp 91-92.
(11) HAUSER, J.R. and CLAUSING, D. (1988) The house of quality, *Harvard business review*, vol. 66 (May-June), pp.63-73
(12) INDUSTRIAL DESIGN SOCIETY OF AMERICA (1992) *Directory of industrial design*, Great falls, IDSA.
(13) KIM, S.H. (1990) *Essence of creativity*, New York, Oxford university press.
(14) KING, B (1989) Better designs in half the time: Implementing quality function deployment in America, Methuen, GOAL/QPC.
(15) LOGAN, R. J. (1997) Research, design, and business strategy. *Design management journal*, Spring, pp.34-39.
(16) LAWSON, B. (1990) *How Designer Thinking: the design process demystified*, Oxford, Butterworth Architecture.
(17) MAIDIQUE, M.A.and ZIRGER, B.J. (1985) The new product learning cycle. *Research policy*, vol.14, pp.299-315.
(18) MONTOYA-WEISS, M.M. AND CALANTONE, R.D. (1994) Determinants of new product performance: A review and meta-analysis. *Journal of product innovation management*, vol.11 (pt.4), pp.394-417.
(19) NEDO (1979) *Product design*, a report by K.G. Corfield, London, National economic development office.
(20) PAHL, G. and BEITZ, W. (1996) Engineering Design, Wallace, K. ed., London, Springer-Verlag.
(21) PUGH, S. (1991) *Total design*, Reading, Addison-Wesley.
(22) ROOZENBURG, N.F.M. and EEKELS, J. (1995) *Product design: Fundamentals and methods*, Chichester, Wiley.
(23) SCRIVENER, S.A.R., CLARK, S.M., CONNOLLY, J., GARNER, S., PALMEN, H., SCHPPO, S. and SMYTH, M.G. (1992) ROCOCO: Phase 1 report. LUTCHI Research Centre, Loughborough University, Leicestershire.
(24) SURI, J.F. (1993) *Observation and Invention: The use of scenarios in interaction design*, CHI '93 and interact '93, Tutorial note 18.
(25) ULRICH, K. T. and EPPINGER, S. D. (1995) *Product design and development*, New York, McGrew-Hill
(26) URBAN , G. L. and von HIPPEL, E. (1988) Lead user analyses for the development of new industrial products, *Management science*, vol.34, pp569-582.
(27) WALSH, V., ROY, R., BRUCE, M., and POTTER, S. (1992) *Winning by design*, Oxford, Blackwell.
(28) WOOD, C. (1996) Vision design: Building the user understanding and experiential lab. *Design management journal*, Fall, pp.24-31.

Experiments in the introduction of shared workspaces within a design environment

A I NEILSON and **W J ION**
Department of Design, Manufacture, and Engineering Management, University of Strathclyde, Glasgow, UK

ABSTRACT

This paper presents the current findings of on-going research[Ψ] which investigates the application and use of shared workspaces. The main objective of this research is to investigate the use of shared workspace technology within the design process and to develop guidelines for its effective introduction within a design environment. The focus of the research is small and medium sized enterprises (SME's) where evidence suggests the greatest resistance to the use of this technology exists, despite the fact that it is currently affordable and available (1). The findings of a series of ongoing experiments involving the use of this technology in live design projects within a company are presented. Emerging guidelines for effective introduction and use of such systems form the conclusion of the paper.

APPLICATION OF SHARED WORKSPACE TECHNOLOGY TO THE DESIGN PROCESS

Video and data conferencing is the terminology commonly adopted for what is described in this paper as shared workspace technology. A shared workspace typically comprises of audio and video conferencing, shared whiteboards, and an application sharing feature.

Today's designers frequently work in geographically distributed teams and typically need to share various types of information throughout the design process. Tools such as electronic-mail,

[Ψ] funded by the Design Council.

faxes and file transfer can be employed to distribute information asynchronously between remote design team members. Although the majority of design communication can be supported asynchronously, in certain instances, synchronous communication is necessary. Normally synchronous communication is achieved through face to face meetings or telephone calls. The latter, however, does not effectively support discussion involving graphical or numerical information. Shared workspaces provide the opportunity for designers who are not co-located to synchronously share and discuss such information in a wide variety if formats such as drawings, calculations, documents.

Research Motivation

A range of benefits are typically achieved by companies adopting shared workspace technology to support the design activity. These range from improvements in work flow to reduction in lead time and a stronger sense of team membership (2&3). Despite these benefits and the fact that the technology is commercially available and affordable the uptake of shared workspace systems within small and medium sized enterprises (SME's) has been low. A number of technological and organizational barriers have been identified as contributing to this (2,3,4 & 5). Additionally, there is a strong need to investigate the use of shared workspace technology within a design context in order to fully understand its full potential and advise on best practice. This is the basis of on-going research at the University of Strathclyde which is funded by the Design Council, part of which is described in this paper.

RESEARCH PROGRAMME

Research Approach

The main objective of the research is to develop a framework for the effective introduction and use of shared workspaces within SME's involved with the design process. The methodology being adopted involves running a number of live experiments within a company. These experiments are being conducted on real design projects in an attempt to obtain results that reflect factors experienced under normal working conditions. A number of experiments are being run, each aimed at collecting different levels of data. The data obtained from these experiments is being validated and extended through a network of partner companies to ensure it is generic and not specific to the company.

The company in which the experiments are being conducted is a mechanical and electrical building services consultancy with a number of offices located throughout the UK, specifically Glasgow, Edinburgh, Inverness, Birmingham and London. The company takes advantage of the widespread location of their offices by adopting what they call 'gateway working'. This involves the delegation of various tasks between two or more design offices throughout the design process depending upon resources and location of the project. This method of working facilitates the company in procuring contracts throughout the UK and allows resources to be used efficiently. Conventionally, the company employs the use of telephone, fax, file transfer, groupware, e-mail, post, and meetings, to support the information sharing demands of gateway working.

Overview of the Experiments

A series of three experiments have been devised and conducted in the company. Each new experiment is based on the findings of the previous and is aimed at generating more detailed data. The first experiment was designed to collect general information such as user reaction and frequency of use following the introduction of a shared workspace system into two of the company offices. Further data was collected during the second experiment allowing information relating to the success of interaction during system usage to be gathered. Experiment three extends data collection to a further dimension providing information on personal, project and cost benefits. Data was captured during the experiments through interviews, questionnaires, observation and system logs. Table 1 summarises the experiments and data captured.

Table 1. Summary of the Experiments

Experiment & Duration	Location & Access. of Shared Workspace System	Number of Projects	Summary of Data Captured
1 6 weeks	Communal area, accessible by all members of staff - separate from the main working area	All projects - between London and Glasgow	The system was used five times during this period. It was typically used to: • discuss project progress; • clarify design details.
2 4 weeks	Communal area, accessible by all members of staff. Trolley mounted for flexibility	Five specific projects between Glasgow and London	Usage increased to nineteen session and included the following activities: • discussion of drawing standards; • discussion of project progress; • clarification of design details; • management meetings.
3 Ongoing	Desk of an individual user within his personal office.	Two specific projects - between Glasgow and Birmingham	Automatic on-line system monitoring is being used to log detail of system component use, providing detailed data for cost/benefit analysis and guideline generation.

Discussion of Results

Experiment 1

From Table 1 it can be seen that the shared workspace was used infrequently during this period. Several reasons could be attributed to this low level of usage:

- Many designers initially felt awkward about using the system mainly due to lack of confidence in using the system and the consciousness of visibility to remote users;
- Access to the communal area could not always be obtained if other meetings were being held there;
- Lack of knowledge or confidence in initiating the system;
- Access was not available from the users' own desks or offices, encouraging use of previously adopted methods of communication, particularly, when available from their own desks;
- System location and availability of staff generally restricted use of the shared workspace system to a pre-arranged basis.

Experiment 2

Usage increased slightly during this second experiment, although remaining at a relatively low level. This can be explained in a number of ways:

- There were still occasions when the shared workspace system was not available when required;
- Some technical problems, associated principally with the quality of audio, discouraged some users;
- Many potential users were working to short deadlines and believed they did not have enough time to learn to use the system to a level at which they felt confident;
- Even though the new location was closer to the main working area users could not work at their own desk;
- Incoming calls were not always received unless prearranged, due to the fact that the shared workspace software was not always left active.

Experiment 3

In this experiment the shared workspace system was installed on the desks of one user in each of the Glasgow and Birmingham offices for use on specific projects. Detailed usage data is being collected through the use of automatic logging of the timing and duration of use of each component in the shared workspace system. This data combined with other quantitative and qualitative data generated by observation and interview will permit a cost/benefit analysis and user guidelines to be produced.

Emerging guidelines

The data collected to date has allowed implementation guidelines to be derived. The key stages in the implementation process have been defined as follows;

- Recognition of need;
- Preparation;
- Introduction;

- Adoption.

The recommended actions and resulting deliverables at each stage in the process are summarised in Table 2.

Table 2. Emerging Guidelines for the Introduction of Shared Workspace Technology

STAGE	ACTIONS	OUTPUT
Stage 1. Recognition of Need	Audit current design procedures to identify: • the key stages of the design process; • the main project participants; • types of information generated and shared at each stage of the design process; • methods currently employed for information sharing; • existing software, hardware and management systems (including network capability)	Development of shared workspace systems specification, including: • location of hardware; • users of the system; • key application areas; • network requirements.
Stage 2. Preparation	• select and procure equipment; • update network infrastructure; • identify pilot projects; • train appropriate staff.	• system installed; • pilot programme staff trained; • pilot projects identified.
Stage 3. Introduction	• use system within pilot programme; • monitor system utilisation and generate cost/benefit data; • develop systems and protocols for integration into existing systems and management procedures.	• identification of key benefits and where they can be achieved in the design process; • protocols for interaction.
Stage 4. Adoption	• review pilot system; • extend utilisation to other projects; • monitor use.	• fully implemented shared workspace system.

CONCLUDING REMARKS

This paper has reviewed on-going research which is investigating the introduction and use of shared workspaces within a design environment. The main thrust of the research addresses management issues associated with shared workspace technology. Similar to other technology (6 & 7), procuring and installing shared workspace hardware and software without addressing the management of the introduction process and system usage tends to lead to under-utilisation and inefficient system use. This has been borne out in the live experiments described in this paper. The research aims to provide guidelines and data that will assist in overcoming these difficulties and lead to more widespread adoption of the technology, where appropriate, to support design communication. An overview of the developing guidelines has been presented in this paper. These will be further developed on the basis of data gathered from the ongoing company experiments and a new series of experiments being initiated in two other companies. The project will at its conclusion provide the following:

- technical and management guidelines for system introduction and use;
- a cost benefit analysis;
- a requirement auditing tool.
- case studies.

This information will assist companies in directing the use of shared workspace technology towards those aspects of the design process where the greatest benefits can be expected. Furthermore, it will assist in the development of management procedures for shared workspace systems ensuring effective integration with existing management systems, infrastructure, and tools. Until full integration is achieved shared workspaces will not be seen as part of 'normal' design practice and consequently their full potential will not be exploited.

REFERENCES

(1) MacCallum, K.J., Ion, W.J. & Fryer, E. 1996. 'Supporting Effective Design Communication', Design Council Co-Partnership Programme Final Report, Dept of Design Manufacture and Engineering Management, University of Strathclyde.

(2) Petrie, C. Madefast Home Page. http://www.madefast.org/

(3) Top Gear. 1996. Ford Motor Company Case Study, Computer Weekly, March 7th

(4) Kumar, V., Glicksman, J and Kramer, G.A. 1994. 'A SHAREd Web to Support Design Teams', Proceedings of 3rd IEEE Workshop on Enabling Technologies: Infrastructure for Collaborative Enterprises.

(5) Chan, B. 1995. 'An Experiment in Supporting Geographically Distributed Engineering Design', EDRC Report No.12-69-95, Carnegie Mellon University.

(6) Munkvold, B.E. 1996. 'Managing IT Implementation in Virtual Enterprises', IEEE International Engineering Management Conference (IEMC96)

(7) Grundin, J. 1989. 'Why Groupware Applications Fail: Problems in Design and Evaluation', Office: Technology and People, 4(3): 245-264.

New product development in SMEs – the mediating effect of behaviour and belief

A O'SHEA, N McBAIN, and **A MORRIS**
Centre for Achievement in Manufacturing and Management, University of Sunderland, UK

Abstract

Manufacturing New Product Development Processes (NPDP) models claiming to represent "best practice" are derived from large manufacturing organisations. It is not axiomatic that they are used by, or suitable for, small and medium sized enterprises (SMEs). We have conducted grounded research, to identify what NPDP SMEs use. We present evidence that a variety of NPDP are knowingly used in preference to best practice. We suggest that choice between NPDPs is not simple and rational. SMEs match NPDP with a perception of self, market and product. A recursive pattern, similar to structuration theory, mediates NPDP (behaviour), and perception (belief).

1.0 INTRODUCTION

New Product Development (NPD) processes have been studied and modelled within academic literature for a considerable period of time. Since 1985 (1) much of this interest has focused upon Japanese "lean production" as best practice (2 and 3). This is not however the only model, or philosophy, that is presented as current "best practice". Academics such as Eisenhardt and Tabrizi present a model of "Experiential Learning" (4); Barclay, et al. propose a "sphenomorphic" model (5), and; Hart, for instance, focuses upon the integration of marketing and design functions in a multiple convergent model (6). Practitioners offer many more examples and descriptions of specific NPD processes that they have personal experience of (for instance, 7 and 8).

There is thus a large amount of published material which describe NPD processes, many of which lay claim to being "best practice". Kessler and Chakrabarti (9) identified over 70 articles published on NPD processes between 1986 and 1995. These articles have concentrated upon empirical or descriptive accounts of NPD processes within large manufacturing organizations. We will argue in this paper that these articles are based upon three paradigms and two major metatheories. We will also argue that they do not address the substantive criticisms of research into innovation and NPD raised by Andrew Van de Ven and Marshall Poole (10) and Richard Wolfe (11). In the light of this we will suggest that *Structuration Theory* (12, 13 and 14) offers a methodology that address these problems and present an empirically based model of NPD process choice based upon this theory. Our argument is based upon empirical evidence gained through grounded research within small manufacturing companies.

2.0 THREE PARADIGMS FOR NPD

A sociological interpretation of the various existing models suggests that they are based upon three paradigmatic views of organizations. The first views organizations as fundamentally controlled by the existing and emerging technologies (technological determinism). The second sees organizations as controlled by, and reactive to, the external environment (environmental determinism). The third suggests that organizations are free to choose their strategies and tactics (social choice).

These paradigms may be further reduced to two sociological metatheories – structural modernism and postmodernism. Both technological and environmental determinism owe much to structural modernism. This metatheory assumes that the external structural features of society shape and control action. Talcott Parsons, for instance, argued that individuals and organizations operated according to normative rules that were themselves determined by the social environment (15). Social choice theory is closer to the postmodernists such as Jean Baudrillard (16) and Michel Foucault (17). Individuals are free to choose their behaviour, strategies and tactics. These choices become contingent upon specific contextual features. Thus there is no general pattern or routine but rather a multiplicity of different and differentiated situations.

Van de Ven and Poole have argued that research into innovation must view innovation, and NPD, as a complex process which involves interaction between the agent (the person, or persons carrying out the NPD), action (what they do, the NPD processes that they use) and structure (organizational, market and social structures) over a period of time (10). They suggest that a theory of innovation, and hence NPD, must address these issues, their interaction and the temporal nature of innovation. Thus we can see that these three paradigms do not adequately reconcile the complex relationship between agency, action and structure. Structural modernism negates the role of the actor to a reaction to system wide requirements; postmodernism negates the role of social structure. We must find some other means of reconciling this paradox.

3.0 STRUCTURATION THEORY

Van de Ven and Poole (10) suggested that agency, action and structure may be resolved by: acknowledging the deterministic and voluntaristic aspects of social systems; clarifying the connections between the various levels of analysis; using time to form a connection between action and structure; developing new theories of action-structure relationships (17). They suggest that structuration theory is a possible means of reconciling this paradox.

Professor Anthony Giddens developed structuration theory (12, 13 and 14) to describe and understand a world where structure exists but is not so deterministic as to preclude deliberate and effective action by people. He separates agency and structure by inserting the notion of system between them. He argues that social systems are constituted by the activities of human agents, who are themselves enabled and constrained by the social properties of these systems. The structures define both the rules (the techniques, norms or procedures) that guide action (13, p. 21), and the resources (both allocative and authorative) that empower action (13, p. 258).

Giddens argues that even though structural properties make action possible, structures themselves have no reality except as they are instantiated in activity or retained mentally as remembered codes of conducts, or rights to resources:

> "According to the notion of the duality of structure, the structural properties of social systems are the medium and outcome to the practices they recursively organize." (13, p. 25).

Moreover, a human's need for ontological security leads to an actor using routine patterns of behaviour that unintentionally reproduce the structure of their world. Humans may, however, still deliberately choose a course of action, and implement it, in defiance of established rules and prevailing powers – i.e. the possibility of purposeful action, or agency, exists:

> "The purposive content of everyday action consists in the continual successful "monitoring" by the actor of her or his own activity; it is indicative of a causal mastery of day-to-day events that actors normally take for granted." (12)

And,

> "It is the specifically reflexive form of the knowledgeability of human agents that is most deeply involved in the recursive ordering of social practice. Continuity of practices presumes reflexivity, but reflexivity in turn is possible only because of the continuity of practices that makes them distinctively "the same" across time and space. "Reflexivity" hence should be understood not merely as "self-consciousness" but as the monitored character of the ongoing flow of social life. ...Purposive action is not composed of an aggregate or series of separate intentions, reasons and motives. Thus it is useful to speak of reflexivity as grounded in the continuous monitoring of action which human beings display and expect others to display. The reflexive monitoring of action depends upon rationalisation, understood here as a process rather than a state and as inherently involved in the competence of agents." (13, p. 3.)

3.1 Structuration, NPD and grounded research

Van de Ven and Poole (10) and Wolfe (11) have suggested that research into innovation within organizations must view innovation as a process with a temporal nature. Whittington (19) has suggested that structuration theory may have a significant role to play in management and organizational studies but at present lacks empirical substantiation. Mestrovic (20) goes further by explicitly criticising Giddens for an *a priori* objectivist use of structuration theory.

Our research, which we will discuss below, has used grounded theory (21 and 22) to identify what NPD processes are used and determine why they are used in preference to others. Grounded theory is a methodology for developing theoretical models based on qualitative inductive research. It is of particular relevance where there is little, if any, substantiated research. NPD within *small* manufacturing firms is one such instance. Previous research has concentrated upon NPD in *large* firms. It is not axiomatic that this is appropriate to small firms.

Grounded theory is based upon building a theoretical model from observed, or reported behaviour, in specific contexts. Testing it within the same, or similar, contexts refines the model. This cycle is repeated until the model is suitably robust for use. The methodology is therefore a process of building and refining a model from empirical data where no *a priori* assumptions may be used.

Our qualitative research has centred upon a sample of SMEs in the NE of England over a period of time. Whilst the sample size is small, we have concentrated upon six SMEs, it should be noted that we are interested in the saturation of specific types of behaviour and contexts rather than identifying generalisable traits. Our analysis leads us to present a possible model of NPD process choice. Our model is similar to structuration theory but has been arrived at by inductive analysis and not an *a priori* attempt to use structuration theory.

3.2 Our Model of NPD Process and choice

We suggest that the organizations we have researched choose to use their respective NPD processes due to a process of rationalising[i]. We will now expand upon this statement.

Giddens' theory of structuration may be represented diagrammatically (Figure 1). This shows both how an agent's reflexive monitoring is connected to social structure *and* how the structural properties of the social system act as both medium and outcome of the agent's actions. Giddens refers to the latter as the *duality of structure*.

Figure 1 Anthony Giddens' Theory of Structuration

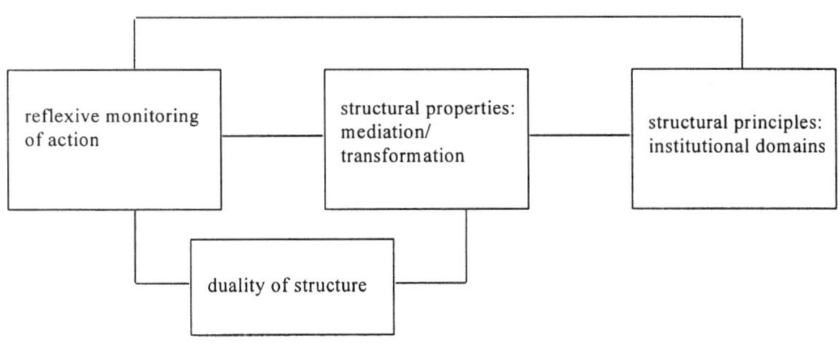

(Adapted from Giddens, (12, p. 191).

In Figure 2, we present our conceptual model of how and why our SMEs choose between NPD processes. It may be seen that this model assumes the basic features of Figure 1.

Figure 2 How SMEs Choose between NPD Processes

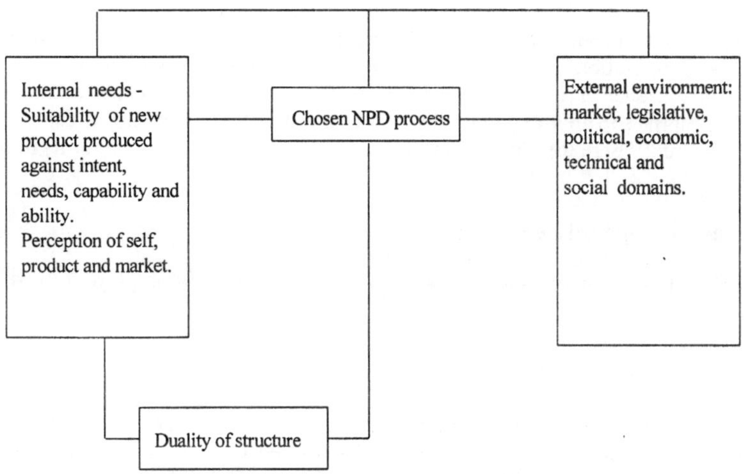

We will now explicate Figure 2 and present empirical data from our research to substantiate it.

4.0 THE EMPIRICAL DATA

Our evidence is based upon on-going longitudinal, grounded methodological research (21 and 22) conducted within six, small manufacturing firms in the NE of England. We will provide evidence here from two of these organizations to support our belief that the NPD is dependent upon reflexive choice. It may be argued that such a small sample lacks statistical validity. Grounded research is however concerned with specification rather than generalisation and particularly the development of a theory that explains behaviour in a specific situation. Generalisability may be tested after theory development and the theory developed, and further generalised, by extension to other situations. We will look at empirical data gathered from two of these SMEs for the reminder of this paper. These SMEs are similar in size, age, turnover and geographical location. They produce the same technical product and sell it to the same customer base.

4.1 The duality of structure and the social system

We suggest in Figure 2 that an organization's internalised beliefs have an important effect upon its choice and use of NPD processes. This interacts with the NPD process by a system of reflexive monitoring to form the basis of the duality of structure. The NPD process is both the medium of reifying the internal beliefs *and* is an outcome.

This duality however is affected by and affects the wider social structure over a long period of time; the *long duree* in Giddens' words (13). Thus a new technology produced by an organization will eventually have an affect upon the external social world. For instance, the invention of the personal computer has had an affect upon, and changed frames such as education, commerce, communications, social interactions, and so on. These different social frames (23) interact with and affect the development, use and perception of the technology.

In our first organization, Company A, the dominant coalition (24) within the organization believe that it operates in a static, unchanging world. It believes that the technology and processes appropriate to this industry have not changed and are unlikely to change in the foreseeable future. Where they have changed it discounts it as inappropriate for its needs and its industry.

> "We're just looking [for] steady, steady progress."

And,

> "I don't think that [their manufacturing process] has changed a great deal since its infancy."

And,

> "The technology that we actually need doesn't really change."

This static world consists of organizations that it believes are similar to itself, share a sense of mutual purpose, and which it both trusts and is trusted by – in effect a form of social "family". It views organizations that do not abide by this family's rules as outsiders who will either change or not survive.

> "We're in it together, for the North East or Durham for that matter. I don't think you will find much infighting with local moulders. You do get new kids on the block, like X in [names a location], who are not very approachable. We've tried and they're too big, or think that they are, or whatever so ...[pause]. But in reality all of us who have been here for a number of years are approachable. It's a way of life really. It's a special little industry that gets along with itself and by itself. That's why we tend to use each other as a sounding board."

And,

> "I suppose we are family orientated, if you hurt one of us you hurt all of us."

It is certainly interesting that the organization's dominant coalition (24) and gatekeepers (25), the people who identify what products to produce and decide how to produce them, have always worked in both this industry and this region and have a limited breadth to their social frames. The educational frame, for instance, is limited to experience of local, NE schools and colleges.

> "[The MD] basically came up through the ranks like the lot of us ... You know who you are working for."

The second SME, Company B has however a rather different view. It believes that its market is changing and is very competitive and that the technology itself has and is changing rapidly. It believes that it must use high technology to differentiate itself from its competitors.

"We want to be first to use new technologies."

And,

"I would say that all the moves [customer driven changes] are much, much quicker nowadays."

And,

"We have recently invested in a novel machine [to the company and the industry]. We've had to rethink the process around it to utilise all of its possibilities. We shouldn't have blinkers around our eyes and stare in one direction. There are always other possibilities."

It believes that employees must be involved in organizational decisions because they will enact the decision. It is able to describe a more sophisticated and complex intra and inter organizational political and social world than Company A.

"...rather than [management imposing] ideas from above, [they] have to come from the organization itself."

Company B was not a NE company originally. Its former parent is Norwegian and many of its senior staff are from continental Europe and have been educated in Europe to post graduate levels. Many of these staff have vocational experience gained from organizations outside of this industry and the NE of England. This change of country has had an effect:

"There were growing pains in the beginning. Things didn't go as well as planned. Then we learned how to do things differently."

4.2 Comments

In the first example, Company A, there is a clear belief in a stable, unchanging environment. This is supported by a limited social system that is established upon and supports a narrow socio-spatial world. It is not surprising then to find that this organization uses an NPD process that supports this. It has two NPD processes, the first is based upon producing new products that rely on past experience. This process results in new products that represent incremental advances upon existing ones. Furthermore, there is no mechanism that exists to transfer experience of producing the new product to existing processes or products. In essence, there is no "double loop learning" (26), no organizational learning, no mechanism to question a belief in a stable world. Its second NPD process is a simple form of copying the actions of others. When the company has to introduce a new process or technology, i.e. where it is required by a major customer, it identifies a local company in its "family" and copies their process verbatim. The company is unlikely to have its vision of a stable world challenged by countervailing information by using this process.

In the second example, Company B, an NPD process exists that is similar to the NPD funnel of Wheelwright and Clark (27 and 28). It differs from this model however in several important ways. First, its NPD process includes a mechanism for "learning by doing". The company does not view the NPD process as limited to a specific time and place (the project world of the compression model). It believes that both a production process and product may be refined with use. Its NPD process includes a mechanism for continual review and change. Second, the company's believes in employee involvement. The NPD process specifically requires employees both to be involved in and to *agree* to an NPD. The NPD process therefore requires that management negotiates with and gains the commitment of its staff. It assumes that the needs, desires and opinions of its staff are different between functions, hierarchical levels and even length of tenure. This NPD process is more complex and socially mediated than those of Company A. It is interesting that Company B has a more complex social system and that one of its first experiences of manufacturing in the UK lead to it deciding that its European history was not entirely applicable to the UK in general and the NE specifically. "There were growing pains in the beginning. Things didn't go as well as planned. Then we learned how to do things differently." The company's formative experience of manufacturing within the UK conflicted with and upset its established Norwegian manufacturing world experience.

Both companies are aware that "best practice" NPD processes have been identified and modelled. Neither however believe that they follow best practice. When Company A was questioned about why it used its NPD processes rather than any form of "best practice" it replied,

> "There are new processes that come onto the market but ... they are not competitive with [what we do]. They never will be, well not in my lifetime I'd imagine. So, well if there was something special we wouldn't get to see it, for a kick-off. I also don't think we are the sort if company that would latch on to it and use it to destroy someone else. I don't think so. *We are not that way inclined.*" (Their emphasis.)

Company A does not believe that "best practice" NPD process models are appropriate to it. Perhaps this not too surprising given its belief in a stable, unchanging world – one in which it has operated successfully for 20 years.

5.0 CONCLUSIONS

Throughout our research the dominant coalitions within the organizations have explained their company's vision and philosophies to us. They have explained how they view their products, markets and what NPD processes they use. The NPD processes are reified examples of action within these companies. In all instances there is a match between the type of action and the dominant coalition's expressed philosophy, vision, etc. This supports the duality of structure that forms the basis of structuration theory.

We believe that evidence from our grounded research shows the importance of socio-technical research. Our work establishes empirical evidence to support a potential theoretical model of structuration at a meso-organizational level. This has important consequences for both academia and industry. For the former it confirms Van de Ven and Poole's assertion (10) that research into innovation processes must resolve the paradox of structure and agency. For the latter it suggests that an organization's ability to identify and implement technologies, processes, operations and practices may depend upon a series of complex, interconnected social issues. We believe that further work is required to substantiate and refine our model both with regard to its specificity and generalisability. Our current research is now focused upon refining its specificity. We hope that future research will test and establish its generalisability.

REFERENCES.

(1) Imai, K., Nonaka, I. and Taeuchi, H. (1985). Managing the new product development process: How Japanese companies learn and unlearn. In, *Readings in the Management of Innovation.* Tushman, M.L. and Moore W.L. (eds.) 2nd edition, Harper Business, 1988.

(2) Lamming, R. (1993). *Beyond Partnership.* New York: Prentice Hall.

(3) Womack, J.P., Jones, D.T. and Roos, D. (1990). *The machine that changed the world.* New York: Harper Perennial.

(4) Eisenhardt, K.M. and Tabrizi, B.N. (1995). Accelerating adaptive processes: Product innovation in the global computer industry. *Administrative Science Quarterly*, 40: pp 84-110.

(5) Barclay, I. Holroyd, P. and Poolton, J. (1994). A spenomorphic model for the management of innovation in a complex environment. *Leadership & Organization Development Journal,* V15 (7): pp 33-44.

(6) Hart, S.J. (1994). The multiple convergent processing model of new product development. *International Marketing Review*, 1 (1): pp 77-92.

(7) Meyer, C. (1993). *Fast cycle time. How to align purpose, strategy, and structure for speed.* New York: The Free Press.

(8) Stalk, G. (1988). Time: The next source of competitive advantage. *Harvard Business Review*, 66 (July-August): pp 41-51.

(9) Kessler, E.H. and Chakrabarti, A.K. (1995). Innovation speed: A conceptual model of context, antecedents, and outcomes. *Academy of Management Review*, 21 (4): pp 1143-1191.

(10) Van de Ven, A., and Poole, M.S. (1989). Methods for studying innovation processes. In, *Research on the management of innovation: The Minnesota studies.* Van de Ven, A.H., Angle, H.L. and Poole, M.S. (Eds.) New York: Harper and Row.

(11) Wolfe, R., (1994). Organizational innovation: Review, critique and suggested research directions. *Journal of Management Studies.* 31(3): 405-431.

(12) Giddens, A. (1993). N*ew rules of the sociological method (2nd. edition). Cambridge: Polity Press.*

(13) Giddens, A. (1984). *The constitution of society.* Cambridge: Polity Press.

(14) Giddens, A. (1992). *The transformation of intimacy: Sexuality, love and eroticism in modern societies.* Cambridge: Polity Press.

(15) Parsons, T., (1971). *The system of modern societies.* Englewood Cliffs: Prentice Hall.

(16) Baudrillard, J., (1983). *Simulations.* New York: Semiotext(e).

(17) Foucault, M. (1975). *Discipline and Punish: The birth of the prison.* Tr. A. Sheridan (1977). London: Penguin Books Ltd, 1991. (Originally, *Surveilleur et punir: Naissance de la prison.* Editions Gallimard.)

(18) Jones, O. (1997). Structuration and management theory. *Working Paper Series.* Aston: University of Aston.

(19) Whittington, R. (1992). Putting Giddens into action: Social systems and managerial agency. *Journal of Management Studies,* **29** (6): pp. 693-712.

(20) Mestrovic, S.G. (1998). *Anthony Giddens. The Last Modernist.* London: Routledge.

(21) Glaser, B.G., and Strauss, A.L. (1967). *The discovery of grounded theory.* Chicago: Aldine.

(22) Strauss, A.L. (1987). *Qualitative analysis for social scientists.* Cambridge: Cambridge University Press.

(23) Goffman, E. (1974). *Frame Analysis.* Cambridge, MA: Harvard University Press.

(24) Prahalad, C.K. and Bettis, R.A. (1986). The dominant logic: A new linkage between diversity and performance. *Strategic Management Journal,* 7: 485-501.

(25) Allen, T. . (1977). *Managing the flow of technology.* Cambridge: MIT Press.

(26) Senge, P.M. (1990). *The fifth discipline: The art and practice of the learning organization.* New York: Doubleday.

(27) Wheelwright, S.C. and Clark, K.B. (1992). *Revolutionising product development.* New York: The Free Press.

(28) Wheelwright, S.C. and Clark, K.B. (1994). Accelerating the design-build-test cycle for effective product development. *International Marketing Review*, 11 (1): pp 32-46.

(29) Aronson, E. (1978). The theory of cognitive dissonance: A current perspective. In, *Cognitive theories in social psychology.* Berkowitz, L. (Ed.) New York: Academic Press.

---

[i] Unlike Giddens we adopt the term "rationalising". We do not assume that an organization's behaviour should be rational in terms of maximising economic return. Instead we adopt the term rationalising from the psychologist Eliot Aronson, who said, "Thus dissonance theory does not rest upon the assumption that man is a rational animal; rather, it suggests that man is a rationalizing animal – that he attempts to appear rational to others and himself." (29).

Integration of fuzzy logic and neural networks to support decision making in engineering design

P Y PAN, K CHENG, and **D K HARRISON**
Department of Engineering, Glasgow Caledonian University, UK

ABSTRACT

Engineering design is a most difficult task which heavily relies on the designer's knowledge and experience. In recent years, AI techniques have been used to assist the design process but mainly rule-based expert systems (RBES) are used. While the RBES approach is effective for engineering design, it has drawbacks such as rigidity and limitation in knowledge representation. This paper proposes a novel approach to automating design decision making using integration of Artificial Neural Networks (ANNs) and Fuzzy Logic (FL). The advantage of this approach is that the system developed on this approach can not only deal with certainty and uncertainty inputs, but also produce an optimal design solution based on the computing intelligence stored in the networks.

1. INTRODUCTION

Engineering design is a most active research area that the researchers have worked on for many decades. Different strategies and mathematical or empirical models for different applications can be found in the literature. In the last two decades, there has been an increasing application of artificial intelligence (AI) for the assistance of engineering design. It is considered that design decision making supported by incorporating AI techniques could be an effective means to meet the challenge of various design requirements(1).

Engineering design, however, is a very complicated process. At the early stages of the design process, it is usually impossible to obtain precise information upon which to make sound straight forward design decisions. The reason is that imprecision is an integral part of the design process, mainly from the intrinsic vagueness of a preliminary, incomplete design description. The design process normally goes on iteratively until the final result is precise. The design iterations, however, sometimes are so complicated that it is too difficult to explicitly describe. In addition, a human is generally ambiguous when he/she makes a decision(2). Thus, there are at least two important issues that should be tackled in order to develop an effective design support system: one is to cope with the certainty and uncertainty inputs, and the other is to implement the iterations in the design process without bias.

In recent years, AI techniques, especially the use of a rule-based expert system (RBES) approach has been strongly recommended by various research workers as a means for computerisation of the design process(3). This is mainly because of the very nature of the design process. Engineering design requires engineering judgement, intuition, experience, and creative abilities in addition to mere number processing routines(3). The RBES approach has the capability to incorporate some of the above-mentioned requirements for the development of a design decision making support system. Many attempts to develop an RBES for this kind of design aid have been reported in recent years. While the RBES approach is an effective AI technique for engineering design, it has drawbacks which are listed as follows(4):

- It lacks the learning capacity and the rules are required to be explicitly stated;
- It requires exhaustive knowledge representation/inference engine programming.
- It is very difficult to incorporate the number processing routines.

Therefore, the need for an approach to overcome these drawbacks without losing the capability of an AI based system is critical. This paper proposes a novel approach to assist the decision making in engineering design using the integration of Artificial Neural Networks (ANNs) and Fuzzy Logic (FL) with conventional design techniques. The architecture of the approach is presented following a discussion of the engineering design processes. The application of the approach for the design and selection of journal bearings is described, which is a design support system under development at Glasgow Caledonian University.

2. ENGINEERING DESIGN PROCESS AND ASSOCIATED COMPUTING INTELLIGENCE SUPPORT

Normally an engineering design process consists of the following design phases as shown in Figure 1, although different applications may have different aspects which should be taken into consideration:

- Clarification of the design
- Conceptual design
- Embodiment design
- Detail design

Clarification of the design

This phase involves the collection of information about the requirements to be embodied in the solution and also about the constraints in a design. It is followed by the drawing up and elaboration of the detailed specification, such as:

- What is the problem really about?
- What implicit wishes and expectations are involved?
- Do the specified constraints actually exist?

Design objectives are established by the constraints and environmental requirements imposed by the application. Obviously, some of the constraints and requirements may be vague and uncertain at this stage. Computer based support to clarify the design specifications is useful especially for a large design project. The support may include design information research through CD-ROM or other electronic media, project verification by using engineering tools and expert systems, etc.

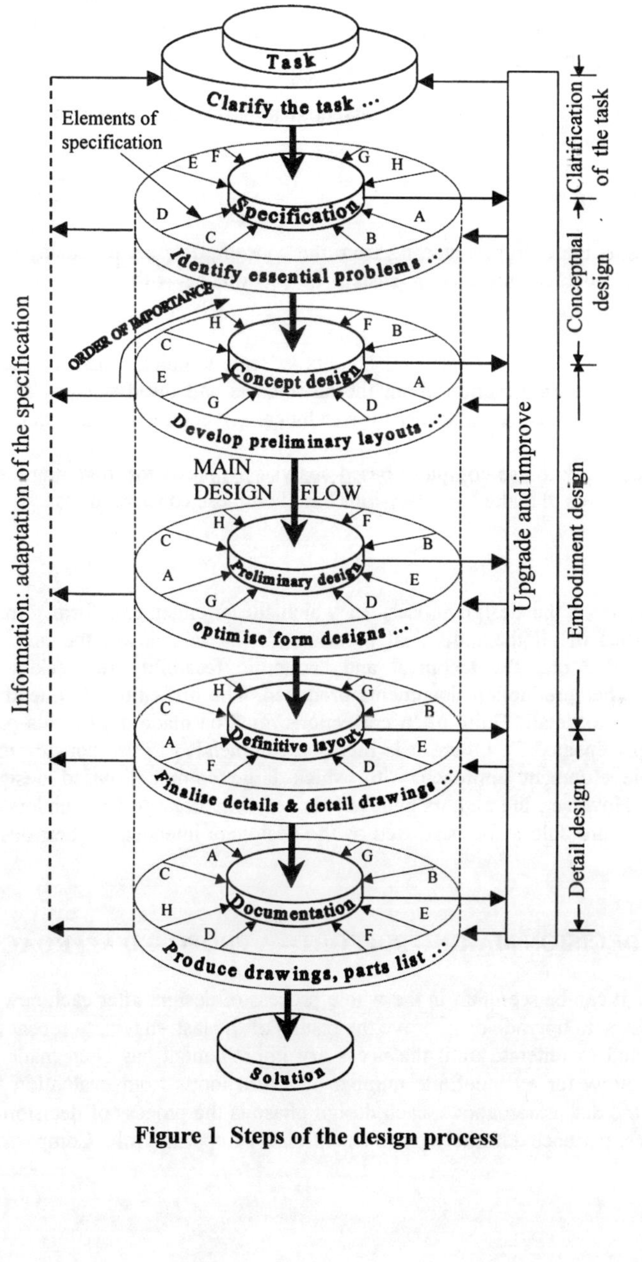

Figure 1 Steps of the design process

Conceptual design

The conceptual design phase involves the establishment of function structures, the search for suitable solution principles and their combination into concept variants. The purpose of this phase is to attempt to avoid the situation where a simple better alternative is overlooked. The concept variants that have been elaborated must now be evaluated. Variants that do not satisfy the demands of the design specification have to be eliminated; the rest must be judged by the methodical application of specific criteria. On the basis of the evaluation the best solution concept can be selected.

Computing support is critical and very essential at this stage, which may include using various analysis/synthesis tools.

Embodiment design

During this phase, the designer, starting from the concept, determines the layout and forms, and develops a technical product or system in accordance with technical and economic considerations.

It is often necessary to produce several layouts to scale simultaneously or successively in order to obtain more information about the advantages and disadvantages of the different design's variants and thus pave the way for a technical and economic evaluation.

Clearly it is necessary to use computer-based analysis/synthesis for modeling the knowledge that enables designers to make such decisions based on these considerations.

Detail design

This is the phase of the design process in which the arrangement, form, dimensions and surface properties of all the individual parts are finally laid down, the precise values of parameters worked out, the technical and economic feasibility re-checked and all the drawings and other production documents produced. The difficulties frequently arise from lack of attention to detail. Quite often corrections must be made during this phase and the preceding steps repeated(5). Often, it is highly recommended to use concurrent design and the product development approach with which large computer based design tools are indispensable. However, the authors believe it is very important to fully understand how the computing tools are able to be best used as the designers intended in their design decision making.

3. DESIGN DECISION MAKING SUPPORT – A PROPOSED APPROACH

From Figure 1 it can be seen that in the whole process of design, after each new step, it may become necessary to upgrade or improve the results of the last; that is, to repeat it at a higher design level, and to reiterate until the necessary improvement has been made. The design process must allow for an indefinite number of return loops from evaluation to synthesis. According to the discussion above, each design phase is the process of decision making and the designer's experience definitely plays an extremely important role. Compared with a new hand, for example, an experienced designer could more easily determine a solution in less time. Thus, the more knowledge and experience a designer has the more successful a solution he/she could obtain. From this point of view, a design decision making process is in principle similar to the way that a neural computing network works.

Artificial Neural Networks (ANNs) have been applied to solve problems with ill-defined, multi-parameter input-output and nonlinear features which are common in design exercises. ANNs have a large number of highly interconnected processing elements (nodes) that demonstrate the ability to learn and generalize from training patterns or data. They, like humans, can perform pattern-matching tasks, while traditional computer architecture, however, is inefficient at coping with these tasks. It has been proven that multilayer feedforward neural networks are universal approximations(6). Recently, neural network models have been developed for multiple criteria decision making(7)(8). The underlying idea of the neural networks approach is to emulate a decision maker's inherent heuristics. In other words, the motivation is to model the decision making underlying preference structures by means of supervised learning based on sampled preference data. ANNs have at least the following advantages compared with rule based reasoning expert systems:

- Knowledge is obtained by learning from examples. Therefore an ANN can be used in problems with poorly understood domains, especially for modeling multivariable, non-linear systems. This feature is particularly relevant to engineering design.

- ANNs can provide solutions to problems which have not previously been experienced as long as the solutions lie within the same domain as the training data.

These characteristics make ANNs potentially useful for modeling the decision making process in engineering design.

Although neural networks have these advantages, they cannot directly deal with linguistic descriptions at the early design stages from the designer. Without full understanding and proper coding of the inputs, ANNs cannot carry out these features automatically. Therefore there needs resort to new AI techniques to cope with such problems.

During the last decade, fuzzy logic has found a variety of applications in various fields ranging from industrial process control to manufacturing and many others. Fuzzy logic systems base their decisions on inputs in the form of

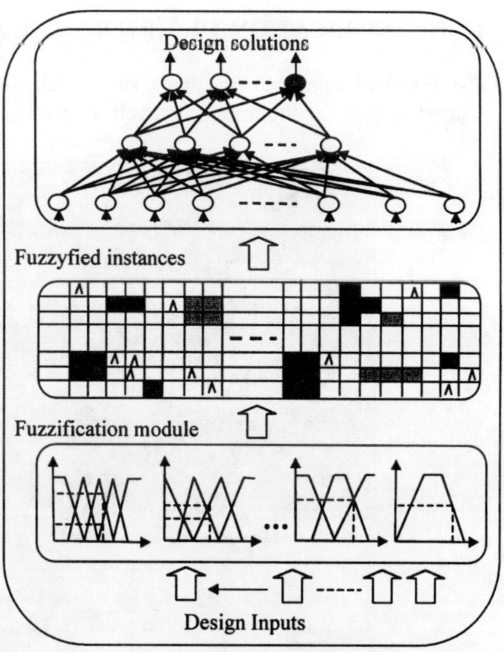

Figure 2 An integrated ANN-FL model for design process

linguistic variables derived from membership functions which are formulas used to determine the fuzzy set to which a value belongs and the degree of membership in that set. The variables are then matched with the preconditions of linguistic IF-THEN rules (fuzzy logic rules), and the response of each rule is obtained through fuzzy implication. To perform compositional rule of inference, the response of each rule is weighted according to the confidence of the degree of membership of its inputs, and the defuzzification procedure is used to generate the appropriate output.

Both neural networks and fuzzy logic systems deal with important aspects of knowledge representation, inference, and learning process but they use different approaches and have their own strengths and weaknesses. ANNs can automatically find the association between the input and the output, i.e. they can learn from sample data automatically, but lack an explanation ability. FLS are aimed at representing human cognitive thinking and perform approximate reasoning, but usually are not self-adaptive. The real power of artificial intelligence lies in the integration of ANNs and FLS(9). Although a number of papers have been written on combined ANNs-FLS techniques(10)(11), ANNs and FLS are commonly used separately for decision making(12).

The proposed integrated architecture of neural networks and fuzzy logic is shown in Figure 2. This model can accept the input of linguistic descriptions and automatically does the iteration in order to find the best solution based on the knowledge stored in the networks.

4. THE APPLICATION OF THE PROPOSED APPROACH

The proposed approach has been successfully used in developing an intelligent design support system for the design and selection of journal bearings. Figure 3 shows the front-

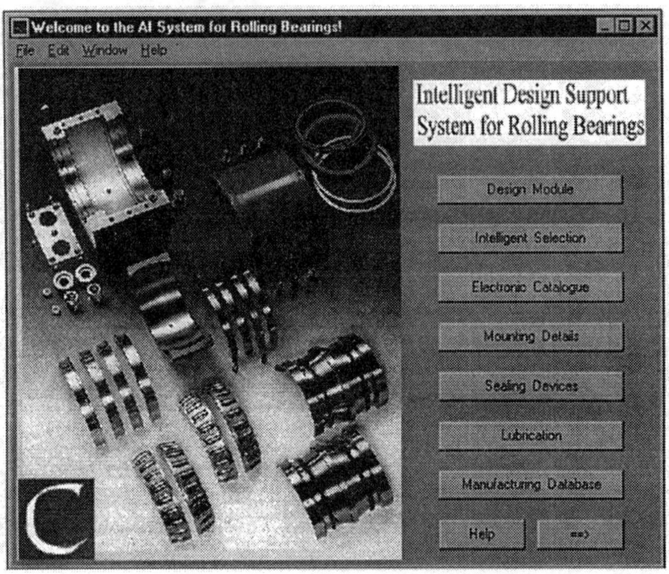

Figure 3 The front-panel of the system

panel of the developed system. In its bearing selection module, the inputs, which may be certainty or uncertainty, will be fuzzified through the fuzzification model as shown in Figure 2 and its output is used directly as the input to the ANN. The trained neural network will find the corresponding selection solution based on the input from the fuzzification model. More details about the system can be found in reference (13).

5. CONCLUDING REMARKS

The main advantages of the proposed approach to the integration of neural networks and fuzzy logic for engineering design can be summarised as follows:

- The fuzzyfication model can deal with certainty and uncertainty inputs from the designer. This feature provides the designer with a very practical and more natural interactive environment at the early design stages.

- The Neural-fuzzy-based structure eliminates the rule-matching process. This feature not only overcomes the potential risks that inference could not be executed without suitable rules existing in a rule-based reasoning expert system, but also greatly facilitates the procedure of implementing the inference engine in a decision making support system for an engineering application.

- This neural-fuzzy-based structure can learn from operating experience if sufficient training examples are provided. It distributively stores knowledge in the connections and link weights. This architecture is a natural structure to perform neural-fuzzy-based learning.

- By integrating ANNs with FL, the weaknesses of FL can be reduced by the strengths of ANNs and the weaknesses of ANNs can be reduced by the strengths of FL.

REFERENCES

1. W.B. Rowe and K. Cheng; Development of a prototype of hypermedia-based engineering design environment - HyperCAD; *Engineering Systems Design and Analysis,* Vol.64-5, ASME 1994, pp.123-129.
2. S. Masui, et al; Decision making support system for house purchase by using fuzzy logic; *Computers Ind. Engng*; Vol.27, 1994, No.1, pp.281-284.
3. H. Adeli and K.V. Balasubramanyam; A novel approach to expert systems for the design of large structures; *AI Magazine*, Vol. 9, 1988, pp.58-63.
4. A. Mukherjee and J.M. Deshpande; Application of artificial neural networks in structural design expert systems; *Computer & Structures*, Vol.54, No.3, 1995, pp.367-375.
5. G. Pahl & W. Beitz, Engineering design – a systematic approach; *The Design Council,* 1988.
6. K. Hornik, M. Stinchcombe and H. White; Multilayer feedforward networks are universal approximators; *Neural Networks*, Vol.2, 1989, pp.359-366.
7. J. Wang and B. Malakooti; Analysis and synthesis of artificial neural networks for modeling complex systems; *In Progress in Neural Networks*, Vol.2, 1994, pp.163-194.

8. J. Wang and B. Malakooti; A feedforward neural network for multiple criteria decision making; *Computers Ops. Res.*, Vol.19, 1992, pp.151-167.
9. Y. Yuan and S. Suarga; On the integration of neural networks and fuzzy logic systems; *Proceedings of the IEEE International Conference on Systems, Man and Cybernetics*, Vol.1, 1995, pp.452-457.
10. N.P. Archer and S. Wang; Fuzzy set representation of neural network classification boundaries; *IEEE Trans. On Systems, Man and Cybernetics*, Vol.21, July/August 1991, No.4, pp.735-742.
11. A Blanco and M. Delgado; A direct fuzzy inference procedure by neural networks; *Fussy Sets and System*, Vol.58, 1991, pp.133-141.
12. Y.H. Chen; Fuzzy rating in mechanical engineering design – application to bearing selection; *Proc Instn Mech Engrs, I.Mech.E.*, Vol.210, 1996, pp.49-53.
13. P.Y. Pan, K. Cheng and D.K. Harrison; A neural-fuzzy intelligent system for the selection of journal bearings; *The Mission of Tribology, I.Mech.E.*, 4 December 1997.

A definition of re-manufacturing within a taxonomy of product end of life scenarios

H J PARKINSON and **G THOMPSON**
Department of Mechanical Engineering, Manchester Metropolitan University, UK

ABSTRACT

Confusion surrounds the terminology associated with product end-of-life processes such as recycling, remanufacture, reuse and refurbishing. In order that product end-of-life strategies can be communicated effectively, an established terminology needs to be developed and adopted. The purpose of this paper is to define these terms, drawing on current literature and case studies in order to establish a taxonomy. Remanufacturing addresses the aims of industrial ecology in a more effective way than material recycling and is a thriving business in many industrial sectors. Before remanufacturing is fully accepted by the public at large, and not just by informed buyers within industry, standards to govern the processes and procedures should be established in order to ensure that the items are as-good-as-new (AGAN). In order to proceed, it is important to define remanufacturing so that it can be differentiated from merely refurbishment and other synonymously used terminology.

1 INTRODUCTION

Consumer pressure, legislation, standards, the need to maintain competitive advantage or the desire to be a good corporate citizen has forced designers and manufacturers to consider the environmental impact of their products. The amenability of a product to remanufacture may mitigate this impact **(1)**. The concept of industrial ecology is based around cyclical reuse of materials at their highest possible purity level **(2)**. The amenability of a product to remanufacture addresses this concept in a more effective way than material recycling. **(3)**. An important part of the research is to define precisely the terminology to be used. Published research and trade literature concerning remanufacture invariably uses terminology interchangeably and imprecisely, leading to possible confusion. This paper will examine how the reuse and reprocessing of products has been defined both in industry and academia. These definitions will be critically examined, in order to establish whether a logical consensus exist.

2.0 PRODUCT END-OF LIFE SCENARIOS

During its life, a product will perform some function and in the case of repairable systems, maintenance takes place. Some products may undergo some form of upgrade. When (for what ever reason) the product ceases to be of further use to the user in its present condition, the product is relinquished. It is at this stage in the product life cycle that this section seeks to address. Figure 1 shows all the likely scenarios for a product or component at the end of its useful life.

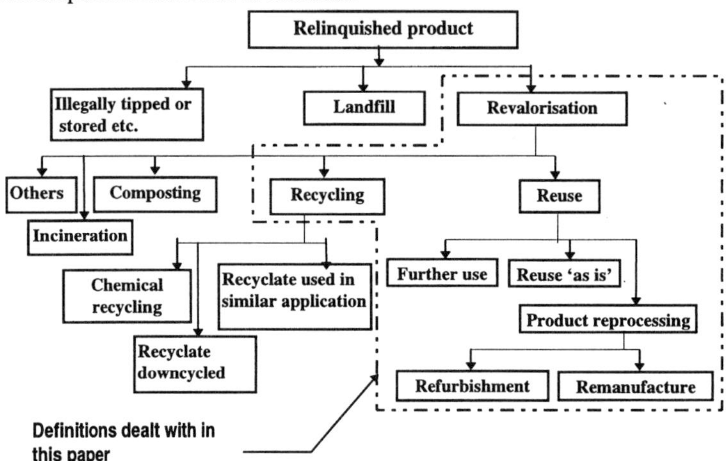

Fig 1. END OF LIFE (E.O.L) SCENARIOS

The product can be landfilled, illegally tipped or stored somewhere *ad infinitum*. These action effectively waste the energy, material and value added through manufacture. Alternatively, the product can be revalorised.

3.0 DEFINITIONS

Table 1 lists the proposed definitions which are subsequently discussed.

DEFINITIONS	
Recycling	Recycling is the process of recovering *material* after a product has been discarded.
Revalorisation	Revalorisation includes any process that seeks to recover any embedded value in a discarded product or material.
Reuse	Reuse means continuing to use an item after it has been relinquished by its previous user, rather than destroying, dumping or recycling it.
Reuse "as is"	Reuse "as is" refers to the reuse of a product with minimal reprocessing.
Further- use	Further-use is the use of a used product for a different purpose than was originally intended.

Term	Definition
Product reprocessing	Product reprocessing is the value adding activity of refurbishment or remanufacture.
Remanufacture	Remanufacture is the reprocessing of products in a such manner, that the product's quality is as good or better than new in terms of appearance, reliability and performance.
Refurbishment	Refurbishing is the reprocessing of products at minimum cost in order to ensure that the performance is within the bounds of what is considered acceptable for resale.
Repair	Repair refers to actions taken in order restore a product or component's functionality.
Reconditioning	Reconditioning is a process within remanufacturing in which used components have their condition restored to within that of the OEM specification.
Upgrade	Upgrade refers to any process that gives a product enhanced functionality
Maintenance	Maintenance is the series of actions taken to enable a product to function at a predetermined level for its intended life-span.

Table 1

3.1 Revalorisation

Revalorisation is a neologism which describes the act of recovering value. It is taken from the French language and has been coined recently in an English language text by Fussler **(4)** to describe any end of life activity which utilises any facet of the value held within a material or item. No English word has been found that has this precise meaning which can collectively describe reuse, recycling, incineration, composting etc. There is currently much research underway into novel ways of profitably disposing of waste through recycling. These methods are outside the scope of this paper, which focuses on the reuse of the product.

3.2 Reuse and recycling

Confusion surrounds the definitions of reuse and recycling. It is important to understand the differences between the two options. Reuse and recycling are often used interchangeably and synonymously. Reuse has been used to indicate all forms of material, product and component re-utilisation., **(6),(18).** The term "reuse" is frequently confused with "recycling". Avoiding such confusion in terminology is important, as the environmental impact of reuse will usually be much less than recycling **(3).**

It has been argued that industry has a vested interest in clouding the issue in order to increase and promote material usage, throughput and profit in a manner that appears environmentally conscious **(7).** Reuse means continuing to use a product, rather than destroying it (as is the case in recycling). When a car is scrapped, the chassis and bodywork are recycled, that is to say melted down into raw steel: the engine or carburettor, however may be salvaged or reused by being placed in another car. Reuse allows the integrity of the item to be retained for either; further use, reuse "as is" or reprocessing, thus recapturing the value added during manufacture. In the recycling process the energy and value added during manufacture are lost. Thus, within the concept of revalorisation, recycling indicates that items are destroyed and the materials from which they are made are used again for similar products or other purposes. This is preferable to landfilling the product, because the energy and resources expended in producing the raw materials are recaptured.

The delineation between reuse and recycling is corroborated by various sources **(1,3,8),** for example:

- "Recycling occurs when machines or subsystems have met the original design's useful life and the machine or subsystem are not remanufacturable or convertible to other useful applications. Reuse is the process of disassembling machines or subsystems to recover usable parts and assemblies for the purpose of utilising them in newly manufactured products"**(9)**.

The following quotations illustrate how confusion can arise:

- "Recycling is defined as the disassembly of the product to recover the material and perhaps components but not normally loosing its function as a system" **(18)**.
- "Recycling: This is the process of *reusing* an asset either by remanufacturing to its original state, converting to a different state, or dismantling to extract its *recyclable* contents. The scope of recycling as it applies within Xerox includes; utilisation of surplus/new assets; equipment remanufacture or conversion; parts repair; asset strip/parts recovery; precious metal recovery and sales ; and environmentally acceptable disposal"**(17)**.

It is interesting to note that reference **(9)** and **(17)** are from the same corporation.

The German document VDI 2243 1993 **(6)** represents the current state of the art in design for product end of life **(15)** and goes into detail regarding definitions, however in the English translation, the terminology refers to recycling as a product and material option "Recycling is the renewed use or utilisation of products in cycles". In this definition, recycling is used to denote all forms of utilising materials and products after they have ceased to be used. The document distinguishes product reuse from material recycling in terms of the former being the "reuse and further use" of products and the latter being the "re-utilisation and further utilisation" of materials. This means that product reuse and material recycling are distinguished by the words "use" and "utilisation". The document, having been translated, may not reflect the German authors exact meaning. It is important to define the two options of recycling and reuse separately in order to avoid confusion. VDI 2243 1993 **(6)** goes on to defines reuse as "Reuse is the renewed use of a used product (old part) for the same purpose as before, using its form without or with limited changes to some parts" which agrees with the definitions put forward in this paper. Recycling is divided into re-utilisation of the material and further utilisation. Re-utilisation is the recycling of materials from a product for use in a similar or identical application. Further utilisation denotes the use of the material in a degraded form in lower performance applications such as plastic battery casing material for car inner-wings, mixed plastics for wall sound installation. Industrial by-products such as tar from coking plants, stamping waste, casting lugs etc. are also referred to within the recycling classification. They can either be recycled for similar applications (casting lugs re-melted and used for production) or made use of elsewhere (coking tar used in asphalt, stamping waste used as ballast).

Other forms of recycling exist such as de-polymerisation, hydrolysis, electrolysis and dissolution. These have been classified as chemical recycling in the taxonomy used in this paper.

3.2.1 Reuse-"as is"

Products can be reused in different ways. Reuse-"as is" refers to the reuse of a product with minimal or no reprocessing. An example of this would be obtaining a used alternator from an automotive breakers.

3.2.2 Further-use

Further-use is the utilisation of a used product for a different purpose than was originally intended. The following are examples of this; the use of a long dead person's tombstones for a paving stone in a church yard, the use of tyres on harbour walls to cushion vessels when they collide with the structure, the use of shopping bags as garbage bags, etc.

4.0 PRODUCT REPROCESSING

If an item is to be reused, but it requires more than superficial work to render it serviceable or in a fit state for resale, then it will be subject to reprocessing. There are two approaches that can be undertaken, refurbishment or remanufacture. This section will attempt to differentiate between the two options, drawing on published studies to substantiate the distinction.

4.1 Remanufacture

The following definitions have been put forward for remanufacture:

- "Remanufacture includes the disassembly, cleaning and inspection of an item. Parts are returned to original equipment manufacturer (OEM) specifications, often through metal adding processes followed by machining. The parts are then reassembled, along with the installation of any engineering changes or improvements that have been released since the item's original manufacture. The assembled product and key sub-assemblies and components are tested, and the item receives a new warranty" **(11).**
- "Remanufacturing is an industrial process in which worn-out products are restored to a like new condition. Through a series of industrial processes in a factory environment, a discarded product is completely disassembled. Usable parts are cleaned, refurbished and put into inventory. Then the new product is reassembled from both old and, where necessary, new parts to produce a unit fully equivalent and sometimes superior to the original new product" **(12).**
- "Remanufacturing is a process in which reasonably large quantities of similar products are brought into a central facility and disassembled. Parts from a specific product are not kept with that product., rather, they are collected by part type, cleaned, inspected for possible repair and reuse. Remanufactured products are then reassembled, usually on an assembly line basis using those recovered parts and new parts where necessary. Remanufacturing then connotes a rather high volume factory arrangement similar to new product manufacturing except the parts flowing to assembly lines are reconditioned parts. The automobile aftermarket ... is an example of remanufacturing as we define it here" **(13).**
- "The returned product is disassembled to an optimal level, certain parts are removed and replaced with new parts (depending on their condition, wear rate), components are cleaned and tested and the product is re-assembled" **(8)** .
- "Remanufacturing involves the reuse of obsolete or obsolescent products by retaining serviceable parts, refurbishing usable parts and introducing replacement components (either identical or upgraded). Such a process is almost always cost effective and almost always environmentally responsible" **(2**) .
- "Remanufacturing....... can be defined as the industrial process where worn out products referred to as cores, are brought back to original specification and condition. In some cases, especially in the remanufacture of OEM automotive parts, remanufactured products exceed original specifications. The reason is that the latest engineering design and specifications, coupled with failure mode counter measures derived from failure analysis, are used instead of the original specification" **(14)** .
- "Remanufacturing is the practice of disassembling, cleaning, refurbishing, replacing parts (as necessary) and reassembling a product in such a manner that the part is at least at least as good as, or better than, new. By remanufacturing a product, the product may be returned to service with a reasonably high degree of confidence that it will endure (at least) another life cycle" **(15)** .
- "Remanufacture has been considered as the transformation of used units, consisting of components and parts, into units which satisfy exactly the same quality and other standards as new units" **(16)** .
- "Remanufacture is defined as the product is recovered and either restored to its original condition (both function and cosmetics) or its function is modified. This includes the reuse of components and materials " **(18)** .

- "Remanufacture transforms durable products which are worn, defective or obsolete to a like new or better condition through a production batch process of disassembly, cleaning, refurbishment and replacement of parts, reassembly and testing" **(19)** .
- "Remanufacturing revives a part to 'like-new' condition if not better in terms of performance and durability. In addition, the part is usually upgraded to match any technological advances **(20)** .
- "To make as close to new as possible. Wearable parts are automatically replaced. The core material is closely inspected and checked against original equipment specifications for tolerances. Replacement parts are new and made in the same production process as original equipment. Testing is performed to manufacturer's specifications and normal production standards"**(21).**
- "Remanufacturing is the process of bringing an assembly to "like-new" condition through reusing, reconditioning and replacing component parts" **(10).**
- "Remanufacture is the process of rebuilding a machine or subsystem. During this process, the machine or subsystem is disassembled, defective components are replaced and the machine or subsystem is reassembled, tested and inspected to ensure that it meets Xerox's newly manufactured machine or subsystem standards" **(9)** .
- "Remanufacturing: Manufacturing process of recycling a machine or part by disassembling and cleaning to a predetermined standard with defective components being replaced with new, or reprocessed components. The equipment is finished and all retrofits deemed as field mandatory are installed. The manufactured machine is inspected and tested to newly manufactured machine test standards" **(17)** .
- "The distinguishing feature of industrial reconditioning * of products as a production line process is the reconditioning of the individual products in large batches as series production. In this process, to some extent the product looses its original identity. The reconditioned products are of a uniform quality equal to that of the new product and also have a uniform appearance" **(6)** .

* This definition is the same as the remanufacturing definition, the word reconditioning has been translated from its German equivalent.

4.2 Refurbishing

- "Refurbishing: Returned products after a certain period of use, are slightly upgraded (i.e. certain parts maybe tested and replaced) to begin a second life" **(8)** .
- "Refurbishing refers to both the repair of damage to the part and the application of protective/aesthetic coatings" **(15)** .
- "Refurbishing is a process where products are usually brought back to some central facility for processing. However, upon disassembly a product's component parts are kept together and after cleaning, inspecting, and replacing with new parts where necessary, the original product is reassembled with most of its original parts. Other words that are sometimes used to connote the same meaning as refurbishing are reworking and rebuilding. This kind of process is not as easily arranged for mass production methods as is the case with remanufacturing. Production volume in refurbishing is usually lower than with remanufacturing. Product characteristics in addition to the volume of products will also help to determine whether refurbishing or remanufacturing is the appropriate mode of operation in a specific instance, The Xerox case ..... is an example of refurbishing" **(13)** .

The following have defined rebuilding, which has entirely the same meaning as refurbishing in this context.

- "Rebuild includes the same disassembly, cleaning and inspection steps that occur during remanufacture. Inspection focuses on identifying defective parts, which are then scrapped and replaced.

Less than original quality parts may be reused (i.e. less than original tolerances or specifications). The item is reassembled and usually tested. It may receive a limited warranty" **(11).**

- "Rebuilding -To recondition by cleaning, inspecting and replacing severely worn or broken parts. Serviceable parts are reused within the manufacturer's acceptable wear limits"**(21)** .
- "Rebuilding follows the same process.(as remanufacture) but its intent is to restore the product to its original mechanical specifications and operating condition. Only parts that are out of specifications or in need of replacement or service are handled"**(20)** .

Lund **(12)** states that in contrast to remanufacture, a rebuilt (or repaired) product normally retains its identity, and only those parts that have failed or are badly worn are replace or serviced. The article goes on to say in remanufacture "the core (the discarded product) parts must be capable of being economically repaired, refurbished or replaced, so that the original function and performance level is restored". Shu **(19)** also describes refurbishment as a stage in the remanufacturing process. This operation is referred to in this paper as either reconditioning if it is within the remanufacturing process or repair if it is part of a refurbishing process.

4.3 Repair

The definition of repair in this paper disguises it from remanufacturing and refurbishing in two ways. It can either be an action taken at the component level within the refurbishing process (see figure 2) or an action taken at product or component level during the use of a product as a maintenance action. The intent of repair in all cases is only to restore the functionality of a product, not necessarily to return its state to that of the original specification. The following definitions substantiate the definition proposed in this paper.

- "Repairing will mean those situations where individual products are brought back to a functioning state usually on a one-at-a-time basis, The process involves trouble-shooting to identify the malfunction and then, appropriate disassembly, repair and reassembly of a single product. We use the term here to mean that activity which takes place in product service centres like T.V. repair shops, automobile garages, etc., and "do-it-yourself" repair by consumers" **(13).**
- "Repairing ... covers all measures to maintain the required condition, to determine and assess the actual condition and to restore the desired condition. Repair measures serve mainly to achieve the intended service life of a product...." "Nevertheless, there is not always a clear separation between reconditioning and repair because the operation performed in each case, as well as the results, are often identical or comparable. It is more appropriate to speak of a relationship .... rather than to seek a strict delineation"**(6)**
- "Repair is usually limited to identifying and replacing defective parts. the amount of disassembly, cleaning and inspection is driven by the repair requested and the assumed or analysed problem to be addressed" **(11)** .

4.4 Traits of Remanufacture and Refurbishment.

The criteria identified in the definitions of remanufacture include; whether the product is changed or updated, a warranty is given, the products loose their identity, a metal adding or removal process is undertaken, a factory environment is used, the product is optimally or fully disassembled, large volumes or batch processes are involved, whether it is a industrial process or the reconditioned parts are put into inventory. The definitions differ regarding the terminology for the reconditioning process, using refurbish, repair and recondition.

The majority of definitions of remanufacture stress the "like new or better" condition or within OEM specification. The product may require some upgrading to compensate for any obsolescence that has occurred since the product first entered that market. The criteria "as-good-as-mew" is *the* critical parameter in defining remanufacture and is the basis of the definition proposed in this paper.

Lund **(12)** distinguishes remanufacture in terms of the quality of the reprocessed product, the scale of the operation and the loss of identity of the parts from their particular product. Another criteria that is stressed is the need for the process to be a batch arrangement **(19)(6)**, where the components lose their identity. These criteria are not essential to the definition of remanufacture. This would mean that machine tool remanufacturing on a one-at-a-time basis to a condition of at least AGAN would be precluded from being termed remanufacture. The definition differentiates on the basis of the process, however the process is a consequence of the economics associate with labour costs, necessitating a mass production approach were the price flexibility is small. For high cost items such as machine tools, there is more flexibility in price, therefore it is viable to remanufacture on a one-at-a-time basis. Research is underway to borrow production systems used in conventional manufacturing in the remanufacturing industry **(24)**. This could mean a move towards just-in-time processes (JIT) and small batch mix model remanufacture driven by subtle changes in the remanufacturing industry **(24)**, where a proliferation of parts (due to increased product variety) is rendering many remanufacturing operations uneconomic. One at a time remanufacturing has the drawback that parts cannibalisation cannot take place unless some form of metal deposition process is available or new parts (expensive) are used. This does not represent an obstacle for OEMs that remanufacture.

Overby **(13)** also differentiates remanufacturing from refurbishing in terms of the loss of identity of the components and the production volumes but no mention is made of the quality or reliability of the remanufactured product. Repair is used interchangeably with reconditioning. Overby's paper goes on to define repair in a different context, as a one-at-a-time action taken to restore the functionality of a product. Different industry sectors have their own terminology for remanufacturing and refurbishing, e.g. overhaul, rebuild, recondition, restoration etc. A range of companies were questioned from various sectors. There was consensus that the distinction between *as good as new* (remanufacturer) and *as good as required* (refurbishment) was a logical demarcation point in product reprocessing **(5)**.

As previously mentioned, the only real criteria which identifies remanufacture from refurbishment is the quality and reliability the consumer can expect. To emphasise the equality or superiority of remanufactured goods compared to brand new, the manufacturer should have guarantees and warranties of a comparable nature. "Warranties offered for remanufactured parts have been offered that are equal to or superior to warranties on brand new equipment" **(23)**. To achieve the necessary levels of quality **(20)**, standards and processes within the remanufacturing facility must be at least as thorough as those of the OEM. The terms "rebuillder" *(refurbisher)* and "remanufacturer" have caused controversy. The original equipment manufacturers (OEMs) claim "only they can be the true remanufacturers" and are trying to create a difference in the two terms even though the United States Federal Trade Commission uses the terms interchangeably. This has caused further problems with patent and copyright laws. So far, it has been upheld in US courts that the automotive rebuilder acquires the owner's right to repair the part upon purchasing the core. The OEMs would prefer to say their parts are subject to copyrights and therefore, only O.E. components should be used **(20)**.

4.5 Reconditioning

- "The task of component reconditioning is to restore the economic value or wear state of used components to that of new components. In many cases the resistance to wear is also increased..." **(6)**.

- Reconditioning is the process of restoring components to a functional and/or satisfactory condition by such methods as: resurfacing, repainting, sleeving etc..**(10).**

As stated previously, in most published work in the field of reuse and recycling, authors have tended to use terminology interchangeably. Reconditioning and refurbishing particularly have been used interchangeably. Other words are also used in an interchangeable manner, e.g. renovate, restore, overhaul etc. Reconditioning has come to be regarded as a step in the remanufacture process along with other processes such as cleaning etc. It is undertaken at the component level. In British trade parlance, remanufacturing is superseding the usage of reconditioning in the automotive industry. This seems to be an American trend, with remanufacture being used to portray a sense of quality. The relationship between component repair, reconditioning, refurbishment and remanufacture are shown in figure 2 below.

5.0 CONCLUSION

This paper has proposed definitions for product revalorisation. Published research has been critically examined and their definitions examined for the key elements which characterise the process. Remanufacture has been characterised by the consensus that it returns products to *AGAN* condition and refurbishing does not. Repair is an action taken either at the product level during maintenance or at the component level during refurbishment. Reconditioning is a process taken during the remanufacture of a product to restore the component to the original specification. The feasibility of returning products to *AGAN* is an area where further research is required

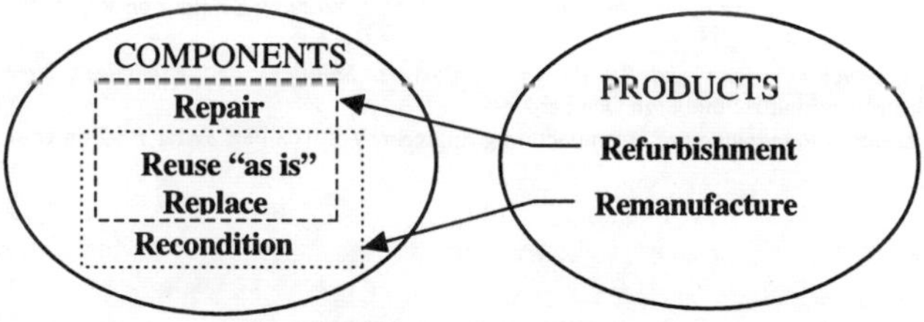

Figure 2 Product Reprocessing

REFERENCES

1. Sweatman, A. and Simon, M. "Design for Environment Tools and Product Innovation" CIRP Seminar "Eco- performance of products" Zurich. March 1996
2. Graedel, T.E and Allenby B.R *"Industrial Ecology"*.. 1995, New Jersey : Prentice Hall
3. Cooper. T *"The reuse of consumer durable in the UK: obstacles and opportunities"*, 1st Int. Working Seminar on Reuse. Eindhoven 1996 Nov. 11-33 pp 63-72
4. Fussler, C.," *Driving Eco-Innovation*" Pitman 1996
5. H. J. Parkinson and G. Thompson *"Product Remanufacture; Practices and Implications for Design"* International Design Conference - Design '98. Dubrovnik, May 19 - 22, 1998.
6. *"Konstruieren Recyclinggerechter Technischer Produkte (Designing Technical Products for Ease of Recycling)"* VDI 2243, VDI-Gesellschaft Entwicklung Konstuction Vertrieb 1993 . English translation BSI Language Services, Milton Keynes UK
7. Failie. S. *"long distance, short life, why big business favours recycling"* The Ecologist, Vol. 22 No 6.
8. Ayrers, R..T., Ferrer, G. and Van Leynseeve, T., *"Eco-Efficiency, Asset Recovery and Remanufacturing"* Working Paper, Centre for the Management of Environmental Resources, INSEAD, Fontainbleau, France, March 1997.

9. Seaver, Wilmer, B *"Design considerations for remanufacturability, recyclability and reusability of user interface modules"* IEEE 1994 Xerox Corporation Inc El Segundo, CA 90245.
10. Amezquita, T., Hammond, R.., and Bras, B., *"Characterising the Remanufacturability of Engineering Systems"* 1995 ASME Advances in Design Automation Conf., DE-Vol. 82, Boston, , USA. 1995, pp 271-278.
11. Levine, L.O., *"Remanufacturing Strategy"* Technical Paper, Society of Manufacturing Eng., MS93-258, 1993.
12. Lund R. T. *"Remanufacturing, United States Experience and implications for developing countries"* The World Bank, Washington D.C. (1984).
13. Overby C. *"Product Design for Recyclability and Life Extension"* Industrial and Engineering Systems Department, Ohio University, Ohio USA ASEE, Annual Conference. 1979. 181-196 pp.
14. Amezquita. T., Hammond, R. C., Bras, B. *"Design for remanufacture"* International Conference on Engineering Design, ICED (Prague 22-24 August 1995.
15. Bras et al *"Design for remanufacturing metrics"* 1st Int. Working Seminar on Reuse. Eindhoven 1996 Nov. 11-33 pp 35-51
16. Daniel, V. et al *"Disassembly operations for recoverable manufacturing systems"* 1st Int. Working Seminar on Reuse. Eindhoven 1996 Nov. 11-33 pp 143-151
17. Berko-Boateng .V. et al *"Asset Recycle Management- A total approach to product design for the environment"* IEEE 1994 Xerox Corporation.
18. Low M.K., Dixon et al.. *"Choice of end of life product management strategy, A case study in alternative telephone concepts."* IEEE 1996.
19. Shu L, and Flowers, W., *"A structured approach to design for remanufacture"* ASME 1993 Winter Annual Meeting. Session on Information Based approaches to Concurrent Engineering
20. Schwartz C., Chairman of Auto Parts Rebuilders Association, the Society of Automotive Engineers (SAE) held a meeting on "Total Life Cycle" in Vienna, Austria. October 1995.
21. Jasper Remanufacturing Article extracted from the Internet http://www.jasperengines.com/rmfg.html
22. North David m., *"Pratt studies more remanufacturing"* Aviation and Space Technology (Nov. 12, 1984) pp 129-133
23. Haynsworth. H. C and Lyons R T. *"Remanufacturing by Design. The missing link"* Production and Inventory Control Management, Second Quarter (1987), pp 25-27
24. Amezquita T *"Lean Remanufacture of an Automobile Clutch"* 1st Int. Working Seminar on Reuse. Eindhoven 1996 Nov. 11-33 pp 5-22 .
25. Steinhilper, R. *"Product Recycling and Eco-design: Challenges, Solutions and Examples"* International IEE/IEEE Conference on Clean Electronics. Edinburgh October 1995 .
26. Sprow, E. ""*Mechanics of Remanufacture*" Manufacturing Engineer, Vol. 108 Part 3 Mar 1992 pp 38-45

A product costing system for a make-to-order manufacturing company

D PRICE and **I McCARTHY**
Department of Mechanical Engineering, University of Sheffield, UK

ABSTRACT

This paper describes a case study of technology transfer, where a bespoke computer aided product costing system is developed for a small make-to-order manufacturing organisation. The paper will describe the existing product costing system within the Company and will evaluate the benefits and disadvantages of the system. Using knowledge elicitation techniques, the process of costing is modelled and this information is used to design and create the new computer aided product costing system.

The paper will describe the creation of this new system in terms of specification, design, validation, training, implementation and system maintenance. A justification of the technology is also given in terms of its technological and operational benefits.

1 TECHNOLOGY TRANSFER

Technology transfer is a complex process because it relies on the interaction of human beings from different organisations who operate within different cultures, have different levels of expertise and resources, are governed by different constraints and have different goals and objectives. Technology transfer is not simply a case of transporting technology from the technology provider (university) and delivering it to the exploiter (receiving organisation). In many cases the technology itself is well established, it is not new or novel, but it is new and novel to the receiving organisation. Thus, the success of the technology transfer project does not always depend on the creation or development of the technology, but rather the process of transferring or implementing the technology. Senker and Senker (1) believe that technology transfer is best characterised as "*a highly interactive and iterative process*". This is confirmed by Lopez-Martinez et al (2) who distinguish between exchange activities (e.g. licensing, etc.) and institutional instruments constructed to enable exchange (e.g. innovation centres, etc.).

The success of technology transfer is dependent on the partnerships between the individuals involved and it is important to distinguish between the acquisition of individual knowledge and the acquisition of knowledge for organisational learning, Senker and Senker (1). Many of the technology transfer projects which the Teaching Company Scheme (TCS) supports require the organisation to undergo a certain degree of organisational learning. This could be the adoption of new management practices in terms of organisational control and operation and thus, it is not sufficient for a firm only to access useful knowledge. It must be able to change by identifying information which can add value and using this information create knowledge which results in organisational learning. This problem of the transfer process and the resulting organisational learning was the one of the founding arguments for setting up the TCS initiative. The TCS was set-up to generate to growth and regeneration in large and small companies, with the overall objective of enhancing the UK economy, but an operational aim was concerned with facilitating the technology transfer, to ensure that the specified technology is appropriate and that the participating organisation has the capability not only to understand the knowledge and technology, but to accept it and integrate it within the organisation.

2 COMPANY BACKGROUND

Mandall Engineering Ltd is a traditional Engineering company with a work force of about 60 people and a turnover of around £3m. Mandall has no products of its own, but specialises in make-to-order metal fabrications to customer drawings, utilising its wide range of fabrication and machining resources. The majority of sales are to the Mining and Hydraulic Pump industries and order values range from £100 to £10'000, in the form of one offs or batches of varying size with small numbers of repeats.

Three years ago the Company, which was struggling financially, was taken over by a new owner following which a major restructuring of management took place. One of the measures introduced to improve the performance of the Company was to employ a graduate engineer via a TCS with the University of Sheffield. The first aim of the scheme was to increase the efficiency of the Production Management System (PMS) by implementing a computer aided system.

The typical route to management within the firm has been up from the shop floor and as a result of a lack of foresight and training over the years the Information Technology skills are not present to implement a Computerised PMS.

Analysis of the existing PMS, and research into the commercial software available revealed two important facts; a commercial package was way out of the allocated budget for the project and a commercial package would force major changes to the satisfactory working methods of the existing PMS, particularly the costing system. This second point would be compounded due to the complete lack of I.T. training among the staff. The decision was made to make a bespoke system in house using a relational database, Microsoft Access. The system would be produced in modules that would integrate together, the first module being Estimation / Quotation which is the basis for this paper.

3 ANALYSIS OF THE ORIGINAL COSTING SYSTEM

The existing costing system was initially analysed by reviewing the procedures in the Company's quality manual, followed by interviewing the Mandall Estimators. From the data collected, flow charts were produced outlining the steps in a concise form giving the basis of the initial technical specification for a computerised system. As well as specifying the quotation procedure it was also necessary to specify other desirable features of a modern computerised system. A problem was apparent here as the Estimators, along with the rest of the staff, had never used and only briefly observed a modern computer system running MS Windows software, which meant they could not ask for what they didn't know was possible. This information had to be planted and then extracted again from the people concerned.

Requests for quotations are sent to Mandall in the form of engineering drawings and specifications. To cost the job the drawings are methodically broken down into raw materials required, that is metal plate and section, fabrication processes, machining processes, bought in items, and subcontract work. The sum of the constituent costs is then found and a percentage added to give the price quoted to the customer. The quote is then filed away along with the drawings and any subcontract quotations obtained, from where it can be retrieved if an order is placed.

The benefits of the original paper based system are that it is both simple to learn and use and that it is reliable. The drawbacks of the system come to light generally when comparing it to the expected benefits of a computerised system; the drawbacks are outlined below:-

- Retrieval of Estimates and Quotations is time consuming. This is not helped by the fact that they are only indexed by quotation number.
- The Estimates are hand written which can lead to errors in communication.
- Once a Quotation is produced there is a delay for typing before it is sent.
- The Estimates and Quotes are stored only on paper with no backup in case of fire or any other disaster.
- The calculations to produce an estimate are done manually which is slow and can lead to unnoticeable errors.

4 SPECIFICATION OF A NEW SYSTEM

The core specification for the new system comes directly from the original tried and tested system in terms of the method of estimation. This will ensure that there are no unnecessary features incorporated in the software to make it more complicated. The system also has to be designed taking into account the fact that it will form the basis for the whole company PMS; customers and suppliers will be stored, along with job estimates which will then be transferred in to orders, the estimated process times being fed in to a production scheduler.

Other features come from the benefits and drawbacks of the original system, such as simple and easy to learn, fast quotation retrieval by multiple fields, automated printing of quotations and estimate details, automatic calculation.

5 DATABASE DESIGN

The basic methodology for database design is described by Cambell et al (3) who describes how to form the structure of your database, deciding on the appropriate tables, fields and relationships between tables. The structure for the Mandall Costing System database is shown in figure 1 and an explanation follows.

The main table in the database is the Quote table which holds the main details of each quote. Each quote has an associated customer, the details of which appear in the Customers table. The two tables, Quote and Customers are linked together via a 'one to many' relationship, that is 'one' customer can have 'many' quotations. Having two tables like this means the customer details are only entered once, rather than entering the details on every quote. For any quote there may be several items that have been requested, these are stored in the QuoteItem table which again is related to the Quote table via a 'one to many' relationship, that is 'one' quote can have 'many' quote items. Similarly every quote item can consist of many plates, many sections, many bought out items, many fabrication processes, many machining processes and many subcontract items, hence tables with one to many relationship with the QuoteItem table were created for each constituent.

A good database structure will make the database efficient in terms of speed and a minimum of duplicated dated will be stored. It was found that it is important to start off the design with the right structure as it becomes increasingly difficult to change as the database complexity increases.

Once an initial structure was formed the programming of the system began. Using Microsoft Access it is possible to make complex stand alone applications without the knowledge of a computer programming language. The main screen is the Quote form (see figure 2) which is either blank ready for a new quote or shows the details of an existing quote. Embedded into the Quote form is the QuoteItem subform which shows the individual quote items for each quote. Clicking on the Plates command button invokes a macro to automatically switch to the Plates form (see figure 3) where the plates associated with the Quote Item are added, similarly with the Sections, Items, Labour (fabrication processes), Machining and Subcontract forms.

The calculations that were previously done manually using a calculator are now done automatically once the relevant data has been inputted; for example costs are calculated and summed, profile burning times are calculated from plate dimensions and welding times are calculated. Data entry is speeded up using drop down boxes with user defined lists as shown in figure 3, and by setting the default value of a field to the most common value. Once all of the details have been added the quote and quote details can be printed at the touch of a button before being sent to the customer.

The database was made into a stand alone application, that can be used without knowledge of MS Access, by adding a Main Menu to navigate through the system and replacing the MS Access toolbars with customised ones.

6 SYSTEM VALIDATION, IMPLEMENTATION AND TRAINING

Once the prototype system was produced one of the Estimators was given basic training on its use and then began to produce estimates on it while still using the original paper based system in parallel as a backup and a check that the calculations were correct.

After various modifications and error 'debugging' the system was networked for multiple users by splitting it into two, an application database and a data only database. The application database was loaded onto each individual computer while the data only database was left on the network server allowing the individual computers to access and store data in the same place.

The simplicity of the application in that it is user friendly and uses an identical method of estimation, meant that users could be trained to use the system easily with only half an hour of tuition and help with any problems if needed.

7 CONCLUSIONS

TCS programmes are successful in both large and small firms, but the structured and monitored process of technology transfer generates greater gains for SMEs, because the competitive success of the transfer usually depends on changing a firm's culture.

The arguments and evidence presented in this paper are intended to support a conclusion that the TCS provides a linkage between SMEs and universities which enhances organisational innovation. In firms which have not previously recruited graduate scientists or engineers, a TCS programme has the potential to put in place an organisational mechanism which initiates knowledge transfer from academia. It can also play a significant role in creating more favourable attitudes to the recruitment of such graduates. In a company which already recruits scientists and engineers, a TCS programme can facilitate transfers of a specific ensemble of scientific, technological or management knowledge which may be critical to the company's future.

Many observers conclude that the success of the TCS is due to its underlying ethos, which creates organisational growth by the mechanism of structured technology transfer, but a key aspect of this process is the role of the Teaching Company Associate (TCA). The TCA is full time member of the SME, but at the same time a university employee. As a full time member, the TCA earns the trust and respect of the decision makers within the company and in many cases will undergo a significant degree of individual learning as a result. This integration into the company is vital for facilitating the technology transfer, because as Senker and Senker (1) conclude the TCAs are change agents within the company, but at the same time they are outsiders guided by outside influences (the university and Teaching Company Directorate).

Few, if any, commercial computerised production management systems cater for make-to-order manufacturing companies who make non-standard products. Implementing a commercial package would require major changes to the PMS of such a company.

The advantages of a bespoke computerised PMS, particularly a system developed in house are valuable. The system can be developed in close collaboration with the end users, making it fit the PMS requirements precisely. The system can be modified, improved and maintained on a continuing basis, quickly and cost effectively.

Microsoft Access is an easy to learn, relational database which can be used to design powerful, stand alone database applications.

The benefits realised since the implementation of the computerised costing system are mainly non tangible and are as follows:-

- Very fast quote retrieval by one or multiple criteria, Customer, Quote Number Part Name, etc.. This produces a man hour saving of approximately 6 hours per week.
- Quotations produced quicker, through faster data entry and elimination of manual calculation.
- Quotes never lost, requiring the job to be re-quoted.
- Quote printed automatically.
- Professional appearance through laser printing.
- Less confusion between departments through untidy handwriting.
- Less calculation error.
- System backed up easily, in case of disaster.
- Multiple users can view or work on a quote at the same time.

8 REFERENCES

1. Senker P and Senker J. (1994), "The Teaching Company Scheme: Transferring Technology and Expertise from Universities to Industry", *TTI Conference*, London, pp 65 - 70, July 1994
2. Johnson S. (1990), "Small firms and the UK labour market: Prospects for the 1990s", in Lopez-Martinez, R. E; Medellin, E.; Scanlon, A. P. and J. L. Solleiro (1994), "Motivations and Obstacles To University-Industry Co-operation (UIC): A Mexican Case", *R&D Management*, vol. 24:1, 1994.
3. Campbell M. (1992), "Microsoft Access Inside & Out", Osborne McGraw-Hill

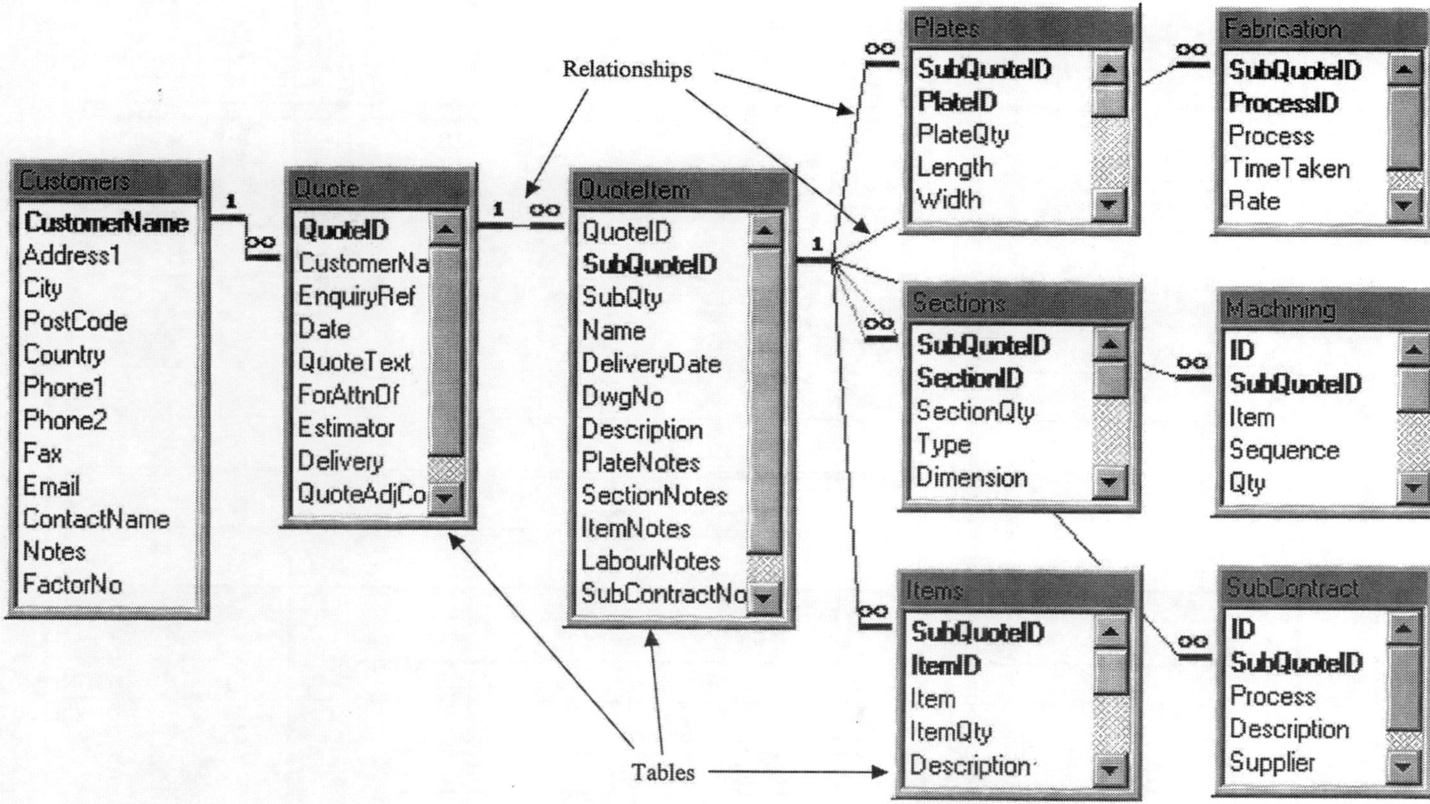

Figure 1
Mandall Costing System structure

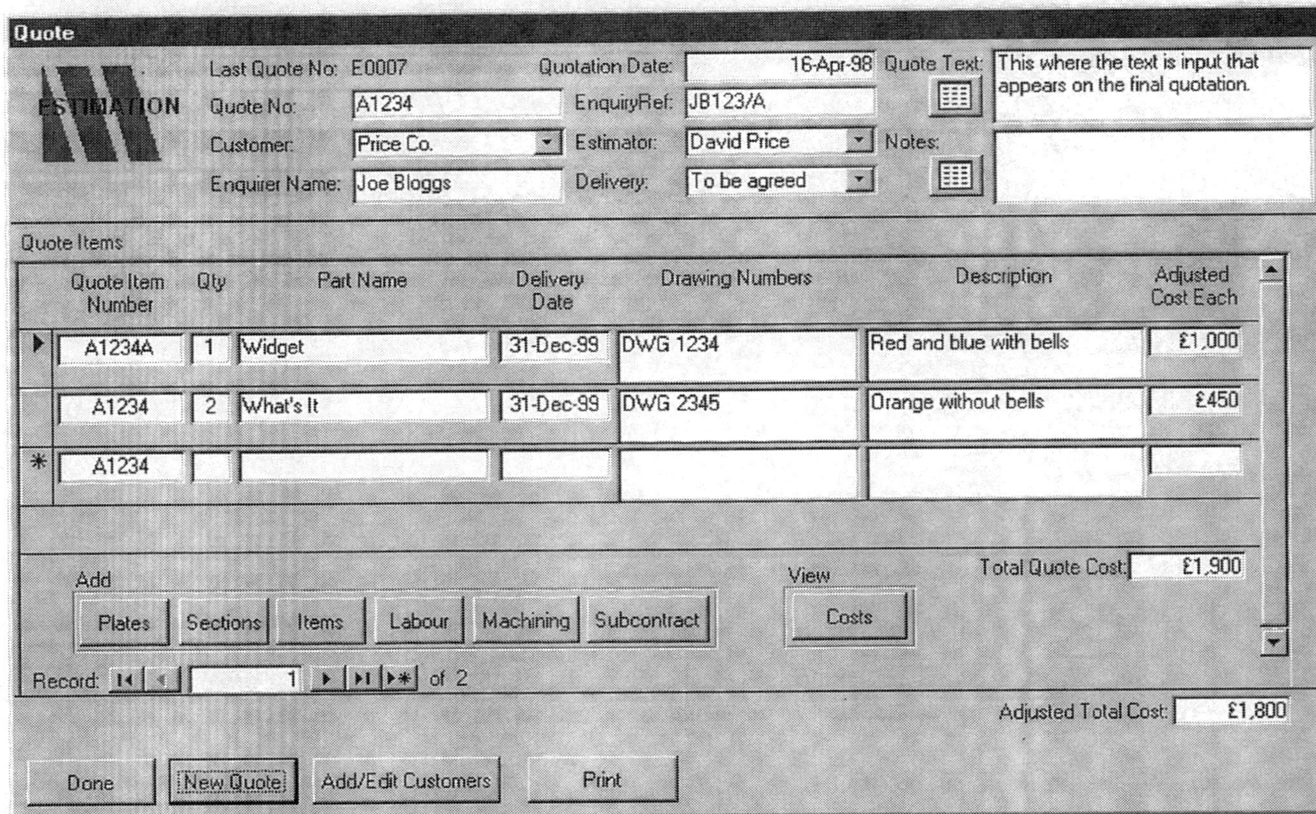

Figure 2
Quote form

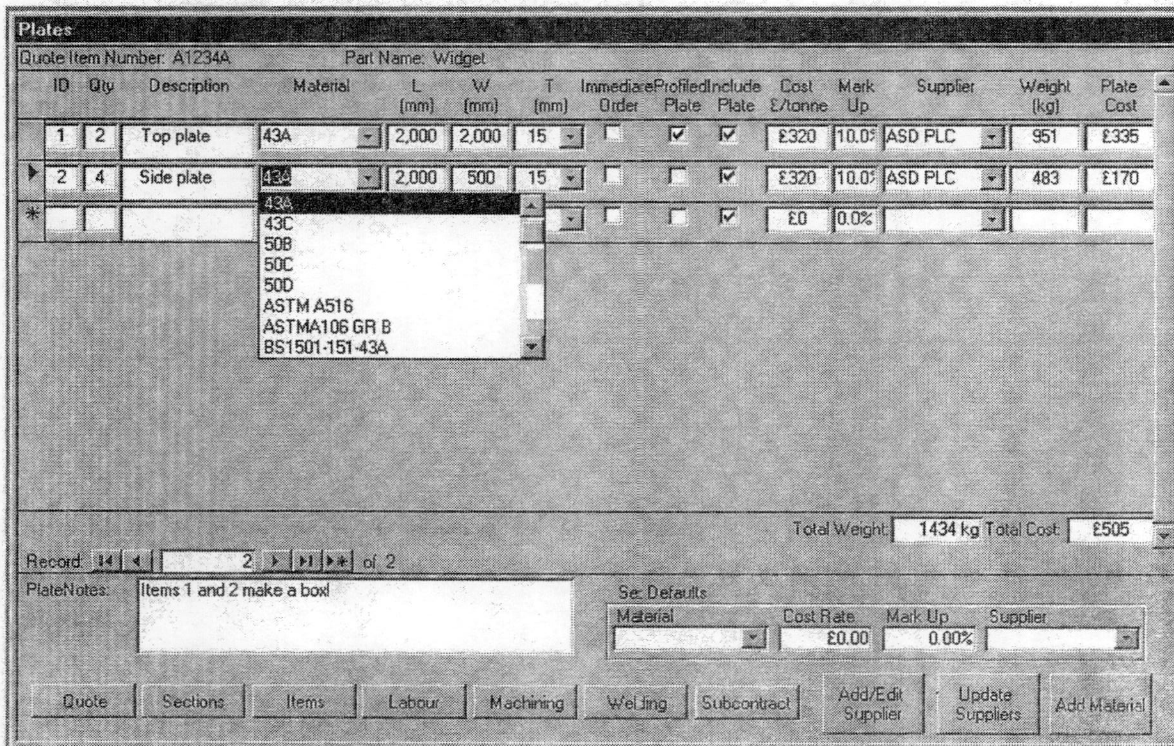

Figure 3
Plate form

Redesign of a product and its manufacturing operations – advanced technologies and technology transfer for commercial advantage

M P ROWLES and **P J FISHER**
School of Manufacturing and Mechanical Engineering, University of Birmingham, Edgbaston, UK

This paper will describe how a small manufacturing enterprise (SME) without any previous experience of advanced manufacturing techniques, gained a major commercial advantage in the redesign of its primary product. Combining the strengths of computer aided design and rapid prototyping, delivered to the company technical and manufacturing benefits and cost reduction allied to high confidence in making capital expenditure.

"If companies do not take up the technology, they will not exist"
Lavern Schmitt, Chrysler Corporation

1. INTRODUCTION

The use of modern methodologies in the rapid development of new products to market is becoming less confined to large companies. Smaller companies are benefiting from the growth of subcontractors providing specialist services at reasonable costs. Selective purchase of such services, without the need for 'in-house' specialisation, provides SMEs with a powerful means to reduce development time and costs. Bringing new products to market quickly is now seen as a major commercial imperative.

2. THE COMPANY

The SME being considered is an established manufacturing company which has been producing variants of its products for 80 years. The manufacturing workforce is divided between high and low skilled people using methodologies which date from the 1930's. Improvements have evolved rather than being initiated and familiarity has bred a high comfort factor. Product technical expertise resided in a single individual who recently retired from the company. No professional manufacturing engineer has ever been employed.

Currently the company is under intense competition both in the home and export markets. The twin problems of high costs and long delivery in a price sensitive, volatile market needed to be addressed. The company has long been recognised in its market for the excellence of its product.

2.1 The product and processes

The product and processes relate to an electromechanical device comprising 32 separate components and weighing a few grammes. The materials employed range from plastic injection mouldings to precision steel pressings and ceramic items. Assembly is extremely labour-intensive and, in part, intricate and highly skilled. No consideration was ever given in the original design to manufacture.

Manufacturing lead time were between 2 and 3 weeks and sub-assemblies were produced in batches, resulting in large quantities of work in progress. Analysis of the manufacturing process revealed a high level of non-productive effort, Amos (1).

3. COMPUTER-AIDED DESIGN (CAD)

The use of CAD, in particular 3-D surface modelling, played an important role within the development program. Mair (2) outlines the advantages of CAD including the reduction in lead-time from concept to design completion, which was a prime consideration.

This reduction is attributable to a combination of many aspects of 3-D CAD. The ability to have high quality visual representation of components during their development increases the confidence of the final assembly and the successful interaction with other components. Several parts were modelled and assembled into their working configuration 'on screen'. By taking sections across the whole mechanism, conflicts between the components were easily highlighted and addressed.

The 3-D models also played an important role in giving people, who were less technically involved with the product, an excellent view of the completed mechanism. This was enhanced with the ability to colour render the components and given them surface textures. The models could then be rotated and examined in any orientation.

The productivity of 3-D design compared to traditional methods is therefore improved. It was possible to make many iterations to the computer models enabling several variations of design to be examined. This was done over a period of days at a very low cost compared with traditional methods, which would take up to several weeks. The result was a wealth of confidence gained in the product prior to the production of tools or hard models.

In creating a 3-D model, the software was able to calculate the volume occupied by the components. With applied values for material density, the mass and centre of mass of the parts was estimated. This not only has implications in the design process, but can also provide the information required for 'quick and nasty' quotations for component manufacture.

The 3-D models can be converted and stored in a standard digital format. This allows interaction with associated CAD tools such as Dynamic Loading (Finite Element Analysis) and Mould Flow Analysis (MFA), the latter being appropriate for this type of development. MFA analyses the material flow path for injection moulded components. It can highlight problem areas for the design such as feed points, difficult to fill sections and built-in stresses.

The digital files provide similar data for the production of manufacturing tools and rapid prototyped models. This eliminates the need for hard copy drawings, thus reducing the associated interpretation problems and reducing lead times considerably. The above features were all used in the design stage of the project.

4. RAPID PROTOTYPING (RP)

The step from the computer generated drawings to rapid prototyping is small. The techniques of RP available are:

Stereolithography
Laminated Object Manufacturing
Solid Ground Curing
Fused Deposition Modelling
Selective Laser Sintering
Direct Shell Production Casting
Photosolidification
Stereophotolithography

The selection of the most suitable technique for this application was dependent on cost, availability of commercial sources and end use of the models. Details of the techniques and materials used in RP are discussed by Reese (3).

The RP models were required for widely different purposes in this project. The important criteria involved in selection were:

- Robustness: Parts need to be strong and durable for use as samples and templates.
- Accuracy: To provide fully working units under electrical power and to verify dimensional integrity of the design.
- Reproducible: To provide masters for vacuum casting process to provide additional components.
- Workable: To withstand repeated assembly/disassembly permitting machining and joining processes.

The above criteria was met by Fused Deposition Modelling using an ABS material.

The ability to have physical representation of the component parts; 60% quicker than by conventional methods such as moulding tool samples, should not be underestimated. The benefits were:

- Verification: Early checking of the robustness of the design and components.
- Integrity: Full fit and function of the components.
- Performance: Indication that the assembly will meet the operational specification.
- Assistance: Providing models for toolmakers in the design/manufacture of injection moulding tools, jigs and fixtures.
- Packaging: Model for specialist packaging for transportation of components and assemblies, both internally and externally.
- Modification: Elimination of expensive post tooling re-design.
- Confidence: Providing real parts for people within the company to examine and comment upon. This was especially important in discussions on radical changes in the manufacturing process. Also an important milestone for management to measure progress

RP models were the single greatest benefit to the company by ensuring that limited investment funds were going to achieve 'right first time' tooling. An example a modelled component is shown in Fig. 1.

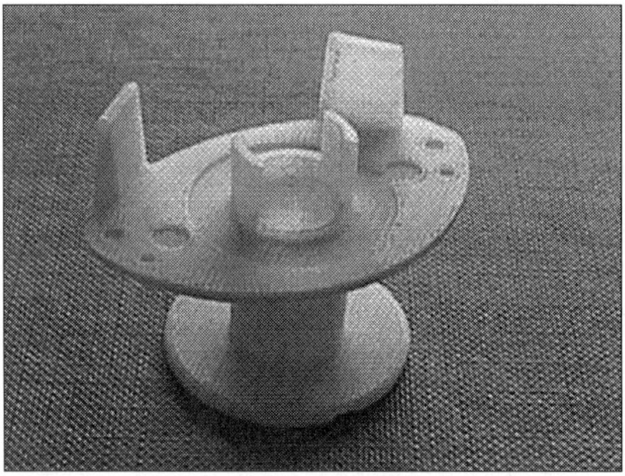

Fig 1. Rapid Prototyped Model (F.D.M.)

5. MANUFACTURING

The redesign of the components has enabled a complete restructuring of the manufacturing process and the introduction of advanced manufacturing techniques. Production has been changed from large batch 'push' systems to a 'pull' process line. Using cellular principles, the batch size has been reduced to a unit of one. This has the effect of reducing the work in progress by 90% aided also by the introduction of KanBan and JIT manufacture.

The new mechanism has been designed for quick and de-skilled assembly. The precision and difficulty is now embedded in both the components and assembly equipment. The simplified build with standardised components combined with unitary batch size has allowed the production line to become flexible enough to handle large variations between order quantities.

The culture and work habits of the existing workforce have been addressed in order to implement and fully realise the benefits of these alternative methods of manufacture.

6. CONCLUSION

Through applying the concept of 'design for manufacture' to the product, the number of components has been reduced from 32 to 16. With innovative methods of assembly introduced into the process, the objective of lead-time production was successfully achieved, thus addressing one of the primary objectives of the project.

The overall project has incorporated the use of several advanced manufacturing technologies which have shortened the lead-time from product conception to manufacturing realisation. Within a timescale of one year the product design was enhanced by 3-D modelling and development accelerated by low-cost accurate prototypes and pre-production models. Included in this timescale was the design and layout of an effective process line.

The result has given a manufacturing lead-time reduction from two to three weeks to same-day despatch and manufacturing cost reduction of 27%. This, whilst still maintaining the company's excellent reputation in the market place for the quality of its product.

By witnessing the impact of new technology and demystifying the process, the company now has plans to apply similar technologies to other products in its catalogue.

REFERENCES

(1) Amos, L. (1997), *Manufacturing process of DIN meters*, M Eng Manufacturing Project School of Manufacturing and Mechanical Engineering, University of Birmingham.

(2) Mair, G. (1994), *Mastering Manufacturing*, Macmillian Master Series, ISBN 0 33 54230 4

(3) Reece, T. (1997), *Rapid Prototyping - How, When, Where.* Rapid News, Vol 5, No 6, pp42 -46.

ACKNOWLEDGMENTS

The authors wish to acknowledge the support given by the company and the Teaching Company Directorate in partially funding this work.

Selection of rapid prototyping manufacturing processes

Y F WANG, R L ROOME, and **R BENHADJ-DJILALI**
School of Mechanical, Aeronautical, and Production Engineering, Kingston University, London, UK

Abstract

This paper describes the formulation and development of an algorithm for the selection of rapid prototyping manufacturing processes based on prototype feature representations.

Rapid prototyping allows the rapid creation of three-dimensional solid objects directly from designs created by CAD, without part-specific tooling or human intervention. Rapid prototyping seeks to reduce lead time and cost, and thereby increase competitiveness.

There are many different rapid prototyping processes available. These have different build methods and use different materials. Lead times and costs vary, and the solid objects made have different accuracy, surface finish and mechanical properties. Which process is the most suitable for a given solid object?

1 INTRODUCTION

The first commercially available rapid prototyping system was Stereolithography, which was launched in 1987 by the 3D Systems Corporation, of Valencia, California [1]. Rapid prototyping facilitates the manufacture of prototypes from CAD models (solid or water-tight surface). The resulting physical prototype conveys more complete information about the product earlier in the development cycle and can aid the production of rapid tooling. The basic process involves building prototypes layer by layer. This is done by taking a CAD modèl of the part and slicing it into a large number of thin horizontal layers. The data are then used to control the rapid prototyping system to form the prototype by sequentially building layer upon layer.

There are now many other such systems commercially available, these include:

- Fused Deposition Modelling (Stratasys)
- Laminated Object Manufacturing (Helisys)
- Selective Laser Sintering (DTM)
- Three Dimensional Plotting (Sanders)

Each rapid prototyping system has its advantages and disadvantages. Time, cost and the size of the prototype and are not the only considerations. Accuracy and surface finish vary between systems. Some prototypes may have thin walls, narrow slots, small diameter holes, blind holes, fine surface detail, interior passages, downward facing surfaces and so on. As different systems use different materials and different build methods, the prototypes have different mechanical properties and are suitable for different applications. Furthermore, the raw prototypes require different post-processing.

Whether the potential customer (novice or expert) is thinking of using a rapid prototyping bureaux or purchasing a rapid prototyping system, the increasing number of systems coming onto the market and the ever improving capabilities of existing systems presents a significant problem in choosing the best system for a particular need.

2 MODEL FEATURE REPRESENTATIONS

Rapid prototyping systems require intermediate processing of the CAD model before it can be used to build a prototype [1]. The current data transfer standard is known as Stereolithography Interface Specification or STL file format [2,3]. This file format, developed for the Stereolithography process, has quickly become a *de facto* standard throughout the rapid prototyping industry, but is not without its drawbacks [4]. The STL file is a faceted representation of the surfaces of the CAD model, which it represents with triangles and their normals. The STL file has two representations, the ASCII representation and the binary representation. The binary representation is more widely used due to its smaller file size, although the ASCII format can easily be read.

By analysing the STL file, the feature representations $f(f_1, f_2, ..., f_n)$ can be extracted. The elements $f_1, f_2, ..., f_n$ may represent part dimensions X, Y, Z, part volume V_m, part outline length L_o, support outline length L_s, surface inclined degree θ, wall thickness W_t, etc.

An STL analysis tool analyses the STL file, extracting its feature representations, then transfers these into rapid prototyping mathematical models to simulate the rapid prototyping processes to find the resulting prototype time, cost, accuracy, surface finish and mechanical properties. Multicriterial optimisation is then used for rapid prototyping system selection based on these simulations. The process which is the most suitable for a given prototype is thus selected.

An overview is shown in Figure 1.

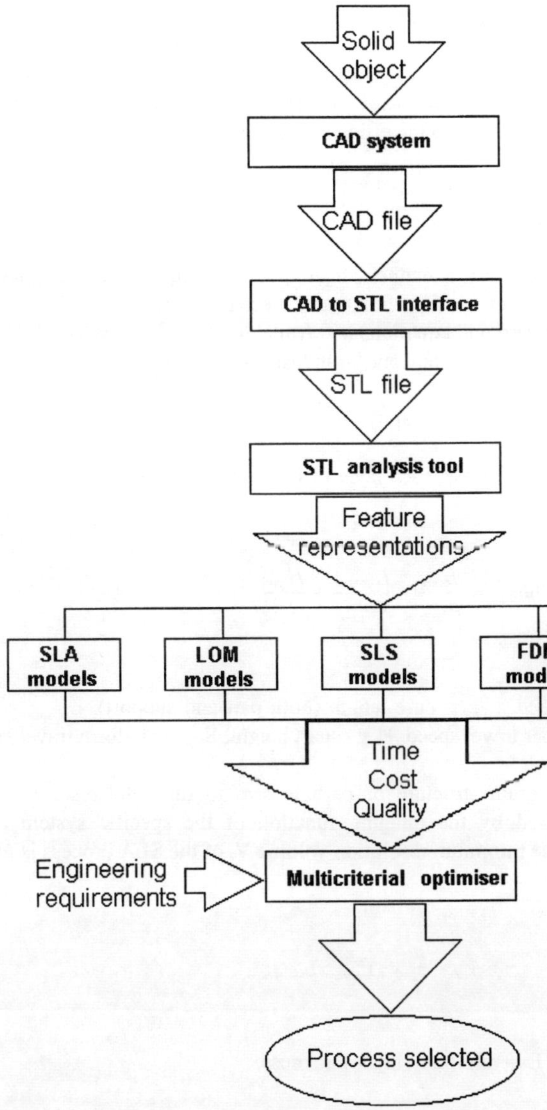

Figure 1 Rapid prototyping manufacturing process selection

The software is being written in Borland C++ and includes Stereolithography, Laminated Object Manufacturing, Selective Laser Sintering, Fused Deposition Modelling and Three-Dimensional Plotting.

"Rapid prototyping technologies are ruled by three principal factors: time, cost and quality" [5]. Each of these factors has several dimensions. Since rapid prototyping aims to reduce lead time, the time factor is of great importance. Quality would include accuracy, surface finish and mechanical properties. Both of these factors together with the requirement to produce a prototype at minimum cost are critical to the efficient use of rapid prototyping.

3 MATHEMATICAL MODELS

For each rapid prototyping process a mathematical model is required which enables the calculation of time, cost, accuracy, surface finish and mechanical properties. In some cases the calculation is straight forward and equations are available or can be developed. However, this is not always the case and experimental and/or industrial data is required.

Examples are afforded by the Stereolithography.

In the Stereolithography (SLA) process, each layer of the prototype is cured with a laser. The run time $Time_{sla}$ is given by:

$$Time_{sla} = \frac{L_{cure}}{S_{laser}} + \frac{L_{scraper}}{S_{scraper}} + \frac{H_{part}}{S_{plat}}$$

where S_{laser} = laser travel speed, L_{cure} = cure length (both part and support), $L_{scraper}$ = scraper travel distance, $S_{scraper}$ = scraper travel speed, H_{part} = part height, S_{plat} = platform travel speed.

Photopolymer resin has no grain structure or particle size to limit the resolution of the process; limits are established by the imaging function of the specific system and the minimum layer thickness. The minimum resolution volume V_r of the SLA process is given by [6]:

$$V_r = \frac{\pi}{4} \times D^2_{laser} x L_t$$

where D_{laser} = laser diameter, L_t= minimum layer thickness.

The accuracy of the SLA process depends upon a number of factors. The following model only considers the laser diameter, shrinkage and the curl distortion [6]:

$$Accuracy_x = \frac{D_{laser}}{2} + \Delta\alpha \qquad Accuracy_y = \frac{D_{laser}}{2} + \Delta\alpha$$

$$Accuracy_z = \frac{a^2}{2\,L_t} \times (\frac{\sigma}{1-\sigma})$$

where D_{laser} = laser diameter, $\Delta\alpha$ =dimensional changes caused by shrinkage, a = length of first unsupported layer, L_t = layer thickness, $\sigma = \Delta\alpha/a$.

The surface finish of the SLA process is dependent on the layer thickness and the angle of the surface:

$$Surfacefinish_{sla} = \frac{L_t \times \cos\theta}{2}$$

where L_t = layer thickness, θ = surface inclined degree.

4 MULTICRITERIAL OPTIMISER

Once the time, cost and quality are calculated for each available rapid prototyping system, the optimum system must be selected. This is not a simple task as the selection criteria may conflict. Work undertaken at the University of Karlsruhe [5] suggests that a multicriterial optimisation algorithm can be used to take account of these competing demands. The algorithm has three target functions:

Time:	min $f_1(x)$ = f(pre-process time, build time, post-process time,.......)
Cost:	min $f_2(x)$ = f(process, material, volume,.......)
Quality:	max $f_3(x)$ = f(accuracy, surface finish, mechanical properties,.......)

If f(x) is the set of all target functions $f_i(x)$:

$f(x) = [f_1(x), f_2(x), f_3(x)]$

Then a value x^* has to be found such that $f(x^*)$ becomes an optimum:

$f(x^*) = \text{opt } f(x)$

This is shown in Figure 2.

The requirements for time, cost and quality can be captured using auxiliary functions $h_i(x)$. The resulting system of equations is then evaluated determining the Pareto optimum solutions. The Pareto optimum solutions are the solutions that meet all of the requirements at the same time. From the set of Pareto optimum solutions the engineer can select the best solution by assigning weights to time, cost and quality.

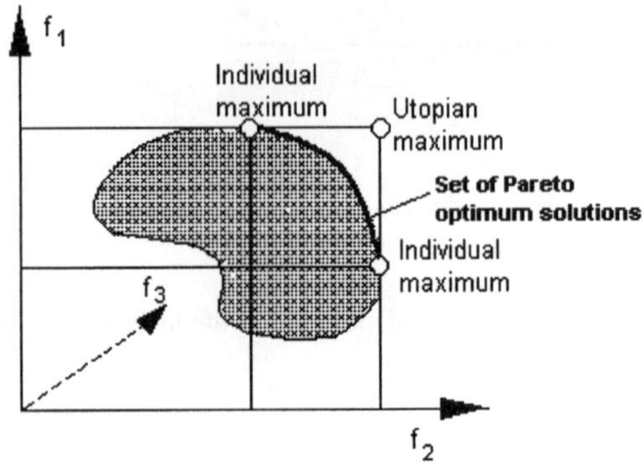

Figure 2 Multicriterial optimisation and Pareto optimum solutions

5 CONCLUSIONS AND FUTURE WORK

The development of a rapid prototyping manufacturing process selector would undoubtedly benefit rapid prototyping system users, both novice and expert. The STL analysis tool and mathematical models are now almost complete and will soon be ready for validation. The STL analysis tool can be expanded to extract information on features such as interior passages and downward facing surfaces which can be problematical on some rapid prototyping systems. The mathematical models can be expanded to include additional rapid prototyping systems and details of rapid prototyping bureaux facilities.

6 REFERENCES

[1] Jacobs, Paul F., "Stereolithography and other Rapid Prototyping Technologies - from Rapid Prototyping to Rapid Tooling", Society of Manufacturing Engineers, 1996.

[2] 3D Systems Inc., "Stereolithography Interface Specification", 3D Systems Publication, Valencia, CA, 1988.

[3] 3D Systems Inc., "SLA-250 Users Reference Manual", 3D Systems Publication, Valencia, CA, 1989.

[4] Dolenc, A. and Makela, I., "Rapid Prototyping from a Computer Scientist's Point of View", Rapid Prototyping Journal, Vol. 2, No. 2, 1996.

[5] Grabowski, J., Erb, J. and Geiger, K., "Choosing the Right Rapid Prototyping Technology for Each Phase of Product Development", Institute for Computer Applications in Planning and Design, University of Karlsruhe, Germany.

[6] Jhonson, J. L., Principles of Computer Automated Fabrication, Palatino Press, 1994.

Creating the conditions for innovation – research programme

A WHITE
International Manufacturing Centre, University of Warwick, Coventry, UK

1.0 ABSTRACT

Innovation is an essential part of manufacturing companies continual need to realign themselves to the dynamic environments in which they operate. The elements of this environment, whether customer needs or new technologies, give rise to the opportunities for companies to gain competitive advantage in their market places. Innovation, be it product or process, are the tangible and non-tangible elements that enable this to happen.
The main objective of the research is to see if there are practices that need to be present within companies internal cultures, environments and procedures that will encourage innovation to take place. These will be identified, prioritised and an attempt made to establish if a "critical mass" of these is needed.

2.0 INTRODUCTION

As already stated innovation is essential to the UK manufacturing industry to remain competitive. This paper takes four distinct views of innovation within companies. The first two of these are literature based, the third is what current UK government policy is doing to assist and promote innovation and finally an outline of the research programme.

This four year research programme is being carried out as part of the Engineering Doctorate programme at the Warwick Manufacturing Group, University of Warwick. Also supporting the programme is the EPSRC, Computer Sciences Corp. and London Taxis International.

3.0 CURRENT LITERATURE

The innovation literature is broad and diverse in its content and scope. This paper objective is to give an overview of the literature and practice from four perspectives. 1) To look at what exactly innovation is defined as. (2) Is there a pattern in terms of when innovation occurs at a national level. (3) The factors that have already been identified as being significant in enhancing the success or failure of innovative practice within organisations. 4)To look at how the economic environment that companies operate in affects it's ability to innovate.

3.1 A Definition of Innovation

The definition innovation must cover three areas. The first is simply what is innovation? There have been several attempts to define this including the DTI's definition (1) which states innovation as being "the successful exploitation of new ideas". Zaltman(2) defines it as being "any idea, practice or material artefact perceived to be new by the relevant unit of adoption" and Kanter (3) as "Innovation is the generation, acceptance and implementation of new ideas, processes, products or services." These three definitions all seem to be pointing towards three elements which must be present if something is to be called an innovation. These are, that it is new to the area in which it is being used, it is successful from generation to implementation and can be in, or an output from, any area within an organisation.

The second area, which was briefly touched on is that of the area within the organisation that the innovation occurs. Damanpour (4) defines three levels, these being technical, administrative and ancillary. Technical is a new idea for a product or service or something associated with production and service operations, administrative are those things which occur within the social system of an organisation and ancillary are those which occur in the organisation - environment interface.

The final area is that of what level the innovation occurs at. Clarke and Staunton (5) define five levels of innovation. The highest level being a generic innovation, this involves radical new "techno-paradigms". An example of this would be discovery and use of electricity. The second level is where a subset of a generic innovation takes place called a epochal innovation. For example the automatic gear-change system in the auto industry. Next comes the altering innovation. This is where something is introduced at the level of the firm. This can be seen when the adoption of a something like sale transactions over the Internet takes place. The next level is entrenching innovations, this is where the same general direction is proceeded but with incremental adjustment. Finally comes the incremental innovation. This is where there is no new input into a system but a reconfiguring of the existing ones. The factors raised in the previous two paragraphs have been placed into table 1 to show how innovation effects various areas within the organisation and at various levels and degrees of impact.

Table 1. Adapted from Clarke and Staunton and Damanpour.

		Level of Innovation				
		Generic	Epochal	Altering	Entrenching	Incremental
Area of Innovation	Technical					
	Administrative					
	Ancillary					

3.2 Pattern of Innovation (6)

The variance in the rate of innovations over time is important to understand in light of the need to plan into the future. In 1925 a Russian economist by the name of Kondratieff published a paper outlining what he saw as a trend that every fifty years the economy went through a full cycle of bust to boom to bust again. Suchumpter in 1939 then built upon this work and the use of the term "business cycles" came into being. This work was built on the pattern which Kondratieff saw. These cycles led to the smelting of iron ore with coal and the mechanisation of the Lancashire coal industry in 1780 to 1842 and the age of the Steam Railway between 1842 and 1897. This outlines the first two cycles in the trend. What we can see from this is that the economy appears to follow a cyclical pattern. Whether this is the long term pattern as outlined here or the shorter ones which we have seen with the recessions and booms of the eighties and nineties. There appears to be times when launching a innovation, particularly product innovation is more likely to succeed due to increased demand in the economy and the higher rate of development of supporting technologies and infrastructure. The phrase "necessity is the mother of inventions" seems to support the above.

3.3 Company Conditions

The study of innovation has long been the focus of many projects. Many of these projects define practices which are essential to innovation (7) (8) (9)(10) and models which facilitate it.

The studies mentioned raise a number of factors which are common to more than one study:

- Clear understanding and identification of customers need. (13)(8)(9)
- Appropriate organisational structures (9)(8)
- Presence of a person in a position of authority who leads the project(8)(9)
- Getting to market quicker than the competition. (10)(13)

In addition to these the following factors are raised:

- An integrated approach of R&D, Design, Production and Marketing. (13)
- Incorporate several times more technologies than their competitors.(10)
- Product superiority through product definition and pre-development activities. (14)

There have also been several models developed which have tried to show how innovation takes place and is facilitated within firms. These are summarised in table 2.(11)

Table 2 Rothwell's five generations of innovation models

Generation	Key features
First / Second	Simply linear models - need pull, technology push.
Third	Coupling model, recognising interaction between different elements and feedback loops between them.
Fourth	Parallel model, integration within the firm, upstream with key suppliers and downstream with demanding and active customers, emphasis on linkages and alliances.
Fifth	Systems integration and extensive networking, flexible and customised response, continuos innovation

This overview at the company level gives insight into the development of the research. The following conclusions can be drawn:

1) There has been a considerable amount of research into the factors that are responsible for the success of innovation.
2) There is a need to develop tools and techniques that will assist the development of these factors within firms.
3) These factors and issues need to be seen within the context and issues facing individual companies and industries.

3.4 Economic Conditions

The correlation between the economic conditions within a country the country's ability to innovate has been established by several writers. Ergas (12) defines three key areas as needing to be present to maximise innovation performance. These have been developed following a study published by the European Policy unit using empirical data from OECD countries. These factors are stated as being:

Demand

"Potential Demand" provides the incentive for firms to invest in R&D by providing a possibility of a pay-back on the money invested. This potential demand is defined in the following terms:

- The present size of the market for a product, or for a product which fulfils broadly the same function;
- The rate of growth of this market;
- The sophistication of the market, in terms of the capacity of customers to change suppliers to acquire an improved product;
- The permeability of the market, in terms of the ease with which buyers obtain information about new products, producers obtain information about customers requirements, and the simplicity and the transparency of the distribution channels linking the two;

Technical Opportunity

The level of demand by itself does not provide incentive to Innovate. The ability of a country's firms to link demand to technological opportunities is largely dependant on trained personnel and systems that provide links to science and research. Therefore a large amount of dependence is placed upon comprehensive educational training which firms need incentives, capacity and facilities in order to deliver.

The Industrial Structure

The role of firms is essential to the "Innovative Performance" of a country as they provide a large proportion of the commercial R&D investment. The factors that effect the conception of these firms and the subsequent behaviour of them from a macro level is intrinsically linked to this process. There are three factors which are identified here:

- Role of new firms.
- Inter firm co-operation .
- Industrial Standardisation.

4.0 UK INNOVATION POLICY

From a history of being recognised for our ability to develop new products but not for our ability to successfully produce and market them, (16) we are now seeing a concentrated effort to address this inconsistency. (15)

4.1 Higher Education / Industry Interface

The knowledge capital contained in many of the UK's universities remains an under capitalised resource. Several initiatives have been started in this area seeking to address this issue, including:

- Effective management of Intellectual Property Rights.
- The development of entrepreneurial and management skills in graduates.
- The development of high quality partnerships.
- Foresight Programme - developing research priorities in the UK.

4.2 Finance

The financing of new R&D and high technology continues to need the development of new and specialised sources of finance. This is primarily due to the level of risk associated with innovative projects and inappropriate structures of normal sources of commercial financing. Several programmes have been developed in this area, including Small Firm Loan Guarantee Scheme, Business Angels, Private Funds and Corporate Venturing.

4.5 DTI

June 1997 saw the launch of Competitiveness UK by the President of the Board of Trade. This programme is seeking to develop partnerships between employees, consumers, government and many private and public bodies which will facilitate the growth and development of, modern, innovative and forward looking businesses. Six working parties have been established, one of which has the remit of answering the question, "How can the "Competitiveness UK" partnership best work with business to encourage more firms to be truly innovative in all aspects of their performance?"
This programme will also see the development of policy for a white paper on competitiveness to be published in 1998. This follows on the work of the DTI's "Winning" report and other initiatives.

5.0 PROPOSED RESEARCH

There is enough evidence to suggest that the next generation of competitive advantage may come from the creation of knowledge and its creative exploitation in convergent and divergent product and technology fields. The following research projects seek to understand what knowledge is needed, the environment that will enable its commercial exploitation to take place, and how it can be managed within the organisation to deliver innovation.

5.1Customer Consciousness

The customers needs, both future and present, must be the main focus of any successful business if its continual growth and survival is to be developed. There has been a large amount of literature and practice showing the importance of this fact in the innovation process. (8)(9)(13) These needs can be most accurately translated into product and service attributes when the whole business is conscious of the customer and therefore successful innovation is produced. The research will seek to develop ways of increasing the customer consciousness and ensuring that the present and future needs are anticipated and met at every level within the organisation.

5.2 Supply Chain Management

With the increased ratio of out-sourced parts to those manufactured in house, the effective management of the supply chain is essential to ensure innovation is comprehensive in all areas of the product and production processes. The objectives of this project will be to develop models of how partnerships can be constructed with suppliers over the life of a product that will facilitate and stimulate innovation in those parts and processes which are to be out-sourced.

5.3 Origins of Innovation

Successful innovation is critical to the growth and development of any business. However the circumstances that surround the conception of these "ideas" are often vague. The factors which enabled the innovation to get to a point where it is brought to reality through a project is generally recognised to be fuzzy. This project will trace the origins of 10 innovations to identify what factors were instrumental in their conception. Then to look at how these factors can be exploited by trying to facilitate and stimulate them within the business. Thus creating a continual stream of new ideas.

5.4 Project Management

The use of detailed project planning has long been accepted as the best way to manage projects. But is it? Does a more "think on your feet" approach enable the creation of an environment where the project meets its targets in the areas of quality cost and time, and allow continual innovation to take place. This project will seek to understand this issue and develop tools to assist in the management of projects in this manner.

6.0 SUMMARY

From the research outlined above we can see that innovation relies on the correct conditions in three areas; the economy, government policy and company practice. What must be understood is that not all these conditions need to be correct before innovation can happen, but when a majority are then there will be a higher occurrence of innovation both at a national and company level when a larger number of them are present. (13) This research is aiming to build on this platform of knowledge by the development of tools and techniques that will transform this knowledge into practices within organisations.

7.0 REFERENCES

(1) DTI. *Innovation- Your Move*. HMSO (1997)

(2) Zaltman, G, et.al. *Innovations and Organisations*. John Wiley and Sons. London. 1973

(3) Moss - Kanter,. R. *The Change Masters.* Simon and Schuster. New York. 1983

(4) West, M. and Farr, J. *Innovation and Creativity at work.* John Wiley and Sons Ltd. London 1990.

(5) Clarke. P, and Staunton. N, *Innovation in Technology and Organisation.* Routledge. 1993

(6) Hall. P, *The Geography of the fifth Kondratieff cycle.* New Society. 1991. Pp535-7

(7) Freeman, et al. *Success and Failure in Industrial Innovation*, Centre for the Study of Industrial Innovation, University of Sussex. 1972

(8) Langrish et al. *Wealth from Knowledge. A Study of Innovation in Industry*. Macmillan 1972.

(9) Peters, T.J. and Waterman, R.H. *In Search of Excellence,* Harper and Row. 1982

(10) Nevens, T.M. et al. *Commercialising Technologies: What Best Companies Do*. Harvard Business Review. May-June 1990 (Report of McKinsey Study)

(11) Rothwell, R, (1992) *Successful Industrial Innovation: Critical Success Factors for the Ninties* . R+D Management 22(3) P221-239

(12) Ergas.H. (1984) *Why do some countries innovate more than others?.* Centre for European Policy Studies.

(13) Twiss, B. *Managing Technological Innovation.* Pitman 4th Ed 1992

(14) Cooper, R.G. and Kleinschmidt, E.J. *New Products: What Separates Winners from Losers.* Journal of Product and Innovation Management, 4, No 3 Sept 1987.

(15) The President of the Board of Trade. *The Innovation Exploitation Barrier. Government Response to the House of Lords Select Committee on Science and Technology, 1996-97 Session.* October 1997. HMSO.

(16) Layton. C, (1972) *Ten Innovations*. Allen and Unwin Ltd. London

Knowledge-based Systems

Development of an expert system for fault diagnosis in compression moulding

M ABURAS, A DAVIES, E JOHN, and **M USENER**
Cardiff School of Engineering, University of Wales, Cardiff, UK

ABSTRACT

This paper describes a project, that was carried out in the interest of a local company, needing to preserve expertise by developing a knowledge based expert system within a KAPPA environment which will be used in their compression moulding plant. It outlines the objectives of the developed expert system, as well as the different parts that constitutes it, and gives a concluding remarks of how useful expert systems are, to the enhancement of the manufacturing process.

1. INTRODUCTION

When the plant is under normal operation, operators are under minimal pressure and control is limited to small number of well acquainted activities. But the moment a fault or failure is experienced, the situation changes radically, and the analysis of the behaviour of the process becomes very difficult.

Experience plays a significant role in this aspect, due to practical difficulties arising in the form of defective products or insufficient resources. Any corrective measure has to be taken promptly and accurately, to bring production up to standard in as short a time as possible. Experience is also required when changing over the production of one product to another.

Although, human experts, can fulfil this requirement, by nature they are subject to stress and fatigue, in such situations, the role of Expert Systems becomes apparent.

2. KAPPA-PC

The KAPPA-PC Applications Development system allows applications to be built in a high-level graphical environment. In KAPPA-PC domain components are represented by structures called objects in a tree hierarchy. Each object has its own characteristics and operations that can be applied to it. The characteristics of an object can be directly inherited from its direct cendant. Objects are defined as either classes or instances. They can represent concrete things like parts or abstract concepts like sale orders.

Object hierarchy represents an important concept. It Defines the relationships among the objects in a model by linking them together into a rank structure that is represented as follows:

Class ——→ Subclass ·········→ Instance

Slots are used in an object to describe all the important properties of the real object. Each slot represents a characteristic of that object in the form of a value assigned to that slot. Objects can have their own slots, inherit slots from ancestors, share slots and communicate with each other using methods.

Kappa uses functions and rule-based systems to process data. There are over 240 pre-defined functions in Kappa with ability to define new functions using the KAL language. Functions are classified as follows: Rule functions, Object functions Block control of expressions, Graphical interface, Logic, string, list and mathematical operators and File management

The Inference Engine is a special program for managing rules and applying them as appropriate. The rule structure in Kappa is as follows:

IF premise **THEN** conclusion.

If car : model #= escort Then Car : make = ford

Rules, use variables, which is an efficient way of codifying information. A single rule with a variable can represent a number of objects. Rules can be assigned priorities over each other. Each rule specifies a set of conditions and a set of conclusions to be made if the conditions are true. The conclusions may represent logical deductions about the knowledge base or specifications of how it changes over time. Each rule is a relatively independent module, so reasoning systems can be built gradually, rule by rule.

3. OBJECTIVES OF THE PROJECT

The main objective of this project was the development of a prototype expert system within a KAPPA environment. This encapsulates important expertise about fault diagnosis in compression moulding and is also able to suggest solutions.

The requirements were as follows:

a) To provide the diagnosis of defects encountered on the products during the compression moulding process, and to demonstrate the ability of suggesting solutions to each of the faults.

b) The expert system should be able to provide the necessary information to start up the production process, to include machine tool setting and operating conditions. In order to preserve data and improve its availability and to save the time and effort that is consumed when using traditional paper forms.

c) Demonstrate the ability to give the necessary information about each fault and ways of avoiding it. This would serve as an educational element provided on line, to make the process operators aware of the various aspects of quality assurance.
d) The system should have a user friendly interface that would enable the efficient use of the expert system by all those concerned in the plant. It should also provide some security features for the protection of the database.

4. THE OVERALL SYSTEM

The overall structure of the system consists of three major parts, which are integrated to form a complete system.

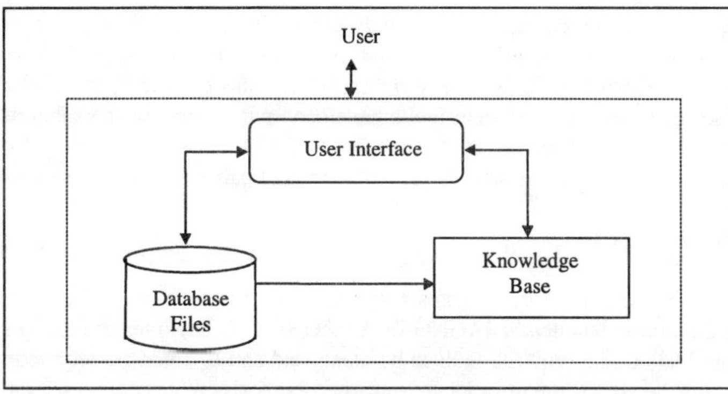

Fig. 1 Over all system structure

5. USER INTERFACE

Since the system is designed to carry out more than one task; it was seen to be appropriate to have a main screen (default), through which the user can gain entry to the more specific screens, and to be able to carry individual tasks independently. All the screens are *instances* of the subclass *Kwindow*, which is provided by KAPPA.

6.1. THE FAULT DIAGNOSTIC PART

This is the main part of the system, which makes the system deserve the name 'Expert system'. Although the finished program was a prototype; efforts were made to make it as practical as possible.

This task was accomplished via the fault diagnosis screen, which is, the main window and holds a list of all the possible faults, by clicking on the arrow in this window the user can select the fault to investigate. The system, then uses pop-up windows to interact with the user on the same screen.

The second window is labelled 'Information' and provides the user with all the necessary information about the selected fault. In this way not only will the user be able to correct the fault, but he will also be aware of the cause of that fault and will gain sufficient knowledge to avoid the re-occurrence of such faults.

6.2. THE HIERARCHY

This is a descriptive structure of the system under consideration, which connects related objects (Classes, subclasses and Instances) together and allow the sharing of attributes through inheritance.

KAPPA provides a parent class called *root*. This class has four default subclasses (Global, Menu, Image, and Kwindow). *Menu, Image and Kwindow* are part of KAPPA's interface, which pass attributes to their respective objects down the hierarchy. When needed, the values of the slots of these subclasses were altered to suit the required application. The subclass 'Global', was used throughout the system, which had slots that are used by functions as wild cards to hold temporary values, *Global* was also used to hold the rule list which is used by the inference engine.

Since this part of the system is concerned with faults, it was natural to create a parent object and give it the name 'fault', which is in turn a subclass of the main class 'root'. All the faults that are considered by the system were made as instances in the parent subclass 'fault', while the factors which produce such faults have been put as slots in the subclass fault. One important slot is *defect,* which stores all faults in its list of allowable values. When the user selects a fault to investigate, the slot *defect* will have that particular fault as its current value.

6.3. THE INTERFACE

As mentioned earlier, the fault diagnosis screen is accessed through the main screen by clicking on the button labelled ' DIAGNOSE A PROBLEM'. The interface enables the user to tell the system, about the required fault to be diagnosed and to enter the required information that the system needs to use in order to start the diagnosis process. The screen Essentially contains two windows through which the user will interact with the system during the diagnostic process, as follows:

The main window : This is a ComboBox Image (provided by Kappa) which is owned by *fault:defect* and hence holds all the instances of the subclass *fault*, representing all the possible defects that can happen in any given defective product, from which the user will be required to select the appropriate fault for investigation. Although the selected fault will inherit all the possible causes of every fault (as empty slots); only the relevant causes to that particular fault would be investigated.

The second window : This is a *Transcript Image* which belongs to the subclass *Image*. *Transcript Images* are used in KAPPA for displaying text, graphs, etc.

Here the transcript image is used to display information about the selected fault. This information contains the important issues that are associated with this fault.

6.4 THE KNOWLEDGE BASE

Two parts constitute the knowledge base of the developed ES:

The first is the database files. Two database files have been created to hold the tool settings and the operating conditions, such as, temperature, cure time , breathe, etc. for a sample of fifteen parts, The tool setting file has been set to hold information about every mould needed to manufacture a particular part which includes: tool number, a list of the accessories required to fix the tool, and location of these parts (Fig. 4).

The contents of these files are used by the inference engine, as a reference to examine the actual settings given by the user, as input.

The second part is the rule set, which contains extracted knowledge about the all possible faults in the compression moulding process within the company. These are formulated in 'IF-THEN' rule form. Each fault is represented by a rule or set of rules. The antecedent part of each rule holds the condition which constitutes a possible causation of a particular fault, while THEN part holds the confirmation of the fault occurrence plus the corrective action to be taken in response. Every fault has a rule for each possible cause. therefore if a defect can happen due to two possible causes, then there will be two rules about this defect. The inference engine will then invoke a search mechanism to reach its decision.

These rules direct the inference engine to the database files to retrieve the ideal settings of the process variables, needed to reason about a given fault, then based on the comparison between these values and the actual operating settings, the rules provide the inference engine with the diagnostic result. Each possible fault in the process is represented by a rule in the knowledge base, for example the fault " Cracks" has the following rule:

```
MakeRule( crack1, [],
   fault:defect #= crack,
   {
   If ( fault:cure_time > DBReadCell( 1, Global:product + 1 ) )
     Then PostMessage( "Cure Time is LONG", " Make Sure That Cure Time is Not Longer
Than  ",
           DBReadCell( 1, Global:product2 + 1 ) )
    Else {
       If ( fault:cure_time < DBReadCell( 1, Global:product
                                         + 1 ))
        Then PostMessage( "Cure Time is SHORT" );
       };
   } );
```

6.5. INPUT AND OUTPUT

It was seen appropriate to make the system prompt the user for the required information rather than making the user input all the information at once. This is to allow the system to be used easily by operators at all levels of skill. and to make the screen neater. The first form of output is pop up windows, which contains messages telling the user about the action that needs to be taken. The second is in the form of text displayed on the *transcript image*, informing the user about the present fault.

6.6. THE INFERENCE MECHANISM

As the user inputs the required information; the system will assign each input as a value to its relevant slot. Once all the necessary information is provided; the inference mechanism starts.

The system then sets its search mode to *depth first*, and the chaining mode is set to *forward chaining*. For faults with more than one cause; more than one rule would fire and because there is no set gaol, the system will search all the eligible rules. The inference engine reaches its conclusion after comparing the data given by the user with the standard values either in the database or in the *allowable values* of the relevant slot, depending on the type of fault. If any of the data given by the user does not match the set criteria, then the system will prompt the user with its conclusion in separate consecutive windows, stating the standard data for each erroneous input.

7.1. THE INFORMATION PART

This constitutes the second main task of the system, which is mainly an input/output medium, through which the user will be able to visualise all the necessary settings of the compression moulding process (Fig. 4).

7.2. INTERFACE

The information screen has two *transcript images*. The top one is for the display of the operating conditions of the press, while the button one is to display the tool setting information. There is also a set of buttons on this screen to activate certain functions, for the purpose of user interaction with the database.

8. DATA RETRIEVAL

KAPPA provides functions for the purpose of accessing database files and spread sheets. In our case the function *DBReadCell* was used to read the stored data, while the functions *formatValue* and *Display Text* were used to display the requested information on the *transcript image*.

9. CONCLUSIONS

Considering the requirements that were set for the fault diagnostic expert system, some conclusions can be drawn:

With regard to the first requirement, the system was found to perform well, proving to be reliable and able to give correct diagnosis to all the faults that are known to it. It was also seen capable of giving solutions to the encountered faults and also providing information regarding selected faults.

In respect of the second requirement, the system was also able to maintain data availability and reliably retrieved all the necessary information regarding any product in the data base. It was also able to efficiently and safely update the database.

As 'user friendly' is a subjective matter, it was hoped that the system's interface is easy to operate and its messages are clear and easy to understand.

The system was not able to give detailed fault rectification or to reason about its decision. This was due to development time restrictions.

References

1. Michael L. Berins; 'Plastics Engineering Handbook'; Chapman and Hall; 1991.
2. M. J. Gordon; 'Total Quality Process Control for Injection Moulding'; Carl Hanser; 1993.
3. KAPPA-PC Reference Manual.

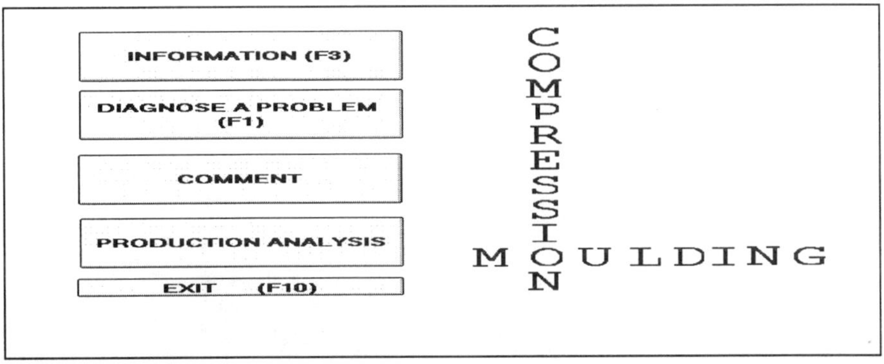

Fig. 2. Main Screen

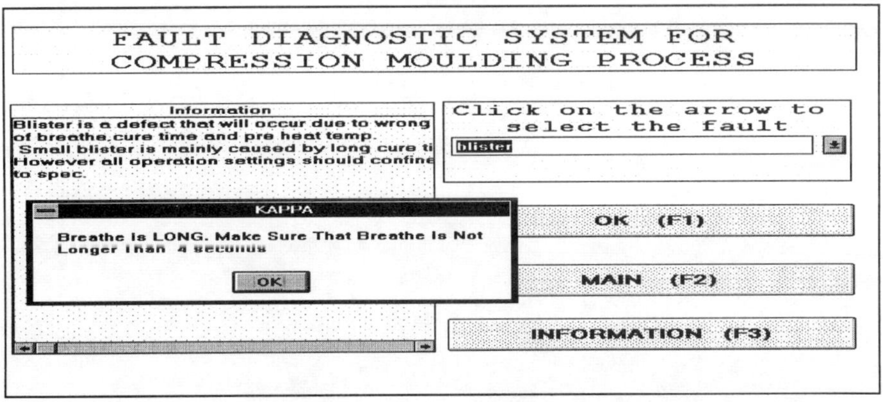

Fig. 3. Fault Diagnosis and online help

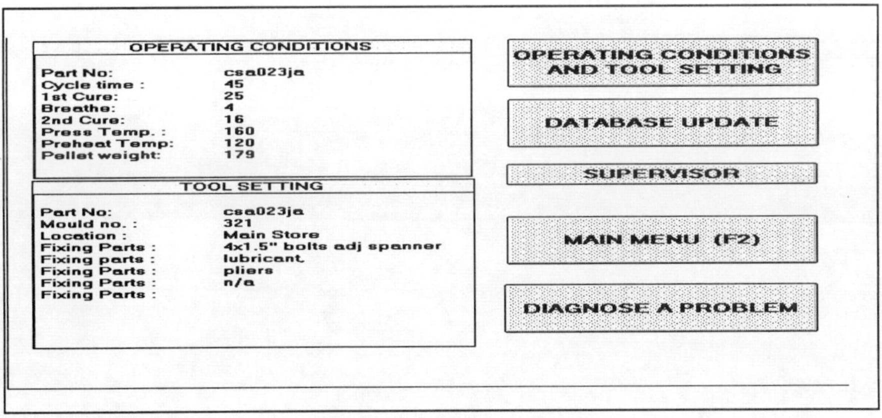

Fig. 4. Start up Information

Optimized multiparameter extraction techniques in sensor array applications

J ANGLESEA, P LOWERY, and P CORCORAN
Sensing and Control Research Group, University of Derby, UK

ABSTRACT

The use of multisensor systems is becoming widespread in a number of applications. Recent developments have led to the identification of multiple parameters that can be extracted from individual sensor elements. However, it is now accepted that the most promising employment of such technology will most probably be in application specific instruments, where a limited set of calibrands would be analysed, based upon a calibration database containing representative parameters of each calibrand. Given the large number of sensors, and the availability of multiple parameters, instrument designers must address the need to maximise the discriminatory information derived from a sensor array. This paper presents a structured approach to selecting the individual array elements (sensors or parameters). Ideally, given a database of the responses of *n* parameters to *p* species, it should be possible to identify a subset of *n* sensor responses that provide a specified level of performance when applied to the classification, or description, of a subset of the *p* species. This paper compares the utility of cost functions and search algorithms that can be applied to the task of optimal sensor array configuration. In particular, the novel use of genetic algorithms when applied to the task of array configuration for gas sensing is described.

1 BACKGROUND

Electronic odour sensing systems are an application of sensor arrays, utilising arrays of gas sensors and sophisticated signal and information processing techniques in order to undertake the tasks of odour detection, classification and description. To increase the reliability and performance of sensor array responses, the trend has been to increase the dimensionality of the information provided from the arrays. This can be achieved by the development of new sensor technologies with smaller physical and power requirements (allowing a greater number of devices to be used), or through the use of multiple parameter extraction techniques in conjunction with novel sensor excitation techniques. These include thermal cycling of operating temperature and analysis of transient sensor response characteristics.

This paper describes the application of optimisation techniques to the configuration of multisensor arrays used in electronic odour sensing classification systems, although the structured approach to optimisation is applicable in many multi-parameter or multi-sensor applications.

2 MULTIPARAMETER EXTRACTION

Historically (1),(2) data collected from multi-sensor systems has generally been acquired in steady state form. This implies a method where some form of difference measure is derived based on a sensors steady state initial baseline, compared with its final value after exposure (Refer to Figure 1 parameter B & Table 1). From this parameter, across the whole of the sensor array, a vector can be derived which can be used as a specific fingerprint for the species being tested. This fingerprint is then stored together with similar acquisitions in a database and used in later stages to aid in pattern recognition.

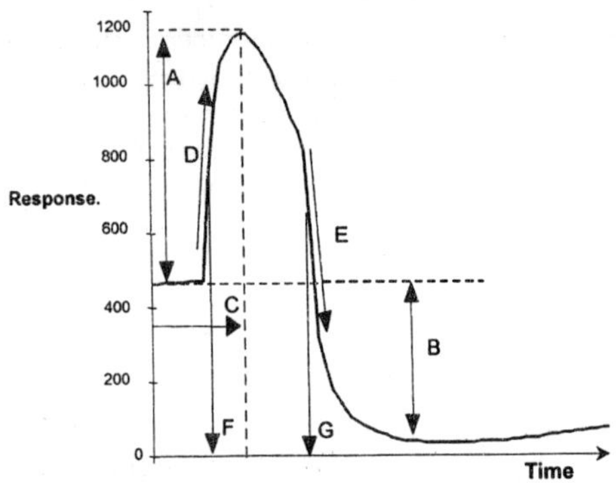

Figure 1 : Extracted parameter metrics

Table 1 : Dynamic response parameters.

Parameter.	Description.
A	Maximum Step Change.
B	Minimum Step Change.
C	Time To Maximum.
D	Maximum Positive Rate Of Change.
E	Maximum Negative Rate Of Change.
F	Time To Maximum Positive Rate Of Change.
G	Time To Maximum Negative Rate Of Change.

If the data sets are analysed over time, it can be seen that each acquisition provides several parameters from each sensor element (Refer to Figure 1 parameters A - G & Table 1). Therefore, the array response can be described in terms of these dynamic parameters.

In steady state analysis, the step response of the sample is taken over some discrete time interval and used as the primary measurement variable. In the dynamic method, the transient signals generated by each sensor are analysed over time and key features are extracted. Studies already completed (3) have shown that there is, at least, similar levels of discriminatory information available in the dynamic signals of reduced dimensionally arrays. These give comparable classification rates to those arrays using standard steady state techniques. Results from these studies have been published previously and the parameter extraction algorithms have been embedded into a commercial electronic nose instrument (3),(4).

3 OPTIMISATION

The optimisation process involves identifying an objective function which provides a measure of the sensor array performance given a particular task (e.g. detection, classification, description). A search technique is then required to explore the search space in order to identify a near-optimal solution without exhaustively searching the total available search space.

3.1 Objective functions

The 'goal' of the optimisation task is determined by the objective (cost) function used. This is a function that takes a subset of the sensors/parameters under consideration, and returns a score that can be compared with other values produced from other subsets of sensors (i.e. it allows the performance of the different sets of sensors/parameters to be compared with each other).

Given a set of classes, each combination of sensor parameters will map the odour space onto parameter space in a unique way. Each class is represented by a 'fingerprint' or array vector, and will be represented in parameter space (given a suitable dataset of results) by a mean location and a spread of points around that mean in n-dimensional space. It should be noted that each fingerprint is normalised. This ensures that a scale invariant pattern is used for analysis (since parameters from different sources can have significant differences in their absolute ranges). Therefore, by finding the set of parameters which provides the greatest distance between calibrands in parameter space will ensure that the task of classification will be simplified.

The representation of distances between classes in parameter space is fundamental to the objective function. This can be undertaken using the Euclidean distance between calibrand means in n-dimensional parameters space. It is important to include some measure of the spread of the responses in the objective function, not just the response means, since this indicates the repeatability of the response in parameter space (i.e. a small spread corresponds to a repeatable response). There are several measure of spread available. These could include the standard deviation, range, and inter-quartile range of the responses. Parameters that provide a large distance between calibrand means in parameter space are not useful if they have such a low repeatability that the different regions overlap in parameter space. The measure of distance between calibrand means in parameter space then becomes:

$$dist_{(Cx,Cy)} = \sqrt{\sum_{n=1}^{no\_of\_sensors} \left[\frac{\left(\mu_{nx} - \mu_{ny}\right)^2}{\sigma_{nx}^2 + \sigma_{ny}^2} \right]}$$

Where:
μ_{nx} is the mean response of the nth sensor to calibrand x
σ_{nx} is the standard deviation of the nth sensor to calibrand x

The computation of the objective function involves calculating this distance function for each pair of calibrands across all the parameters currently being considered to provide a matrix of distance values.

3.2 Search algorithms

A search algorithm is a technique used to investigate a space defined by some objective function. A search algorithm uses two techniques to find a global maximum (a minimisation function can always be expressed as a maximisation function): exploration to investigate novel areas of the search space, and exploitation to make use of previous knowledge already gained about the search space in regions already visited (5). Since these two requirements are contradictory, a good search algorithm must demonstrate a trade-off between the two.

There are three main types of search algorithm. The three types are enumerative techniques, calculus-based search techniques and random search techniques.

An enumerative approach would examine every location defined in the search space, and is therefore guaranteed to find the maximum value of any defining function. However, an example dataset used to describe the response of an electronic odour sensing system with 64 parameters produces a search space with over 1.8 x 10^{19} individual locations. An exhaustive search of such a large number of locations is impractical. This type of search is explorative, but exhibits no exploitation of information previously obtained about the nature of the search space.

A calculus-based search technique, also known as hill-climbing search, is initiated at a random location in the search space and performs an analysis of the surrounding locations. The search selects the direction that shows the greatest increase in objective function value, and repeats this procedure in order to travel to a region of greater fitness. The main problem associated with calculus-based search techniques is that they are prone to the effects of local minima. They are local in scope, and hence can only find the best position in the search space within a small area around the current search point. This problem can be alleviated by performing several searches, although the technique is not suited to noisy, multi-modal search spaces.

3.3 Genetic algorithm search

A genetic algorithm search is modeled on the biological process of natural selection. This is often referred to as 'survival of the fittest'. In order for a genetic algorithm search to be applied to a particular task, the variables that define a solution to the objective function must be encoded into a string, known in genetic terms as a chromosome. This chromosome in fact represents a particular point in the n-dimensional search space, where n is the length of the chromosome (6). Each chromosome is a valid solution to the problem and the 'fitness' of each chromosome is computed by the objective function.

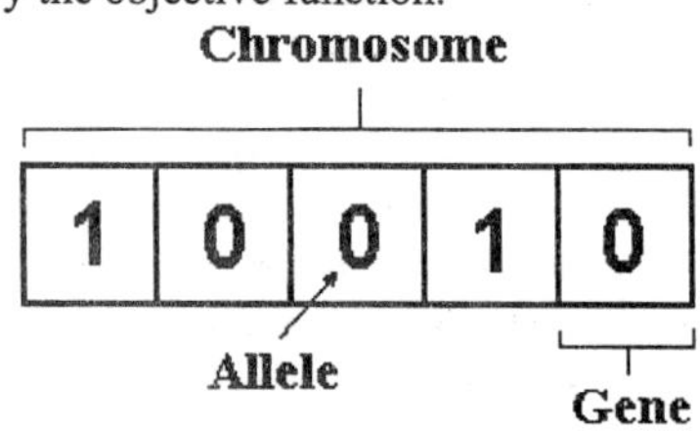

Figure 2: Chromosome representation in genetic algorithms

The principal of operation of a genetic algorithm is to maintain a population of strings (chromosomes), and through a three stage process (selection, crossover and mutation) breed successive generations of strings in order to arrive at some optimal or near optimal solution for the particular problem domain.

In the selection process, pairs of chromosomes are selected at random to produce 'offspring'. The likelihood of a chromosome being selected for mating is related to the 'fitness' of that chromosome. i.e. chromosomes with higher fitnesses (as governed by the objective function) are more likely to be selected for mating, and hence their genes have a greater influence on the next generation of chromosomes.

Once enough pairs of strings have been selected (matching the population size), there is a chance that pairs will mate, and that 'crossover' will occur. This is effectively a transfer of genetic information from one chromosome's child to the other in order to produce new offspring that inherit some genetic information from the previous generation. However, it can also introduce novelty into the population by incorporating previously untried combinations of alleles (gene values). The chance of a crossover occurring is a parameter of the genetic algorithm search which is set heuristically (typically in the range 40%-60% chance of crossover occurring).

Since the positional representation of parameters in the chromosomes was arbitrary, a novel template crossover technique was used. Each gene was given a 50% probability that crossover was to occur. If crossover was to take place within a particular gene, then the children of the two strings would contain reversed alleles at that location as shown in Figure 3.

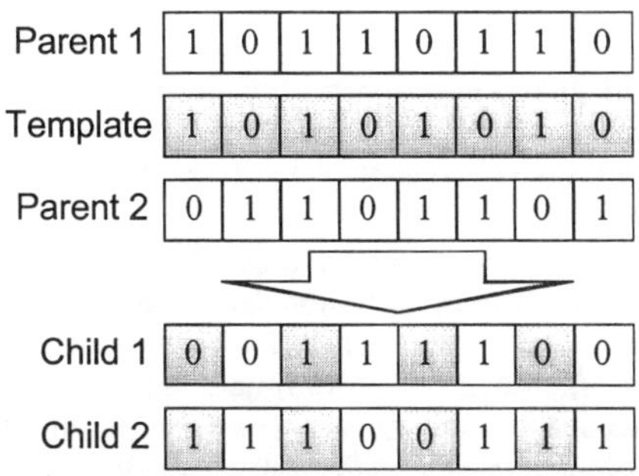

Figure 3: Template crossover genetic operator

The other mechanism by which alleles can change is by mutation. This is very much a secondary mechanism compared to the main crossover activity, as is the case in the biological systems on which genetic algorithms are modelled. In each generation, there is a small probability (in the order of 10^{-3}) that each gene will mutate. The main purpose of the mutation operation is to help escape local minima within the search space by exposing (at least theoretically) all possible areas of the search space to the search algorithm.

4 EXPERIMENTAL METHOD

The transformation from the parameter sets to the desired outputs of a pattern classifier is both non-linear and unknown, hence traditional multivariate statistical techniques are of limited use. However, artificial neural networks more readily accommodate these non-linearities and are well suited to information processing in electronic odour sensing systems utilising such parameters.(7) An artificial neural network was used to classify various parameter sets extracted using two different multi-parameter extraction techniques; Dynamic Parameter Method (DPM) and Total Signature Difference (TSD) along with three GA optimised sets of parameters which were subsets of the DPM parameters. For comparison, the standard Single Temperature Method (STM) was used to extract a set of reference parameters. The datasets were generated from three types of brewed tea headspaces, generated on a static headspace sampling electronic nose system.

To test the validity of the neural network solutions, prior to training each neural network four samples were randomly removed from each tea class for cross-validation. This process was repeated four times to generate four separate training data sets and four corresponding test data sets. Each training set (48 samples in total) was used to train a 3-layer neural network using the back-propagation technique. The number of output units was always 3 (one per class), with the number of input units being equal to the number of input variables, giving 8 for the STM network, 64 for the DPM network, 208 for the TSD network and a variable number for the GA configurations depending on the number of input parameters selected. The number of hidden units were 5 (STM), 20 (DPM) and 80 (TSD) respectively, and 13, 15 and 15 for the GA1, GA2 and GA3 data sets respectively.

Data Set	Number of Parameters	Classification Rate - Overall Data	Classification Rate - Unseen Data
STM	8	75	69
DPM	64	90	81
TSD	208	95	90
GA1	13	93	83
GA2	15	91	75
GA3	15	93	67

Table 2 : Classification rates for various sets of parameters.

5 RESULTS

The results are shown in Table 2. The highest classification rate was achieved using the 208 parameter set, resulting in 90% correct classification of unseen samples and an overall classification rate of 95%. The 64 parameter set also showed an improvement (81%) over the 8-parameter technique (69%) for unseen data.

The three optimised configurations were then tested using one of the four training and test data sets. This indicated that the optimisation technique had identified a reduced set of parameters (13-15 parameters), approximately 7% of the size of the set used in the TSD approach (208 parameters), which exhibit a near equivalent level of success in the classification of both seen and unseen samples. The three optimised configurations each have overall correct classification rates greater than 90%, with the best unseen classification rate being 83%, equal to that achieved using the Dynamic Parameter Method. Two of the GA search solutions for unseen data were better or equivalent to the both DPM and STM solutions for unseen data.

6 DISCUSSION

The work presented demonstrates that improved levels of discrimination between calibrands can be achieved by the introduction of multi-parameter extraction techniques, providing increased sensor array sensitivity and repeatability by utilising novel parameters that are more reliable than those extracted from steady-state techniques. The optimisation work shows that by identifying an optimised set of parameters, the dimensionality of the array fingerprint can be reduced, and yet near-equivalent levels of classification can be achieved. It would seem that increases in array sensitivity should be pursued through the identification of key sets of parameters that facilitate improvements in classification rates. This will provide useful guidance in the design of sensors and sensor systems.

7 CONCLUSIONS

The use of multi-parameter extraction techniques in conjunction with a genetic algorithm approach to parameter optimisation, can be used to identify near-optimal parameter configurations for classifying multivariable fingerprints from sensor arrays. This has necessitated the development of novel representation schemes and reproduction algorithms. The technique is applicable to any multisensor/parameter system and future work will concentrate on applying these optimisation techniques to other multisensor applications.

REFERENCES

[1] K.C Persuad, G Dodd. 1982. "Analysis of Discrimination Mechanisms In The Mammalian Olfactory System Using A Model Nose". Nature. Vol 299 pp 352 - 355.

[2] K.C Persuad, P Pelosi. 1985. "An Approach to an Artificial Nose". Artificial Internal Organs. Vol 31 pp 297-300.

[3] T.D Gibson, O Prosser, J.N Hulbert, R.W Marshall, P Corcoran, P Lowery, E.A Ruck-Keene. 1996. "Detection And Simultaneous Identification Of Micro-organisms From Headspace Samples Using An Electronic Nose". EuroSensors

[4] P.Lowery, P.Corcoran. 1997. "The Enhanced Classification Of Loose Leaf Teas Using A Thermally Cycled Gas Sensor Array And Neural Network Pattern Recognition". Eurosensors XI Conference, Poland, September 97. ISBN 83-908335-0-6

[5] D. Beasley, D.R. Bull and R.R. Martin, An Overview of Genetic Algorithms: Part 1, Fundamentals, *University Computing 15(2)*, (1993) p58-69

[6] L. Davis, Handbook of Genetic Algorithms, *Van Norstrand Reinhold* (1991)

[7] P. Corcoran and P. Lowery, Neural processing in an electronic odour sensing system, *Proc. 4th Int. Conf. on Artificial Neural Networks, IEE*, (1995) p415-420

A prototype knowledge-based system for bore quality

S ARUNACHALAM, S M NGUM, and **J M O'SULLIVAN**
Division of Manufacturing System Engineering, University of Hertfordshire, Hatfield, UK
K F GILL
Department of Mechanical Engineering, University of Leeds, UK

ABSTRACT

To date, abrasive reaming has received very little attention towards technical advancements, in comparison with other forms of hole-finishing techniques such as numerically controlled grinding or honing machines. This paper presents the development of a prototype knowledge-based system for abrasive reaming process. The main function of the system is to provide users, with selection of reaming parameters at a given set of condition which will enable the users to obtain a better overall bore quality required by the product specification. The reaming parameters considered in the KBS is derived from Taguchi based experimental analysis.

1. INTRODUCTION

Abrasive reaming involves many variables, governing the bore quality (1). It is essential to identify the effects of these parameters and their interactions, for maximum benefit. Currently, the success of the process is largely dependent on skill and very manually oriented. This is very time consuming and prone to high scrap rates and low productivity (2),(3). It is expected that a knowledge-based system with suitable machining parameters shall significantly optimise the process leading to better bore quality, higher production rates and higher efficiency.

A knowledge-based system is a knowledge intensive program capable of producing solutions to a range of problems, which generally require the knowledge of a human expert to provide an acceptable solution (4),(5). The prototype system discussed here is capable of assisting the reaming operator to identify the process related problems such as out-of-control situations and to suggest alternative actions to solve the problems.

2. CRITICAL PROCESS PARAMETER SUPPORTED BY EXPERIMENTAL DATA

The tooling in abrasive reaming consists of a barrel shaped sleeve, electroplated with cubic boron nitride or natural diamond, mounted on a tapered mandrel. The process uses a floating system to eliminate misalignment between the bore and the reamer axis (6). Bore quality produced by abrasive reaming is assessed by the accuracy of internal diameter, roundness, cylindricity, straightness and surface finish.

The overall bore geometry determines the functional features of reamed bores; so its control and predictability are of great importance. There are several problems likely to be encountered in abrasive reaming such as, production of out of roundness bores, barrel shapes, metallurgical damage of surface layer, etc. It is therefore important to identify the influence of process variables and their interactions for improving the bore geometry. Figures 1-5 show the key parameters influencing the reaming behaviour and the bore quality characteristics.

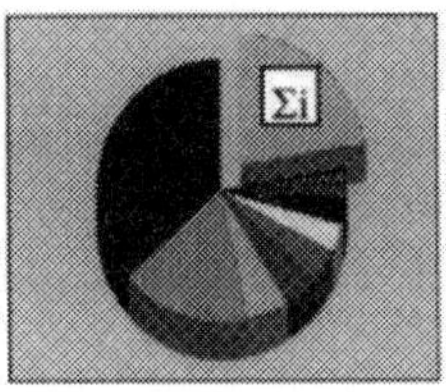

Figure 1: Significant Factor influences for surface finish

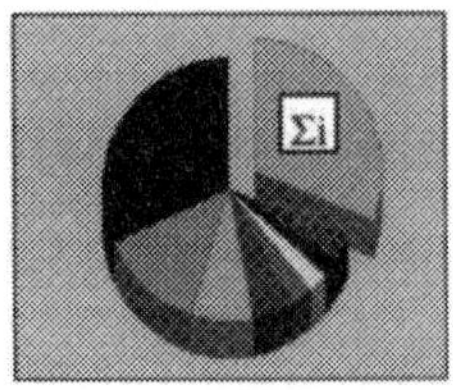

Figure 2: Significant factor influences for bore diameter

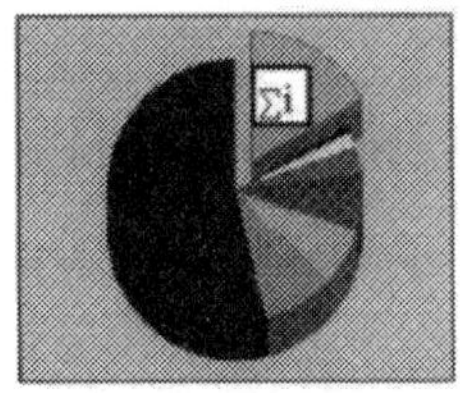

Figure 3: Significant factor influence for roundness

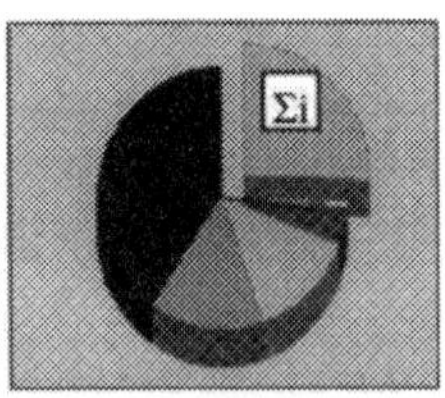

Figure 4: Significant factor influence for cylindricity

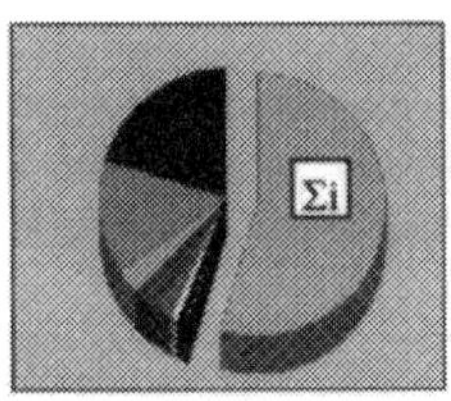

Figure 5: Significant factor influence for straightness

Σi- Interactions, others & error
- Grit size
- Stock removal
- Dressing
- Reamer speed
- Feed rate
- External geometry
- Gimbals centre
- Plain or grooved
- Sleeve parallel length

3. CONCEPTUALISATION OF KNOWLEDGE-BASE

The model of Figure 1 represents a lower level decomposition of the conceptual model for the superabrasive reaming (SAR) KBS. In principle, the model highlights the knowledge and data relationships - inputs, outputs, controls and mechanisms, governing the engineering of the proposed system (7), (8).

Each box represents a relevant system property loosely called an activity which SAR must consider to achieve an optimum solution. Derived knowledge (A31), for example, examines data which, has initially been acted upon by the knowledge base and rejected by phase one of a solution process. Rules are then fired which attempt to study the nature of this data (Table 1) and derive some relevant process by a different set of reasoning mechanisms (9). It may, of course, fail to yield any useful data. Hence, the feedback loop ensures that all possible perspectives of a rule are examined. This is the basis of reasoning agents in this context. A successful derivation results in an output classed as 'data relationship & constraints'. All these operations are heavily monitored by the nature of the database for reaming - hence the control factor (C1). The development environment for implementation of the model constitutes the mechanism (M1).

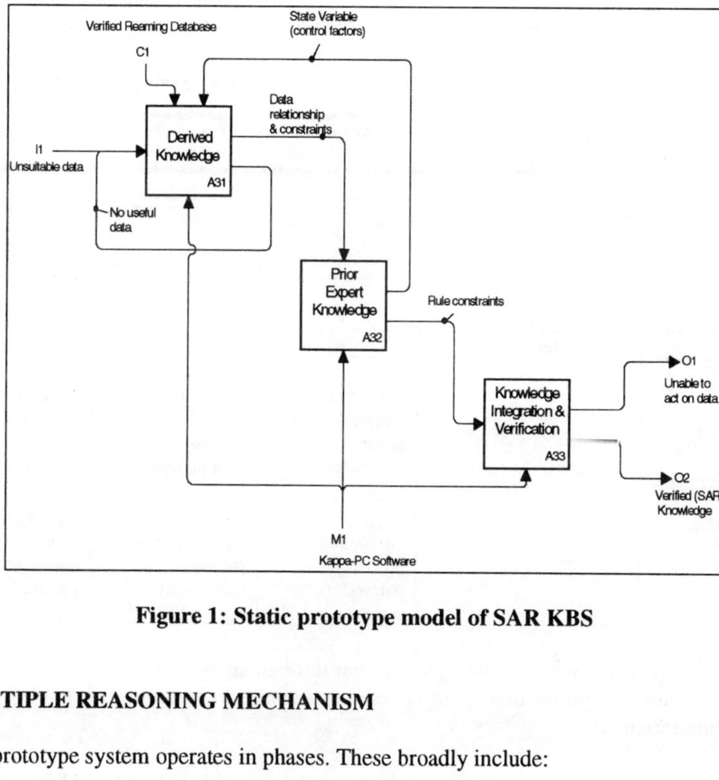

Figure 1: Static prototype model of SAR KBS

4. MULTIPLE REASONING MECHANISM

The prototype system operates in phases. These broadly include:

PHASE 1	• characterising the behaviour and features of SAR, so that we can formulate the SAR problems
PHASE 2	• formalising the problem and the problem-solving elements (agents)
PHASE 3	• computerising the various pieces of information regarding the problems and their solutions, so that a decision support can be developed.

Table 1 below illustrates the primary considerations by the SAR Knowledge-Base System. It is derived from a Taguchi-based experimental analysis at 95% & 99% confidence levels (10).

This allows only the critical factors such as reamer rotational speed, feeds, nature of dressing etc. to be included in the control system and as a result improve on the succinctness of the database of facts. Each agent, as illustrated in the table below, monitors a manageable range of these factors. For example, Agent 1 will base its decision(s) on factors that result in the optimisation of surface finish of a bore. Agent 2, on the other hand, though using some common factors will be responsible for factors that influence cylindricity of the bore. Thus, an agent is a problem-solver in a constrained problem domain as illustrated in Table 1.

Table 1: Example of process considerations for an example agent.

		SYSTEM CONTROL AGENTS		
MODEL FRAMEWORK	**Examples of Influential Factors**	**AGENT 1**	**AGENT 2**	**AGENT 3**
INPUT	• Tool geometry			
	• Reamer speed	✔		
	• Depth of cut	✔		✔
	• Fixture configuration			✔
	• Nature of abrasive		✔	
	• Specimen	✔	✔	✔
	• Feed rate	✔	✔	
CONTROL		Database for Maximum allowable initial process parameters.	Database for Maximum allowable initial process parameters.	Database for Maximum allowable initial process parameters.
MECHANISM		Kappa-Pc software	Kappa-Pc software	Kappa-Pc software
OUTPUT		Optimum process parameters for surface finish	Optimum process parameters for cylindricity	Optimum process parameters for roundness

The results of these monitoring is brought together through an overall co-operation mechanism for achieving process settings that result in optimum reaming conditions. This the essence of a rule structured such as:

IF (*'Agent 1' & 'Agent 2'*) // results of rules fired by agents 1&2

THEN

(*Abrasive type is.......*) // suggested optimum conditions
(*Optimum speed is....*)
(*Optimum feed is......*)

5. CONCLUSION

This paper has highlighted the importance that a Knowledge-Based System can contribute in the selection of optimum parameters to produce a well defined bore geometry. The SAR prototype system assists the operator in selecting the most appropriate reaming conditions to achieve the most optimum bore quality based on learnt facts about previous reaming experiments. As stated earlier, the validity of the learnt facts, and as such the knowledge base, has been developed based on a Taguchi experimental analysis.

6. FURTHER WORK

Feedback is also being sought from experts in the field of abrasive reaming in association with the limited research findings available(1). Given the demands currently being placed on the prototype system, further proofing of the database of facts is required. Current efforts are being directed towards engineering the acquired knowledge as discussed in this paper. This will be the subject of a subsequent paper in the very near future.

7. ACKNOWLEDGEMENT

This project is supported by the International Scientific Co-operation of the European Commission. The authors duly acknowledge the assistance of DK Holdings Limited, Kent.

REFERENCES

1. S.Arunachalam and J.M.O'Sullivan, *Superabrasive Reaming - An Innovative Technique for Bore Finishing,* Industrial Diamond Review, Vol.57, No.574, pp.76-80, 1997
2. J.Miller, *Superabrasive Reaming of Bores In Cast Iron,* Industrial Diamond Review, Vol.56, No.2, pp.52-57, 1996
3. S.Arunachalam, I. McAndrew and J.M.O'Sullivan, *Barriers to Implementing New Technology Within a QS-9000 Framework*, International Journal of Technology Management, Oct. 1997 (to be published)
4. Pham, D.T. (ed)., *Expert Systems in Engineering,* IFS (Bedford) and Springer Verlog (Berlin), 1988
5. A.R. Mirzai,. and J.R.Leigh, *An Overview of the Application of Neural Networks in Process Engineering,* Computing and Control Engineering Journal, pp. 105-108, 1992
6. S.Arunachalam, J.M.O'Sullivan and J.Miller, *Design of Equipment for Experimental Investigation of Superabrasive Reaming Process,* 13th NCMR, Glasgow, pp.1-5, 1997
7. A.H. Kobbacy, *The Use of Knowledge-based Systems in Evaluation and Enhancement of Maintenance Routines,* International Journal of Production Economics, Vol.57, No574, pp.76-80,1997
8. G.Russel, *Application of Knowledge-based Expert System to the Design of VLSI Circuits,* Computer-Aided Engineering Journal, pp.83-90, 1991
9. S.M.Ngum and K.F.Gill, *A Methodology for Shrinkage Predictions in Powder Pressed Parts Using a KBS,* Matodor Conference, UMIST, PP.146-151, April 1996
10. K.R.Ranjit, *A Primer on the Taguch Method,* Van Nostrand Reinhold, 1990

Reasoning about conditional effects in control systems via directed domain constraints

A AYESH and **G KELLEHER**
CMS School, Liverpool John Moores University, UK

Abstract

In this paper a manufacturing system will illustrate the need to represent conditional effects in control systems. Control system operators will be described showing the need to represent and reason about conditional effects in order for the control system to perform the required tasks, as well as having a reasonable way of defining actions. An automated generation process [1, 2] is used to derive interesting facts about the actions' consequences and represent them as Directed Domain Constrains (DDC) [2]. Applying this automated reasoning method to the prototype system will highlight its benefits in reasoning about control systems with conditional effects.

1. Introduction.

The manufacturing system that is used here is a prototype that represents systems where there are different objects with different sizes and shapes that need to be placed in different places. It is assumed that there is a robotic arm and a control system for the arm that selects actions from a given action repertoire. The description of the actions used shows the need to represent and reason about conditional effects in order for the control system to perform the required tasks as well as having a reasonable way of defining the actions. An automated generation process [1, 2] is used to derive interesting facts about the actions' consequences and represent them as Directed Domain Constrains (DDC) [2]. The original theoretical account of DDC that is based on classical STRIPS [3] will be presented. The modification to that model, as well as the expansion of STRIPS representational language to cope with conditional effects will be explained. Applying this automated reasoning method to the prototype system will highlight the benefits of using such a method in reasoning about control systems with conditional effects.

2. Background - STRIPS and Directed Domain Constraints.

STRIPS is a well known action description language from the planning literature. A STRIPS-based planning system has two main components which need to be represented: world models (states) and actions. World models are represented by sets of first order sentences [4]. Actions represent the changes that are possible within the domain and the situations in which they can be applied. The actions (also termed operators) are represented as a triple (P, D, A) where: P is the set of preconditions that has to be true in order to execute the operator. A is the add list which consists of the conditions that are going to be true (added to world model state). D is the delete list which consists of the conditions that are going to be deleted from the world model state.

Directed Domain Constraints [1], [5] and [2] take the form of implications and express the idea that given some state of the world it will always be true that in any world containing that state some other state will also be present (but not necessarily vice versa). They are typically intended to represent the basic "laws" of a domain. For example they provide information such as the fact that (under a standard blocks world ontology) a block cannot be both clear and have something on top of it. This could be represented as (On a b) -> (not (Clear b)).

The DDC conditions that are provided in [1] describe a set of conditions which, if satisfied by a set operators, allow the DDC to be inferred. These conditions are:

Directed Mutual Persistence (DMP) : This definition is to ensure that when two states are introduced into a world model the second state persists for at least as long as the first. This is achieved by defining two conditions for elimination which are Mutual Elimination (ME) (both conditions e.g. X Y are deleted at the same time) and Elimination Blocking (EB) (the deletion of the first condition e.g. X is required in the pre-conditions of the operator that deletes the second condition e.g. Y).

DMP is denoted by (π S1 S2) to mean there is DMP relationship from S1 to S2.

$$(\forall x)(\forall y)(\forall o)\ ((\pi\ x\ y) \Leftrightarrow ((\text{operator } o) \Rightarrow\ ((\neg\ y \in (\text{post-conditions } o)) \Rightarrow ((\neg\ x \in (\text{post-conditions } o)) \vee ((\neg\ x \in (\text{preconditions } o))).$$

Intuitively it is obvious that, provided we can guarantee DMP from one state to another, if we can ensure that every world which might have the first state introduced into it also has the second state, then there will be a directed domain constraint from the first state to the second. The means by which this is ensured is by requiring one of the following two conditions to hold.

Persistent Precondition (PP) : If the first state is introduced then the second must already be extant. The PP definition as follows (denoted Ψ):

$$(\forall y)(\forall o)\ ((\Psi\ y\ o) \Leftrightarrow (((\text{operator } o) \wedge (\ y \in (\text{preconditions } o))) \Rightarrow\quad (\neg\ y \notin (\text{post-conditions } o))).$$

Mutual Assertion (MA) : Covers the case where, if the first state is introduced into a world model then the second is also introduced.. MA will be denoted by (μ S1 S2 Op). The definition of MA as Kelleher provided is :

$$(\forall x)(\forall y)(\forall o)\ ((\mu\ x\ y\ o) \Leftrightarrow (((\text{operator } o) \wedge (x{\neq}y) \wedge ((x \in (\text{post-}\quad \text{conditions } o))) \Rightarrow\quad (y \in (\text{post-conditions } o)\ ))).$$

Formalisation of Directed Domain Constraints : DDC formalism is the conjunction of DMP condition an either of PP or MA conditions [2, p9].

$$(\forall x)(\forall y)(\forall o)\ ((\text{operator } o) \wedge (x \in (\text{post-conditions } o)) \wedge ((\mu\ x\ y\ o) \vee (\Psi\ y\ o)) \wedge (\pi\ x\ y))) \Rightarrow\quad (\delta\ x\ y)).$$

Definition: DDC Generation Process.

Providing two properties X and Y it can be said there is DDC between from X to Y iff there is PP or MA (establishing conditions) from X to Y to establish DDC and there is DMP (maintaining conditions) from X to Y to maintain the DDC, then DDC will be correctly generated in the format $X \Rightarrow Y$.

3. Conditional Effects: Formalism.

To formalise conditional effects, the idea of primary and secondary pre- and post- conditions needs to be introduced [6, p96]. The main idea that there are primary pre-conditions that should be always satisfied to execute the operator, and primary post-conditions that should be always asserted after the operator's execution. The secondary pre-conditions on the other hand may or may not satisfied, thus the secondary post-conditions may or may not asserted. Moreover the assertion of the secondary post-conditions depends on which of the pre-conditions are satisfied.

Conditional effects could be represented formally as disjunction between the primary pre-/post- conditions and the primary pre-/post- conditions and secondary ones. Thus an operator with conditional effects can generally be represented by

4. Manufacture System for Stocking Components

In computer manufacturing systems, there are typically machines that fit components to the motherboard circuits. These parts (gates, etc.) have different shapes, sizes and locations on the motherboard. The robotics hand(s) should be adjustable to cope with the differences in shapes and sizes. At the same time, the robotics arm(s) should fit the parts in the right places on the motherboard according to part that is currently held by the robotic arm. As the details of the system is unimportant to the communication of the ideas here we will concentrate on the pedagogic example.

There are three type of objects, big boxes, small boxes and balls. There is a robotic arm that can pick up one object at the time. The robotic hand can adjust itself according to the object measures. However, the user is allowed to assert Default for the default size of the objects.

The substitution of variables in this system is based on the returned facts from another system, such as visual system or sensing system. Giving the robotics hand the size it should adjust itself in case the size is not the default one. The visual system gives the status of the object as well as the size, if the object is damaged, then the object should goes either to the store4 or to any other default bin store.

R = Robotic arm that can be adjusted as well as given a default size.

Obj. = Objects $\in$ {Big-box, Small-box, Ball}.

The actions, that are expected from this system to be able to perform, as they are represented in STRIPS-like format thus:

(Detect-Obj.
Pre-conditions: Obj.-available
Post-conditions:

(Adjust
Pre-conditions: $\neg$ Default - measures, R-available
Post-conditions: R-adjusted,
(When big-box then big-box-adj.)
(When small-box then small-box-adj.)
(When ball then ball-adj.)).

(Pick-up
Pre-conditions: R-available, R-adjusted, Obj-available

Post-conditions: ¬ R-available, R-hold-Obj.)
(When ¬ Obj-Damaged then R-status-store)
(When Obj-Damaged then R-status-bin)).

(Store
Pre-conditions: R-hold-Obj., R-status-store
Post-conditions: R-available,
(When big-box then In-store1)
(When small-box then In-store2)
(When ball then In-store3)).

(Bin
Pre-conditions: R-hold-Obj., R-status-bin
Post-conditions: R-available, (In-store4 OR In-default-bin)).

5. Dealing with Conditional Effects

The operators 'Adjust', 'Pickup' and 'Store' have conditional effects the "when-clause". Since the original DDC conditions and algorithms cannot deal with operators with conditional effects, conditional effects should be dealt with before the DDC algorithms can be applied. The proposed solution is to look at conditional effects as a disjunction of pairs where there the first element of the pair is the condition in the when-clause and the second is the effect of when-clause. Once they come to be disjunctive sentences the decomposition process of disjunctive sentences that is described in earlier work [7] can be applied. The result of the decomposition process will be conjunction of pairs. Each conjunction of pairs will be added to a generated operator. The condition part of each pair will be included in the pre-conditions of the operator that contains that pair, so the effect part will be added to the post-conditions. By this method a set of standard STRIPS-like operators will be generated which represent the non-standard operator in terms of effects (reachable states). At the same time the original DDC algorithms can be applied. As an example consider the pickup operator.

(Pick-up
Pre-conditions: R-available, R-adjusted, Obj-available
Post-conditions: ¬ R-available, R-hold-Obj.)
(When ¬ Obj-Damaged then R-status-store)
(When Obj-Damaged then R-status-bin)).

There are two pairs in a disjunctive sentence ((¬ Obj-Damaged R-status-store) OR (Obj-Damaged R-status-bin)). This produce the following combination via the disjunction re-writing process.

(¬ Obj-Damaged R-status-store)
(Obj-Damaged R-status-bin)
((¬ Obj-Damaged R-status-store) and (Obj-Damaged R-status-bin))

This leads to the creation of four operators

(Pick-up-1
Pre-conditions: R-available, R-adjusted, Obj-available
Post-conditions: ¬ R-available, R-hold-Obj).
(Pick-up-2
Pre-conditions: R-available, R-adjusted, Obj-available, ¬ Obj-Damaged
Post-conditions: ¬ R-available, R-hold-Obj, R-status-store).
(Pick-up-3

Pre-conditions: R-available, R-adjusted, Obj-available, Obj-Damaged
Post-conditions: ¬ R-available, R-hold-Obj, R-status-bin).
(Pick-up-4
Pre-conditions: R-available, R-adjusted, Obj-available, ¬Obj-Damaged, Obj-Damaged
Post-conditions: ¬R-available, R-hold-Obj, R-status-store, R-status-bin).

6. Generating DDC from The Previous Operators.
As an example of generating DDC, let us take the Pick-up operator. There is a potential DDC between ¬R-available and R-adjusted since they satisfy the Persistent Pre-conditions condition (PP). Looking through the other operators there is no operator that denies R-adjusted without denying ¬R-available. In fact there is no operator that denies R-adjusted. Thus the DDC from ¬R-available to R-adjusted can be established.

The other example is a potential DDC between ¬R-available and R-hold-Obj since they satisfy the Mutual Assertion (MA) condition. However, this DDC is denied by two operators (store and bin) because ¬R-available is denied in those two operators while R-hold-Obj is not. At the same time this DDC represents a natural fact with regards to the given operators, when the robot is not available then the robot should hold an object. This implies an error in the operators description which can be found in the fact that store and bin operators do not assert ¬R-hold-Obj which will make the system stop operating after the first stacking of an object because it will never be able to execute Pick-up operator again. This discussion leads us to the next section in which the use of DDC as a reasoning and analysis tool will be discussed.

7. Using DDC as a Reasoning Tool.
A potential and probably expected DDC has failed the DDC conditions in the previous section because of incorrect operator description. Once this DDC is not generated the correctness of the operators' description would be questioned. This shows the potential of DDC as a reasoning and analysis tool that enables us to debug the system [8] and determine its behaviour in advance. In addition it enables us to predict deadlocks, where the system stops operating because of an error occurred (lake of resources, fault in design, etc.), in the system as it is the case in the previous section.

It can help to determine, by generating DDC, whether the system matches the required specifications or not since DDC shows the possible states (behaviour) of the system given the specifications. However only the generated DDC can be used, since the current algorithms do not show all the potential DDC; whereas some failed potential DDC may indicate errors in the description in a way that is useful as the actually generated DDCs. At the same time the generated DDC would be more useful if it was known where they are generated from so the operator that does not match the given specifications can be pursued.
We propose here to create two tables that hold information about the generation of DDC. The first table contains information about the generated DDC, while the other contains information about the potential failed DDC.

The first table provides information about the operators that establish the DDC (i.e. either PP or MA conditions is satisfied) and the operators that maintain the DDC (i.e. DMPR is satisfied). The second table provides information about the operators that establish the potential DDC (i.e. either PP or MA conditions is satisfied) and the operators that make the potential DDC fails (i.e. DMPR is not satisfied). By maintaining these tables, DDC can be used as a reasoning and analysis tool during the design of the manufacturing system where some of the operators that cause the deadlock or incorrectness of the system (with regards to given specifications) can be identified and dealt with.

8. Conclusion and Future Work.

In this paper, the use of DDC as a reasoning tool for manufacturing systems has been reviewed. The need to reason about conditional effects has been discussed as well as a method of dealing with conditional effects. Since the original DDC generating algorithms do not deal with conditional effects an approach to re-write non-standard operators to standard operators has been demonstrated. The generation of some DDCs have been demonstrated and the motivations for using DDCs as an analysis tool to reason about manufacturing systems have been discussed. A proposal has been made to keep information about the generation of DDC that enable us to trace the incorrect operators. As future work the creation of these tables may enhance the DDC generating algorithms in aim to support a simulation system to be used as computer-aided design (CAD) tool for manufacturing systems.

9. References

[1] G. Kelleher and A. G. Cohn, "Automatically Synthesising Domain Constraints from Operator Descriptions," presented at Proceedings European Conference on Artificial Intelligence, Chester, 1992.

[2] G. Kelleher, "Determining General Consequences of Sets of Actions," Liverpool John Moores University, Liverpool 1995.

[3] R. E. Fikes and N. J. Nilsson, "STRIPS: A New Approach to the Application of Theorem Proving to Problem Solving," in *Readings in Planning*, J. Allen, J. Hendler, and A. Tate, Eds. San Mateo, CA, USA: Morgan Kaufmann Publishers, INC., 1990, pp. 88-97.

[4] V. Lifschitz, "On the Semantics of STRIPS,", *Readings in Planning*, J. Allen, J. Hendler, and A. Tate, CA, USA: Morgan Kaufmann Publishers, INC., 1990, pp. 523-530.

[5] G. Kelleher, "Planning for the Support of Computer Users," *Int. J. Man-Machine Studies*, pp. 925-964, 1993.

[6] A. Gerevini and L. Schubert, "Computing Parameter Domains as an Aid to Planning," presented at The Third International Conference on Artificial Intelligence Planning Systems (AIPS 96), Edinburgh, Scotland, 1996.

[7] A. Ayesh and G. Kelleher, "An Introduction to The Generation of Directed Domain Constraints (DDC) From more Expressive STRIPS-like Models I: Disjunctive Sentences,", The 15th Workshop of the UK Planning & Scheduling SIG, Liverpool, UK, 1996.

[8] A. Ayesh and G. Kelleher, "Aspects of Temporal Usability Via Constraint-Based Reasoning: The Use of Constraints as a Debugging Mechanism Within General Descriptions of The User's Possible Actions,", The IASTED International Conference on Artificial Intelligence and Soft Computing., Cancun, Mexico, 1998.

Integration of machining analysis with knowledge-based design

S C COOPER and **I-S FAN**

The majority of product cost although realised in production is determined during design. There is a need for tools that allow a better understanding of the implications of design on manufacturing. This paper briefly reviews research and currently available aids to support design for machining. An architecture is then proposed for a design for machining approach that exploits knowledge based engineering technology. Parts of this architecture are currently being realised in collaborative research with the UK civil aircraft industry.

1 INTRODUCTION

The decisions taken during the design process have the greatest influence upon product cost for the least investment (1). Design for Manufacture and Assembly (DFM/A) tools aim to facilitate the consideration, during design, of the implications that the design of a product has upon the cost of the manufacture and assembly of its parts. The need for such tools is accentuated by the current trend for Original Equipment Manufacturers (OEMs) to focus predominantly upon design and assembly whilst subcontracting much of their piece part manufacture and sub-assembly work to an extended supply chain.

The work presented here has been completed through a research project involving members of the UK civil aircraft industry. This industry is moving towards a mode of operation where an increasing proportion of the manufacture is subcontracted (2). In addition, factors including risk sharing partnerships, industrial off-set agreements and US dollar trading encourage the development of international, rather than local, supply chains (2, 3). In such an environment the designer in the OEM is remote from the manufacturing site and multi-functional teams are more difficult to assemble - although in practice they are formed if the contract in question is sufficiently large. A survey of 40 companies across Europe, North America and South-East Asia found the use of current DFM/A systems to be at a low level (4).

The work presented in this paper focuses particularly upon Design for Machining (DFM-Machining), machining being the predominant process used in the production of civil aircraft structures. A method is proposed by which DFM-Machining may be supported in the civil aircraft industry through exploitation of the Knowledge Based Engineering (KBE) systems that are being increasingly used in the sector. The next section of this paper will briefly review recent work in DFM-Machining and section three will then introduce the basic concepts of KBE systems. The fourth section will propose a framework for integrating DFM-Machining with KBE design applications. The fifth section then discusses the practical limitations of such an application, finally leading to brief conclusions.

2 RELATED WORK

A range of tools and methods are available to support DFM-Machining. Design guidelines are widely published and define basic good practice, backed up by graphical and tabular data to support the selection of appropriate materials, tolerances and surface finishes. Parametric cost estimation tools, which are constructed by analysis of previous designs to identify correlation between manufacturing cost and parameters such as material, production process and part size, are also available. However, the former of these approaches is unable to provide the designer with a cost estimate of a part and neither can adequately indicate how costs are associated with individual features.

In order to support DFM-Machining at a detailed level it is necessary to construct a process plan. The activities entailed in doing this are listed below.

- Selection of machine tools
- Selection of machining operations
- Sequencing of machining operations
- Determination of set-up requirements
- Design of jigs and fixtures
- Selection of cutting tools
- Calculation of cutting parameters
- Tool path planning

Commercially available DFM-Machining tools require the user to provide most of this information. They contain routines that can estimate the costs of individual cuts but it is left to the user to devise an economical means to machine the part. Similarly, although contemporary CAM systems include powerful routines to generate tool paths based upon the CAD model, it is still the user who must determine how the part should be cut. Both approaches require the user to be sufficiently knowledgeable of the machining process to be able to devise an appropriate process plan. However, detailed knowledge of how a part may be machined is not always available during design.

A large amount of research has been completed in generative Computer Aided Process Planning (CAPP) in an attempt to automate more of the process planning activities. However, as recently noted by Shah and Mäntylä (5), there remain several problem areas. The first step in deriving how a part may be machined entails obtaining its machining features. The recognition of feature interactions remains a problem however this is achieved and yet is particularly important for process planning, where relative orientations and specified tolerances can affect the definition of set-ups and operation sequencing. In linking machining features to operations, it is possible to use a direct association, with final selection based upon parameters such as tolerances and dimensions. However the more general approach, which considers the volume that can be removed by a cutter swept along a path is rather more complex and is not easily generalised to all types of features and processes. McMahon and Browne (6) further note that there can be problems of combinatorial explosion in machining sequence generation. They also report that although a large number of generative process planning systems have been reported in the research literature, few are in use in industry.

Through their work with the civil aircraft industry the authors have found that in many cases it may be possible to apply a plan based DFM-Machining approach that avoids many of the difficulties associated with generative CAPP. Although technically advanced, much of the structure of a civil aircraft is conceptually mature. Parts having basically similar geometry with varying dimensions and details are repeated throughout the structure and, with a little more variation, across aircraft models. The rules that determine the basic geometry of these parts are well known and the industry has exploited this by using KBE systems, such as Concentra's ICAD (7), to partially automate the design of such items. Similarly, although there are some differences subject to the eventual supplier, the machining of these items is also relatively well known. Given that the geometry created by the KBE design application is in a known format, there is an opportunity to make use of the machining knowledge applied in the planning of previous parts to analyse how new items may be produced.

3 KNOWLEDGE BASED ENGINEERING DESIGN

KBE systems are used to automate the design of certain components in various industries including aerospace and automotive. They may be used when the geometry of the part is defined by a set of known rules. Considerable savings in design time can be achieved by application to families of parts which have geometry that is subject to generic rules or if the design is likely to proceed through a series of iterations during development. The general architecture of a KBE design application is shown in Figure 1. The model of the generic part consists of a series of objects arranged in a hierarchical tree structure. If the part being modelled can be divided into a series of similar objects then it will normally be the case that a single, low level, generic object will be coded and its parent constructed so as to create multiple instances of the child. It is often the case that these low-level objects can be coded to be generic across a range of different models, hence reducing the coding effort. A specific instance of a part is created by specifying the inputs of the top-level part. This then enables the specification of the inputs to the objects in each successive level of the tree with multiple instances of lower level objects being created as necessary.

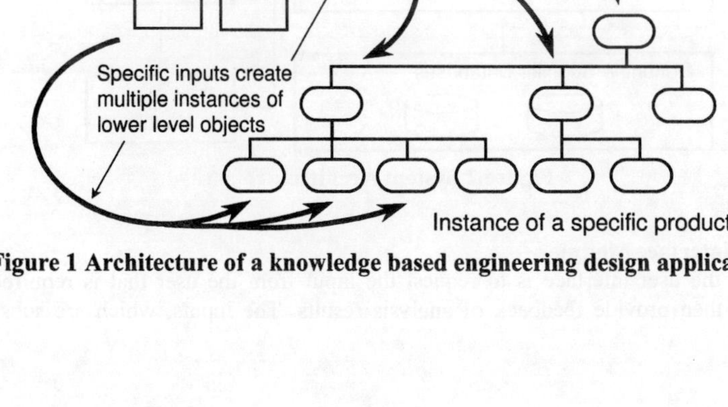

Figure 1 Architecture of a knowledge based engineering design application

4 KNOWLEDGE BASED DFM-MACHINING

A DFM-Machining system using KBE technology and integrated with a KBE design application is proposed. The architecture of the system is illustrated in Figure 2.

4.1 Design Model

The design model provides the input to the system, rather than being an element of the system itself. It is being developed independently by the collaborating OEM and its role in this application is to provide access to the part geometry in a known format. It is assumed that the constraints incorporated in the design model ensure the parts it generates can be machined.

4.2 Machining Feature Specification (MFS) Module

The MFS module takes the geometrical model and generates a representative set of machining elements. A modular implementation is used, to allow maximum reusability of low-level objects. The part level object is dedicated to a particular part family. The feature set level objects represent commonly occurring feature sets, such as a group of nested pockets. The feature level objects are generic machining features such as pockets, holes and slots. The machining elements generate the paths of individual cuts required to machine each feature. Sequencing constraints are passed down the object tree as the part is decomposed. The interactions between the MFS module and other parts of the system will be explained in the following descriptions of the other system modules.

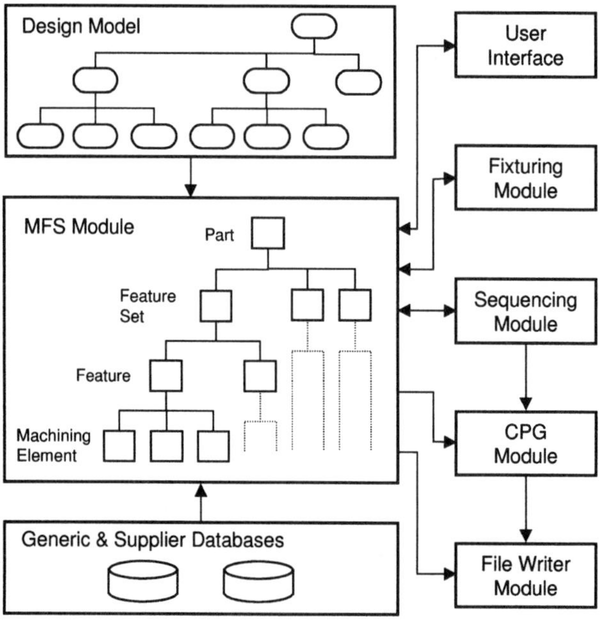

Figure 2 System Architecture

4.3 User Interface Module

The role of the user interface is to request the input from the user that is required by the system and then provide feedback of analysis results. The inputs, which are subsequently provided to other system modules, include the specification of the part, billet, machine tool and machine origin. The feedback provided includes details of the fixturing points and elements, a list of required cutters, cutting time for the whole part and individual features, and representations of relative cutter usage and feature machining time.

4.4 Fixturing Module

The parts considered in this work are predominantly fixtured by bolting the workpiece to a base plate at points which are either outside of the finished part or inside large areas of waste within the part. During machining, channels are cut between the part and the waste areas to leave them attached only by small tie bars. The fixturing module contains rules based upon existing practice to calculate the location of these bolts and tie bars. This then enables the MFS module to ensure that the generated features will not interfere with fixturing elements.

4.5 Sequencing Module

The role of the sequencing module is to specify the order in which to machine the machining elements. Ordering is primarily determined from the precedence constraints defined during the instantiation by the MFS module. These precedence constraints relate to access requirements, avoidance of component deformation and current practice. Remaining sequencing aims to minimise cutter changes and non-cutting movement. As sequencing is completed there is interaction with the machining elements, to access the cutter position at the end of each cut and to provide the start point for the next cut.

4.6 Cutter Path Generation (CPG) Module

The role of the CPG module is to stitch together and process the cutter location, speed and feed data generated by the machining elements, in the order specified by the sequencing module, to produce a complete representation of the cutter path. This includes the insertion of appropriate tool change and rapid feed commands between the machining element paths.

4.7 File Writer Module

This module provides a permanent record of the analysis results, including the user inputs and the generated cutter path.

4.8 Generic and Supplier Databases

The generic and supplier databases contain details of machines and cutters for use at various points during the analysis. The generic database provides a default source of information on cutting tools and machining parameters. The supplier specific database provides details of specific machine types and capabilities, and cutting tools. Machine information includes the available axes of movement and spindle power, which can affect machining element definition. The cutter data is used during cutter selection and the specification of cutting parameters.

5 DISCUSSION AND CONCLUSIONS

The architecture presented above considers many of the activities involved in process planning. One activity that is left solely to the user, is the selection of the machine tool. The authors have found that in practice, machine tool selection is based primarily on component size and available plant capacity. The second of these factors may not made available as a matter of course by potential suppliers. Hence it was considered preferable to allow the user to specify the machine tool after discussion with the supplying company.

Several other activities are completed by the system but are left for the user to optimise. For example, cutter selection is completed at the feature level and as such leaves open the possibility for reducing machining cost by optimising the choice of cutters across the whole part. However, given the complexity of implementing an automated routine which fully optimises cutter selection with respect to machining cost it was considered preferable to leave this task to the skilled process planner. The restricted range of standard cutters means that in any case the benefits achievable from optimisation could be limited. A similar philosophy pervades in the sequencing of the process plan and the determination of cutter parameters. In both cases the choices which the system must make are restricted based upon current practice. The consequence is that although the resulting plan may not be optimal, it should be realistic. This is sufficient to provide DFM-Machining feedback to the designer. Once design is complete, the process plan can be refined using the expertise of a process planner. The aim in this work has not been to completely automate the manufacturing planning, only its most repetitive tasks, firstly to provide feedback to the designer and secondly to allow the process planner to focus their expertise on how to produce an optimum plan.

In conclusion, an architecture has been presented for a DFM-Machining approach which exploits KBE technology to allow analysis of the members of part families. The realisation of a system based on this architecture is a major task. Current work is focusing on a family parts used in civil aircraft structures, analysing their machining in consort with their suppliers to identify how the design can be decomposed into its machining elements. The use of supplier data in the architecture has been described. Another stream of work has produced a mechanism for accessing supplier capability information using Internet technology. Further details of this part of the research are provided by Roy *et al* (8).

6 ACKNOWLEDGEMENTS

This research is supported by the UK Engineering and Physical Sciences Research Council (EPSRC) under grant GR/K95758. The authors also acknowledge the valuable assistance given in this work by the collaborating companies.

REFERENCES

1 **Ullman, D.G.** *The Mechanical Design Process*, 1992 (McGraw Hill).
2 **Eriksson, S.** *Global Shift in the Aircraft Industry: A Study of Airframe Manufacturing with Special Reference to the Asian NIEs*, 1995 (University of Gothenburg).
3 **Hayward, K.** *The World Aerospace Industry: Collaboration and Competition*, 1994 (Duckworth and RUSI, London)
4 **Fan, I.S., Cooper, S., Sehdev, K,** and **Williams, G.** Design for Manufacture in a Multi-Tier Acquisition Environment. *Proceedings of CALS Europe '96*, Paris, 29-31 May 1996.
5 **Shah, J.** and **Mäntylä, M.** *Parametric and Feature-Based CAD/CAM: Concepts, Techniques and Applications*, 1995 (John Wiley & Sons).
6 **McMahon, C.** and **Browne, J.** *CADCAM: Principles, Practice and Manufacturing Management*, 2nd ed, 1998 (Addison-Wesley).
7 **ICAD** Concentra Corporation, Cambridge, Massachusetts
8 **Roy, R., Cooper, S.C., Fan, I.S., Sehdev, K.** and **Ward, N.J.** Management of supplier capability information in the extended enterprise using Internet technology. *5th ISPE International Conference on Concurrent Engineering*, Tokyo, 15-17 July 1998.

The integration of a knowledge-based system and a Gantt chart scheduling application for solving the machine shop scheduling problem

S LOKE, S ARUNACHALAM, and **T PAGE**
Manufacturing System Engineering Department, University of Hertfordshire, Hatfield, UK

ABSTRACT

In many companies, the scheduling process is usually complex, time consuming, and only understood by a limited number of personnel intimately involved with the manual task and thereby effecting productivity. This paper discusses the application of Artificial Intelligence to improve the efficiency in Job Shop Scheduling. The work involves the development of a knowledge-based system which is then integrated with a Project Management program for the graphical scheduling process. The knowledge-based system will declare machine utilisation for a given machine-part code, derived from the information system database. Data from the Information system could be exported to the Gantt Chart scheduling program and also produce job cards and reports when required.

1. INTRODUCTION

A common problem for scheduling systems in the production industry is the assignment of customer orders to shop floor processing (i.e. deciding the sequence of orders to be carried out). As part of cost reduction program, management are implementing knowledge-based production control system. This would enable the manual scheduling procedures to be carried out more quickly and efficiently. The development of the knowledge-based system (KBS) discussed in this paper offers a user-friendly interface, which will result in improved flexibility enabling shop-floor staffs to make instant decisions on schedules assisted by the database. KBS assisted scheduling system also provides management with a visual tool for costing and analysis of relevant data [1].

2. JOB SHOP SCHEDULING

In a job shop, there is a range of jobs to be processed, each of which may have a different routing among departments or machines in the shop. The two important decisions an operations manager must make involve scheduling and sequencing the jobs. Technically speaking, scheduling is the process of assigning starting and completion times to jobs. Sequencing involves the determination of the order in which the jobs are processed. In practice, this distinction is not often made, however, and the term scheduling is used to mean both the timing and sequencing of jobs. In this instance, we are attempting to perform both sequencing and scheduling of a job shop with the aid of a knowledge-based system. The data extracted from the knowledge-based system (KBS) is then fed into a Project Management program, which generates Gantt charts and printouts. Reports and/or job cards could also be produced from the KBS.

Scheduling for shop floor operations can be described as the task of organising (n) operations to be executed on (m) machines, so that a global performance criteria (e.g. makespan) is optimised, and manufacturing constrained such as due date, processing time etc. (with known deterministic values) are satisfied. Previous works on this area showed that under certain conditions, some rules performed better than others indicating the inadequacy of classical off-line scheduling methods [2][3].

In many companies, the scheduling procedure applied is usually complex, and only understood by the personnel fully involved [4]. The scheduler determines when and what resources are required for the jobs to be accomplished. Learning curve for a new scheduler to acquire the expertise and understand the production environment is normally longer. The experienced scheduler generally retains the expertise in memory. This affects the efficiency of the productivity if the experienced scheduler was unavailable.

3. KNOWLEDGE-BASED SYSTEM

The following are the basic components of knowledge-based system namely, inference engine, knowledge base, working memory, knowledge acquisition module, and the user interface module. A knowledge base could not be used unless it had a good inference mechanism enabling it to use the knowledge in the system. Different types of inference engine are possible, rule-based depending on the type of knowledge representation scheme adopted. In a knowledge-based system, the inference engine examines facts and executes rules contained in the knowledge base according to the inference and control procedures selected.

Domain knowledge typically has many forms, including descriptive definitions of domain-specific terms, description of individual objects, classes of objects and their inter relationships, and criteria for making decisions. The power of a knowledge-based system lies in the specific knowledge of the problem domain, with potentially the most power system being the ones containing the most knowledge. The most popular approach to represent the domain knowledge needed for a knowledge is by production rules. Production rules are, in effect a subset of predicate calculus with an added prescriptive component indicating how the information in the rules is to be used during reasoning. The following is an example of production rule [5].

IF part P_i is to be dispatched to machine M_a that is occupied by another part P_j
THEN check availability of an alternative machine M_b cf100
where cf refers to the certainty factor of 100%.

4. IMPLEMENTATION OF AI TO JOB-SHOP SCHEDULING

The aim of all scheduling strategy is to achieve one or more of the following:

1. Completion of work by due-date
2. Maximum material utilisation
3. Maximum product output capacity
4. Minimum setting cost

However, before the problem definition can be addressed, a number of criteria need to be satisfied. These criteria would help us determine how well we are meeting the strategy that we have mapped out. The two principal criteria used to evaluate schedules for job shop problems are:

(a) Flowtime - time a job spends in a shop
(b) Lateness - the difference between completion time and due date.

Minimising flowtime reduces work-in-progress and inventory. Effect of late deliveries is difficult to quantify but the result could be detrimental to acceptance of goods, payment and goodwill. In any case, customer service is the prime consideration and the meeting of deadlines for delivery of finished goods is of paramount importance. Therefore, the selection of the appropriate criteria in setting up the conditions for the inference engine would enable us to manipulate the desired outcome of a schedule to meet the objective of our choice. This is available as items of the menu in the conditions module.

5. The paradigm of scheduling n parts over m machines.

A scheduling model is developed and defined by system constraints. A mixed-integer programming model for the problem of scheduling *n* operations with precedence constraints on *m* machines is presented. The scheduling problem considered is a generalisation of the scheduling problem presented in [7].

The main goal of the scheduling model is to minimise the total completion time of all components.

$$\text{Min} \sum_{i=1}^{n} x_{[i]}$$

subject to

$$x_{il} - x_{ik} \geq t_{il} \qquad \text{for all } [k, l] \in O_i, \text{ for all } i \qquad (1)$$

$$x_{il} - x_{ik} + M(1 - y_{kl}) \geq t_{il} \quad \text{for all } [k, l] \in O_i, \text{ for all } i \qquad (2)$$

$$x_{ik} - x_{il} + My_{kl}) \geq t_{ik} \qquad (3)$$

$$x_{jl} - x_{ik} + M(1 - z_{kl}) \geq t_{jl} \quad \text{for all } [k, l] \in N_p, \text{ for all } p \qquad (4)$$

$$x_{ik} - x_{jl} + Mz_{kl}) \geq t_{jk} \qquad i \neq j \qquad (5)$$

$$x_{ik} \geq t_{jk} \qquad \text{for all } [i, k] \in I_i, i = 1, \ldots, n \quad (6)$$

$$x_{ik} \geq 0 \qquad \text{for all other } i, k \quad (7)$$

$$y_{kl} = 0, 1 \qquad \text{for all } [k, l] \in Q_i, \text{ for all } i \quad (8)$$

$$z_{kl} = 0, 1 \qquad \text{for all } [k, l] \in N_p, \text{ for all } p \quad (9)$$

where: n = number of parts

n_i = number of operations of part O_i, $i = 1, \ldots, n$

m = number machines

R_i = set of pairs of operations $[k, l]$ for part Pi, where operation k precedes operation l, $i = 1, \ldots, n$

Q_i = set of pairs of operations $[k, l]$ for part Pi, where k and l can be performed in any order, $i = 1, \ldots$

I_i = set of operations without precedence constraints, $i = 1, \ldots, n$

N_p = set of operations to be performed on machine p, $p = 1, \ldots, m$

t_{il} = processing time of operation l of part P_i, $l = 1, \ldots, n_i$, $i = 1, \ldots, n_i$,

M = an arbitrary large positive number

x_{ik} = completion time of operation k of part Pi, $k = 1, \ldots, n_i$, $i = 1, \ldots, n$

$x_{[i]}$ = completion time of the root operation (the last in the sequence) of part P_i, $i = 1, \ldots, n$

$$y_{kl} = \begin{cases} 1 & \text{if operation } k \text{ precedes operation } l \\ 0 & \text{otherwise } [k, l] \in Q_i, i = 1, \ldots, n \end{cases}$$

$$z_{kl} = \begin{cases} 1 & \text{if operation } k \text{ precedes operation } l \\ 0 & \text{otherwise } [k, l] \in N_p, p = 1, \ldots, m \end{cases}$$

Expression (1) is a constraint imposing that the operations of each part are processed according to the precedence required. Expressions (2) and (3) ensure that any two operations belonging to the same part cannot be processed at the same time. Expressions (4) and (5) ensure that a machine cannot process more than one part at the same time. Constraints (6) and (7) impose non-negativity and (8) and (9) impose integrity.

6. IMPLEMENTATION OF KNOWLEDGE-BASED SYSTEM FOR JOB SHOP SCHEDULING

To build our knowledge base, a database is constructed using Microsoft Access. Production rules are used to represent knowledge in the knowledge-based system developed. A eight-digit numerical coding system was used to identify the various parts for machining. This is an identifying key in the tables and acts as the machine-part code for each part. Tables are developed to classify the parts according to the coding nomenclature e.g. if a part is tubular, rectangular etc. The main criteria for classification is based on the most appropriate machining method to be employed in due course and this is reflected in the part code. The assumption is made that 'if two parts have the same shape, they will be made on the same machines'. Initial data entry and classification has to be done manually by a knowledgeable person. Correspondingly, tables are developed for individual machines and related to many machines table enabling machine selection according to characteristics and functions available. To avoid duplication of data and to enable selection of control parameters e.g. manufacturing tolerances, resources, and materials, common components of similar tables are placed in one table and linked by relationships.

The scheduling model and selection criteria are then written as independent modules in Visual Basic to work as the main inference engine for the knowledge-based system. Indeed these modules are aided by numerous queries and special macros written to manipulate and collate the data in the different tables. Once the module for scheduling has been executed for a job, the results are stored or appended to a table in an appropriate format that could be opened as a data file in Microsoft Project. This data file is then uploaded into Microsoft Project by selecting Tools and open database file. Gantt charts of the scheduling and sequencing decisions are produced and further analysis to the charts could be done with modifications to parameters such as resources. On acceptance of the conditions produced, job cards with bar codes are generated as reports from Ms Access.

During production, the state of progress of jobs is monitored by installed bar-code readers next to the workstation. Data fed back to the main database is stored and used as knowledge for future scheduling and sequencing exercises. The availability of this data allows us to have at hand valuable information on the actual time needed for the various tasks being scheduling for future repetitive jobs.

7. DISCUSSION AND CONCLUSION

Although the method described might seem a long-winded way of achieving scheduling, it has been shown that the task can be done in this manner and with the aid of a knowledge-based system. In addition, a company with this system could eventually build up a very useful database of the operating capabilities of their machinery and a knowledge base of parts classification. Initial data entry would still need to be done manually by a person with knowledge of machining. A notable reduction in the number of people involved in the actual job of scheduling and the recurring of mistakes has been reduced considerably. These two factors have helped improved working conditions in the company.

A number of problems were identified, for example, the assumption made about any two parts having the same shape being made on the same machines is not entirely true. The two parts may be identical in shape, but could be made rightly on entirely different machines, because: they differ greatly in size, or one has much closer manufacturing tolerances than other, or one is required in much greater quantities than other, or because again, one has a special feature such as a keyway, which is not needed on the others. However, to simplify the model for the current moment of development this has been adopted for the current prototype.

Further improvement to the model is desired and is currently been worked on. The knowledge-based system developed is continuously tested and verified with the help of the production scheduler. It provides a daily feasible schedule automatically depending on the resources and manufacturing constraints. The developed knowledge-based system is a good decision aid to the production scheduler. In addition, the database being developed currently would become the cornerstone of the production information system of the company.

REFERENCES

1. Mason, D, Vinodrai, K.P, *Development of a computerised job-shop scheduling system for use in a small screen printing company*, proc. of the 13th national conf. on manufacturing, pp.311-315, 1997
2. Browne, J. (1986) *Production planning and control, simulation and AI technique*, proc.2nd advanced manufacturing seminar, Coventry Polytechnic, May 1986
3. Wbester,M, Alder, C, Hierarchical scheduling in a distributed electronics manufacturing environment, *Advanced manufacturing processes, systems and technologies*, pub.Mechanical Engineering, pp.549-558, 1996
4. Stecke,K.E. (1984), *Design, planning, scheduling and control of flexible manufacturing systems*, proc. TIMS-ORSA conf. on FMS
5. Pham, D.T (ed), Expert systems in engineering, IFS (Bedford) and Springer Verlag (Berlin) 1998
6. Kusiak, A. (1987). "Artificial intelligence and operations research in flexible manufacturing systems, "INFOR, Vol.25, No.1, pp2-12.
7. Baker, K.R. (1974). Introduction to Sequencing and Scheduling, John Wiley, New York

Gas/solid separation using a novel compact in-line vortex separator

T WHITE
School of Mechanical Engineering, Cranfield University, UK (also at BHR Group Limited, Cranfield, UK)
R THORN
School of Engineering, University of Derby, UK
S SASHAR
CALtec, BHR Group Limited, Cranfield, UK

SYNOPSIS

The need to separate solids from gas flows is a common problem in a number of industries. This research is concerned with the design and optimisation of a compact vortex separator for gas/solid separation duties. In-line vortex separators have an inherent reliability due to their simple design. They have been proven in demanding environments such as the nuclear, mining and automotive industries. To exploit further the advantages of the vortex separator, collaborative research has been underway to develop a novel gas/solid inline separator for use in natural gas production systems.

The starting point for work has been CALtec's oil/gas inline separator known as WELLSEP. Preliminary tests have been performed using air/sand mixture in the laboratory, and so far encouraging results have been obtained. Results from recent gas/solid separation experiments are presented, and the current status of the CFD (Computational Fluid Dynamics) modelling work is briefly described.

NOMENCLATURE

Eu	Euler Number (dimensionless)
ΔP	Static pressure drop across separator (Pa)
V	Characteristic velocity (m/s)
ρ	Gas density (kg/m^3)

1. INTRODUCTION

Presence of sand in well fluids is a major threat to equipment throughout the production chain and could become a more wide spread problem as producing wells age. Multiphase separators, typically in the form of conventional gravity separators, are utilized within the oil and gas industries for removing solids from production systems. However these are typically bulky in size and their associated installation and cleaning costs make them costly and unsuitable for most subsea operations.

Extensive research has been conducted at BHR Group's oil and gas division, CALtec, to develop a new generation of compact separators suitable for both offshore and subsea applications which are more efficient, compact and simpler to operate, with minimum maintenance requirement.

The concept of rapid phase separation by subjecting a multiphase flow to high centrifugal forces within a separation chamber has remained one of CALtec 's core research activities over recent years. This has led to the development of a compact inline cyclone separator called WELLSEP, which has potential for use in a variety of applications. WELLSEP is very compact in comparison with conventional gravity separators. The main reason for this is that it does not rely on normal gravity separation and high residence time to achieve phase separation.

The general configuration is shown in figure 1. The separator consists of a number of sections which includes a static device for inducing a spin in the multiphase flow, a dynamic separation chamber were most of the phase separation takes place, an outlet section for the sand phase and an outlet section (vortex finder) for the gas phase. The outlets for the sand and gas phase have particular features to help in collecting the separated phase efficiently and to achieve pressure recovery.

1.1 Cyclone Separators

Structurally, cyclone separators can be divided into two groups. In the first and more common group, known as reverse flow cyclones as shown in figure 2, the sand laden gas enters tangentially near one end to generate high 'g' forces. The mixture spins down the collector near the wall, then, still rotating, the cleaned gas returns upwards to emerge centrally at the top end. For this reason they are called "return flow" cyclones or "reverse flow" cyclones. The sand is extracted at the opposite end.

In the second group, as in figure 3 (which includes the WELLSEP separator), the sand laden gas enters at one end and spins through the cyclone chamber. Both phase components leave at the other end; the clean gas flows along the central core and the solids or the denser phase flows near the wall. This second group is generically known as "straight through", or "uniflow" or "axial flow" cyclones.

CALtec's WELLSEP separator has several advantages over the conventional return flow cyclones. In a return flow cyclone, sand particles entering through the inlet can be picked up by the exiting gas since the inlet is adjacent to the gas exit, resulting in a reduction in the efficiency. The fact that the cleaned gas is forced to reverse direction creates additional problems, which are particularly worsened if there are fluctuations in the flow and the operating pressure. Previous studies (1,2), have suggested that the gas reversal creates an additional pressure drop, so that a reverse flow cyclone has a typically higher pressure drop than the WELLSEP. Also, the diameter of the return flow cyclone body must be larger than that of the WELLSEP design to accommodate the flow of the returning gas. Therefore, for given operational duties, WELLSEP will be more compact. In addition the axially opposite inlet and outlet configuration of WELLSEP makes the layout and arrangement of interconnecting piping much simpler than that for return flow cyclones with tangential inlets.

2. PRELIMINARY TESTING

The WELLSEP separator has proven its ability to operate effectively as an integral gas/liquid separator for the successful WELLCOM system (3), where the main requirement was to have the separated liquid phase free of gas. Its potential as a unit for use in other separation duties, although recognized, needed to be fully examined. Indeed, its original design satisfied the limited requirements of the WELLCOM system for gas/liquid separation duties but this was not necessarily an optimal design for gas/solid separation duties. Therefore the purpose of this preliminary study was to evaluate the separation performance of several design configurations so as to determine the optimum designs of the compact inline separator for gas/solid separation duties.

2.1 Experimental test flowloop

A two-phase flowloop was constructed for this study. It was designed to reduce the quantity of sand particles sedimenting onto the pipework and/or collection of particles at obstruction points along the flow path of the fluid. Therefore pipe bore reducers and orifice plates, where necessary, were manufactured using an eccentric formation and intrusive devices, such as control valves, were located downstream of the filter boxes.

The gas/solid test flowloop, shown in figure 4, consisted of a 11-kW rotodynamic fan capable of generating approx. 0.3 m^3/s of air at 0.2 barg pressures, supporting pipework, the axial flow cyclone separator, filter boxes (each holding a polypropylene 1 μm aperture sock filter), pressure/flow control gate valves and supporting instrumentation. The use of air as the transporting medium permitted venting of the gas to the atmosphere once cleaned.

Sand was introduced into the gas stream via an injection arrangement at the fan inlet. This mixture then passed through a series of pipe bore reducers, during which stream flow data was collected, before entering the horizontally positioned compact inline separator. Once in the separator, the clean gas stream left the separator via the axial outlet, while the sand laden gas stream was ejected via the tangential outlet. A "safety" filter box was placed downstream of the axial outlet to ensure a safe clean gas discharge to the atmosphere. Downstream of the tangential outlet a "primary" filter box was placed to capture sand particles. Furthermore, an additional "safety" filter box was installed further downstream, to ensure that no sand particles were released to atmosphere in case of the filter bag failure in the primary filter box. The control valve located on the tangential outlet was used to regulate the amount of gas flow purged through the outlet. The filters were weighed before and after each test run to determine the amount of sand collected.

2.2 Separation Performance Criteria

The performance of the test separator is indicated in terms of the total collection efficiency of the test unit, i.e. the weight ratio of the sand collected by the tangential line filter units and the total amount fed into the system and the pressure drop across the separator. Although each test run was performed at a gas velocity exceeding the saltation velocity of the sand, the gas flowrate was increased after each test run so as to flush the test rig and remove any sedimented sand within the test rig.

In order to obtain more general information regarding the pressure drop characteristics across the separator, a dimensionless group which characterises the performance of cyclone separators was used. This axial loss coefficient, is defined as the static pressure drop ΔP across the separator's inlet and gas outlet, divided by the dynamic pressure calculated from the characteristic velocity V, and is called the Euler number (Eu). It is mathematically expressed as:

$$Eu = \frac{\Delta P}{1/2 \rho V^2} \qquad \text{(EQ1)}$$

2.3 Test Programme

Before any experimental tests were run, the key variables which influence the performance of an axial flow cyclone were identified through a literature survey (1,2,5). The high 'g' generator design, settling chamber length, and the purge flowrate were chosen as independent parameters to be varied to determine their effect on the two main performance variables; the total collection efficiency and the axial loss coefficient. The test sand sample used was the ISO standard AC coarse test dust. For each test run, a mean gas inlet velocity of 10.0 m/s and a solid loading less than 5.0 g/m^3 was chosen.

The preliminary test programme was divided into two parts. In the 1st stage, an examination of the effects of the purge flowrates on the separation performance was performed. For the 2nd stage, the effects of varying the settling chamber lengths and high 'g' generator design on the overall performance were performed under predetermined flowrate conditions.

3. RESULTS AND DISCUSSION

3.1 Effects of the purge flow rate

The results obtained, illustrated in figures 5 and 6, show that by increasing the purge flowrate the collection efficiency of the separator increases and the axial loss coefficient decreases . The results also showed that the separator continues to separate and collect particles from a gas stream without the need for a purge flow to drive the particles through the outlet but with reduced efficiency. These results are in agreement with the findings of the previous studies (5). It has been found that an increase in the collection efficiency is achieved with an increase in the purge flowrate and it has been concluded that presence of a purge flow enhances the separator's 'natural' collection efficiency.

Reviewing the results obtained for 0% purge, one could suggest that it is the particle's momentum which drives it through the 'solids' outlet into the horizontal tangential outlet. Once outside the influence of a gas flow, gravitational effects will finally bring the particle to rest onto the bottom of the collection device.

3.2 Effects of the changing the settling chamber length and high 'g' generator geometry

Based on the results obtained from the purge flowrate experiments, it was decided that the remaining preliminary experiments be performed under the most robust flowrate conditions, so as to illustrate fully any performance variations obtained from altering the chosen structural parameters. Therefore the volumetric purge flowrate was set at 0 %, as this gave the lowest collection efficiency for the tested operating duties.

The increase in the settling chamber length, as shown in figure 7, proves that an increase the collection efficiency is achieved as the separation chamber length increased to an optimum limit. The length of the chamber ranged from 1.5 through to 4.1 ID. It is expected that increasing the time spent by the particles under the influence of the vortex field encourages a greater migration of the particles to the separator wall, where they can be collected, and hence increases the collection efficiency. However a settling chamber length beyond a certain limit appears to impose a hindering effect; thereby reducing the collection efficiency of the test separator. This trend has also been observed from previous axial flow and return flow cyclone studies (1,2,6). These studies concluded that the combination of the settling chamber length exceeding the natural turning length of the generated vortex and the influence of particle re-entrainment from the cyclone wall impedes the collection capability of the cyclones.

Further studies have been planned including a full factorial study on several structural parameters and high pressurised natural gas trials to demonstrate the performance of the separator over a great range of operating duties.

4. COMPUTATIONAL MODELLING

In conjunction with the experimental investigations, modelling work has taken place to develop a computational fluid dynamics (CFD) model of the compact separator using a commercial software package CFX-4. A 3-dimensional 50,000 cell geometry of the separator's settling chamber, involute chamber and gas outlet have been modelled, figure 8. Profiles of the axial and tangential components of the inlet gas vortex velocities generated from a laser Doppler anemometry (LDA) study were used to set up the inlet boundary condition of the new cyclone performance model. This eliminated the need to model the complex geometry of the inlet device. Inside the settling chamber, comparisons of the experimentally measured tangential and axial velocities with the CFD predictions were made.

The turbulent model currently employed is the k-epsilon turbulent model, however the generated results were not encouraging and computational runs using the RNG k-epsilon model and the Reynold Stress model are now planned.

Once good agreement is reached between the LDA and CFD velocities, particle tracking incorporating an inelastic bounce routine will be performed. Grade efficiency curves will be developed and the calculated total collection efficiency will be compared with the total collection efficiency of the laboratory tests. The use of computational modelling is expected to help to optimise the design and some of the key dimensions of the separator.

5. CONCLUSIONS

The work presented has shown that CALtec's WELLSEP is capable of operating as a gas/solid separator for a wide range of operation and duties. More detailed tests have been planned to investigate further the influence of selected parameters. Its compactness, lack of moving parts and requiring no active control are the key benefits of this separator which makes it ideal for applications in not only oil and gas industries but in other industries where gas/solid separation duties are involved.

6. REFERENCES

1. Daniels, T.C. 1957, 'Investigation of a vortex air cleaner', *The engineer*, no. 203, March, pp 358-362

2. Jackson, R. 1963, *Mechanical equipment for removing grit and dust from gases*, The British Coal Utilisation Research Association, Leatherhead.

3. Sarshar, M. et al., 1997 'Field testing and applications of WELLCOM system using jet pump technology', *Proceedings of the 8th International Conference on Multiphase 97*,Kluwer Academic Publishers, 18-20 June

4. Arato, E.G & Barnes N.D. 1992, 'Application of a novel in-line free vortex separator used for gas/liquid separation within a two-phase pumping system', *Proceeding of the 4th International Conference on Hydrocyclone Conference; Analysis and Applications*, ed. Svarovsky, L & Thew, M.T, Kluwer Academic Publishers, vol. 12, 23-25 Sept., pp 377-396.

5. Stenhouse, J.I.T & Trow, M. 1985, 'Particle separation efficiency of uniflow cyclones', *Proceeding of Filtration Society meeting on Gas Cyclones*, Filtration Society, London, England, 4 Jun., pp 1-10.

6. Hoffman, A.C. et al., 1995, 'Evidence of the "natural vortex length" and its effects on the separation efficiency of gas cyclones', *Filtration and Separation Journal*, Sept., pp34-40

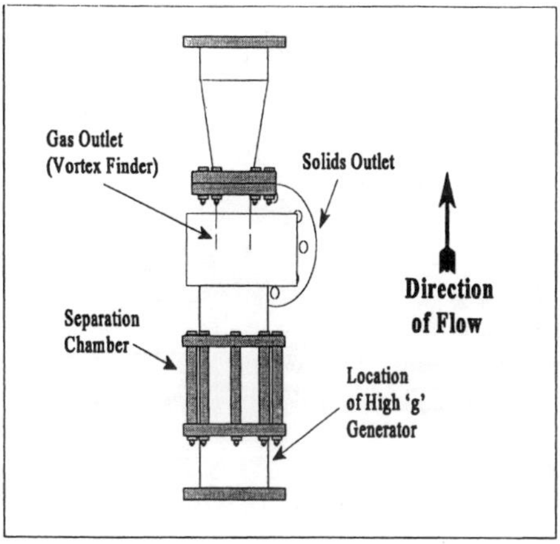

Figure 1: General configuration of WELLSEP

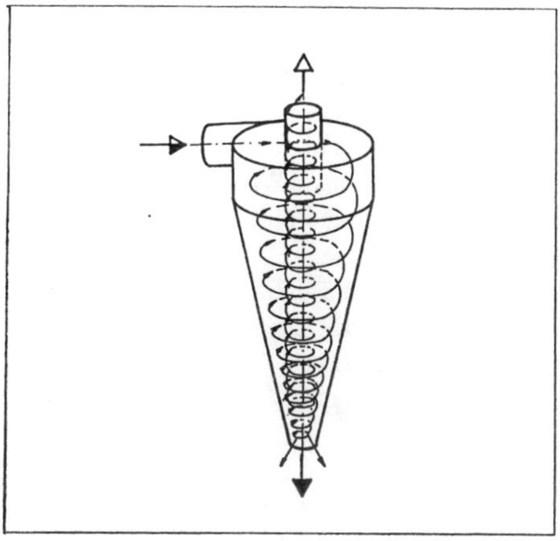

Figure 2: Illustration of the gas flows inside a typical Return Flow Cyclone separator (taken from (4))

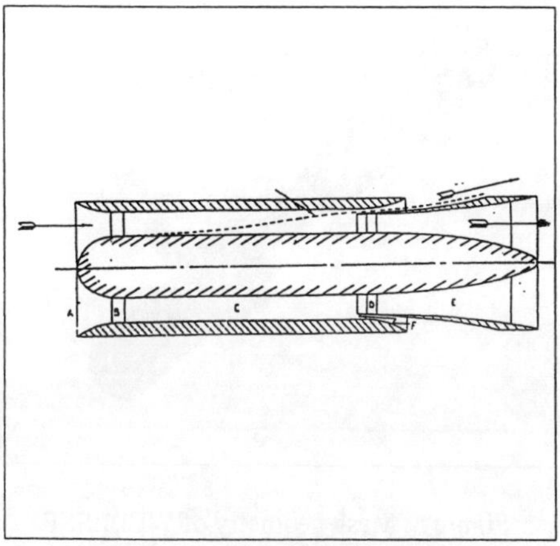

Figure 3: Illustration of the gas flows inside an Axial Flow Cyclone separator (taken from (2))

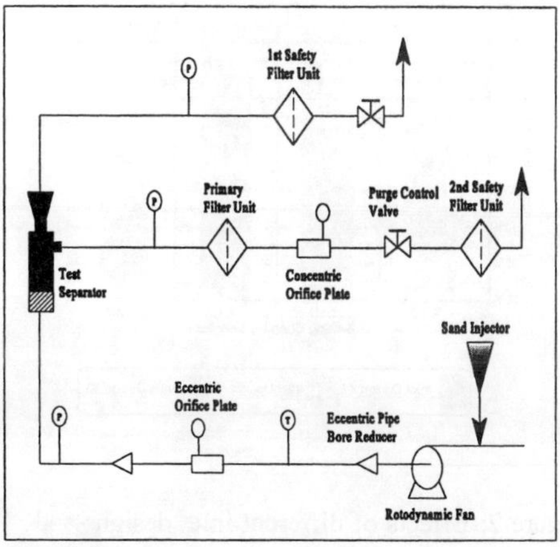

Figure 4: Test flowloop at CALtec Laboratory

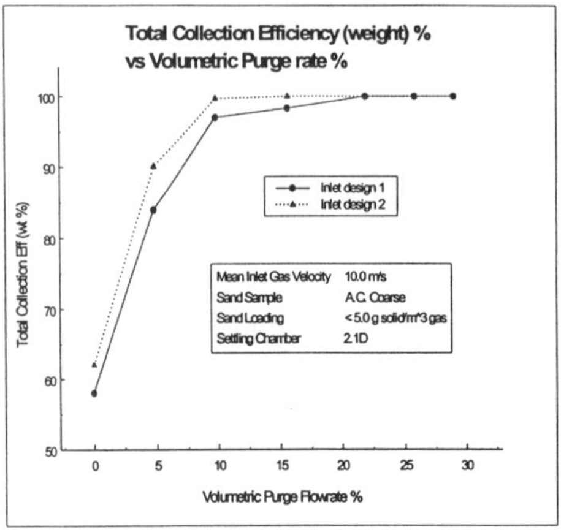

Figure 5: Effect of purge flowrate on collection efficiency (including 0 % purge flowrate)

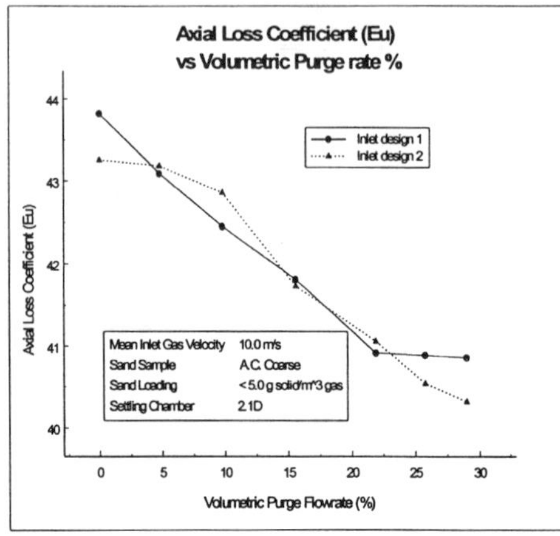

Figure 6: Effect of purge flowrate on axial loss coefficient (including 0 % purge flowrate)

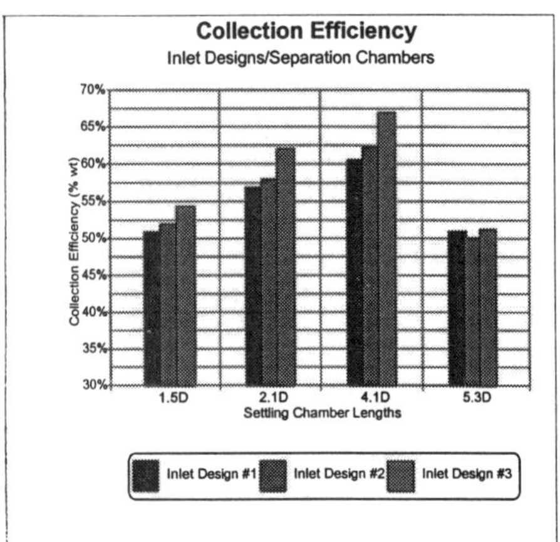

Figure 7: Effects of different inlet designs and separation chamber lengths on collection efficiency

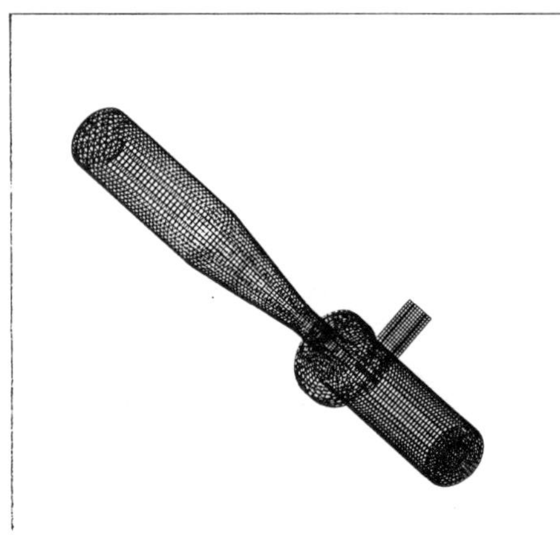

Figure 8: Mesh geometry of WELLSEP

Knowledge-based executive car evaluation using the evidential reasoning approach

J-B YANG
Manchester School of Management, UMIST, Manchester, UK
D-L XU
REL Instrumentation Limited, Manchester, UK

ABSTRACT

This paper presents the application of the Evidential Reasoning (ER) approach to an executive car selection problem. Both quantitative data and qualitative judgements are involved in evaluating each car on individual attributes defining the attractiveness of a car. What would be of interest to both customers and manufacturers is to find a framework in which to combine different types of knowledge for generating an overall picture for each car and furthermore a clear ranking of all cars in question. The ER approach provides a general and flexible framework for such synthesis in a rational way. Details are provided for modelling the car selection problem. Analytic results are generated and discussed using the ER approach.

1. INTRODUCTION

Most real-world decision problems in design of complex engineering products or manufacturing systems involve multiple performance attributes, including both quantitative attributes and qualitative attributes of subjective nature. The evaluation of a design option on a qualitative attribute is inevitably associated with many complex interrelated factors and uncertain subjective judgements. To assess competitive design options, problem often arises as to how numerical numbers and subjective judgements with uncertainty could be used in a consistent and rational way in order to generate a clear ranking order with the decision makers' (designers' or customers') opinions being properly taken into account.

One simple and commonly used way of dealing with the above decision problems is to directly quantify subjective judgements using some numerical scale. Following such quantification, traditional decision making methods like the additive utility function approach (1) may be employed to evaluate different design options. However, it is rather difficult to use such a direct quantification method to treat both inaccurate subjective judgements and precise numerical data in a rational and consistent manner. Furthermore, uncertainty and vagueness common in human knowledge cannot be properly catered for in such a method.

In the recent years, an Evidential Reasoning (ER) approach has been developed for dealing with the above complex decision problems in engineering design and in management alike (3), (4). This paper presents the application of the ER approach to the evaluation of several competitive executive cars, including ACURA 3.2TL Premium, BMW 325I, Mercedes Bens C280 and Mazda Mellenia S, to name a few.

The evaluation was conducted by investigating the different achievement levels of each model on many assessment attributes such as *price*, *general dimension*, *performance, chassis* and *general finish*. Most of such generic attributes are associated with many other more specific attributes, which are easier to assess directly. For example, *performance* may be assessed through *fuel economy*, *ride quality*, *acceleration*, *braking*, *handling* and so on. Some of these attributes such as *fuel economy* and *acceleration* could be measured by numbers. Others such as *ride quality* and *handling* may be assessed using subjective judgements. The question is then how to combine the mixed numerical and judgmental knowledge to generate an overall ***rational*** evaluation for each model in question.

In this paper, the modelling of the car selection problem is first investigated. The evaluation procedure of the ER approach and the synthesis results for the above executive car selection problem will then be reported in some detail.

2. CAR EVALUATION MODELS

2.1 Performance Criteria

Selecting a car may not be a simple or an easy decision for many people to make. It is equally important for manufacturers to know why customers buy or do not buy their cars. Some people may buy a car simply because they have used the same type of cars for years. Many people, especially first time buyers, however, may choose a car based on the evaluations and comparisons of different types of cars available in the markets.

In evaluation of cars, many factors need to be taken into account apart from *price*, such as *general finish, performance, chassis* and *general dimension*. These factors are referred to as attributes in this paper. Some attributes such as *price* may be measured directly by numbers. The assessment of some other attributes such as *general finish* may not be so straightforward. They are qualitative in nature and are associated with other factors that are easier to measure. For instance, the evaluation of *general finish* is related to factors like *styling*, *ergonomics*, *noise isolation* and *interior comfort*. In general, a hierarchy of factors can be established for evaluation of qualitative attributes.

Figure 1 shows a typical evaluation hierarchy for car selection. In the hierarchy, factors relevant to car selection are broken down to different levels. At the top level (level 1) is the ranking of cars based on the evaluations of attributes at a level immediately below (level 2). Factors at levels 3 detail relevant attributes at level 2. These factors may be further broken down to lower-level factors if they are still difficult to be measured directly. Factors that are directly measurable are referred to as basic factors in the paper.

In the hierarchy, not all factors at a single level play the same role in the evaluation of a factor associated at a level immediately above. The decision maker may have different views on the importance of factors in evaluation. Among the five attributes at level 2, for example, a customer may consider *performance* as the most important attribute, followed by *price* and *chassis*. *General dimension* and *general finish* may be regarded as relatively less important attributes. This is, however, the customer's individual preference. Another customer may have an entirely different view. Similarly, the relative importance of factors at level 3 associated with the same upper-level attribute also needs to be taken into account.

The relative weight of each attribute is shown by a number in the bracket. For example, the relative weights of *price*, *general dimension, performance, chassis* and *general finish* are 0.15, 0.1, 0. 5, 0.15 and 0.1 respectively. The relative weights of factors associated with *general dimension* are 0.3 for *cargo capacity*, 0.2 for *fuel capacity*, 0.4 for *weigh/power ratio* and 0.1 for *dimension.* Note that weight indicates relative importance and therefore the numbers are only relatively meaningful. Also note that the weights could be normalised. In other words, the total weight of all factors associated with the same upper level attribute is summed to one. The weights in Figure 1 represent the customer's personal preference.

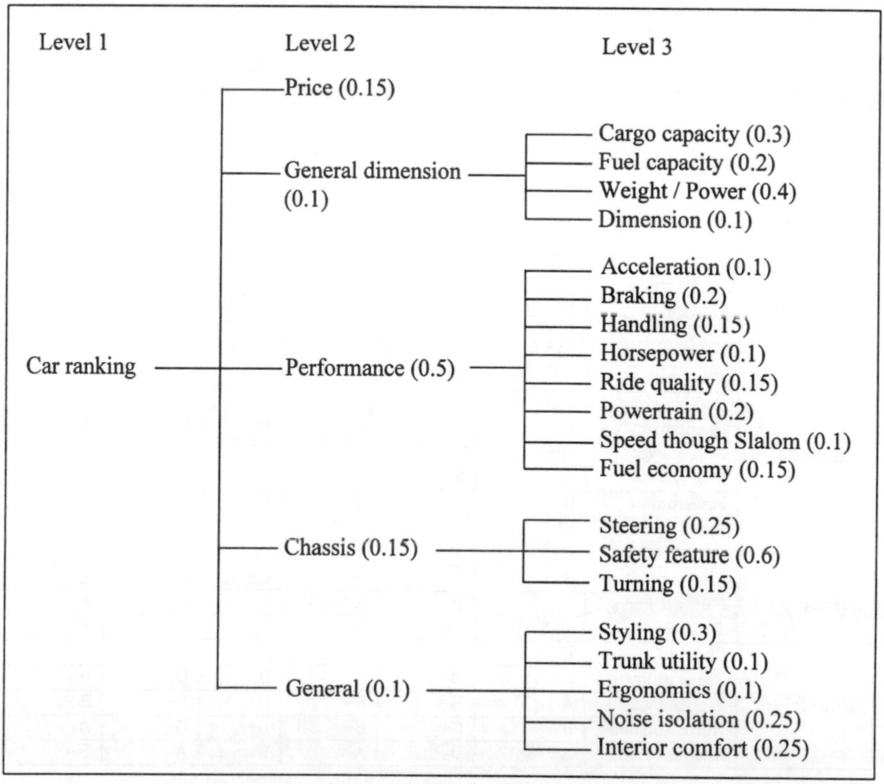

Figure 1 A Hierarchy of Factors for Car Selection

2.2 Performance Evaluation and Data Collection

The attributes shown in Figure 1 need to be defined and measured properly. Some of the attributes can be defined numerically and measured using numbers in certain units. For instance, *price* is defined as total purchasing price and measured in pounds. *Cargo capacity* is defined as the size of the luggage compartment measured in cubic feet. Other quantitative attributes include *fuel capacity, weight/power ratio, dimension, acceleration, braking, horsepower, fuel economy,* and *turning*. They can all be measured numerically.

Qualitative attributes are those difficult to measure by numbers in the first instance. They are normally evaluated using human judgements. *Powertrain*, for example, is related to the performance of an engine. It is evaluated on the basis of the quietness and smoothness of the engine. *Steering* is another qualitative attribute. It is related to the manoeuvrability of a car and is assessed based on the tightness and accuracy of steering. Other qualitative attributes include *handling, ride quality, speed through slalom, safety features, styling, trunk utility, ergonomics, noise isolation*, and *interior comfort*.

Data related to individual car evaluation could be collected from different sources such as magazines, motor show and World Wide Web. In this study, six cars are selected for analysis and they are ACURA 3.2TL Premium, BMW 325I, Infinity I30t, Lexus ES300, Mercedes Bens C280 and Mazda Mellenia S. The evaluation data were collected from BBC Top Gear Magazine, Motor Trend Performance Cars, Motor Show 1996 at the National Exhibition Centre in Birmingham, and World Wide Web sites for the manufacturers producing these cars. Table 1 shows the original assessment data collected for the six cars.

Table 1 Original Car Evaluation Data Collected

		Acura 3.2 TL Premium	BMW 3251	Infinity I30t	Lexus ES300	Mazda Mellenia	Mercedes Benz
Price ($)		36020	36420	34200	36453	33760	39332
Dimensions	Cargo cap.	14.1	10.3	14.1	14.3	13.3	13.7
	Fuel cap.	17.2	17.2	18.5	18.5	18.0	16.4
	Weight/power	17.6	16.7	16.9	18.0	16.2	17.0
	Dimension	12.23	10.47	12.06	11.61	11.90	11.04
Performance	Acceleration	8.8	8.0	7.7	8.4	8.0	7.9
	Braking	128	124	127	134	135	126
	Handling	B	A	B	B-	B+	A
	Horsepower	196	152	182	183	138	171
	Ride quality	A-	B-	B	B+	B+	A-
	Powertrain	B	B+	A	B	A-	A
	Speed th. Sla.	63.3	61.4	62.0	63.1	65.8	63.2
	Fuel economy	20	20	21	20	19	20
Chassis	Steering	B	A	B-	C+	B	A-
	Safety features	A	A	A	A	A	A+
	Turning	34.8	34.1	34.8	36.7	37.4	35.2
General	Styling	B-	B+	C-	B	A-	B+
	Trunk utility	A-	B-	A	B	B	B+
	Ergonomics	A	B-	B+	B+	B+	B+
	Noise isolation	B+	C+	B+	B+	B+	A-
	Interior comfort	A	B-	A-	B+	A-	B+

2.3 Model Building

Data in Table 1 are in different units and not ready for comparative analysis. They can be projected to the normalised utility space (4), (5) by defining utility grades, such as typical five grades named by E, A, B, C and P. The utilities of these grades may be given by u(E)=1, u(A)=0.8, u(B)=0.5, u(C)=0,2 and u(P)=0, respectively. The higher the utility of a grade, the better is the grade. The quantitative data and qualitative judgements may be transformed to the utility space using utility assessment methods (6).

A piecewise linear utility function could be established for a quantitative attribute by identifying the figures having the utilities of 1, 0.8, 0.5, 0.2 and 0. Take *price* for example. The acceptable *price* range is assumed to be between $30,000 and $40,000. This means that $30,000 is the lowest possible price and $40,000 the highest acceptable price. The utilities of $30,000 and $40,000 are then assessed to one and zero, respectively. Suppose the utilities of $32,000, $35,000 and $38,000 are assessed to be 0.8, 0.5 and 0.2 respectively. Then, the utility of any other price between $30,000 and £40,000 may be calculated using a piecewise linear utility function (5).

Table 2 Equivalent Standardised Car Evaluation Data

		Acura 3.2 TL	BMW 3251	Infinity I30t	Lexus ES300	Mazda Mellenia	Mercedes Benz
Price ($)		C(0.34) B(0.66)	C(0.47) B(0.53)	B(0.73) A(0.27)	C(0.48) B(0.52)	B(0.59) A(0.41)	P(0.67) C(0.33)
Dimension	Cargo cap.	B(0.40) A(0.60)	P(0.75) C(0.25)	B(0.40) A(0.60)	B(0.27) A(0.73)	B(0.83) A(0.17)	B(0.60) A(0.40)
	Fuel cap.	C(0.33) B(0.67)	C(0.33) B(0.67)	A(0.85) E(0.15)	A(0.85) E(0.15)	B(0.43) A(0.57)	P(0.35) C(0.65)
	Weight/power	C(0.50) B(0.50)	B(0.73) A(0.27)	B(0.93) A(0.07)	C(0.83) B(0.17)	B(0.33) A(0.67)	B(1)
	Dimension	A(0.95) E(0.05)	C(0.43) B(0.57)	B(0.13) A(0.87)	B(0.50) A(0.50)	B(0.26) A(0.74)	B(0.97) A(0.03)
Performance	Acceleration	C(1)	C(0.23) B(0.77)	B(0.96) A(0.04)	C(0.63) B(0.37)	C(0.23) B(0.77)	C(0.13) B(0.87)
	Braking	B(0.67) A(0.33)	A(1)	B(0.50) A(0.50)	C(0.67) B(0.33)	C(0.83) B(0.17)	B(0.33) A(0.67)
	Handling	B(1)	A(1)	B(1)	C(0.33) B(0.67)	B(0.67) A(0.33)	A(1)
	Horsepower	A(0.25) E(0.75)	C(0.33) B(0.67)	B(0.07) A(0.93)	B(0.03) A(0.97)	C(0.93) B(0.07)	B(0.53) A(0.47)
	Ride quality	B(0.33) A(0.67)	C(0.33) B(0.67)	B(1)	B(0.67) A(0.33)	B(0.67) A(0.33)	B(0.33) A(0.67)
	Powertrain	B(1)	B(0.67) A(0.33)	A(1)	B(1)	B(0.33) A(0.67)	A(1)
	Speed through slalom	C(0.30) B(0.70)	P(0.10) C(0.90)	C(0.83) B(0.17)	C(0.37) B(0.63)	B(0.26) A(0.74)	C(0.33) B(0.67)
	Fuel economy	B(0.33) A(0.67)	B(1)	B(0.67) A(0.33)	C(0.33) B(0.67)	C(0.67) B(0.33)	B(1)
Chassis	Steering	B(1)	A(1)	B(0.67) A(0.33)	C(0.67) B(0.33)	B(1)	B(0.33) A(0.67)
	Safety features	A(1)	A(0.35) E(0.65)	A(1)	A(1)	A(1)	A(0.50) E(0.50)
	Turning	B(0.67) A(0.33)	B(0.33) A(0.67)	B(0.67) A(0.33)	C(0.57) B(0.43)	C(0.90) B(0.10)	B(0.87) A(0.13)
General	Styling	C(0.33) B(0.67)	B(0.67) A(0.33)	P(0.50) C(0.50)	B(1)	B(0.33) A(0.67)	B(0.85) A(0.15)
	Trunk utility	B(0.33) A(0.67)	C(0.33) B(0.67)	A(1)	B(1)	B(1)	B(0.67) A(0.33)
	Ergonomics	A(1)	C(0.33) B(0.67)	B(0.67) A(0.33)	B(0.67) A(0.33)	B(0.67) A(0.33)	B(0.67) A(0.33)
	Noise isolation	B(0.67) A(0.33)	C(0.67) B(0.33)	B(0.67) A(0.33)	B(0.67) A(0.33)	B(0.67) A(0.33)	B(0.33) A(0.67)
	Interior comfort	A(1)	C(0.33) B(0.67)	B(0.33) A(0.67)	B(0.67) A(0.33)	B(0.33) A(0.67)	B(0.67) A(0.33)

The utilities of qualitative attributes can also be assessed in a similar way. Take *handling* for example. The evaluation of *handling* may be classified into several categories. Without loss of generality and to simply the analysis, define five evaluation categories equivalent to the five utility grades. Suppose the utility grade E is equivalent to a *handling* evaluation category of "excellent control in cruising at high speed", A to "good control in cruising at high speed but soft wobbling felt sometimes", B to "good control but moderate wobbling felt sometimes", C to "average control and tight felt", and P to "difficult to control when cruising at high speed". By equivalent it is meant that the utility grade and the evaluation category in question have the same utility value. Since the number of evaluation categories for a qualitative attribute is limited, their utilities can be assessed using a utility assessment method.

Having obtained the utility values for all the data as shown in Table 1, the technique introduced in reference (6) may be employed to convert Table 1 to Table 2. The details about the conversion are available in reference (6).

The data displayed in Table 2 are equivalent to the corresponding ones shown in Table 1 in terms of utility equivalence. The data in Table 2 are organised in a consistent judgmental format. Take for example a quantitative attribute the *price* of Acura 3.2 TL Premium. It reads that the price of Acura 3.2 TL Premium is between the utility grades C and B with the degrees of belief of 0.34 and 0.66, respectively. The utility of this price is therefore equal to the following expected utility: $0.34u(C) + 0.66u(B) = 0.34\times0.2 + 0.66\times0.5 = 0.398$. On the other hand, the utility of the price $36,020 was directly assessed to be 0.398.

A qualitative attribute is measured in a similar way in Table 2. For instance, the *ride quality* of Acura 3.2 TL Premium is between the utility grades B and A with the degrees of belief of 0.33 and 0.67. The utility of this judgement is therefore equal to: $0.33u(B) + 0.67u(A) = 0.34\times0.5 + 0.67\times0.8 = 0.701$. As shown in Table 1, the original judgement about the *ride quality* of Acura 3.2 TL Premium is A- whose utility was assessed to be 0.701.

Table 2 presents a decision model for the car selection problem. In this model, both quantitative data (numbers) and qualitative (subjective) judgements are measured in the same manner which is consistent and rational in terms of utility equivalence. Another feature of this model is that uncertain judgements can also be treated in the format without converting to certain numbers.

3. THE EVIDENTIAL REASONING APPROACH AND CAR SELECTION

3.1 Evidential Reasoning for Car Evaluation

The evaluation problem is to provide a clear ranking of these six executive cars for the customer on the basis of the evaluation hierarchy and weights shown in Figure 1 and the evaluation data given in Table 2 (or Table 1). This requires the rational synthesis of the evaluations of each car on all the 20 basic factors at level 3 in Figure 1, resulting in the propagation of the evaluations to the 5 attributes at level 2. The generated evaluations for the 5 attributes at level 2 are then combined to provide an assessment for each car. A ranking of the cars should result from such assessments. The difficulty in dealing with such a decision problem lies in the fact that both quantitative data and qualitative judgements need to be taken into account simultaneously and rationally.

The Evidential Reasoning (ER) approach provides a flexible yet rational way of dealing with such uncertain synthesis problems by means of evidence combination for multiple attributes. The ER approach can model the narrowing of individual evaluations and provide a combined distributive evaluation for each car. It can also generate a clear ranking for the cars based on the final expected utilities obtained. A brief description of the ER approach is provided in the rest of this section. More details about the ER approach can be found in references (2), (3), (4) and (6).

Suppose an attribute y_k is associated with L_k factors $e_1 \ldots e_i \ldots e_{L_k}$. Let H_n be the nth utility grade, H the whole utility space and N the number of utility grades defined. Define $m_{k,i}^{n}$ as a probability mass indicating the degree to which the ith factor e_i supports the hypothesis that the attribute y_k should be assessed to a utility grade H_n. $m_{k,i}^{n}$ can be generated from the given degrees of belief and weights (3), (4).

Let m_k^n be a probability mass measuring the degree to which y_k is evaluated to H_n. m_k^n is obtained by combining all factors associated with y_k. If $m_{I_k,(i+1)}^{n}$ is defined as the degree to which the first $(i+1)$ factors support the hypothesis that y_k should be assessed to a utility grade H_n, then $m_{I_k,(i+1)}^{n}$ can be generated using the following recursive evidential reasoning algorithm (3), (4)

$$\{H_n\}:\ m_{I_k(i+1)}^{n} = K_{I_k(i+1)}(m_{I_k(i)}^{n}m_{k,i+1}^{n} + m_{I_k(i)}^{n}m_{k,i+1}^{H} + m_{I_k(i)}^{H}m_{k,i+1}^{n}),\ n = 1,\cdots,N$$

$$\{H\}:\ m_{I_k(i+1)}^{H} = K_{I_k(i+1)}m_{I_k(i)}^{H}m_{k,i+1}^{H}$$

$$K_{I_k(i+1)} = \left[1 - \sum_{\tau=1}^{N}\sum_{\substack{j=1 \\ j\neq\tau}}^{N} m_{I_k(i)}^{\tau}m_{k,i+1}^{j}\right]^{-1}$$

$$i = 1,\cdots,L_k - 1$$

Note that in the above algorithm $m_{I_k,(1)}^{n} = m_{k,1}^{n}$ and $m_k^n = m_{I_k,(L_k)}^{n}$. Once m_k^n is obtained, the state of the attribute y_k can be evaluated in terms of the defined utility grades as follows

$$S(y_k) = \{(m_k^n,\ H_n),\ \text{for } n = 1,\cdots,\text{N}\}$$

which means that the attribute y_k is evaluated to H_n with a degree of belief of m_k^n, $n=1, \ldots, N$.

3.2 Executive Car Selection

Applying the ER approach to combining the evaluations on the basic factors at level 3 leads to the combined evaluations of each car on the attributes at level 2, as shown in Table 3. A distributive evaluation is thus generated for each car on every attribute. By combining all the evaluations on the four basic factors at level 3, for example, the *general dimension* of Acura

3.2 TL Premium is assessed to utility grades C, B, A and E with the degrees of belief of 0.2578, 0.6438, 0.0655 and 0.0005, respective. This combined evaluation reads that the *general dimension* of Acura 3.2 TL Premium is to a large extent (more than 90%) assessed to the utility grades C and B. The rationale of this synthesis process is evident by looking at the relevant data in Table 2 and the relative weights of these four basic factors in Figure 1.

Table 3 Combined Evaluations at Level 2

	Acura 3.2 TL Premium	BMW 3251	Infinity I30t	Lexus ES300	Mazda Mellenia	Mercedes Benz
Price ($)	C(0.3400) B(0.6600)	C(0.4700) B(0.5300)	B(0.7300) A(0.2700)	C(0.4800) B(0.5200)	B(0.5900) A(0.4100)	P(0.6700) C(0.3300)
Dimension	C(0.2578) B(0.6438) A(0.0655) E(0.0005)	P(0.0700) C(0.0524) B(0.7022) A(0.1340)	B(0.7062) A(0.2590) E(0.0040)	C(0.4985) B(0.2313) A(0.2135) E(0.0065)	B(0.4678) A(0.5087)	P(0.0073) C(0.0136) B(0.9324) A(0.0223)
Performance	C(0.0275) B(0.7812) A(0.1310) E(0.1447)	P(0.0026) C(0.0735) B(0.5159) A(0.3456)	C(0.0173) B(0.6083) A(0.3251)	C(0.1512) B(0.7743) A(0.0327)	C(0.2670) B(0.4685) A(0.2006)	C(0.0090) B(0.2766) A(0.6700)
Chassis	B(0.0200) A(0.9610)	B(0.0017) A(0.5240) E(0.4589)	B(0.0121) A(0.9723)	C(0.0153) B(0.0083) A(0.9554)	C(0.0061) B(0.0156) A(0.9573)	B(0.0096) A(0.5912) E(0.3825)
General	C(0.0717) B(0.4808) A(0.3985)	C(0.1345) B(0.7867) A(0.0468)	P(0.1483) C(0.1483) B(0.2279) A(0.3586)	B(0.9430) A(0.0395)	B(0.4720) A(0.4992)	B(0.7744) A(0.1996)

Table 4 Final Evaluations and Ranking of the Cars

	Acura 3.2 TL Premium	BMW 3251	Infinity I30t	Lexus ES300	Mazda Mellenia	Mercedes Benz
Evaluation	C(0.0299) B(0.7752) A(0.1419) E(0.0084)	P(0.0026) C(0.0704) B(0.5748) A(0.2840) E(0.0107)	P(0.0016) C(0.0116) B(0.5815) A(0.3603)	C(0.1392) B(0.7663) A(0.0502)	C(0.1626) B(0.4715) A(0.3082)	P(0.0143) C(0.0141) B(0.2983) A(0.6137) E(0.0080)
Expected utility	05378	05682	06039	04733	05437	06767
Ranking	5	3	2	6	4	1

Using the ER approach for the attributes at level 2, the evaluations shown in Table 3 can be further synthesised to generate a final evaluation for each car, as shown in the second row of Table 4. Such an evaluation takes into account both the original data of Table 1 and the customer's weights of Figure 1 in a systematic way and thus provides a rational basis for generating a ranking of the cars.

An expected utility for each car is generated as the sum of the utility grades times the corresponding degrees of belief obtained, as shown in the third row of Table 4. The final ranking of the cars is produced on the basis of the magnitude of the final expected utilities generated.

As a result of this analysis, Mercedes Benz C280 is found to best suit the customer, as he preferred a high performance mid-sedan with good safety features. Infinity I30t is also a very good choice, followed by BMW 325I. The customer is advised not to choose Lexus ES300 if his preferences are properly represented by the weights shown in Figure 1.

4. CONCLUDING REMARKS

Real world assessment problems are often associated with both quantitative data and qualitative judgements of subjective nature. It is crucial to all the parties involved to synthesise different types of decision knowledge in order to arrive at a consensus view in a consistent and rational manner. The ER approach provides a general and flexible framework for such complex decision analysis. The application of the ER approach to the above executive car selection problem, as presented in this paper, shows its potential to help deal with a wide range of complex decision analysis problems of similar nature.

ACKNOWLEDGEMENTS

Mr Y.K. Cheong collected and analysed the data as part of his final year research project that was conducted at the University of Birmingham.

REFERRENCES

(1) Hwang, C.L. and Yoon, K. (1981) *Multiple Attribute Decision Making Methods and Applications*, Springer-Verlag, Berlin.

(2) Zhang, Z.J., Yang, J.B. and Xu, D.L.(1990) A hierarchical analysis model for multiobjective decision making. In Analysis, Design and Evaluation of Man-Machine System 1989. Selected Papers from the 4th IFAC/IFIP/ IFORS/IEA Conference, Xian, P.R. China, September 1989, Oxford, UK: Pergamon, pp.13-18.

(3) Yang, J.B. and Singh, M.G. (1994) An evidential reasoning approach for multiple attribute decision making with uncertainty. IEEE Transactions on Systems, Man, and Cybernetics, 24(1), pp.1-18.

(4) Yang, J.B. and Sen, P. (1994) A general multi-level evaluation process for hybrid MADM with uncertainty. IEEE Transactions on Systems, Man, and Cybernetics, 24(10), pp.1458-1473.

(5) Yang, J.B. and Sen, P. (1996) "Preference modelling by estimating local utility functions for multiobjective optimisation", European Journal of Operational Research, 95, pp.115-138.

(6) Yang, J.B. and Sen, P. (1997) Multiple attribute design evaluation of complex engineering products using the evidential reasoning approach. Journal of Engineering Design, 8(3), pp.211-230.

Quality Control Management

which the ACI best practice indicators are determined. More importantly, such measures should be central to the management system on any site, and as such easy to obtain.

Traditionally, on-site measurement of progress and delay has been deemed central to the operations of construction firms. The highly fragmented and thus competitive construction market has induced the cutting of profit margins to a minimum for all parties to a construction project. The measurement of work done has subsequently been essential for contractors and sub-contractors alike, as a means for justifying claims for work performed over and above that defined in the original contract.

Recent developments in construction project procurement (PFI, DBFO, etc) have resulted in a relative reduction in terms of on-site measurement. The lack of a "claims culture" upon which the traditional approach to project management has been based is seen as negating the need to measure.

In establishing a benchmarking process, an important issue has therefore been the re-educating of site personnel in relation to the need for measurement - from a purely management perspective. This has been achieved through a combination of mechanisms across the range of projects being benchmarked. The type of mechanism(s) used was principally based upon the size and type of project, and the level of resistance to change demonstrated by site personnel. On all projects a combination of site based tutorials, workshops and seminars were held. The need for benchmarking, from both a project (operational) and company (strategic) perspective formed the basis of each of these interactions.

In all cases, top-down management support was obtained. In practical terms this was demonstrated by the hands-on involvement of a senior management 'champion' in the benchmarking initiative. It is important to note that the effective transmission of the support down through the project organisation varied according to the levels of resistance to change demonstrated by the site management function.

In broad terms this resistance to change has been classified by Baylis (6) as follows:

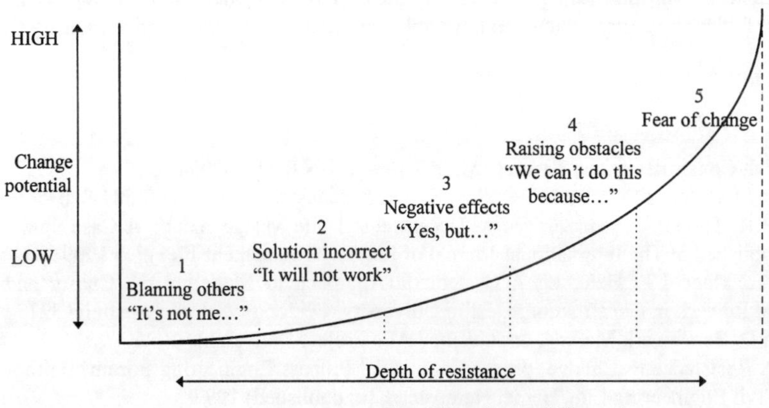

Fig. 2 Baylis model.

This model (fig. 2) is supported by Weiss (7) who asserts that an individual's resistance to change is dependent upon "their need for security, their fear of the unknown, no felt need for change, having a vested interest threatened by the change...". In addition, an individuals own interpretation (which may differ or contradict the official view), the timing of the change and the availability of necessary resources, are also noted as having influence.

Both Plant (8) and King & Anderson (9) suggest that the best means of overcoming such resistance in through the involvement and informing of those that the change directly effects. Hence, the approach taken to introduce the ACI benchmarking initiative may be seen as a potential lesson in successful change management.

4.4 Initial Outcomes

To date, the ACI benchmarking initiative has been implemented on a total of 6 construction projects across the UK. Analysis of the collected data after just three months has resulted in the identification of several areas of potential best practice in site operations which are currently being case studied.

The benchmark metrics and best practice analyses in turn will feed the development of an Agile Construction Database and Agile Construction Model, respectively. Given the confidential nature of the information held on the database, no further discussion of this element of the work is permissible. The Agile Construction Model, however, in its demonstration format can be viewed on the Agile wed site (10).

5.0 Conclusion

The work conducted by the ACI in developing and implementing the performance and quality monitoring measures for construction, have shown, how it is not possible to just simply apply methods that have been developed in other sectors to the industry, as the complex contractual arrangements and general lack of record keeping mean that new means of obtaining the data have to be developed. With the methods developed and means of collecting data established the outcomes of the performance and quality monitoring methods showed that considerable improvements in both performance and quality are available.

The ACI is now developing ways of improving the performance and quality through the areas identified in the study. The improvement methods will again draw on the experiences of other industries which will have to be modified for the construction industry.

6.0 REFERENCES

(1) Latham M, Constructing The Team, H.M.S.O. 1994 ISBN 0-11-752994-X

(2) Womack J P & Jones D T, Lean Thinking, Simon & Schuster 1996 ISBN 0-684-81035-2

(3) Barber P R, Tomkins C and Graves A, Decentralised Site Management - A Case Study, To be Published in The International Journal of Project Management 17(1)Feb 1999

(4) Love P E, Mandel P, Heng Li, A Systematic Approach to Modeling the Cuases and Effects of Rework in Construction, Deakin University, Geelong, Victoria, Australia 3217

(5) Madigan D, Benchmark Method, unpublished ACI Project Deliverable 1996

(6) Baylis P, Resistance to Change, proceedings of 3rd Process Engineering Forum, Balfour Beatty Civil Engineering Ltd., Hemel Hempstead, (unpublished) 1997

(7) Weiss J W, Organisational behaviour and change: managing diversity, cross-cultural dynamics, and ethics, West Publishing Company, 1996 ISBN 0-314-06929-1

(8) Plant R, Managing change and making it stick, Gower 1987 ISBN 0-566-02736-4

(9) King N & Anderson N, Innovation and Change in Organisations, Routledge 1995

(10) (URL) http://www.bath.ac.uk/Departments/Management/research/agile/acm/home.htm

Refinement and modelling of CMM data

R BARDELL, K SIVAYOGANATHAN, and **V BALENDRAN**
Department of Mechanical and Manufacturing Engineering, Nottingham Trent University, UK

ABSTRACT

The Co-ordinate Measuring Machine (CMM), is a Cartesian robot used to collect relevant data from physical parts. Representing this point data as a surface model is a major step in reverse engineering. However, surface model generation can cause inaccuracies, originating from both the data collection and modelling stages.

This paper presents a method of model refinement, based on trimming planes, using the ACIS® Toolkit as the geometric engine to perform this task. 2D trim planes are presented and a method of 3D boundary trimming is proposed.

1 INTRODUCTION

In its initial stages, the Reverse Engineering (RE) industry relies heavily on the accurate surface data acquisition of a physical prototype. The Co-ordinate Measuring Machine (CMM) has proved to be successful at this task, collecting point data to an accuracy of one micron (1). This 3D data is then applied to generating a surface model, in a Part-to-CAD stage. The accuracy of this model is paramount, relating to areas such as physical reproduction of CAD models, and accuracy analysis.

The accuracy of a created surface model can be improved, by implementing surface refinement methods as a means of reducing modelling errors. This in turn reduces machining errors.

2 REVERSE ENGINEERING

Conventional engineering involves the stages of concept design, using Computer Aided Design (CAD), and results in the creation of a physical part, via the creation of a machining part program, utilising Computer Numerically Controlled (CNC) machining.

RE is the reverse process, and can be defined as: 'Systematic evaluation of a product with the purpose of replication. This involves either direct copies or adding improvements to existing design' (2). RE comprises of three main areas, outlined in the following sections.

2.1 Data Collection using the CMM

3D point data is usually collected automatically, using either point to point or scanning methods. Scanning methods are usually preferred, having improved accuracy and speed due to the fast pecking action of the probe on the surface. CMM scanning takes place uni-directionally or bi-directionally across a surface, physically touching the part in question with a spherical probe.

A problem with contact probe data collection is the issue of probe compensation. Despite the fact that the probe makes contact with the surface on its circumference, the recorded point is taken at the centre of the probe. Thus the point on the actual surface differs from the recorded point by the probe radius. In the generation of a surface model, it is essential that this is taken into account by offsetting the surface by this probe radius.

In most cases, the collection of data involves placing the object on a planar 'bed'. This is due to most scanning algorithms requiring a regular quadrilateral boundary within which data is collected.

2.2 Data Modelling

The modelling of CMM data in this work is achieved using ACIS®, a CAD development tool which is interfaced using the C++ language. This results in a CAD surface model, based on the collected points. This process forms a number of stages.

Uni-directional scanning results in a series of parallel scan lines made up of dense data points. These scan points are interpolated, forming uni-directional spline curves as shown in figure 1. The quadrilateral boundaries as mentioned previously can be seen here by the extents of the excess data which is not needed.

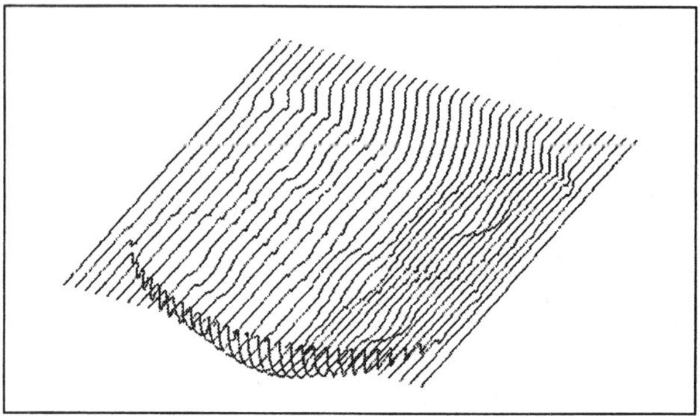

Figure 1 Uni-directional Scan Data Lines

(data collected by LK Ltd.)

Knot points are introduced at intervals along these splines in order to form perpendicular spline curves necessary for the generation of a spline network. From this spline network, patch boundary curves and tangent vectors are used to create a composite surface made up of tangent continuous (C^1) Gregory/Charrot patches.

The point to point method of data collection has the advantage of allowing the direct generation of a surface model from data points, without knot insertion (3). However this limits continuity conditions.

From a surface model it is possible to create machining paths in a CAD-to-Part stage, as well as analyse the generated surface for accuracy.

2.3 Machining

CAD surface information can be used to create part programs for CNC machines, causing the original measured part to be reverse engineered. As well as creating a physical model, this stage also forms an additional error determination stage, as inspection of this new model can take place.

It is important to assess a surface, as its accuracy cannot be assumed merely because accurate point data is used. Thus an accuracy analysis stage completes the RE process.

2.4 Accuracy Analysis

Surface assessment is a necessary step, both for aesthetic and functional properties (1). This usually takes the form of physical assessment, with contact inspection based on a reverse engineered model. However, assessing the CAD model can save time and money in the RE process. Basic numerical analysis involves positional and tangency continuity constraints, examining both inter-point and inter-patch continuity.

Deviation analysis allows the comparison of a point on the fitted model with the corresponding (closest) digitised point. The close correlation between original data points and points on a generated surface must be maintained, within defined tolerances, with the aim of estimating the surface model deviation. This stage also encompasses comparative analysis, where an optimum level of accuracy should be established, based on parameter optimisation.

3 MODEL REFINEMENT

It is not known if a set of data points will generate an adequate surface model until the modelling stage is complete. In some cases, if the data collected is inadequate, further data collection can take place. Often, however, it is unfeasible to return to a physical part once data collection has taken place. Where this is the case, a degree of manipulation of the data may be necessary to improve the surface accuracy.

Model refinement is possible at object boundaries, aiming to improve the accuracy of the model, at areas prone to errors. This involves the trimming and extension of surfaces to reduce the errors in modelling which occur due to high curvature change. These problems occur in both convex and concave areas on a surface, but in particular where data cannot be collected at concave corners approaching a right angle. This is due to the spherical nature of the probe. An additional problem is caused by the continuous nature of spline interpolation (4). This problem is highlighted in figure 2, where the surface deviates from its desired position at areas of high curvature change, resulting in a smoothing of the corner at the base of the object. Here all the collected data is wanted, thus it is necessary to refine the surface model at these areas to improve accuracy.

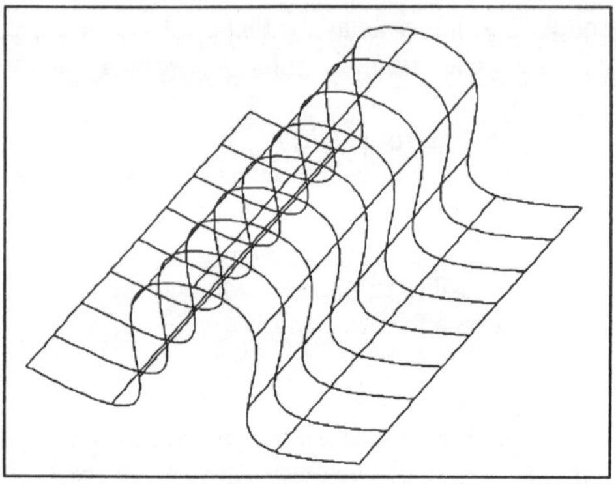

Figure 2 Surface Deviation

3.1 2D Boundary Trimming

There are two main applications for planar (2D) trimming.

1) In areas where minimal data was collected due to probing constraints at concave areas, causing surface deviations, as seen in figure 2.
2) Where the data collection stage requires quadrilateral boundaries around the object to be inspected, causing excess data, as seen in figure 1.

2D boundary trimming involves the automatic creation of a trimming plane at or near the base plane of the object. This refines the surface model by improving the surface accuracy at concave areas on the surface model where minimal data was collected. This reduces the surface deviation, improving the accuracy of the refined model, as compared with the original model.

This form of model refinement allows discontinuities such as sharp corners to be modelled, as seen in figure 3, where data is manipulated and remodelled.

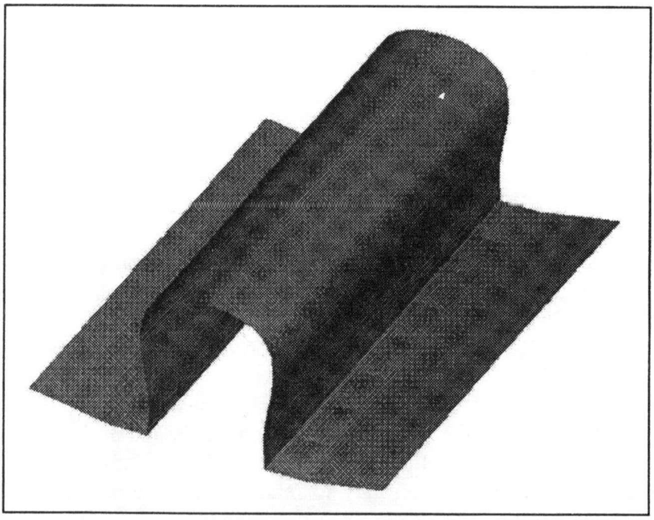

Figure 3 Refined Surface Model with Improved Data Fitting

In some cases excess data is modelled. This due to the nature of artificially creating quadrilateral boundaries around objects, as required for the data collection stage. This causes continuity problems as seen in figure 2, which would be reduced if data was trimmed to where data is actually needed. Figure 4 shows a refined surface model, where trimmed data is discarded, reducing the surface area. This greatly benefits the speed of accuracy analysis stages, as only wanted points are analysed for deviation.

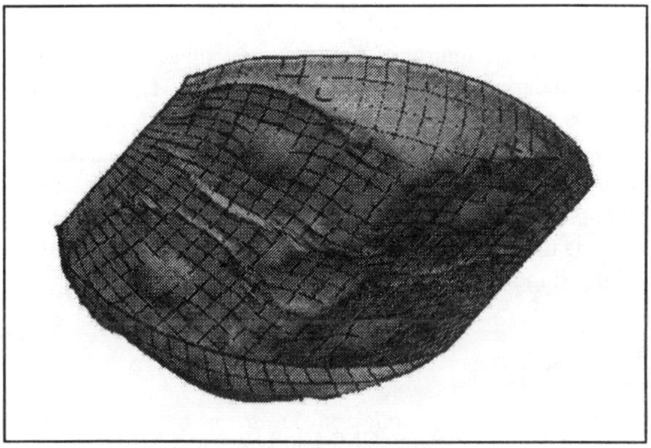

Figure 4 Refined Surface Model, with Discarded Excess Data

2D trimming is only effective with surfaces with a planar boundary. This is the case with most CMM scanned external boundaries, due to the need for a planar 'bed'. However, internal boundaries of features will inevitably have 3D boundaries.

3.2 3D Boundary Trimming

Creating 3D boundaries around internal features is proposed as a method of improving the accuracy of the overall model. This relies on segmenting the surface, creating 3D sub-surface trim boundaries. Each sub-surface can be trimmed along its corresponding boundary and then be modelled separately.

This uses the methods of Gaussian and Mean curvature based decomposition to create 3D boundaries from point data. Each data point is associated with one of eight possible surface type labels (5). These are given as peak, pit, ridge, valley, saddle ridge, saddle valley, flat and minimal (6). Automatic edge detection is then possible, and from this, surface refinement can occur.

4 CONCLUSION AND FUTURE WORK

Errors are expected at the data collection stage, and in most cases these are inevitable, such as collecting data from concave corners. In some objects, corners are present, but in many cases, corners are introduced into a data set due to the functionality of scanning routines. This involves unnecessary corners between the wanted object and the planar 'bed' necessary for scanning routines.

This paper shows how 2D trim planes can be used to improve the accuracy of a surface model by modifying data and introducing corners in areas where inadequate data was collected. 2D trim planes can also be used to discard unnecessary data which is unavoidably collected by the CMM contact measuring method of unidirectional scan lines. This results in a reduction of the data set for processing tasks, such as accuracy analysis.

Gaussian and Mean curvature based methods aim to define internal boundaries, improving model accuracy. Fitting sub-surfaces within defined internal boundaries may lead to greater overall surface accuracy.

This model refinement stage improves the accuracy of surface models at areas of high curvature, such as boundaries between differing surface types, as well as offering a computationally effective means of obtaining smaller data files.

Improvements to surface accuracy can occur by the use of bi-directional scan lines. This would more accurately define both 2D and 3D boundaries. The data collection stage can be further enhanced by implementing methods of using a 3D boundary, bounding the object to be measured, rather than the usual 2D quadrilateral boundary. This external 3D boundary can then be used in the modelling stage.

5 ACKNOWLEDGEMENTS

Acknowledgements to LK Ltd for existing CMM software and hardware.

6 REFERENCES

1 Smith, G., Hill, T., Rockliffe, S., and Harris, J., 1994, 'Geometric Assessment of Sculptured Features', *Proceedings of the Tenth National Conference on Manufacturing Research,* pp. 663-667.

2 Aronson, R., 1996, 'Forward Thinkers take to RE', *Man Engineer*, Nov ed., pp. 34-44.

3 Bardell, R., Balendran, V., and Sivayoganathan, K., 1997, 'Representing Surfaces from CMM Data,' *Proceedings of the Thirteen National Conference on Manufacturing Research*, pp.22-26.

4 Sarraga, R., 1990, 'Computer Modelling of Surfaces with Arbitrary Shapes', *IEEE Computer Graphics & Applications,* vol. 10, no. 2, pp. 67-77.

5 Balendran, V., Sivayoganathan, K., and Howarth, M., 1992, 'Sensor Aided Fettling', *Proceedings of the Eighth National Conference on Manufacturing Research*, pp.132-136.

6 Besl, P., and Jain, R., 1988, *'Range Image Segmentation'*, Algorithms, Architectures and Systems, ed. H. Freeman, Academic Press.

Analysing quality data to facilitate continuous improvement

A J BECKETT, D BANCE, and C E R WAINWRIGHT
CIM Institute, Cranfield University, UK

ABSTRACT

The growth of modern inspection technology has led to increasingly automated inspection processes, which are aimed at achieving more rapid and reliable fault detection. A by-product of this new technology has been to allow the collection of large amounts of data about manufactured product at little or no time penalty to production.

This paper proposes an approach to implementing a system for analysing this production inspection data, in order to assist continuous improvement of production processes, product design, development and new research.

This is achieved by creating a knowledge management framework, which allows both reactive problem solving approaches and proactive data driven approaches.

1. INTRODUCTION

In recent years, large numbers of 'total engineering' techniques have been developed with the basic aim of increasing the quality, reliability and efficiency of manufactured products. These techniques, such as total quality approaches (1,2) and statistical process controls (3,4), aim to provide effective control of the manufacturing process, instead of relying solely upon product inspection, which is seen as an activity that adds no value to the product. Concurrently, the growth of modern inspection technology has led to increasingly automated inspection processes, which are aimed at achieving rapid and reliable fault detection.

A by-product of this new technology has been to allow the collection of large amounts of data about manufactured product at little or no time penalty to production (5). This has created the problem of what to actually do with this data, after the primary quality yield information has been extracted from it. This paper proposes that modern inspection technology can in fact facilitate the extraction of more fundamental information about the production processes, the product design and can further prompt areas for new research. This information can in turn lead to a greater understanding of the key drivers of product and manufacturing performance.

2. CONTINUOUS IMPROVEMENT AND KNOWLEDGE CREATION

The principle of Continuous Improvement centres upon the idea that products and processes may be improved by analysing them, in order to identify areas that need to be addressed. This assumes that with knowledge of how a product or process operates, it may be made to operate better.

It would be useful at this stage to define the route by which this knowledge is generated. Nonaka (6) and Newman (7) have defined knowledge as the identification, within information, of an opportunity that may be exploited for competitive advantage. Knowledge is thus a nebulous commodity, which only has worth to the extent that it may be made use of. What constitutes knowledge at one instant may degrade into information when the opportunity for exploiting that knowledge passes. Information, on the other hand, is much more clearly definable as the identification of patterns within data. The knowledge obtained is finally exploited by a technology, such as a design improvement or process change.

This process of knowledge creation, the DIKT model (7) is shown in Figure 1.

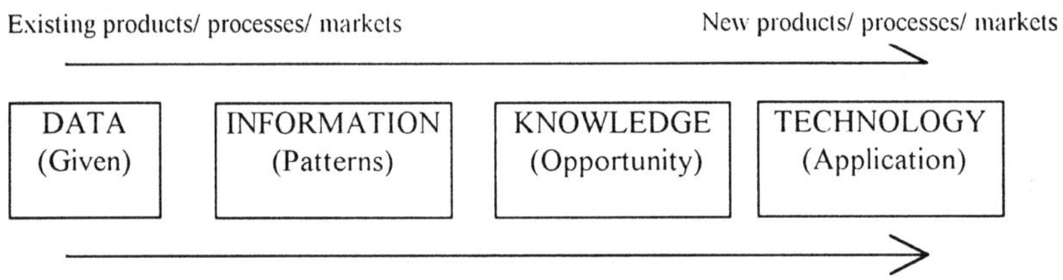

Figure 1: The DIKT model

3. APPROACHES TO DATA ANALYSIS

The DIKT model itself assumes that the process of knowledge creation starts with the data and 'works upwards' to supply knowledge. In practise, this is only one aspect of the process. Often, the absence of a technology drives the search for knowledge, which in turn generates information and data requirements, and the model could be said to work in reverse. Therefore, there are two basic approaches that may be adopted to extract information from large amounts of data, which represent the two extremes on a continuum. These are proactive and reactive approaches.

3.1 The proactive approach

This may be termed 'data led reasoning' (8). It centres upon the principle of subjecting the data to all available analysis techniques, and reviewing the results. The main advantage of this approach is that because the analysis is impartial, results can be obtained that were not necessarily anticipated. This could illuminate issues that were not suspected previously. In addition, maintaining the analysis over an extended period can show the trends of the measured

properties over time, which contributes to the accuracy of statistical techniques, and may lead to the anticipation of and prevention of problems.
The main disadvantage of this approach is that it is time consuming and therefore expensive to analyse all the available data. Furthermore, because the benefits of the analyses can only be measured after the fact, it is very difficult to justify to management, since effectively the analysis is being carried out in the 'hope' of useful information being obtained. For this reason, the proactive approach is commonly confined to information that has previously proved to be necessary, such as periodic quality reports.

In summary, the proactive approach is rigorous, but is expensive, with any benefits unclear at the outset.

3.2 The reactive approach

This approach is sometimes termed 'expectation led reasoning' (8) or 'fire-fighting'. Most of the common problem solving techniques (9) use this approach, which is based upon the principle of firstly defining the information required, and conducting the necessary analyses to acquire the desired information from the available data. The principle advantage of this approach is that the value of the analysis is easy to define, since it may be judged whether the analysis has successfully yielded the necessary information. Also, it is straightforward to justify to management, because the value of the expected results can be accurately anticipated.

However, the approach is adopted in response to a specific issue, and timescales are often very tight indeed. This precludes some types of exhaustive analyses and data collection exercises. In addition, problem solving exercises tend to be 'snapshots' of the production process at specific times, whereas a more continuous study might reveal more relevant information.

In summary, reactive approaches are a necessary part of an organisation's ability to rectify failures of its systems, but suffer significantly from time constraints and often only delay the onset of further, often more significant, issues.

3.3 Resolving the two approaches

A dilemma that often occurs in industry is the issue of how much resources should be dedicated to reactive problem solving, and how much to proactive knowledge creation. Because of the complex nature of the business environment, there are a multitude of factors which influence an organisation's position on this proactive/reactive continuum, and indeed this position will shift as the business and manufacturing environment changes. It may be hypothesised that there is no analytical solution to this problem, and that organisations 'iterate' to a position on the continuum that suits their environment..

4. A KNOWLEDGE MANAGEMENT FRAMEWORK

The aim of this paper was to show how an organisation could use its production quality data to institute continuous improvements. An organisation must therefore be able to stabilise its position on the proactive/reactive continuum, and so manage the process by which its data is transformed, via the two approaches, into product and process knowledge.

Given that an iterative solution is required, we can decompose both approaches to their constituent parts of the DIKT model, and create a flexible feedback control system around the DIKT principles. This framework is shown in figure 2.

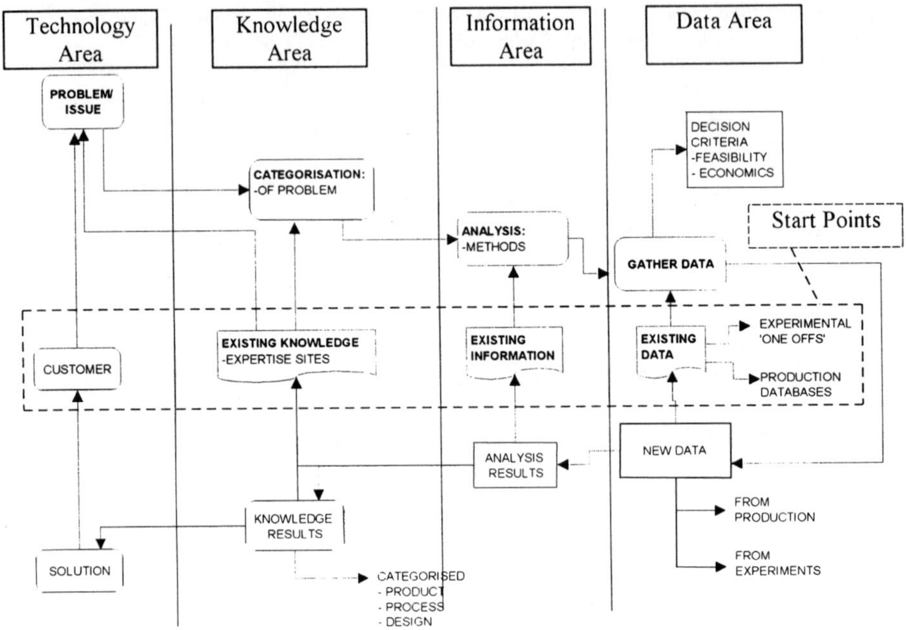

Figure 2: The knowledge management framework

As can be seen, the system is based around organisational mechanisms for the storage of data, information, and knowledge. Both proactive and reactive approaches are incorporated. For example, a common trigger for a reactive investigation is an issue identified by the customer concerning the product (in DIKT terminology, the organisations present 'technology'). This causes the creation of a knowledge requirement, which can either be satisfied by present knowledge within the organisation or can lead to the search for information around that issue. This in turn may be satisfied by the existing information store, or generates a requirement for data.

Conversely, the proactive approach is catered for by the existing set of information analyses being fed relevant data. This is communicated to the knowledge workers within the organisation, who can in turn use it to drive the creation of new technologies.

5. IMPLEMENTING THE FEEDBACK SYSTEM.

Nonaka has proposed (6) that organisational knowledge is of a different nature to knowledge held by individuals within the organisation. Indeed, if that knowledge is restricted to a few key personnel, its overall value to the organisation will be much less. Thus the data, information and knowledge gained by individual members has to be accessed and made explicit within the organisation as a whole.

To achieve this, the four aspects of the DIKT model must be managed. Management of the technology of an organisation has been the traditional emphasis of manufacturing industry, since it concentrates on how the existing design is manufactured efficiently, how build standards are controlled, whether customers are happy with the product, etc.

The three other main management areas must be introduced in addition to the technology management already in operation. These are detailed in the following sections.

5.1 Data management

The data produced by automatic test equipment (ATE) has to be collected and controlled. Most pieces of ATE output raw data in different formats, which have to be translated for importation into a central set of databases. Since flexibility in the system is required to facilitate reactive data requirements, a map of production processes is desirable, showing current data measures, potential data measures and also which measures may be used as proxies for measures that are not possible.

Most of the critical issues concern the data format and access to the data. Once set up, these operations can be supervised with minimal management involvement, and can be easily built into company procedures in accordance with ISO 9000 requirements.

5.2 Information management

The information management area involves the generation of information from the data, and the collation, storage and transmission of that information around the organisation. For proactive measures, like data management, this requires minimal effort once set up initially. Reports on the state and trends of specific parameters may be quickly directed throughout the organisation to the relevant knowledge areas. Also, using a common access area, such as shared drives or more easily an intranet site, access to current and historical information can be facilitated, which makes reactive enquiries a great deal easier. This can be tied to other design information such as technical documents, part specifications and process instructions.

Often, a situation will arise that requires a new piece of information, and the manager of the information area will have to search for relevant data to enable its creation.

5.3 Knowledge management

The purpose of the knowledge management area is primarily to guide the proactive research effort of the organisation and to act as a resource for reactive investigations.
Unlike the previous two areas, it is not advisable to automate, since it involves judgement as to what research is worthwhile, that is, the knowledge that is required by the organisation. This definition will be closely linked to the strategy and purpose of the organisation, which changes over time. Instead, the knowledge management area consists of a research steering committee, which can translate the current organisational strategy into a set of concrete objectives and research areas. This approach is consistent with the 'middle-up-down' management style proposed by Nonaka (6) where middle managers interpret upper management strategy statements and resolve those against lower management operational requirements and ideas.

6. INDUSTRIAL APPLICATION

This research is sponsored by Domino UK Ltd, a leading manufacturer of industrial ink jet printing machines. Domino require a system by which they can analyse production quality and reliability data, and introduce the resultant knowledge into the design process in order to consistently improve delivered quality at a lower overall cost. To achieve this, it has been proposed to implement the knowledge management framework in a number of discrete, easily managed stages. These stages will be based around ATE currently being installed on the production line. The first phase is to take one stand-alone piece of ATE, and sequentially apply Data management and Information management mechanisms to it. Having proven the principle, these two steps will be performed on other pieces of ATE as they become part of an integrated production inspection system. In parallel, the research will integrate with an embryonic research steering group to provide the knowledge management aspect of the system. This approach allows progress to be monitored by the sponsor, to ensure that Domino is receiving concrete benefits from the project.

Once fully installed, Domino will have an interactive system, which manages production data to provide a continuous improvement mechanism for its products and processes.

7. CONCLUSIONS

This paper has described a system by which an organisation can manage the creation of knowledge from production quality data, resolving both proactive and reactive knowledge creation requirements. This translates to, in effect, a three part management system which integrates data, information, and knowledge management requirements.

Implementing this system at the sponsoring company will enable inspection data to be made greater use of, not only providing a historical record but providing a mechanism by which product and processes may be continuously improved.

8. REFERENCES

1) Deming, W.E. *Out of the crisis.* 1986. Cambridge University Press. ISBN 0-58-130553-5
2) Juran, J.M. *Quality control handbook.* 1989. 4th Ed. McGraw-Hill. ISBN 0-07-033176-6
3) Shewhart, W.A. *Economic control of quality of manufactured product.* 1931. Van Nostrand
4) Taguchi, G. *Quality engineering in production systems.* 1989. McGraw-Hill.
5) Tannock, J.D.T. *Automating quality systems.* 1992. Chapman & Hall. ISBN 0412-40910-0
6) Nonaka, I. *The knowledge creating company.* 1991. Harvard Business Review, Nov-Dec Issue, p96-104.
7) Newman, V. *Redefining knowledge management to deliver competitive advantage.* 1997. Vol.1, No.2, p123-128.
8) Rogers, E. *Cognitive co-operation through visual interaction.* 1995. Knowledge based systems, Vol.8, No.2-3, p117-125
9) Kepner, C; Tregoe, B. *The rational manager.* 1965. McGraw-Hill.

The application of Taguchi to wafer plate development

A L CRUICKSHANK
Sasib Bakeries UK Limited, Newton-le-Willows, UK
C E R WAINWRIGHT
The CIM Institute, Cranfield University, UK

1. THE APPLICATION OF TAGUCHI TO WAFER PLATE DEVELOPMENT.

2. ABSTRACT

The high volume manufacture of biscuit wafers is achieved using automated machinery applying traditional processes which have been developed heuristically over many decades. The wafer plate is a critical element of wafer making machines and operates in a highly hostile environment, subject to great mechanical and thermal stresses on a repetitive basis. Positioning of the wafer plate are crucial to the process consistency and are measured in microns. Previous designs have evolved heuristically with the consequence that no scientific theory has been developed relating to the manner in which a wafer plate performs during operation. Similarly, the hostile environment of the process is such that monitoring is not possible. Therefore the design performance and limitations of the wafer plate are unknown.

During wafer production there are a great number of variables which change dynamically, with many process properties being disturbed by measurement. Hence, given the many variables and their interactions which exist gives rise to the need to carry out a detailed Taguchi based 'design of experiments' (DoE) activity to enable the controlling of the process to be understood with the optimum use of resources.

Therefore this paper describes the DoE activities undertaken and presents the results obtained. The paper concludes with a description of how the results will influence the future design and subsequent manufacture of wafer plates.

3. INTRODUCTION

Data acquisition in one form or another is associated with all experiments, the most common approach adopted is to adjust each variable one at a time. This is regarded as a very inefficient method by Taguchi. Taguchi's 'Design of experiments' is applied to the development of wafer plates so as to further work carried out from a food technology perspective and integrate it with engineering related data. Work previously carried out measured the effect of differing ingredients with relation to the final wafer product. These variables were adjusted individually in the original work and during duplication are omitted or were adjusted to either maximum or minimum values as outlined in the principles of Taguchi. This decision to maintain these parameters as constants or as variables was based in the conclusions found by Pritchard et al, 1975(1).

The most significant of the ingredients adjusted in the above work which has an effect from an engineering perspective was the aerating ingredient as this affects the pressure developed and thus the plate response to the process. This was deemed as a variable which must be adjusted and responses recorded. The remaining additives were concluded to have organoleptic effects as opposed to physical. The work also highlighted that solids content is the most critical factor in the production of wafer sheets. This being due to the effect which it has on the viscosity of the batter. This is further supported by further work undertaken by Cruickshank, 1997, (2)on the analysis of wafer plate design in which batter viscosity is a critical factor in determining the dispersion of the batter prior to 'going off'.

Further process variable were determined as critical by Pritchard et al, 1973, (3)These include deposit time, volume of batter and deposit pattern. These were carried out with a fixed batter recipe and viscosity so as to reduce variables. This again is against Taguchi principles. Of the above variables, the batter volume was deemed most appropriate to vary.

4. BASICS OF TAGUCHI

With the use of Taguchi, many of these variables are adjusted simultaneously as determined by using two level coding to give the design units values. Orthogonality is also applied to carry out Taguchi on several variables at once to map the interactions. The number of observations is determined by the number of variables and the two levels as highlighted in deriving two level coding for the design units.

This is given as $tc = 2^k$

Where tc = Number of observations
k = Number of factors.

Thus for 3 factors 8 runs must be carried out. 4 factors gives 16 runs. It is also regarded that adjusting more than 6 factors in an experiment as the data becomes unmanageable. The use of computers aids in this issue and is not ignored. A 8 channel data recorder was used to give dynamic data capture as well as aiding in plotting the monitored variables. The values monitored in these experiments were 4 temperature (2 per plate) and a pressure reading. For this paper the pressure is the resultant data under analysis.

5. CASE STUDY

So as to apply Taguchi, experiments were carried out on Standard sized wafer plates. These being cast iron plates with dimensions nominally 550mm * 350mm * 25mm thick. These are supported by a carrier assembly. The total mass of the complete assembly being 140 Kg. Ovens contain upwards of 45 pairs of assemblies, thus mass is a criteria to reduce wear, initial costs and subsequent running costs. A stand alone test rig was used to simulate the oven conditions and allow instrumentation access.

Standard batter mix is as follows:

Water 4500 ml
Flour 3500 g
Salt 11g
Bicarbonate of Soda 11g
Lecithin 28g
Groundnut oil 54g

Extremes of the following variables to give satisfactory wafers was determined. The quantities of the remaining ingredients remained the constant:

Baking temperatures.	150°C	190°C	(Manually adjusted)
Viscosity*	'Thick'	'Runny'	
Batter volume	200 ml	300ml	
Aerator Quantity	11g	40g	

* Thick = 65 Seconds Ford Cup No.5B. Runny = 25 Seconds Ford Cup No.5B. No SI unit comparison determined.

As can be seen the variables give rise to the need to produce 8 batches (tc = 2^3 = 8). The remaining variable being manually adjusted within each batch.

Duplication of aspects of previous work was required so as to gain data which was of interest from the engineering perspective. Thus enabling the range of conditions to which the plates would be subjected, to be determined. Again using the Taguchi principles; the quantities used the extreme values as shown above, to obtain correlation and interaction relationships between the batter properties and the wafer plate performance

6. RESULTS

For simplicity a Graph of all values against the extremes of temperatures is now shown (this is selected as the process variables are time and temperature, in commercially available ovens):

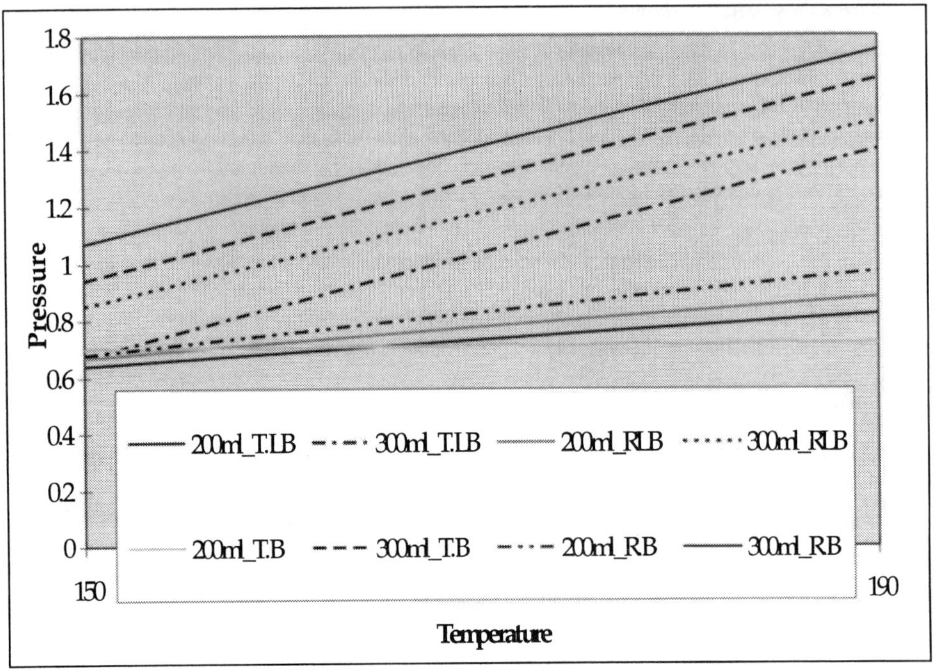

Figure 1:Graph of Pressure against Temperature for differing batter variables

6.1 Data processing

It is instructive to note that an immediate correlation can be drawn between Temperature and pressure. The difference between 200 ml and 300 ml is also apparent. The correlation between variables must now be quantified. This is started with 'Yates order' to place the test into a set sequence. This separates the highs and lows of each variable. Using above data gives the pressures relating to each variable:

Bicarbonate of Soda High = 8.21	Bicarbonate of Soda low = 6.15
Viscosity High =7.55	Viscosity Low = 7.49
Volume High = 9.98	Volume Low = 5.06
Temperature High = 8.81	Temperature Low = 6.23

Taguchi requires that the statistical interactions are assessed. Using the four variables to determine the effect on pressure gives rise to 6 pairs of interactions.
The most surprising of the interactions is that of viscosity and Bicarbonate of soda.

The Iso-level chart is shown:

		Viscosity -	Viscosity +
Bicarbonate	+	1.11	1.00
	-	1.17	0.89

Figure 2: An Iso-level chart showing 2 variable interaction

From this data it can be seen that an increase in viscosity reduces the pressure developed when there is a reduced quantity of Sodium bicarbonate in the batter mix. The effect is reduced when the sodium bicarbonate levels are high. This supports initial statements relating to the importance of determining viscosity in the production of wafer and the design of the wafer plates, as for any given recipe, the viscosity can greatly affect the properties of the final wafer product.

7. CONCLUSIONS

The application of Taguchi has shown that the effects of adjusting all variables within a set of experiments can be carried out in an efficient manner gaining much information which can be of great use with subsequent variable isolation and interaction procedures. Also highlighted are the compound effects of these changes. This was carried out in considerably less time that would be required for adjustments of one factor at a time. It is also recognised that the issues would not have been achievable without the use of the automated data acquisition software and further information technology in respect to the next stage of development in the production of enlarged wafer plates. The results have also given rise to the extremes to which the plates will be subjected to in normal operating life allowing the design criteria for pressure and temperature issues to be verified.

With the information gained from the tests, an optimised design based on the process requirements can be achieved. Without this work, the assumptions in place which have been used during the previous heuristic designs would continue to have built on excesses with regards to the mass and thermal properties of the plates. By feeding this information back into the design process it is now possible to reduce the plate thickness which in turn will allow a carrier of reduced dimensions and mass. This will in turn reduce the costs, and extend bearing life as previous stated.

This leads to the next stage or work to be carried out in the development of wafer plate technology which is to be completed with the use of Finite element design in a 3D cad environment.

References

(1)Pritchard et al. 1975. "The influence of ingredients on the properties of wafer sheets". Report No. 66 FMBRA.

(2)Cruickshank & Wainwright, 1997. "Technical analysis of wafer plate design". NCMR XIII, Glasgow. PP. 93-97.

(3) Pritchard et al. 1973. "The influence of process variables on the properties of wafer sheets". Report No. 56 FMBRA

A dimensional control methodology for the variation analysis of aircraft structure assembly

D J JEFFREYS and **P G LEANEY**
Department of Manufacturing Engineering, Loughborough University, UK
G WOOD
British Aerospace Airbus Limited, Bristol, UK

ABSTRACT

This paper describes the methodology used in a dimensional control analysis case study undertaken in conjunction with British Aerospace Airbus Ltd (BAAL). The methodology was applied for the assembly variation modelling of a carbon fibre composite (CFC) wing box structure developed for future civil large aircraft projects. With the planned introduction of large scale CFC components into future aircraft structures, BAAL are using dimensional control analysis to investigate and evaluate new manufacturing and assembly techniques and processes. Focusing on tolerancing and assembly operations, both components and fixturing have been investigated to identify areas where dimensional quality could be improved further. The methodology is closely supported through the use of an advanced simulation software tool and requires a systematic approach from design to manufacture. The assembly variation studies are currently assisting BAAL with both manufacturing and assembly process evaluation leading to the design optimisation of both component and tooling entities.

1. INTRODUCTION

Sources of variation (dimensional noise) need to be identified and managed if aircraft manufacturing organisations wish to become more competitive in the global market place. Variation can be found in manufactured components and subassemblies, assembly tooling and fixturing, the design/manufacturing/assembly/inspection processes, and human intervention, (1).

Traditionally, design engineers are primarily concerned with issues of style, function, and structural integrity. The communication of tolerancing, datuming and location schemes of the product then becomes a secondary concern, (2). An important part of the design process requires the determination of nominal dimensions and the application of tolerances. These need to be administered such that each individual part will meet the performance intent of the design and requirements of interchangeability. Dimensions therefore not only specify the size and shape of an object, but they maintain the design integrity of the part during manufacturing and assembly, (3,4). Many of these issues are covered by blanket company and international standards and are then left to various interpretations by the down stream activities of manufacturing, tooling and assembly building in unnecessary cost into some areas. This has led to the existence of the 'hidden factory' (5), where even at the late stages of production operators and fitters are "adjusting" out variational problems. The result of this is wasted time and money, as well as compromising product quality.

The introduction of dimensional variation from upstream manufacturing processes in addition to poorly defined location and datuming schemes may produce major problems for an assembly organisation. The discipline of dimensional control is an engineering methodology that manages variation from the early product definition stages through to full production. Computer 3D simulation tools are used to predict how much and where variation will occur in an assembly due to component variation, assembly methods, and assembly sequences, (6). A major objective of a dimensional control methodology is to expose the hidden factory to the whole organisation creating a network of cross functional activities that actively promotes engineering concurrency.

The aim of this paper is to report a particular methodology adopted for a geometric dimensional analysis study on a development wing box structure, to highlight why the assembly analysis was necessary, and what was achieved. The assembly analysis technique was undertaken to investigate and evaluate dimensional control issues surrounding the use of new production techniques, technologies and processes in the application of large CFC components into future large civil aircraft structures at British Aerospace Airbus Ltd, (BAAL).

2. THE ASSEMBLY VARIATION MODELLING APPROACH

The appropriation and preparation of data necessary to facilitate the assembly variation modelling requires a systematic approach for robust results. In this case study the nature and the source of the data are outlined and key points are highlighted. The acquisition of appropriate data was time consuming and iterative but became a major contributor in helping focus in on factors potentially effecting dimensional quality prior to the assembly variation simulation and analysis. Due to the amount of preparation time required to undertake an assembly analysis, a structured approach in the investigation, preparation and presentation of data will save time, reduce errors and produce a clear data audit trail. Both manufacturing and assembly process knowledge (process limits and the expected variation between these levels) needed to be captured for components and tooling at each related assembly stage.

The assembly modelling, simulation and analysis were undertaken using the Variation Systems Analysis VSA-3D (7) software product embedded in the CATIA Computer Aided Engineering (CAE) system and operated on an IBM RS/6000 AIX platform. The VSA-3D product has several sections which relies on CATIA's Functional Dimensional and Tolerancing (FD&T) module to generate feature, datum and tolerance data.

The development CFC wing box programme at BAAL is at an advanced stage of development. Physical as well as functional detail architecture has been defined for component and fixture geometry. As a consequence, the following data was available for the assembly simulations:

- Detailed 3D CAD models were made available for use in CATIA for all geometry with assembly sequencing defined for the development structure;
- The development CFC wing box structure detailed product dimensional objectives were known (targets and metrics).

In the early stages the process is sequential but later becomes more concurrent and iterative. The assembly modelling process was divided into the following twelve steps:

(i) Define the engineering challenges

The challenges were highlighted and documented by the composite wing programme engineering team resulting mainly from work undertaken previously on prototype builds and from experimentation. Defining the engineering challenges allowed the formulation of aims and objectives for the analysis. This was an important stage which determined the 'focus' and 'context' for the analysis.

(ii) Define aims/objectives of the assembly analysis study

From the documented challenges a set of analysis aims and objectives were developed. These defined the analysis boundary (both physical and functional) not allowing any one model to become over complex. The challenges were grouped and several analysis studies were planned, each having its own set of aims and objectives.

(iii) Determine the analysis zone

Physical zones were created for the analysis and only the components having an effect on the dimensional quality within that zone were included. The boundaries of the zones were determined by the aims and objectives of the analysis, but encompassed all physical components and elements of manufacturing and assembly tooling potentially having an effect on the 'focus' of the investigation.

(iv) Identify all the key features for components and tooling

These were the physical features (areas of surface geometry) owned by components, manufacturing, inspection and assembly tooling that influence the dimensional quality aspects of the zone under analysis. Component features can be split into the following categories:

- Features that are associated with the manufacture of the component part;
- Features associated with the location system;
- Features associated with component and fixture interfaces;
- Features associated with product dimensional quality control;
- Features to be used as component and fixture datums.

Some features shared more than one category. An example of this is when a plane, a hole and a slot in a component shares the status of being both a locating and datuming feature set.

(v) Quantify manufacturing and assembly process variation
This was undertaken for all component parts and assembly tooling. For this to be achieved, knowledge on all manufacturing processes used to produce the individual component and assembly tooling key features was required. This data was made available from manufacturing and assembly development process inspection data.

Assembly simulation incorporating unrealistic statistical formulae of tolerance variation can give misleading results. It is important to recognize that steps need to be taken to adequately ascertain manufacturing process data through current SPC programmes or other inspection data systems. Manufacturing and assembly processes at BAAL were being measured for capability limits (Cp), how well the process is under control (Cpk), and how adequate samples are being distributed within those limits (distribution type).

The use of the Geometric Dimensioning & Tolerancing (GD&T) standard Y14.5M-1994, ASME (8), allowed each component to be described in clear and concise geometric and mathematical terms avoiding ambiguous tolerance and datuming communication. The FD&T fuction in CATIA automatically determines the feature type and only allows the placement of datums and tolerances in accordance to a GD&T standard. Datums and component tolerances are defined in conjunction with feature definitions ensuring that no datum and tolerance call outs can be created violating a given GD&T standard.

(vi) Determine the sequence of component assembly
For each analysis study a sequence was defined representing in which order components were assembled together. The components were taken from a parts list defined at an earlier stage and were represented diagramatically as an assembly 'tree,' with the components and subassemblies diverging into the total assembly for that stage. The assembly tree defined whether components were located directly through jigs and fixtures (independent location, via holes, pins, offsets), located via other component parts (kit build, via component interfaces, offsets) or located via a combination of the two. The assembly tree allowed tractability at every stage and provided a tolerance map highlighting a 'variation route' progressing through to the final assembly via component and fixturing features.

(vii) Define assembly method (assembly moves) for each component part
To ensure a dimensionally stable build each component was constrained in all six degrees of freedom (DOF), three in translation and three in rotation. Each component was analysed for conditions of over constraint (more than six DOF) or under constraint (less than six DOF). If either condition existed the assembly 'philosophy' for that component was investigated and revised where ever possible to ensure the correct constraint status in 3D space.

VSA-3D was used to facilitate the component to fixture assembly moves using the following software templates:

- Datum target based moves: this allows the placement of an unlimited number of datum target points on component and fixtures to facilitate the assembly of curved and complex features;
- Assembly feature based moves: allows the assembly of feature combinations such as hole/hole (via pin), hole/pine, pine/slot, plane/plane, or any combination of these.

For any type of assembly move the normal convention references the component being moved as the 'Object' and the component/fixture into which it is to be moved as the 'Target'.

(viii) Define the analysis measurements to be undertaken
The VSA-3D software was used to measure the following entities of the development wing box:

- Gap clearance: the gap condition between rib bottom edge and skin surface to facilitate mechanical fastener, film and paste adhesive bonding assembly processes;
- Virtual hole condition: the virtual condition between skin and rib holes to facilitate mechanical fasteners.

(ix) Undertake tolerance analysis simulation
When all the proceeding steps had been completed the assembly of the wing box section was simulated 2,000 times for each scenario. Three types of tolerance analysis were undertaken for each measurement:

- Monte Carlo: a random number generator which is related to a distribution type;
- High-Low-Medium: this simulation holds each individual tolerance in turn at its high, low and medium level while alternating all remaining tolerances through all combinations of high, low and medium levels;
- Extreme case: simulates the build setting all variation at the widest level, i.e. the worst combination to produce maximum variation in an assembly.

(x) Analyse the simulation results
When the simulation was completed a report was compiled by the VSA-3D software for each measurement analysis. Each type of analysis undertaken was presented in its relevant report format. The Monte Carlo (MC) simulation presented the results in a distribution and provided information on the magnitude of variation associated with each measurement. The High-Low-Median (HLM) results were presented as a Pareto analysis indicating the percentage contribution to each tolerance in the analysis in relation to the variation in the output measured. MC and HLM analysis were undertaken for each measurement.

(xi) Revise parameters and repeat process for further investigation
Several analysis studies were undertaken of defined assembly zones. Subsequent studies of each zone were undertaken with the different tolerance parameters revised to explore the resultant effect upon the assembly structure. This process of rerunning the analysis models with different tolerance parameters allowed the assessment of major contributors to geometric variation. It also helps in defining the most adequate tolerances required to achieve a high level of confidence of assemblability through the relaxation of tolerances that have little contribution and tightening of tolerances that have a major contribution.

(xii) Document the analysis results
The final stage was to document the analysis results of the different zones in a formal report. The report should include clear data trails of where the required information was gained for each step of the approach, therefore in the event of inconclusive results these data sources may be revisited. A typical analysis report should contain information such as objectives, assumptions, modelling methodology, drawing/CAD data, results and conclusions.

3. CONCLUSION

This paper outlines a systematic approach used for the modelling of manufacturing, tooling and assembly issues used in support of the composite wing programme at BAAL. These projects continues to ensure the successful development and productionisation of future large civil aircraft and provides a base for the development of advanced manufacturing and assembly technologies. The dimensional control approach of assembly analysis is currently assisting in identifying, quantifying and managing sources of variation potentially introduced to CFC wing box structures.

Assembly variation modelling required both product and process knowledge most of which had been characterized. Such data is forming the basis for a better understanding of the products control and development resulting in improved robustness. The credibility of any simulation result to be used for component, subassembly and tooling design or redesign purpose relies upon model validation. This in turn will be dependent upon clear audit trails in the acquisition of data.

The case study experience also points to the need for a multifunctional team approach for effective product and process design and revision. The aim of introducing the assembly variation modelling process at BAAL is to characterize process and product capability and to make this information available to all engineering functions in so far as this will enable variation management to be exercised at all stages of aircraft development and production. This dimensional control methodology is ensuring that tolerance verses variation related data can be fed back to the design, manufacturing, assembly and inspection areas of the organisation. This will provide essential information at all stages of product development necessary to ensure total aircraft robustness.

REFERENCES

1. McCuistion P. J., 1994, "The role of dimensional analysis within concurrent engineering", Ohio University, USA (Conference paper)
2. Craig M., 1996, "Limits of tolerance", Manufacturing Engineer, Vol:75, No:3, 139-143
3. Henzold G., 1995, "Handbook of geometric tolerancing - Design, manufacture and inspection", John Wiley & Sons, England, UK, ISBN 0-471-94816-0
4. Spotts M. F., 1983, "Dimensioning and tolerancing for quality production", Prentice-Hall, New Jersey, ISBN 0-13-214684-3
5. Leaney P G, 1996, "Design for dimensional control", in Design for X - Concurrent engineering imperatives, Ed. G Q Huang, Chapman & Hall, UK, ISBN 0-412-78750-4
6. Craig M, 1992, "Controlling the variation", Manuf. Breakthrough, Vol:1, No:6, pp343-348
7. Variation Simulation Analysis Ltd, 1993, "VSA-3D Analysis Reference", St. Clair Shores, MI, USA (Product brochure)
8. ASME, 1995, "ASME Y14.5M-1994, Dimensioning and Tolerancing", The American Society of Mechanical Engineers, New York, USA, ISBN 0-8273-4694-8

Summary

The purpose of this paper is to relate the experience of a case study being undertaken at British Aerospace Airbus Ltd (BAAL) and outline the use of a particular dimensional control (DC) methodology as part of a product development philosophy. The need for dimensional variation (noise) modelling and tolerance analysis is the basis for the emerging discipline of DC. The control of dimensional variation needs to be addressed throughout the design, manufacture and assembly phases of product development and production. Dimensional variation in the build of an aircraft structure will occur due to variation in component manufacture (manufacturing process noise) and the assembly processes (assembly process noise) including tools, fixtures, fasteners, as well as the procedures adopted. If uncontrolled, this variation will effect aircraft dimensional quality characteristics as well as increasing production cost and lead-time.

The study proposes a particular methodology for the DC of an assembly structure. The key points are highlighted, for example, the acquisition of appropriate data is a time consuming and iterative process which, by itself, can focus on the factors affecting dimensional variation before an assembly model can be simulated and analysed. Dimensional variation modelling requires process knowledge sometimes never before characterized, however, such data will form the basis for better process control and a more robust design. In addition, the credibility of any variation results being used for design evaluation purposes means that assembly models must be validated and this in turn depends on clear audit trails in data acquisition. The case study experience also highlights the need for product and process data to be extracted from a range of engineering disciplines. This can become the catalyst for an effective multifunctional team approach with regard to integrated product and process engineering. The ultimate aim is to characterize process capability and make it available to product engineers in so far as this enables the DC methodology to be exercised from the earliest stages of product design and development.

Corresponding author: D.J. Jeffreys
Department of Manufacturing Engineering
Loughborough University
Loughborough, Leics.
LE11 3TU
Tel: 01509 228254
Fax: 01509 267725
E-mail: d.j.jeffreys@lboro.ac.uk

Relationships between capability indices and signal-to-noise ratios for use in the manufacturing industry

L KOUNIS, J M O'SULLIVAN, and **I R McANDREW**
Faculty of Engineering and Information Science, University of Hertfordshire, Hatfield, UK

ABSTRACT

This paper reports on the implications when using capability indices and signal-to-noise ratios for continuous improvements. The aim was to address the problems that are created when trying to use both at the same time, and forms part of a wider study into their implication. When robust designs are required on processes that have adequate capability indices it was found that in many cases the new parameters can reduce the capability index. Likewise, when trying to improve capability indices on robust designs the signal-to-noise ratio will no longer be at its optimum. Reviewing these two situations shows that for achieving robustness and acceptable capability a compromise must be achieved.

INTRODUCTION

The last decade has seen major changes in the ways that industry has viewed quality, its importance and place in the development procedure. The introduction of BS EN ISO 9000 in 1987 and, now QS 9000, have directed the attention towards the need for continuous improvement if quality management system are to take a proactive role. Indeed, QS 9000 is directing companies towards even higher levels of quality and advanced quality planning to satisfy the ever increasing demand of quality by customers. To meet the needs of a quality system companies are having to use techniques to ensure the predicted quality levels cannot only be met but can be further improved, e.g. SPC, FMEA's, QFD and Taguchi. Not only are these techniques accepted as the way forward, but as an essential to survive.

Within the area of Statistical Process Control it is generally accepted that capability indices are a key indicator to predict reject levels (Besterfield, 1991). Furthermore, Motorola's Six-Sigma approach has focused the aims of reaching Cpk indices of at least 2.0 if reject levels are to be accepted by all involved. This, in its own right, is a difficult task and might not be possible if tolerances are set at unachievable levels. In many ways the subject of tolerance design has been discussed by Taguchi, but companies are more interested in the robustness of a product than individual elements (Logothetis, 1992).

For many manufacturing processes capability indices are stated at the contract stage and subsequent evaluation may require proof of this achievement. Now, with the advent of QS 9000, robustness of designs and processes is also a requirement for these processes. Continuous improvement will result in a company trying to increase the Cpk values while endeavouring to find settings for robustness. The most used approach is to measure the Signal-to-Noise values and subsequently select the parameters to this value.

Capability indices and robustness have separate requirements which may result in contradictions with each other ; particularly in cases where maximising one has detrimental affects on the other. If compromises are required then it is necessary to establish a feasible target.

COMPARISON OF TECHNIQUES

Signal-to-Noise ratios depend upon the application and it is generally agreed that there are three principal ones; although there have been many proposed by others. This paper is primarily concerned with: minimal-the-best, nominal-the-best and maximum-the-best. Peace (1996) discussed the concerns of each and showed that the method should be to centralise the process and then reduce the variability. This in its own right is practical but the application falls short due to commercial constraints. McAndrew and O'Sullivan (1995) found that the use of Signal-to-Noise ratios were normally used by industry to show a form of robustness in design and process but seldom as a long-term commitment to achieving optimum settings. Thus, robustness is more likely to produce situations where target values are met but no real reduction in variation is achieved.

For Signal-to-Noise ratios to be used successfully it is important that it takes the form of an iterative stage, otherwise the improvements in robustness are not at their optimum. To centralise the average to a target can be achieved by many other methods and if Taguchi's approach is used it will probably be an ineffective use of time and effort. Indeed, Jones (1996) in his research showed that as many as 80% of companies that used Taguchi's approach believed it to be a technique that they would be unlikely to use again.

The alternative direction to robustness for ensuring continuous improvement has been to use capability indices as a way of reducing the likelihood of rejects. This technique also focuses on the target value and variations that exist; however, the ends of the variation are related to the specification limits. The most widely used index is the Cpk value which is determined from the following formula:

$$\text{Cpk [min]:} \quad USL - \mu / 3\sigma \quad \text{or} \quad \mu - LSL / 3\sigma$$

As it is impossible to establish, μ, without measuring the complete population the sample average is used in its place. By using the 3σ for each side 99.73% of the population variation can be explained with respect to the specification limits (USL and LSL). With Cpk's of greater than 1 the likelihood of a reject is less than the 4 per 1000 level, likewise this probability decreases as higher values are achieved. Processes that have Cpk values of 2.0 are only expected to produce 3.4 rejects per million

Clearly, as Cpk's can be used to measure the probability of rejects they are an interest to all in a supply-customer chain. Wierda (1992) has argued that Cpk's of 2.0 are now becoming accepted as a base value not as a long-term target. Given this situation, the central position of a process and its variation is secondary in its importance to the absolute value of the capability index.

Since the mid-1990's QS 9000 is being imposed on many manufacturing companies that work within the automotive sector. Other sectors are also following, or intend to follow QS 9000. One of the principal difference between QS 9000 and ISO 9000 is that the former requires designs and processes to be both capable and robust. For a company to become accredited with QS 9000 there must be clear evidence that capability and robustness are being achieved and on a continuous basis.

To achieve capability and robustness companies are determining capability indices and robustness by determining the Cpk and Signal-to-Noise ratios for designs and processes alike. Naturally, not all values achieved are acceptable and research has shown that attention is being directed to improve the worst situations not continuous improvement in all areas. It may be that if all designs and processes reach a minimum accepted level, which has not been defined, real continuous improvement will occur.

It is when both capability indices and Signal-to-Noise ratios are performed on the same process that concerns can and do arise. As discussed earlier, they both focus on the target value and variation of the process, except that capability indices also relate to the specifications.

A Signal-to-Noise ratio is expressed as a function of log to the base 10 in the following form for nominal-the-best:

$$-10 \log_{10} (y^2 / \sigma^2)$$

The first part of this formula is to expand the differences of the part within the brackets. It is the parts within the brackets where detailed comparisons are made. Taguchi (1990) has argued that specifications should be ignored in preference to minimising the loss; usually expressed as Loss to Society. Therefore, the most suitable setting will be focused towards the targeted value. In effect, this formula expresses the square of the average divided by its variance. A smaller variance, less variation, will result in larger values; although these are in a negative form. Thus, a value with the highest number will result in the desired setting, i.e. the most robust situation. No account is made for the positions from the ends of the variation to the specifications.

The above is in contrast to the approach used with capability indices; although a central average will always produce the highest value. What both of these approaches cannot ensure is how to target the central position, they only allow a comparison of how effective such changes can be in practice.

Below in figure 1 is a hypothetical representation of the most suitable outputs from both approaches showing how they relate to specification limits.

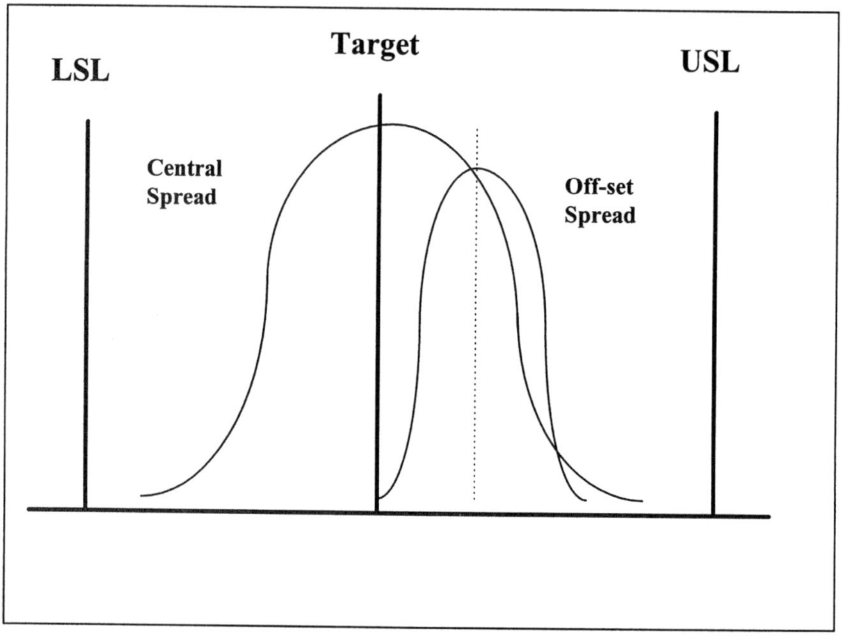

Figure 1, signal-to-noise v. capability indices.

Figure 1 clearly shows the two different situation from these approaches. The off-set spread may have a smaller variation than the central spread but the average is not in the most important situation if the loss is to minimised. For these situations the central spread would give the highest Signal-to-Noise ratio whilst the off-set would result in the highest capability index as the ends of the curves are further way from the specifications.

In such a situation as discussed above there would be no clear answer on which setting to use when both approaches are required, as in QS 9000. To choose one would result in the other being compromised. It could be argued that which ever one was selected would be wrong as the other is reduced. To this end it is not possible to make recommendations that everyone would agree was correct. How could you improve robustness while increasing the likelihood of rejects being produced? In such a situation there would have to be a compromise and this requires a general approach or set of guidelines to be made. At present there is neither and individual judgements need to be made.

ANALYSIS

For a continuous improvement environment both robustness and capability are necessary. If such a situation occurs as in figure 1, a solution would be to carry out further investigations to determine possible interactions, changes to inputs and process integrity checks to try to obtain settings that satisfy both approaches. This would assume that the process is always subjected to continuous improvements regardless of how high are the capability and Signal-to-noise ratios.

A short-term solution is that providing the capability is at least 2.0 for the Signal-to-noise ratio the highest Signal-to-noise ratio can be selected regardless as rejects will be less than 3.4 per million. If, for figure 1, it was not nominal-the-best but minimum-the-best the conclusions would still be different. The robustness would still hold true as the best selection but the capability is still best achieved by the off-set situation. However, for maximum-the-best the off-set situation would be the best for Signal-to-Noise ratio and capability index. Thus, each of the three Signal-to-Noise ratios require individual compromises.

There is clearly a need to clarify this situation and determine parameters that allow for a balanced answer. Whilst a confusion can occur then further barriers exist to adopting and maintaining a continuous environment.

IMPLICATIONS

Followers of Taguchi's philosophy will reject the capability index as it does not minimise the loss to society. This may very well be true but commercial decisions are based on engineering specification. While designers include specifications there will always be checks to determine if rejects exist. Capability indices are always going to be necessary whilst rejects are possible. Companies are still and, likely in the future, to be selected upon their performance to meeting specifications. Unless the need to remove specification occurs, which introduces new problems, there will always be a potential need to perform capability investigations.

CONCLUSIONS

Continuous improvement can be shown by using either Signal-to-Noise ratios or capability indices. However, when using both, as recommended by QS 9000, situations can occur where conflicting answers to problems exist. Selecting which one, or a compromise needs to be taken on a statistical basis, manufacturing requirements and the product. If the present situation does not change then opponents of modern quality techniques will argue that this potential concern is a reason to ignore such investigations. Whilst the authors would not denounce robustness as a technique, it is not acceptable to reduce the capability and, hence increase rejects, as a method of reducing robustness. The long-term aim must be to establish methods that promote both.

REFERENCES

Besterfield, D. (1991) Quality Control, 2nd Ed. Prentice Hall.

Jones, H. J. (1996) Developing Designed Experiments, *Inter. Journal of Quality and Reliability,* Vol. **25**, pp 87-95.

Logothetis, N. (1992) Managing for Total Quality, Prentice Hall.

McAndrew, I. R. and O'Sullivan, J. M. (1995) Designed Experiments for Assembling CPR, *11th NCMR,* De Montford.

Peace, G. (1996) Taguchi Methods, Addison Wesley

Taguchi, G. (1990) Controlling Proceses for high Quality, Low Cost, *8th symposium on Tagughi Methods,* MI Dearborn, American Supply Institute.

Wierda, S. J. (1992) A multi-variant process capability index, *Proc. 9th Inter. Conf. Israel Soc. Qual. Assur.* Jerusalem, Israel.

Authors' Index